The Finite Element Method
Third edition

The Finite Element Method

(The third, expanded and revised edition of *The Finite Element Method in Engineering Science*)

O. C. Zienkiewicz, FRS

Professor of Civil Engineering and
Head of the Department of Civil Engineering,
University of Wales, Swansea

McGRAW-HILL Book Company (UK) Limited

London · New York · St Louis · San Francisco · Auckland
Bogotá · Guatemala · Hamburg · Johannesburg · Lisbon
Madrid · Mexico · Montreal · New Delhi · Panama · Paris
San Juan · São Paulo · Singapore · Sydney · Tokyo · Toronto

Published by

McGRAW-HILL Book Company (UK) Limited
MAIDENHEAD · BERKSHIRE · ENGLAND

First edition 1967
Reprinted 1968
Reprinted 1969
Reprinted 1970
Second revised edition 1971
Reprinted 1972
Reprinted 1973
Reprinted 1975
Third revised edition 1977
Reprinted 1978
Reprinted 1979
Reprinted 1981
Reprinted 1982

British Library Cataloguing in Publication Data

Zienkiewicz, Olgierd Cecil

The finite element method. — 3rd expanded and revised ed.
1. Finite element method 2. Engineering mathematics
I. Title II. Finite element method in engineering science
620'.001'515353 TA347.F5 77–30152

ISBN 0–07–084072–5

6 7 8 9 MP 8 4 3 2

Printed and bound in the United States of America

To my wife and to my mother

Preface

The present volume may be regarded as the third edition of the *Finite Element Method on Structural and Continuum Mechanics* first published in 1967. Although the size is now some three times that of the original edition—it is written with identical objectives; first to teach and second to provide a 'state of the art' reference base of the subject—which is now recognized as one of considerable importance to both practicing engineers and physicists as well as researchers.

Since the first volume was written the number of research publications on the finite element method has been increasing almost exponentially. Close to 8000 references are recorded and many more are available as internal reports, etc.† While in early days the contributors have been almost exclusively engineers, today a large number of these come from the field of mathematics which has now adopted the method and made a great contribution to the understanding of finite element method. Clearly, at this stage, a book doing justice to all points of view is impracticable—and in this volume much selection and filtering had to be done representing the viewpoints of the author. This acknowledges both the mathematical basis and the need for intuitive creative thought. Thus although the book starts with the basis of a physical discrete system—and introduces the finite element approximations via well understood elasticity examples—in Chapter 3 the concepts of fundamental mathematical approximation are presented (in a manner avoiding, with apology to mathematicians, some of the jargon and pedantry so as to make it suitable for engineers or physicists). In some later chapters we show, however, how some of the usually accepted criteria can be modified and violated with success. In particular Chapter 11 provides some of the recent developments in this context, showing how a cancellation of errors can occur through inexact integration etc.

† An excellent bibliography compiled by D. Norrie and G. de Vries (IFI/PLENUM 1976) shows the following rate of publication with figures in parentheses giving number of papers in the year: 1961 (10); 1962 (15); 1963 (25); 1964 (33); 1965 (67); 1966 (134); 1967 (162); 1968 (303); 1969 (531); 1970 (510); 1971 (844); 1972 (1004); 1973 (1169); 1974 (1377); 1975 (880 incompl.).

The general definitions of the finite element method can today be made so wide (vide Chapter 3) as to include other useful approximation processes. In particular finite difference methods will now be recognized as a subclass of the procedure and (with some imagination) the boundary integral method, which has been lately used with much success for certain classes of problem, can be brought under the general definition. This generalization is made with two-fold object. First to improve our understanding—second to incorporate selective advantages of the alternatives in a unified manner. Chapter 23 is devoted to a recent development by which the boundary integral and finite element methodologies are combined.

The application of the finite element method is today so wide that it is impossible to present an exhaustive picture in one volume. The reader will find, however, that the main fields of solid mechanics both in linear and non-linear phases, fluid mechanics, heat transfer and electro-magnetism have received some attention—and depending on his interest he can direct his selection appropriately. Clearly the study of the complete volume in one course is not recommended and a teacher using the text will make an appropriate selection of chapters. It is hoped, however, that the wider coverage will prove its use by providing a reasonably self-contained reference to many fields of activity into which sooner or later every one of us is thrown. The notes of the text have been successfully used at many levels of presentation, ranging in Chapters 1–3 from undergraduate courses through postgraduate teaching to courses including practitioners involved in development of the method. The prerequisite knowledge of mathematics and mechanics does not go much beyond a reasonable undergraduate engineering or physics course and some more abstract topics—such as matrices and vectors—are expounded in appendices.

The finite element process is essentially dependent for its success on skillful use of computers and efficient numerical techniques. Emphasis on the latter is made throughout the book but in the concluding chapter, written by Professor R. L. Taylor, much of the programming experience of the University of California, Berkeley, and of the University of Wales College, Swansea, is incorporated in a fairly complete computer system which the reader can use immediately for a variety of problems or extend readily to suit his own needs. For simplicity the system is limited in capacity. This at the same time avoids machine dependency—but its expansion to a larger size can readily be made.

Acknowledgements

To many friends in this field all over the world who, sharing the author's enthusiasm, have contributed through discussions and their own researches to many of the ideas here reported.

To my colleagues and research students at Swansea without whose effort this book could not have been written.

To the innumerable sponsoring agencies for supporting students and research. Here particular thanks go to the Science Research Council of U.K. who have over the year provided the base for much of the work.

To my wife for her help and forbearance.

Contents

List of symbols

Below a list of principal symbols used in this book is presented for easy reference, although all are defined in the text as they occur. On many occasions, additional ones have to be used in a minor context and a *non-uniqueness* arises. It is hoped that appropriate text explanation will avoid confusion.

The symbols are listed roughly in the order of occurrence in chapter sequence.

Matrices and column vectors are denoted by bold symbols, e.g., **K** and **a** and $\mathbf{K}^{\mathrm{T}}$ stands for transpose of **K**. Dots are used to denote differentiation with respect to one variable, e.g., $\frac{d}{dt} \equiv \dot{\mathbf{a}}$, etc.

Chapter	*Symbol*	
1	$\mathbf{a}_i$, $\mathbf{a}$	nodal or global displacements
	$\mathbf{q}_i^e$	nodal force at i due to element e
	$\mathbf{K}^e$, $\mathbf{K}$	stiffness matrix (element/global)
	$\mathbf{f}_{pi}^e$	nodal element force at i due to p, etc.
	$\mathbf{r}_i$	external nodal force
	$\boldsymbol{\sigma}$	stress (vector)
	$\mathbf{L}$, $\mathbf{T}$	transformation matrices
	$\mathbf{b}$	alternative parameters
	$\mathbf{u}$	displacement vector (components u, v and w)
2, 4, 5, 6	$\boldsymbol{\varepsilon}$	strain (vector)
	$\mathbf{L}$	strain operator
	$\mathbf{N}$	(displacement) shape function
	$\mathbf{B} = \mathbf{LN}$	strain shape function
	$\mathbf{D}$	elasticity matrix
	$\mathbf{b}$	body force (vector)
	E	Young's modulus
	ν	Poisson's ratio
	$\boldsymbol{\varepsilon}_0$, $\boldsymbol{\sigma}_0$	initial strain or stress

Chapter	Symbol	Meaning
	$\mathbf{t}$	boundary traction
	b_x, t_x, etc.,	x – components of body forces and tractions
	ε_x, γ_{xy}, σ_x, τ_{xy}	x – components of direct and shear strain or stress
	U	strain energy
	W	potential energy of loads
	Π	total potential energy
	$\mathbf{I}$	identity matrix
	h	representative element dimension
	ϕ	body force potential (or other scalar function)
	$\boldsymbol{\phi}$	body force potential nodal values
	$\mathbf{m}^T = [1, 1, 0]$ or $[1, 1, 1, 0, 0, 0]$	matrix equivalent of Kronecker delta for two or three dimensional strain/stress vectors
	$x, y, z, x', y', z', r, z, \theta$	Cartesian or cylindrical co-ordinates
3	$\mathbf{A}(\mathbf{u})$, $\mathbf{B}(\mathbf{u})$, etc.	operators defining governing differential equations and boundary conditions
	$\mathbf{u}$, $\boldsymbol{\phi}$, ϕ	unknown function
	$\mathbf{v}$	'test' function
	$\mathbf{a}$, $\mathbf{b}$, etc.	nodal (or other) parameters defining the trial expansion $\mathbf{u} \simeq \mathbf{Na}$
	$\mathbf{w}_j$	'weight' function
	Π	a stationary functional
	$\mathbf{L}$	a linear differential operator
	$\mathbf{C}(\mathbf{u})$	constraint condition on $\mathbf{u}$
	$\boldsymbol{\lambda}$	Lagrangian multiplier
	$\mathbf{n}^T = [n_x, n_y, n_z]$	vector normal to boundary
	α	penalty number
	∇	gradient operator $= \left[\frac{\partial}{\partial x}, \frac{\partial}{\partial y}, \frac{\partial}{\partial z}\right]^T$
7, 8, 9	l_k^n	Lagrange polynomials
	ξ, η, (ζ)	element, curvilinear, coordinates two and three dimensions
	L_1, L_2, (L_3)	triangular (area) or tetrahedral (volume) coordinates
	$\mathbf{J}$	Jacobian matrix
	H_i, w_i	quadrature weights
10	w	plate deflection

Chapter	Symbol	Meaning
	M_x, M_y, M_{xy}	generalized stress components (moments)
	$\theta_{xi} \theta_{yi}$	rotations
	H^n_{mi}	Hermitian polynomials
	t	plate thickness
11	K, G	bulk and shear moduli
12	$\mathbf{G}$	operator linking stresses and tractions on boundary
13	$\mathbf{K}^b, \mathbf{K}^p$	stiffness matrices in bending and in-plane action respectively
	$\lambda_{x'y}$, etc.	direction cosines of between x' and y axes, etc.
	$\mathbf{V}_{ij}$	vector connecting point i to j
	l_{ij}	length of vector $\mathbf{V}_{ij}$
14	ϕ	angle of tangent to shell and Z axis
	R_s and r	radii of curvature
17	$\mathbf{k}, k$	permeability matrix or coefficient
	$\mathbf{H}$	discretized problem matrix
	p	pressure
	ϕ	potential
18, 19	$\mathbf{\Psi}(\mathbf{a})$	non-linear discrete equation operator
	$\mathbf{K}_T$	tangent matrix
	F	yield function
	Q	plastic potential
	$\mathbf{K}_\sigma$	initial stress matrix
20, 21	$\mathbf{M}$	mass matrix
	$\mathbf{C}$	damping matrix
	$\omega_i, \bar{\mathbf{a}}_i$	i-th eigenvalue or eigenvector
	ω	frequency
	y_i	mode participation factor
	λ	characteristic number
	$\mathbf{u}$	velocity vector
22	μ	viscosity
	ρ	density
	R_n	Reynolds number
	α	upwinding parameter
23	H_0	Hankel function
	$K_{\mathrm{I}}, K_{\mathrm{II}}, K_{\mathrm{III}}$	stress intensity factors

1. Some Preliminaries: The Standard Discrete System

1.1 Introduction

The limitations of the human mind are such that it cannot grasp the behaviour of its complex surroundings and creations in one operation. Thus the process of subdividing all systems into their individual components or 'elements', whose behaviour is readily understood, and then rebuilding the original system from such components to study its behaviour is a natural way in which the engineer, the scientist, or even the economist proceeds.

In many situations an adequate model is obtained using a finite number of well-defined components. Such problems we shall term *discrete*. In others the subdivision is continued indefinitely and the problem can only be defined using the mathematical fiction of an infinitesimal. This leads to differential equations or equivalent statements which imply an infinite number of elements. Such systems we shall term *continuous*.

With the advent of digital computers, *discrete* problems can generally be solved readily even if the number of elements is very large. As the capacity of all computers is finite, *continuous* problems can only be solved exactly by mathematical manipulation. Here, the available mathematical techniques usually limit the possibilities to oversimplified situations.

To overcome the intractability of the realistic type of continuum problem, various methods of *discretization* had from time to time been proposed both by engineers and mathematicians. All involve an *approximation* which, hopefully, is of such a kind that it approaches, as closely as desired, the true continuum solution as the number of discrete variables increases.

The discretization of continuum problems has been approached differently by mathematicians and engineers. The first have developed general techniques applicable directly to differential equations governing the problem, such as finite difference approximations,[1,2] various weighted residual procedures,[3,4] or approximate techniques of determining the stationarity of properly defined 'functionals'. The engineer, on the other hand, often approaches the problem more intuitively by creating an analogy between real discrete elements and finite portions of a continuum domain. For instance, in the field of solid mechanics McHenry,[5] Hrenikoff,[6] and Newmark[7] have, in the early 1940s, shown that reasonably good solutions to a continuum problem can be obtained by substituting small portions of the continuum by an arrangement of simple elastic bars. Later, in the same context, Argyris[8] and Turner *et al.*[9] showed that a more direct, but no less intuitive, substitution of properties can be made much more directly by considering that small portions or 'elements' in a continuum behave in a simplified manner.

It is from the engineering 'direct analogy' view that the term 'finite element' has been born. Clough[10] appears to be the first to use this term, which implies in it a direct use of *standard methodology applicable to discrete systems*. Both conceptually and from the computational viewpoint, this is of the utmost importance. The first allows an improved understanding to be obtained; the second the use of a unified approach to the variety of problems and the development of standard computational procedures.

Since the early 1960s much progress has been made, and today the purely mathematical and 'analogy' approaches are fully reconciled. It is the object of this text to present a view of the finite element method as *a general discretization procedure of continuum problems posed by mathematically defined statements*.

In the analysis of problems of a discrete nature, a standard methodology has been developed over the years. The civil engineer, dealing with structures, first calculates his force–displacement relationships for each element of the structure and then proceeds to assemble the whole following a well-defined procedure of establishing local equilibrium at each 'node' or connecting point of the structure. From such equations the solution of the unknown displacements becomes possible. Similarly, the electrical or hydraulic engineer, dealing with a network of electrical components (resistors, capacitances, etc.) or hydraulic conduits, first establishes a relationship between currents (flows) and potentials for individual elements and then proceeds to assemble the system by ensuring continuity of flows.

All such analyses follow a standard pattern which is universally adaptable to discrete systems. It is thus possible to define a *standard discrete system*, and this chapter will be primarily concerned with establishing the processes applicable to such systems. Much of what is presented here will be known to engineers, but some reiteration is at this stage advisable. As the treatment of elastic, solid structures has been the most developed area of activity this will be introduced first, followed by examples from other fields, before attempting a complete generalization.

The existence of a unified treatment of 'standard discrete problems' leads us to the first definition of the finite element process as a method of approximation to continuum problems such that

(*a*) the continuum is divided into a finite number of parts (elements), the behaviour of which is specified by a finite number of parameters, and

(*b*) the solution of the complete system as an assembly of its elements follows precisely the same rules as those applicable to *standard discrete problems*.

It will be found that numerous classical mathematical procedures of approximation fall into this category—as well as the various direct approximations used in engineering. It is thus difficult to determine the origins of the finite element method and the precise moment of its invention.

Table 1.1 shows the process of evolution which led to the present-day concepts of finite element analysis. Chapter 3 will give, in more detail, the mathematical basis which evolved from the classical landmarks.[11-20]

1.2 The Structural Element and System

To introduce the reader to the general concept of the discrete system we shall first consider a structural engineering example of linear elasticity.

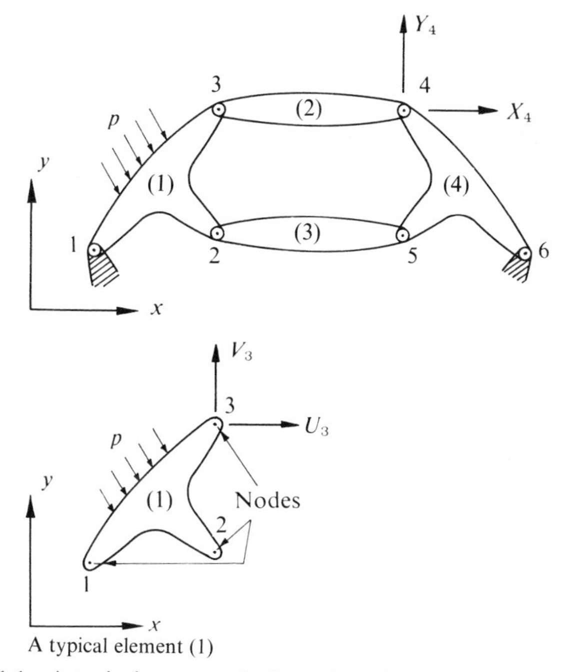

Fig. 1.1 A typical structure built up from interconnected elements

TABLE 1.1
FAMILY TREE OF FINITE ELEMENT METHODS

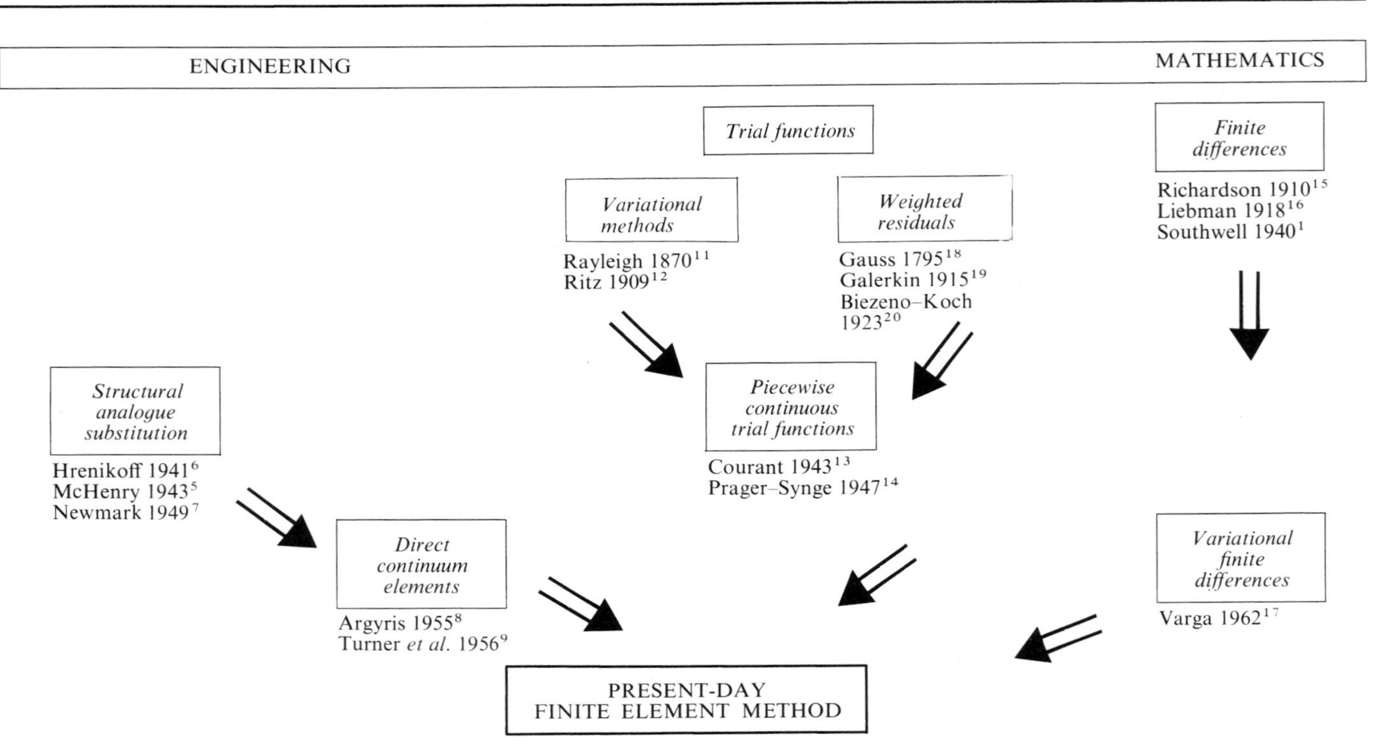

Let Fig. 1.1 represent a two-dimensional structure assembled from individual components and interconnected at the nodes numbered 1 to n. The joints at the nodes, in this case, are pinned so that moments cannot be transmitted.

As a starting point it will be assumed that by separate calculation, or for that matter from the results of an experiment, the characteristics of each element are precisely known. Thus, if a typical element labelled (1) and associated with nodes 1, 2, 3 is examined, the forces acting at the nodes are uniquely defined by the displacements of these nodes, the distributed loading acting on the element (p), and its initial strain. The last may be due to temperature, shrinkage, or simply an initial 'lack of fit'. The forces and the corresponding displacements are defined by appropriate components (U, V and u, v) in a common co-ordinate system.

Listing the forces acting on all the nodes (three in the case illustrated) of the element (1) as a matrix† we have

$$\mathbf{q}^1 = \begin{Bmatrix} \mathbf{q}_1^1 \\ \mathbf{q}_2^1 \\ \mathbf{q}_3^1 \end{Bmatrix}; \quad \mathbf{q}_1^1 = \begin{Bmatrix} U_1 \\ V_1 \end{Bmatrix}, \text{ etc.} \tag{1.1}$$

and for the corresponding nodal displacements

$$\mathbf{a}^1 = \begin{Bmatrix} \mathbf{a}_1^1 \\ \mathbf{a}_2^1 \\ \mathbf{a}_3^1 \end{Bmatrix}; \quad \mathbf{a}_1^1 = \begin{Bmatrix} u_1 \\ v_1 \end{Bmatrix}, \text{ etc.} \tag{1.2}$$

Assuming linear elastic behaviour of the element, the characteristic relationship will always be of the form

$$\mathbf{q}^1 = \mathbf{K}^1\mathbf{a}^1 + \mathbf{f}_p^1 + \mathbf{f}_{\varepsilon 0}^1 \tag{1.3}$$

in which $\mathbf{f}_p^1$ represents the nodal forces required to balance any distributed loads acting on the element, and $\mathbf{f}_{\varepsilon 0}^1$ the nodal forces required to balance any initial strains such as may be caused by temperature change if the nodes are not subject to any displacement. The first of the terms represents the forces induced by displacement of the nodes.

Similarly, the preliminary analysis or experiment will permit a unique definition of stresses or internal reactions at any specified point or points of the element in terms of the nodal displacements. Defining such stresses by a matrix $\boldsymbol{\sigma}^1$ a relationship of the form

$$\boldsymbol{\sigma}^1 = \mathbf{S}^1\mathbf{a}^1 + \boldsymbol{\sigma}_p^1 + \boldsymbol{\sigma}_{\varepsilon 0}^1 \tag{1.4}$$

† A limited knowledge of matrix algebra will be assumed throughout this book. This is necessary for reasonable conciseness and forms a convenient book-keeping form. For readers not familiar with the subject a brief appendix is included in which sufficient principles of matrix algebra are given to follow intelligently the development. Matrices (and vectors) will be distinguished by bold print throughout.

is obtained in which the last two terms are simply the stresses due to the distributed element loads or initial stresses respectively when no nodal displacement occurs.

The matrix $\mathbf{K}^e$ is known as the element stiffness matrix and the matrix $\mathbf{S}^e$ as the element stress matrix for an element (e).

Relationships Eqs. (1.3) and (1.4) have been illustrated on an example of an element with three nodes and with the interconnection points capable of transmitting only two components of force. Clearly, the same arguments and definitions will apply generally. An element (2) of the hypothetical structure will possess only two points of interconnection, others may have quite a large number of such points. Similarly, if the joints were considered as rigid, three components of generalized force and of generalized displacement would have to be considered, the last corresponding to a moment and a rotation respectively. For a rigidly jointed, three-dimensional structure the number of individual nodal components would be six. Quite generally therefore—

$$\mathbf{q}^e = \begin{Bmatrix} \mathbf{q}_1^e \\ \mathbf{q}_2^e \\ \vdots \\ \mathbf{q}_m^e \end{Bmatrix} \quad \text{and} \quad \mathbf{a}^e = \begin{Bmatrix} \mathbf{a}_1 \\ \mathbf{a}_2 \\ \vdots \\ \mathbf{a}_m \end{Bmatrix} \tag{1.5}$$

with each $\mathbf{q}_i$ and $\mathbf{a}_i$ possessing the same number of components or *degrees of freedom*.

The stiffness matrices of the element will clearly always be square and of the form

$$\mathbf{K}^e = \begin{bmatrix} \mathbf{K}_{ii}^e & \mathbf{K}_{ij}^e & \dots & \mathbf{K}_{im}^e \\ \vdots & \vdots & & \vdots \\ \mathbf{K}_{mi}^e & \dots & \dots & \mathbf{K}_{mm}^e \end{bmatrix} \tag{1.6}$$

in which $\mathbf{K}_{ii}^e$, etc., are submatrices which are again square and of the size $l \times l$, where l is the number of force components to be considered at the nodes.

As an example, the reader can consider a pin-ended bar of a uniform section A and modulus E in a two-dimensional problem shown in Fig. 1.2. The bar is subject to a uniform lateral load p and a uniform thermal expansion strain

$$\varepsilon_0 = \alpha T.$$

If the ends of the bar are defined by the co-ordinates x_i, y_i and x_n, y_n its length can be calculated as

$$L = \sqrt{\{(x_n - x_i)^2 + (y_n - y_i)^2\}}$$

and its inclination from the horizontal as

$$\alpha = \tan^{-1} \frac{y_n - y_i}{x_n - x_i}.$$

Only two components of force and displacement have to be considered at the nodes.

The nodal forces due to the lateral load are clearly

$$\mathbf{f}_p^e = \begin{Bmatrix} U_i \\ V_i \\ U_n \\ V_n \end{Bmatrix}_p = -\begin{Bmatrix} -\sin\alpha \\ \cos\alpha \\ -\sin\alpha \\ \cos\alpha \end{Bmatrix} \cdot \frac{pL}{2}$$

and represent the appropriate components of simple beam reactions, $pL/2$. Similarly, to restrain the thermal expansion ε_0 an axial force $(E\alpha TA)$ is needed, which gives the components

$$\mathbf{f}_{\varepsilon 0}^e = \begin{Bmatrix} U_i \\ V_i \\ U_n \\ V_n \end{Bmatrix}_{\varepsilon 0} = -\begin{Bmatrix} -\cos\alpha \\ -\sin\alpha \\ \cos\alpha \\ \sin\alpha \end{Bmatrix} (E\alpha TA).$$

Finally, the element displacements

$$\mathbf{a}^e = \begin{Bmatrix} u_i \\ v_i \\ u_n \\ v_n \end{Bmatrix}$$

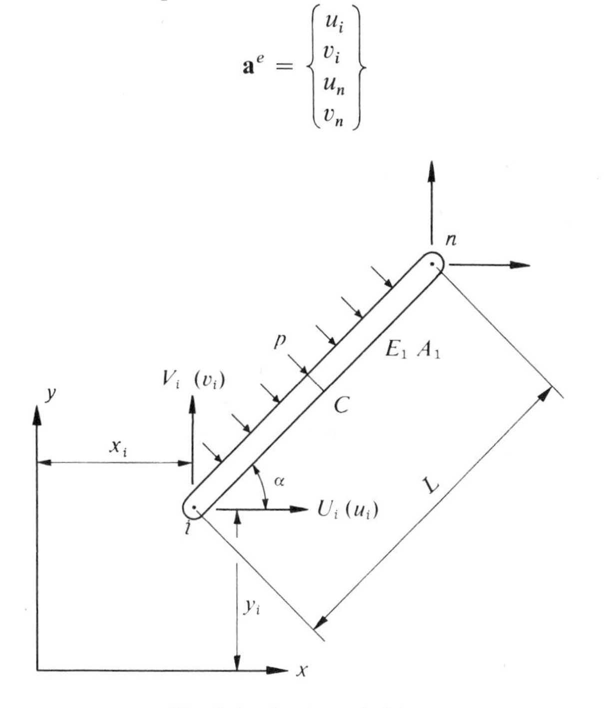

Fig. 1.2 A pin-ended bar

will cause an elongation $(u_n - u_i)\cos\alpha + (v_n - v_i)\sin\alpha$. This, when multiplied by EA/L, gives the axial force whose components can again be found. Rearranging these in the standard form gives

$$\mathbf{K}^e\mathbf{a}^e = \begin{Bmatrix} U_i \\ V_i \\ U_n \\ V_n \end{Bmatrix}_\delta =$$

$$= \frac{EA}{L}\left[\begin{array}{cc|cc} \cos^2\alpha & \sin\alpha\cos\alpha & -\cos^2\alpha & -\sin\alpha\cos\alpha \\ \sin\alpha\cos\alpha & \sin^2\alpha & -\sin\alpha\cos\alpha & -\sin^2\alpha \\ \hline -\cos^2\alpha & -\sin\alpha\cos\alpha & \cos^2\alpha & \sin\alpha\cos\alpha \\ -\sin\alpha\cos\alpha & -\sin^2\alpha & \sin\alpha\cos\alpha & \sin^2\alpha \end{array}\right]\begin{Bmatrix} u_i \\ v_i \\ u_n \\ v_n \end{Bmatrix}$$

The components of the general Eq. (1.3) have thus been established for the elementary case discussed. It is again quite simple to find the stresses at any section of the element in the form of relation Eq. (1.4). For instance, if attention is focused on the mid-section C of the beam the extreme fibre stresses determined from the axial tension to the element and the bending moment can be shown to be

$$\boldsymbol{\sigma}_C^e = \begin{Bmatrix} \sigma_1 \\ \sigma_2 \end{Bmatrix}_C = \frac{E}{L}\begin{bmatrix} -\cos\alpha, & -\sin\alpha, & \cos\alpha, & \sin\alpha \\ -\cos\alpha, & -\sin\alpha, & \cos\alpha, & \sin\alpha \end{bmatrix}\mathbf{a}^e$$

$$+ \begin{Bmatrix} 1 \\ -1 \end{Bmatrix}\frac{pL^2}{8}\frac{d}{I} - \begin{Bmatrix} 1 \\ 1 \end{Bmatrix}E\alpha T$$

in which d is the half depth of the section and I its second moment of area. All the terms of Eq. (1.4) can now be easily recognized.

For more complex elements more sophisticated procedures of analysis are required but the results are of the same form. The engineer will readily recognize that the so-called 'slope–deflection' relations used in analysis of rigid frames are only a special case of the general relations.

It may perhaps be remarked, in passing, that the complete stiffness matrix obtained for the simple element in tension turned out to be symmetric (as indeed was the case with some submatrices). This is by no means fortuitous but follows from the principle of energy conservation and from its corollary—the well-known Maxwell–Betti reciprocal theorem.

The element properties were assumed to follow a simple linear relationship. In principle, similar relationships could be established for non-linear materials, but discussion of such problems will be held over at this stage.

1.3 Assembly and Analysis of a Structure

Consider again the hypothetical structure of Fig. 1.1. To obtain a complete solution the two conditions of

(*a*) displacement compatibility, and

(*b*) equilibrium

have to be satisfied throughout.

Any systems of nodal displacements **a**

$$\mathbf{a} = \begin{Bmatrix} \mathbf{a}_1 \\ \vdots \\ \mathbf{a}_n \end{Bmatrix} \tag{1.7}$$

listed now for the whole structure in which all the elements participate, automatically satisfies the first condition.

As the conditions of overall equilibrium have already been satisfied *within* an element all that is necessary is to establish equilibrium conditions at the nodes of the structure. The resulting equations will contain the displacements as unknowns, and once these have been solved the structural problem is determined. The internal forces in elements, or the stresses, can easily be found by using the characteristics established *a priori* for each element by Eq. (1.4).

Consider the structure to be loaded by external forces **r**

$$\mathbf{r} = \begin{Bmatrix} \mathbf{r}_1 \\ \vdots \\ \mathbf{r}_n \end{Bmatrix} \tag{1.8}$$

applied at the nodes in addition to the distributed loads applied to the individual elements. Again, any one of the forces r_i must have the same number of components as that of the element reactions considered. In the example in question

$$\mathbf{r}_i = \begin{Bmatrix} X_i \\ Y_i \end{Bmatrix} \tag{1.9}$$

as the joints were assumed pinned, but at this stage a generality with an arbitrary number of components will be assumed.

If now the equilibrium conditions of a typical node, i, are to be established, each component of $\mathbf{r}_i$ has, in turn, to be equated to the sum of the component forces contributed by the elements meeting at the node. Thus, considering *all* the force components we have:

$$\mathbf{r}_i = \sum_{e=1}^{m} \mathbf{q}_i^e = \mathbf{q}_i^1 + \mathbf{q}_i^2 + \cdots \tag{1.10}$$

in which $\mathbf{q}_i^1$ is the force contributed to node i by element 1, $\mathbf{q}_i^2$ by element 2, etc. Clearly, only the elements which include point i will contribute non-zero forces but for tidiness all the elements are included in the summation.

Substituting from the definition (1.3) the forces contributed to node i are noting that nodal variables $\mathbf{a}_i$ are common and thus omitting the superscript e we have

$$\mathbf{r}_i = \left(\sum_{e=1}^{m} \mathbf{K}_{i1}^e\right)\mathbf{a}_1 + \left(\sum_{e=1}^{m} \mathbf{K}_{i2}^e\right)\mathbf{a}_2 + \cdots + \mathbf{f}_i^e \tag{1.11}$$

where

$$\mathbf{f}^e = \mathbf{f}_p^e + \mathbf{f}_{\varepsilon 0}^e$$

The summation again only concerns the elements which contribute to node i. If all such equations are assembled we have simply

$$\mathbf{Ka} = \mathbf{r} - \mathbf{f} \tag{1.12}$$

in which the submatrices are

$$\begin{aligned} \mathbf{K}_{im} &= \sum_{e=1}^{m} \mathbf{K}_{im}^e \\ \mathbf{f}_i &= \sum_{e=1}^{m} \mathbf{f}_i^e \end{aligned} \tag{1.13}$$

with summations including all elements. This simple rule for assembly is very convenient because, as soon as a coefficient for a particular element is found it can be put immediately into the appropriate 'location' specified in the computer. *This general assembly process can be found to be the common and fundamental feature of all finite element calculations and should be well understood by the reader.*

If different types of structural elements are used and are to be coupled it must be remembered that the rules of matrix summation permit this to be done only if these are of identical size. The individual submatrices to be added have therefore to be built up of the same number of individual components of force or displacement. Thus, for example, if a member capable of transmitting moments to a node is to be coupled at that node to one which in fact is hinged, it is necessary to complete the stiffness matrix of the latter by insertion of appropriate (zero) coefficients in the rotation or moment positions.

1.4 The Boundary Conditions

The system of equations resulting from Eq. (1.12) can be solved once the prescribed support displacements have been substituted. In the example of Fig. 1.1, where both components of displacement of nodes 1 and 6 are zero, this will mean the substitution of

$$\mathbf{a}_1 = \mathbf{a}_6 = \begin{Bmatrix} 0 \\ 0 \end{Bmatrix}$$

which is equivalent to reducing the number of equilibrium equations (in this instance twelve) by deleting the first and last pairs and thus reducing the total number of unknown displacement components to eight. It is, nevertheless, always convenient to assemble the equation according to relation Eq. (1.12) so as to include all the nodes.

Clearly, without substitution of a minimum number of prescribed displacements to prevent rigid body movements of the structure, it is impossible to solve this system, because the displacements cannot be uniquely determined by the forces in such a situation. This physically obvious fact will mathematically be interpreted in the matrix $\mathbf{K}$ being singular, i.e., not possessing an inverse. The prescription of appropriate displacements after the assembly stage will permit a unique solution to be obtained by deleting appropriate rows and columns of the various matrices.

If all the equations of a system are assembled, their form is

$$\begin{aligned} \mathbf{K}_{11}\mathbf{a}_1 + \mathbf{K}_{12}\mathbf{a}_2 + \cdots &= \mathbf{r}_1 - \mathbf{f}_1 \\ \mathbf{K}_{21}\mathbf{a}_1 + \mathbf{K}_{22}\mathbf{a}_2 + \cdots &= \mathbf{r}_2 - \mathbf{f}_2 \\ \text{etc.} \end{aligned} \tag{1.14}$$

and it will be noted that if any displacement, such as $\mathbf{a}_1 = \bar{\mathbf{a}}_1$, is prescribed then the external 'force' $\mathbf{r}_1$ cannot be prescribed and remains unknown. The first equation could then be *deleted* and substitution of known values of $\mathbf{a}_1$ made in the remaining equations. This process is computationally cumbersome and the same objective is served by adding a large number, $\alpha\mathbf{I}$, to the coefficient $\mathbf{K}_{11}$ and replacing the right-hand side, $\mathbf{r}_1 - \mathbf{f}_1$, by $\bar{\mathbf{a}}_1\alpha$. If α is very much larger than other stiffness coefficients this alteration effectively replaces the first equation by the equation

$$\alpha\mathbf{a}_1 = \alpha\bar{\mathbf{a}}_1 \tag{1.15}$$

that is, the required prescribed condition, but the whole system remains symmetric and minimal changes are necessary in the computation sequence. A similar procedure will apply to any other prescribed displacement. The above artifice has been introduced by Payne and Irons.[21] An alternative procedure avoiding the assembly of equations corresponding to nodes with prescribed boundary values will be presented in Chapter 24.

When all the boundary conditions are inserted the equations of the system can be solved for the unknown displacements and stresses, and internal forces in each element obtained.

1.5 Electrical and Fluid Networks

Identical principles of deriving element characteristics and of assembly will be found in many non-structural fields. Consider for instance an assembly of electrical resistances shown in Fig. 1.3.

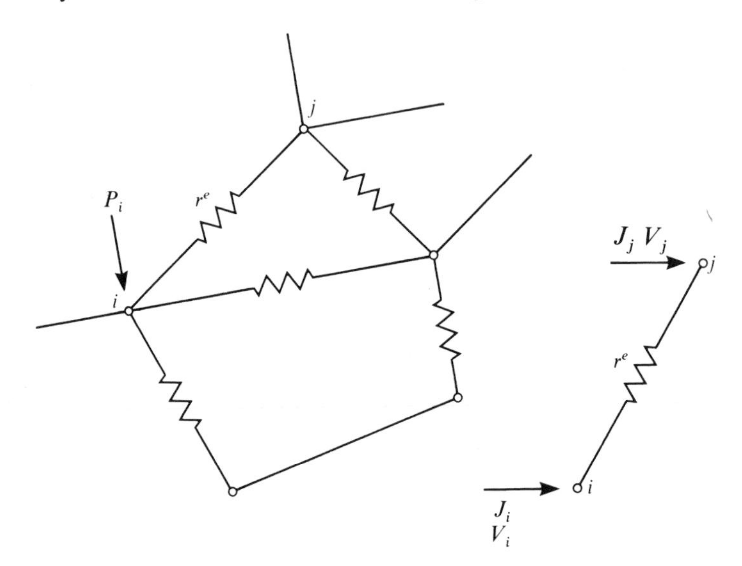

Fig. 1.3 A network of electrical resistances

If a typical resistance-element, *ij*, is isolated from the system we can write by Ohm's law the relation between the currents *entering* the element at the ends and the end voltages as

$$J_i^e = \frac{1}{r^e}(V_i - V_j)$$

$$J_j^e = \frac{1}{r^e}(V_j - V_i)$$

or in matrix form

$$\begin{Bmatrix} J_i^e \\ J_j^e \end{Bmatrix} = \frac{1}{r^e}\begin{bmatrix} 1 & -1 \\ -1 & 1 \end{bmatrix}\begin{Bmatrix} V_i \\ V_j \end{Bmatrix}$$

which in our standard form is simply

$$\mathbf{I}^e = \mathbf{K}^e\mathbf{V}^e \tag{1.16}$$

This form clearly corresponds to the stiffness relationship (1.3), indeed if external current were supplied along the length of the element the element 'force' terms could also be found.

To assemble the whole network the continuity of potential at nodes is assumed and a current balance imposed there. If P_i now stands for an external imput of current at node i we must have with complete analogy to Eq. (1.11)

$$P_i = \sum_{m=1}^{m=h} \sum K_{im}^e V_m \tag{1.17}$$

where the second summation is over all 'elements', and once again for all the nodes

$$\mathbf{P} = \mathbf{KV} \tag{1.18}$$

in which

$$K_{ij} = \sum_{e=1}^{m} K_{ij}^e$$

Matrix notation in the above has been dropped as the quantities such as voltage and current and hence also the coefficients of the 'stiffness' matrix are scalars.

If the resistances were replaced by fluid-carrying pipes in which a laminar regime pertained, identical formulation would once again result with V standing for the hydraulic head and I for flow.

For pipe networks usually encountered, however, the linear laws are in general not valid. Typically the flow-head relationship is of a form

$$I_i = c(V_i - V_j)^\gamma \tag{1.19}$$

where the index γ lies between 0·5 and 0·7. Even now it would still be possible to write relationships in for the form (1.16) noting, however, that the matrices $\mathbf{K}^e$ are no longer an array of constants but are known functions of $\mathbf{V}$. The final equations can once again be assembled but their form will be non-linear and in general iterative techniques of solution will be needed.

Finally it is perhaps of interest to mention the more general form of an electric network subject to an alternating current. It is customary to write the relationships between the current and voltage in *complex form* with the resistance being replaced by complex impedance. Once again the standard forms of (1.16) to (1.18) will be obtained but with each quantity divided into real and imaginary parts.

Identical solution procedures can be used if the equality of the real and imaginary quantities is considered at each stage. Indeed with modern digital computers it is possible to use standard programming making use of facilities available for dealing with complex numbers. Reference to some problems of this class will be made in a later chapter dealing with vibration problems.

1.6 The General Pattern

To consolidate the concepts discussed in this chapter an example will be considered. This is shown in Fig. 1.4(*a*) where five discrete elements are inter-connected. These may be of structural, electrical, or any other linear type. In the solution:

The first step is the determination of element properties from the geometric material and loading data. For each element the stiffness matrix as well as the corresponding 'nodal loads' are found in the form of Eq. (1.3). Each element has its own identifying number and specified nodal connection. For example:

element		connection			
1		1	3	4	
2		1	4	2	
3		2	5		
4		3	6	7	4
5		4	7	8	5

Assuming that properties are found in global co-ordinates we can enter each 'stiffness' or 'force' component in its position of the global matrix as shown in Fig. 1.4(*b*). Each shaded square represents a single coefficient or a submatrix of type $\mathbf{K}_{ij}$ if more than one quantity is being considered at the nodes. Here, for each element, its separate contribution is shown and the reader can verify the position of the coefficients. Note that various types of 'elements' considered here present no difficulty in specification. (All 'forces', including nodal ones, are here associated with elements for simplicity.)

The second step is the assembly of the final equations of type given by Eq. (1.12). This is simply accomplished according to the rule of Eq. (1.13) by *simple addition* of all numbers in the appropriate space of the global matrix. The result is shown in Fig. 1.4(*c*) where the non-zero coefficients are indicated by shading.

As the matrices are symmetric only the half above the diagonal shown needs, in fact, to be found.

All the non-zero coefficients are confined within a *band* or *profile* which can be calculated *a priori* for the nodal connections. Thus in computer programs only the storage of the elements within the upper half of the profile is necessary as shown in Fig. 1.4(*c*).

The third step is the insertion of prescribed boundary conditions into the final assembled matrix as discussed in Section 1.3. This is followed by

The final step of solving the resulting equation system. Here many different methods can be employed, some of which will be discussed in Chapter 24.

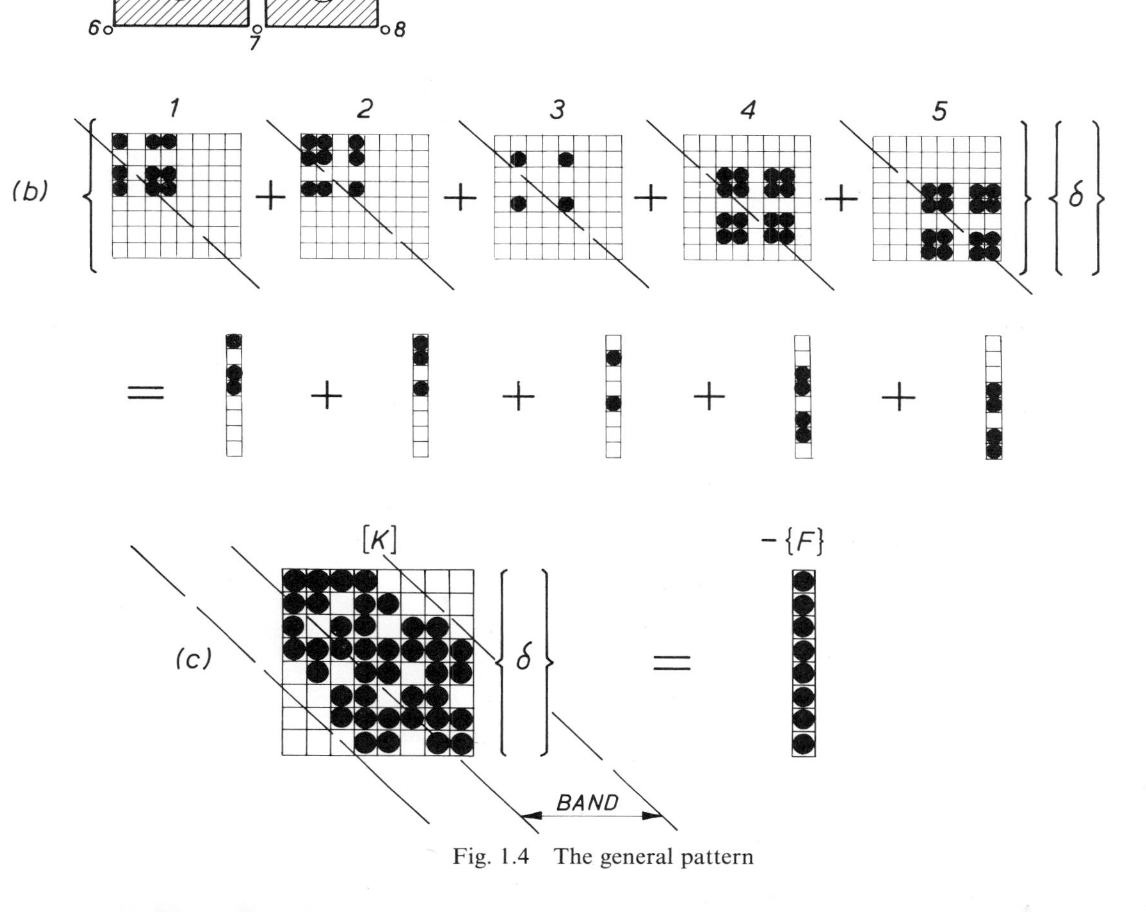

Fig. 1.4 The general pattern

Indeed the general subject of equation solving, though extremely important, is in general beyond the scope of this book.

The final step discussed above will be followed by substitution to obtain stresses, currents or other desired *output* quantities.

All operations involved in the structural or other network analysis are thus of an extremely simple and repetitive kind.

We can now define *the standard discrete system* as one in which such conditions prevail.

1.7 The Standard Discrete System

In the *standard discrete system*, whether it is structural or any other kind, we find that:

(1) A set of discrete parameters, say $\mathbf{a}_i$, can be identified which describes simultaneously the behaviour of each element, e, and of the whole system. We shall call these the *system parameters*.

(2) For each element a set of quantities $\mathbf{q}_i^e$ can be computed in terms of the system parameters $\mathbf{a}_i$. The general function relationship can be non-linear

$$\mathbf{q}_i^e = \mathbf{q}_i^e(\mathbf{a}) \tag{1.20}$$

but in many cases a linear form also exists giving

$$\mathbf{q}_i^e = \mathbf{K}_{i1}^e \mathbf{a}_1 + \mathbf{K}_{i2}^e \mathbf{a}_2 + \cdots + \mathbf{f}_i^e. \tag{1.21}$$

(3) The *system equations* are obtained by a simple addition

$$\mathbf{r}_i = \sum_{e=1}^{m} \mathbf{q}_i^e \tag{1.22}$$

where $\mathbf{r}_i$ are system quantities (often prescribed as zero).

In the linear case this results in a system of equations

$$\mathbf{Ka} + \mathbf{f} = \mathbf{r} \tag{1.23}$$

such that

$$\mathbf{K}_{ij} = \sum_{e=1}^{m} \mathbf{K}_{ij}^e \qquad \mathbf{f}_i = \sum_{e=1}^{m} \mathbf{f}_i^e \tag{1.24}$$

from which the solution for the system variables $\mathbf{a}$ can be found.

The reader will observe that this definition includes the structural and electrical examples already discussed. However, it is broader. In general neither linearity nor symmetry of matrices must exist—although in many problems this will naturally arise. Further, the narrowness of inter-connections existing in usual elements is not essential.

While much further detail could be discussed (we refer the reader to specific books for more exhaustive studies in the structural context[22–24]), we feel that the general exposé given here should suffice for further study of this book.

Only one further matter relating to the change of discrete parameters need be mentioned here. The process of so-called transformation of co-ordinates is vital in many contexts and must be fully understood.

1.8 Transformation of Co-ordinates

It is often convenient to establish the characteristics of an individual element in a co-ordinate system which is different from that in which the external forces and displacements of the assembled structure or system will be measured. A different co-ordinate system may, in fact, be used for every element, to ease the computation. It is a simple matter to transform the co-ordinates of the displacement and force components of Eq. (1.3) to any other co-ordinate system. Clearly, it is necessary to do so before an assembly of the structure can be attempted.

Let the local co-ordinate system in which the element properties have been evaluated be denoted by prime suffix, and the common co-ordinate system necessary for assembly be not annotated. The displacement components can be transformed by a suitable matrix of direction cosines $\mathbf{L}$ as

$$\mathbf{a}' = \mathbf{L}\mathbf{a} \tag{1.25}$$

As the corresponding force components must perform the same amount of work in either system†

$$\mathbf{q}^{\mathrm{T}}\mathbf{a} = \mathbf{q}'^{\mathrm{T}}\mathbf{a}' \tag{1.26}$$

and inserting (1.25) we have

$$\mathbf{q}^{\mathrm{T}}\mathbf{a} = \mathbf{q}'^{\mathrm{T}}\mathbf{L}\mathbf{a}$$

or

$$\mathbf{q} = \mathbf{L}^{\mathrm{T}}\mathbf{q}' \tag{1.27}$$

The set of transformations given by (1.25) and (1.27) is called *contragradient*.

To transform 'stiffnesses' which may be available in local co-ordinates to global ones note that if we write

$$\mathbf{q}' = \mathbf{K}'\mathbf{a}' \tag{1.28}$$

† With $(\ )^{\mathrm{T}}$ standing for transpose of the matrix.

then by (1.27), (1.28), and (1.25)

$$\mathbf{q} = \mathbf{L}^{\mathrm{T}}\mathbf{K}'\mathbf{L}\mathbf{a}$$

or in global co-ordinates

$$\mathbf{K} = \mathbf{L}^{\mathrm{T}}\mathbf{K}'\mathbf{L} \tag{1.29}$$

The reader can verify the usefulness of the above transformations by re-working the sample example of the pin-ended bar. In many complex problems an external constraint of some kind may be imagined enforcing the requirement (1.25) with the number of degrees of freedom of **a** and **a**′ being quite different. Even in such instances the relations (1.26) and (1.27) continue to be valid.

An alternative and more general argument can be applied to many other situations of discrete analysis. We wish to replace a set of parameters **a** in which the system equations have been written by another one related to it by a transformation matrix **T** as

$$\mathbf{a} = \mathbf{T}\mathbf{b}. \tag{1.30}$$

In a linear case the system equations are of the form

$$\mathbf{K}\mathbf{a} = \mathbf{r}-\mathbf{f} \tag{1.31}$$

and on substitution we have

$$\mathbf{K}\mathbf{T}\mathbf{b} = \mathbf{r}-\mathbf{f}. \tag{1.32}$$

The new system can be premultipled simply by $\mathbf{T}^{\mathrm{T}}$ yielding

$$(\mathbf{T}^{\mathrm{T}}\mathbf{K}\mathbf{T})\mathbf{b} = \mathbf{T}^{\mathrm{T}}\mathbf{r}-\mathbf{T}^{\mathrm{T}}\mathbf{f} \tag{1.33}$$

which will preserve the symmetry of equations if matrix **K** is symmetric. However, occasionally the matrix **T** is not square and expression (1.30) represents in fact *an approximation* in which a larger number of parameters **a** is *constrained*. Clearly the system of Eq. (1.32) gives more equations than necessary for a solution of the reduced set of parameters **b**, and the final expression (1.33) presents a reduced system which in some sense approximates to the original one.

We have thus introduced the basic idea of approximation, which will be the subject of subsequent chapters where infinite sets of quantities are reduced to finite sets.

References

1. R. V. SOUTHWELL, *Relaxation Methods in Theoretical Physics*, Clarendon Press, 1946.
2. D. N. DE G. ALLEN, *Relaxation Methods*, McGraw-Hill, 1955.
3. S. H. CRANDALL, *Engineering Analysis*, McGraw-Hill, 1956.

4. B. A. FINLAYSON, *The Method of Weighted Residuals and Variational Principles*, Academic Press, 1972.
5. D. MCHENRY, 'A lattice analogy for the solution of plane stress problems', *J. Inst. Civ. Eng.*, **21**, 59–82, 1943.
6. A. HRENIKOFF, 'Solution of problems in elasticity by the framework method', *J. Appl. Mech.*, **A8**, 169–75, 1941.
7. N. M. NEWMARK, 'Numerical methods of analysis in bars plates and elastic bodies' in *Numerical Methods in Analysis in Engineering* (ed. L. E. Grinter), Macmillan, 1949.
8. J. H. ARGYRIS, *Energy Theorems and Structural Analysis*, Butterworth, 1960 (reprinted from *Aircraft Eng.*, 1954–55).
9. M. J. TURNER, R. W. CLOUGH, H. C. MARTIN and L. J. TOPP, 'Stiffness and deflection analysis of complex structures', *J. Aero. Sci.*, **23**, 805–23, 1956.
10. R. W. CLOUGH, 'The finite element in plane stress analysis', *Proc. 2nd A.S.C.E. Conf. on Electronic Computation*, Pittsburgh, Pa., Sept. 1960.
11. LORD RAYLEIGH (J. W. STRUTT), 'On the theory of resonance', *Trans. Roy. Soc. (London)*, **A161**, 77–118, 1870.
12. W. RITZ, 'Über eine neue Methode zur Lösung gewissen Variations—Probleme der mathematischen Physik', *J. Reine Angew. Math.*, **135**, 1–61, 1909.
13. R. COURANT, 'Variational methods for the solution of problems of equilibrium and vibration', *Bull. Am. Math. Soc.*, **49**, 1–23, 1943.
14. W. PRAGER and J. L. SYNGE, 'Approximation in elasticity based on the concept of function space', *Q. J. Appl. Math.*, **5**, 241–69, 1947.
15. L. F. RICHARDSON, 'The approximate arithmetical solution by finite differences of physical problems', *Trans. Roy. Soc. (London)*, **A210**, 307–57, 1910.
16. H. LIEBMAN, 'Die angenäherte Ermittlung: harmonischen, functionen und konformer Abbildung', *Sitzber. Math. Physik Kl. Bayer Akad. Wiss. München*, **3**, 65–75, 1918.
17. R. S. VARGA, *Matrix Iterative Analysis*, Prentice-Hall, 1962.
18. C. F. GAUSS, See *Carl Friedrich Gauss Werks*, Vol. VII, Göttingen, 1871.
19. B. G. GALERKIN, 'Series solution of some problems of elastic equilibrium of rods and plates' (Russian), *Vestn. Inzh. Tech.*, **19**, 897–908, 1915.
20. C. B. BIEZENO and J. J. KOCH, 'Over een Nieuwe Methode ter Berekening van Vlokke Platen', *Ing. Grav.*, **38**, 25–36, 1923.
21. N. A. PAYNE and B. M. IRONS, Private communication, 1963.
22. R. K. LIVESLEY, *Matrix Methods in Structural Analysis*, 2nd ed., Pergamon Press, 1975.
23. J. S. PRZEMIENIECKI, *Theory of Matrix Structural Analysis*, McGraw-Hill, 1968.
24. H. C. MARTIN, *Introduction to Matrix Methods of Structural Analysis*, McGraw-Hill, 1966.

2. Finite Elements of an Elastic Continuum—Displacement Approach

2.1 Introduction

In many phases of engineering the solution of stress and strain distributions in elastic continua is required. Special cases of such problems may range from two-dimensional plane stress or strain distributions, axi-symmetrical solids, plate bending, and shells, to fully three-dimensional solids. In all cases the number of interconnections between any 'finite element' isolated by some imaginary boundaries and the neighbouring elements is infinite. It is therefore difficult to see at first glance how such problems may be discretized in the same manner as was described in the preceding chapter for simpler structures. The difficulty can be overcome (and the approximation made) in the following manner:

(*a*) The continuum is separated by imaginary lines or surfaces into a number of 'finite elements'.

(*b*) The elements are assumed to be interconnected at a discrete number of nodal points situated on their boundaries. The displacements of these nodal points will be the basic unknown parameters of the problem, just as in the simple, discrete, structural analysis.

(*c*) A set of functions is chosen to define uniquely the state of displacement within each 'finite element' in terms of its nodal displacements.

(*d*) The displacement functions now define uniquely the state of strain within an element in terms of the nodal displacements. These strains, together with any initial strains and the constitutive properties of the material will define the state of stress throughout the element and, hence, also on its boundaries.

(*e*) A system of forces concentrated at the nodes and equilibrating the boundary stresses and any distributed loads is determined, resulting in a stiffness relationship of the form of Eq. (1.3).

Once this stage has been reached the solution procedure can follow the standard discrete system pattern described earlier.

Clearly a series of approximations has been introduced. Firstly, it is not always easy to ensure that the chosen displacement functions will satisfy the requirement of displacement continuity between adjacent elements. Thus, the compatibility condition on such lines may be violated (though within each element it is obviously satisfied due to uniqueness of displacements implied in their continuous representation). Secondly, by concentrating the equivalent forces at the nodes, equilibrium conditions are satisfied in the overall sense only. Local violation of equilibrium conditions within each element and on its boundaries will usually arise.

The choice of element shape and of the form of the displacement function for specific cases leaves much choice to ingenuity and skill of the engineer, and obviously the degree of approximation which can be achieved will much depend on these factors.

The approach outlined here is known as the displacement formulation.[1,2]

So far, the process described is justified only intuitively, but what in fact has been suggested is equivalent to the minimization of the total potential energy of the system in terms of a prescribed displacement field. If this displacement field is defined in a suitable way, then convergence to the correct result must occur. The process is then equivalent to the well-known Ritz procedure. This equivalence will be proved in a later section of this chapter where also a discussion of the necessary convergence criteria will be made.

The recognition of the equivalence of the finite element method with a minimization process was late.[3,2] However, Courant in 1943[4]† and Prager and Synge[5] in 1947 proposed methods in essence identical.

This broader basis of the finite element method allows it to be extended to other continuum problems where a variational formulation is possible. Indeed, general procedures are now available for a finite element discretization of any problem defined by a properly constituted set of differential equations. Such generalizations will be discussed in the next chapter, and throughout the book application to non-structural problems will be made. It will be found that the processes described in this chapter are essentially an application of trial-function and Galerkin-type approximations to a particular case of solid mechanics.

† It appears that Courant had anticipated the essence of the finite element method in general, and of a triangular element in particular, as early as 1923 in a paper entitled *On a convergence principle in calculus of variations,* Kön. Gesellschaft der Wissenschaften zu Göttingen, Nachrichten, Berlin 1923. He states: 'We imagine a mesh of triangles covering the domain . . . the convergence principles remain valid for each triangular domain.'

2.2 Direct Formulation of Finite Element Characteristics

The 'prescriptions' for deriving the characteristics of a 'finite element' of a continuum, which were outlined in general terms, will now be presented in more detailed mathematical form.

It is desirable to obtain results in a general form applicable to any situation, but to avoid introducing conceptual difficulties the general relations will be illustrated with a very simple example of plane stress analysis of a thin slice. In this a division of the region into triangular-shaped elements is used as shown in Fig. 2.1. Relationships of general validity will be underlined. Again, matrix notation will be implied.

2.2.1 *Displacement function.* A typical finite element, e, is defined by nodes, i, j, m, etc., and straight line boundaries. Let the displacements $\mathbf{u}$ at any point within the element be approximated as a column vector, $\hat{\mathbf{u}}$

$$\underline{\underline{\mathbf{u} \approx \hat{\mathbf{u}} = \sum \mathbf{N}_i \mathbf{a}_i^e = [\mathbf{N}_i, \mathbf{N}_j, \ldots] \begin{Bmatrix} \mathbf{a}_i \\ \mathbf{a}_j \\ \vdots \end{Bmatrix}^e = \mathbf{N}\mathbf{a}^e}} \tag{2.1}$$

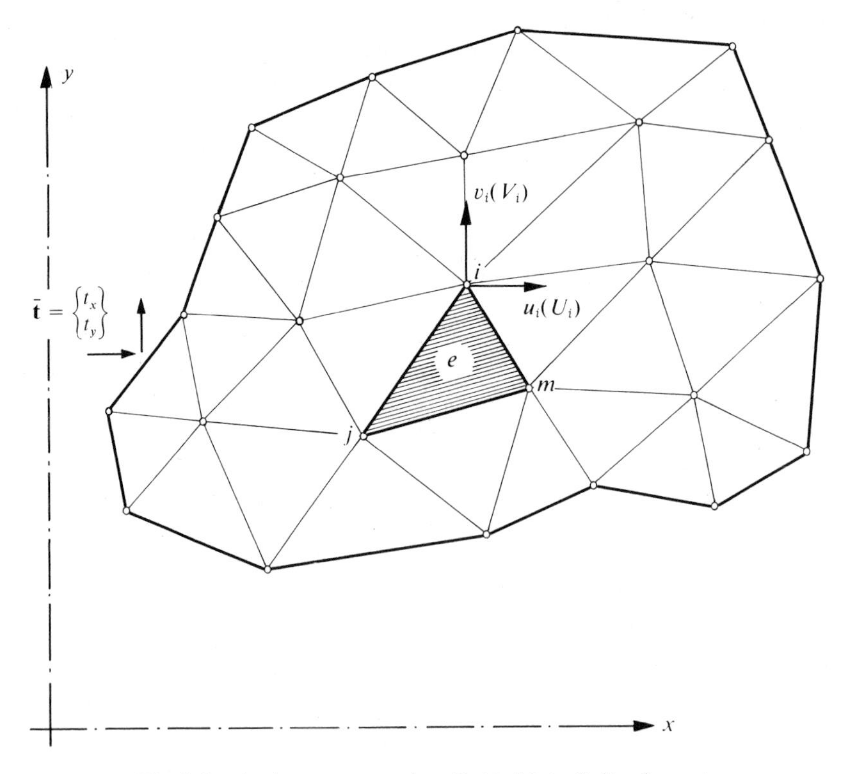

Fig. 2.1 A plane stress region divided into finite elements

in which the components of $\mathbf{N}$ are prescribed functions of position and $\mathbf{a}^e$ represents a listing of nodal displacements for a particular element.

In the case of plane stress for instance

$$\mathbf{u} = \begin{Bmatrix} u(x, y) \\ v(x, y) \end{Bmatrix}$$

represents horizontal and vertical movements of a typical point within the element and

$$\mathbf{a}_i = \begin{Bmatrix} u_i \\ v_i \end{Bmatrix}$$

the corresponding displacements of a node i.

The functions $\mathbf{N}_i$, $\mathbf{N}_j$, $\mathbf{N}_m$ have to be so chosen as to give appropriate nodal displacements when the co-ordinates of the appropriate nodes are inserted in Eq. (2.1). Clearly, in general

$$\mathbf{N}_i(x_i, y_i) = \mathbf{I} \quad \text{(identity matrix)},$$

while

$$\mathbf{N}_i(x_j, y_j) = \mathbf{N}_i(x_m, y_m) = 0 \text{ etc.},$$

which is simply satisfied by suitable linear functions of x and y.

If both the components of displacement are interpolated in an identical manner then we can write

$$\mathbf{N}_i = N_i\mathbf{I}$$

and obtain N_i from Eq. (2.1) by noting that $N_i = 1$ at x_i and y_i but zero at other vertices.

The most obvious linear interpolation in the case of a triangle will yield the shape of N_i of the form shown in Fig. 2.2. Detailed expressions for such a linear interpolation are given in Chapter 4, but at this stage can be readily derived by the reader.

The functions $\mathbf{N}$ will be called *shape functions* and will be seen later to play a paramount role in finite element analysis.

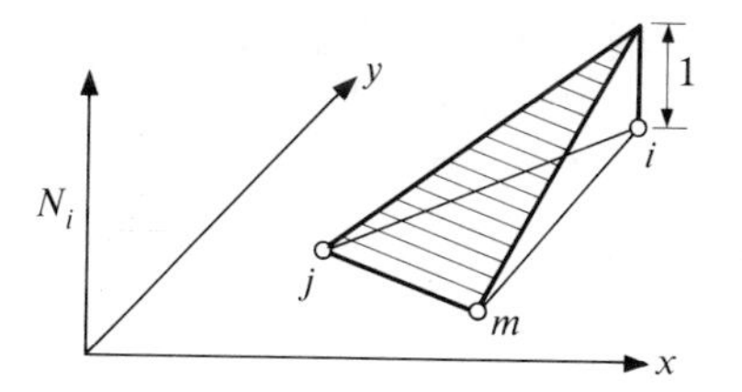

Fig. 2.2

2.2.2 *Strains.* With displacements known at all points within the element the 'strains' at any point can be determined. These will always result in a relationship which can be written in matrix notation as

$$\underline{\underline{\boldsymbol{\varepsilon} = \mathbf{Lu}}} \tag{2.2}$$

where $\mathbf{L}$ is a suitable linear operator. Using Eq. (2.1), the above equation can be approximated as

$$\underline{\underline{\hat{\boldsymbol{\varepsilon}} = \mathbf{Ba}}} \tag{2.3}$$

with

$$\underline{\underline{\mathbf{B} = \mathbf{LN}.}} \tag{2.4}$$

For the plane stress case the relevant strains of interest are those occurring in the plane and are defined in terms of the displacements by well-known relations[6] which define the operator $\mathbf{L}$:

$$\boldsymbol{\varepsilon} = \begin{Bmatrix} \varepsilon_x \\ \varepsilon_y \\ \gamma_{xy} \end{Bmatrix} = \begin{Bmatrix} \dfrac{\partial u}{\partial x} \\ \dfrac{\partial v}{\partial y} \\ \dfrac{\partial u}{\partial y} + \dfrac{\partial v}{\partial x} \end{Bmatrix} = \begin{bmatrix} \dfrac{\partial}{\partial x}, & 0 \\ 0\ , & \dfrac{\partial}{\partial y} \\ \dfrac{\partial}{\partial y}, & \dfrac{\partial}{\partial x} \end{bmatrix} \begin{Bmatrix} u \\ v \end{Bmatrix}$$

With the shape functions $\mathbf{N}_i$, $\mathbf{N}_j$, and $\mathbf{N}_m$ already determined, the matrix $\mathbf{B}$ will easily be obtained. If the linear form of these functions is adopted then, in fact, the strains will be constant throughout the element.

2.2.3 *Stresses.* In general, the material within the element boundaries may be subjected to initial strains such as may be due to temperature changes, shrinkage, crystal growth, and so on. If such strains are denoted by $\boldsymbol{\varepsilon}_0$ then the stresses will be caused by the difference between the actual and initial strains.

In addition it is convenient to assume that at the outset of analysis the body is stressed by some known system at initial residual stresses $\boldsymbol{\sigma}_0$ which for instance could be measured but the prediction of which without the full knowledge of the material's history is impossible. These stresses can simply be added on to the general definition. Thus, assuming general linear elastic behaviour, the relationship between stresses and strains will be linear and of the form

$$\underline{\underline{\boldsymbol{\sigma} = \mathbf{D}(\boldsymbol{\varepsilon} - \boldsymbol{\varepsilon}_0) + \boldsymbol{\sigma}_0}} \tag{2.5}$$

where $\mathbf{D}$ is an elasticity matrix containing the appropriate material properties.

Again, for the particular case of plane stress three components of stress corresponding to the strains already defined have to be considered. These are, in familiar notation

$$\boldsymbol{\sigma} = \begin{Bmatrix} \sigma_x \\ \sigma_y \\ \tau_{xy} \end{Bmatrix}$$

and the **D** matrix may be simply obtained from the usual isotropic stress–strain relationship[6]

$$\varepsilon_x - (\varepsilon_x)_0 = \frac{1}{E}\sigma_x - \frac{\nu}{E}\sigma_y,$$

$$\varepsilon_y - (\varepsilon_y)_0 = -\frac{\nu}{E}\sigma_x + \frac{1}{E}\sigma_y,$$

$$\gamma_{xy} - (\gamma_{xy})_0 = \frac{2(1+\nu)}{E}\tau_{xy},$$

i.e., on solving,

$$\mathbf{D} = \frac{E}{1-\nu^2}\begin{bmatrix} 1 & \nu & 0 \\ \nu & 1 & 0 \\ 0 & 0 & (1-\nu)/2 \end{bmatrix}$$

2.2.4 *Equivalent nodal forces.* Let

$$\mathbf{q}^e = \begin{Bmatrix} \mathbf{q}_i^e \\ \mathbf{q}_j^e \\ \vdots \end{Bmatrix}$$

define the nodal forces which are equivalent statically to the boundary stresses and distributed loads on the element. Each of the forces $\mathbf{q}_i^e$ must contain the same number of components as the corresponding nodal displacement $\mathbf{a}_i$ and be ordered in the appropriate, corresponding directions.

The distributed loads **b** are defined as those acting on a unit volume of material within the element with directions corresponding to those of the displacements **u** at that point.

In the particular case of plane stress the nodal forces are, for instance,

$$\mathbf{q}_i^e = \begin{Bmatrix} U_i \\ V_i \end{Bmatrix}$$

with components U and V corresponding to the directions of u and v displacements, and the distributed load is

$$\mathbf{b} = \begin{Bmatrix} b_x \\ b_y \end{Bmatrix}$$

in which b_x and b_y are the 'body force' components.

To make the nodal forces statically equivalent to the actual boundary stresses and distributed loads, the simplest procedure is to impose an arbitrary (virtual) nodal displacement and to equate the external and internal work done by the various forces and stresses during that displacement.

Let such a virtual displacement be $\delta \mathbf{a}^e$ at the nodes. This results, by Eqs. (2.1) and (2.2), in displacements and strains within the element equal to

$$\delta \mathbf{u} = \mathbf{N}\, \delta \mathbf{a}^e \quad \text{and} \quad \delta \boldsymbol{\varepsilon} = \mathbf{B}\, \delta \mathbf{a}^e \tag{2.6}$$

respectively.

The work done by the nodal forces is equal to the sum of the products of the individual force components and corresponding displacements, i.e., in matrix language

$$\delta \mathbf{a}^{e\mathrm{T}} \mathbf{q}^e \tag{2.7}$$

Similarly, the internal work per unit volume done by the stresses and distributed forces is

$$\delta \boldsymbol{\varepsilon}^{\mathrm{T}} \boldsymbol{\sigma} - \delta \mathbf{u}^{\mathrm{T}} \mathbf{b} \tag{2.8}$$

or†

$$\delta \mathbf{a}^{\mathrm{T}} (\mathbf{B}^{\mathrm{T}} \boldsymbol{\sigma} - \mathbf{N}^{\mathrm{T}} \mathbf{b}) \tag{2.9}$$

Equating the external work with the total internal work obtained by integrating over the volume of the element, V^e, we have

$$\delta \mathbf{a}^{e\mathrm{T}} \mathbf{q}^e = \delta \mathbf{a}^{e\mathrm{T}} \left(\int_{V^e} \mathbf{B}^{\mathrm{T}} \boldsymbol{\sigma}\, d(\mathrm{vol}) - \int_{V^e} \mathbf{N}^{\mathrm{T}} \mathbf{b}\, d(\mathrm{vol}) \right). \tag{2.10}$$

As this relation is valid for any value of the virtual displacement, the equality of the multipliers must exist. Thus

$$\mathbf{q}^e = \int_{V^e} \mathbf{B}^{\mathrm{T}} \boldsymbol{\sigma}\, d(\mathrm{vol}) - \int_{V^e} \mathbf{N}^{\mathrm{T}} \mathbf{b}\, d(\mathrm{vol}). \tag{2.11}$$

This statement is valid quite generally for any stress–strain relations. With the linear law of Eq. (2.5) we can write

$$\mathbf{q}^e = \mathbf{K}^e \mathbf{a}^e + \mathbf{f}^e \tag{2.12}$$

† Note that by rules of matrix algebra for transpose of products

$$(\mathbf{AB})^{\mathrm{T}} = \mathbf{B}^{\mathrm{T}} \mathbf{A}^{\mathrm{T}}.$$

where

$$\mathbf{K}^e = \int_{V^e} \mathbf{B}^{\mathrm{T}}\mathbf{D}\mathbf{B}\,\mathrm{d(vol)} \tag{2.13a}$$

and

$$\mathbf{f}^e = -\int_{V^e} \mathbf{N}^{\mathrm{T}}\mathbf{b}\,\mathrm{d(vol)} - \int_{V^e} \mathbf{B}^{\mathrm{T}}\mathbf{D}\boldsymbol{\varepsilon}_0\,\mathrm{d(vol)} + \int_{V^e} \mathbf{B}^{\mathrm{T}}\boldsymbol{\sigma}_0\,\mathrm{d(vol)}. \tag{2.13b}$$

In the last equation the three terms represent forces due to body forces, initial strain, and initial stress respectively. The relations have the characteristics of the discrete structural elements described in Chapter 1.

If the initial stress system is self-equilibrating, as must be the case with normal residual stresses, then the forces given by the initial stress term of Eq. (2.13b) are identically zero after assembly. Thus frequently evaluation of this force component is omitted. However, if for instance a machine part is manufactured out of a block in which residual stresses are present or if an excavation is made in rock where known tectonic stresses exist a removal of material will cause a force imbalance which results from the above term.

For the particular example of the plane stress triangular element these characteristics will be obtained by the appropriate substitution. It has already been noted that the **B** matrix in that example was not dependent on the co-ordinates, hence the integration will become particularly simple.

The interconnection and solution of the whole assembly of the elements follow the simple structural procedures outlined in Chapter 1. In general, external concentrated forces may exist at the nodes and the matrix

$$\mathbf{r} = \begin{Bmatrix} \mathbf{r}_1 \\ \mathbf{r}_2 \\ \vdots \\ \mathbf{r}_n \end{Bmatrix} \tag{2.14}$$

will be added to the consideration of equilibrium at the nodes.

A note should be added here concerning elements near the boundary. If, at the boundary, displacements are specified, no special problem arises. Consider, however, the boundary as subject to a distributed external loading, say $\bar{\mathbf{t}}$ per unit area. A loading term on the nodes of the element which has a boundary face A^e will now have to be added. By the virtual work consideration, this will simply result in

$$-\int_{A^e} \mathbf{N}^{\mathrm{T}}\bar{\mathbf{t}}\,\mathrm{d(area)} \tag{2.15}$$

with the integration taken over the boundary area of the element. It will be noted that $\bar{\mathbf{t}}$ must have the same number of components as $\mathbf{u}$ for the above expression to be valid.

Such a boundary element is shown again for the special case of plane stress in Fig. 2.1. An integration of this type is sometimes not carried out explicitly. Often by 'physical intuition' the analyst will consider the boundary loading to be represented simply by concentrated loads acting on the boundary nodes and calculate these by direct static procedures. In the particular case discussed the results will be identical.

Once the nodal displacements have been determined by solution of the over all 'structural' type equations, the stresses at any point of the element can be found from the relations in Eqs. (2.2) and (2.3) giving

$$\boldsymbol{\sigma} = \mathbf{DBa}^e - \mathbf{D}\boldsymbol{\varepsilon}_0 + \boldsymbol{\sigma}_0 \tag{2.16}$$

in which the typical terms of the relationship of Eq. (1.4), p. 5, will be immediately recognized, the element stress matrix being

$$\mathbf{S}^e = \mathbf{DB}. \tag{2.17}$$

To this the stresses

$$\boldsymbol{\sigma}_{\varepsilon_0} = -\mathbf{D}\boldsymbol{\varepsilon}_0 \quad \text{and} \quad \boldsymbol{\sigma}_0 \tag{2.18}$$

have to be added.

The absence of the term of stresses due to distributed loading $\boldsymbol{\sigma}_p^e$ needs a comment. It is due to the fact that the internal equilibrium within any element has not been considered, and only overall equilibrium conditions were established.

2.2.5 *Generalized nature of displacements, strains, and stresses.* The meaning of displacements, strains, and stresses in the illustrative case of plane stress was obvious. In many other applications, shown later in this book, this terminology may be applied to other, less obvious, quantities. For example, in considering plate elements the 'displacement' may be characterized by the lateral deflection and the slopes of the plate at a particular point. The 'strains' will then be defined as the curvatures of the middle surface and the 'stresses' as the corresponding internal bending moments.

All the expressions derived here are generally valid provided the sum product of displacement and corresponding load components represents truly the external work done, while that of the 'strain' and corresponding 'stress' components results in the total internal work.

2.3 Generalization to the Whole Region—Internal Nodal Force Concept Abandoned

In the preceding section the virtual work principle was applied to a single element and the concept of equivalent nodal force was retained. The assembly principle thus followed the conventional, direct equilibrium, approach.

The idea of nodal forces contributed by elements replacing the continuous interaction is a conceptual difficulty although it has a considerable appeal to 'practical' engineers and does at times allow an interpretation which otherwise would not be obvious to the more rigorous mathematician. There is, however, no need to consider each element individually and the reasoning of the previous section may be applied directly to the whole continuum.

Equation (2.1) can be interpreted as applying to the whole structure, that is,

$$\mathbf{u} = \overline{\mathbf{N}}\mathbf{a} \tag{2.19}$$

in which $\mathbf{a}$ lists all the nodal points and

$$\overline{\mathbf{N}}_i = \mathbf{N}_i^e \tag{2.20}$$

when the point concerned is within a particular element e and i is a point associated with that element. If point i does not occur within the element

$$\overline{\mathbf{N}}_i = 0. \tag{2.21}$$

Matrix $\overline{\mathbf{B}}$ can be similarly defined and we shall drop the bar superscript considering simply that the shape functions, etc., are defined over the whole region V.

For any virtual displacement $\delta\mathbf{a}$ we can now write the sum of internal and external work for the whole region as

$$-\delta\mathbf{a}^{\mathrm{T}}\mathbf{r} = \int_V \delta\mathbf{u}^{\mathrm{T}}\mathbf{b}\,\mathrm{d}V + \int_A \delta\mathbf{u}^{\mathrm{T}}\bar{\mathbf{t}}\,\mathrm{d}A - \int_V \delta\boldsymbol{\varepsilon}^{\mathrm{T}}\boldsymbol{\sigma}\,\mathrm{d}V \tag{2.22}$$

Substituting Eqs. (2.19), (2.3), and (2.5) we have once again a system of ordinary equations:

$$\underline{\underline{\mathbf{K}\mathbf{a} + \mathbf{f} = \mathbf{r}}} \tag{2.23}$$

where

$$\underline{\underline{\mathbf{K}} = \int_V \mathbf{B}^{\mathrm{T}}\mathbf{D}\mathbf{B}\,\mathrm{d}V} \tag{2.24a}$$

$$\underline{\underline{\mathbf{f} = -\int_V \mathbf{N}^{\mathrm{T}}\mathbf{b}\,\mathrm{d}V - \int_A \mathbf{N}^{\mathrm{T}}\bar{\mathbf{t}}\,\mathrm{d}A - \int_V \mathbf{B}^{\mathrm{T}}\mathbf{D}\boldsymbol{\varepsilon}_0\,\mathrm{d}V + \int_V \mathbf{B}^{\mathrm{T}}\boldsymbol{\sigma}_0\,\mathrm{d}V}} \tag{2.24b}$$

The integrals are taken over the whole volume V and over the whole surface area A on which the tractions are given.

It is immediately obvious from the above that

$$\underline{\underline{\mathbf{K}_{ij} = \sum \mathbf{K}_{ij}^{e}}} \qquad \underline{\underline{\mathbf{f}_i = \sum \mathbf{f}_i^{e}}} \tag{2.25}$$

by virtue of the property of definite integrals requiring that the total be the sum of the parts:

$$\int_V (\quad) \, dV = \sum \int_{V^e} (\quad) \, dV \tag{2.26}$$

The same is obviously true for the surface integrals in Eq. (2.25). We see thus that the 'secret' of the approximation possessing the required behaviour of a 'standard discrete system of Chapter 1' lies simply in the requirement of writing the approximation in an integral form.

The assembly rule as well as the whole derivation has been achieved without involving the concept of 'interelement forces'. In the remainder of this chapter the element superscript will be dropped unless specifically needed. Also no differentiation between element and system shape functions will be made.

However, an important point arises immediately. In considering the virtual work for the whole system (Eq. (2.22)) and equating this to the sum of the element contributions it is implicitly assumed that no discontinuity between adjacent elements develops. If such a discontinuity developed a contribution equal to the work done by the stresses in the separations would have to be added.

Thus the displacement field defined by the shape functions has to be such that only finite strains exist on the interfaces, i.e., displacement continuity must exist to make the general equations valid. More will be said about this necessary condition later.

2.4 Displacement Approach as a Minimization of Total Potential Energy

The principle of virtual displacements used in the previous sections ensured satisfaction of equilibrium conditions within the limits prescribed by the assumed displacement pattern. Only if the virtual work equality for all, arbitrary, variations of displacement was ensured (prescribing only the boundary conditions) would the equilibrium be complete.

If the number of parameters of $\mathbf{a}$ which prescribes the displacement increases without limit then ever closer approximation of all equilibrium conditions can be ensured.

The virtual work principle as written in Eq. (2.22) can be restated in a different form if the virtual quantities $\delta\mathbf{a}$, $\delta\mathbf{u}$, and $\delta\boldsymbol{\varepsilon}$ are considered as *variations* (or differentials) of the real quantities.

Thus, for instance, we can write

$$\delta\left(+\mathbf{a}^{\mathrm{T}}\mathbf{r}+\int_V \mathbf{u}^{\mathrm{T}}\mathbf{b}\,\mathrm{d}V+\int_A \mathbf{u}^{\mathrm{T}}\bar{\mathbf{t}}\,\mathrm{d}A\right) = -\delta W \tag{2.27}$$

for the first three terms of Eq. (2.22), where W *is the potential energy of the external loads*. The above is certainly true if $\mathbf{r}$, $\mathbf{b}$, and $\bar{\mathbf{t}}$ are conservative (or independent of displacement).

The last term of Eq. (2.22) can, for certain materials, be written as

$$\delta U = \int_V \delta\boldsymbol{\varepsilon}^{\mathrm{T}}\boldsymbol{\sigma}\,\mathrm{d}V \tag{2.28}$$

where U is the 'strain energy' of the system. For the elastic, linear material described by Eq. (2.5) the reader can verify that

$$U = \tfrac{1}{2}\int_V \boldsymbol{\varepsilon}^{\mathrm{T}}\mathbf{D}\boldsymbol{\varepsilon}\,\mathrm{d}V+\int_V \boldsymbol{\varepsilon}^{\mathrm{T}}\mathbf{D}\boldsymbol{\varepsilon}_0\,\mathrm{d}V-\int_V \boldsymbol{\varepsilon}^{\mathrm{T}}\boldsymbol{\sigma}_0\,\mathrm{d}V \tag{2.29}$$

will, after differentiation, yield the correct expression providing $\mathbf{D}$ is a symmetric matrix. (This is indeed a necessary condition for single-valued U to exist.)

Thus instead of Eq. (2.22) we can write simply

$$\delta(U+W) = \delta(\Pi) = 0 \tag{2.30}$$

in which the quantity Π is called the *total potential energy*.

The above statement means that for equilibrium to be ensured the *total potential energy must be stationary* for variations of admissible displacements. The finite element equations derived in the previous section (Eqs. (2.23)–(2.25)) are simply the statements of this variation with respect to displacements constrained to a finite number of parameters $\mathbf{a}$ and could be written as

$$\frac{\partial\Pi}{\partial\mathbf{a}} = \begin{Bmatrix} \dfrac{\partial\Pi}{\partial\mathbf{a}_1} \\ \dfrac{\partial\Pi}{\partial\mathbf{a}_2} \\ \vdots \end{Bmatrix} = 0. \tag{2.31}$$

It can be shown that in elastic situations the total potential energy is not only stationary but is a minimum.[7] *Thus the finite element process seeks such a minimum within the constraint of an assumed displacement pattern.*

The greater the degrees of freedom, the more closely will the solution approximate to the true one ensuring complete equilibrium, providing the true displacement can, in the limit, be approximated. The necessary convergence conditions for the finite element process could thus be derived. Discussion of these will, however, be deferred to a later section.

It is of interest to note that if true equilibrium requires an absolute minimum of the total potential energy, Π, an approximate finite element solution by displacement approach will always provide an approximate Π greater than the correct one. *Thus a bound on the value of the total potential energy is always achieved.*

If the function Π could be specified, *a priori*, then the finite element equations could be derived directly by differentiation specified by Eq. (2.31).

The well-known Rayleigh[8]–Ritz[9] process of approximation frequently used in elastic analysis uses precisely this approach. The total potential energy expression is formulated and the displacement pattern is assumed to vary with a finite set of undetermined parameters. A set of simultaneous equations minimizing the total potential energy with respect to these parameters is set up. Thus the finite element process as described so far is identically the Rayleigh–Ritz procedure. The difference is only in the manner in which the displacements are prescribed. In the Ritz process traditionally used these are usually given by expressions valid throughout the whole region thus leading to simultaneous equations in which no banding occurs and the coefficient matrix is full. In the finite element process this specification is usually piecewise, each nodal parameter influencing only adjacent elements, and thus a sparse and usually banded matrix of coefficients is found.

By its nature the conventional Ritz process is limited to relatively simple geometrical shapes of the total region while this limitation only occurs in finite element analysis in the element itself. Thus complex, realistic, configurations can be assembled from relatively simple element shapes.

A further difference in kind is in the usual association of the undetermined parameter with a particular nodal displacement. This allows a simple physical interpretation invaluable to an engineer. Doubtless much of the popularity of the finite element process is due to this fact.

2.5 **Convergence Criteria**

The assumed shape functions limit the infinite degrees of freedom of the system, and the true minimum of the energy may never be reached, irrespective of the fineness of subdivision. To ensure convergence to the correct result certain simple requirements have to be satisfied. Obviously, for instance, the displacement function should be able to represent the true displacement distribution as closely as possible. It will be found that this is not so if the chosen functions are such that straining of an element is possible when this is subject to rigid body displacements. Thus, the first criterion the displacement function must obey is as follows.

Criterion 1. The displacement function chosen should be such that it does not permit straining of an element to occur when the nodal displacements are caused by a rigid body displacement.

This self-evident condition can be violated easily if certain types of function are used; care must therefore be taken in the choice of displacement functions.

A second criterion stems from the same requirements. Clearly, as elements get smaller nearly constant strain conditions will prevail in them. If, in fact, constant strain conditions exist, it is most desirable for good accuracy that a finite size element is able to reproduce these exactly. It is possible to formulate functions which satisfy the first criterion but at the same time require a strain variation throughout the element when the nodal displacements are compatible with a constant strain solution. Such functions will, in general, not show a good convergence to an accurate solution and cannot, even in the limit, represent the true strain distribution. The second can therefore be formulated as

Criterion 2. The displacement function has to be of such a form that if nodal displacements are compatible with a constant strain condition such constant strain will in fact be obtained. (In this context again a generalized 'strain' definition is implied.)

It will be observed that Criterion 2 in fact incorporates the requirement of Criterion 1, as rigid body displacements are a particular case of constant strain—with a value of zero. This criterion was first stated by Bazeley *et al.*[10] in 1965. Strictly both criteria need only be satisfied in the limit as the size of the element tends to zero. However, the imposition of these criteria on elements of finite size leads to improved accuracy.

Lastly, as already mentioned in section 2.3, it is implicitly assumed in the derivation presented that no contribution to the virtual work arises at element interfaces. It therefore appears necessary that the following criterion be included:

Criterion 3. The displacement functions should be so chosen that the strains at the interface between elements are finite (even though indeterminate).

This criterion implies a certain continuity of displacements between elements. In the case of strains being defined by first derivatives, as in the plane example quoted here, the displacements only have to be continuous. If, however, as in the plate and shell problems, the 'strains' are defined by second derivatives of deflections, first derivatives of these have also to be continuous.[2]

The above criteria are mathematically included in a statement of 'functional completeness' and the reader is referred for full mathematical discussion elsewhere.[11–16] The 'heuristic' proof of the convergence requirements given here is sufficient for practical purposes in all but the most pathological cases and we shall generalize all of the above criteria in Chapter 3.

2.6 Discretization Error and Convergence Rate

In the foregoing section we have assumed that the approximation to the displacement as represented by Eq. (2.1) will yield the exact solution in the limit at the size h, if elements decrease. The arguments for this are simple: as the expansion is capable, in the limit, of reproducing exactly any displacement form conceivable in the continuum, then as the solution of each approximation is unique it must approach in the limit of $h \to 0$, the unique exact solution. In some cases the exact solution is indeed obtained with a finite number of subdivisions (or even with one element only) if the *polynomial expansion used in that element can fit exactly the correct solution*. Thus, for instance, if the exact solution is of the form of a quadratic polynomial *and* the shape functions include all the polynomials of that order, the approximation will yield the exact answer.

The last argument helps in determining the order of convergence of the finite element procedure as the exact solution can always be expanded in the vicinity of any point (or node) i as a polynomial

$$\mathbf{u} = \mathbf{u}_i + \left(\frac{\partial \mathbf{u}}{\partial x}\right)_i x + \left(\frac{\partial \mathbf{u}_i}{\partial y}\right)_i y + \cdots \tag{2.32}$$

If within an element of 'size' h a polynomial expansion of order p is employed, this can fit locally the Taylor expansion up to that order and, as x and y are of the order of magnitude h, the error in $\mathbf{u}$ will be of the order $O(h^{p+1})$. Thus, for instance, in the case of the plane elasticity problem discussed, we used a linear expansion and $p = 1$. We should therefore expect a *convergence* rate of order $O(h^2)$, i.e., the error in displacement being reduced to $\frac{1}{4}$ for a halving of the mesh spacing.

By a similar argument the strains (or stresses) which are given by mth derivatives of displacement should converge with an error of $O(h^{p+1-m})$, i.e., as $O(h)$ in the example quoted, where $m = 1$. The strain energy being given by the square of stresses will show an error of $O(h^{2(p+1-m)})$ or $O(h^2)$ in the plane stress example.

The arguments given here are perhaps a trifle 'heuristic' from a mathematical viewpoint—they are, however, true[16] and give correctly the orders of convergence. Much more elaborate mathematical analyses have been frequently carried out attempting not only to determine the order of convergence but also error bounds. None of these has to date proved particularly useful, as generally these are given in terms of quantities unknown *a priori*. Further, the mere determination of the order of convergence often suffices to extrapolate the solution to the correct result. Thus, for instance, if the displacement converges at $O(h^2)$ and we have two approximate solutions u^1 and u^2 obtained with meshes of size h and $h/2$, we can write with u being the exact solution

$$\frac{u^1 - u}{u^2 - u} = \frac{O(h^2)}{O(h/2)^2} = 4 \tag{2.33}$$

From the above an (almost) exact solution u can be predicted. This type of extrapolation was first introduced by Richardson[17] and is of use if convergence is monotonic.

The discretization error is not the only one possible in the finite element computation. In addition to obvious mistakes which can occur when using computers, errors due to *round-off* are always possible. With the computer operating on numbers rounded off to a finite number of digits, a reduction of accuracy occurs every time differences between 'like' numbers are being formed. In the process of equation-solving many subtractions are necessary and accuracy decreases. Problems of matrix conditioning, etc., enter here and the user of the finite element method must at all times be aware of accuracy limitations which simply do not allow the exact solution ever to be obtained. Fortunately in many computations, by using modern machines which carry a large number of significant digits, these errors are often small!

2.7 Displacement Functions with Discontinuity between Elements—Non-conforming Elements and the Patch Test

In some cases considerable difficulty is experienced in finding displacement functions for an element which will automatically be continuous along the whole interface between adjacent elements.

As already pointed out, the discontinuity of displacement will cause infinite strains at the interfaces—a factor ignored in the formulation presented because the energy contribution is limited to the elements themselves.

However, if, in the limit, as the size of the subdivision decreases continuity is restored, then the formulation already obtained will still tend to the correct answer. This condition is always reached if

(*a*) a constant strain condition automatically ensures displacement continuity,
(*b*) the constant strain criterion of the previous section is satisfied.

To test that such continuity is achieved for any mesh configuration when using such *non-conforming* elements it is necessary to impose, on an arbitrary patch of elements, nodal displacements corresponding to any state of constant strain. *If nodal equilibrium is simultaneously achieved without the imposition of external, nodal, forces and if a state of constant stress is obtained, then clearly no external work has been lost through inter-element discontinuity.*

Elements which pass such a *patch test* will converge, and indeed at times non-conforming elements will show a superior performance to conforming elements.

The patch test was first introduced by Irons[10] and has since been demonstrated to give a sufficient condition for convergence.[16, 18, 19]

We shall return to the subject of non-conforming elements in Chapter 11. In some of the problems dealt with in this book 'discontinuous' displacement functions of this type will be used with success. In such elements however, bounds on the functional will no longer be available.

2.8 **Bound on Strain Energy in a Displacement Formulation**

While the approximation obtained by the finite element displacement approach always overestimates the true value of Π, the total potential energy (the absolute minimum corresponding to the exact solution), this is not directly useful in practice. It is, however, possible to obtain a more useful limit in special cases.

Consider in particular the problem in which no 'initial' strains or initial stresses exist. Now by principle of energy conservation the strain energy will be equal to the work done by the external loads which increase uniformly from zero.[20] This work done is equal to $-\frac{1}{2}W$ where W is the potential energy of the loads.

Thus

$$U+\tfrac{1}{2}W = 0 \tag{2.34}$$

or

$$\Pi = U+W = -U \tag{2.35}$$

whether an exact or approximate displacement field is assumed.

Thus in the above case the approximate solution always *underestimates* the value of U and a displacement solution is frequently referred to as the *lower bound solution.*

If only one external concentrated load R is present the strain energy bound immediately informs us that the deflection under this load has been underestimated (as $U = -\frac{1}{2}W = \frac{1}{2}\mathbf{r}^T\mathbf{a}$). In more complex loading cases the usefulness of this bound is limited as neither deflections nor stresses, i.e., the quantities of real engineering interest, can be bounded.

It is important to remember that this bound on strain energy is only valid in the absence of any initial stresses or strains.

The expression for U in this case can be obtained from Eq. (2.29) as

$$U = \frac{1}{2}\int_V \boldsymbol{\varepsilon}\mathbf{D}\boldsymbol{\varepsilon}^T \mathrm{d(vol)} \tag{2.36}$$

which becomes by Eq. (2.2) simply

$$U = \frac{1}{2}\mathbf{a}^T\left[\int_V \mathbf{B}^T\mathbf{D}\mathbf{B}\,\mathrm{d(vol)}\right]\mathbf{a} = \frac{1}{2}\mathbf{a}^T\mathbf{K}\mathbf{a} \tag{2.37}$$

a 'quadratic' matrix form in which $\mathbf{K}$ is the 'stiffness' matrix previously discussed.

The above energy expression is always positive from physical considerations. It follows therefore that the matrix $\mathbf{K}$ occurring in all the finite element assemblies is not only symmetric but is 'positive definite' (a property defined in fact by the requirements that the quadratic form should always be greater than or equal to zero).

This feature is of importance when the numerical solution of the simultaneous equations involved is considered as simplifications arise in the case of 'symmetric positive definite' equations.

2.9 Direct Minimization

The fact that the finite element approximation reduces to the problem of minimizing the total potential energy Π defined in terms of a finite number of nodal parameters led us to formulation of the simultaneous set of equations given symbolically by Eq. (2.31). This is the most usual and convenient approach especially in linear solutions but other search procedures, now well developed in the field of optimization, could be used to estimate the lowest value of Π. In this text we shall continue with the simultaneous equation process but the interested reader could well bear the alternative possibilities in mind.[21, 22]

2.10 An Example

The concepts discussed and the general formulation cited are a little abstract and the reader may at this stage seek to test his grasp of the nature of the approximations derived. While detailed computations of a two-dimensional element system are best left to the computer, we can perform a simple hand calculation on a one-dimensional finite element of a beam.

P

k i j

x w

L L L

i j

$$\mathbf{d}_i = \begin{Bmatrix} w_i \\ w_{xi} \end{Bmatrix} \equiv \begin{pmatrix} w_i \\ \theta_i \end{pmatrix} \qquad \mathbf{d}_j = \begin{Bmatrix} w_j \\ w_{xj} \end{Bmatrix} \equiv \begin{pmatrix} w_j \\ \theta_j \end{pmatrix}$$

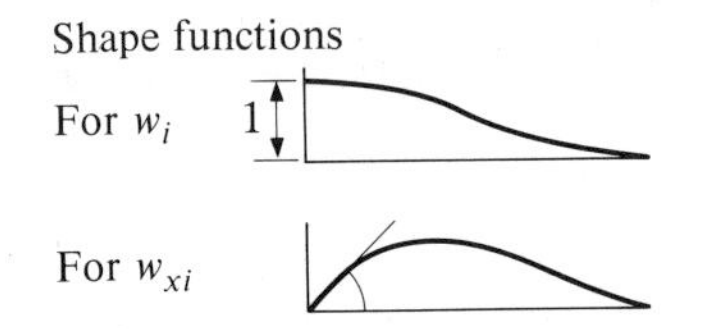

Fig. 2.3 A beam element and its shape functions

Indeed, this example will allow us to introduce the concept of generalized stresses and strains in a simple manner.

Consider a beam shown in Fig. 2.3. The generalized 'strain' here is the curvature. Thus we have

$$\boldsymbol{\varepsilon} \equiv \kappa = -\frac{\mathrm{d}^2 w}{\mathrm{d}x^2}$$

where w is the deflection, which is the basic unknown. The generalized stress (in the absence of shear deformation) will be the bending moment M, which is related to the 'strain' as

$$\boldsymbol{\sigma} \equiv M = -EI\frac{\mathrm{d}^2 w}{\mathrm{d}x^2}.$$

Thus immediately we have, using the general notation of previous sections,

$$\mathbf{D} \equiv EI.$$

If the displacement w is discretized we can write

$$w = \mathbf{Na}$$

for the whole system or, for an individual element, ij.

In this example the strains are expressed as the second derivatives of displacement and it is necessary to ensure that both w and its slope

$$w_x \equiv \frac{\mathrm{d}w}{\mathrm{d}x} = \theta$$

be continuous between elements. This is easily accomplished if the nodal parameters are taken as the values of w and the slope, w_x. Thus,

$$\mathbf{a}_i = \begin{Bmatrix} w \\ w_x \end{Bmatrix}_i = \begin{Bmatrix} w_i \\ \theta_i \end{Bmatrix}.$$

The shape functions now will be derived. If we accept that in an element two nodes (i.e., four variables) define the deflected shape we can assume this to be given by a cubic

$$w = \alpha_1 + \alpha_2 x + \alpha_3 x^2 + \alpha_4 x^3.$$

This will define the shape functions corresponding to w_i and w_{xi} by taking for each a cubic giving unity for the appropriate points ($x = 0;\ L$) and zero for other quantities, as shown in Fig. 2.3.

The expressions for the shape function can be written for the element shown as

$$\begin{matrix} \mathbf{N}_i = \\ \mathbf{N}_j = \end{matrix} \begin{bmatrix} 1-3(x/L)^2+2(x/L)^3, & L(x/L-2(x/L)^2+(x/L)^3) \\ 3(x/L)^2-2(x/L)^3, & L(-(x/L)^2+(x/L)^3) \end{bmatrix}$$

Immediately we can write

$$\mathbf{B}_i = -\frac{d^2}{dx^2}\mathbf{N}_i = [6-12(x/L), (4-6(x/L))L]/L^2$$

$$\mathbf{B}_j = -\frac{d^2}{dx^2}\mathbf{N}_j[-6+12(x/L), (2-6(x/L))L]/L^2$$

and the stiffness matrices for the element can be written as

$$\mathbf{K}_{ij}^e = \int_0^L \mathbf{B}_i^{\mathrm{T}}\, EI\, \mathbf{B}_j\, dx$$

We shall leave the detailed calculation of this and the 'forces' corresponding to a uniformly distributed load p (assumed constant) on ij and zero elsewhere to the reader. He will observe that the finally assembled equations for a node i are of the form linking three nodal displacements i, j, k. Explicitly these equations are for elements of equal length L.

$$EI\begin{bmatrix} -12/L^3, & -6/L^2 \\ 6/L^2, & 2/L \end{bmatrix}\begin{Bmatrix} w_k \\ \theta_k \end{Bmatrix} + EI\begin{bmatrix} 24/L^3, & 0 \\ 0, & 8/L \end{bmatrix}\begin{Bmatrix} w_i \\ \theta_i \end{Bmatrix} +$$

$$+ EI\begin{bmatrix} -12/L^3, & +6/L^2 \\ -6/L^2, & 2/L \end{bmatrix}\begin{Bmatrix} w_j \\ \theta_j \end{Bmatrix} + \begin{Bmatrix} pL/2 \\ -pL^2/12 \end{Bmatrix} = 0$$

It is of interest to compare these with the *exact* form represented by the so-called 'slope–deflection' equations which can be found in standard texts.

Here it will be found that the finite element approximation has achieved the exact solution as the cubic polynomial was capable of representing it for a uniform load. For other distributed loads it is easy to show that the difference between the approximation and the exact equations decreases as the length of elements tends to zero.

2.11 Concluding Remarks

The 'displacement' approach to analysis of elastic solids is still undoubtedly the most popular and easily understood procedure. In many of the following chapters we shall use the general formulae developed here in the context of linear elastic (Chapters 4, 5, 6, 10, 13, 14) or non-linear analysis (Chapters 18, 19), the main variants being the definitions of the stresses, generalized strains, and other associated quantities. It is thus convenient to summarize the essential formulae, and this is done in Appendix 2.

In the next chapter we shall show that the procedures developed here are but a particular case of finite element discretization applied to the governing equilibrium equations written in terms of displacements. Clearly, alternative starting points are possible. Some of these will be mentioned in Chapter 12.

References

1. R. W. CLOUGH, 'The finite element in plane stress analysis', *Proc. 2nd A.S.C.E. Conf. on Electronic Computation*, Pittsburgh Pa., Sept. 1960.
2. R. W. CLOUGH, 'The finite element method in structural mechanics', Chapter 7 of *Stress Analysis* (eds. O. C. Zienkiewicz and G. S. Holister), Wiley, 1965.
3. J. SZMELTER, 'The energy method of networks of arbitrary shape in problems of the theory of elasticity', *Proc. I.U.T.A.M., Symposium on Non-Homogeneity in Elasticity and Plasticity* (ed. W. Olszak), Pergamon Press, 1959.
4. R. COURANT, 'Variational methods for the solution of problems of equilibrium and vibration', *Bull. Am. Math. Soc.*, **49**, 1–23, 1943.
5. W. PRAGER and J. L. SYNGE, 'Approximation in elasticity based on the concept of function space', *Quart. Appl. Math.*, **5**, 241–69, 1947.
6. S. TIMOSHENKO and J. N. GOODIER, *Theory of Elasticity*, 2nd ed., McGraw-Hill, 1951.
7. K. WASHIZU, *Variational Methods in Elasticity and Plasticity*, 2nd ed., Pergamon Press, 1975.
8. J. W. STRUTT (Lord Rayleigh), 'On the theory of resonance', *Trans. Roy. Soc. (London)*, **A161**, 77–118, 1870.
9. W. RITZ, 'Über eine neue Methode zur Lösung gewissen Variations—Probleme der mathematischen Physik', *J. Reine angew. Math.*, **135**, 1–61, 1909.
10. G. P. BAZELEY, Y. K. CHEUNG, B. M. IRONS and O. C. ZIENKIEWICZ, 'Triangular elements in bending—conforming and non-conforming solutions', *Proc. Conf. Matrix Methods in Structural Mechanics*, Air Force Inst. Tech., Wright-Patterson A.F. Base, Ohio, 1965.

11. S. C. MIKHLIN, *The Problem of the Minimum of a Quadratic Functional,* Holden-Day, 1966.
12. M. W. JOHNSON and R. W. MCLAY, 'Convergence of the finite element method in the theory of elasticity', *J. Appl. Mech., Trans. Am. Soc. Mech. Eng.*, 274–8, 1968.
13. S. W. KEY, *A convergence investigation of the direct stiffness method*, Ph.D. Thesis, Univ. of Washington, 1966.
14. T. H. H. PIAN and PING TONG, 'The convergence of finite element method in solving linear elastic problems', *Int. J. Solids Struct.*, **3**, 865–80, 1967.
15. E. R. DE ARRANTES OLIVEIRA, 'Theoretical foundations of the finite element method', *Int. J. Solids Struct.*, **4**, 929–52, 1968.
16. G. STRANG and G. J. FIX, *An Analysis of the Finite Element Method*, p. 106, Prentice-Hall, 1973.
17. L. F. RICHARDSON, 'The approximate arithmetical solution by finite differences of physical problems', *Trans. Roy. Soc.* (*London*), **A210**, 307–57, 1910.
18. B. M. IRONS and A. RAZZAQUE, 'Experience with the patch test' in *Mathematical Foundations of the Finite Element Method*', pp. 557–87 (ed. A. R. Aziz), Academic Press, 1972.
19. B. FRAEIJS DE VEUBEKE, 'Variational principles and the patch test', *Int. J. Num. Meth. Eng.*, **8**, 783–801, 1974.
20. B. FRAEIJS DE VEUBEKE, 'Displacement and equilibrium models in the finite element method', Chapter 9 of *Stress Analysis* (ed. O. C. Zienkiewicz and G. S. Holister), Wiley, 1965.
21. R. L. FOX and E. L. STANTON, 'Developments in structural analysis by direct energy minimization', *J.A.I.A.A.*, **6**, 1036–44, 1968.
22. F. K. BOGNER, R. H. MALLETT, M. D. MINICH and L. A. SCHMIT, 'Development and evaluation of energy search methods in non-linear structural analysis', *Proc. Conf. Matrix Methods in Structural Mechanics*, Air Force Inst. Tech., Wright-Patterson A.F. Base, Ohio, 1965.

3. Generalization of the Finite Element Concepts—Weighted Residual and Variational Approaches

3.1 **Introduction**

We have so far dealt with one possible approach to the approximate solution of the particular problem of linear elasticity. Many other continuum problems arise in engineering and physics and usually these problems are posed by appropriate differential equations and boundary conditions to be imposed on the unknown function or functions. It is the object of this chapter to show that all such problems can be dealt with by the finite element method.

Posing the problem to be solved in most general terms we find that we seek an unknown function **u** such that it satisfies a certain differential equation set

$$\mathbf{A}(\mathbf{u}) = \begin{Bmatrix} A_1(\mathbf{u}) \\ A_2(\mathbf{u}) \\ \vdots \end{Bmatrix} = 0 \qquad (3.1)$$

in a 'domain' (volume, area, etc.) Ω, Fig. 3.1, together with certain boundary conditions

$$\mathbf{B}(\mathbf{u}) = \begin{Bmatrix} B_1(\mathbf{u}) \\ B_2(\mathbf{u}) \\ \vdots \end{Bmatrix} = 0 \qquad (3.2)$$

on the boundaries Γ of the domain (Fig. 3.1).

The function sought may be a scalar quantity or may represent a vector of several variables. Similarly, the differential equation may be a single one or a set of simultaneous equations. It is for this reason that we have resorted in the above to matrix notation.

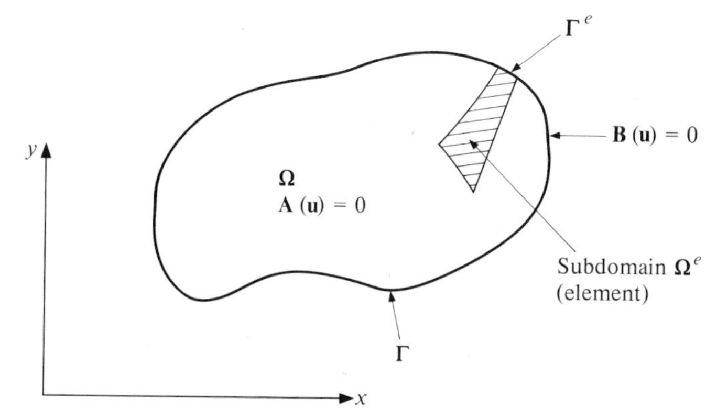

Fig. 3.1 Problem domain Ω and Γ boundary

The finite element process, being one of approximation, will seek the solution in the approximate form

$$\mathbf{u} \approx \hat{\mathbf{u}} = \sum_{1}^{r} \mathbf{N}_i \mathbf{a}_i = \mathbf{N}\mathbf{a} \tag{3 3}$$

where $\mathbf{N}_i$ are shape functions prescribed in terms of independent variables (such as the coordinates x, y, etc.) and all or some of the parameters $\mathbf{a}_i$ are unknown.

We have seen that precisely the same form of approximation was used in the displacement approach to elasticity problems in the last chapter. We also noted there that (*a*) the shape functions were usually defined locally for elements or subdomains and (*b*) the properties of discrete systems were recovered if the approximating equations were cast in *an integral form*, viz. Eqs. (2.22)–(2.26).

With this object in mind we shall seek to cast the equation from which the unknown parameters $\mathbf{a}_i$ are to be obtained in an integral form

$$\int_{\Omega} \mathbf{G}_j(\hat{\mathbf{u}})\, d\Omega + \int_{\Gamma} \mathbf{g}_j(\hat{\mathbf{u}})\, d\Gamma = 0 \quad (j = 1-n). \tag{3.4}$$

in which $\mathbf{G}_j$ and $\mathbf{g}_j$ prescribe known functions or operators.

These integral forms will permit the approximation to be obtained element by element and an assembly to be achieved by the use of procedures developed for *standard discrete systems* in Chapter 1, as, providing the functions $\mathbf{G}_j$ and $\mathbf{g}_j$ are integrable, we have

$$\int_{\Omega} \mathbf{G}_j\, d\Omega + \int_{\Gamma} \mathbf{g}_j\, d\Gamma = \sum_{e=1}^{m} \left(\int_{\Omega^e} \mathbf{G}_j\, d\Omega + \int_{\Gamma^e} \mathbf{g}_j\, d\Gamma \right) \tag{3.5}$$

where Ω^e is the domain of each element and Γ^e its part of boundary.

Two distinct procedures are available for obtaining the approximation in such integral forms. The first is the *method of weighted residuals*, the second the determination of *variational functionals* for which stationarity is sought. We shall deal with both approaches in turn.

If the differential equations are linear, i.e., if we can write (3.1) and (3.2) as

$$\mathbf{A}(\mathbf{u}) \equiv \mathbf{L}\mathbf{u}+\mathbf{p} = 0 \quad \text{in } \Omega \tag{3.6}$$

$$\mathbf{B}(\mathbf{u}) \equiv \mathbf{M}\mathbf{u}+\mathbf{t} = 0 \quad \text{on } \Gamma \tag{3.7}$$

then the approximating equation system (3.4) will yield a set of linear equations of the form

$$\mathbf{K}\mathbf{a}+\mathbf{f} = 0 \tag{3.8}$$

with

$$\mathbf{K}_{ij} = \sum_{e=1}^{m} \mathbf{K}_{ij}^{e}; \qquad \mathbf{f}_i = \sum_{e=1}^{m} \mathbf{f}_i^{e} \tag{3.9}$$

The reader not used to abstraction may well by now be lost in the meaning of the various terms. We shall introduce here some typical sets of differential equations for which we will seek solutions (and which may make the problems a little more definite).

Example 1. Steady-state heat conduction equations in a two-dimensional domain

$$A(\phi) = \frac{\partial}{\partial x}\left(k\frac{\partial \phi}{\partial x}\right)+\frac{\partial}{\partial y}\left(k\frac{\partial \phi}{\partial y}\right)+Q = 0$$

$$\begin{aligned} B(\phi) &= \phi-\bar{\phi} = 0 \quad \text{on } \Gamma_\phi \\ &= k\frac{\partial \phi}{\partial n}-\bar{q} = 0 \quad \text{on } \Gamma_q \end{aligned} \tag{3.10}$$

where (with the n direction normal to Γ) $\phi \equiv \mathbf{u}$ indicates temperature, k is conductivity, and $\bar{\phi}$ and $\bar{q}$ are the prescribed values of temperature and heat flow on the boundaries.

In the above problem k and Q can be functions of position and, if the problem is non-linear, of ϕ or its derivatives.

Example 2. Steady-state heat conduction–convection equation in two dimensions

$$\begin{aligned} A(\phi) &= \frac{\partial}{\partial x}\left(k\frac{\partial \phi}{\partial y}\right)+\frac{\partial}{\partial y}\left(k\frac{\partial \phi}{\partial y}\right)+ \\ &+ u\frac{\partial \phi}{\partial x}+v\frac{\partial \phi}{\partial y}+Q = 0 \end{aligned} \tag{3.11}$$

with boundary conditions as in the first example. Here u and v are known functions of position and represent velocities of the fluid through which heat transfer occurs.

Example 3. A system of three equations equivalent to problem of example 1 with $Q = 0$

$$\mathbf{A}(\mathbf{u}) = \begin{Bmatrix} \dfrac{\partial}{\partial x}(kq_x)+\dfrac{\partial}{\partial y}(kq_y) \\ q_x-\dfrac{\partial \phi}{\partial x} \\ q_y-\dfrac{\partial \phi}{\partial y} \end{Bmatrix} = 0 \tag{3.12}$$

in Ω and

$$\mathbf{B}(\mathbf{u}) = \begin{matrix} \phi-\bar{\phi} = 0 & \text{on } \Gamma_\phi \\ \left.\begin{matrix} q_x-\bar{q}_x = 0 \\ q_y-\bar{q}_y = 0 \end{matrix}\right\} & \text{on } \Gamma_q. \end{matrix}$$

Here the unknown function vector $\mathbf{u}$ corresponds to the set

$$\mathbf{u} = \begin{Bmatrix} \phi \\ q_x \\ q_y \end{Bmatrix}.$$

In Chapter 17 we shall return to detailed examples of the above field and other examples will be introduced throughout the book. The three sets of problems will, however, be useful in their full form or reduced to one dimension (by suppressing the y variation) to illustrate the various approaches used in this chapter.

WEIGHTED RESIDUAL METHODS

3.2 Integral or 'Weak' Statements Equivalent to the Differential Equations

As the set of differential equations (Eqs. (3.1) has to be zero at each point of the domain Ω, it follows that

$$\int \mathbf{v}^{\mathrm{T}}\mathbf{A}(\mathbf{u})\,\mathrm{d}\Omega \equiv \int (v_1 A_1(\mathbf{u})+v_2 A_2(\mathbf{u})+\cdots)\,\mathrm{d}\Omega \equiv 0 \tag{3.13}$$

where

$$\mathbf{v} = \begin{Bmatrix} v_1 \\ v_2 \\ \vdots \end{Bmatrix} \tag{3.14}$$

is a set of arbitrary functions equal in number to the number of equations (or components of $\mathbf{u}$) involved.

The statement is however more powerful. *We can assert that if (3.13) is satisfied for any* $\mathbf{v}$ *then the differential equations (3.1) must be satisfied at all points of the domain.* The proof of the validity of this statement is obvious if one considers the possibility that $\mathbf{A}(\mathbf{u}) \neq 0$ at any point or part of the domain. Immediately, a function $\mathbf{v}$ can be found which makes the integral of (3.13) non-zero, and hence the point is proved.

If the boundary condition (3.12) are to be simultaneously satisfied, then either we ensure such satisfaction by the choice of a function $\hat{\mathbf{u}}$ or require that

$$\int_{\Gamma} \mathbf{v}^{\mathrm{T}}\mathbf{B}(\mathbf{u})\, \mathrm{d}\Gamma \equiv \int (v_1 B_1(\mathbf{u}) + v_2 B_2(\mathbf{u}) + \cdots)\, \mathrm{d}\Gamma = 0 \qquad (3.15)$$

for any set of functions $\mathbf{v}$.

Indeed, the integral statement that

$$\int_{\Omega} \mathbf{v}^{\mathrm{T}}\mathbf{A}(\mathbf{u})\, \mathrm{d}\Omega + \int_{\Gamma} \bar{\mathbf{v}}^{\mathrm{T}}\mathbf{B}(\mathbf{u})\, \mathrm{d}\Gamma = 0 \qquad (3.16)$$

is satisfied for all $\mathbf{v}$ and $\bar{\mathbf{v}}$ is equivalent to the satisfaction of the differential equations (3.1) and their boundary conditions (3.2).

In the above discussion it was implicitly assumed that integrals such as those in Eq. (3.16) are capable of being evaluated. This places certain restrictions on the possible families to which the functions $\mathbf{v}$ or $\mathbf{u}$ must belong. *In general we shall seek to avoid functions which result in any term in the integrals becoming infinite.*

Thus, in Eq. (3.16) we limit the choice of $\mathbf{v}$ and $\bar{\mathbf{v}}$ to single, finite value functions without restricting the validity of previous statements.

What restrictions need to be placed on the functions $\mathbf{u}_1$, $\mathbf{u}_2$, ..., etc.? The answer depends obviously on the order of differentiation implied in the equations $\mathbf{A}(\mathbf{u})$ (or $\mathbf{B}(\mathbf{u})$). Consider, for instance, a function $\mathbf{u}$ which is continuous but has a discontinuous slope in the x direction, as shown in Fig. 3.2. We imagine this discontinuity to be replaced by a continuous variable in a very small distance Δ and study the behaviour of the derivatives. It is easy to see that although the first derivative is not defined here, it can be integrated but the second derivative tends to infinity. The function illustrated would be a suitable choice for $\mathbf{u}$ if only first derivatives occurred in the differential equation. *Such a function is said to be* C_0 *continuous.*

In a similar way it is easy to see that if nth order derivatives occur in any term of $\mathbf{A}$ or $\mathbf{B}$ then the function has to be such that its $n-1$ derivatives are continuous (C_{n-1} continuity).

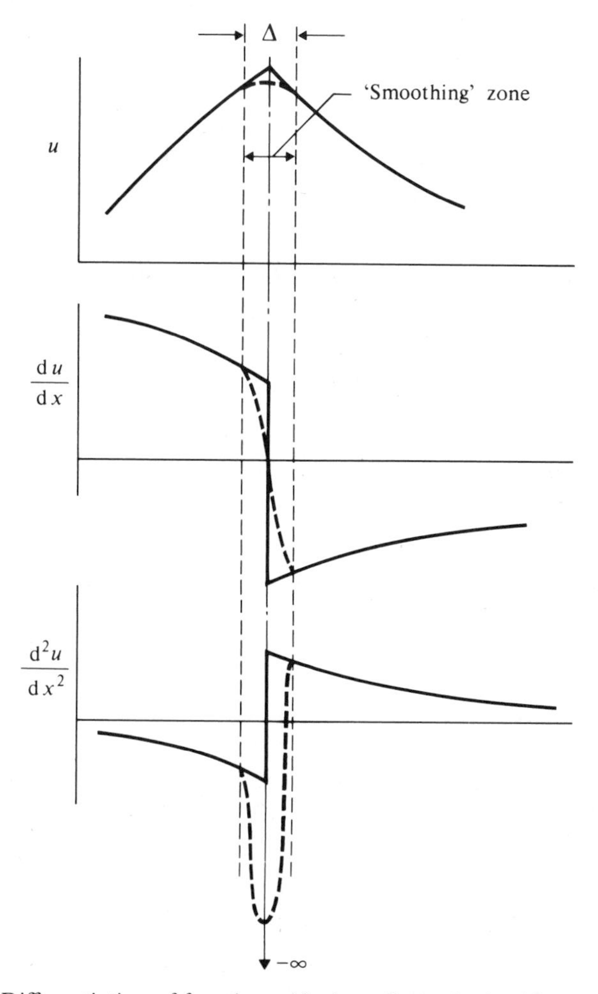

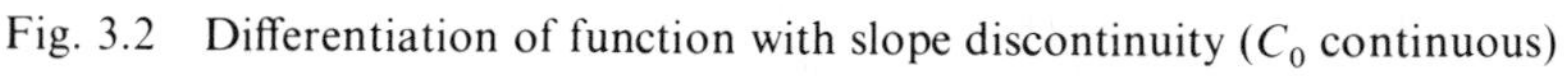
Fig. 3.2 Differentiation of function with slope discontinuity (C_0 continuous)

On many occasions it is possible to perform an integration by parts on Eq. (3.16) and replace it by an alternative statement of the form

$$\int_{\Omega} \mathbf{C}(\mathbf{v})^{\mathrm{T}}\mathbf{D}(\mathbf{u})\, \mathrm{d}\Omega + \int_{\Gamma} \mathbf{E}(\bar{\mathbf{v}})^{\mathrm{T}}\mathbf{F}(\mathbf{u})\, \mathrm{d}\Gamma = 0. \tag{3.17}$$

In this the operators **C** to **F** contain lower order derivatives than those occurring in operators **A** and **B**. Now a lower order of continuity is required in the choice of the **u** function at a price of higher continuity for **v** and $\bar{\mathbf{v}}$.

The statement (3.17) is now more 'permissive' than the original problem posed by Eqs. (3.1), (3.2), or (3.16) and is called a *weak form* of these equations. It is a somewhat surprising fact that often this weak form is more realistic physically than the original differential equation which implied an excessive 'smoothness' of the true solution.

Integral statements of the form of (3.16) and (3.17) will form the basis of finite element approximations, and we shall discuss them later in fuller detail. Before doing so we shall apply the new formulation to an example.

3.3 Weak Form of the Heat Conduction Equation—Forced and Natural Boundary Conditions

Consider now the integral form of Eq. (3.10). We can write the statement (3.16) as

$$\int_\Omega v\left[\frac{\partial}{\partial x}\left(k\frac{\partial\phi}{\partial x}\right)+\frac{\partial}{\partial y}\left(k\frac{\partial\phi}{\partial y}\right)+Q\right]\mathrm{d}x\,\mathrm{d}y + \\ +\int_{\Gamma_q}\bar{v}\left[k\frac{\partial\phi}{\partial n}-\bar{q}\right]\mathrm{d}\Gamma = 0 \qquad (3.18)$$

noting that v and $\bar{v}$ are scalar functions and presuming that one of the boundary conditions, i.e.,

$$\phi-\bar{\phi} = 0$$

is automatically satisfied by the choice of the functions ϕ.

Equation (3.18) can now be integrated by parts to obtain a weak form similar to Eq. (3.17). We shall make use here of general formulae for such integration (Green's formulae) which we derive in Appendix 3 and which will, on many occasions, be useful, i.e.,

$$\int_\Omega v\frac{\partial}{\partial x}\left(k\frac{\partial\phi}{\partial x}\right)\mathrm{d}x\,\mathrm{d}y \equiv -\int_\Omega \frac{\partial v}{\partial x}\left(k\frac{\partial\phi}{\partial x}\right)\mathrm{d}x\,\mathrm{d}y+\oint_\Gamma v\left(k\frac{\partial\phi}{\partial x}\right)n_x\,\mathrm{d}\Gamma \\ \int v\frac{\partial}{\partial y}\left(k\frac{\partial\phi}{\partial y}\right)\mathrm{d}x\,\mathrm{d}y \equiv -\int \frac{\partial v}{\partial y}\left(k\frac{\partial\phi}{\partial y}\right)\mathrm{d}x\,\mathrm{d}y+\oint v\left(k\frac{\partial\phi}{\partial y}\right)n_y\,\mathrm{d}\Gamma. \qquad (3.19)$$

We have thus

$$-\int_\Omega\left(\frac{\partial v}{\partial x}k\frac{\partial\phi}{\partial x}+\frac{\partial v}{\partial y}k\frac{\partial\phi}{\partial y}-Q\,v\right)\mathrm{d}x\,\mathrm{d}y+\oint_\Gamma vk\left(\frac{\partial\phi}{\partial x}n_x+\frac{\partial\phi}{\partial y}n_y\right)\mathrm{d}\Gamma + \\ +\int_{\Gamma_q}\bar{v}\left[k\frac{\partial\phi}{\partial n}-\bar{q}\right]\mathrm{d}\Gamma = 0. \qquad (3.20)$$

Noting that the derivative along the normal is given as

$$\frac{\partial\phi}{\partial n} \equiv \frac{\partial\phi}{\partial x}n_x+\frac{\partial\phi}{\partial y}n_y \qquad (3.21)$$

and, further, making

$$v = -\bar{v} \tag{3.22}$$

without loss of generality (as both functions are arbitrary), we can write Eq. (3.20) as

$$\int_{\Omega} \nabla^{\mathrm{T}} v \, k \nabla \phi \, \mathrm{d}\Omega - \int_{\Omega} vQ \, \mathrm{d}\Omega - \int_{\Gamma_q} v\bar{q} \, \mathrm{d}\Gamma - \int_{\Gamma_\phi} vk \frac{\partial \phi}{\partial n} \mathrm{d}\Gamma = 0 \tag{3.23}$$

where the operator ∇ is simply

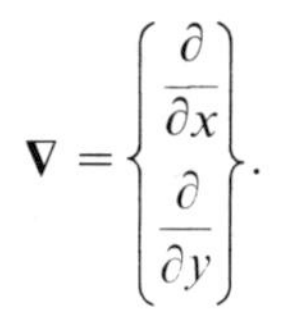

$$\nabla = \begin{Bmatrix} \dfrac{\partial}{\partial x} \\ \dfrac{\partial}{\partial y} \end{Bmatrix}.$$

We note

(*a*) that the variable ϕ has disappeared from the integrals taken along the boundary Γ_q and that the boundary conditions

$$B(\phi) = k\frac{\partial \phi}{\partial n} - \bar{q} = 0$$

on that boundary is automatically satisfied. Such a condition is known as a *natural boundary* condition; and

(*b*) that if the choice of ϕ is so restricted as to satisfy the *forced boundary conditions* $\phi - \bar{\phi} = 0$, we can omit the last term of Eq. (3.23) by restricting the choice of v to functions which give $v = 0$ on Γ_ϕ.

The form of Eq. (3.23) is the *weak form* of the heat conduction statement equivalent to Eq. (3.17). It admits discontinuous conductivity coefficients k and temperature ϕ which show discontinuous first derivatives —a real possibility not admitted in the differential form.

3.4 Approximation to Integral Formulations: the Weighted Residual Method

If the unknown function $\mathbf{u}$ is approximated by the expansion (3.3), i.e.,

$$\mathbf{u} \approx \hat{\mathbf{u}} = \sum_{1}^{r} \mathbf{N}_i \mathbf{a}_i = \mathbf{Na} \tag{3.3}$$

then it is clearly impossible to satisfy both the differential equation and the boundary conditions in a general case. The integral statements (3.16) or (3.17) allow an approximation to be made if, in place of *any function*, $\mathbf{v}$ we put a finite set of prescribed functions

$$\mathbf{v} = \mathbf{w}_j; \qquad \bar{\mathbf{v}} = \bar{\mathbf{w}}_j \qquad (j = 1-n) \tag{3.24}$$

where n is the number of unknown parameters $\mathbf{a}_i$ entering the problem $(n \leqslant r)$.

Equations (3.16) and (3.17) thus yield a set of ordinary equations from which parameters $\mathbf{a}$ can be determined, i.e., for Eq. (3.16) we have a set

$$\int_\Omega \mathbf{w}_j^\mathrm{T} \mathbf{A}(\mathbf{Na})\, \mathrm{d}\Omega + \int_\Gamma \overline{\mathbf{w}}_j^\mathrm{T} \mathbf{B}(\mathbf{Na})\, \mathrm{d}\Gamma = 0 \qquad (j = 1-n) \tag{3.25}$$

or, from Eq. (3.17),

$$\int_\Omega \mathbf{C}(\mathbf{w}_j)^\mathrm{T} \mathbf{D}(\mathbf{Na})\, \mathrm{d}\Omega + \int_\Gamma \mathbf{E}(\overline{\mathbf{w}}_j)^\mathrm{T} \mathbf{F}(\mathbf{Na})\, \mathrm{d}\Gamma = 0 \qquad (j = 1-n). \tag{3.26}$$

If we note that $\mathbf{A}(\mathbf{Na})$ represents the *residual or error* obtained by substitution of the approximation into the differential equation (and $\mathbf{B}(\mathbf{Na})$ the residual of the boundary conditions), then Eq. (3.25) is a *weighted integral of such residuals.* The approximation thus may be called the *method of weighted residuals.*

In its classical sense it was first described by Crandall,[1] who points out the various forms used since the end of the last century. More recently a very full exposé of the method has been given by Finlayson.[2] Clearly, almost any set of independent functions $\mathbf{w}_j$ could be used for the purpose of weighting and, according to the choice of function, a different name can be attached to each process. Thus the various common choices are:

(*a*) *Point collocation*[3]

$\mathbf{w}_j = \boldsymbol{\delta}_j$, where $\boldsymbol{\delta}_j$ is such that for $x \neq x_j;\ y \neq y_j,\ \ \mathbf{w}_j = 0$ but $\int_\Omega \mathbf{w}_j \,\mathrm{d}\Omega = I$ (unit matrix). This procedure is equivalent to simply making the residual zero at n points within the domain and integration is 'nominal' (incidentally although $\mathbf{w}_j$ defined here does not satisfy our integrability criteria of section 3.2, it is nevertheless admissible).

(*b*) *Subdomain collocation*[4]

$\mathbf{w}_j = I$ in Ω_j and zero elsewhere. This essentially makes the integral of the error zero over the specified subdomain of the domain.

(*c*) *The Galerkin method*[5,6] (Bubnov-Galerkin)

$\mathbf{w}_j = \mathbf{N}_j$. Here simply the original shape (or basis) functions are used as weighting. This method, as we shall see, leads frequently (but by no means always) to symmetric matrices and for this and other reasons will be adopted in our finite element work almost exclusively.

The name of 'weighted residuals' is clearly much older than that of the 'finite element method'. The latter uses mainly locally based (element) functions in the expansion of Eq. (3.3) but the general procedures are identical. As the process leads always to equations which, being of integral form, can be obtained by summation of contributions from various subdomains, we choose to embrace all weighted residual approximations under the finite element name. Frequently, simultaneous use of both local and 'global' trial functions will be found to be useful.

In mathematical literature the names of Petrov-Galerkin[6] are often associated with the use of weighting functions such that $\mathbf{w}_j \neq \mathbf{N}_j$.

3.5 **Examples**

To illustrate the procedure of weighted residual approximation and its relation to the finite element process let us consider some specific examples.

Example 1. One-dimensional equation of heat conduction (Fig. 3.3). The problem here will be a one-dimensional representation of the heat conduction equation Eq. (3.10) with unit conductivity. (This problem could equally well represent many other physical situations, e.g., deformation of a loaded string.) Here we have

$$A(\phi) = \frac{\mathrm{d}^2\phi}{\mathrm{d}x^2}+Q = 0 \quad (0 \leqslant x \leqslant L) \tag{3.27}$$

with $Q = Q(x)$ given by $Q = 1$ $(0 \leqslant x < L/2)$ and $Q = 0$ $(L/2 < x \leqslant L)$. The boundary conditions assumed will be simply $\phi = 0$ at $x = 0$ and $x = L$.

In the first case we shall consider a one- or two-term approximation of the Fourier series form, i.e.,

$$\phi \approx \hat{\phi} = \sum a_i \sin\frac{\pi x i}{L}; \quad N_i = \sin\frac{\pi x i}{L} \tag{3.28}$$

with $i = 1$ and $i = 1$ and 2. These satisfy the boundary conditions exactly and are continuous throughout the domain. We can thus use either Eq. (3.16) or Eq. (3.17) for approximation with equal validity. We shall use the former, which allows various weighting functions to be adopted. In Fig. 3.3 we present the problem and its solution using point collocation, subdomain collocation, and the Galerkin method.†

As the chosen expansion satisfies *a priori* the boundary conditions there is no need to introduce them into the formulation, which is given simply by

$$\int_0^L w_j\left[\frac{\mathrm{d}^2}{\mathrm{d}x^2}\left(\sum N_i a_i\right)+Q\right]\mathrm{d}x = 0. \tag{3.29}$$

The full working out of this problem is left as an exercise to the reader. Of more interest to the true finite element field is the use of piecewise

† In the case of point collocation using $i = 1$, $x_i = L/2$, a difficulty arises about the value of Q (as this is either zero or one). The value of $\frac{1}{2}$ was therefore used for the example.

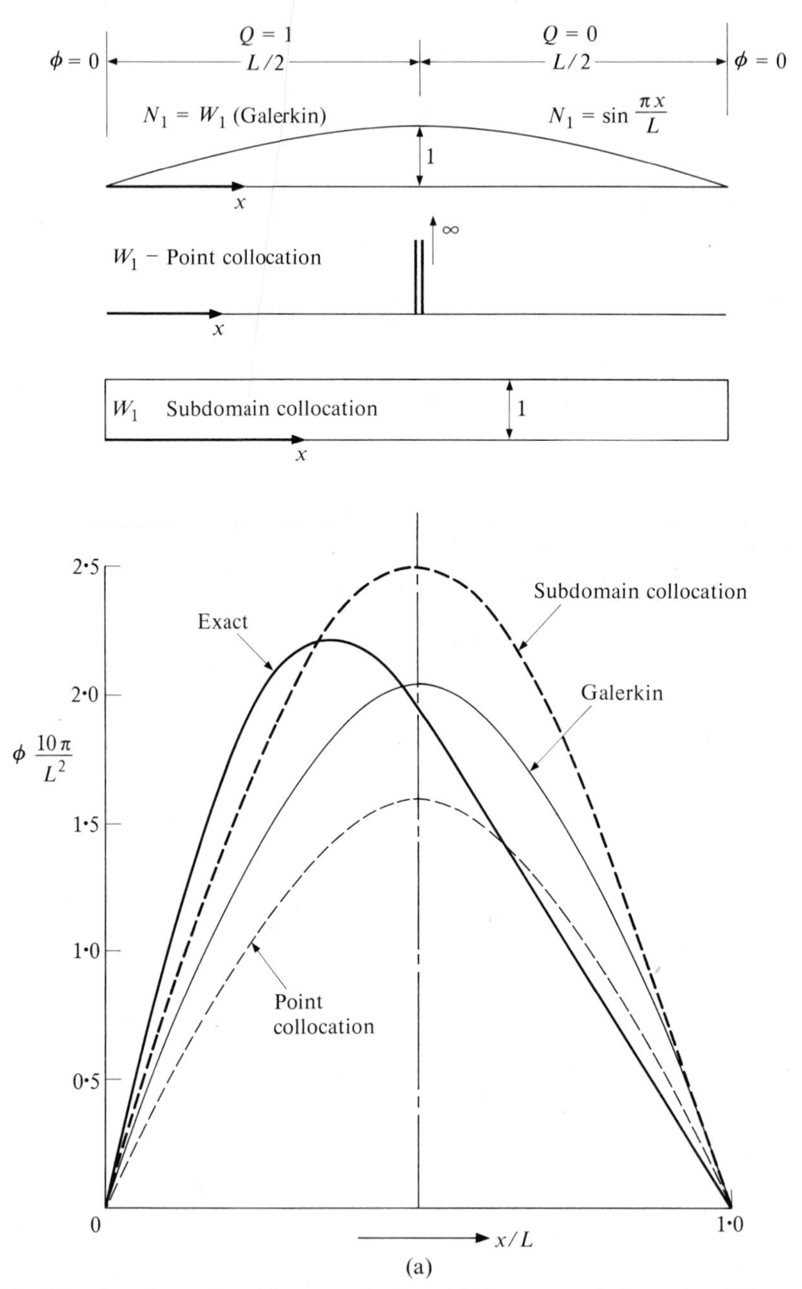

Fig. 3.3 One-dimensional heat conduction. (*a*) One-term solution using different weighting procedures. (*b*) Two-term solutions using different weighting procedures

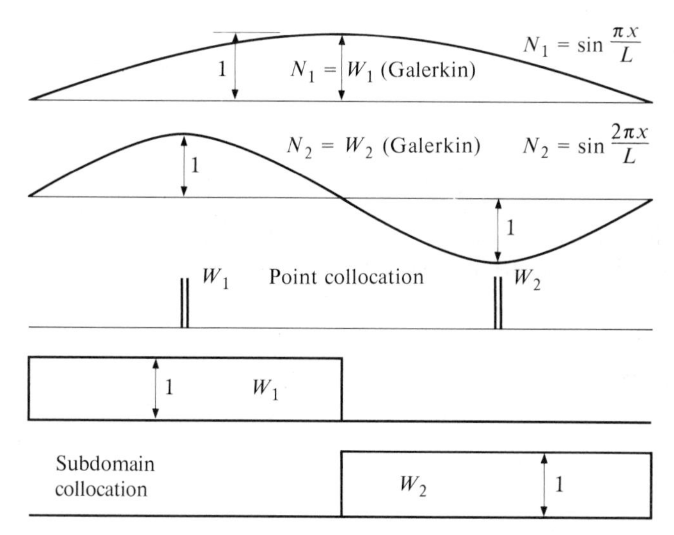

$N_1 = \sin \frac{\pi x}{L}$
1
$N_1 = W_1$ (Galerkin)
$N_2 = W_2$ (Galerkin)
$N_2 = \sin \frac{2\pi x}{L}$
1
1
W_1
Point collocation
W_2
1
W_1
Subdomain collocation
W_2
1

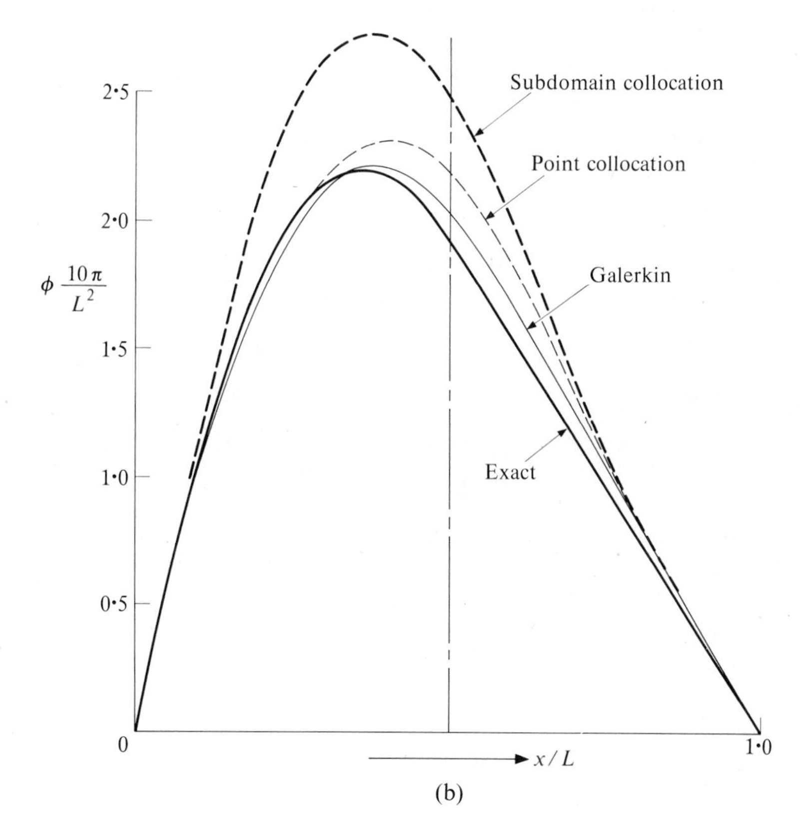

Subdomain collocation
Point collocation
Galerkin
Exact
2·5
2·0
1·5
1·0
0·5
0
1·0
$\phi \frac{10\pi}{L^2}$
x/L

(b)

defined (locally based) functions in place of the global functions of Eq. (3.28). Here, to avoid imposing slope continuity, we shall use the equivalent of Eq. (3.17) obtained by integrating Eq. (3.29) by parts. This yields

$$\int_0^L \left[\frac{\mathrm{d}w_j}{\mathrm{d}x} \frac{\mathrm{d}}{\mathrm{d}x} \sum N_i a_i - w_j Q \right] \mathrm{d}x = 0. \tag{3.30}$$

The boundary terms disappear identically if $w_j = 0$ at the two ends.

The above equations can be written as

$$\mathbf{Ka} + \mathbf{f} = 0 \tag{3.31}$$

where for each 'element' of length L^e

$$\begin{aligned} K_{ji}^e &= \int_0^{L^e} \frac{\mathrm{d}w_j}{\mathrm{d}x} \frac{\mathrm{d}N_i}{\mathrm{d}x} \mathrm{d}x \\ f_j^e &= -\int_0^{L^e} w_j Q \, \mathrm{d}x \end{aligned} \tag{3.32}$$

with the usual rules of addition pertaining, i.e.,

$$K_{ji} = \int_0^L \frac{\mathrm{d}w_j}{\mathrm{d}x} \frac{\mathrm{d}N_i}{\mathrm{d}x} \mathrm{d}x; \quad f_j^e = -\int_0^L w_j Q \, \mathrm{d}x \tag{3.33}$$

In the computation we shall use the Galerkin procedure, i.e., $w_j = N_j$, and the reader will observe that the matrix $\mathbf{K}$ is then symmetric, i.e., $K_{ij} = K_{ji}$.

As the shape functions need only be of C_0 continuity, a piecewise linear approximation is conveniently used, as shown in Fig. 3.4. Considering a typical element ij shown, we can write (moving the origin of x to point i)

$$N_j = x/L^e \qquad N_i = (L^e - x)/L^e \tag{3.34}$$

giving for a typical element

$$\begin{aligned} K_{ij}^e = K_{ji}^e &= -1/L^e; \quad K_{ii} = K_{jj} = 1/L^e \\ f_j^e &= -QL^e/2 = f_i^e \end{aligned} \tag{3.35}$$

Assembly of a typical equation at a node i is left to the reader, who is well advised to carry out the calculations leading to the results shown in Fig. 3.4 for a two- and four-element subdivision.

Some points of interest immediately arise if results of Figs. 3.3 and 3.4 are compared. With smooth global shape functions the Galerkin method gives better overall results than those achieved for the same number of unknown parameters $\mathbf{a}$ with locally based functions. This we shall find to be the general case with higher order approximations yielding a better

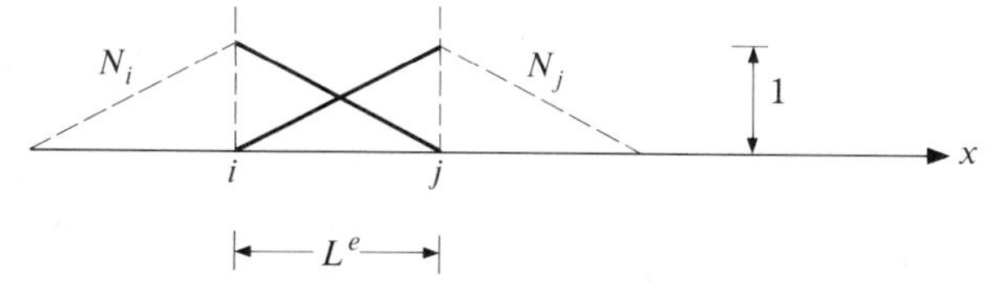

Locally based linear shape functions

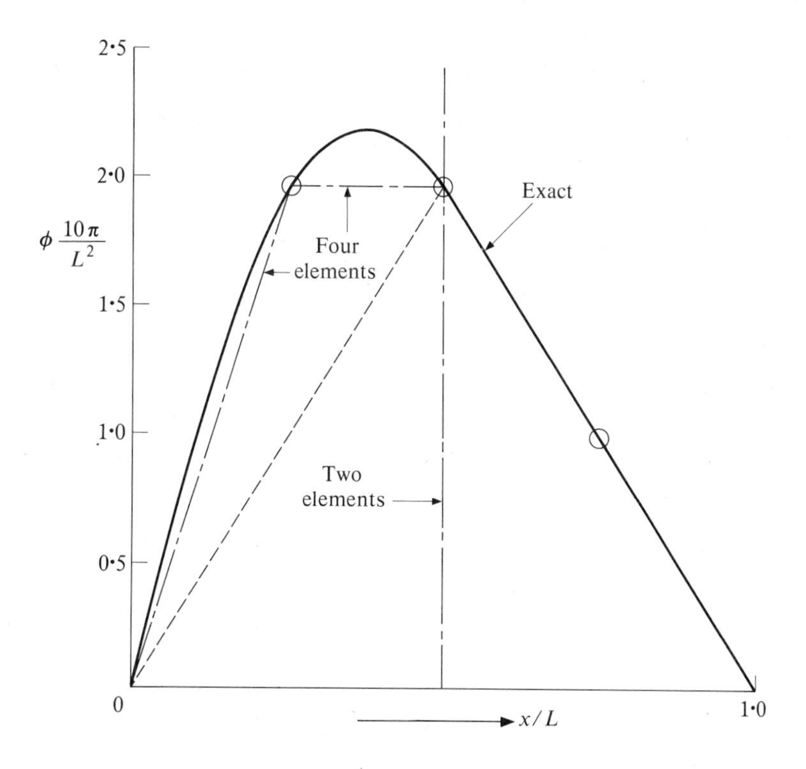

Fig. 3.4 Galerkin–finite element solution of problem of Fig. 3.3 using linear locally based shape functions

accuracy. Further, it will be observed that the linear approximation has given the exact answers at the nodal points. This is a property of the particular equation being solved and unfortunately does not carry over to the general problems.[7] Lastly, the reader will observe how easy it is to create equations with any degree of subdivision once the element properties Eq. (3.35) have been derived. This is not the case with global approximation where new integrations have to be carried out for each new parameter introduced. It is this repeatability feature that is one of the finite element advantages.

Example 2. Steady-state heat conduction–convection in two dimensions. The Galerkin formulation. We have already introduced the problem in section 3.1 and defined it by Eq. (3.11) with appropriate boundary conditions. The equation differs only in the convective terms from that of simple heat conduction for which the weak form has already been obtained in Eq. (3.23). We can write the weighted residual equation immediately from this, substituting $v = w_j$ and adding the convective terms. Thus we have

$$\int_\Omega \nabla^T w_j\, k \nabla \hat{\phi}\, d\Omega - \int w_j \left(u \frac{\partial \hat{\phi}}{\partial x} + v \frac{\partial \hat{\phi}}{\partial y} \right) d\Omega - \int_\Omega w_j Q\, d\Omega - \int_{\Gamma_q} w_j \bar{q}\, d\Gamma = 0 \tag{3.36}$$

with $\hat{\phi} = \sum N_i a_i$ being such that the prescribed values of $\bar{\phi}$ are given on the boundary $\Gamma = \phi$ and that $w_j = 0$ on that boundary.

Specializing to the Galerkin approximation, i.e., putting $w_j = N_j$, we have immediately a set of equations of the form

$$\mathbf{Ka} + \mathbf{f} = 0 \tag{3.37}$$

with

$$K_{ji} = \int_\Omega \nabla^T N_j k \nabla^T N_i\, d\Omega - \int_\Omega \left(N_j u \frac{\partial N_i}{\partial x} + N_j v \frac{\partial N_i}{\partial y} \right) d\Omega$$
$$= \int_\Omega \left(\frac{\partial N_j}{\partial x} k \frac{\partial N_i}{\partial x} + \frac{\partial N_j}{\partial y} k \frac{\partial N_i}{\partial y} \right) d\Omega + - \int_\Omega \left(N_j u \frac{\partial N_i}{\partial x} + N_j v \frac{\partial N_i}{\partial y} \right) d\Omega \tag{3.38a}$$

$$f_j = -\int_\Omega N_j Q\, d\Omega - \int_{\Gamma_q} N_j \bar{q}\, d\Gamma \tag{3.38b}$$

Once again the components K_{ji} and f_j can be evaluated for a typical element or subdomain and systems equations built up by standard methods.

At this point it is important to mention that to satisfy the boundary conditions some of the parameters $\mathbf{a}_i$ have to be prescribed and approximation equations must be in number equal only to the unknown parameters. It is nevertheless convenient to form all equations for all parameters and prescribe the fixed values at the end using precisely the same techniques as we have described in Chapter 1 for the insertion of prescribed boundary conditions in standard discrete problems.

A further point concerning the coefficients of matrix **K** should here be noted. The first part, corresponding to the pure heat conduction equation, is symmetric ($K_{ij} = K_{ji}$) but the second is not and thus a system of non-symmetric equations needs to be solved. There is a basic reason for such non-symmetries which will be discussed in section 3.3.

To make the problem concrete consider the domain Ω to be divided into regular square elements of side h (Fig. 3.5). To preserve C_0 continuity with nodes placed at corners, shape functions given as the product of the linear expansions can be written. For instance, for node i, as shown in Fig. 3.5,

$$N_i = \frac{x}{h} \cdot \frac{y}{h}$$

and for node j,

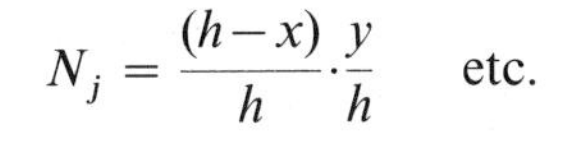

$$N_j = \frac{(h-x)}{h} \cdot \frac{y}{h} \qquad \text{etc.}$$

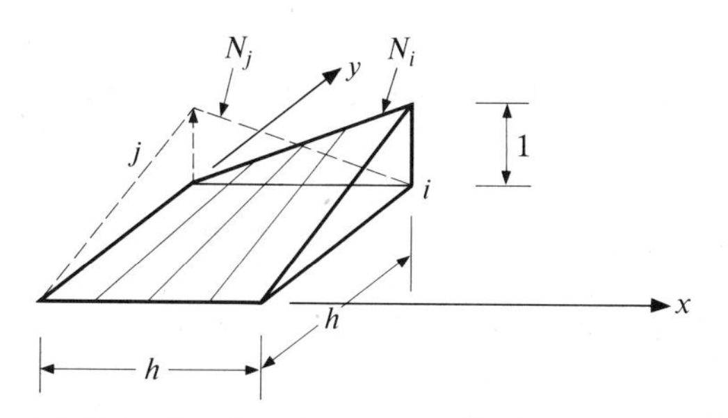

(a) Shape functions for a square C_0 element

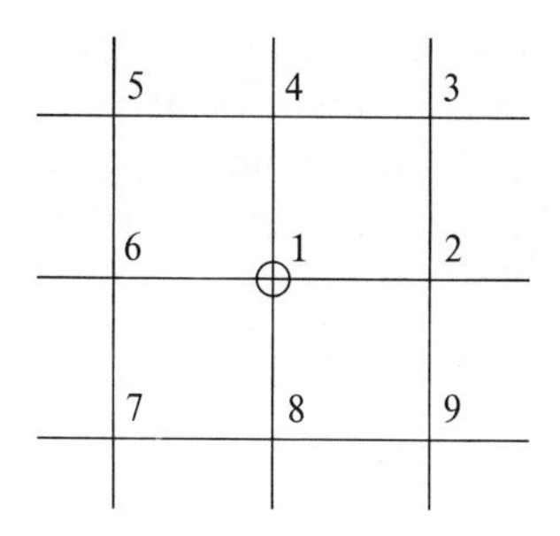

(b) Nodes 'connected' by equation for node 1

Fig. 3.5 A linear square element of C_0 continuity. (a) Shape functions for a square element. (b) 'Connected' equation for node 1

With these shape functions the reader is invited to evaluate typical element contributions and to assemble the equations for point 1 of the mesh numbered as shown in Fig. 3.5. The result will be (if no boundary of type Γ_q is present and Q is assumed to be constant)

$$\frac{8}{3}a_1-\left(\frac{1}{3}-\frac{uh}{3k}-\frac{vh}{6k}\right)a_2-\left(\frac{1}{3}-\frac{uh}{12k}-\frac{vh}{12k}\right)a_3-\left(\frac{1}{3}-\frac{uh}{6k}-\frac{vh}{3k}\right)a_4$$
$$-\left(\frac{1}{3}+\frac{uh}{12k}-\frac{vh}{12k}\right)a_5-\left(\frac{1}{3}+\frac{uh}{3k}-\frac{vh}{6k}\right)a_6-\left(\frac{1}{3}+\frac{uh}{12k}+\frac{vh}{6k}\right)a_7$$
$$-\left(\frac{1}{3}-\frac{uh}{6k}+\frac{vh}{3k}\right)a_8-\left(\frac{1}{3}+\frac{uh}{12k}+\frac{vh}{12k}\right)a_9 = 4h^2Q \tag{3.39}$$

This equation is similar to those which would be obtained by using finite difference approximations to the same equations in a fairly standard manner.[8,9] In the example discussed some difficulties arise when the convective terms are large. In such cases the Galerkin weighting is not acceptable and other forms have to be used. This is discussed in detail in Chapter 22 (section 22.8).

3.6 Virtual Work as the 'Weak Form' of Equilibrium Equations for Analysis of Solids or Fluids

In the previous chapter we introduced the finite element by the way of an application to the solid mechanics problem of elasticity. The integral statement necessary for formulation in terms of finite element approximation was supplied via the principle of *virtual work*, which was assumed to be so basic as not to merit proof. Indeed, to many this is so, and the virtual work principle is considered as a statement of mechanics more fundamental than the traditional equilibrium conditions of Newton's laws of motion. Others will argue with this view and will point out that all work statements are derived from the classical laws pertaining to the equilibrium of the particle. We shall therefore show in this section that the virtual work statement is simply a 'weak form' of equilibrium equations.

In a general three-dimensional continuum the equilibrium equations of an elementary volume can be written in terms of the components of the symmetric Cartesian stress tensor as[10]

$$\left\{\begin{matrix} \dfrac{\partial\sigma_{xx}}{\partial x}+\dfrac{\partial\tau_{xy}}{\partial y}+\dfrac{\partial\tau_{xz}}{\partial z} \\ \dfrac{\partial\sigma_{y}}{\partial y}+\dfrac{\partial\tau_{yx}}{\partial x}+\dfrac{\partial\tau_{yz}}{\partial z} \\ \dfrac{\partial\sigma_{z}}{\partial z}+\dfrac{\partial\tau_{xz}}{\partial x}+\dfrac{\partial\tau_{yz}}{\partial y} \end{matrix}\right\}+\left\{\begin{matrix} b_x \\ b_y \\ b_z \end{matrix}\right\}=0 \tag{3.40}$$

where $\mathbf{b}^{\mathrm{T}} = [b_x, b_y, b_z]$ stands for the forces acting per unit volume (which may well include the acceleration effects).

In solid mechanics the six stress components will be some general functions of the components of the displacement

$$\mathbf{u}^{\mathrm{T}} = [u, v, w] \tag{3.41}$$

and in fluid mechanics of the velocity vector $\mathbf{u}$, which has similar components. Thus Eq. (3.40) can be considered as a general equation of form (3.1), i.e., $\mathbf{A}(\mathbf{u}) = 0$. To obtain a weak form we shall proceed as before, introducing an arbitrary weighting function vector $\delta\mathbf{u}$, defined as

$$\delta\mathbf{u}^{\mathrm{T}} = [\delta u, \delta v, \delta w]. \tag{3.42}$$

We can write now the integral statement of Eq. 3.13 as

$$\int_V \delta\mathbf{u}^{\mathrm{T}}\mathbf{A}(\mathbf{u})\,\mathrm{d}V = \int_V \left[\delta u\left(\frac{\partial \sigma_x}{\partial x}+\frac{\partial \tau_{xy}}{\partial y}+\frac{\partial \tau_{xz}}{\partial z}+b_x\right)+ \right. \\ \left. +\delta v(\ldots)+\delta w(\ldots)\right]\mathrm{d}V \tag{3.43}$$

where V, the volume, is the problem domain.

Integrating each term by parts and rearranging we can write above as

$$-\int_V \left[\sigma_x\frac{\partial}{\partial x}(\delta u)+\tau_{xy}\left(\frac{\partial}{\partial x}(\delta u)+\frac{\partial}{\partial y}(\delta v)\right)+\cdots \right. \\ \left. -\delta u b_x-\delta v b_y-\delta w b_z\right]\mathrm{d}V \tag{3.44}$$
$$+\int_A [\delta u(\sigma_x n_x+\tau_{xy}n_y+\tau_{xz}n_z)+\cdots \\ +\delta v(\ldots)+\delta w(\ldots)]\,\mathrm{d}A = 0$$

where A is the surface area of solid (here again Green's formulae of Appendix 3 are used).

In the first set of bracketed terms we can recognize immediately the small strain operators acting on $\delta\mathbf{u}$—which can be termed a virtual displacement (or virtual velocity). We can therefore introduce virtual strain (or strain rate) defined as

$$\delta\boldsymbol{\varepsilon} = \begin{Bmatrix} \dfrac{\partial}{\partial x}(\delta u) \\ \dfrac{\partial}{\partial y}(\delta v) \\ \dfrac{\partial}{\partial z}(\delta w) \\ \vdots \end{Bmatrix} \tag{3.45}$$

Similarly, the terms in the second integral will be recognized as forces $\mathbf{t}$:

$$\mathbf{t} = [t_x, t_y, t_z] \tag{3.46}$$

acting per unit area of the surface A. Arranging the six stress components in a vector $\boldsymbol{\sigma}$ and similarly the six virtual strain (or rate of virtual strain) components in a vector $\delta\boldsymbol{\varepsilon}$, we can write Eq. (3.44) simply as

$$\int_V \delta\boldsymbol{\varepsilon}^{\mathrm{T}}\boldsymbol{\sigma}\,\mathrm{d}V - \int_V \delta\mathbf{u}^{\mathrm{T}}\mathbf{b}\,\mathrm{d}V - \int_\Gamma \delta\mathbf{u}^{\mathrm{T}}\mathbf{t}\,\mathrm{d}\Gamma = 0 \tag{3.47}$$

which is the virtual work statement used in Eqs. (2.10) and (2.22) of Chapter 2.

We see from above that the virtual work statement is precisely the weak form of the equilibrium equations and is valid for non-linear as well as linear stress–strain (or stress–rate of strain) relations.

The finite element approximation which we have derived in Chapter 2 *is in fact a Galerkin formulation of the weighted residual process applied to the equilibrium equation.* Thus, if we take $\delta\mathbf{u}$ as the shape function

$$\delta\mathbf{u} = \mathbf{N} \tag{3.48}$$

where the displacement field is discretized, i.e.,

$$\mathbf{u} = \sum \mathbf{N}_i\mathbf{a}_i \tag{3.49}$$

together with the constitutive relation of Eq. (2.5), we shall determine once again all the basic expressions of Chapter 2 which are so essential to the solution of elasticity problems.

Similar expressions will be vital to the formulation of equivalent fluid mechanics problems, as will be shown in Chapter 22.

3.7 Partial Discretization

In the approximation to the problem of solving the differential equation (Eqs. (3.1)) by an expansion of the standard form of Eq. (3.3), we have assumed that the shape functions $\mathbf{N}$ included in them *all* the independent co-ordinates of the problem and that $\mathbf{a}$ was simply a set of constants. The final approximation equations were thus always of an algebraic form, from which a unique set of parameters could be determined.

In some problems it is convenient to proceed differently. Thus, for instance, if the independent variables are x, y, and z we could allow the parameters $\mathbf{a}$ to be functions of z and do the approximate expansion only in the domain of x, y, say $\bar{\Omega}$. Thus, in place of Eq. (3.3) we would have

$$\begin{aligned} \mathbf{u} &= \mathbf{Na} \\ \mathbf{N} &= \mathbf{N}(x, y) \\ \mathbf{a} &= \mathbf{a}(z) \end{aligned} \tag{3.50}$$

Clearly the derivatives of $\mathbf{a}$ with respect to z will remain in the final discretization and the result will be a set of ordinary differential equations with z as the independent variable. In linear problems such a set will have the appearance

$$\mathbf{Ka}+\mathbf{C\dot{a}}+\cdots+\mathbf{f}=0 \tag{3.51}$$

where $\dot{\mathbf{a}} \equiv (\mathrm{d}/\mathrm{d}t)\mathbf{a}$, etc.

Such partial discretization can obviously be used in different ways, but is particularly useful when the domain $\overline{\Omega}$ is not dependent on z, i.e., when the *problem is prismatic*. In such a case the coefficients of the ordinary differential Eq. (3.51) are independent of z and the solution of the system can frequently be carried out efficiently by standard analytical methods.

This type of partial discretization has been applied extensively by Kantorovitch[11] and is frequently known by his name. In Chapter 15 we shall discuss such semi-analytical treatments in the context of prismatic solids where the final solution is obtained in terms of Fourier series. The most frequently encountered 'prismatic' problem is one involving the time variable, where the space domain $\overline{\Omega}$ is not subject to change. We shall discuss such problems in detail in Chapters 20 and 21, but it is convenient by way of illustration to consider here the heat conduction in a two-dimensional equation in its transient state. This is obtained from Eq. (3.10) by addition of the heat storage term $c(\partial\phi/\partial t)$, where c is the specific heat. We now have a problem posed in a domain $\Omega(x, y, t)$ in which the following equation holds

$$A(\phi) \equiv \frac{\partial}{\partial x}\left(k\frac{\partial\phi}{\partial x}\right)+\frac{\partial}{\partial y}\left(k\frac{\partial\phi}{\partial y}\right)+Q-c\frac{\partial\phi}{\partial t}=0 \tag{3.52}$$

with boundary conditions identical to those of Eq. (3.10).

Taking

$$\phi \approx \hat{\phi} = \sum N_i a_i \tag{3.53}$$

with $a_i = a_i(t)$ and $N_i = N_i(x, y)$ and using the Galerkin weighting procedure we follow precisely the steps outlined in Eqs. (3.36)–(3.38) and arrive at a system of ordinary differential equations

$$\mathbf{Ka}+\mathbf{C}\frac{\mathrm{d}\mathbf{a}}{\mathrm{d}t}+\mathbf{f}=0 \tag{3.54}$$

Here the expression for K_{ij} is identical with that of Eq. (3.38a) (convective terms neglected), f_i identical to Eq. (3.38b), and the reader can verify that the matrix $\mathbf{C}$ is defined by

$$C_{ij} = \int_{\Omega} N_i c N_j \,\mathrm{d}x\,\mathrm{d}y \tag{3.55}$$

Once again the matrix **C** can be assembled from its element contribution. We shall return to the details of the possible analytical or numerical solution of such ordinary equation systems in Chapters 21 and 22. However, to illustrate the detail and the possible advantage of the process of partial discretization, we shall consider a very simple problem.

Example. Consider a square prism of size L in which the transient heat conduction equation [Eq. (3.52)] applies and assume that the rate of heat generation varies with time as

$$Q = Q_0 e^{-\alpha t} \tag{3.56}$$

(this approximates a problem of heat development due to hydration of concrete). We assume that at $t = 0$, $\phi = 0$ throughout. Further, we shall take $\phi = 0$ on all boundaries throughout all times.

As a first approximation a shape function for a one-parameter solution is taken:

$$\phi = N_1 a_1;$$
$$N_1 = \cos\frac{\pi x}{L}\cos\frac{\pi y}{L} \tag{3.57}$$

with x and y measured from the centre (Fig. 3.6).

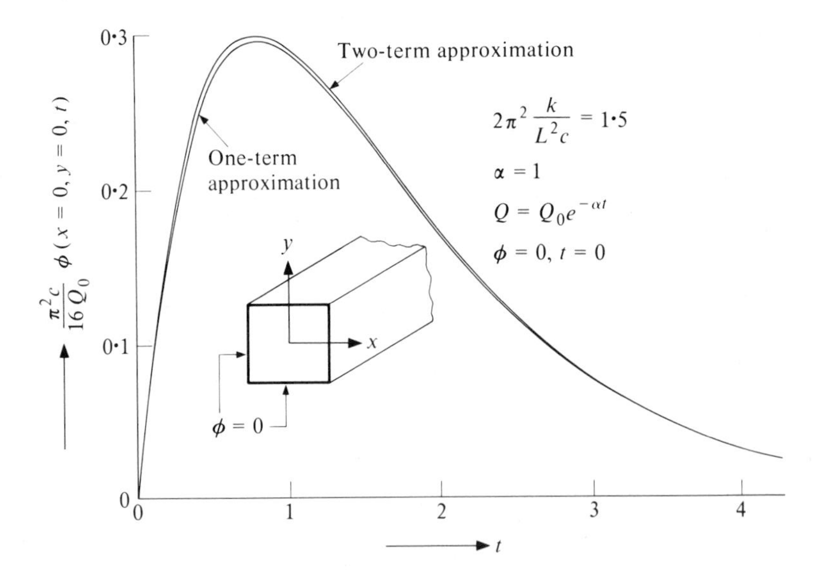

Fig. 3.6 Two-dimensional transient heat development in a square prism—plot of temperature at centre

Evaluating the coefficients, we have

$$K_{11} = \int_{-L/2}^{L/2}\int_{-L/2}^{L/2} k\left(\frac{\partial N_1}{\partial x}\right)^2 + k\left(\frac{\partial N_1}{\partial y}\right)^2 \mathrm{d}x\,\mathrm{d}y = \frac{\pi^2 k}{2}$$

$$C_{11} = \int_{-L/2}^{L/2}\int_{-L/2}^{L/2} cN_1\,\mathrm{d}x\,\mathrm{d}y = \frac{L^2 c}{4} \tag{3.58}$$

$$f_1 = \int_{-L/2}^{L/2}\int_{-L/2}^{L/2} N_1 Q_0\,\mathrm{e}^{-\alpha t}\,\mathrm{d}x\,\mathrm{d}y = \frac{4Q_0 L^2}{\pi^2}\mathrm{e}^{-\alpha t}$$

This leads to an ordinary differential equation with one parameter a_1

$$C_{11}\frac{\mathrm{d}a_1}{\mathrm{d}t} + K_{11}a_1 + f_1 = 0 \tag{3.59}$$

with $a_1 = 0$; $t = 0$. The exact solution of this is easy to obtain, as is shown in Fig. 3.6 for specific values of the parameters α and k/L^2c.

On the same figure we show a two-parameter solution with

$$N_2 = \cos\frac{3\pi x}{L}\cos\frac{3\pi y}{L} \tag{3.60}$$

which the reader can pursue to test his grasp of the problem. The second component of the Fourier series is here omitted due to the required symmetry of solution.

The remarkable accuracy of the one-term approximation in this example should be noted.

3.8 **Convergence**

In the previous sections we have discussed how approximate solutions can be obtained by use of an expansion of the unknown function in terms of trial or shape functions. Further, we have stated the necessary conditions that such functions have to fulfil in order that the various integrals can be evaluated over the domain. Thus if various integrals contain only the values of N or its first derivatives, then N has to be C_0 continuous. If second derivatives are involved, C_1 continuity is needed, etc. The problem to which we have not yet addressed ourselves are the questions of *just how good is the approximation*? and *how it can be systematically improved to approach the exact answer*? The first question is almost impossible to answer and presumes the knowledge of the exact solution. The second is more rational and can be answered if we consider some systematic way in which the number of parameters **a** in the standard expansion of Eq. (3.3)

$$\hat{\mathbf{u}} = \sum_{1}^{r} \mathbf{N}_i \mathbf{a}_i \tag{3.3}$$

is presumed to increase.

In some of the examples (e.g., pp. 51 and 62) we have assumed, in effect, a trigonometric Fourier type series limited to a finite number of terms with a single form of trial function assumed over the whole domain. Here addition of new terms would be simply an extension of the number of the terms in the series included in the analysis, and as the Fourier series is known to be able to represent any desired function within any accuracy desired as the number of terms increases, we can talk about *convergence* of the approximation to the true solution as the number of terms increases.

In other examples of this chapter we have used locally based functions which are fundamental in the finite element analysis. Here we have tacitly assumed that *convergence occurs as the size of elements decreases and, hence, the number of* **a** *parameters specified at nodes increases.* It is with such convergence that we have to be concerned and we have already discussed this in the context of the analysis of elastic solids in Chapter 2 (section 2.6).

Clearly we have now to determine

(*a*) that as the number of elements increases, the unknown functions can be approximated as closely as required, and

(*b*) how the error decreases with the size, h, of the element subdivisions. (h could be here some typical dimension of an element.)

The first problem is that of *completeness* of the expansion and we shall here assume that all trial functions are polynomials (or at least include certain terms of a polynomial expansion).

Clearly, as the approximation discussed here is to the weak, integral form typified by Eqs. (3.13) or (3.17) it is necessary that every term occurring under the integral be in the limit capable of being approximated as nearly as possible and, in particular, giving a single constant value over an infinitesimal part of the domain Ω.

If a derivative of order m exists in any such terms, then it is obviously necessary for the local polynomial to be at least of the order m so that, in the limit, such a constant value be obtained.

We will thus state that a necessary condition for the expansion to be convergent is the *criterion of completeness*: that a constant value of mth derivative be attainable in the element domain (if mth derivatives occur in the integral form) when the size of any element tends to zero.

This criterion is automatically ensured if the polynomials used in the shape function N are complete to the mth order. This criterion is also equivalent to the one of constant strain postulated in Chapter 2 (section 2.5).

If the actual order of a complete polynomial used in the finite element expansion is $p \geqslant m$, then *the order of convergence* can be ascertained by seeing how closely such a polynomial can follow the local Taylor expansion of the unknown $\mathbf{u}$. Clearly the order of error will be simply $O(h^{p+1})$ since only terms of order p can be rendered correctly.

The knowledge of the order of convergence helps in ascertaining how good the approximation is if studies on several decreasing mesh sizes are conducted. Once again we have re-established some of the conditions discussed in Chapter 2.

We shall not discuss, at this stage, approximations which do not satisfy the postulated continuity requirements except to remark that once again, in many cases, convergence and indeed improved results can be obtained. (vide Chapter 11.)

VARIATIONAL PRINCIPLES

3.9 What are 'Variational Principles'?

What are variational principles and how can they be useful in the approximation to continuum problems? It is to these questions that the following sections are addressed.

First a definition: a 'variational principle' specifies a scalar quantity (functional) Π, which is defined by an integral form

$$\Pi = \int_{\Omega} F\left(\mathbf{u}, \frac{\partial}{\partial x}\mathbf{u}, \ldots\right) \mathrm{d}\Omega + \int_{\Gamma} E\left(\mathbf{u} + \frac{\partial}{\partial x}\mathbf{u}, \ldots\right) \mathrm{d}\Gamma \qquad (3.61)$$

in which $\mathbf{u}$ is the unknown function and F and E are specified operators. The solution to the continuum problem is a function $\mathbf{u}$ which make Π stationary with respect to small changes $\delta\mathbf{u}$. Thus, for a solution to the continuum problem, the 'variation' is

$$\delta\Pi = 0. \qquad (3.62)$$

If a 'variational principle' can be found, then immediately means are established for obtaining approximate solutions in the standard, integral form suitable for finite element analysis.

Assuming a trial function expansion in the usual form

$$\mathbf{u} \approx \hat{\mathbf{u}} = \sum \mathbf{N}_i \mathbf{a}_i \qquad (3.3)$$

we can insert this into Eq. (3.61) and write

$$\delta\Pi = \frac{\partial \Pi}{\partial \mathbf{a}_1}\delta\mathbf{a}_1 + \frac{\partial \Pi}{\partial \mathbf{a}_2}\delta\mathbf{a}_2 + \cdots = \frac{\partial \Pi}{\partial \mathbf{a}}\delta\mathbf{a}. \qquad (3.63)$$

This being true for any variations $\delta \mathbf{a}$ yields a set of equations

$$\frac{\partial \Pi}{\partial \mathbf{a}} = \begin{Bmatrix} \dfrac{\partial \Pi}{\partial \mathbf{a}_1} \\ \vdots \\ \dfrac{\partial \Pi}{\partial \mathbf{a}_n} \end{Bmatrix} = 0 \tag{3.64}$$

from which parameters $\mathbf{a}_i$ are found. The equations are of an integral form necessary for the finite element approximation as the original specification of Π was given in terms of domain and boundary integrals.

The process of finding stationarity with respect to trial function parameters $\mathbf{a}$ is an old one and is associated with the names of Rayleigh[12] and Ritz.[13] It has become extremely important in finite element analysis which, to many investigators, is typified as a 'variational process'.

If the functional Π is 'quadratic', i.e., if the function $\mathbf{u}$ and its derivatives occur in powers not exceeding 2, then Eq. (3.64) reduces to a standard linear form similar to Eq. (3.8), i.e.,

$$\frac{\partial \Pi}{\partial \mathbf{a}} \equiv \mathbf{K}\mathbf{a} + \mathbf{f} = 0. \tag{3.65}$$

It is easy to show that the matrix $\mathbf{K}$ will now always be symmetric. To do this let us consider a variation of the vector $\partial \Pi / \partial \mathbf{a}$ generally. This we can write as

$$\delta\left(\frac{\partial \Pi}{\delta \mathbf{a}}\right) = \begin{bmatrix} \dfrac{\partial}{\partial \mathbf{a}_1}\left(\dfrac{\partial \Pi}{\partial \mathbf{a}_1}\right)\delta \mathbf{a}_1, & \dfrac{\partial}{\partial \mathbf{a}_2}\left(\dfrac{\partial \Pi}{\partial \mathbf{a}_1}\right)\delta \mathbf{a}_2, & \ldots \\ \vdots & & \end{bmatrix} \equiv \mathbf{K}_{\mathrm{T}}\, \delta \mathbf{a} \tag{3.66}$$

in which $\mathbf{K}_{\mathrm{T}}$ is generally known as the tangent matrix of significance in non-linear analysis (see Chapter 18). Now it is easy to see that

$$\mathbf{K}_{\mathrm{T}ij} = \frac{\partial^2 \Pi}{\partial \mathbf{a}_i\, \partial \mathbf{a}_j} = \mathbf{K}_{\mathrm{T}ji}^{\mathrm{T}} \tag{3.67}$$

hence $\mathbf{K}_{\mathrm{T}}$ is symmetric.

For a quadratic functional we have, from Eq. (3.65),

$$\delta\left(\frac{\partial \Pi}{\partial \mathbf{a}}\right) = \mathbf{K}\, \delta \mathbf{a} \qquad \text{or} \qquad \mathbf{K} = \mathbf{K}^{\mathrm{T}} \tag{3.68}$$

and hence symmetry must exist.

The fact that *symmetric matrices will arise whenever a variational principle exists is one of the most important merits of variational approaches for discretization.*

How then do 'variational principles' arise and is it always possible to construct these for continuous problems?

To answer the first part of the question we note that frequently the physical aspects of the problem can be stated directly in a variational principle form. Such theorems as minimization of total potential energy to achieve equilibrium in mechanical systems, least energy dissipation principles in viscous flow, etc., may be known to the reader and are considered by many as the basis of formulation. We have already referred to the first of these in section 2.4 of Chapter 2.

Variational principles of this kind are 'natural' ones but unfortunately they do not exist for all continuum problems for which well-defined differential equations may be formulated.

However, there is another category of variational principles which we may call 'contrived'. Such contrived principles can always be constructed for any differentially specified problems either by extending the number of unknown functions **u** by additional variables known as Lagrange multipliers, or by procedures imposing a higher degree of continuity requirements such as least square problems. In subsequent sections we shall discuss, respectively, such 'natural' and 'contrived' variational principles.

Before proceeding further it is worth noting that, in addition to symmetry arising in equations derived by variational means, sometimes further motivation arises. When 'natural' variational principles exist the quantity Π may be of specific interest itself. If this arises a variational approach possesses the merit of easy evaluation of this functional.

The reader will observe that if the functional is 'quadratic' and yields Eq. (3.65), then we can write the approximate 'functional' Π simply as

$$\Pi = \tfrac{1}{2}\mathbf{a}^{\mathrm{T}}\mathbf{K}\mathbf{a} + \mathbf{a}^{\mathrm{T}}\mathbf{f}. \tag{3.69}$$

That this is true the reader can observe by simple differentation.†

† Observe that

$$\delta\Pi = \tfrac{1}{2}\delta(\mathbf{a}^{\mathrm{T}})\mathbf{K}\mathbf{a} + \tfrac{1}{2}\mathbf{a}^{\mathrm{T}}\mathbf{K}\,\delta\mathbf{a} + \delta\mathbf{a}^{\mathrm{T}}\mathbf{f}.$$

As **K** is symmetric

$$\delta\mathbf{a}^{\mathrm{T}}\mathbf{K}\mathbf{a} \equiv \mathbf{a}^{\mathrm{T}}\mathbf{K}\,\delta\mathbf{a}$$

hence

$$\delta\Pi = \delta\mathbf{a}^{\mathrm{T}}(\mathbf{K}\mathbf{a} + \mathbf{f}) = 0$$

gives

$$\mathbf{K}\mathbf{a} + \mathbf{f} = 0.$$

3.10 'Natural' Variational Principles and their Relation to Governing Differential Equations

3.10.1 *Euler equations.* If we consider the definitions of Eqs. (3.61) and (3.62) we observe that for stationarity we can write, after performing some differentiations,

$$\delta\Pi = \int_\Omega \delta\mathbf{u}^{\mathrm{T}}\mathbf{A}(\mathbf{u})\,\mathrm{d}\Omega + \int_\Gamma \delta\mathbf{u}^{\mathrm{T}}\mathbf{B}(\mathbf{u})\,\mathrm{d}\Gamma = 0. \tag{3.70}$$

As the above has to be true for any variations $\delta\mathbf{u}$, we must have

$$\mathbf{A}(\mathbf{u}) = 0 \quad \text{in } \Omega \tag{3.71}$$

and

$$\mathbf{B}(\mathbf{u}) = 0 \quad \text{on } \Gamma.$$

If $\mathbf{A}$ corresponds precisely to the differential equations governing the problem and $\mathbf{B}$ to its boundary conditions, then the variational principle is a *natural* one. Equations (3.70)–(3.71) are known as the Euler differential equations corresponding to the variational principle requiring the stationarity of Π. It is easy to show that for any variational principle a corresponding set of Euler equations can be established. The reverse is unfortunately not true, i.e., only certain forms of differential equations are Euler equations of a variational functional. In the next section we shall consider the conditions necessary for the existence of variational principles and give a prescription for the establishment of Π from a set of suitable linear differential equations. In this section we shall continue to assume that the form of the variational principle is known.

To illustrate the process let us consider now a specific example:

Suppose we specify the problem by requiring the stationarity of a functional

$$\Pi = \int_\Omega \left[\frac{1}{2}k\left(\frac{\partial\phi}{\partial x}\right)^2 + \frac{1}{2}k\left(\frac{\partial\phi}{\partial y}\right)^2 - Q\phi\right]\mathrm{d}\Omega - \int_{\Gamma_q} \bar{q}\phi\,\mathrm{d}\Gamma \tag{3.72}$$

in which k and Q depend only on position and $\delta\phi$ is such that $\delta\phi = 0$ on Γ_ϕ where Γ_ϕ and Γ_q are bounding the domain Ω.

We now perform the variation. This can be written following rules of differentiation as

$$\delta\Pi = \int_\Omega \left[k\frac{\partial\phi}{\partial x}\,\delta\left(\frac{\partial\phi}{\partial x}\right) + k\frac{\partial\phi}{\partial y}\,\delta\left(\frac{\partial\phi}{\partial y}\right) - Q\,\delta\phi\right]\mathrm{d}\Omega - \int_{\Gamma_q} (\bar{q}\,\delta\phi)\,\mathrm{d}\Gamma \tag{3.73}$$

As

$$\delta\left(\frac{\partial\phi}{\partial x}\right) = \frac{\partial}{\partial x}(\delta\phi) \tag{3.74}$$

we can integrate by parts (as in section 3.3) and, noting that $\delta\phi = 0$ on Γ_ϕ, obtain

$$\delta\Pi = -\int_\Omega \delta\phi\left[\frac{\partial}{\partial x}\left(k\frac{\partial\phi}{\partial x}\right)+\frac{\partial}{\partial y}\left(k\frac{\partial\phi}{\partial y}\right)+Q\right]\mathrm{d}\Omega+\int_{\Gamma_q}\delta\phi\left(k\frac{\partial\phi}{\partial n}-\bar{q}\right)\mathrm{d}\Gamma = 0. \tag{3.75a}$$

This is of the form of Eq. (3.70) and we immediately observe that the Euler equations are

$$\mathbf{A}(\phi) = \frac{\partial}{\partial x}\left(k\frac{\partial\phi}{\partial y}\right)+\frac{\partial}{\partial y}\left(k\frac{\partial\phi}{\partial y}\right)+Q \quad \text{in } \Omega$$

$$\mathbf{B}(\phi) = k\frac{\partial\phi}{\partial n}-\bar{q} = 0 \quad \text{on } \Gamma_q. \tag{3.75b}$$

If ϕ is so prescribed that $\phi = \bar{\phi}$ on Γ_ϕ and $\delta\phi = 0$ on that boundary, then the problem is precisely the one we have already discussed in section 3.3 and the functional (3.72) specifies the *two-dimensional heat conduction* problem in an alternative way.

In this case we have 'guessed' the functional but the reader will observe that the variation operation could have been carried out for any functional specified and corresponding *Euler* equations could have been established.

Let us continue the problem to obtain an approximate solution of the linear heat conduction problem.

Taking, as usual,

$$\phi \approx \hat{\phi} = \sum N_i a_i = \mathbf{N}\mathbf{a} \tag{3.76}$$

we substitute this approximation into the expression for the functional Π (Eq. (3.72)) and obtain

$$\Pi = \int_\Omega \frac{1}{2}k\left(\sum\frac{\partial N_i}{\partial x}a_i\right)^2 \mathrm{d}\Omega+\int_\Omega \frac{1}{2}k\left(\sum\frac{\partial N_i}{\partial y}a_i\right)^2 \mathrm{d}\Omega$$
$$-\int_\Omega Q\sum N_i a_i\,\mathrm{d}\Omega-\int_{\Gamma_q}\bar{q}\sum N_i a_i\,\mathrm{d}\Gamma \tag{3.77}$$

On differentiation with respect to a typical parameter a_j we have

$$\frac{\partial\Pi}{\partial a_j} = \int_\Omega k\left(\sum\frac{\partial N_i}{\partial x}a_i\right)\frac{\partial N_j}{\partial x}\mathrm{d}\Omega+\int k\left(\sum\frac{\partial N_i}{\partial y}a_i\right)\frac{\partial N_j}{\partial y}\mathrm{d}\Omega$$
$$-\int_\Omega Q\,N_j\,\mathrm{d}\Omega-\int_{\Gamma_q}\bar{q}N_j\,\mathrm{d}\Gamma \tag{3.78}$$

and a system of equations for solution of the problem is

$$\mathbf{K}\mathbf{a}+\mathbf{f} = 0 \tag{3.79}$$

with

$$K_{ij} = K_{ji} = \int_{\Omega} k\left(\frac{\partial N_i}{\partial x}\frac{\partial N_j}{\partial x}\right) \mathrm{d}\Omega + \int_{\Omega} k\frac{\partial N_i}{\partial y}\frac{\partial N_j}{\partial y}\,\mathrm{d}\Omega$$

$$f_j = -\int_{\Omega} N_j Q\,\mathrm{d}\Omega - \int_{\Gamma_q} N_j \bar{q}\,\mathrm{d}\Gamma. \tag{3.80}$$

The reader will observe that the approximation equations are here identical with those obtained in section 3.5 for the same problem using the Galerkin process. No special advantage accrues to the variational formulation here—and indeed we can predict now that Galerkin and variational procedures must give the same answer for cases where natural variational principles exist.

3.10.2 *Relation of the Galerkin method to approximation via variational principles.* In the preceding example we have observed that the approximation obtained by the use of a natural variational principle and by the use of the Galerkin weighting process proved identical. That this is the case follows directly from Eq. (3.70), in which the variation was derived in terms of the original differential equations and the associated boundary conditions.

If we consider the usual trial function expansion

$$\mathbf{u} \approx \hat{\mathbf{u}} = \mathbf{N}\mathbf{a} \tag{3.3}$$

we can write the variation of this approximation as

$$\delta\hat{\mathbf{u}} = \mathbf{N}\,\delta\mathbf{a} \tag{3.81}$$

and inserting the above into (3.70) yields

$$\delta\Pi = \delta\mathbf{a}^{\mathrm{T}}\int_{\Omega} \mathbf{N}^{\mathrm{T}}\mathbf{A}(\mathbf{N}\mathbf{a})\,\mathrm{d}\Omega + \delta\mathbf{a}^{\mathrm{T}}\int_{\Gamma} \mathbf{N}^{\mathrm{T}}\mathbf{B}(\mathbf{N}\mathbf{a})\,\mathrm{d}\Gamma.$$

The above form being true of all $\delta\mathbf{a}$ requires that the expression under the integrals should be zero. The reader will immediately recognize this as simply the Galerkin form of the weighted residual statement discussed earlier (Eq. (3.25)), and identity is hereby proved.

We need to underline, however, that this is only true if the Euler equations of the variational principle coincide with the governing equations of the original problems. The Galerkin process thus retains its greater range of applicability.

At this stage another point must be made, however. If we consider a *system* of governing equations

$$\mathbf{A}(\mathbf{u}) = \begin{Bmatrix} A_1(\mathbf{u}) \\ A_2(\mathbf{u}) \\ \vdots \end{Bmatrix} = 0 \tag{3.1}$$

with $\hat{\mathbf{u}} = \mathbf{N}\mathbf{a}$, the Galerkin weighted residual equation becomes (disregarding the boundary conditions)

$$\int_\Omega \mathbf{N}^{\mathrm{T}} \mathbf{A}(\hat{\mathbf{u}})\, \mathrm{d}\Omega = 0. \tag{3.83}$$

This form is not unique as the equation systems $\mathbf{A}$ can be ordered in a number of ways. Only one such ordering will correspond precisely with the Euler equations of a variational principle (if this exists) and the reader can verify that for an equation system weighted in the Galerkin manner at best only one arrangement of the vector $\mathbf{A}$ results in a symmetric set of equations.

As an example, consider for instance the one-dimensional heat conduction problem (Example 1, section 3.5) redefined as an equation system with two unknowns, ϕ being the temperature and q the heat flow. Disregarding at this stage the boundary conditions we can write these equations as

$$\mathbf{A}(\mathbf{u}) = \begin{Bmatrix} q - \dfrac{\mathrm{d}\phi}{\mathrm{d}x} \\ \dfrac{\mathrm{d}q}{\mathrm{d}x} \end{Bmatrix} + \begin{Bmatrix} 0 \\ Q \end{Bmatrix} = 0 \tag{3.84}$$

or as a linear equation system,

$$\mathbf{A}(\mathbf{u}) \equiv \mathbf{L}\mathbf{u} + \mathbf{b} = 0$$

in which

$$\mathbf{L} \equiv \begin{bmatrix} 1, & -\dfrac{\mathrm{d}}{\mathrm{d}x} \\ \dfrac{\mathrm{d}}{\mathrm{d}x}, & 0 \end{bmatrix}; \quad \mathbf{b} = \begin{Bmatrix} 0 \\ Q \end{Bmatrix}; \quad \mathbf{u} = \begin{Bmatrix} q \\ \phi \end{Bmatrix}. \tag{3.85}$$

Writing the trial function in which a different interpolation is used for each function

$$\mathbf{u} = \sum \mathbf{N}_i \mathbf{a}_i; \quad \mathbf{N}_i = \begin{bmatrix} N_i^1 & 0 \\ 0 & N_i^2 \end{bmatrix}$$

and applying the Galerkin process, we arrive at a usual linear equation system with

$$K_{ij} = \int_{\Omega} N_i^{\mathrm{T}} \mathbf{L} N_j \, \mathrm{d}x = \int_{\Omega} \begin{bmatrix} N_i^1 N_j^1, & -N_i^1 \dfrac{\mathrm{d}}{\mathrm{d}x} N_j^2 \\ N_i^2 \dfrac{\mathrm{d}}{\mathrm{d}x} N_i^1, & 0 \end{bmatrix} \mathrm{d}x \tag{3.86}$$

This form yields a symmetric equation† system after integration by parts, and

$$K_{ij} = K_{ji}. \tag{3.87}$$

If the order of equations were simply reversed, i.e., using

$$\mathbf{A}(\mathbf{u}) = \begin{bmatrix} \dfrac{dq}{dx} \\ q - \dfrac{\mathrm{d}\phi}{\mathrm{d}x} \end{bmatrix} + \begin{Bmatrix} Q \\ 0 \end{Bmatrix} = 0 \tag{3.88}$$

application of the Galerkin process would now lead to non-symmetric equations quite different from those arising using the variational principle. The second type of Galerkin approximation would clearly be less desirable due to loss of symmetry in the final equations. It is easy to show that the first system corresponds precisely to Euler equations of a variational functional.

3.11 Establishment of Natural Variational Principles for Linear, Self-adjoint Differential Equations

3.11.1 *General theorems.* General rules for deriving natural variational principles from non-linear differential equations are complicated and even the tests necessary to establish the existence of such variational principles are not simple. Much mathematical work has been done, however, in this context by Veinberg,[14] Tonti,[15] Oden,[16] and others.

For linear differential equations the situation is much simpler and a thorough study is available in the works of Mikhlin,[17,18] and in this section a brief presentation of this is given.

We shall consider here only the establishment of variational principles for linear system of equations with *forced* boundary conditions, implying

† As

$$\int N_i^1 \frac{\mathrm{d}}{\mathrm{d}x} N_j^2 \, \mathrm{d}x \equiv -\int N_j^2 \frac{\mathrm{d}}{\mathrm{d}x} N_i^1 \, \mathrm{d}x + \text{boundary terms}$$

only variation of functions which yield $\delta\mathbf{u} = 0$ on their boundaries. The extension to include natural boundary conditions is simple and will be omitted.

Writing a linear system of differential equations as

$$\mathbf{A}(\mathbf{u}) \equiv \mathbf{Lu}+\mathbf{b} = 0 \tag{3.89}$$

in which $\mathbf{L}$ is a linear, differential, operator it can be shown that natural variational principles require that the operator $\mathbf{L}$ be such that

$$\int_\Omega \boldsymbol{\psi}^{\mathrm{T}}\mathbf{L}\boldsymbol{\gamma}\, \mathrm{d}\Omega = \int \boldsymbol{\gamma}^{\mathrm{T}}\mathbf{L}\boldsymbol{\psi}\, \mathrm{d}\Omega + \text{b.t.} \tag{3.90}$$

for any two function sets ψ and γ. In the above, 'b.t.' stands for boundary terms which we disregard in the present context. The property required in the above operator is called one of *self-adjointness* or *symmetry*.

If the operator $\mathbf{L}$ is self-adjoint, the variational principle can be written immediately as

$$\Pi = \int_\Omega \left[\frac{1}{2}\mathbf{u}^{\mathrm{T}}\mathbf{Lu}+\mathbf{u}^{\mathrm{T}}\mathbf{b}\right] \mathrm{d}\Omega + \text{b.t.} \tag{3.91}$$

To prove the veracity of the last statement a variation needs to be considered. We write thus

$$\delta\Pi = \int_\Omega \left[\frac{1}{2}\delta\mathbf{u}^{\mathrm{T}}\mathbf{Lu}+\frac{1}{2}\mathbf{u}^{\mathrm{T}}\delta(\mathbf{Lu})+\delta\mathbf{u}^{\mathrm{T}}\mathbf{b}\right] \mathrm{d}\Omega + \text{b.t.} \tag{3.92}$$

Noting that for any linear operator

$$\delta(\mathbf{Lu}) \equiv \mathbf{L}\,\delta\mathbf{u} \tag{3.93}$$

and that $\mathbf{u}$ and $\delta\mathbf{u}$ can be treated as any two independent functions, by identity (3.90), we can write Eq. (3.92) as

$$\delta\Pi = \int_\Omega \delta\mathbf{u}^{\mathrm{T}}(\mathbf{Lu}+\mathbf{b})\, \mathrm{d}\Omega + \text{b.t.} \tag{3.94}$$

We observe immediately that the term in the brackets, i.e., the Euler equation of the functional, is identical with the original equation postulated, and therefore the variational principle is verified.

The above gives a very simple test and a prescription for the establishment of natural variational principles for differential equations of the problem.

Consider, for instance, two examples.

Example 1: This is a problem governed by the differential equation similar to the heat conduction equation, e.g.,

$$\nabla^2\phi + c\phi + Q = 0 \tag{3.95}$$

with c and Q being dependent on position only.

The above can be written in the general form of Eq. (3.89), with

$$\mathbf{L} \equiv \left[\frac{\partial^2}{\partial x^2}, \frac{\partial^2}{\partial y^2}, c\right]; \mathbf{b} \equiv Q. \tag{3.96}$$

Verifying that self-adjointness applies (which we leave to the reader as an exercise), we immediately have a variational principle

$$\Pi = \int_\Omega \left\{\frac{1}{2}\phi\left[\frac{\partial^2 \phi}{\partial x^2} + \frac{\partial^2 \phi}{\partial y^2} + c\phi\right] + Q\phi\right\} \mathrm{d}x\,\mathrm{d}y \tag{3.97}$$

with ϕ satisfying the forced boundary condition, i.e., $\delta\phi = 0$ on Γ. Integration by parts of the first two terms results in

$$\Pi = -\int_\Omega \left[\frac{1}{2}\left(\frac{\partial \phi}{\partial x}\right)^2 + \frac{1}{2}\left(\frac{\partial \phi}{\partial y}\right)^2 - \frac{1}{2}c\phi^2 - Q\phi\right] \mathrm{d}x\,\mathrm{d}y \tag{3.98}$$

on noting that boundary terms with prescribed ϕ do not alter the principle.

Example 2: This problem concerns the equation system discussed in the previous section (Eqs. (3.84)–(3.85)). Again self-adjointness of the operator can be tested—and found satisfied. We now write the functional as

$$\Pi = \int_\Omega \left(\frac{1}{2}\begin{Bmatrix} q \\ \phi \end{Bmatrix}^{\mathrm{T}} \begin{bmatrix} 1, & -\frac{\mathrm{d}}{\mathrm{d}x} \\ \frac{\mathrm{d}}{\mathrm{d}x}, & 0 \end{bmatrix} \begin{Bmatrix} q \\ \phi \end{Bmatrix} + \begin{Bmatrix} q \\ \phi \end{Bmatrix}^{\mathrm{T}} \begin{Bmatrix} 0 \\ q \end{Bmatrix}\right) \mathrm{d}x$$

$$= \int_\Omega \left(q^2 - q\frac{\mathrm{d}\phi}{\partial x} + \phi\frac{\mathrm{d}q}{\mathrm{d}x} + \phi q\right) \mathrm{d}x. \tag{3.99}$$

The verification of the correctness of the above, by executing a variation, is left to the reader.

These two examples illustrate the simplicity of application of the general expressions. The reader will observe that self-adjointness of the operator will generally exist if even orders of differentiation are present. For odd orders the self-adjointness is only possible if the operator is a 'skew'-symmetric matrix such as occurs in the second example.

3.11.2 *Adjustment for self-adjointness.* On occasion a linear operator which is not self-adjoint can be adjusted so that self-adjointness is achieved without altering the basic equation. Consider, for instance, a problem governed by the following differential equation of a standard linear form:

$$\frac{\mathrm{d}^2\phi}{\mathrm{d}x^2} + \alpha\frac{\mathrm{d}\phi}{\mathrm{d}x} + \beta\phi + q = 0. \tag{3.100}$$

In this equation α and β are some functions of x. It is easy to see that the operator **L** is now a scalar

$$L \equiv \left[\frac{d^2}{dx^2}+\alpha\frac{d}{dx}+\beta\right] \tag{3.101}$$

and is not self-adjoint.

Let p be some, as yet undetermined, function of x. We shall show that it is possible to convert Eq. (3.100) to a self-adjoint form by multiplying it by this function. The new operator becomes

$$\bar{L} = pL. \tag{3.102}$$

To test for symmetry with any two functions ψ and γ we write

$$\int_\Omega \psi(pL\gamma)\,dx = \int_\Omega \left[\psi p\frac{d^2\gamma}{dx^2}+\psi p\alpha\frac{d\gamma}{dx}+\psi p\beta\gamma\right] dx. \tag{3.103}$$

On integration of the first term, by parts, we have

$$\int_\Omega \left(-\frac{d(\psi p)}{dx}\frac{d\gamma}{dx}+\psi p\alpha\frac{d\gamma}{dx}+\beta\psi p\gamma\,dx\right) dx + \text{b.t.}$$

$$= \int\left(-\frac{d\psi}{dx}p\frac{d\gamma}{dx}+\psi\frac{d\gamma}{dx}\left(p\alpha-\frac{dp}{dx}\right)+\psi p\beta\gamma\right) dx + \text{b.t.} \tag{3.104}$$

Symmetry (and therefore self-adjointness) is now achieved in the first and last terms. The middle term will only be symmetric if it disappears, i.e., if

$$p\alpha - \frac{dp}{dx} = 0 \tag{3.105}$$

or

$$\frac{dp}{p} = \alpha\,dx$$

$$p = e^{\int \alpha\,dx}. \tag{3.106}$$

By using this value of p the operator is made self-adjoint and a variational principle for the problem of Eq. (3.100) is easily found.

Procedure of this kind has been used by Guyman[19] *et al.* to derive variational principles for a convective diffusion equation which is not self-adjoint. (We have noted such lack of symmetry in the equation in section 3.5, Example 2.)

A similar method for creating variational functionals can be extended to a special case of non-linearity of Eq. (3.89) when

$$\mathbf{b} = \mathbf{b}(\mathbf{u}, x, \ldots) \tag{3.107}$$

If Eq. (3.92) is inspected we note that we could write

$$\delta(\mathbf{u}^{\mathrm{T}}\mathbf{b}) = \delta(\mathbf{g}) \tag{3.108}$$

if

$$\mathbf{g} = \int \mathbf{b}^{\mathrm{T}}\, \mathrm{d}\mathbf{u}.$$

This integration is generally quite easy to accomplish.

3.12 Maximum, Minimum, or a Saddle Point?

In discussing variational principles so far we have assumed simply that at the solution point $\delta\Pi = 0$, or that the functional is stationary. It is often desirable to know whether Π is at a maximum, minimum, or simply at a 'saddle point'. If a maximum or a minimum is involved, then the approximation will always be 'bounded', i.e., will provide approximate values of Π which are either smaller or larger than the correct ones. This in itself may be of practical significance.

When, in elementary calculus, we consider a stationary point of a function Π of one variable a, we investigate the rate of change of $\mathrm{d}\Pi$ with $\mathrm{d}a$ and write

$$\mathrm{d}(\mathrm{d}\Pi) = \mathrm{d}\left(\frac{\partial\Pi}{\partial a}\,\mathrm{d}a\right) = \frac{\partial^2\Pi}{\partial a^2}(\mathrm{d}a)^2. \tag{3.109}$$

The sign of the second derivative determines whether Π is a minimum, maximum, or simply stationary (saddle point), as shown in Fig. 3.7. By analogy in the calculus of variation we shall consider changes of $\delta\Pi$.

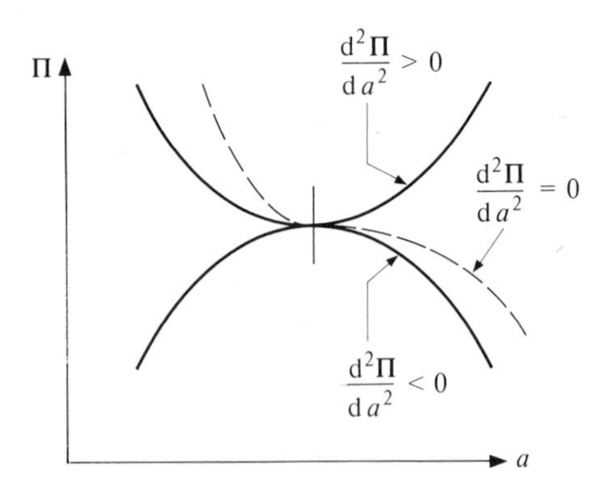

Fig. 3.7 Maximum, minimum and a 'saddle' point for a functional Π of one variable.

Noting the general form of this quantity given by Eq. (3.63) and the notion of the second derivative of Eq. (3.66) we can write, in terms of discrete parameters,

$$\begin{aligned}\delta(\delta\Pi) &\equiv \delta\left(\left(\frac{\partial \Pi}{\partial \mathbf{a}}\right)^{\mathrm{T}} \delta\mathbf{a}\right) \\ &= \delta\mathbf{a}^{\mathrm{T}}\,\delta\left(\frac{\partial \Pi}{\partial \mathbf{a}}\right) \\ &= \delta\mathbf{a}^{\mathrm{T}}\mathbf{K}_{\mathrm{T}}\,\delta\mathbf{a}\end{aligned} \tag{3.110}$$

If, in the above, $\delta(\delta\Pi)$ is always negative then Π is obviously reaching a maximum, if it is always positive then Π is a minimum, but if the sign is indeterminate this shows only the existence of a saddle point.

As $\delta\mathbf{a}$ is an arbitrary vector this statement is equivalent in requiring the matrix $\mathbf{K}_{\mathrm{T}}$ to be negative definite for a maximum *or* positive definite for a minimum. The form of the matrix $\mathbf{K}_{\mathrm{T}}$ (or in linear problems of $\mathbf{K}$ which is identical to it) is thus of great importance in variational problems.

3.13 Constrained Variational Principles. Lagrange Multipliers and Adjoint Functions

3.13.1 *Lagrangian multipliers.* Consider the problem of making a functional Π stationary, subject to the unknown $\mathbf{u}$ obeying some set of additional differential relationships

$$\mathbf{C}(\mathbf{u}) = 0 \quad \text{in } \Omega. \tag{3.111}$$

We can introduce this constraint by forming another functional

$$\bar{\Pi} = \Pi + \int_{\Omega} \boldsymbol{\lambda}^{\mathrm{T}}\mathbf{C}(\mathbf{u})\,\mathrm{d}\Omega \tag{3.112}$$

in which $\boldsymbol{\lambda}$ is some set of functions of the independent co-ordinates in domain Ω known as the *Lagrangian multipliers.* The variation of the new functional is now

$$\delta\bar{\Pi} = \delta\Pi + \int_{\Omega} \delta\boldsymbol{\lambda}^{\mathrm{T}}\mathbf{C}(\mathbf{u})\,\mathrm{d}\Omega + \int \boldsymbol{\lambda}^{\mathrm{T}}\,\delta\mathbf{C}(\mathbf{u})\,\mathrm{d}\Omega \tag{3.113}$$

and this is zero providing $\mathbf{C}(\mathbf{u}) = 0$ (and hence $\boldsymbol{\delta}\mathbf{C} = 0$) and, simultaneously,

$$\delta\Pi = 0. \tag{3.114}$$

In a similar way, constraints can be introduced at some points or over boundaries of the domain. For instance, if we require that **u** obey

$$\mathbf{E}(\mathbf{u}) = 0 \quad \text{on } \Gamma \tag{3.115}$$

we would add to the original functional the term

$$\int_{\Gamma} \boldsymbol{\lambda}^{\mathrm{T}} \mathbf{E}(\mathbf{u})\, \mathrm{d}\Gamma \tag{3.116}$$

with $\boldsymbol{\lambda}$ now being an unknown function defined only on Γ. Alternatively, if the constraint **C** is applicable at one or more points of the system, then the simple addition of $\boldsymbol{\lambda}^{\mathrm{T}}\mathbf{C}(\mathbf{u})$ at these points to the general functional Π will introduce the constraints.

It appears, therefore, possible to introduce always additional functions $\boldsymbol{\lambda}$ and modify a functional to include any prescribed constraints. In the 'discretization' process we shall now have to use trial functions to describe both **u** and $\boldsymbol{\lambda}$.

Writing, for instance,

$$\hat{\mathbf{u}} = \sum \mathbf{N}_i \mathbf{a}_i = \mathbf{N}\mathbf{a}; \quad \hat{\boldsymbol{\lambda}} = \sum \tilde{\mathbf{N}}_i \mathbf{b}_i = \tilde{\mathbf{N}}\mathbf{b} \tag{3.117}$$

we shall obtain a set of equations

$$\frac{\partial \Pi}{\partial \mathbf{c}} = \begin{Bmatrix} \dfrac{\partial \Pi}{\partial \mathbf{a}} \\ \dfrac{\partial \Pi}{\partial \mathbf{b}} \end{Bmatrix} = 0; \qquad \mathbf{c} = \begin{Bmatrix} \mathbf{a} \\ \mathbf{b} \end{Bmatrix} \tag{3.118}$$

from which both the sets of parameters **a** and **b** can be obtained. It is somewhat paradoxical that the 'constrained' problem has resulted in a larger number of unknown parameters than the original one and, indeed, complicated the solution. We shall, nevertheless, find practical use for Lagrangian multipliers in formulating some physical variational principles, and will make use of these in a more general context in Chapter 12.

The point about increasing the number of parameters to introduce a constraint may perhaps be best illustrated in a simple algebraic situation in which we require a stationary value of a quadratic function of two variables a_1 and a_2.

$$\Pi = 2a_1^2 - 2a_1 a_2 + a_2^2 + 18a_1 + 6a_2 \tag{3.119}$$

subject to a constraint

$$a_1 - a_2 = 0 \tag{3.120}$$

The obvious way to proceed would be to insert about equality 'constraint' and obtain

$$\Pi = a_1^2 + 24a_1 \tag{3.121}$$

and write, for stationarity,

$$\frac{\partial \Pi}{\partial a_1} = 0 = 2a_1 + 24; \quad a_1 = a_2 = -12. \tag{3.122}$$

Introducing a Langrangian multiplier λ we can alternatively find the stationarity of

$$\bar{\Pi} = 2a_1^2 - 2a_1a_2 + a_2^2 + 18a_1 + 6a_2 + \lambda(a_1 - a_2) \tag{3.123}$$

and write *three* simultaneous equations

$$\frac{\partial \bar{\Pi}}{\partial a_1} = 0; \qquad \frac{\partial \bar{\Pi}}{\partial a_2} = 0; \qquad \frac{\partial \bar{\Pi}}{\partial \lambda} = 0. \tag{3.124}$$

The solution of the above system yields again the correct answer

$$a_1 = a_2 = -12; \qquad \lambda = 6$$

but at considerably more effort. Unfortunately, in most continuum problems the direct elimination of constraints cannot be so simply accomplished.†

Before proceeding further it is of interest to investigate the form of equations resulting from the modified functional Π of Eq. (3.112). If the original functional Π gave as its Euler equations a system

$$\mathbf{A}(\mathbf{u}) = 0 \tag{3.125}$$

then we have

$$\delta\bar{\Pi} = \int_\Omega \delta\mathbf{u}^{\mathrm{T}}\mathbf{A}(\mathbf{u})\,\mathrm{d}\Omega + \int_\Omega \delta\boldsymbol{\lambda}^{\mathrm{T}}\mathbf{C}(\mathbf{u})\,\mathrm{d}\Omega + \int_\Omega \boldsymbol{\lambda}^{\mathrm{T}}\,\delta\mathbf{C}\,\mathrm{d}\Omega. \tag{3.126}$$

Substituting the trial functions (3.117) we can write, if the constraints are a linear set of equations.

$$\mathbf{C}(\mathbf{u}) = \mathbf{L}_1\mathbf{u} + \mathbf{C}_1$$

$$\delta\bar{\Pi} = \delta\mathbf{a}^{\mathrm{T}}\int_\Omega \mathbf{N}^{\mathrm{T}}\mathbf{A}(\hat{\mathbf{u}})\,\mathrm{d}\Omega + \delta\mathbf{b}^{\mathrm{T}}\int_\Omega \tilde{\mathbf{N}}^{\mathrm{T}}(\mathbf{L}_1\hat{\mathbf{u}} + \mathbf{C}_1)\,\mathrm{d}\Omega + \delta\mathbf{a}^{\mathrm{T}}\int_\Omega (\mathbf{L}_1\mathbf{N})^{\mathrm{T}}\bar{\boldsymbol{\lambda}}\,\mathrm{d}\Omega = 0. \tag{3.127}$$

As this has to be true for all variations $\delta\mathbf{a}$ and $\delta\mathbf{b}$, we have a system of equations

$$\int_\Omega \mathbf{N}^{\mathrm{T}}\mathbf{A}(\hat{\mathbf{u}})\,\mathrm{d}\Omega + \int_\Omega (\mathbf{L}_1\mathbf{N})^{\mathrm{T}}\bar{\boldsymbol{\lambda}}\,\mathrm{d}\Omega = 0.$$

$$\int_\Omega \tilde{\mathbf{N}}^{\mathrm{T}}(\mathbf{L}_1\hat{\mathbf{u}} + \mathbf{C}_1)\,\mathrm{d}\Omega = 0 \tag{3.128}$$

† In the finite element context, Szabo *et al.*[20] use such direct elimination; however, this involves considerable algebraic manipulation.

For linear equations **A** the first term of the first equation is precisely the ordinary, unconstrained, variational approximation

$$\mathbf{K}\mathbf{a}+\mathbf{f} \tag{3.129}$$

and inserting again the trial functions (3.117) we can write the approximated Eq. (3.128) as a linear system.

$$\mathbf{K}_C\mathbf{c} = \begin{bmatrix} \mathbf{K}, & \mathbf{K}_{ab} \\ \mathbf{K}_{ab}^{\mathrm{T}}, & 0 \end{bmatrix} \begin{Bmatrix} \mathbf{a} \\ \mathbf{b} \end{Bmatrix} + \begin{Bmatrix} \mathbf{f} \\ \mathbf{o} \end{Bmatrix} = 0 \tag{3.130}$$

with

$$\mathbf{K}_{ab}^{\mathrm{T}} = \int_\Omega \tilde{\mathbf{N}}^{\mathrm{T}}\mathbf{L}_1\mathbf{N}\,\mathrm{d}\Omega. \tag{3.131}$$

Clearly the equation system is symmetric but now possesses zeros on the diagonal, and therefore the variational principle Π is merely stationary. Further, computational difficulties may be encountered unless the solution process allows for zero diagonal terms.

3.13.2 *Identification of Lagrangian multipliers. Forced boundary conditions and modified variational principles.* Although the Lagrangian multipliers were introduced as a mathematical fiction necessary for the enforcement of certain external constraints required to satisfy the original variational principle, we shall find that in most physical situations they can be identified with certain physical quantities of importance to the original mathematical model. Such an identification will follow immediately from the definition of the variational principle established in Eq. (3.112) and through the second of the Euler equations corresponding to it. The variation $\delta\bar{\Pi}$, written in Eq. (3.113), supplies through its first two terms the original Euler equation of the problem corresponding to the functional Π and the constraint equation. The last term can be always rewritten as

$$\int \boldsymbol{\lambda}^{\mathrm{T}}\,\delta\mathbf{C}(\mathbf{u})\,\mathrm{d}\Omega \equiv \int \delta\mathbf{u}^{\mathrm{T}}\mathbf{R}(\boldsymbol{\lambda}, \mathbf{u})\,\mathrm{d}\Omega + \text{b.t.} \tag{3.132}$$

imposing the requirements that

$$\mathbf{R}(\boldsymbol{\lambda}, \mathbf{u}) = 0. \tag{3.133}$$

This supplies the identification of $\boldsymbol{\lambda}$.

In the literature of variational calculation such identification arises frequently and the reader is referred to the excellent text by Washizu[21] for numerous examples.

Here we shall introduce this identification by means of the example considered in section 3.10.1. As we have noted the variational principle of Eq. (3.72) established the governing equation and the natural boundary conditions of the heat conduction problem providing the forced boundary condition

$$\mathbf{C}(\phi) = \phi - \bar{\phi} = 0 \tag{3.134}$$

was satisfied on Γ_ϕ in the choice of trial function for ϕ.

The above, forced boundary, condition can however be considered as a constraint on the original problem. We can write the constrained variational principle as

$$\bar{\Pi} = \Pi + \int_{\Gamma_\phi} \lambda(\phi - \bar{\phi}) \, \mathrm{d}\Gamma \tag{3.135}$$

where Π is given by Eq. (3.72).

Performing the variation we have

$$\delta\bar{\Pi} = \delta\Pi + \int_{\Gamma_\phi} \delta\lambda(\phi - \bar{\phi}) + \int_{\Gamma_\phi} \delta\phi\lambda \, \mathrm{d}\Gamma \tag{3.136}$$

$\delta\Pi$ is now given by the expression (3.75a) augmented by an integral

$$\int_{\Gamma_\phi} \delta\phi k \frac{\partial\phi}{\partial n} \, \mathrm{d}\Gamma \tag{3.137}$$

which previously was disregarded (as we have assumed $\delta\phi = 0$ on Γ_ϕ). In addition to the conditions of Eq. (3.75b), we now require that

$$\int_{\Gamma_\phi} \delta\lambda(\phi - \bar{\phi}) \, \mathrm{d}\Gamma + \int_{\Gamma_\phi} \delta\phi\left(\lambda + k\frac{\partial\phi}{\partial n}\right) \mathrm{d}\Gamma = 0 \tag{3.138}$$

which must be true for all variations $\delta\lambda$ and $\delta\phi$. The first simply reiterates the constraint

$$\phi - \bar{\phi} = 0 \quad \text{on } \Gamma_\phi \tag{3.139}$$

The second *defines* λ as

$$\lambda = -k\frac{\partial\phi}{\partial n}. \tag{3.140}$$

Noting that $k(\partial\phi/\partial n)$ is equal to the flux $-q$ on the boundary Γ_ϕ, a physical identification of the multiplier has been achieved.

The identification of the Lagrangian variable leads to the possible establishment of a modified variational principle in which λ is replaced by the identification.

We could thus write a new principle for the above example:

$$\bar{\bar{\Pi}} = \Pi + \int_{\Gamma_\phi} k\frac{\partial\phi}{\partial n}(\phi - \bar{\phi}) \, \mathrm{d}\Gamma \tag{3.141}$$

in which once again Π is given by the expression (3.72) but ϕ is not constrained to satisfy any boundary conditions. Use of such modified variational principles can be made to restore inter-element continuity and appears to have been introduced for that purpose first by Kikuchi and Ando.[22] We shall discuss such applications in that context in Chapter 12.

A further extension of such principles has been made use of by Chen and Mei[23] and Zienkiewicz.[24] Washizu[21] discusses many such applications in the context of structural mechanics. The reader can verify that the variational principle expressed in Eq. (3.141) leads to an automatic satisfaction of all the necessary boundary conditions in the example considered.

The use of modified variational principles restores the problem to the original number of unknown functions or parameters and is thus computationally advantageous.

3.13.3 *A general variational principle: adjoint functions and operators.* The Lagrangian multiplier procedure leads to an obvious procedure of 'creating' a variational principle for any set of equations

$$\mathbf{A}(\mathbf{u}) = 0. \tag{3.142}$$

Treating all the above equations as a set of constraints we can obtain such a general variational functional for Eq. (3.57) simply by putting $\Pi = 0$ and writing

$$\bar{\Pi} = \int_{\Omega} \boldsymbol{\lambda}^{\mathrm{T}} \mathbf{A}(\mathbf{u}) \, \mathrm{d}\Omega \tag{3.143}$$

now requiring stationarity for all variations of $\delta\boldsymbol{\lambda}$ and $\delta\mathbf{u}$. The new variational principle has, however, been introduced at the expense of doubling the number of variables in the discretized situation. Treating the case of linear equations only, i.e.,

$$\mathbf{A}(\mathbf{u}) = \mathbf{L}\mathbf{u} + \mathbf{g} = 0 \tag{3.144}$$

and discretizing we note, going through the steps involved in Eqs. (3.126)–(3.130), that the final system of equations now takes the form

$$\begin{bmatrix} 0, & \mathbf{K}_{ab} \\ \mathbf{K}_{ab}^{\mathrm{T}} & 0 \end{bmatrix} \begin{Bmatrix} \mathbf{a} \\ \mathbf{b} \end{Bmatrix} + \begin{Bmatrix} 0 \\ \mathbf{f} \end{Bmatrix} = 0 \tag{3.145}$$

with

$$\begin{aligned} \mathbf{K}_{ab}^{\mathrm{T}} &= \int_{\Omega} \tilde{\mathbf{N}} \mathbf{L}^{\mathrm{T}} \mathbf{N} \, \mathrm{d}\Omega \\ \mathbf{f} &= \int_{\Omega} \tilde{\mathbf{N}}^{\mathrm{T}} \mathbf{g} \, \mathrm{d}\Omega. \end{aligned} \tag{3.146}$$

Equations are completely decoupled and the second set can be solved independently for all the parameters **a** describing the unknowns in which we were originally interested without consideration of parameters **b**. It will be observed that this second set of equations is *identical with an, apparently arbitrary, weighted residual process*. We have thus completed the full circle and obtained the weighted residual forms of section 3.4 from a general variational principle.

The function $\boldsymbol{\lambda}$ which appears in the variational principle of Eq. (3.143) is known as the *adjoint function to* **u**.

By performing a variation on Eq. (3.143) it is easy to show that the Euler equations of the principle are such that

$$\mathbf{A}(\mathbf{u}) = 0 \tag{3.147}$$

and

$$\mathbf{A}^*(\mathbf{u}) = 0 \tag{3.148}$$

where the operator $\mathbf{A}^*$ is such that

$$\int \boldsymbol{\lambda}^{\mathrm{T}}\, \delta(\mathbf{A}\mathbf{u})\, \mathrm{d}\Omega = \int \delta\mathbf{u}^{\mathrm{T}} \mathbf{A}^*(\boldsymbol{\lambda})\, \mathrm{d}\Omega. \tag{3.149}$$

The operator $\mathbf{A}^*$ is known as the adjoint operator and will exist only in linear problems.

For the full significance of the adjoint operator the reader is advised to consult mathematical texts.[25]

3.14 Constrained Variational Principles. Penalty Functions and the Least Square Method

3.14.1 *Penalty functions.* In the previous section we have seen how the process of introducing Lagrange multipliers allows constrained variational principles to be obtained at the expense of increasing the total number of unknowns. Further, we have shown that even in linear problems the algebraic equations which have to be solved are now complicated by having zero diagonal terms. In this section we shall consider an alternative procedure of introducing constraints which does not possess these drawbacks.

Considering once again the problem of obtaining stationarity of Π with a set of const aint equations $\mathbf{C}(\mathbf{u}) = 0$ in domain Ω, we note that the product

$$\mathbf{C}^{\mathrm{T}}\mathbf{C} = C_1^2 + C_2^2 + \cdots \tag{3.150}$$

where

$$\mathbf{C}^{\mathrm{T}} = [C_1, C_2, \ldots]$$

must always be a quantity which is positive or zero. Clearly, the latter value is found when the constraints are satisfied and clearly the variation

$$\delta(\mathbf{C}^T\mathbf{C}) = 0 \tag{3.151}$$

as the product reaches that minimum.

We can now immediately write a new functional

$$\bar{\bar{\Pi}} = \Pi + \alpha \int \mathbf{C}^T(\mathbf{u})\mathbf{C}(\mathbf{u})\, d\Omega \tag{3.152}$$

in which α is a 'penalty number' and require the stationarity for the constrained solution. If Π itself is a minimum of the solution then α should be a positive number. The solution obtained by the stationarity of the functional $\bar{\Pi}$ will satisfy the constraints only approximately. The larger the value of α the better will the constraints be achieved. Further, it seems obvious that the process is best suited to cases where Π is a minimum (or maximum) principle—but success can be obtained even with purely saddle point problems. The process is equally applicable to constraints applied on boundaries or simple discrete constraints. In this latter case integration is dropped.

To clarify ideas let us once again consider the algebraic problem of section 3.13, in which the stationarity of a functional given by Eq. (3.119) was sought subject to a constraint. With the penalty function approach we now could seek the minimum of a functional

$$\bar{\bar{\Pi}} = 2a_1^2 - 2a_1a_2 + a_2^2 + 18a_1 + 6a_2 + \alpha(a_1 - a_2)^2 \tag{3.153}$$

with respect to the variation of both parameters a_1 and a_2. Writing the two simultaneous equations

$$\frac{\partial\bar{\bar{\Pi}}}{\partial a_1} = 0; \qquad \frac{\partial\bar{\bar{\Pi}}}{\partial a_2} = 0 \tag{3.154}$$

we find that as α is increased we approach the correct solution. In Table 3.1 the results are set out demonstrating the convergence.

TABLE 3.1

$\alpha =$	1	2	6	10	100
$a_1 =$	−12·00	−12·00	−12·00	−12·00	−12·00
$a_2 =$	−13·50	−13·00	−12·43	−12·78	−12·03

The reader will observe that in a problem formulated in above manner the constraint introduces no additional unknown parameters—but neither does it decrease their original number. The process will always result in strongly positive definite matrices if the original variational principal is one of a minimum.

In practical application the method of penalty functions has proved quite effective,[26] and indeed is often introduced intuitively. One such 'intuitive' application was already made when we enforced the value of boundary parameters in the manner indicated in Chapter 1, section 1.4.

In the example presented there (and frequently practiced in the real assembly of discretized finite element equations), the forced boundary conditions are not introduced *a priori* and the problem gives, on assembly, a singular system of equations

$$\mathbf{Ka}+\mathbf{f} = 0 \tag{3.155}$$

which can be obtained from a functional (providing $\mathbf{K}$ is symmetric)

$$\Pi = \frac{1}{2}\mathbf{a}^{\mathrm{T}}\mathbf{Ka}+\mathbf{a}^{\mathrm{T}}\mathbf{f}. \tag{3.156}$$

Introducing a prescribed value of a_1, i.e., writing

$$a_1-\bar{a}_1 = 0 \tag{3.157}$$

the functional can be modified to

$$\bar{\bar{\Pi}} = \Pi+\alpha(a_1-\bar{a}_1)^2 \tag{3.158}$$

yielding

$$\bar{\bar{\mathbf{K}}}_{11} = K_{11}+2\alpha, \qquad \bar{\bar{\mathbf{f}}}_1 = f_1-2\alpha\bar{a}_1 \tag{3.159}$$

and giving no change in any of the other matrix coefficients. This is precisely the procedure adopted in Chapter 1 for modifying the equations, to introduce prescribed values of a_1 (2α here replacing α, the 'large number' of section 1.4). Many applications of such a 'discrete' kind are discussed by Campbell.[27]

In the second example we shall consider the problem of beam deflection discussed in Chapter 2 (section 2.10). This problem can be stated as the minimization of total potential energy given by

$$\Pi = \int_0^L EI\left(\frac{\mathrm{d}^2w}{\mathrm{d}x^2}\right)^2 \mathrm{d}x-\int_0^L wq\,\mathrm{d}x. \tag{3.160}$$

As the above formulation requires w to be modelled with C_1 continuity it is of interest to investigate a possibility of a reformulation imposing only C_0 continuity. Such an alternative form would be to require the minimization of

$$\Pi = \int_0^L \frac{1}{2}EI\left(\frac{\mathrm{d}\theta}{\mathrm{d}x}\right)^2 \mathrm{d}x-\int_0^L wq\,\mathrm{d}x \tag{3.161}$$

subject to the constraint

$$C \equiv \frac{\mathrm{d}w}{\mathrm{d}x} - \theta = 0. \tag{3.162}$$

θ here is obviously the slope approximation and Π is now a function of two variables, θ and w, which can be interpolated with C_0 continuity.

A modified variational principle using the penalty function can now be introduced:

$$\bar{\bar{\Pi}} = \Pi + \alpha \int_0^L \left(\frac{\mathrm{d}w}{\mathrm{d}x} - \theta\right)^2 \tag{3.163}$$

where α is a large number.

The structural engineer will immediately recognize that the physical meaning of α is that of shear rigidity, i.e.,

$$\alpha = \frac{1}{2} GA \tag{3.164}$$

and that the formulation presented is simply that for a beam in which the slopes and rotations of sections vary independently and the additional term stands for the strain energy absorbed in shear.

The thick shell and plate elements discussed in Chapter 11 are but an extension of the process here given.

It is easy to show in another context[26,28] that the use of a high Poisson's ratio ($\nu \rightarrow 0{\cdot}5$) for the study of incompressible solids or fluids is in fact equivalent to the introduction of a penalty term to suppress any compressibility allowed by an arbitrary displacement variation.

The use of the penalty function in the finite element context presents certain difficulties.

First, the constrained functional of Eq. (3.152) leads to equations of the form

$$(\mathbf{K}_1 + \alpha \mathbf{K}_2)\mathbf{a} + \mathbf{f} = 0$$

where $\mathbf{K}_1$ derives from the original functions and $\mathbf{K}_2$ from the constraints. As α increases the above equation degenerates

$$\mathbf{K}_2 \mathbf{a} = -\mathbf{f}/\alpha \rightarrow 0$$

and $\mathbf{a} = 0$ unless the matrix $\mathbf{K}_2$ is singular. This singularity does not always arise and we shall discuss means of its introduction in Chapter 11.

Second, with large but finite values of α numerical difficulties will be encountered. Noting that discretization errors can be of comparable magnitude to those due to not *satisfying* the constraint, we can make

$$\alpha = \text{const.}\ (1/h)^n$$

ensuring a limiting convergence to the correct answer.

Fried[29,30] discusses this problem in detail, and again we shall return to it in Chapter 11.

A more general discussion of the whole topic is given in reference 31.

3.14.2 *Least square approximation.* In section 3.13.3 we have shown how a constrained variational principle procedure could be used to construct a general variational principle if the constraints become simply the governing equations of the problem

$$\mathbf{C}(\mathbf{u}) = \mathbf{A}(\mathbf{u}). \tag{3.166}$$

Obviously the same procedure can be used in the context of the penalty function approach by setting $\Pi = 0$ in Eq. (3.152). We thus can write a 'variational principle'

$$\bar{\bar{\Pi}} = \int_\Omega (A_1^2 + A_2 + \cdots) \, d\Omega = \int_\Omega \mathbf{A}^\mathrm{T}(\mathbf{u})\mathbf{A}(\mathbf{u}) \, d\Omega \tag{3.167}$$

for any set of differential equations. In the above equation the boundary conditions are assumed to be satisfied by $\mathbf{u}$ (forced boundary condition) and the parameter α is dropped as it becomes simply a multiplier.

Clearly, the above statement is simply a requirement that the sum of the squares of the residuals of the differential equations should be a minimum at the correct solution. This minimum is obviously zero at that point, and the process is simply the well-known *least square method* of approximation.

It is equally obvious that we could obtain the correct solution by minimizing any functional of the form

$$\bar{\bar{\Pi}} = \int_\Omega (p_1 A_1^2 + p_2 A_2^2 + \cdots) \, d\Omega = \int_\Omega \mathbf{A}^\mathrm{T}(\mathbf{u})\mathbf{p}\mathbf{A}(\mathbf{u}) \, d\Omega \tag{3.168}$$

in which $p_1, p_2, \ldots$, etc. are positive valued functions or constants and $\mathbf{p}$ is a diagonal matrix

$$\mathbf{p} = \begin{bmatrix} p_1 & & 0 \\ & p_2 & \\ 0 & & \ddots \end{bmatrix} \tag{3.169}$$

The above alternative form is sometimes convenient as it puts different importance to the satisfaction of individual components of the equation and allows additional freedom in the choice of the approximate solution. Once again this weighting function could be so chosen as to ensure a constant ratio of terms contributed by various elements—although this has not yet been put into practice.

Least square methods of the kind shown above are a very powerful alternative procedure for obtaining integral forms from which an approximate solution can be started, and have recently been used with considerable success.[32, 33] As the least square variational principles can be written for *any* set of differential equations without introducing additional variables, we may well enquire what is the difference between these and the *natural variational principles* discussed previously. On performing a variation in a specific case the reader will find that the Euler equations which are obtained no longer give the original differential equations but give higher order derivatives of these. This introduces a possibility of spurious solutions if incorrect boundary conditions are used. Further, higher order continuity of trial function is now generally needed. This may be a serious drawback but frequently can be by-passed by stating the problem originally as a set of lower order equations.

We shall now consider the general form of discretized equations resulting from the least square approximation for linear equation sets (again neglecting boundary conditions which are enforced). Thus, if we take

$$\mathbf{A(u)} = \mathbf{Lu}+\mathbf{b} \tag{3.170}$$

and take the usual trial function approximation

$$\hat{\mathbf{u}} = \mathbf{Nu} \tag{3.171}$$

we can write, substituting into (3.168),

$$\bar{\bar{\Pi}} = \int [(\mathbf{LN})\mathbf{a}+\mathbf{b}]^{\mathrm{T}}\,\mathbf{p}[(\mathbf{LN})\mathbf{a}+\mathbf{b}]\,\mathrm{d}\Omega \tag{3.172}$$

and

$$\delta\bar{\bar{\Pi}} = \int_\Omega \delta\mathbf{a}^{\mathrm{T}}(\mathbf{LN})^{\mathrm{T}}\mathbf{p}[(\mathbf{LN})\mathbf{a}+\mathbf{b}]\,\mathrm{d}\Omega + \int_\Omega ((\mathbf{LN})\mathbf{a}+\mathbf{b})^{\mathrm{T}}\,\mathbf{p}(\mathbf{LN})\,\delta\mathbf{a}\,\mathrm{d}\Omega \tag{3.173}$$

or, as $\mathbf{p}$ is symmetric,

$$\delta\bar{\bar{\Pi}} = \delta\mathbf{a}^{\mathrm{T}}\left[(2\int_\Omega (\mathbf{LN})^{\mathrm{T}}\,\mathbf{p}(\mathbf{LN})\,\mathrm{d}\Omega)\mathbf{a}+\int_\Omega (\mathbf{LN})^{\mathrm{T}}\,\mathbf{pb}\,\mathrm{d}\Omega\right]. \tag{3.174}$$

This yields immediately the approximation equation in the usual form

$$\mathbf{Ka}+\mathbf{f} = 0 \tag{3.175}$$

and the reader can observe that the matrix $\mathbf{K}$ is symmetric and positive definite.

To illustrate an actual example, consider a problem governed by Equation (3.95) of this chapter for which we have obtained already a *natural* variational principle (Eq. (3.98)) in which only first derivatives were involved requiring C_0 continuity for **u**. Now, if we use the operator **L** and term **b** defined by Eq. (3.96), we have a set of approximating equations with

$$K_{ij} = 2\int_{\Omega} (\nabla^2 N_i + cN_i)\,\mathbf{p}(\nabla^2 N_j + cN_j)\,\mathrm{d}x\,\mathrm{d}y$$

$$f_i = \int_{\Omega} (\nabla^2 N_i + cN_i) Q\,\mathrm{d}x\,\mathrm{d}y. \tag{3.176}$$

The reader will observe that now a C_1 continuity is needed for the trial functions **N**.

An alternative avoiding this difficulty is to write Eq. (3.95) as a first order system. This can be written as

$$\mathbf{A}(\mathbf{u}) = \begin{Bmatrix} \dfrac{\partial \phi_x}{\partial x} + \dfrac{\partial \phi_y}{\partial y} + c\phi + Q \\ \dfrac{\partial \phi}{\partial x} - \phi_x \\ \dfrac{\partial \phi}{\partial y} - \phi_y \end{Bmatrix} = 0 \tag{3.177}$$

or, introducing a vector **u**,

$$\mathbf{u}^{\mathrm{T}} = [\phi, \phi_x, \phi_y] = (\mathbf{Na})^{\mathrm{T}} \tag{3.178}$$

as the unknown we can write the standard linear form (3.170) as

$$\mathbf{Lu} + \mathbf{b} = 0$$

$$\mathbf{L} = \begin{bmatrix} c, & \dfrac{\partial}{\partial x}, & \dfrac{\partial}{\partial y} \\ \dfrac{\partial}{\partial x}, & 1, & 0 \\ \dfrac{\partial}{\partial y}, & 0, & 1 \end{bmatrix}; \qquad \mathbf{b} = \begin{Bmatrix} Q \\ 0 \\ 0 \end{Bmatrix} \tag{3.179}$$

The reader can now perform the substitution into Eq. (3.174) to obtain the approximation equations in a form requiring only C_0 continuity—introduced, however, at the expense of additional variables. Use of such forms has been made extensively in the finite element context.[32, 33]

3.15 Concluding Remarks

This very extensive chapter presents the general possibilities of using the finite element processes in almost any mathematical or mathematically modelled physical problem. The essential approximation processes have been given in as simple a form as possible, at the same time presenting a fully comprehensive picture which should allow the reader to understand much of the literature and indeed to experiment with new permutations. In the chapters that follow we shall apply to various physical problems only a limited selection of the methods to which allusion has been made. In some we shall show, however, that certain extensions of the process are possible (Chapter 12) and in another (Chapter 11) how a violation of some of the rules here expounded can be accomplished with benefit.

The numerous approximation procedures discussed fall into several categories. To remind the reader of these, we present in Table 3.2 a comprehensive catalogue of the methods used here and in Chapter 2.

TABLE 3.2
FINITE ELEMENT APPROXIMATION

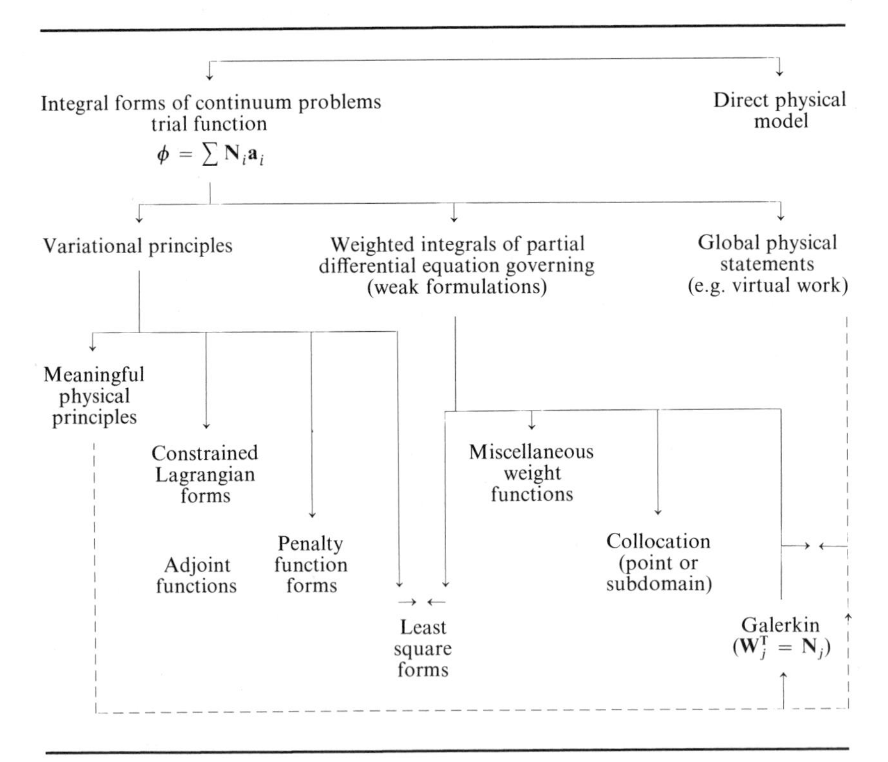

The only aspect of the finite element process mentioned in that table that has not been discussed here is that of a *direct physical method.* In such models an 'atomic' rather than continuum concept is the starting point. While much interest exists in the possibilities offered by such models, their discussion is outside the scope of this book.

In all the continuum processes discussed the first step is always the choice of suitable shape or trial functions. A few simple forms of such functions have been introduced as the need demanded it. In later chapters a fuller discussion of shape function applications will be given. The reader who has mastered the essence of the present chapter will have little difficulty in following the remainder of the book.

References

1. S. H. CRANDALL, *Engineering Analysis*, McGraw-Hill, 1956.
2. B. A. FINLAYSON, *The Method of Weighted Residuals and Variational Principles*, Academic Press, 1972.
3. R. A. FRAZER, W. P. JONES, and S. W. SKEN, *Approximations to functions and to the solutions of differential equations*, Aero. Research Committee Report 1799, 1937.
4. C. B. BIEZENO and R. GRAMMEL, *Technische Dynamik*, p. 142, Springer-Verlag, 1933.
5. B. G. GALERKIN, 'Series solution of some problems of elastic equilibrium of rods and plates' (Russian), *Vestn. Inzh. Tech.*, **19**, 897–908, 1915.
6. Also attributed to Bubnov, 1913: see S. C. MIKHLIN, *Variational Methods in Mathematical Physics*, Macmillan, 1964.
7. I. CHRISTIE, D. F. GRIFFITHS, A. R. MITCHELL, and O. C. ZIENKIEWICZ, 'Finite element methods for second order equations with significant first derivatives', *Int. J. Num. Meth. Eng.*, **10**, 1389–96, 1976.
8. R. V. SOUTHWELL, *Relaxation Methods in Theoretical Physics*, Clarendon Press, 1946.
9. R. S. VARGA, *Matrix Iterative Analysis*, Prentice-Hall, 1962.
10. S. TIMOSHENKO and J. N. GOODIER, *Theory of Elasticity*, 2nd ed., McGraw-Hill, 1951.
11. L. V. KANTOROVITCH and V. I. KRYLOV, *Approximate Methods of Higher Analysis*, Wiley (International), 1958.
12. J. W. STRUTT (Lord Rayleigh), 'On the theory of resonance', *Trans. Roy. Soc. (London)*, **A161**, 77–118, 1870.
13. W. RITZ, 'Über eine neue Methode zur Lösung gewissen Variations—Probleme der mathematischen Physik', *J. Reine angew. Math.* **135**, 1–61, 1909.
14. M. M. VEINBERG, *Variational Methods for the Study of Nonlinear Operators*, Holden-Day, 1964.
15. E. TONTI, 'Variational formulation of non-linear differential equations', *Bull. Acad. Roy. Belg. (Classe Sci.)*, **55**, 137–65 and 262–78, 1969.
16. J. T. ODEN, 'A general theory of finite elements—I: Topological considerations', pp. 205–21, and 'II: Applications', pp. 247–60. *Int. J. Num. Meth. Eng.*, **1**, 1969.
17. S. C. MIKHLIN, *Variational Methods in Mathematical Physics*, Macmillan, 1964.

18. S. C. MIKHLIN, *The Problems of the Minimum of a Quadratic Functional*, Holden-Day, 1965.
19. G. L. GUYMON, V. H. SCOTT, and L. R. HERRMANN, 'A general numerical solution of the two-dimensional differential–convection equation by the finite element method', *Water Res. Res.*, **6**, 1611–15, 1970.
20. B. A. SZABO and T. KASSOS, 'Linear equation constraints in finite element approximations', *Int. J. Num. Meth. Eng.*, **9**, 563–80, 1975.
21. K. WASHIZU, *Variational Methods in Elasticity and Plasticity*, 2nd ed., Pergamon Press, 1975.
22. F. KIKUCHI and Y. ANDO, 'A new variational functional for the finite element method and its application to plate and shell problems', *Nucl. Eng. Des.*, **21**, 95–113, 1972.
23. H. S. CHEN and C. C. MEI, *Oscillations and water forces in an offshore harbour*, Ralph M. Parsons Laboratory for Water Resources and Hydrodynamics, Report 190, Cambridge, Mass., 1974.
24. O. C. ZIENKIEWICZ, D. W. KELLY and P. BETTESS, 'The coupling of the finite element method and boundary solution procedures', *Int. J. Num. Meth. Eng.*, **11**, 355–75, 1977.
25. I. STAKGOLD, *Boundary Value Problems of Mathematical Physics*, MacMillan, 1967.
26. O. C. ZIENKIEWICZ, 'Constrained variational principles and penalty function methods in the finite element analysis', *Lecture Notes in Mathematics*, No. 363, pp. 207–214, Springer-Verlag, 1974.
27. J. CAMPBELL, *A finite element system for analysis and design*, Ph.D. Thesis, Swansea, 1974.
28. D. J. NAYLOR, 'Stresses in nearly incompressible materials for finite elements with application to the calculation of excess pore pressures', *Int. J. Num. Meth. Eng.*, **8**, 443–60, 1974.
29. I. FRIED, 'Finite element analysis of incompressible materials by residual energy balancing', *Int. J. Solids Struct.*, **10**, 993–1002, 1974.
30. I. FRIED, 'Shear in C^0 and C^1 bending finite elements', *Int. J. Solids Struct.*, **9**, 449–60, 1973.
31. O. C. ZIENKIEWICZ and E. HINTON. 'Reduced integration, function smoothing and non-conformity in finite element analysis', *J. Franklin Inst.*, **302**, 443–61, 1976.
32. P. P. LYNN and S. K. ARYA, 'Finite elements formulation by the weighted discrete least squares method', *Int. J. Num. Meth. Eng.*, **8**, 71–90, 1974.
33. O. C. ZIENKIEWICZ, D. R. J. OWEN, and K. N. LEE, 'Least square finite element for elasto-static problems—use of reduced integration', *Int. J. Num. Meth. Eng.*, **8**, 341–58, 1974.

4. Plane Stress and Plane Strain

4.1 Introduction

Two-dimensional elastic problems were the first successful examples of the application of the finite element method.[1,2] Indeed, we have already used this situation to illustrate the basis of the finite element formulation in Chapter 2 where the general relationships were derived. These basic relationships are given in Eqs. (2.1)–(2.5), and (2.23), (2.24) which for quick reference are summarized in Appendix 2.

In this chapter the particular relationships for the problem in hand will be derived in more detail, and illustrated by suitable practical examples, a procedure that will be followed throughout the remainder of the book.

Only the simplest, triangular, element will be discussed in detail but the basic approach is general. More elaborate elements to be discussed in later chapters would be introduced to the same problem in an identical manner.

The reader not familiar with the applicable basic definitions of elasticity is referred to elementary texts on the subject, in particular to the text by Timoshenko and Goodier,[3] whose notation will be widely used here.

In both problems of plane stress and plane strain the displacement field is uniquely given by the u and v displacements in directions of the cartesian, orthogonal x and y axes.

Again, in both, the only strains and stresses that have to be considered are the three components in the x–y plane. In the case of *plane stress*, by definition, all other components of stress are zero and therefore give no contribution to internal work. In *plane strain* the stress in a direction perpendicular to the x–y plane is not zero. However, by definition, the strain in that direction is zero, and therefore no contribution to internal work is made by this stress, which can in fact be explicitly evaluated from the three main stress components, if desired, at the end of all computation.

4.2 Element Characteristics

4.2.1 *Displacement functions.* Figure 4.1 shows the typical triangular element considered, with nodes i, j, m numbered in an anti-clockwise order.

The displacements of a node have two components

$$\mathbf{a}_i = \begin{Bmatrix} u_i \\ v_i \end{Bmatrix} \tag{4.1}$$

and the six components of element displacements are listed as a vector

$$\mathbf{a}^e = \begin{Bmatrix} \mathbf{a}_i \\ \mathbf{a}_j \\ \mathbf{a}_m \end{Bmatrix}. \tag{4.2}$$

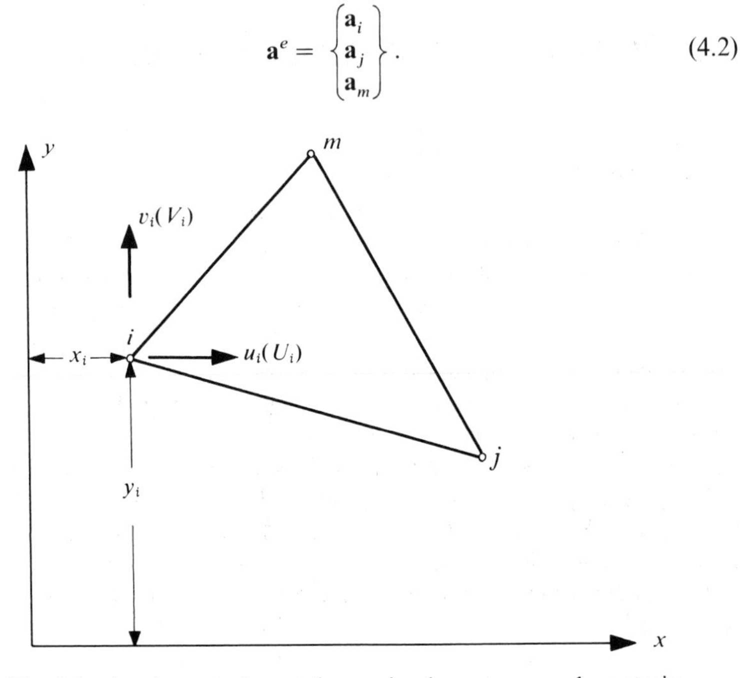

Fig. 4.1 An element of a continuum in plane stress or plane strain

The displacements within an element have to be uniquely defined by these six values. The simplest representation is clearly given by two linear polynomials

$$\begin{aligned} u &= \alpha_1 + \alpha_2 x + \alpha_3 y, \\ v &= \alpha_4 + \alpha_5 x + \alpha_6 y. \end{aligned} \tag{4.3}$$

The six constants α can be evaluated easily by solving the two sets of three simultaneous equations which will arise if the nodal co-ordinates are inserted and the displacements equated to the appropriate nodal displacements. Writing, for example,

$$\begin{aligned} u_i &= \alpha_1 + \alpha_2 x_i + \alpha_3 y_i \\ u_j &= \alpha_1 + \alpha_2 x_j + \alpha_3 y_j \\ u_m &= \alpha_1 + \alpha_2 x_m + \alpha_3 y_m \end{aligned} \tag{4.4}$$

we can easily solve for α_1, α_2, and α_3 in terms of the nodal displacements u_i, u_j, u_m and obtain finally

$$u = \frac{1}{2\Delta}\{(a_i+b_ix+c_iy)u_i+(a_j+b_jx+c_jy)u_j+(a_m+b_mx+c_my)u_m\} \tag{4.5a}$$

in which

$$\begin{aligned} a_i &= x_jy_m-x_my_j \\ b_i &= y_j-y_m = y_{jm} \\ c_i &= x_m-x_j = x_{mj} \end{aligned} \tag{4.5b}$$

with the other coefficients obtained by a cyclic permutation of subscripts in the order, i, j, m, and where †

$$2\Delta = \det \begin{vmatrix} 1 & x_i & y_i \\ 1 & x_j & y_j \\ 1 & x_m & y_m \end{vmatrix} = 2x \text{ (area of triangle } ijm). \tag{4.5c}$$

As the equations for the vertical displacement v are similar we also have

$$v = \frac{1}{2\Delta}\{(a_i+b_ix+c_iy)v_i+(a_j+b_jx+c_jy)v_j+(a_m+b_mx+c_my)v_m\}. \tag{4.6}$$

Though not strictly necessary at this stage we can represent the above relations Eqs. (4.5a) and (4.6) in the standard form of Eq. (2.1)

$$\mathbf{u} = \begin{Bmatrix} u \\ v \end{Bmatrix} = \mathbf{N}\mathbf{a}^e = [\mathbf{I}N_i, \mathbf{I}N_j, \mathbf{I}N_m]\mathbf{a}^e \tag{4.7}$$

with $\mathbf{I}$ a two by two identity matrix, and

$$N_i = (a_i+b_ix+c_iy)/2\Delta \text{ etc.} \tag{4.8}$$

The chosen displacement function automatically guarantees continuity of displacements with adjacent elements because the displacements vary linearly along any side of the triangle and, with identical displacement imposed at the nodes, the same displacement will clearly exist all along an interface.

4.2.2 *Strain* (*total*). The total strain at any point within the element can be defined by its three components which contribute to internal work.

† *Note:* if co-ordinates are taken from the centroid of the element then

$$x_i+x_m+x_j = y_i+y_j+y_m = 0 \quad \text{and} \quad a_i = 2\Delta/3 = a_j = a_m.$$

See also Appendix 4 for a summary of integrals for a triangle.

Thus

$$\boldsymbol{\varepsilon} = \begin{Bmatrix} \varepsilon_x \\ \varepsilon_y \\ \gamma_{xy} \end{Bmatrix} = \begin{bmatrix} \dfrac{\partial}{\partial x}, & 0 \\ 0, & \dfrac{\partial}{\partial y} \\ \dfrac{\partial}{\partial y}, & \dfrac{\partial}{\partial x} \end{bmatrix} \begin{Bmatrix} u \\ v \end{Bmatrix} = \mathbf{Lu}. \tag{4.9}$$

Substituting Eq. (4.7) we have

$$\boldsymbol{\varepsilon} = \mathbf{Ba}^e = [B_i, B_j, B_m] \begin{Bmatrix} \mathbf{a}_i \\ \mathbf{a}_j \\ \mathbf{a}_m \end{Bmatrix} \tag{4.10a}$$

with a typical matrix $\mathbf{B}_i$ given by

$$\mathbf{B}_i = \mathbf{L}\mathbf{I}N_i = \begin{bmatrix} \dfrac{\partial N_i}{\partial x}, & 0 \\ 0, & \dfrac{\partial N_i}{\partial y} \\ \dfrac{\partial N_i}{\partial y}, & \dfrac{\partial N_i}{\partial x} \end{bmatrix} = \frac{1}{2\Delta} \begin{bmatrix} b_i, & 0 \\ 0, & c_i \\ c_i, & b_i \end{bmatrix} \tag{4.10b}$$

This defines matrix **B** of Eq. (2.2) explicitly.

It will be noted that in this case the **B** matrix is independent of the position within the element, and hence the strains are constant throughout it. Obviously, the criterion of constant strain mentioned in Chapter 2 is satisfied by the shape functions.

4.2.3 *Initial strain* (*thermal strain*). 'Initial' strains, that is strains which are independent of stress, may be due to many causes. Shrinkage, crystal growth or, most frequently, temperature changes will, in general, result in an initial strain vector.

$$\boldsymbol{\varepsilon}_0 = \begin{Bmatrix} \varepsilon_{x0} \\ \varepsilon_{y0} \\ \gamma_{xy0} \end{Bmatrix}. \tag{4.11}$$

Although this initial strain may, in general, depend on the position within the element, it will usually be defined by average, constant, values. This is consistent with the constant strain conditions imposed by the prescribed displacement function.

Thus, for the case of *plane stress* in an isotropic material in an element subject to a temperature rise θ^e with a coefficient of thermal expansion α, we will have, for instance,

$$\boldsymbol{\varepsilon}_0 = \begin{Bmatrix} \alpha\theta^e \\ \alpha\theta^e \\ 0 \end{Bmatrix} \tag{4.12}$$

as no shear strains are caused by a thermal dilatation.

In *plane strain* the situation is more complex. The presumption of plane strain implies that stresses perpendicular to the x–y plane will develop due to thermal expansion even without the three main stress components, and hence the initial strain will be affected by the elastic constants.

It can be shown that in such a case

$$\boldsymbol{\varepsilon}_0 = (1+\nu) \begin{Bmatrix} \alpha\theta^e \\ \alpha\theta^e \\ 0 \end{Bmatrix} \tag{4.13}$$

where ν is the Poisson's ratio.

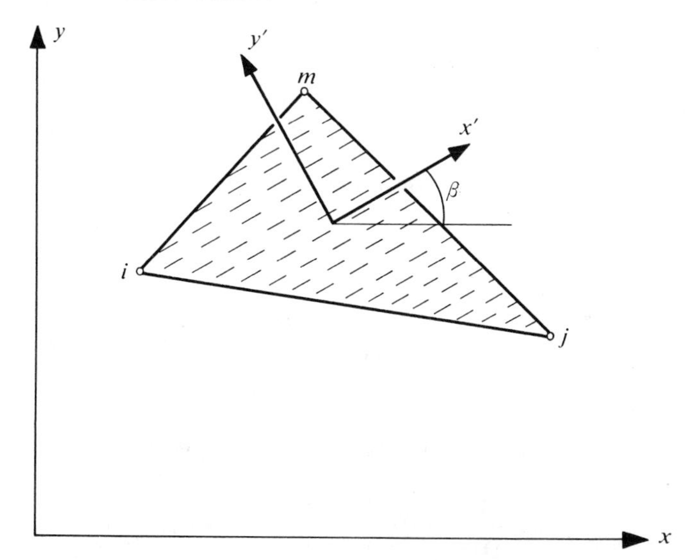

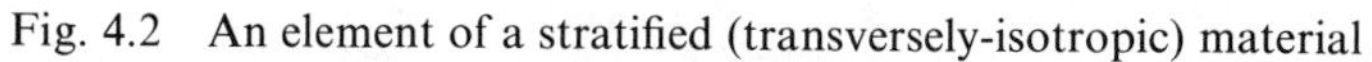
Fig. 4.2 An element of a stratified (transversely-isotropic) material

Anisotropic materials present special problems, since the coefficients of thermal expansion may vary with direction. Let x' and y' in Fig. 4.2 show the principal directions of the material. The initial strain due to thermal expansion becomes, with reference to these co-ordinates for plane stress

$$\boldsymbol{\varepsilon}_0' = \begin{Bmatrix} \varepsilon_{x'0} \\ \varepsilon_{y'0} \\ \gamma_{x'y'0} \end{Bmatrix} = \begin{Bmatrix} \alpha_1\theta^e \\ \alpha_2\theta^e \\ 0 \end{Bmatrix} \tag{4.14}$$

where α_1 and α_2 are the expansion coefficients referred to the x' and y' axes respectively.

To obtain the strain components in the x, y system it is necessary to use an appropriate strain transformation matrix $\mathbf{T}$ giving

$$\boldsymbol{\varepsilon}_0' = \mathbf{T}^T \boldsymbol{\varepsilon}_0 \tag{4.15}$$

With the β as defined in Fig. 4.2 it is easily verified that

$$\mathbf{T} = \begin{bmatrix} \cos^2 \beta & \sin^2 \beta & -2 \sin \beta \cos \beta \\ \sin^2 \beta & \cos^2 \beta & 2 \sin \beta \cos \beta \\ \sin \beta \cos \beta & -\sin \beta \cos \beta & \cos^2 \beta - \sin^2 \beta \end{bmatrix}$$

Thus, $\boldsymbol{\varepsilon}_0$ can be simply evaluated. It will be noted that no longer is the shear component of strain equal to zero in the x–y co-ordinates.

4.2.4 *Elasticity matrix.* The matrix $\mathbf{D}$ of the relation Eq. (2.5)

$$\boldsymbol{\sigma} = \begin{Bmatrix} \sigma_x \\ \sigma_y \\ \tau_{xy} \end{Bmatrix} = \mathbf{D} \left(\begin{Bmatrix} \varepsilon_x \\ \varepsilon_y \\ \gamma_{xy} \end{Bmatrix} - \boldsymbol{\varepsilon}_0 \right) \tag{4.16}$$

can be explicitly stated for any material (excluding here $\boldsymbol{\sigma}_0$ which is simply additive).

Plane stress—isotropic material. For plane stress in an isotropic material we have, by definition,

$$\begin{aligned} \varepsilon_x &= \sigma_x/E - \nu\sigma_y/E + \varepsilon_{x0} \\ \varepsilon_y &= -\nu\sigma_x/E + \sigma_y/E + \varepsilon_{y0} \\ \gamma_{xy} &= 2(1+\nu)\tau_{xy}/E + \varepsilon_{xy0}. \end{aligned} \tag{4.17}$$

Solving the above for the stresses, we obtain matrix $\mathbf{D}$ as

$$\mathbf{D} = \frac{E}{1-\nu^2} \begin{bmatrix} 1 & \nu & 0 \\ \nu & 1 & 0 \\ 0 & 0 & (1-\nu)/2 \end{bmatrix} \tag{4.18}$$

in which E is the elastic modulus and ν is the Poisson's ratio.

Plane strain—isotropic material. In this case a normal stress σ_z exists in addition to the three other stress components. For the special case of isotropic thermal expansion we have

$$\begin{aligned} \varepsilon_x &= \sigma_x/E - \nu\sigma_y/E - \nu\sigma_z/E + \alpha\theta^e \\ \varepsilon_y &= -\nu\sigma_x/E + \sigma_y/E - \nu\sigma_z/E + \alpha\theta^e \\ \gamma_{xy} &= 2(1+\nu)\tau_{xy}/E. \end{aligned} \tag{4.19}$$

but in addition

$$\varepsilon_z = 0 = -\nu\sigma_x/E - \nu\sigma_y/E + \sigma_z/E + \alpha\theta^e.$$

On eliminating σ_z and solving for the three remaining stresses we obtain the previously quoted expression for the initial strain Eq. (4.13), and by comparison with Eq. (4.16), the matrix **D** is

$$\mathbf{D} = \frac{E(1-\nu)}{(1+\nu)(1-2\nu)} \begin{bmatrix} 1 & \nu/(1-\nu) & 0 \\ \nu/(1-\nu) & 1 & 0 \\ 0 & 0 & (1-2\nu)/2(1-\nu) \end{bmatrix}. \tag{4.20}$$

Anisotropic materials. For a completely anisotropic material, 21 independent elastic constants are necessary to define completely the three-dimensional stress–strain relationship.[4,5]

If two-dimensional analysis is to be applicable a symmetry of properties must exist, implying at most six independent constants in the **D** matrix. Thus, it is always possible to write

$$\mathbf{D} = \begin{bmatrix} d_{11} & d_{12} & d_{13} \\ & d_{22} & d_{23} \\ \text{(sym)} & & d_{33} \end{bmatrix} \tag{4.21}$$

to describe the most general two-dimensional behaviour. (The necessary symmetry of the **D** matrix follows from the general equivalent of the Maxwell–Betti reciprocal theorem and is a consequence of invariant energy irrespective of the path taken to reach a given strain state.)

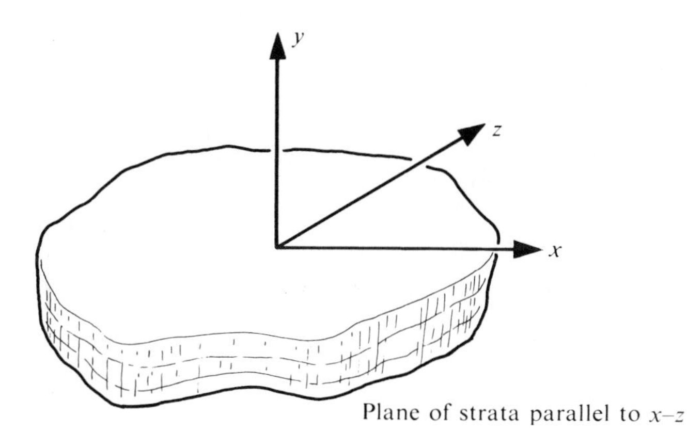

Fig. 4.3 A stratified (transversely-isotropic) material

A case of particular interest in practice is that of a 'stratified' or transversely isotropic material in which a rotational symmetry of properties exists within the plane of the strata. Such a material possesses only five independent elastic constants.

The general stress–strain relations give in this case, following the notation of Lekhnitskii,[4] and taking now the y axis as perpendicular to the strata (neglecting initial strain), Fig. 4.3

$$\begin{aligned}
\varepsilon_x &= \sigma_x/E_1 - \nu_2\sigma_y/E_2 - \nu_1\sigma_z/E_1 \\
\varepsilon_y &= -\ \nu_2\sigma_x/E_2 + \sigma_y/E_2 - \nu_2\sigma_z/E_2 \\
\varepsilon_z &= -\nu_1\sigma_x/E_1 - \nu_2\sigma_y/E_2 + \sigma_z/E_1 \\
\gamma_{xz} &= \{2(1+\nu_1)/E_1\}\tau_{xz} \\
\gamma_{xy} &= \frac{1}{G_2}\tau_{xy} \\
\gamma_{yz} &= \frac{1}{G_2}\tau_{yz}
\end{aligned} \tag{4.22}$$

in which the constants E_1, ν_1 (G_1 is dependent) are associated with the behaviour in plane of the strata and E_2, G_2, ν_2 with a direction normal to these.

The **D** matrix in two dimensions becomes now, taking

$$\frac{E_1}{E_2} = n \quad \text{and} \quad \frac{G_2}{E_2} = m$$

$$\mathbf{D} = \frac{E_2}{(1-n\nu_2^2)} \begin{bmatrix} n & n\nu_2 & 0 \\ n\nu_2 & 1 & 0 \\ 0 & 0 & m(1-n\nu_2^2) \end{bmatrix} \tag{4.23}$$

for plane stress, or

$$\mathbf{D} = \frac{E_2}{(1+\nu_1)(1-\nu_1-2n\nu_2^2)} \times$$

$$\times \begin{bmatrix} n(1-n\nu_2^2) & n\nu_2(1+\nu_1) & 0 \\ n\nu_2(1+\nu_1) & (1-\nu_1^2) & 0 \\ 0 & 0 & m(1+\nu_1)(1-\nu_1-2n\nu_2^2) \end{bmatrix} \tag{4.24}$$

for plane strain.

When, as in Fig. 4.2, the direction of strata is inclined to the x axis then to obtain the **D** matrices in the universal co-ordinates a transformation is necessary. Taking $\mathbf{D}'$ as relating the stresses and strains in the inclined co-ordinate system (x', y') it is easy to show that

$$\mathbf{D} = \mathbf{T}\mathbf{D}'\mathbf{T}^{\mathrm{T}} \tag{4.25}$$

where $\mathbf{T}$ is the same as given in Eq. (4.15)

If the stress systems $\boldsymbol{\sigma}'$ and $\boldsymbol{\sigma}$ correspond to $\boldsymbol{\varepsilon}'$ and $\boldsymbol{\varepsilon}$ respectively then by equality of work

$$\boldsymbol{\sigma}'^{\mathrm{T}}\boldsymbol{\varepsilon}' = \boldsymbol{\sigma}^{\mathrm{T}}\boldsymbol{\varepsilon}$$

or

$$\boldsymbol{\varepsilon}'^{\mathrm{T}}\mathbf{D}'\boldsymbol{\varepsilon}' = \boldsymbol{\varepsilon}^{\mathrm{T}}\mathbf{D}\boldsymbol{\varepsilon}$$

from which Eq. (4.25) follows on substitution of Eq. (4.15). (*See also* Chapter 1.)

4.2.5 *The stiffness matrix.* The stiffness matrix of the element ijm is defined from the general relationship Eq. (2.13) by coefficient

$$\mathbf{K}_{ij}^{e} = \int \mathbf{B}_i^{\mathrm{T}}\mathbf{D}\mathbf{B}_j t \,\mathrm{d}x\,\mathrm{d}y \tag{4.26}$$

where t is the thickness of the element and the integration is taken over the area of the triangle. If the thickness of the element is assumed to be constant, an assumption convergent to the truth as size of elements decreases, then, as neither of the matrices contains x or y we have, simply

$$\mathbf{K}_{ij}^{e} = \mathbf{B}_i^{\mathrm{T}}\mathbf{D}\mathbf{B}_j t\Delta \tag{4.27}$$

where Δ is the area of the triangle (defined already by Eq. (4.5)). This form is now sufficiently explicit for computation with the actual matrix operations being left to the computer.

4.2.6 *Nodal forces due to initial strain.* These are given directly by the expression Eq. (2.13) which, on performing the integration, become

$$(\mathbf{f}_i)_{\varepsilon 0}^{e} = -\mathbf{B}_i^{\mathrm{T}}\mathbf{D}\boldsymbol{\varepsilon}_0 t\Delta, \text{ etc.} \tag{4.28}$$

These 'initial strain' forces are contributed to the nodes of an element in an unequal manner and require precise evaluation. Similar expressions are derived for initial stress forces.

4.2.7 *Distributed body forces.* In the general case of plane stress or strain each element of unit area in the x–y plane is subject to forces

$$\mathbf{b} = \begin{Bmatrix} b_x \\ b_y \end{Bmatrix}$$

in the direction of the appropriate axes.

Again, by Eq. (2.13), the contribution of such forces to these at each node is given by

$$\mathbf{f}_i^{e} = -\int N_i \begin{Bmatrix} b_x \\ b_y \end{Bmatrix}$$

or by Eq. (4.7)

$$\mathbf{f}_i^e = -\begin{Bmatrix} b_x \\ b_y \end{Bmatrix} \int N_i \, dx \, dy, \text{ etc.} \tag{4.29}$$

if the body forces b_x and b_y are constant. As N_i is no longer constant the integration has to be carried out explicitly. Some general integration formulae for a triangle are given in Appendix 3.

In this special case the calculation will be simplified if the origin of co-ordinates is taken at the centroid of the element. Now

$$\int x \, dx \, dy = \int y \, dx \, dy = 0$$

and on using Eq. 4.8)

$$\mathbf{f}_i^e = -\begin{Bmatrix} b_x \\ b_y \end{Bmatrix} \int a_i \, dx \, dy/2\Delta = -\begin{Bmatrix} b_x \\ b_y \end{Bmatrix} a_i/2 = -\begin{Bmatrix} b_x \\ b_y \end{Bmatrix} \Delta/3 \tag{4.30}$$

by relations noted on p. 95.

Explicitly, for the whole element

$$\mathbf{f}^e = \begin{Bmatrix} f_i^e \\ f_j^e \\ f_m^e \end{Bmatrix} = -\begin{Bmatrix} b_x \\ b_y \\ b_x \\ b_y \\ b_x \\ b_y \end{Bmatrix} \Delta/3 \tag{4.31}$$

which means simply that the total forces acting in x and y direction due to the body forces are distributed to the nodes in three equal parts. This fact corresponds with physical intuition, and was often assumed implicitly.

4.2.8 *Body force potential.* In many cases the body forces are defined in terms of a body force potential ϕ as

$$b_x = -\frac{\partial \phi}{\partial x}, \qquad b_y = -\frac{\partial \phi}{\partial y} \tag{4.32}$$

and this potential, rather than the values of b_x and b_y is known throughout the region and is specified at nodal points. If $\boldsymbol{\phi}^e$ lists the three values of the potential associated with the nodes of the element, i.e.,

$$\boldsymbol{\phi}^e = \begin{Bmatrix} \phi_i \\ \phi_j \\ \phi_m \end{Bmatrix} \tag{4.33}$$

and has to correspond with constant values of b_x and b_y, ϕ must vary linearly within the element. The 'shape function' of its variation will obviously be given by a procedure identical to that used in deriving Eqs. (4.4) to (4.6), and yields

$$\phi = [N_i, N_j, N_m]\boldsymbol{\phi}^e. \tag{4.34}$$

Thus,

$$b_x = -\frac{\partial\phi}{\partial x} = -[b_i, b_j, b_m]\boldsymbol{\phi}^e/2\Delta$$

and

$$b_y = -\frac{\partial\phi}{\partial y} = -[c_i, c_j, c_m]\boldsymbol{\phi}^e/2\Delta. \tag{4.35}$$

The vector of nodal forces due to the body force potential will now replace Eq. (4.31) by

$$\mathbf{f}^e = \frac{1}{6}\begin{bmatrix} b_i, & b_j, & b_m \\ c_i, & c_j, & c_m \\ b_i, & b_j, & b_m \\ c_i, & c_j, & c_m \\ b_i, & b_j, & b_m \\ c_i, & c_j, & c_m \end{bmatrix}\boldsymbol{\phi}^e \tag{4.36}$$

4.2.9 *Evaluation of stresses.* The formulae derived enable the full stiffness matrix of the structure to be assembled, and a solution for displacements to be obtained.

The stress matrix given in general terms in Eq. (2.16) is obtained by the appropriate substitutions for each element.

The stresses are, by the basic assumption, constant within the element. It is usual to assign these to the centroid of the element, and in most of the examples in this chapter this procedure is followed. An alternative consists of obtaining stress values at the nodes by averaging the values in the adjacent elements. Some 'weighting' procedures have been used in this context on an empirical basis but their advantage appears small.

It is usual to arrange for the computer to calculate the principal stresses and their directions of every element.

4.3 **Examples—An Assessment of Accuracy**

There is no doubt that the solution to plane elasticity problems as formulated in Section 4.2 is, in the limit of subdivision, an exact solution.

Indeed at any stage of a finite subdivision it is an approximate solution as, say, a Fourier series solution with a limited number of terms.

As already explained in Chapter 2 the total strain energy obtained during any stage of approximation will be below the true strain energy of the exact solution. In practice it will mean that the displacements, and hence also the stresses, will be underestimated by the approximation in its *general picture*. However, it must be emphasized that this is not necessarily true at every point of the continuum individually; hence the value of such a bound in practice is not great.

What is important for the engineer to know is the order of accuracy achievable in typical problems with a certain fineness of element subdivision. In any particular case the error can be assessed by comparison with known, exact, solutions or by a study of the convergence, using two or more stages of subdivision.

With the development of experience the engineer can assess *a priori* the order of approximation that will be involved in a specific problem tackled with a given element subdivision. Some of this experience will perhaps be conveyed by the examples considered in this book.

In the first place attention will be focused on some simple problems for which exact solutions are available.

Uniform stress field. If the exact solution is in fact that of a uniform stress field then, whatever the element subdivision, the finite element solution will coincide exactly with the exact one. This is an obvious corollary of the formulation, nevertheless it is useful as a first check of written computer programs.

Linearly varying stress field. Here, obviously, the basic assumption of constancy of stress within elements means that solution will be approximate only. In Fig. 4.4 a simple example of a beam subject to constant bending moment is shown with a fairly coarse subdivision. It is readily seen that the axial (σ_y) stress given by the element 'straddles' the exact values and, in fact, if the constant stress values are associated with centroids of the elements and plotted, the best 'fit' line represents the exact stresses. [See Ch. 11 for optimal sampling points].

The horizontal and shear stress components differ again from the exact values (which are simply zero). Again, however, it will be noted that they oscillate by equal, small amounts around the exact values.

At internal nodes, if the average of stresses of surrounding elements is taken it will be found that the exact stresses are very closely represented. The average at external faces is not, however, so good. The overall improvement in representing the stresses by nodal averages, as shown on Fig. 4.4, is often used in practice for improvement of the approximation.

A weighting of averages near the faces of the structure can further be used for refinement. Without being dogmatic on this point, it seems

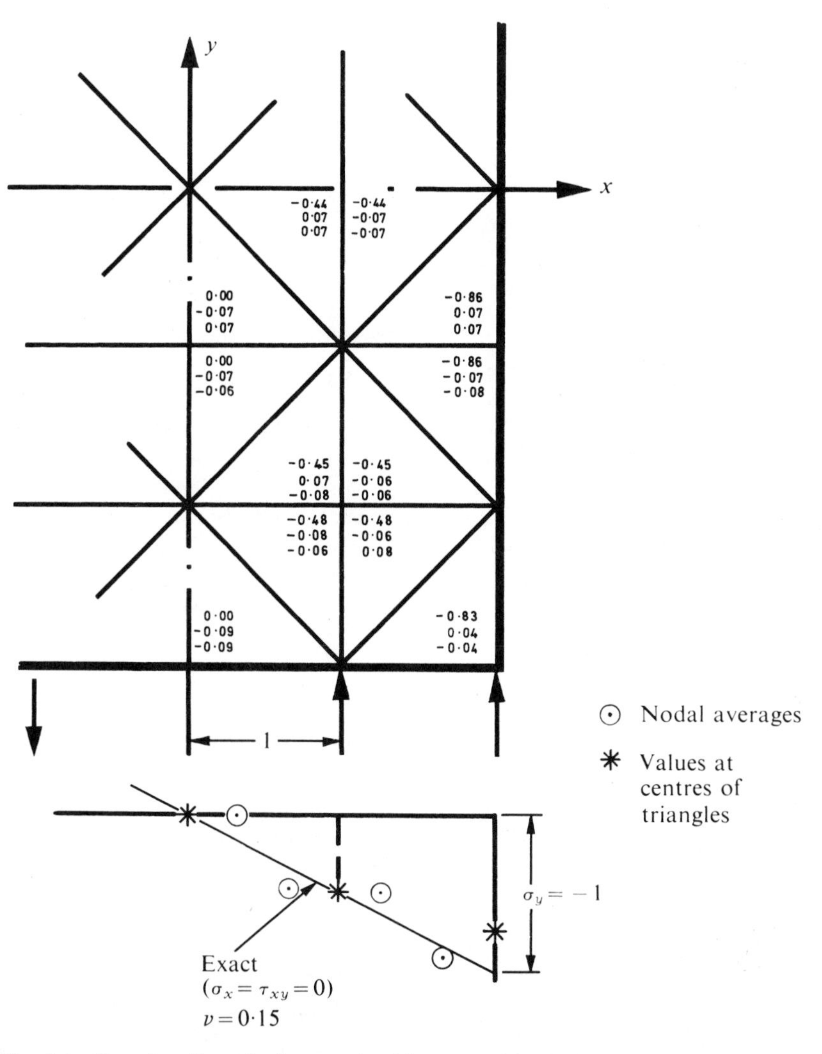

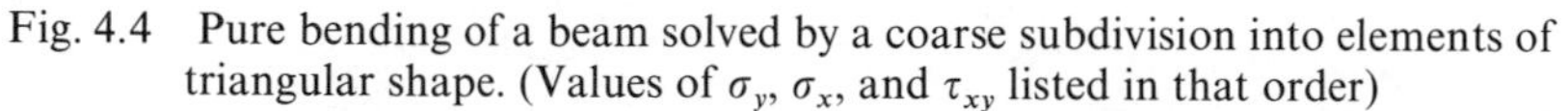

Fig. 4.4 Pure bending of a beam solved by a coarse subdivision into elements of triangular shape. (Values of σ_y, σ_x, and τ_{xy} listed in that order)

preferable, when accuracy demands this, simply to use a finer mesh subdivision.

Stress concentration. A more realistic test problem is shown in Figs. 4.5 and 4.6. Here the flow of stress around a circular hole in an isotropic and in an anisotropic stratified material is considered when the stress conditions are uniform.[6] A graded division into elements is used to allow a more detailed study in the region where high stress gradients are expected.

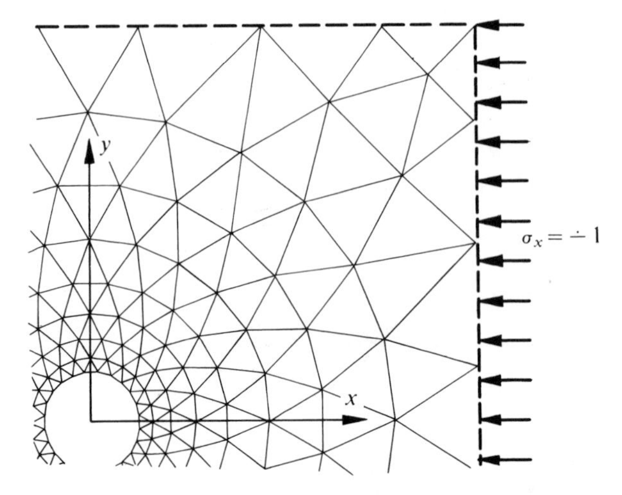

Fig. 4.5 A circular hole in a uniform stress field. (a) isotropic material; (b) stratified (orthotropic) material; $E_x = E_1 = 1$, $E_y = E_2 = 3$, $\nu_1 = 0{\cdot}1$, $\nu_2 = 0$, $G_{xy} = 0{\cdot}42$

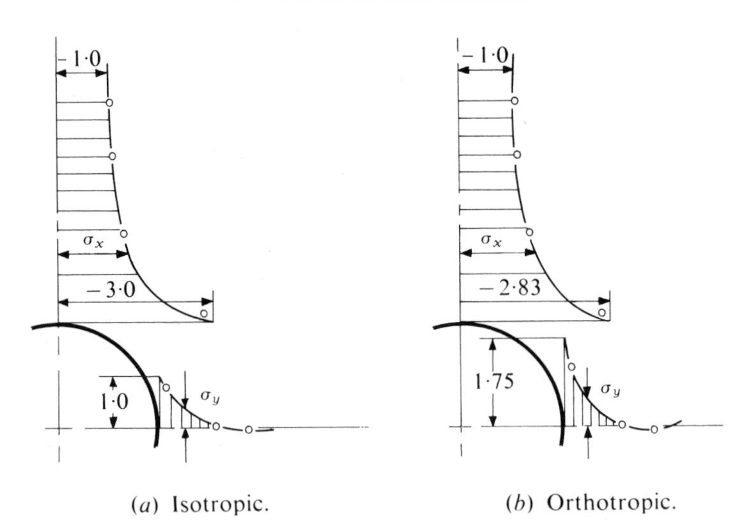

Fig. 4.6 Comparison of theoretical and finite element results for cases (*a*) and (*b*) of Fig. 4.5

The high degree of accuracy achievable can be assessed from Fig. 4.6 where some of the results are compared against exact solutions.[3,7]

In later chapters we shall see that even more accurate answers can be obtained with the use of more elaborate elements; however, the principles of the analysis remain identical.

4.4 Some Practical Applications

Obviously, the practical applications of the method are limitless, and the finite element method has superseded experimental technique for plane problems because of its high accuracy, low cost, and versatility. The ease of treatment of material anisotropy, thermal stresses, or body force problems add to its advantages.

A few examples of actual applications to complex problems of engineering practice will now be given.

Stress flow around a reinforced opening (Fig. 4.7). In steel pressure vessels or aircraft structures, openings have to be introduced in the stressed skin. The penetrating duct itself provides some reinforcement round the edge and, in addition, the skin itself is increased in thickness to reduce the stresses due to the concentration effects.

Analysis of such problems treated as cases of plane stress presents no difficulties. The elements are so chosen as to follow the thickness variation, and appropriate values of this are assigned.

The narrow band of thick material near the edge can be represented either by special beam-type elements, or more easily in a standard program by very thin triangular elements of the usual type, to which appropriate thickness is assigned. The latter procedure was used in the problem shown in Fig. 4.7 which gives some of the resulting stresses near the opening itself. The fairly large extent of the region introduced in the analysis and the grading of the mesh should be noted.

An anisotropic valley subject to tectonic stress[6] (Fig. 4.8). A symmetrical valley subject to a uniform horizontal stress is considered. The material is stratified, hence is 'transversely isotropic', and the direction of strata varies from point to point.

The stress plot shows the tensile region that develops. This phenomenon is of considerable interest to geologists and engineers concerned with rock mechanics.

A dam subject to external and internal water pressures[8,9] (Fig. 4.9). A buttress dam on a somewhat complex rock foundation is here analysed. The heterogeneous foundation region is subject to plane strain conditions while the dam itself is considered as a plate (plane stress) of variable thickness.

With external and gravity loading no special problems of analysis arise,

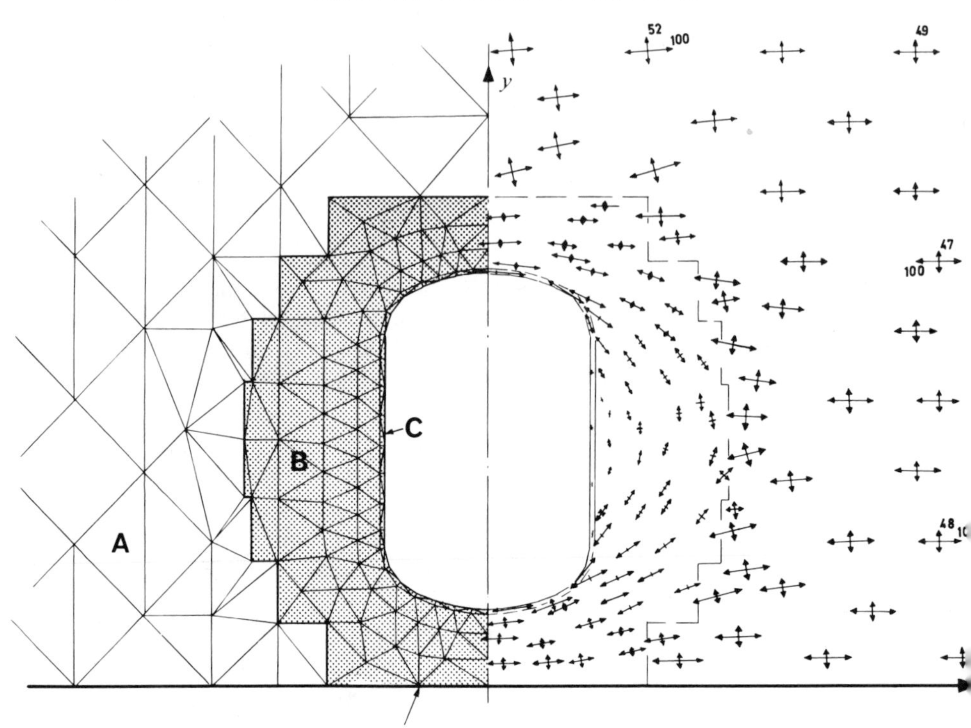

Restrained in y direction from movement.

Fig. 4.7 A reinforced opening in a plate. Uniform stress field at a distance from opening $\sigma_x = 100$, $\sigma_y = 50$. Thickness of plate regions A, B, and C is in the ratio of 1:3:23

though perhaps it should be mentioned that it was found worth while to 'automatize' the computation of gravity nodal loads.

When pore pressures are considered, the situation, however, requires perhaps some explanation.

It is well known that in a porous material the water pressure is transmitted to the structure as a *body force* of magnitude

$$b_x = -\frac{\partial p}{\partial x}, \qquad b_y = -\frac{\partial p}{\partial y} \tag{4.37}$$

and that now the external pressure need not be considered.

The pore pressure p is, in fact, now a body force potential, as defined in Eq. (4.32). Figure 4.9 shows the element subdivision of the region and the outline of the dam. Figure 4.10(a) and (b) show the stresses resulting from gravity (applied to the dam only) and due to water pressure assumed to be acting as an external load or, alternatively, as an internal pore pressure. Both solutions indicate large tensile regions, but the increase of stresses due to the second assumption is important.

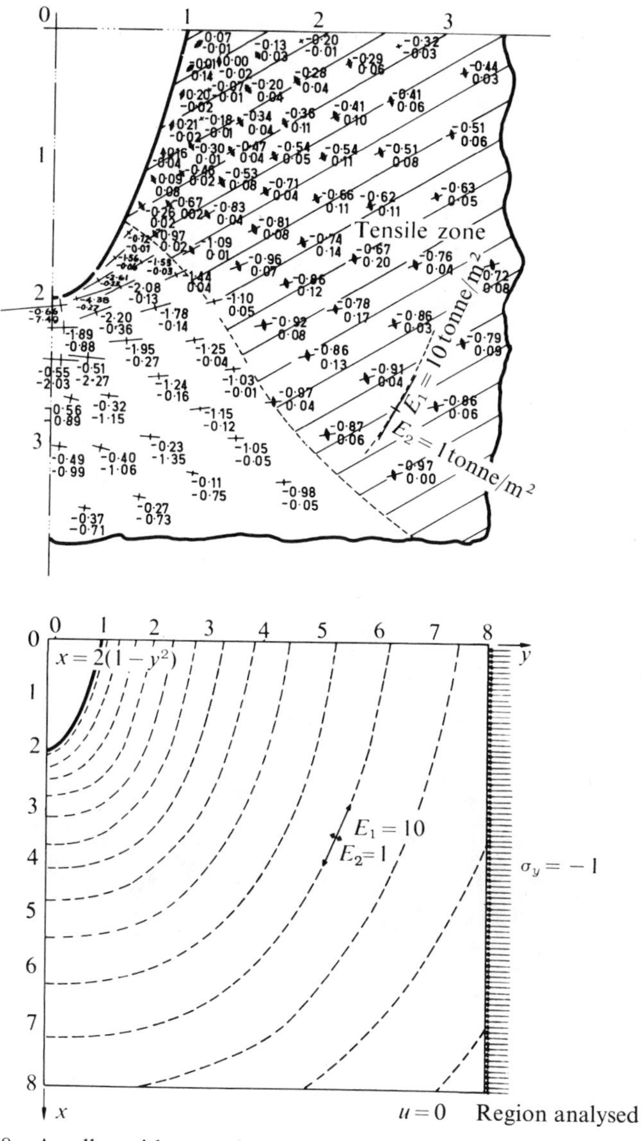

Fig. 4.8 A valley with curved strata subject to a horizontal tectonic stress (plane strain 170 nodes, 298 elements)

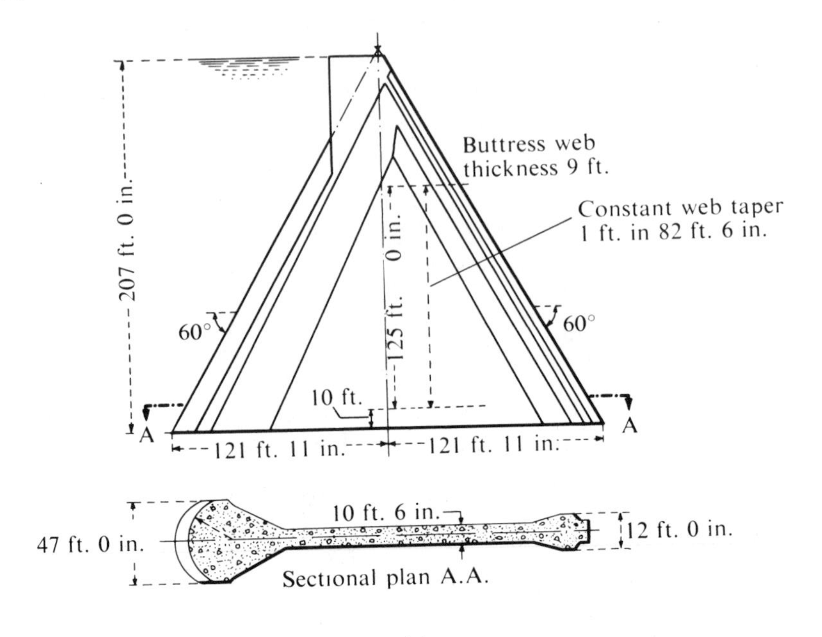

(*a*)

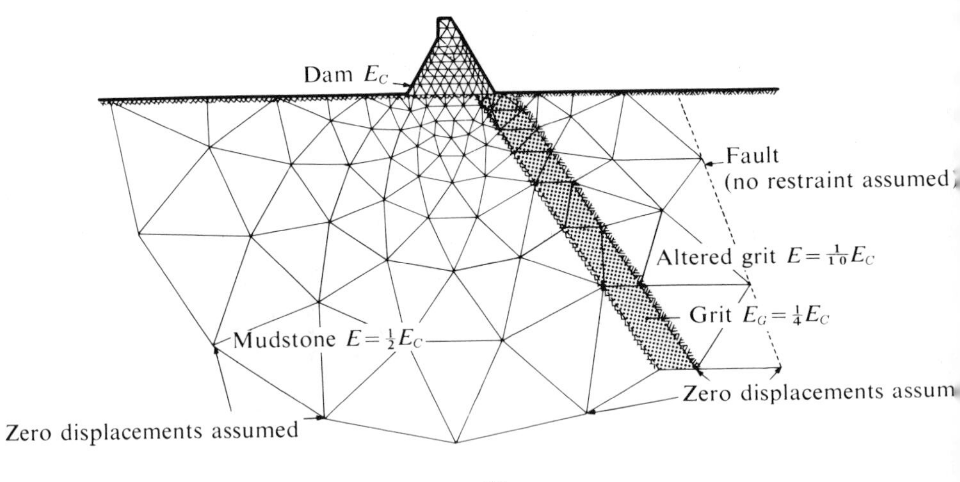

(*b*)

Fig. 4.9 Stress analysis of a buttress dam. Plane stress condition assumed in a dam and plane strain in foundation. (*a*) The buttress section analysed. (*b*) Extent of foundation considered and division into finite elements

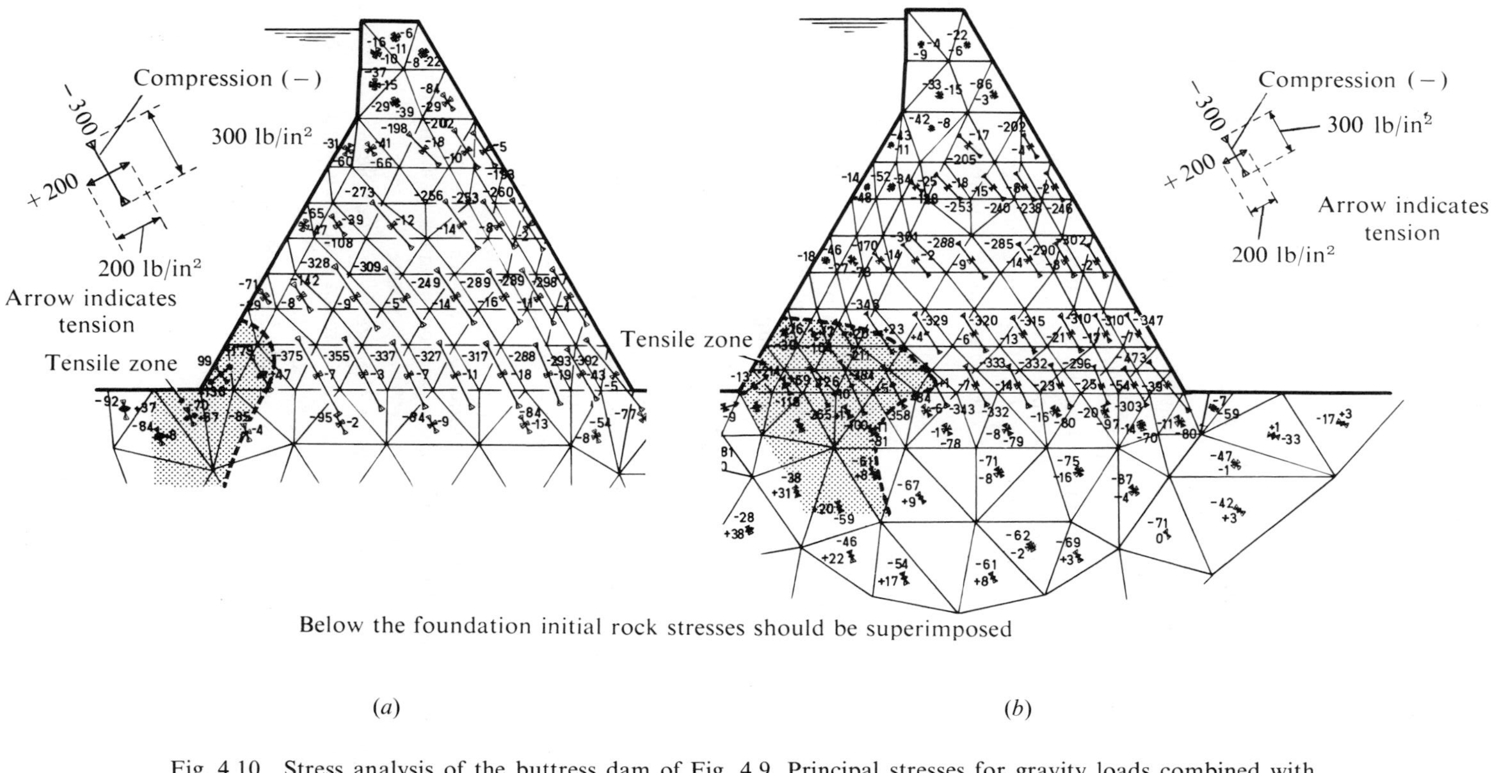

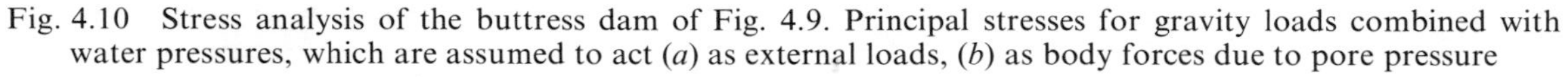

Fig. 4.10 Stress analysis of the buttress dam of Fig. 4.9. Principal stresses for gravity loads combined with water pressures, which are assumed to act (*a*) as external loads, (*b*) as body forces due to pore pressure

The stresses calculated here are the so-called 'effective' stresses. These represent the forces transmitted between the solid particles and are defined in terms of the *total* stresses $\boldsymbol{\sigma}$ and the pore pressures p by

$$\boldsymbol{\sigma}' = \boldsymbol{\sigma} + \mathbf{m}p \qquad \mathbf{m}^{\mathrm{T}} = [1, 1, 0] \tag{4.38}$$

i.e., simply by removing the hydrostatic pressure component from the *total* stress.[10]

The effective stress is of particular importance in the mechanics of porous media such as occur in the study of soils, rocks, or concrete. The basic assumption in deriving the body forces of Eq. (4.37) is that only the effective stress is of any importance in deforming the solid phase. This leads immediately to another possibility of formulation.[11] If we examine the equilibrium conditions of Eq. (2.10) we note that this is written in terms of total stresses. Writing the constitutive relation, Eq. (2.5), in terms of effective stresses, i.e.,

$$\boldsymbol{\sigma}' = \mathbf{D}'(\boldsymbol{\varepsilon} - \boldsymbol{\varepsilon}_0) + \boldsymbol{\sigma}_0' \tag{4.39}$$

and substituting into the equilibrium equation [Eq. (2.10)] we find that Eq. (2.12) is again obtained with the stiffness matrix using the matrix $\mathbf{D}'$ and the force terms of Eq. (2.13b) being augmented by an additional force

$$-\int_{V^e} \mathbf{B}^{\mathrm{T}}\mathbf{m}p \, \mathrm{d(vol)} \tag{4.40}$$

or, if p is interpolated by shape functions N_i', the force becomes

$$-\int_{V^e} \mathbf{B}^{\mathrm{T}}\mathbf{m}\mathbf{N}' \, \mathrm{d(vol)}\mathbf{P}^e. \tag{4.41}$$

This alternative form of introducing the pore pressure effects allows a discontinuous interpolation of p to be used (as in Eq. (4.40) no derivatives occur) and this is now frequently used in practice.

Cracking. The tensile stresses in the previous example will doubtless cause the rock to crack. If a stable situation can develop when such a crack spreads then the dam can be considered safe.

Cracks can be introduced very simply into the analysis by assigning zero elasticity values to chosen elements. An analysis with a wide cracked wedge is shown in Fig. 4.11, where it can be seen that with the extent of the crack assumed no tension within the dam body develops.

A more elaborate procedure for following crack propagation and resulting stress redistribution can be developed and will be discussed later (*see* Chapter 18).

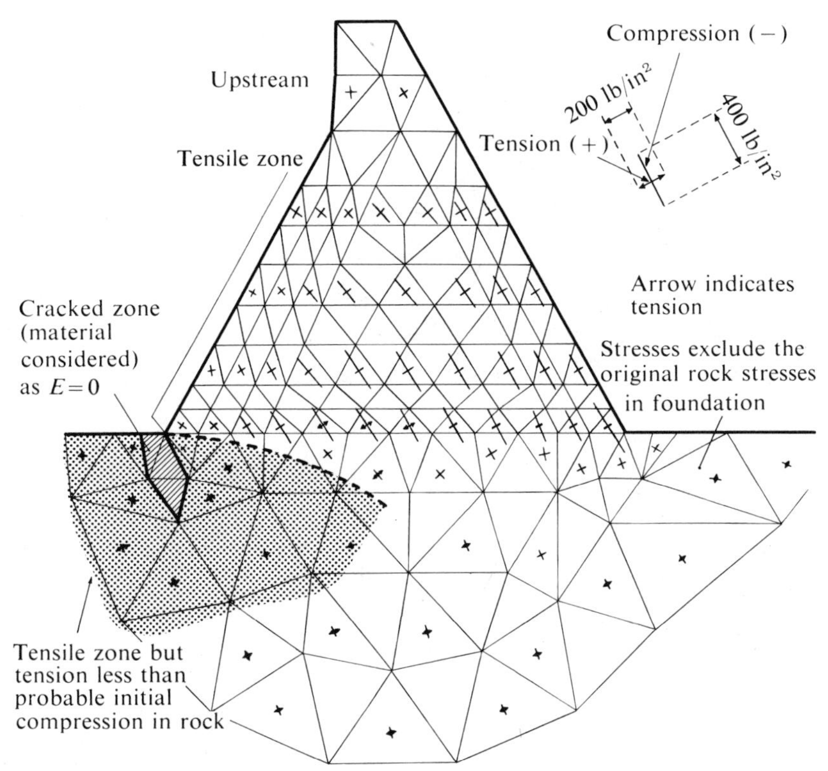

Fig. 4.11 Stresses in a buttress dam. An introduction of a 'crack' modifies stress distribution (same loading as Fig. 4.10(*b*))

Thermal stresses. As an example of thermal stress computation the same dam is shown under simple temperature distribution assumptions. Results of this analysis are given in Fig. 4.12.

Gravity dams. A buttress dam is a natural example for the application of finite element methods. Other types, such as gravity dams with or without piers and so on, can also be simply treated. Figure 4.13 shows an analysis of a large dam with piers and crest gates.

In this case an approximation of assuming a two-dimensional treatment in the vicinity of the abrupt change of section, i.e., where the piers join the main body of the dam, is clearly involved, but this leads to localized errors only.

It is important to note here how, in a single solution, the grading of element size is used to study concentration of stress at the cable anchorages, the general stress flow in the dam, and the foundation behaviour.

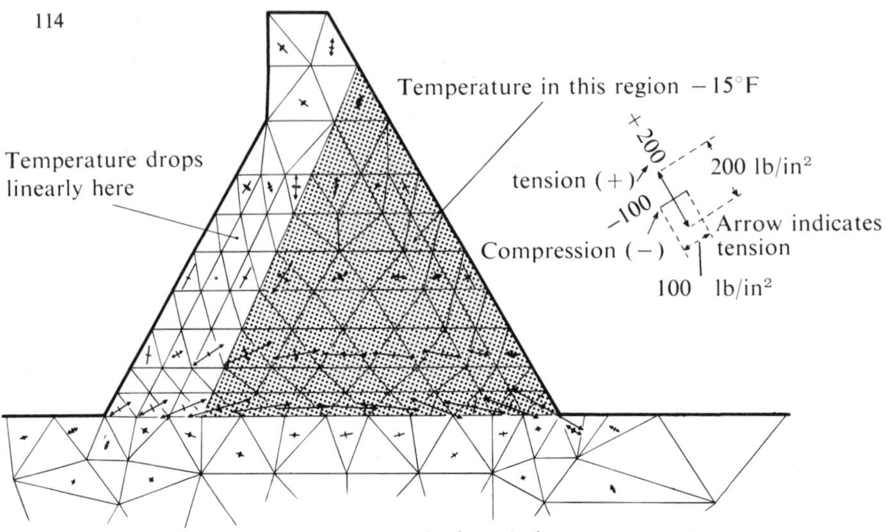

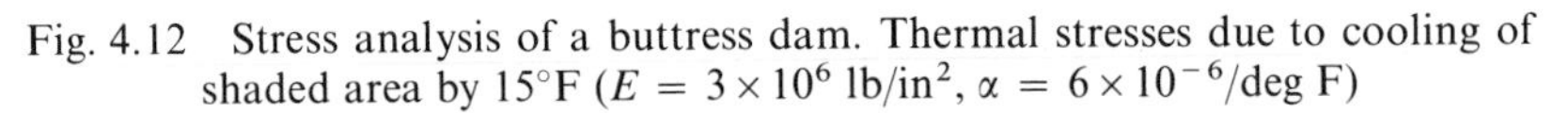

Fig. 4.12 Stress analysis of a buttress dam. Thermal stresses due to cooling of shaded area by 15°F ($E = 3 \times 10^6$ lb/in², $\alpha = 6 \times 10^{-6}$/deg F)

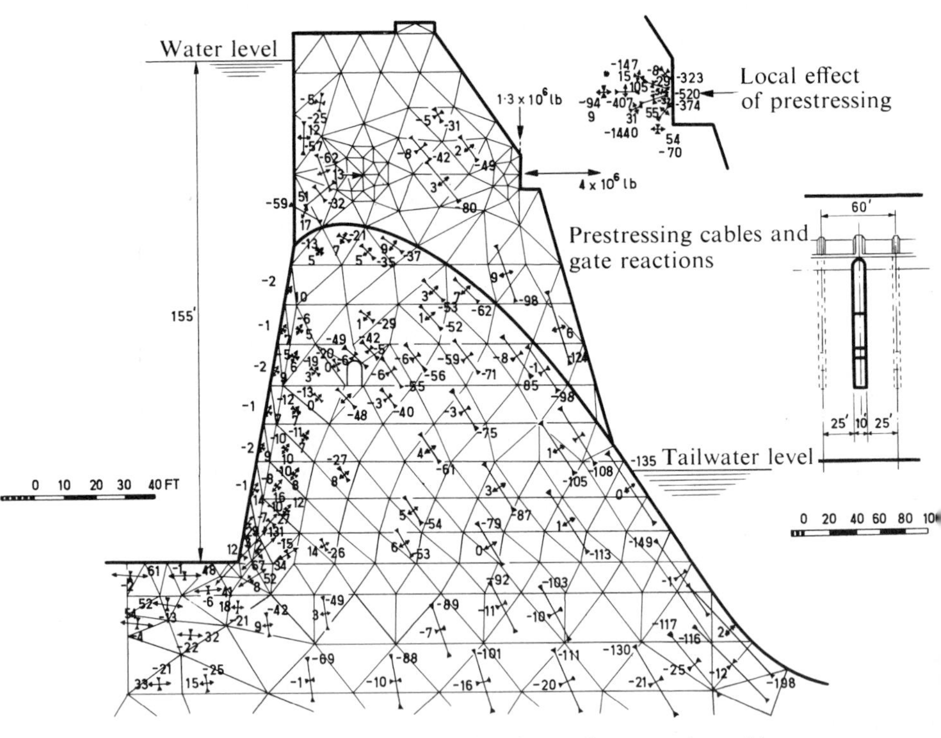

Fig. 4.13 A large barrage with piers and prestressing cables

The linear ratio of size of largest to smallest elements is of the order of 30 to 1 (the largest elements occurring in the foundation are not shown in the figure).

Underground power station. This last example illustrated in Figs. 4.14 and 4.15 shows an interesting large-scale application. Here principal stresses are plotted automatically. In this analysis very many different components of $\boldsymbol{\sigma}_0$, the initial stress, were used due to uncertainty of knowledge about geological conditions. The rapid solution and plot of many results enabled the limits within which stresses vary to be found and an engineering decision arrived at.

4.5 Special Treatment of Plane Strain with an Incompressible Material

It will have been noted that the relationship Eq. (4.20) defining the elasticity **D** matrix for an isotropic material breaks down when the Poisson's ratio reaches a value of 0·5 as the factor in the parentheses becomes infinite. A simple way of side-stepping the difficulty presented is to use values of Poisson's ratio approximating to 0·5 but not equal to it. Experience shows, however, that if this is done the solution deteriorates unless special formulations such as those discussed in Chapter 11 are used. An alternative procedure has been suggested by Hermann.[12] This involves the use of a different variational formulation, and will be discussed in Chapter 12.

References

1. M. J. TURNER, R. W. CLOUGH, H. C. MARTIN, and L. J. TOPP, 'Stiffness and deflection analysis of complex structures', *J. Aero. Sci.*, **23**, 805–23, 1956.
2. R. W. CLOUGH, 'The finite element in plane stress analysis', *Proc. 2nd A.S.C.E. Conf. on Electronic Computation*, Pittsburgh, Pa., Sept. 1960.
3. S. TIMOSHENKO and J. N. GOODIER, *Theory of Elasticity*, 2nd ed., McGraw-Hill, 1951.
4. S. G. LEKHNITSKII, *Theory of Elasticity of an Anisotropic Elastic Body*, Translation from Russian by P. Fern, Holden Day, San Francisco, 1963.
5. R. F. S. HEARMON, *An Introduction to Applied Anisotropic Elasticity*, Oxford Univ. Press, 1961.
6. O. C. ZIENKIEWICZ, Y. K. CHEUNG, and K. G. STAGG, 'Stresses in anisotropic media with particular reference to problems of rock mechanics', *J. Strain Analysis*, **1**, 172–82, 1966.
7. G. N. SAVIN, *Stress Concentration Around Holes*, Pergamon Press, 1961. (Translation from Russian.)
8. O. C. ZIENKIEWICZ and Y. K. CHEUNG, 'Buttress dams on complex rock foundations', *Water Power*, **16**, 193, 1964.
9. O. C. ZIENKIEWICZ and Y. K. CHEUNG, 'Stresses in buttress dams', *Water Power*, **17**, 69, 1965.
10. K. TERZHAGI, *Theoretical Soil Mechanics*, Wiley, 1943.

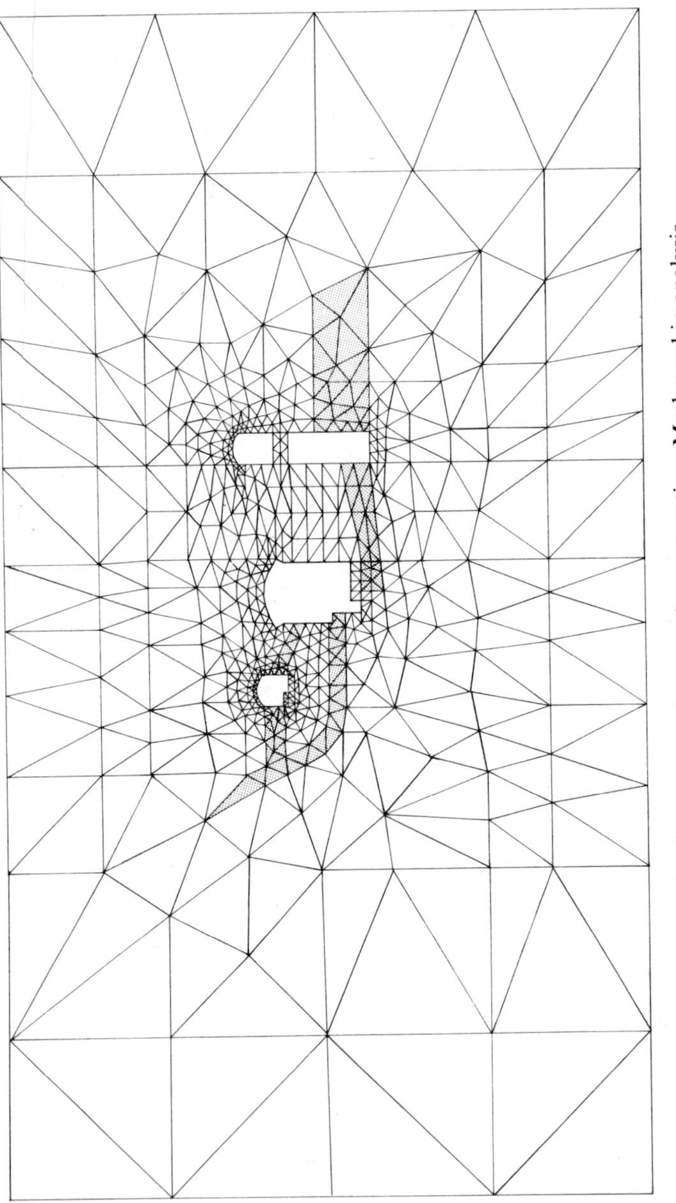

Fig. 4.14 An underground power station. Mesh used in analysis.

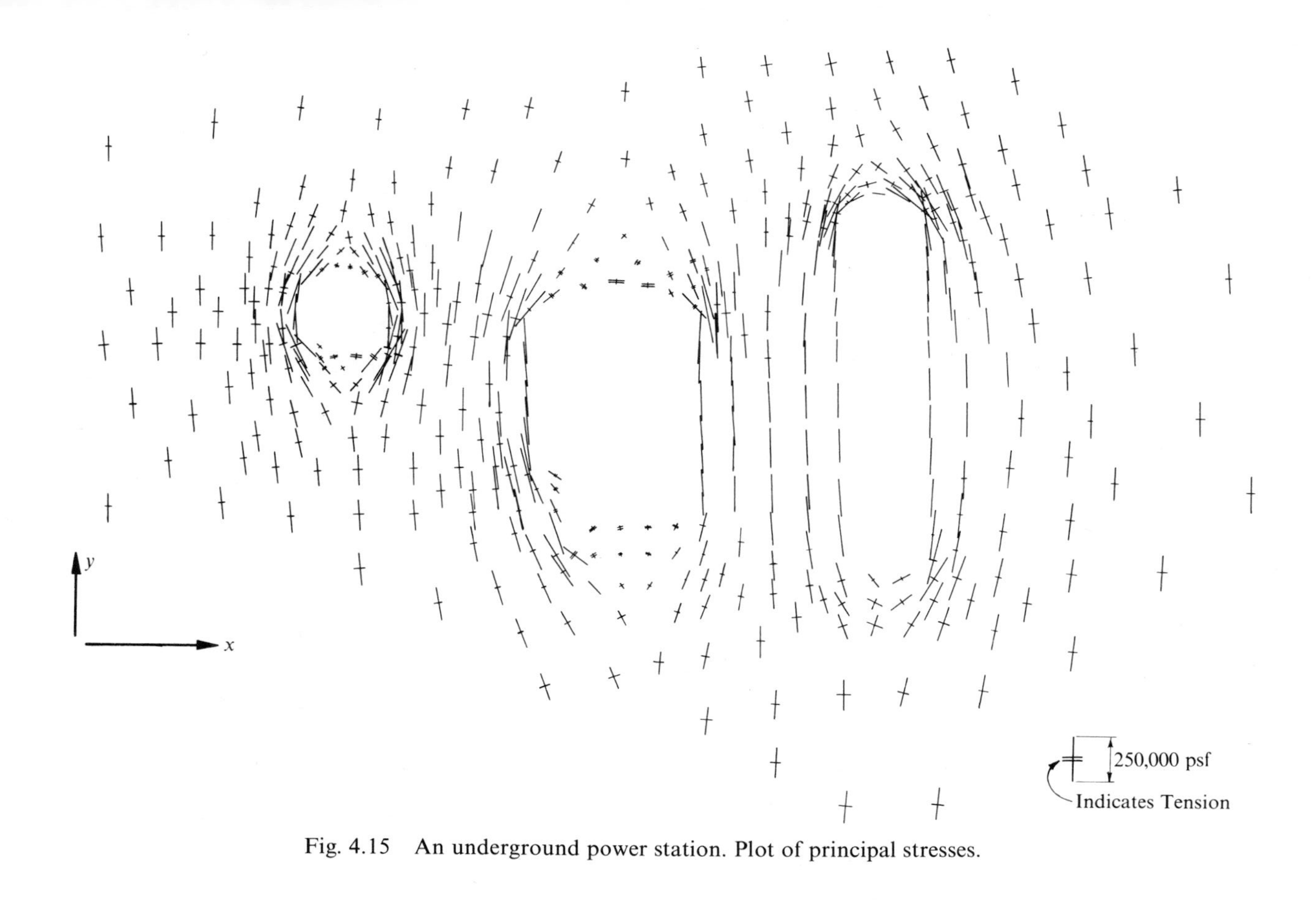

Fig. 4.15 An underground power station. Plot of principal stresses.

11. O. C. Zienkiewicz, C. Humpheson, and R. W. Lewis, 'A unified approach to soil mechanics problems, including plasticity and visco-plasticity', *Int. Symp. on Numerical Methods in Soil and Rock Mechanics*, Karlsruhe, 1975. See also Ch. 4 pp. 151–178 of *Finite Elements in Geomechanics*, ed. G. Gudehus, Wiley, 1977.
12. L. R. Herrmann, 'Elasticity equations for incompressible, or nearly incompressible materials by a variational theorem', *J.A.I.A.A.*, **3**, 1896, 1965.

5. Axi-Symmetric Stress Analysis

5.1 Introduction

The problem of stress distribution in bodies of revolution (axi-symmetric solids) under axi-symmetric loading is of considerable practical interest. The mathematical problems presented are very similar to those of plane stress and plane strain as, once again, the situation is two-dimensional.[1,2] By symmetry, the two components of displacements in any plane section of the body along its axis of symmetry define completely the state of strain and, therefore, the state of stress. Such a cross-section is shown in Fig. 5.1. If r and z denote respectively the radial and axial co-ordinates of a point, with u and v being the corresponding displacements, it can readily be seen that precisely the same displacement functions as those

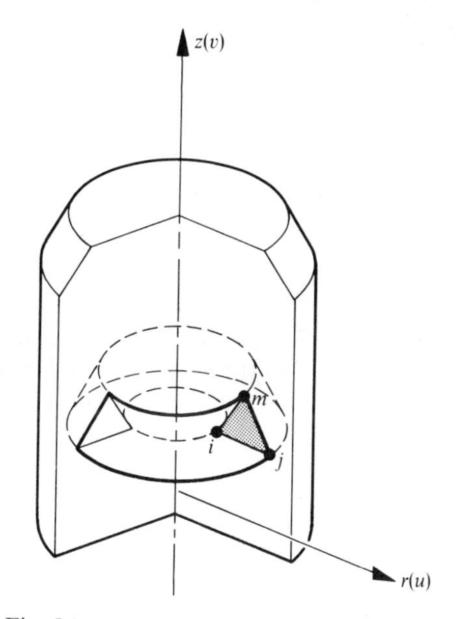

Fig. 5.1 Element of an axi-symmetric solid

used in Chapter 4 can be used to define the displacements within the triangular element i, j, m shown.

The volume of material associated with an 'element' is now that of a body of revolution indicated on Fig. 5.1, and all integrations have to be referred to this.

The triangular element is again used mainly for illustrative purposes, the principles developed being completely general.

In plane stress or strain problems it was shown that internal work was associated with three strain components in the co-ordinate plane, the stress component normal to this plane not being involved due to zero values of either the stress or the strain.

In the axi-symmetrical situation any radial displacement automatically induces a strain in the circumferential direction, and as the stresses in this direction are certainly non-zero, this fourth component of strain and of the associated stress has to be considered. Here lies the essential difference in the treatment of the axi-symmetric situation.

The reader will find the algebra involved in this chapter somewhat more tedious than that in the previous one but, essentially, identical operations are once again involved, following the general formulation of Chapter 2.

5.2 Element Characteristics

5.2.1 *Displacement function.* Using the triangular shape of element (Fig. 5.1) with the nodes i, j, m numbered in the anti-clockwise sense, we define the nodal displacement by its two components as

$$\mathbf{a}_i = \begin{Bmatrix} u_i \\ v_i \end{Bmatrix} \tag{5.1}$$

and the element displacements by the vector

$$\mathbf{a}^e = \begin{Bmatrix} \mathbf{a}_i \\ \mathbf{a}_j \\ \mathbf{a}_m \end{Bmatrix}. \tag{5.2}$$

Obviously, as in section 4.2.1, a linear polynomial can be used to define uniquely the displacements within the element. As the algebra involved is identical to that of Chapter 4 it will not be repeated here. The displacement field is now given again by Eq. (4.7).

$$\mathbf{u} = \begin{Bmatrix} u \\ v \end{Bmatrix} = [\mathbf{I}N_i, \mathbf{I}N_j, \mathbf{I}N_m]\mathbf{a}^e \tag{5.3}$$

with

$$N_i = (a_i + b_i r + c_i z)/2\Delta, \text{ etc.}$$

and **I** a two-by-two identity matrix. In the above

$$
\begin{aligned}
a_i &= r_j z_m - r_m z_j \\
b_i &= z_j - z_m = z_{jm} \\
c_i &= r_m - r_j = r_{mj}
\end{aligned}
\tag{5.4}
$$

etc., in cyclic order. Once again Δ is the area of the element triangle.

5.2.2 *Strain* (*total*). As already mentioned, four components of strain have now to be considered. These are, in fact, all the non-zero strain components possible in an axi-symmetric deformation. Figure 5.2 illustrates and defines these strains and the associated stresses.

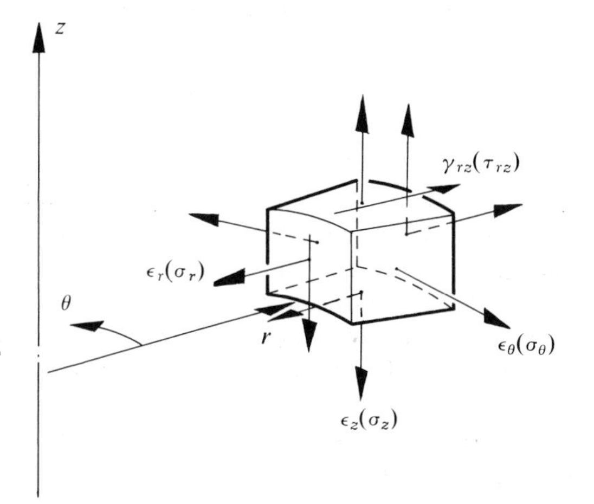

Fig. 5.2 Strains and stresses involved in the analysis of axi-symmetric solids

The strain vector defined below lists the strain components involved and defines them in terms of the displacements of a point. The expressions involved are almost self-evident and will not be derived here. The interested reader can consult a standard elasticity textbook[3] for the full derivation. We have thus

$$
\boldsymbol{\varepsilon} = \begin{Bmatrix} \varepsilon_z \\ \varepsilon_r \\ \varepsilon_\theta \\ \gamma_{rz} \end{Bmatrix} = \begin{Bmatrix} \dfrac{\partial v}{\partial z} \\[2mm] \dfrac{\partial u}{\partial r} \\[2mm] \dfrac{u}{r} \\[2mm] \dfrac{\partial u}{\partial z} + \dfrac{\partial v}{\partial r} \end{Bmatrix} = \mathbf{L}\mathbf{u}
\tag{5.5}
$$

Using the displacement functions defined by Eqs. (5.3) and (5.4) we have

$$\boldsymbol{\varepsilon} = \mathbf{B}\mathbf{a}^e = [\mathbf{B}_i, \mathbf{B}_j, \mathbf{B}_m]\mathbf{a}^e$$

in which

$$\mathbf{B}_i = \begin{bmatrix} 0 & , \dfrac{\partial N'_i}{\partial z} \\ \dfrac{\partial N'_i}{\partial r} , & 0 \\ \dfrac{1}{r}N'_i , & 0 \\ \dfrac{\partial N'_i}{\partial z} , & \dfrac{\partial N'_i}{\partial r} \end{bmatrix} = \frac{1}{2\Delta}\begin{bmatrix} 0 & , & c_i \\ b_i & , & 0 \\ a_i/r + b_i + c_i z/r & , & 0 \\ c_i & , & b_i \end{bmatrix} \text{etc.} \tag{5.6}$$

With the **B** matrix now involving the co-ordinates r and z, the strains are no longer constant within an element as in the plane stress or strain case. This strain variation is due to the ε_θ term. If the imposed nodal displacements are such that u is proportional to r then indeed the strains will all be constant. As this is the only state of displacement coincident with a constant strain condition it is clear that the displacement function satisfies the basic criterion of Chapter 2.

5.2.3 *Initial strain* (*thermal strain*). In general, four independent components of initial strain vector can be envisaged

$$\boldsymbol{\varepsilon}_0 = \begin{Bmatrix} \varepsilon_{z0} \\ \varepsilon_{r0} \\ \varepsilon_{\theta 0} \\ \gamma_{rz0} \end{Bmatrix}. \tag{5.7}$$

Although this can, in general, be variable within the element, it will be convenient to take the initial strain as constant there.

The most frequently encountered case of initial strain will be that due to a thermal expansion. For an isotropic material we shall have then

$$\boldsymbol{\varepsilon}_0 = \begin{Bmatrix} \alpha\theta^e \\ \alpha\theta^e \\ \alpha\theta^e \\ 0 \end{Bmatrix} \tag{5.8}$$

where θ^e is the average temperature rise in an element and α is the coefficient of thermal expansion.

A general case of anisotropy need not be considered since axial symmetry would be impossible to achieve under such circumstances. A case of some interest in practice is that of a 'stratified' material, similar to the one discussed in Chapter 4, in which the plane of isotropy is normal to the axis of symmetry (Fig. 5.3). Here, two different expansion coefficients are possible; one in the axial direction α_z and another in the plane normal to it, α_r.

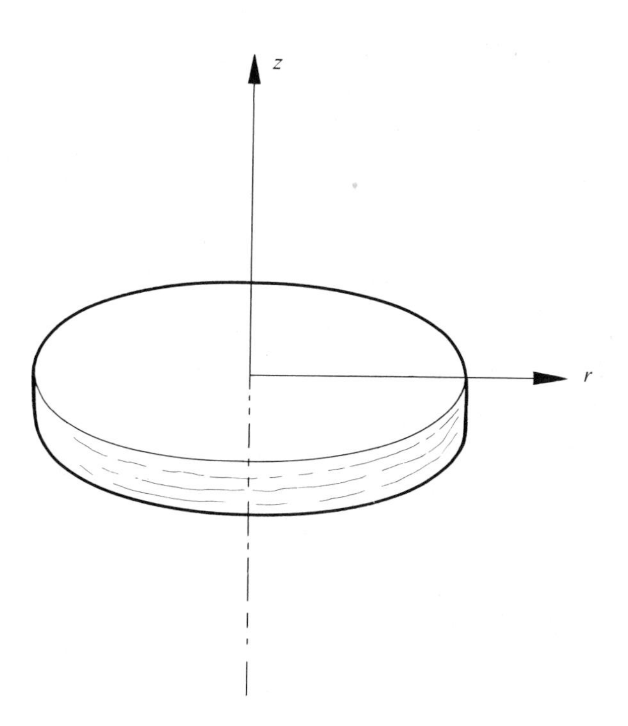

Fig. 5.3 Axi-symmetrically stratified material

Now the initial thermal strain becomes

$$\boldsymbol{\varepsilon}_0 = \begin{Bmatrix} \alpha_z \theta^e \\ \alpha_r \theta^e \\ \alpha_r \theta^e \\ 0 \end{Bmatrix}. \tag{5.9}$$

Practical cases of such 'stratified' anisotropy often arise in laminated or fibreglass construction of machine components.

5.2.4 *Elasticity matrix*. The elasticity matrix **D** linking the strains $\boldsymbol{\varepsilon}$

and the stresses $\boldsymbol{\sigma}$ in the standard form (Eq. 2.5)

$$\boldsymbol{\sigma} = \begin{Bmatrix} \sigma_z \\ \sigma_r \\ \sigma_\theta \\ \tau_{rz} \end{Bmatrix} = \mathbf{D}(\boldsymbol{\varepsilon}-\boldsymbol{\varepsilon}_0)+\boldsymbol{\sigma}_0$$

needs now to be derived.

The anisotropic, 'stratified' material will be first considered as the isotropic case can be simply presented as a special case.

Anisotropic, stratified, material (Fig. 5.3). With the z axis representing the normal to the planes of stratification we can rewrite Eqs. (4.22) (again ignoring the initial strains and stresses for convenience) as:

$$\begin{aligned} \varepsilon_z &= \sigma_z/E_2 - \nu_2\sigma_r/E_2 - \nu_2\sigma_\theta/E_2 \\ \varepsilon_r &= -\nu_2\sigma_z/E_2 + \sigma_r/E_1 - \nu_1\sigma_\theta/E_1 \\ \varepsilon_\theta &= -\nu_2\sigma_z/E_2 - \nu_1\sigma_r/E_1 + \sigma_\theta/E_1 \\ \gamma_{zr} &= \tau_{zr}/G_2. \end{aligned} \tag{5.10}$$

Writing again

$$\frac{E_1}{E_2} = n \quad \text{and} \quad \frac{G_2}{E_2} = m$$

we have, on solving for the stresses that:

$$\mathbf{D} = \frac{E_2}{(1+\nu_1)(1-\nu_1-2n\nu_2^2)}$$

$$\begin{bmatrix} 1-\nu_1^2 & n\nu_2(1+\nu_1), & n\nu_2(1+\nu_1) & , & 0 \\ & n(1-n\nu_2^2), & (\nu_1+n\nu_2^2)n & , & 0 \\ & & n(1-n\nu_2^2) & , & 0 \\ \text{symmetric} & & , & m(1+\nu_1) & \times(1-\nu_1-2n\nu_2^2) \end{bmatrix} \tag{5.11}$$

Isotropic material. For an isotropic material we can obtain the **D** matrix by taking

$$E_1 = E_2 = E \quad \text{or} \quad n = 1$$

and

$$\nu_1 = \nu_2 = \nu$$

and using the well-known relationship between isotropic elastic constants

$$\frac{G_2}{E_2} = \frac{G}{E} = m = \frac{1}{2(1+\nu)}.$$

Substituting in Eq. (5.11) we have now

$$\mathbf{D} = \frac{E(1-\nu)}{(1+\nu)(1-2\nu)} \begin{bmatrix} 1, & \frac{\nu}{1-\nu}, & \frac{\nu}{1-\nu}, & 0 \\ & 1, & \frac{\nu}{1-\nu}, & 0 \\ & & 1, & 0 \\ & \text{symmetric} & & \frac{1-2\nu}{2(1-\nu)} \end{bmatrix}. \tag{5.12}$$

5.2.5 *The stiffness matrix.* The stiffness matrix of the element *ijm* can now be computed according to the general relationship Eq. (2.13). Remembering that the volume integral has to be taken over the whole ring of material we have

$$\mathbf{K}_{ij}^e = 2\pi \int \mathbf{B}_i^{\mathrm{T}} \mathbf{D} \mathbf{B}_j r \, \mathrm{d}r \, \mathrm{d}z \tag{5.13}$$

with $\mathbf{B}$ given by Eq. (5.6) and $\mathbf{D}$ by either Eq. (5.11) or Eq. (5.12), depending on the material.

The integration cannot now be performed as simply as was the case in the plane stress problem because the $\mathbf{B}$ matrix depends on the co-ordinates. Two possibilities exist: the first that of numerical integration and the second of an explicit multiplication and term-by-term integration.

The simplest approximate procedure is to evaluate $\overline{\mathbf{B}}$ for a centroidal point

$$\bar{r} = (r_i + r_j + r_m)/3$$

and

$$\bar{z} = (z_i + z_j + z_m)/3.$$

In this case we have simply as a first approximation

$$\mathbf{K}_{ij}^e = 2\pi \overline{\mathbf{B}}_i^{\mathrm{T}} \mathbf{D} \overline{\mathbf{B}}_j \bar{r} \Delta \tag{5.14}$$

with Δ being the triangle area.

More elaborate numerical integration schemes could be used by evaluating the integrand at several points of the triangle. Such methods will be discussed in detail in Chapter 8. However, it can be shown that if the numerical integration is of such an order that the volume of the element is exactly determined by it, then in the limit of subdivision, the solution will converge to the exact answer.[4] The 'one point' integration suggested here is of such a type, as it is well known that the volume of a body of revolution is given exactly by the product of the area and the path swept around by its centroid. With the simple triangular element used here a fairly fine subdivision is in any case needed for accuracy and most practical programs use the simple approximation which, surprisingly perhaps, is in fact usually superior to the exact integration (*vide* Chapter 11). One reason for this is the occurrence of logarithmic terms in the exact formulation. These involve ratios of the type r_i/r_m and, when the element is at a large distance from the axis, such terms tend to unity and evaluation of the logarithm is inaccurate.

5.2.6 *External nodal forces.* In the case of two-dimensional problems of the previous chapter the question of assigning of the external loads was so obvious as not to need further comment. In the present case, however, it is important to realize that the nodal forces represent a combined effect of the force acting along the whole circumference of the circle forming the element 'node'. This point was already brought out in the integration of the expressions for the stiffness of an element, such integrations being conducted over the whole ring.

Thus, if $\bar{R}$ represents the radial component of force per unit length of the circumference of a node or a radius r, the external 'force' which will have to be introduced in the computation is

$$2\pi r\bar{R}.$$

In the axial direction we shall, similarly, have

$$2\pi r\bar{Z}$$

to represent the combined effect of axial forces.

5.2.7 *Nodal forces due to initial strain.* Again, by Eq. (2.13)

$$\mathbf{f}^e = -2\pi \int \mathbf{B}^{\mathrm{T}}\mathbf{D}\boldsymbol{\varepsilon}_0 r \, \mathrm{d}r \, \mathrm{d}z \tag{5.15}$$

or partitioning and noting that $\boldsymbol{\varepsilon}_0$ is constant

$$\mathbf{f}_i^e = -2\pi \left(\int \mathbf{B}_i^{\mathrm{T}} r \, \mathrm{d}r \, \mathrm{d}z \right) \mathbf{D}\boldsymbol{\varepsilon}_0. \tag{5.16}$$

The integration can be performed in a similar manner to that used in the determination of the stiffness.

It will be readily seen that, again, an approximate expression using a centroidal value is

$$\mathbf{f}_i^e = -2\pi \bar{\mathbf{B}}_i^{\mathrm{T}}\mathbf{D}\boldsymbol{\varepsilon}_0 \bar{r}\Delta \tag{5.17}$$

Initial stress forces are treated in an identical manner.

5.2.8 *Distributed body forces.* Distributed body forces such as those due to gravity (if acting along the z axis), centrifugal force in rotating machine parts or pore pressure, often occur in axi-symmetric problems.

Let such forces be denoted by

$$\mathbf{b} = \begin{Bmatrix} b_r \\ b_z \end{Bmatrix} \tag{5.18}$$

per unit volume of material in directions of r and z respectively. By the general Eq. (2.13) we have

$$\mathbf{f}_i^e = -2\pi \int \mathbf{I} N_i \begin{Bmatrix} b_r \\ b_z \end{Bmatrix} r \, dr \, dz \tag{5.19}$$

Using a co-ordinate shift similar to that of section 4.2.7 it is easy to show that the first approximation, if the body forces are constant, results in

$$\mathbf{f}_i^e = -2\pi \begin{Bmatrix} b_r \\ b_z \end{Bmatrix} \bar{r} \Delta/3. \tag{5.20}$$

Although this is not exact the error term will be found to decrease with reduction of element size and, as it is also self-balancing, it will not introduce inaccuracies. Indeed, as will be shown in Ch. 11, the convergence rate is maintained.

If the body forces are given by a potential similar to that defined in section 4.2.8, i.e.,

$$b_r = -\frac{\partial \phi}{\partial r}, \qquad b_z = -\frac{\partial \phi}{\partial z} \tag{5.21}$$

and if this potential is defined linearly by its nodal values, an expression equivalent to Eq. (4.36) can again be used with the same degree of approximation.

In many problems the body forces vary, proportionately to r. For example in rotating machinery we have centrifugal forces

$$b_r = \omega^2 \rho r \tag{5.22}$$

where ω is the angular velocity and ρ the density of the material.

5.2.9 *Evaluation of stresses.* The stresses now vary throughout the element as will be appreciated from Eqs. (5.5) and (5.6). It is convenient now to evaluate the average stress at the centroid of the element. The stress matrix resulting from Eqs. (5.6) and (2.3) gives there, as usual

$$\bar{\boldsymbol{\sigma}}^e = \mathbf{D}\bar{\mathbf{B}}\mathbf{a}^e - \mathbf{D}\boldsymbol{\varepsilon}_0 + \boldsymbol{\sigma}_0. \tag{5.23}$$

It will be found that a certain amount of oscillation of stress values between elements occurs and better approximation can be achieved by averaging nodal stresses.

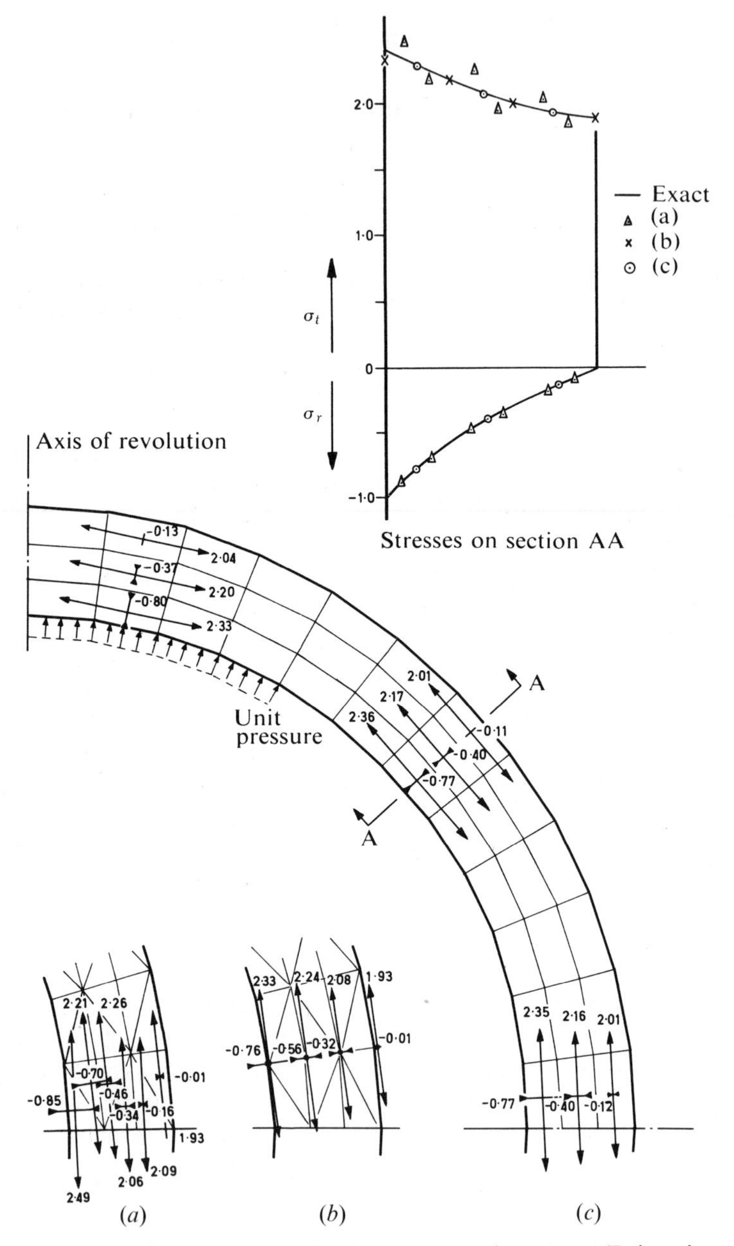

Fig. 5.4 Stresses in a sphere subject to an internal pressure (Poisson's ratio $\nu = 0{\cdot}3$). (*a*) Triangular mesh—centroidal values; (*b*) Triangular mesh—nodal averages; (*c*) Quadrilateral mesh obtained by averaging adjacent triangles

5.3 Some Illustrative Examples

Test problems such as those of a cylinder under constant axial or radial stress give, as indeed would be expected, solutions which correspond to exact ones. This is again an obvious corollary of the ability of the displacement function to reproduce constant strain conditions.

A problem for which an exact solution is available and in which almost linear stress gradients occur is that of a sphere subject to internal pressure. Figure 5.4(*a*) shows the centroidal stresses obtained using rather a coarse mesh, and the stress oscillation around the exact values should be noted. (This oscillation becomes even more pronounced at larger values of Poisson's ratio although the exact solution is independent of it.) In Fig. 5.4(*b*) the very much better approximation obtained by averaging the stresses at nodal points is shown, and in Fig. 5.4(*c*) a further improvement is given by element averaging. The close agreement with exact solution even for the very coarse subdivision used here shows the accuracy achievable. The displacements at nodes compared with the exact solution are given in Fig. 5.5.

In Fig. 5.6 thermal stresses in the same sphere are computed for a steady-state temperature variation shown. Again, excellent accuracy is demonstrated by comparison with the exact solution.

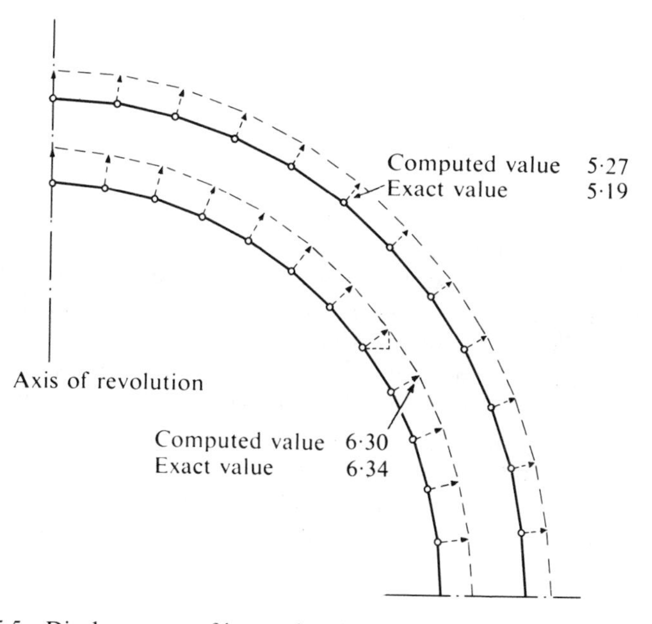

Fig. 5.5 Displacements of internal and external surfaces of sphere under loading of Fig. 5.4

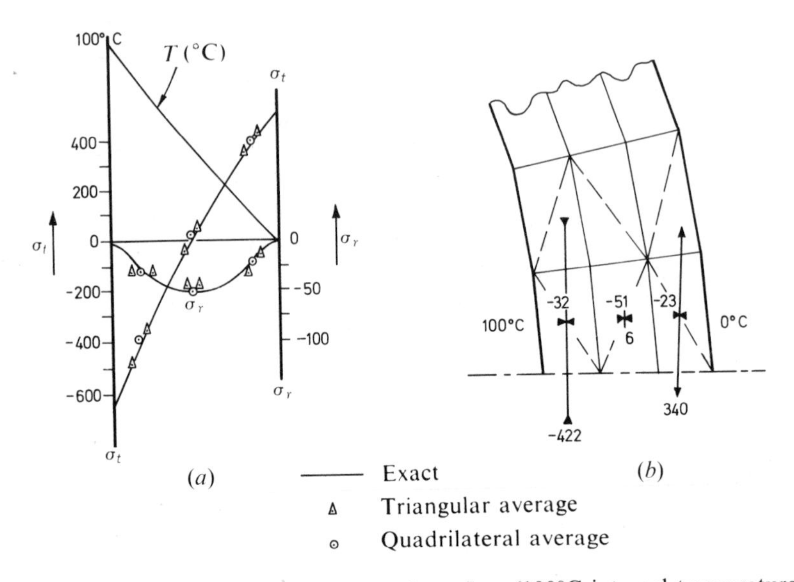

Fig. 5.6 Sphere subject to steady-state heat flow (100°C internal temperature, 0° external temperature). (*a*) Temperature and stress variation on radial section; (*b*) 'Quadrilateral' averages

5.4 **Practical Applications**

Two examples of practical application of the programs available for axi-symmetrical stress distribution are given here.

A prestressed concrete reactor pressure vessel. Figure 5.7 shows the stress distribution in a relatively simple prototype pressure vessel. Due to symmetry only one-half of the vessel is analysed, the results given here referring to the components of stress due to an internal pressure. Similar results due to the effect of prestressing cables are readily obtained by putting in the appropriate nodal loads due to these cables.

In Fig. 5.8 contours of equal major principal stresses caused by temperature are shown. The thermal state is due to a steady-state heat conduction and itself was found by the finite element method in a way described in Chapter 17.

Foundation pile. Figure 5.9 shows the stress distribution around a foundation pile penetrating two different strata. This non-homogeneous problem presents no difficulties and is treated by the standard program.

5.5 Non-symmetrical Loading

The method described in the present chapter can be extended to deal with non-symmetrical loading. If the circumferential loading variation is expressed in circular harmonics then it is still possible to focus attention on one axial section although the degree of freedom is now increased to three.

Some details of this process are described in Chapter 15. For full description reference 5 should be consulted.

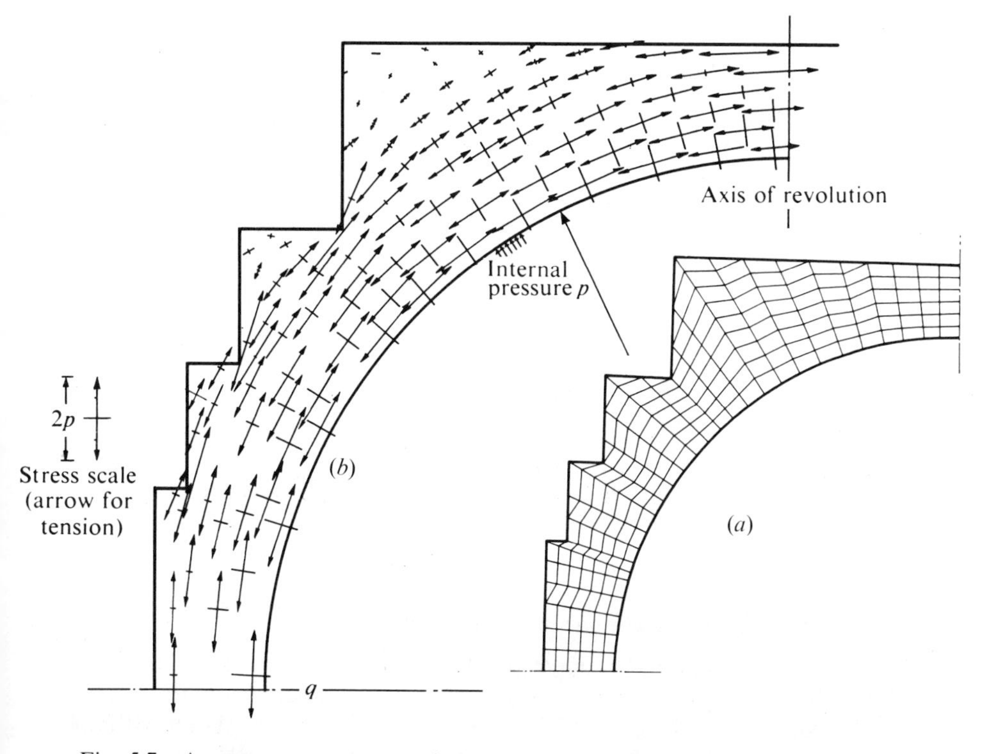

Fig. 5.7 A reactor pressure vessel. (*a*) 'Quadrilateral' mesh used in analysis; this was generated automatically by a computer. (*b*) Stresses due to a uniform internal pressure (automatic computer plot). Solution based on quadrilateral averages. Poisson's ratio $\nu = 0{\cdot}15$

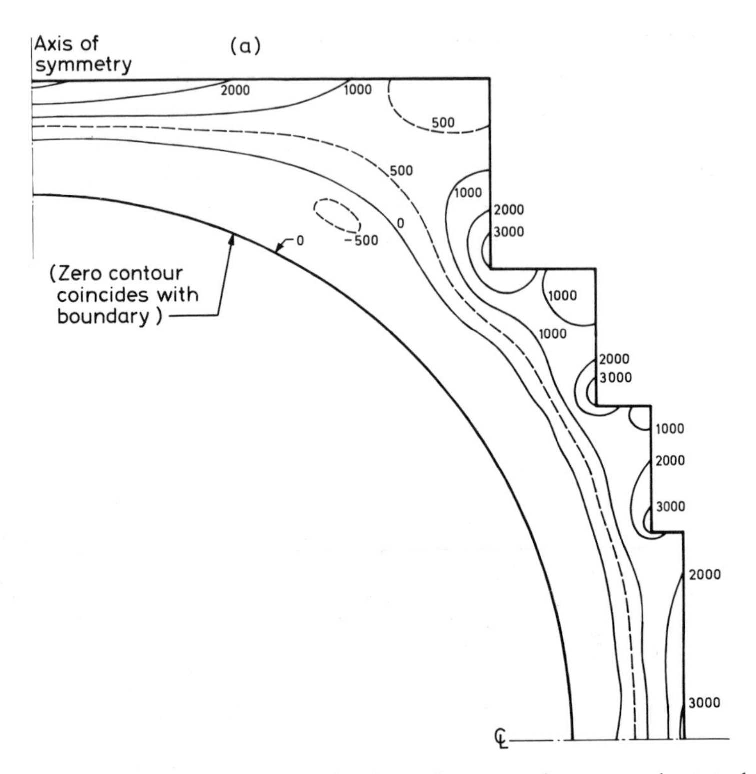

Fig. 5.8 A reactor pressure vessel. Thermal stresses due to steady-state heat conduction. Contours of major principal stress in lb/in^2.
(Interior temperature 400°C, Exterior temperature 0°C, $\alpha = 5 \times 10^{-6}/°C$. $E = 2{\cdot}58 \times 10^6$ lb/in, $\nu = 0{\cdot}15$)

5.6 **Axi-symmetry, plane strain and plane stress**

In the previous chapter we have noted that plane stress and strain analysis was done in terms of three stress and strain components and, indeed, both cases would be generally incorporated in a single program with an indicator changing appropriate constants in the matrix **D**. Doing this loses track of the σ_z component in the plane strain case which has to be separately evaluated. Further, special expressions (viz Eq. 4.13) had to be used to introduce initial strains. This is inconvenient especially when non-linear constitutive laws are used (to be discussed in Chapter 18) and an alternative of writing the plane strain case in terms of four stress/strain components as a special case of axi-symmetric analysis is highly recommended.

If the axi-symmetric strain definition of Eq. (5.5) is examined, we note

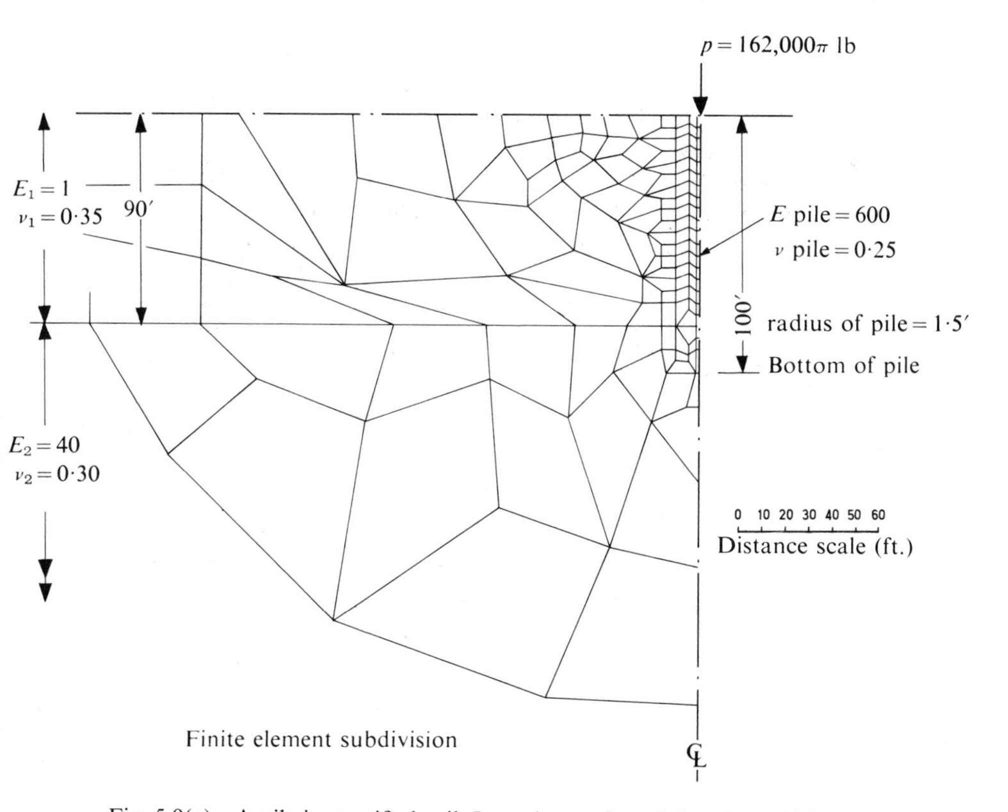

Fig. 5.9(*a*) A pile in stratified soil. Irregular mesh and data for problem

that $r = \infty$ gives $\varepsilon_0 \equiv 0$. Thus plane strain conditions are obtained. If we replace the coordinates

$$r \text{ and } z \quad \text{by} \quad x \text{ and } y$$

and further change in the stiffness expressions the volume of integration

$$2\pi r \text{ to } 1$$

the axi-symmetric formulation becomes available from plane strain directly.

Plane stress conditions can similarly be incorporated requiring in addition substitution of the axi-symmetric **D** matrix by the Eqs. (4.18) or (4.23) augmented by an appropriate zero row and column. Thus, at the cost of an additional storage of the fourth stress and strain component, all the cases discussed can be incorporated in a single format.

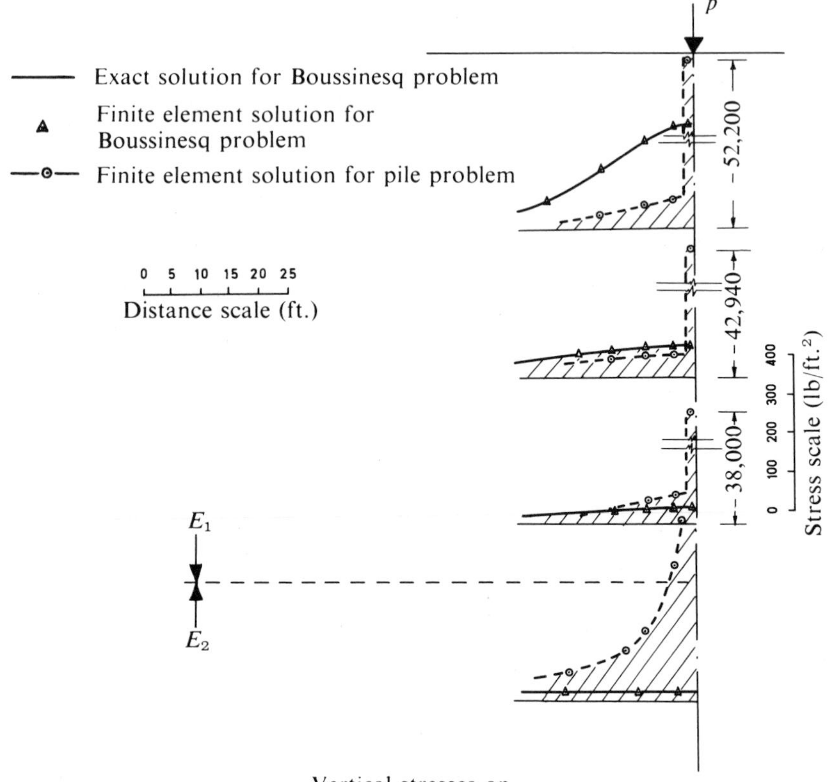

Fig. 5.9(*b*) A pile in stratified soil. Plot of vertical stresses on horizontal sections. Solution also plotted for Boussinesq problem obtained by making $E_1 = E_2 = E_{\text{pile}}$, and this is compared with exact values

References

1. R. W. Clough, chapter 7, *Stress Analysis*, ed. O. C. Zienkiewicz and G. S. Holister, Wiley, 1965.
2. R. W. Clough and Y. R. Rashid, 'Finite element analysis of axi-symmetric solids', *Proc. A.S.C.E.*, **91**, EM.1, 71, 1965.
3. S. Timoshenko and J. N. Goodier, *Theory of Elasticity*, 2nd ed., McGraw-Hill, 1951.
4. B. M. Irons, Comment on 'Stiffness matrices for sector element' by I. R. Raju and A. K. Rao, *J.A.I.A.A.*, **7**, 156–7, 1969.
5. E. L. Wilson, 'Structural analysis of axisymmetric solids', *J.A.I.A.A.*, **3**, 2269–74, 1965.

6. Three-Dimensional Stress Analysis

6.1 Introduction

It will have become obvious to the reader by this stage of the book that there is but one further step to apply the general finite element procedure to fully three-dimensional problems of stress analysis. Such problems embrace clearly all the practical cases, though for some, the various two-dimensional approximations give an adequate and more economical 'model'.

The simplest two-dimensional continuum element was a triangle. In three dimensions its equivalent is a tetrahedron, an element with four nodal corners—and this chapter will deal with the basic formulation of such an element. Immediately, a difficulty not encountered previously is presented. It is one of ordering of the nodal numbers and, in fact, of a suitable representation of a body divided into such elements.

The first suggestions for use of the simple tetrahedral element appear to be those of Gallagher *et al.*[1] and Melosh.[2] Argyris[3,4] elaborated further on the theme and Rashid[5] has shown that with the largest modern computers such a formulation can still be applied to realistic problems.

It is immediately obvious, however, that the number of simple tetrahedral elements which has to be used to achieve a given degree of accuracy has to be very large. This will result in very large numbers of simultaneous equations in practical problems, which may place a severe limitation on the use of the method in practice. Further the band width of the resulting equation system becomes large leading to big computer storage requirements.

To realize the order of magnitude of the problems presented let us assume that the accuracy of triangle in two-dimensional analysis is comparable to that of a tetrahedron in three dimensions. If an adequate stress analysis of a square, two-dimensional region requires a mesh of some $20 \times 20 = 400$ nodes, the total number of simultaneous equations is around 800 given two displacement variables for node. (This is a fairly realistic figure.) The band width of the matrix involves 20 nodes (see chapter on computation), i.e., some 40 variables.

An equivalent three-dimensional region is that of a cube with $20 \times 20 \times 20 = 8000$ nodes. The total number of simultaneous equations is now some 24,000 as three displacement variables have to be specified. Further, the band width involves now an interconnection of some $20 \times 20 = 400$ nodes or 1200 variables.

Given that with usual solution techniques the computation effort is roughly proportional to the number of equations and to the square of the band width, the magnitude of the problems can be appreciated. It is not surprising therefore that efforts to improve accuracy by use of complex elements with many degrees of freedom have been strongest in the area of three-dimensional analysis.[6,7,8,9,10] The use and practical application of such elements will be described in the following chapters. However, the presentation of this chapter gives all the necessary ingredients of formulation for three-dimensional elastic problems and so follows directly from the previous ones. Extension to more elaborate elements will be self evident.

6.2 Tetrahedral Element Characteristics

6.2.1 *Displacement functions.* Figure 6.1 illustrates a tetrahedral element i, j, m, p in space defined by the x, y, and z co-ordinates.

The state of displacement of a point is defined by three displacement components, u, v, and w in directions of the three co-ordinates x, y, and z. Thus

$$\mathbf{u} = \begin{Bmatrix} u \\ v \\ w \end{Bmatrix}. \tag{6.1}$$

Just as in a plane triangle where a linear variation of a quantity was defined by its three nodal values, here a linear variation will be defined by the four nodal values. In analogy to Eq. (4.3) we can write, for instance

$$u = \alpha_1 + \alpha_2 x + \alpha_3 y + \alpha_4 z. \tag{6.2}$$

Equating the values of displacement at the nodes we have four equations of the type

$$u_1 = \alpha_1 + \alpha_2 x_i + \alpha_3 y_i + \alpha_4 z_i \text{ etc.} \tag{6.3}$$

from which α_1 to α_4 can be evaluated.

Again, it is possible to write this solution in the form similar to that of

Eq. (4.5) by using a determinant form, i.e.

$$u = \frac{1}{6V}\Big\{(a_i + b_i x + c_i y + d_i z)u_i - (a_j + b_j x + c_j y + d_j z)u_j + (a_m + b_m x + c_m y + d_m z)u_m - (a_p + b_p x + c_p y + d_p z)u_p\Big\} \tag{6.4}$$

with

$$6V = \det\begin{vmatrix} 1 & x_i & y_i & z_i \\ 1 & x_j & y_j & z_j \\ 1 & x_m & y_m & z_m \\ 1 & x_p & y_p & z_p \end{vmatrix} \tag{6.5a}$$

in which, incidentally, the value V represents the volume of the tetrahedron. By expanding the other relevant determinants into their co-factors

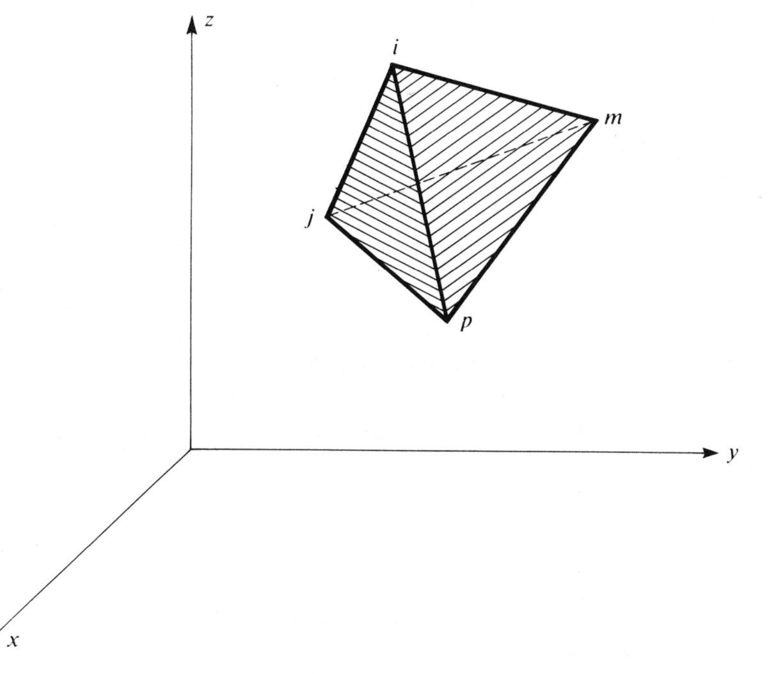

Fig. 6.1 A tetrahedral volume. (Always use a consistent order of numbering, e.g., starting with p count the other nodes in an anti-clockwise order as viewed from p—$pijm$ or $mipj$ etc.)

we have

$$a_i = \det \begin{vmatrix} x_j & y_j & z_j \\ x_m & y_m & z_m \\ x_p & y_p & z_p \end{vmatrix}, \qquad b_i = -\det \begin{vmatrix} 1 & y_j & z_j \\ 1 & y_m & z_m \\ 1 & y_p & z_p \end{vmatrix},$$

$$c_i = -\det \begin{vmatrix} x_j & 1 & z_j \\ x_m & 1 & z_m \\ x_p & 1 & z_p \end{vmatrix}, \qquad d_i = -\det \begin{vmatrix} x_j & y_j & 1 \\ x_m & y_m & 1 \\ x_p & y_p & 1 \end{vmatrix} \tag{6.5b}$$

with the other constants defined by cyclic interchange of the subscripts in the order p, i, j, m.

The ordering of nodal numbers p, i, j, m must follow a 'right-hand' rule obvious from Fig. 6.1. In this the first three nodes are numbered in an anti-clockwise manner when viewed from the last one. (See Appendix 5.)

The element displacement is defined by the twelve displacement components of the nodes as

$$\mathbf{a}^e = \begin{Bmatrix} \mathbf{a}_i \\ \mathbf{a}_j \\ \mathbf{a}_m \\ \mathbf{a}_p \end{Bmatrix} \tag{6.6}$$

with

$$\mathbf{a}_i = \begin{Bmatrix} u_i \\ v_i \\ w_i \end{Bmatrix} \text{ etc.}$$

We can write the displacements of an arbitrary point as

$$\mathbf{u} = [\mathbf{I}N_i, \mathbf{I}N_j, \mathbf{I}N_m, \mathbf{I}N_p]\mathbf{a}^e \tag{6.7}$$

with shape functions defined as

$$N_i = (a_i + b_i x + c_i y + d_i z)/6V \text{ etc.} \tag{6.8}$$

and $\mathbf{I}$ being a three by three identity matrix.

Once again the displacement functions used will obviously satisfy continuity requirements on interfaces between various elements. This fact is a direct corollary of the linear nature of the variation of displacements.

6.2.2 *Strain matrix.* Six strain components are relevant in full three-dimensional analysis. The strain matrix can now be defined as

$$\boldsymbol{\varepsilon} = \begin{Bmatrix} \varepsilon_x \\ \varepsilon_y \\ \varepsilon_z \\ \gamma_{xy} \\ \gamma_{yz} \\ \gamma_{zx} \end{Bmatrix} = \begin{Bmatrix} \dfrac{\partial u}{\partial x} \\ \dfrac{\partial v}{\partial y} \\ \dfrac{\partial w}{\partial z} \\ \dfrac{\partial u}{\partial y} + \dfrac{\partial v}{\partial x} \\ \dfrac{\partial v}{\partial z} + \dfrac{\partial w}{\partial y} \\ \dfrac{\partial w}{\partial x} + \dfrac{\partial u}{\partial z} \end{Bmatrix} = \mathbf{L}\mathbf{u} \tag{6.9}$$

following the standard notation of Timoshenko's elasticity text. Using Eqs. (6.4) to (6.7) it is an easy matter to verify that

$$\boldsymbol{\varepsilon} = \mathbf{B}\mathbf{a}^e = [\mathbf{B}_i, \mathbf{B}_j, \mathbf{B}_m, \mathbf{B}_p]\mathbf{a}^e \tag{6.10}$$

in which

$$\mathbf{B}_i = \begin{bmatrix} \dfrac{\partial N_i'}{\partial x}, & 0, & 0 \\ 0, & \dfrac{\partial N_i'}{\partial y}, & 0 \\ 0, & 0, & \dfrac{\partial N_i'}{\partial z} \\ \dfrac{\partial N_i'}{\partial y}, & \dfrac{\partial N_i'}{\partial x}, & 0 \\ 0, & \dfrac{\partial N_i'}{\partial z}, & \dfrac{\partial N_i'}{\partial y} \\ \dfrac{\partial N_i'}{\partial z}, & 0, & \dfrac{\partial N_i'}{\partial x} \end{bmatrix} = \frac{1}{6V} \begin{bmatrix} b_i, & 0, & 0 \\ 0, & c_i, & 0 \\ 0, & 0, & d_i \\ c_i, & b_i, & 0 \\ 0, & d_i, & c_i \\ d_i, & 0, & b_i \end{bmatrix} \tag{6.11}$$

with other submatrices obtained in a similar manner simply by interchange of subscripts.

Initial strains, such as those due to thermal expansion, can be written in the usual way as a six-component vector which, for example, in an isotropic thermal expansion is simply

$$\boldsymbol{\varepsilon}_0 = \begin{Bmatrix} \alpha\theta^e \\ \alpha\theta^e \\ \alpha\theta^e \\ 0 \\ 0 \\ 0 \end{Bmatrix} \tag{6.12}$$

with α being the expansion coefficient and θ^e the average element temperature rise.

6.2.3 *Elasticity matrix.* With complete anisotropy the **D** matrix relating the six stress components to the strain components can contain 21 independent constants (*vide* section 4.2.4).

In general, thus,

$$\boldsymbol{\sigma} = \begin{Bmatrix} \sigma_x \\ \sigma_y \\ \sigma_z \\ \tau_{xy} \\ \tau_{yx} \\ \tau_{zx} \end{Bmatrix} = \mathbf{D}(\boldsymbol{\varepsilon}-\boldsymbol{\varepsilon}_0)+\boldsymbol{\sigma}_0. \tag{6.13}$$

Although no difficulty presents itself in computation when dealing with such materials, since the multiplication will never be carried out explicitly, it is convenient to recapitulate here the **D** matrix for an isotropic material. This, in terms of the usual elastic constants E (modulus) and ν (Poisson's ratio), can be written as

$$\mathbf{D} = \frac{E(1-\nu)}{(1+\nu)(1-2\nu)} \times \begin{bmatrix} 1, & \nu/(1-\nu), & \nu/(1-\nu), & 0, & 0, & 0 \\ & 1, & \nu/(1-\nu), & 0, & 0, & 0 \\ & & 1, & 0, & 0, & 0 \\ & \text{symmetric} & & \dfrac{(1-2\nu)}{2(1-\nu)}, & 0, & 0 \\ & & & & \dfrac{(1-2\nu)}{2(1-\nu)}, & 0 \\ & & & & & \dfrac{(1-2\nu)}{2(1-\nu)} \end{bmatrix}. \tag{6.14}$$

6.2.4 *Stiffness, stress, and load matrices.* The stiffness matrix defined by the general relationship Eq. (2.10) can be now explicitly integrated since the strain and stress components are constant within the element.

The general ij submatrix of the stiffness matrix will be a three by three matrix defined as

$$\mathbf{K}_{ij}^{e} = \mathbf{B}_i^{\mathrm{T}}\mathbf{D}\mathbf{B}_j V^e \tag{6.15}$$

where V^e represents the volume of the elementary tetrahedron.

The nodal forces due to the initial strain become, similarly to Eq. (4.28)

$$\mathbf{f}_i^e = -\mathbf{B}_i^{\mathrm{T}}\mathbf{D}\boldsymbol{\varepsilon}_0 V^e \tag{6.16}$$

with similar expression for forces due to initial stresses.

In fact, the similarity with the expressions and results of Chapter 4 is such that further explicit formulation is unnecessary. The reader will find no difficulty in repeating the various steps needed for the formulation of a computer program.

Distributed body forces can once again be expressed in terms of their b_x, b_y, and b_z components or in terms of the body-force potential. Not surprisingly, it will once more be found that if the body forces are constant the nodal components of the total resultant are distributed in four equal parts (*vide* Eq. (4.30)).

6.3 Composite Elements with Eight Nodes

The division of a space volume into individual tetrahedra sometimes presents difficulties of visualization and could easily lead to errors in nodal

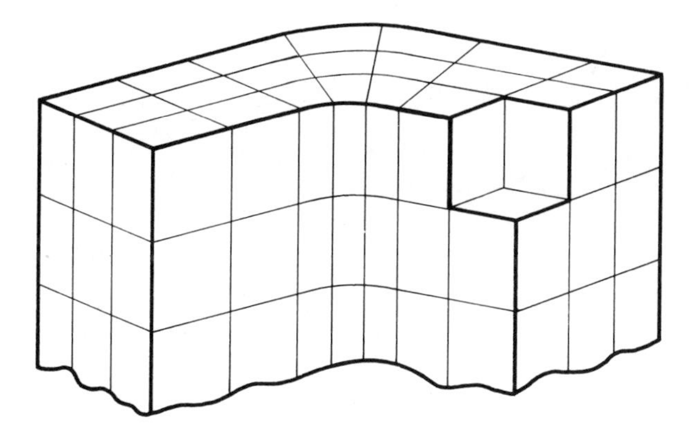

Fig. 6.2 A systematic way of dividing a three-dimensional object into "brick"-type elements

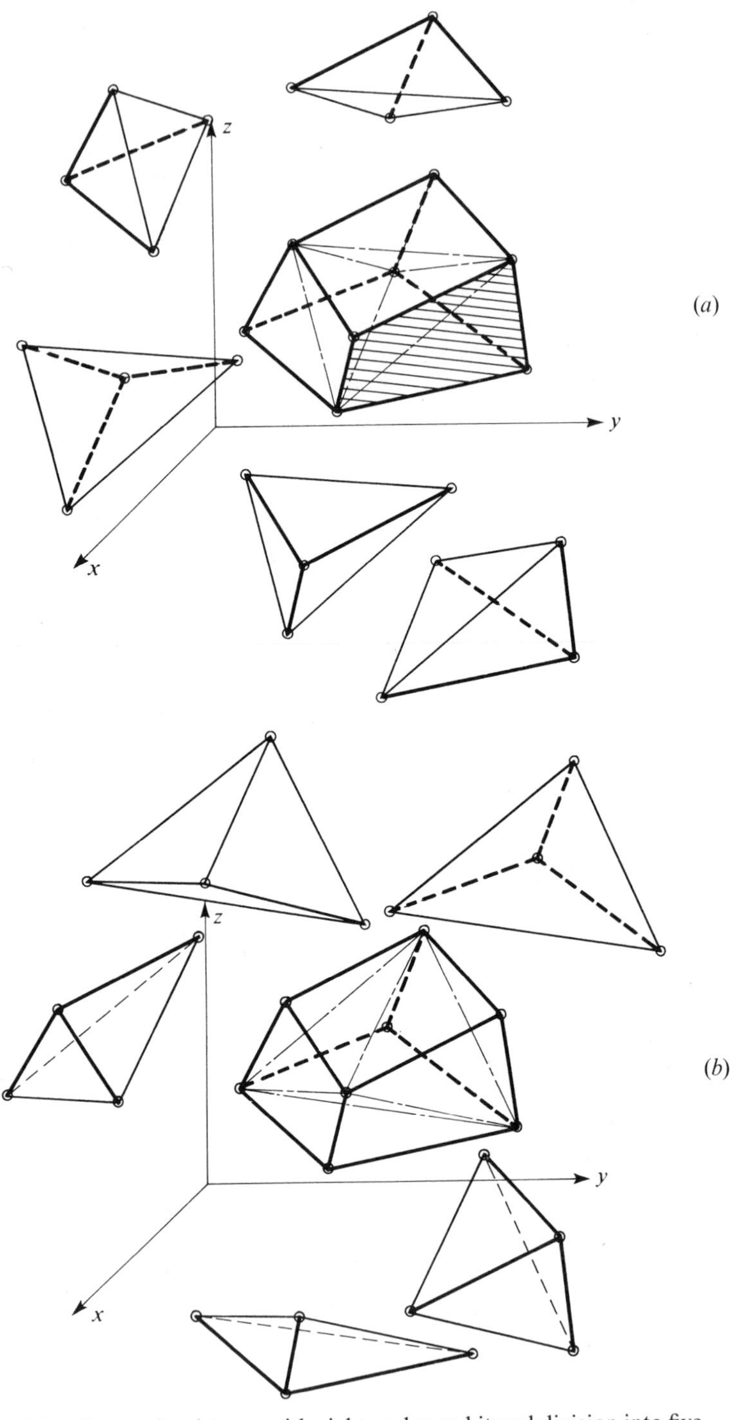

Fig. 6.3 Composite element with eight nodes and its subdivision into five tetrahedra by alternatives (*a*) or (*b*)

numbering, etc. A more convenient subdivision of space is into eight-cornered brick elements. By sectioning a three-dimensional body parallel sections can be drawn and, each one being subdivided into quadrilaterals a systematic way of element definition could be devised as in Fig. 6.2.

Such elements could be assembled automatically from several tetrahedra and the process of creating these tetrahedra left to a simple logical program. For instance Fig. 6.3 shows how a typical brick can be divided into five tetrahedra in two (and only two) distinct ways. Indeed by averaging the two types of subdivision a slight improvement of accuracy can be obtained. Stresses could well be presented as averages for a whole brick-like element.

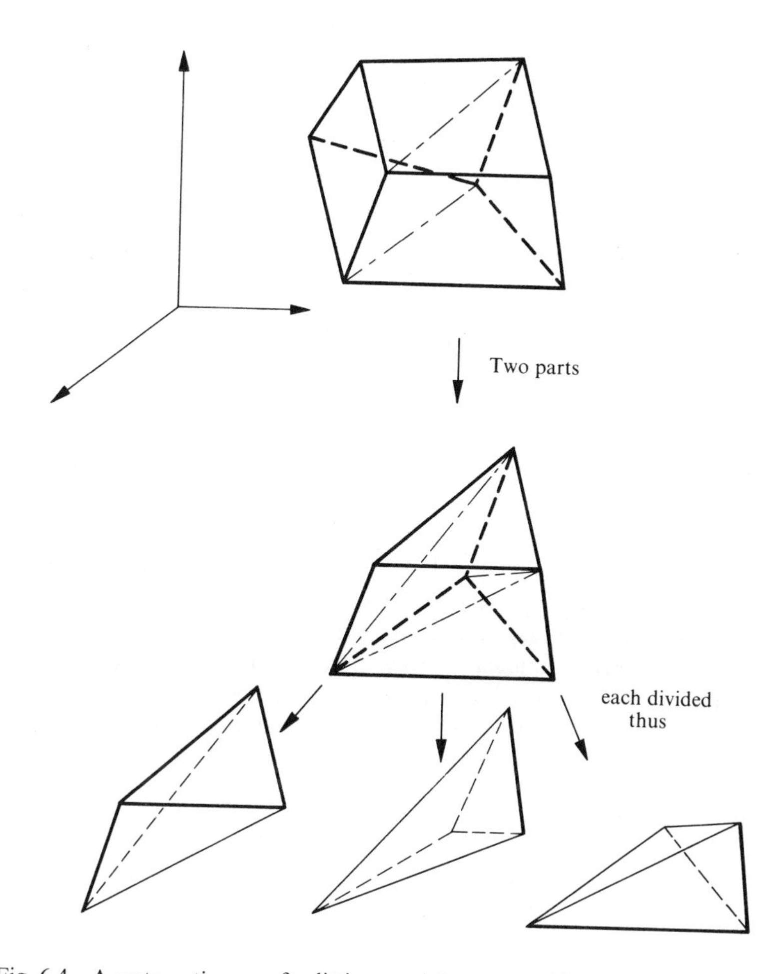

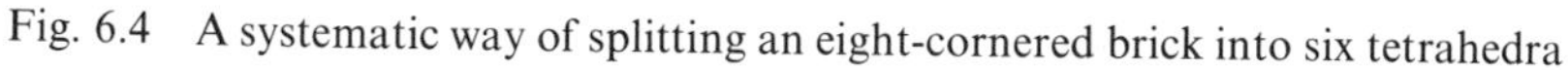

Fig. 6.4 A systematic way of splitting an eight-cornered brick into six tetrahedra

In Fig. 6.4 an alternative subdivision of a brick into six tetrahedra is shown. Here obviously the number of alternatives is very great.

In later chapters it will be seen how the basic bricks can be obtained directly with more complex types of shape functions.

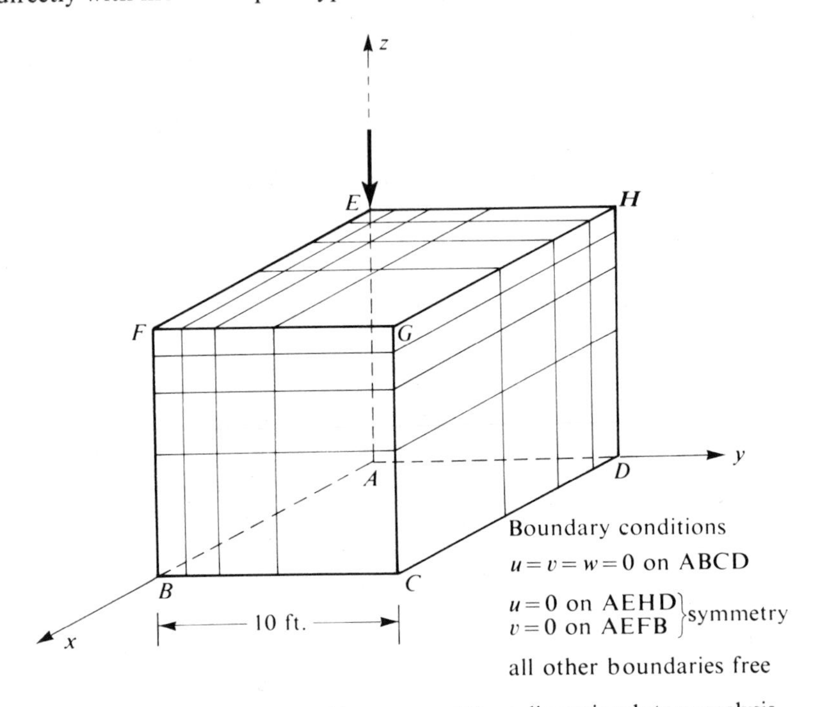

Fig. 6.5 The Boussinesq problem as one of three-dimensional stress analysis

6.4 Examples and Concluding Remarks

A simple, illustrative example of application of simple, tetrahedral, elements is shown in Figs. 6.5 and 6.6. Here the well-known Boussinesq problem of an elastic halfspace with a point load is approximated to by analysing a cubic volume of space. Use of symmetry is made to reduce the size of the problem and the boundary displacements are prescribed in a manner shown in Fig. 6.5.[11] As zero displacements were prescribed at a finite distance below the load a correction obtained from the exact expression was applied before executing the plots shown in Fig. 6.6. Comparison of both stresses and displacement appears reasonable although it will be appreciated that the division is very coarse. However, even this trivial problem involved the solution of some 375 equations. More ambitious problems treated with simple tetrahedra are given in references 5 and 11. Figure 6.7, taken from the former, illustrates an analysis of a complex

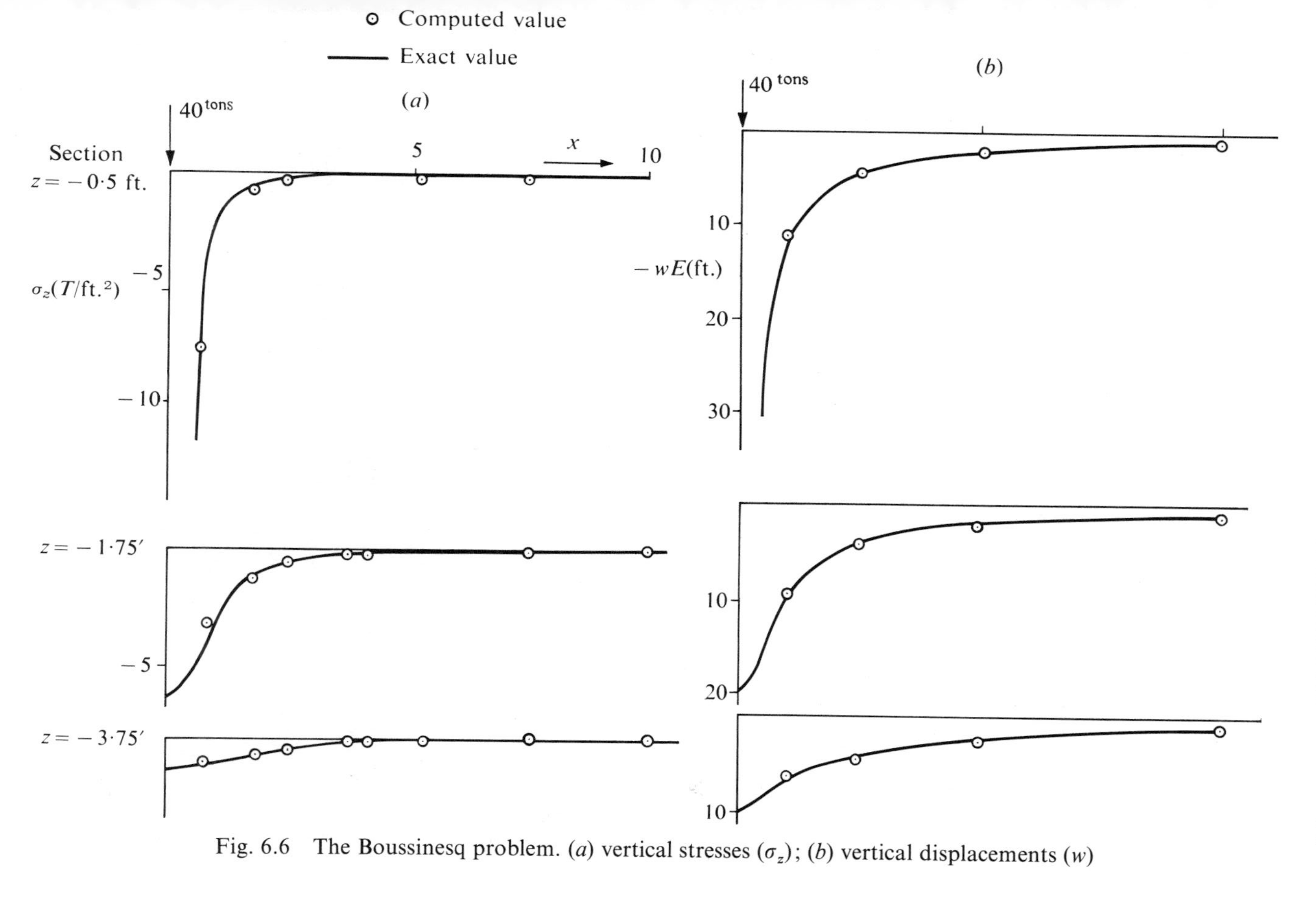

Fig. 6.6 The Boussinesq problem. (*a*) vertical stresses (σ_z); (*b*) vertical displacements (w)

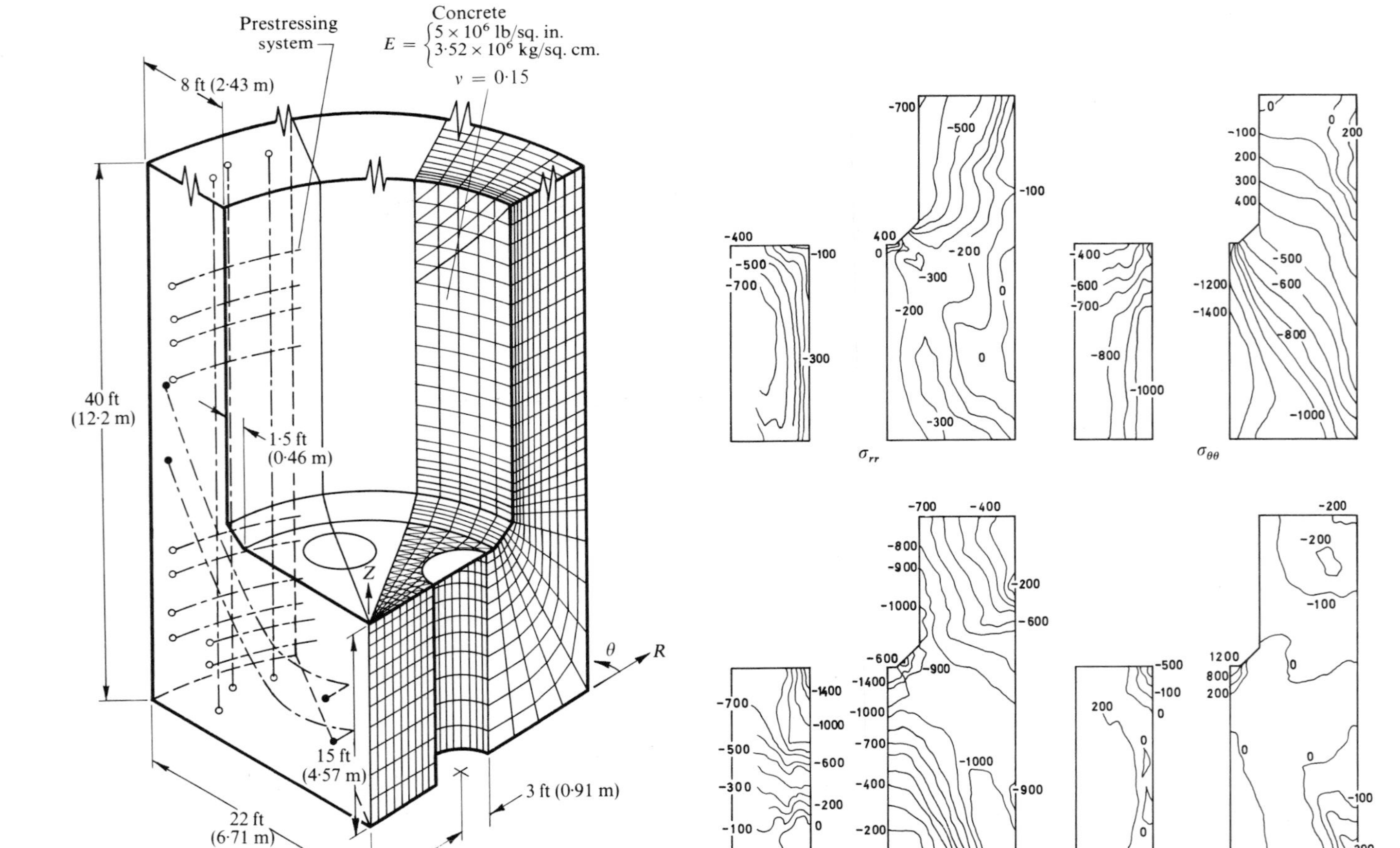

Fig. 6.7 A nuclear pressure vessel analysis using simple tetrahedral elements.[5] Geometry, subdivision and some stress results

pressure vessel. Some 10,000 degrees of freedom are involved in this analysis. In Chapter 9 it will be seen how the use of complex elements permit a sufficiently accurate analysis to be performed with a much smaller total number of degrees of freedom for a very similar problem.

References

1. R. H. GALLAGHER, J. PADLOG, P. P. BIJLAARD, 'Stress analysis of heated complex shapes', *A.R.S. Journal*, 700–7, 1962.
2. R. J. MELOSH, 'Structural analysis of solids', *Proc. Amer. Soc. Civ. Eng.*, **S.T.4**, 205–23, Aug. 1963.
3. J. H. ARGYRIS, 'Matrix analysis of three-dimensional elastic media—small and large displacements', *J.A.I.A.A.*, **3**, 45–51, Jan. 1965.
4. J. H. ARGYRIS, 'Three-dimensional anisotropic and inhomogeneous media—matrix analysis for small and large displacements', *Ingenieur Archiv.*, **34**, 33–55, 1965.
5. Y. R. RASHID and W. ROCKENHAUSER, 'Pressure vessel analysis by finite element techniques', *Proc. Conf. on Prestressed Concrete Pressure Vessels*, Inst. Civ. Eng., 1968.
6. J. H. ARGYRIS, 'Continua and Discontinua', *Proc. Conf. Matrix Methods in Structural Mechanics*, Wright Patterson Air Force Base, Ohio, Oct. 1965.
7. B. M. IRONS, 'Engineering applications of numerical integration in stiffness methods', *J.A.I.A.A.*, **4**, 2035–7, 1966.
8. J. G. ERGATOUDIS, B. M. IRONS, and O. C. ZIENKIEWICZ, 'Three dimensional analysis of arch dams and their foundations', *Proc. Symp. Arch Dams*, Inst. Civ. Eng., 1968.
9. J. H. ARGYRIS and J. C. REDSHAW, 'Three dimensional analysis of two arch dams by a finite element method', *Proc. Symp. Arch Dams*, Inst. Civ. Eng., 1968.
10. S. FJELD, 'Three dimensional theory of elastics', *Finite Element Methods in Stress Analysis* (eds. I. Holand and K. Bell), Tech. Univ. of Norway, Tapir Press, Trondheim, 1969.
11. J. OLIVEIRA PEDRO, Thesis 1967, Laboratorio Nacional de Engenharia Civil, Lisbon.

7. Element Shape Functions—Some General Families of C_0 Continuity

7.1 Introduction

In the three previous chapters the reader was shown in some detail how linear elasticity problems could be formulated and solved using very simple finite element forms. Although the detailed algebra was concerned with shape functions which arose from triangular and tetrahedral shapes only it should by now be obvious that other element forms could equally well be used. Indeed, once the element and the corresponding shape functions are determined, subsequent operations follow a standard, well-defined path which could be entrusted to an algebraist not familiar with the physical aspects of the problem. It will be seen later that in fact it is possible to program a computer to deal with wide classes of problems by specifying the shape functions only. The choice of these is, however, a matter to which intelligence has to be applied and in which the human factor remains paramount. In this chapter some rules for generation of several families of one-, two-, and three-dimensional elements will be presented.

In the problems of elasticity illustrated in Chapters 4, 5, and 6 the displacement variable was a vector with two or three components and the shape functions were written in a matrix form. They were, however, derived for each component separately and in fact the matrix expressions in these were derived by multiplying a scalar function by an identity matrix (e.g., Eqs. (4.7), (5.3), and (6.7)). We shall therefore concentrate in this chapter on the scalar shape function forms omitting the prime and calling these simply N_i.

The shape functions used in the displacement formulation of elasticity problems were such that to satisfy the convergence criteria of Chapters 2 and 3

(*a*) the continuity of the unknown only had to occur between elements (i.e., slope continuity is not required), or C_0 continuity.

(*b*) the function has to allow any arbitrary linear form to be taken so that the constant strain (constant first derivative) criterion could be observed.

The shape functions described in this chapter will only require the satisfaction of these two criteria. They will be thus applicable to all the problems of the preceding chapters and also to other problems which require only these conditions to be obeyed. For instance all problems of Chapter 17 can use the forms here determined. Indeed they are applicable to any situation where the functional Π (see Chapter 3) is defined by derivatives of the first order only.

The element families discussed will progressively have an increasing number of degrees of freedom. The question may well be asked as to whether any economic or other advantage is gained by increasing thus the complexity of an element. The answer here is not an easy one although it can be stated as a general rule that as the order of an element increases so the total number of unknowns in a problem can be reduced for a given accuracy of representation. Economic advantage requires, however, a reduction of total computations and data preparation effort and this does not follow automatically for a reduced number of total variables as, though equation solving times may be reduced, the time required for element formulation increases.

An overwhelming economic advantage in case of three-dimensional analysis has been already hinted at in the previous chapter on three-dimensional analysis.

The same kind of advantage arises on occasion in other problems but in general the optimum element may have to be determined from case to case.

In section 2.6 of Chapter 2 (and more generally in section 3.8 of Chapter 3) we have shown that the order of error in the approximation is $O(h^{p+1})$, where h is the element 'size' and p is the complete polynomial present in the expansion. Clearly, as the element shape functions increase in order so will the order of error increase, and convergence to the exact solution become more rapid. While this says nothing about the magnitude of error at a particular subdivision, it is clear that we should seek element shape functions with the highest complete polynomial for a given number of degrees of freedom.

TWO-DIMENSIONAL ELEMENTS

7.2 **Rectangular Elements—Some Preliminary Considerations**

Conceptually (especially if the reader is conditioned by education to thinking in the Cartesian co-ordinate system) the simplest element form is that of a rectangle with sides parallel to x and y axes. Consider for instance a rectangle shown in Fig. 7.1 with nodal points numbered 1 to 8, located as shown, and at which the values of the unknown function ϕ form the element parameters. How can suitable shape functions for this element be determined?

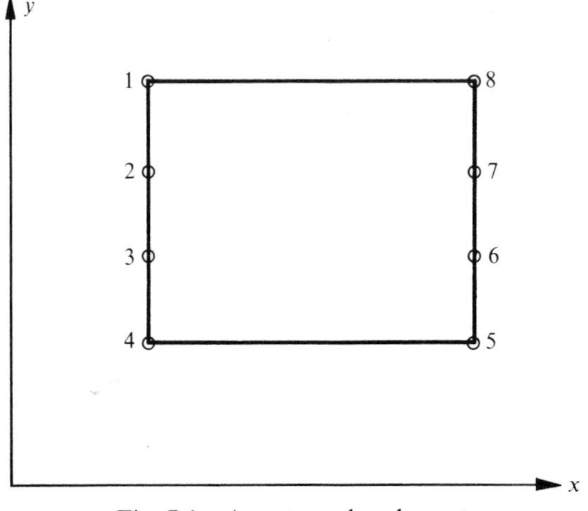

Fig. 7.1 A rectangular element

Let us first assume that this is expressed in a polynomial form in x and y. To ensure inter-element continuity of ϕ along the top and bottom sides the variation must be linear. Two points at which the function is common between elements lying above or below exist, and as two values determine uniquely a linear function, its identity all along these sides is ensured with that given by adjacent elements. Use of this fact was already made in specifying linear expansions for a triangle.

Similarly if a cubic variation along the vertical sides is assumed continuity will be preserved there as four values determine a unique cubic expansion. Conditions for satisfying the first criterion are now obtained.

To ensure the existence of arbitrary values of first derivative all that is necessary is that all the linear terms of the expansion be retained.

Finally, as eight points are to determine uniquely the variation of the function only eight coefficients of the expansion can be retained and thus we could write

$$\phi = \alpha_1 + \alpha_2 x + \alpha_3 y + \alpha_4 xy + \alpha_5 y^2 + \alpha_6 xy^2 + \alpha_7 y^3 + \alpha_8 xy^3. \tag{7.1}$$

The choice can in general be made unique by retaining the lowest possible expansion terms though in this case apparently no such choice arises.†

† Retention of a higher order term of expansion, ignoring one of lower order, will usually lead to a poorer approximation though still retaining convergence.[1]

The reader will easily verify that all the requirements have now been satisfied.

Substituting co-ordinates of the various nodes a set of simultaneous equations will be obtained.

This can be written in exactly the same manner as was done for a triangle in Eq. (4.4) as

$$\begin{Bmatrix} \phi_1 \\ \vdots \\ \phi_8 \end{Bmatrix} = \begin{bmatrix} 1, x_1, y_1, x_1 y_1, y_1^2, x_1 y_1^2, y_1^3, x_1 y_1^3 \\ \cdot \quad \cdot \quad \cdot \quad \cdot \quad \cdot \quad \cdot \quad \cdot \quad \cdot \quad \cdot \quad \cdot \quad \cdot \quad \cdot \\ \cdot \quad \cdot \quad \cdot \quad \cdot \quad \cdot \quad \cdot \quad \cdot \quad \cdot \quad \cdot \quad \cdot \quad \cdot \quad \cdot \end{bmatrix} \begin{Bmatrix} \alpha_1 \\ \vdots \\ \alpha_8 \end{Bmatrix} \tag{7.2}$$

or simply as

$$\boldsymbol{\phi}^e = \mathbf{C}\boldsymbol{\alpha}. \tag{7.3}$$

Formally

$$\boldsymbol{\alpha} = \mathbf{C}^{-1}\boldsymbol{\phi}^e \tag{7.4}$$

and we could write Eq. (7.1) as

$$\phi = \mathbf{P}\boldsymbol{\alpha} = \mathbf{P}\mathbf{C}^{-1}\boldsymbol{\phi}^e \tag{7.5}$$

in which

$$\mathbf{P} = [1, x, y, xy, y^2, xy^2, y^3, xy^3]. \tag{7.6}$$

Thus the shape functions for the element defined by

$$\phi = \mathbf{N}\boldsymbol{\phi}^e = [N_1, N_2, \ldots, N_8]\boldsymbol{\phi}^e \tag{7.7}$$

can be found from

$$\mathbf{N} = \mathbf{P}\mathbf{C}^{-1}. \tag{7.8}$$

This process, frequently used in practice as it does not involve much ingenuity, has, however, some considerable disadvantages. Occasionally an inverse of $\mathbf{C}$ may not exist[1,2] and *always* considerable algebraic difficulty is experienced in obtaining an inverse in general terms suitable for all element geometries. It is therefore worth while to consider whether shape functions $N_i(x, y)$ can be written down directly. Before doing this some general properties of these functions have to be mentioned.

Inspection of the defining relation, Eq. (7.7) reveals immediately some important characteristics. First, as this expression is valid for all components of $\boldsymbol{\phi}^e$

$$N_i = 1$$

at node i and is equal to zero at all other nodes. Further, the basic type of variation along boundaries defined for continuity purposes (e.g., linear in x and cubic in y in the above example) must be retained. The typical form of the shape functions for the elements considered is illustrated for two typical nodes isometrically in Fig. 7.2. It is clear that these could have been written down directly as a product of a suitable linear function in x with a cubic in y. The easy solution of this example is not always as obvious but given sufficient ingenuity, a direct derivation of shape function is always recommended.

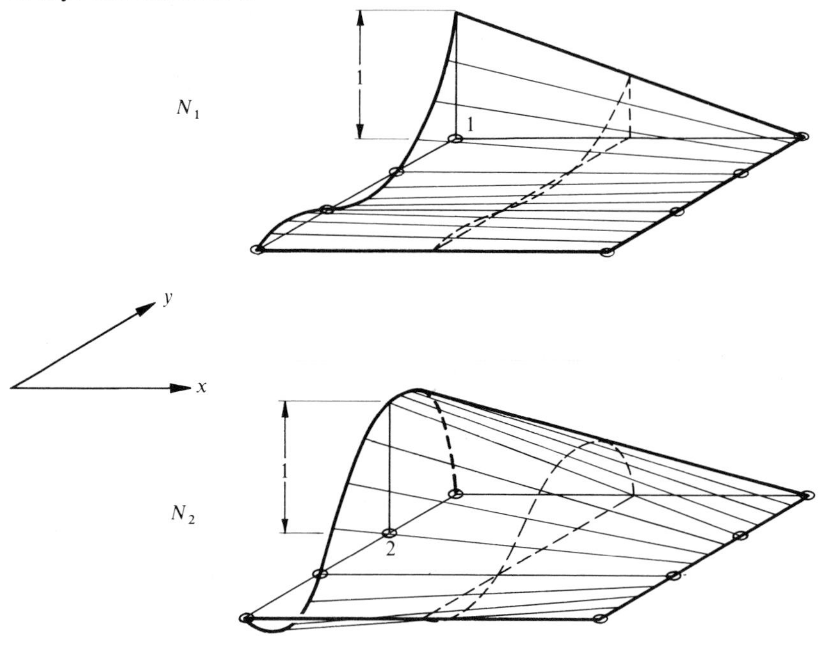

Fig. 7.2 Shape functions for elements of Fig. 7.1

It will be convenient to use normalized co-ordinates in further investigation. Such normalized co-ordinates are shown in Fig. 7.3 and are so chosen that on the faces of the rectangle their values are ± 1.

$$\begin{aligned} \xi &= (x-x_c)/a, \qquad & d\xi &= dx/a, \\ \eta &= (y-y_c)b, \qquad & d\eta &= dy/b. \end{aligned} \tag{7.9}$$

Once the shape functions are known in the normalized co-ordinates, translation into actual co-ordinates or transformation of the various expressions occurring, for instance, in stiffness derivation is trivial and work can be carried out using these.

7.3 **Completness of Polynomials**

The shape function derived in the previous section was of a rather special form [*vide* Eq. (7.1)]. Only a linear variation with the co-ordinate x was

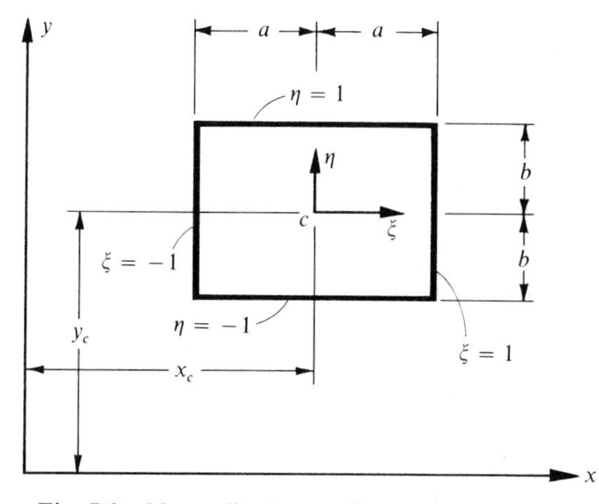

Fig. 7.3 Normalized co-ordinates for a rectangle

permitted, while in y a full cubic was available. The complete polynomial contained in it was thus of order 1 and in general use a convergence order corresponding to a linear variation would occur despite an increase of the total number of variables. Only in situations where the linear variation in x corresponded closely to the exact solution would a higher order of convergence occur—and for this reason elements with such 'preferential' directions should be restricted to special use, e.g., in narrow beams or strips. In general, we shall seek element expansions which possess the highest order of a complete polynomial for a minimum of degrees of freedom. In this context it is useful to recall the Pascal triangle (Fig. 7.4) from which the number of terms occurring in a polynomial in two variables x, y can be readily ascertained. For instance, first order polynomial require three terms, second order six terms, third order ten terms, etc.

7.4 **Rectangular Elements—Lagrange Family**[3,4,5,6]

An easy and systematic method of generating shape functions of any order can be achieved by simple products of appropriate polynomials in the two co-ordinates. Consider an element shown in Fig. 7.5 in which a series of nodes, external and internal, is placed on a regular grid. It is required to determine a shape function for the point indicated by the heavy circle. Clearly a product of a fifth order polynomial in ξ which has a value of unity at points of the second column of nodes and zero elsewhere with that of a fourth order polynomial in η having unity on the co-ordinate corresponding to the top row of nodes and zero elsewhere satisfies all the inter-element continuity conditions and gives unity at the nodal point concerned.

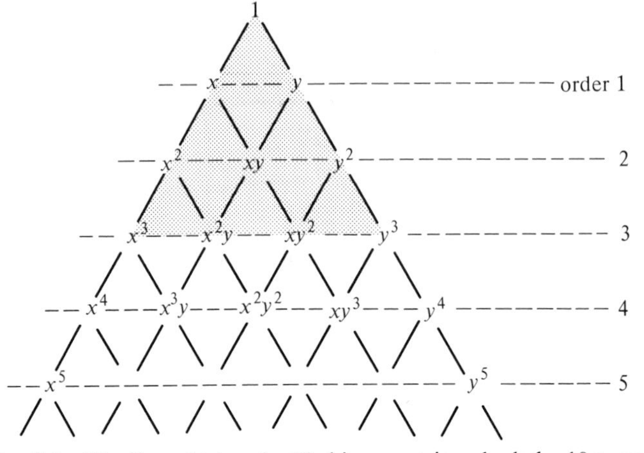

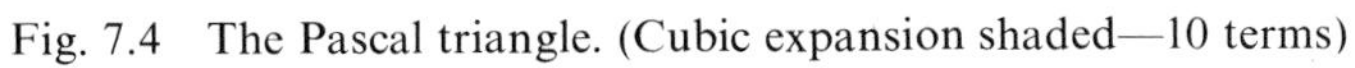

Fig. 7.4 The Pascal triangle. (Cubic expansion shaded—10 terms)

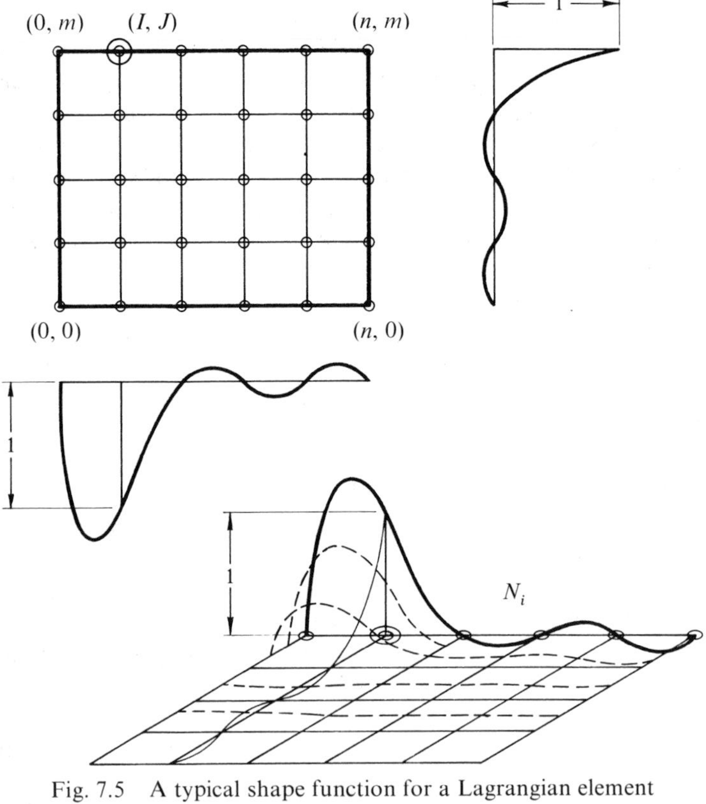

Fig. 7.5 A typical shape function for a Lagrangian element

$$\begin{pmatrix} n = 5 & I = 1 \\ m = 4 & J = 4 \end{pmatrix}$$

Polynomials in one co-ordinate having this property are known as Lagrange polynomials and can be written down directly as

$$l_k^n(\xi) = \frac{(\xi-\xi_0)(\xi-\xi_1)\ \dots(\xi-\xi_{k-1})(\xi-\xi_{k+1})\dots(\xi-\xi_n)}{(\xi_k-\xi_0)(\xi_k-\xi_1)\dots(\xi_k-\xi_{k-1})(\xi_k-\xi_{k+1})\dots(\xi_k-\xi_n)} \tag{7.10}$$

giving unity at ξ_k and passing through n points.

Thus in two dimensions, if we label the node by its column and node number, I, J, we have

$$N_i \equiv N_{IJ} = l_I^n(\xi) l_J^m(\eta) \tag{7.11}$$

where n and m stand for the number of subdivisions in each direction.

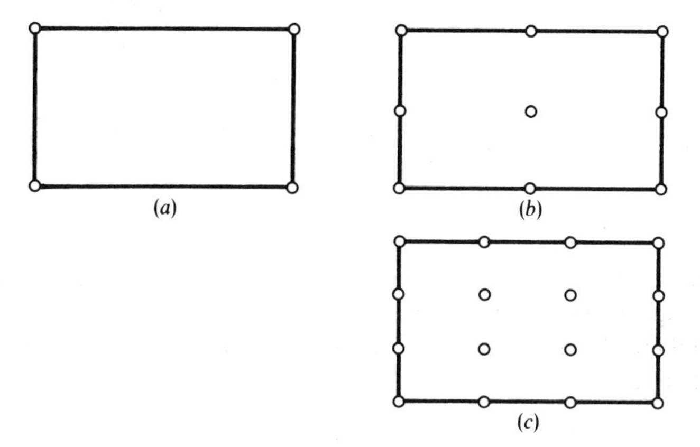

Fig. 7.6 Three elements of the Lagrange family (*a*) linear, (*b*) quadratic, and (*c*) cubic

Figure 7.6 shows a few members of this unlimited family. Though it is easy to generate, the usefulness of this family is limited not only due to a large number of internal nodes present but also due to the poor curve-fitting properties of the higher order polynomials. It will be noticed that the expressions of shape function will contain some very high order terms while omitting some lower ones.

Indeed if we examine the polynomial terms present in a situation where $n = m$ we observe in Fig. 7.7, based on the Pascal triangle, that a very large number of *parasitic* polynomial terms is present.[7]

7.5 Rectangular Elements—'Serendipity' Family [3, 4]

It is usually most convenient to make the functions dependent on nodal values placed on the element boundary. Consider for instance the first three elements of Fig. 7.8 (page 157). In each a progressively increasing and equal number of nodes is placed on the element boundary. The variation of

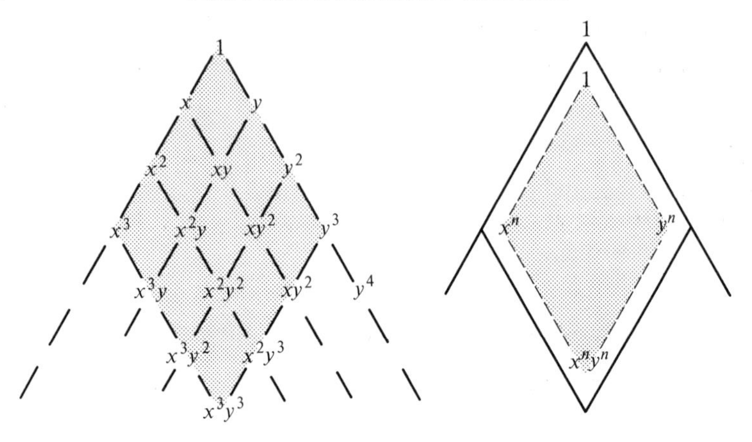

Fig. 7.7 Terms generated by a Lagrangian expansion of order 3×3 (or $n \times n$). Complete polynomials of order 3 (or n)

function on the edges to ensure continuity is linear, parabolic, and cubic in increasing element development.

To achieve the shape function for the first element it is obvious that a product of the form

$$\tfrac{1}{4}(\xi+1)(\eta+1) \tag{7.12}$$

gives unity at top right corners where $\xi = \eta = 1$ and zero at all the other corners. Further, a linear variation of the shape function of all sides exists and hence continuity is satisfied. Indeed this element is identical to the Lagrangian one with $n = 1$.

Introducing new variables

$$\xi_0 = \xi\xi_i, \qquad \eta_0 = \eta\eta_i \tag{7.13}$$

the form

$$N_i = \tfrac{1}{4}(1+\xi_0)(1+\eta_0) \tag{7.14}$$

allows all shape functions to be written down in one expression.

As a linear combination of these shape functions yields any arbitrary linear variation of ϕ the second convergence criterion is satisfied.

The reader can verify that the following functions satisfy all the necessary criteria for quadratic and cubic members of the family.

'Quadratic' element

Corner nodes

$$N_i = \tfrac{1}{4}(1+\xi_0)(1+\eta_0)(\xi_0+\eta_0-1). \tag{7.15}$$

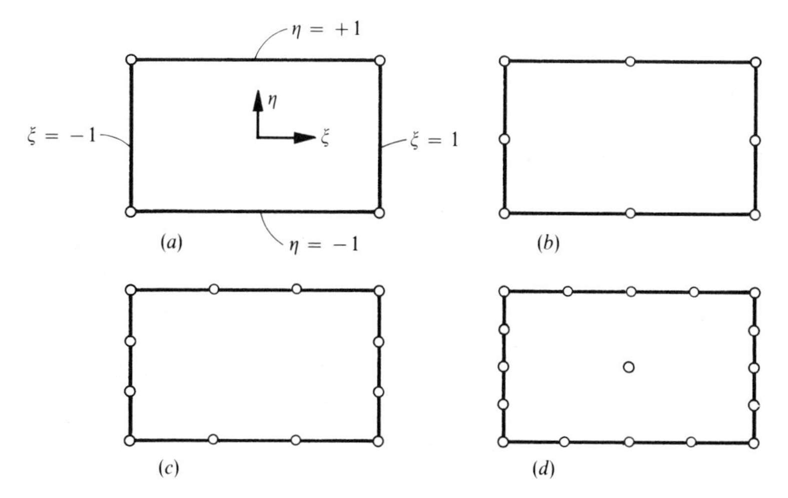

Fig. 7.8 Rectangles of boundary node (serendipity) family (*a*) linear, (*b*) Quadratic, (*c*) cubic, (*d*) quartic

Mid-side nodes

$$\xi_i = 0, \qquad N_i = \tfrac{1}{2}(1-\xi^2)(1+\eta_0),$$
$$\eta_i = 0, \qquad N_i = \tfrac{1}{2}(1+\xi_0)(1-\eta^2).$$

'Cubic' element

Corner nodes

$$N_i = \tfrac{1}{32}(1+\xi_0)(1+\eta_0)[-10+9(\xi^2+\eta^2)]. \tag{7.16}$$

Mid-side nodes

$$\xi_i = \pm 1 \qquad \text{and} \qquad \eta_i = \pm\tfrac{1}{3}$$
$$N_i = \tfrac{9}{32}(1+\xi_0)(1-\eta^2)(1+9\eta_0)$$

with the remaining mid-side node expression obtained by changing variables.

In the next, quartic, member[8] of this family a central node is added so that all terms of a complete fourth order expansion would be available. This central node adds a shape function $(1-\xi^2)(1-\eta^2)$ which is zero on all outer boundaries.

The above functions have been originally derived by inspection, and progression to yet higher members is difficult and requires some ingenuity. It was therefore appropriate to name this family 'Serendipity' after the famous princes of Serendip noted for their chance discoveries (Horace Walpole, 1754).

However, a quite systematic way of generating the 'serendipity' shape functions can be devised, which becomes apparent from Fig. 7.9 where the generation of a quadratic shape function is presented.[9]

As a starting point we observe that for *mid-side* nodes a Lagrangian interpolation of a quadratic × linear type sufficies to determine N_i at nodes 5 to 8. N_5 and N_8 are shown at Fig. 7.9(*a*) and (*b*). For a *corner* node, such as Fig. 7.9(c), we start with a bilinear $\hat{N}_1$ and note immediately that while $\hat{N}_1 = 1$ at node 1, it is not zero at nodes 5 or 8 (Step 1). Successive substraction of $\frac{1}{2}N_5$ (Step 2) and $\frac{1}{2}N_8$ (Step 3) ensures that a zero value is obtained at these nodes. The reader can verify that the expressions obtained coincide with those of Eqs. (7.15) and (7.16).

Indeed it should now be obvious that for all higher order elements the *mid-side* and *corner shape* functions can be generated by an identical process. For the former a simple multiplication of *m*th order and first order Lagrangian interpolations suffices. For the latter a combination of

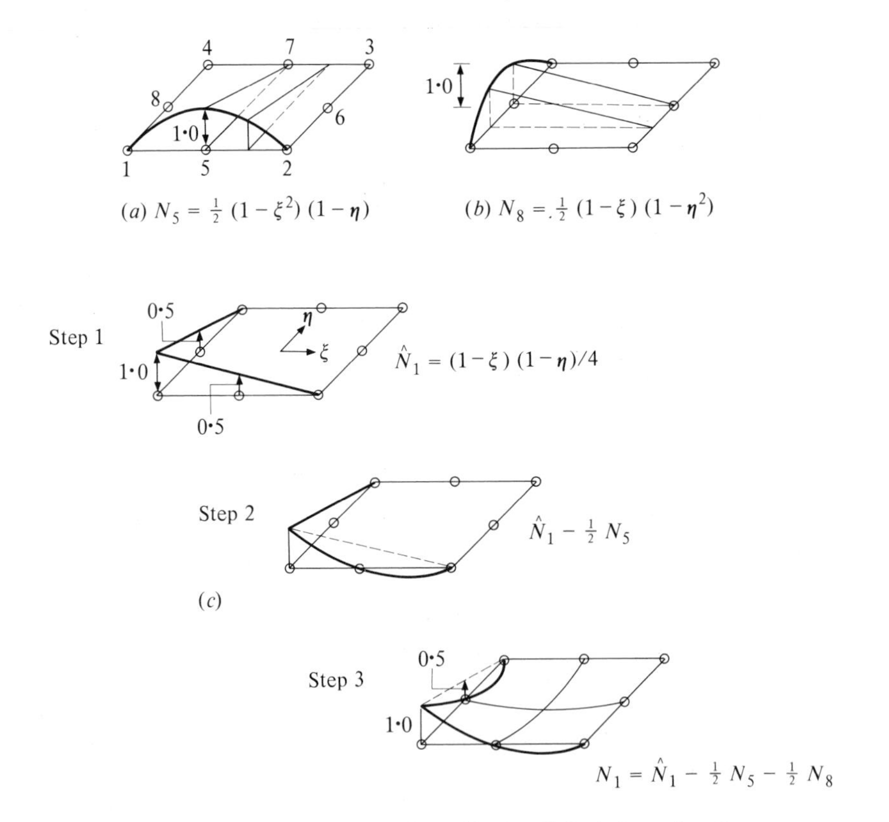

Fig. 7.9 Systematic generation of 'serendipity' shape functions

bilinear corner functions, together with appropriate fractions of mid-side shape functions to ensure zero at appropriate nodes, is necessary.

Indeed it is quite easy to generate shape functions for elements with different number of nodes along each side by a systematic algorithm. This may be very desirable if a transition between elements of different order is to be achieved enabling a different order of accuracy in separate sections of a large problem to be studied. Figure 7.10 illustrates the necessary shape functions for a cubic/linear transition. Use of such mixed elements was first introduced in reference 9, but the simpler formulation used here is that of reference 7.

With the mode of generating shape functions for this class of elements available it is immediately obvious that fewer degrees of freedom are now necessary for a given complete polynomial expansion. Figure 7.11 shows this for a cubic element where only two surplus terms arise (as compared with six surplus terms in a Lagrangian of same order).

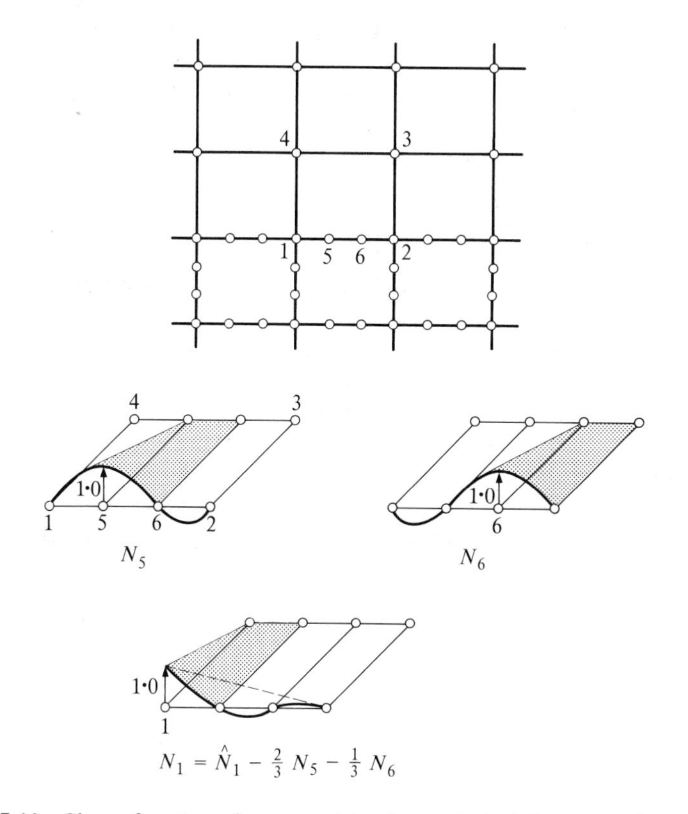

Fig. 7.10 Shape functions for a transition 'serendipity' element, cubic/linear

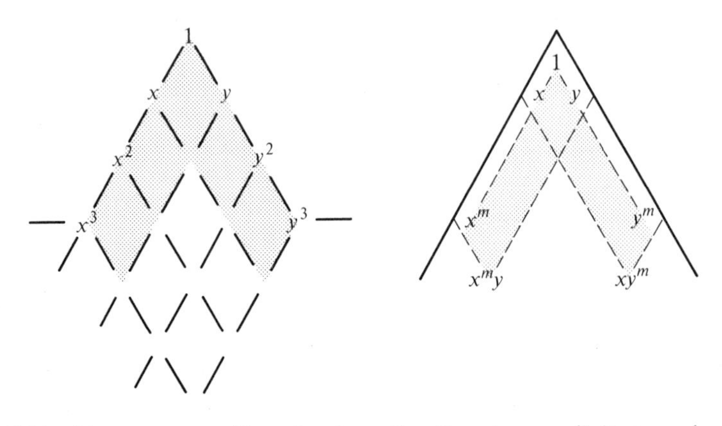

Fig. 7.11 Terms generated by edge shape functions in serendipity type elements (3×3 and $n \times n$)

It is immediately evident, however, that the functions generated by nodes placed only along the edges will not generate complete polynomials beyond the cubic. For higher order ones it is necessary to supplement the expansion by internal nodes (as was done in the quartic element of Fig. 7.8) or by use of 'nodeless' variables (discussed in the next section) which contain the appropriate polynomial terms.

7.6 **Internal Nodes and 'Nodeless' Variables**

It is instructive to note that both 'serendipity' and 'Lagrangian' elements of Figs. 7.6, and 7.8 are identical in their linear form but differ in the existence of the central node in the quadratic form. The shape functions for the two types of quadratic elements are shown in Fig. 7.12.

On the boundaries of the element the function is uniquely determined by boundary nodes only and hence (although the actual shapes of the first shape function differ internally), on the boundary both are identical. The additional degree of freedom of the Lagrangian type element is represented by adding multiples of the shape which has zero values along all the boundaries. The parameter multiplying that shape is in fact the value of ϕ at the central node.

Now, clearly it would be possible to achieve precisely the same degree of freedom of the first, 'serendipity' type, element by adding the additional shape function which has zero value on all boundaries multiplied by some parameter a^* associated with the element. All the shapes available in the Lagrangian element would again be available but now the multiplying factor obviously does not correspond to any nodal value of ϕ. a^* may be termed a *nodeless variable* associated with the element.

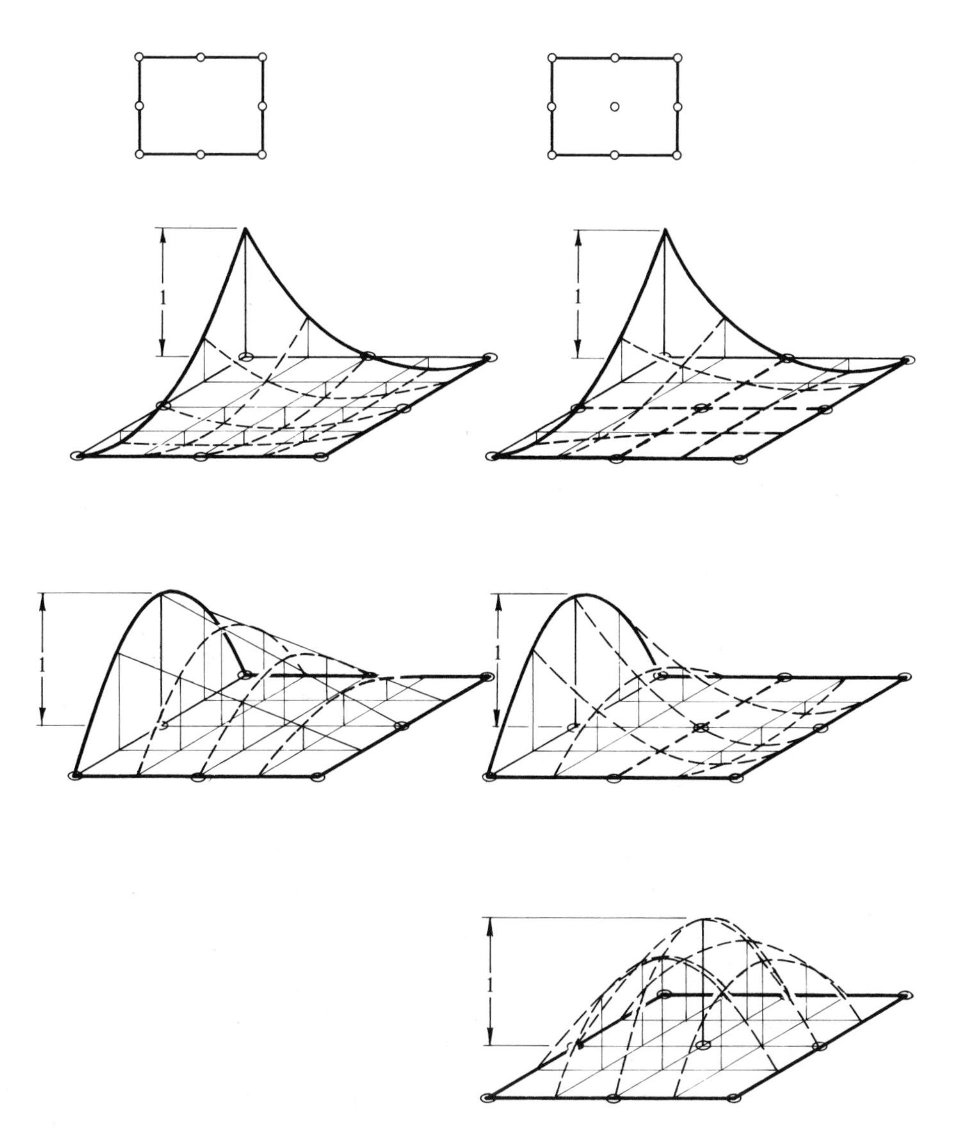

Fig. 7.12 Shape functions for quadratic elements of 'serendipity' and 'Lagrange' families

The minimization of the functional with respect to that variable can be treated in precisely the same way as if it were an internal node—but the *physical interpretation* of the quantities such as nodal forces, etc., is no longer obvious. Several such nodeless variables can be associated with any element if desired.

There is usually little advantage in doing this as the additional freedom of the function choosing its best form is constrained on the boundaries.

In the expressions followed so far polynomial terms were used exclusively. This has many advantages—in particular that the linear terms necessary for representing the constant derivative requirements are available. There is no necessity, however, to be so constrained when considering the higher, additional, freedoms.

For instance a function of form

$$\cos \pi\xi/2 \cos \pi\eta/2 \tag{7.17}$$

would be equally suitable in the preceding example giving identically zero on the boundaries and not altering the degree of complete polynomials.

The shape functions necessary in serendipity type elements of quartic or higher orders to supplement the incomplete expansions are most conveniently introduced by such 'nodeless' variables.

7.7 **Elimination of Internal Variables before Assembly—Substructures**

Internal nodes and nodeless variables yield in the usual way the element properties (Chapters 2 and 3)

$$\frac{\partial \Pi^e}{\partial \mathbf{a}^e} = \mathbf{K}^e\mathbf{a}^e + \mathbf{f}^e. \tag{7.18}$$

As $\mathbf{a}^e$ can be subdivided into parts which are common with other elements, $\bar{\mathbf{a}}^e$, and others which occur in the particular element only, $\bar{\bar{\mathbf{a}}}^e$ we can immediately write

$$\frac{\partial \Pi}{\partial \bar{\bar{\mathbf{a}}}^e} = \frac{\partial \Pi^e}{\partial \bar{\bar{\mathbf{a}}}^e} = 0$$

and eliminate $\bar{\bar{\mathbf{a}}}^e$ from further consideration. Writing Eq. (7.18) in a partitioned form we have

$$\frac{\partial \Pi^e}{\partial \mathbf{a}^e} = \begin{Bmatrix} \dfrac{\partial \Pi^e}{\partial \bar{\mathbf{a}}^e} \\ \dfrac{\partial \Pi^e}{\partial \bar{\bar{\mathbf{a}}}^e} \end{Bmatrix} = \begin{bmatrix} \bar{\mathbf{K}}^e & \hat{\mathbf{K}}^e \\ \hat{\mathbf{K}}^{e\mathrm{T}} & \bar{\bar{\mathbf{K}}}^e \end{bmatrix} \begin{Bmatrix} \bar{\mathbf{a}}^e \\ \bar{\bar{\mathbf{a}}}^e \end{Bmatrix} + \begin{Bmatrix} \bar{\mathbf{f}}^e \\ \bar{\bar{\mathbf{f}}}^e \end{Bmatrix} \tag{7.19}$$

$$= \begin{Bmatrix} \dfrac{\partial \Pi^e}{\partial \bar{\mathbf{a}}^e} \\ 0 \end{Bmatrix}.$$

From the second set of equations given in above we can write

$$\bar{\bar{\mathbf{a}}}^e = -(\bar{\bar{\mathbf{K}}}^e)^{-1}(\hat{\mathbf{K}}^{e\mathrm{T}}\bar{\mathbf{a}}^e + \bar{\bar{\mathbf{f}}}^e) \tag{7.20}$$

which on substitution yields

$$\frac{\partial \Pi^e}{\partial \bar{\mathbf{a}}^e} = \mathbf{K}^{*e}\bar{\mathbf{a}}^e + \mathbf{f}^{*e} \tag{7.21}$$

in which

$$\begin{aligned} \mathbf{K}^{*e} &= \bar{\mathbf{K}}^e - \hat{\mathbf{K}}^e\bar{\bar{\mathbf{K}}}^{e-1}\hat{\mathbf{K}}^{e\mathrm{T}} \\ \mathbf{f}^{*e} &= \bar{\mathbf{f}}_e - \hat{\mathbf{K}}^e(\bar{\bar{\mathbf{K}}}^e)^{-1}\bar{\bar{\mathbf{f}}}^e. \end{aligned} \tag{7.22}$$

Assembly of the total region then follows, only considering the element boundary variables, thus giving a considerable saving in the equation-solving effort at the expense of a few additional manipulations carried out at the element stage.

Perhaps a structural interpretation of this elimination is desirable. What in fact is involved is the separation of a part of the structure from its surroundings and determination of its solution separately for any prescribed displacements at the interconnecting boundaries. $\mathbf{K}^{*e}$ is now simply the overall stiffness of the separated structure and $\mathbf{f}^{*e}$ the equivalent set of nodal forces.

If the triangulation of Fig. 7.13 is interpreted as an assembly of pin-jointed bars the reader will recognize immediately the well-known device of 'substructures' used frequently in structural engineering.

Such a substructure is in fact simply a complex element from which the internal degrees of freedom have been eliminated.

Immediately a new possibility for devising more elaborate, and presumably more accurate, elements is presented.

Let Fig. 7.13(a) be interpreted as a continuum field subdivided into triangular elements. The substructure results in fact in one complex element shown in Fig. 7.13(b) with a number of boundary nodes.

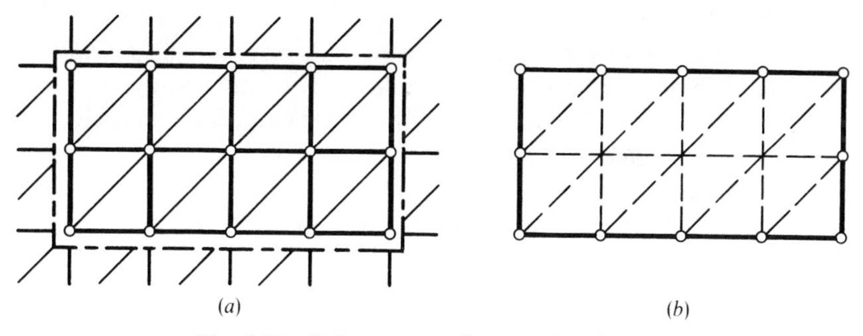

Fig. 7.13 Substructure of a complex element

The only difference from elements derived in previous section is the fact that the unknown ϕ is now not approximated internally by one set of smooth shape functions but by a series of piecewise approximations. This presumably results in a slightly poorer approximation but an economic advantage may arise if the total computation time for such an assembly is saved.

Substructuring is an important device in complex problems particularly where a repetition of complicated components arises.

In simple, small-scale finite element analysis, much improved use of simple triangular elements was found by the use of simple sub-assemblies of the triangles (or indeed tetrahedra). For instance a quadrilateral based on four triangles from which the central node is eliminated was found to give an economic advantage over a direct use of simple triangles, Fig. 7.14. This and other sub-assemblies based on triangles are discussed in detail by Doherty *et al.*[10]

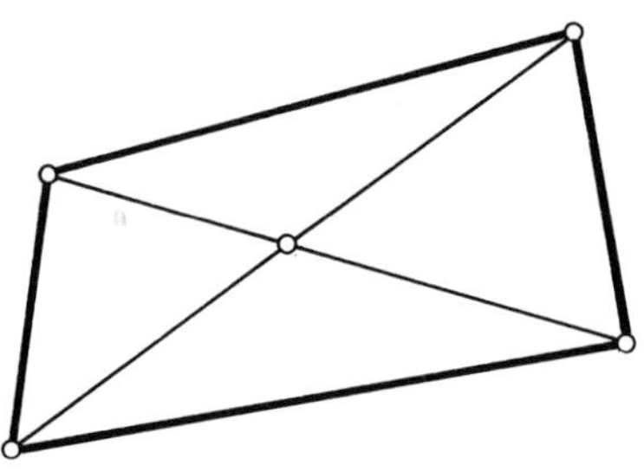

Fig. 7.14 A quadrilateral made up by four simple triangles

7.8 Triangular Element Family

The advantage of an arbitrary triangular shape in approximating to any boundary shape has been amply demonstrated in earlier chapters. Its apparent superiority here over the rectangular shapes needs no further discussion. The question of generating more elaborate elements needs to be further developed.

Consider a series of triangles generated on a pattern indicated in Fig. 7.15. The number of nodes in each member of the family is now such that a complete polynomial expansion, of the order needed for inter-element compatibility, is ensured. This follows by comparison with the Pascal triangle of Fig. 7.4 in which we see the number of nodes coinciding exactly with the number of polynomial terms required. This particular feature puts the triangle family in a special, privileged position, in which the inversion of the **C** matrices of Eq. (7.3) will always exist.[2] However, once again a direct generation of shape functions will be preferred—and indeed will be shown to be particularly easy.

Before proceeding further it is convenient to define a special set of normalized co-ordinates for a triangle.

7.8.1 *Area co-ordinates.* While Cartesian directions parallel to the sides of a rectangle were a natural choice for that shape, in the triangle these are not convenient.

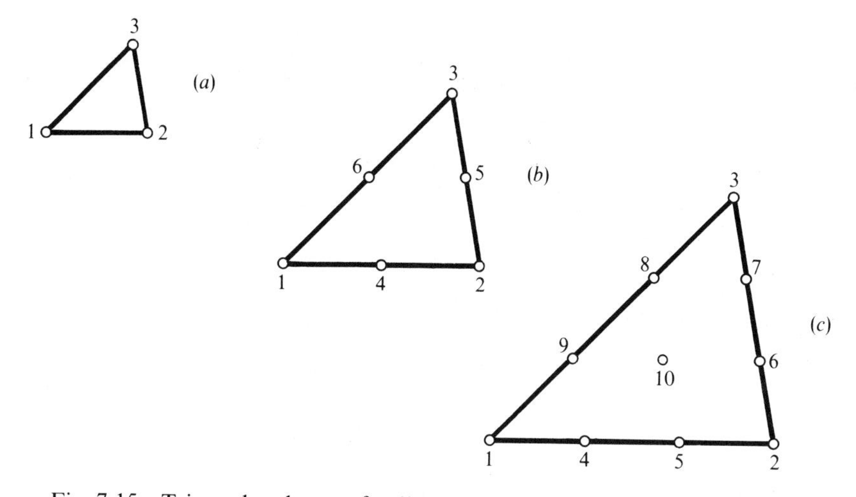

Fig. 7.15 Triangular element family (*a*) linear, (*b*) quadratic, and (*c*) cubic

A convenient set of co-ordinates, L_1, L_2, and L_3 for a triangle 1, 2, 3, Fig. 7.16, is defined by the following linear relation between these and the Cartesian system:

$$\begin{aligned} x &= L_1 x_1 + L_2 x_2 + L_3 x_3 \\ y &= L_1 y_1 + L_2 y_2 + L_3 y_3 \\ 1 &= L_1 + L_2 + L_3. \end{aligned} \tag{7.23}$$

To every set, L_1, L_2, L_3 (which are not independent, but are related by the third equation) corresponds a unique set of Cartesian co-ordinates. At point 1, $L_1 = 1$ and $L_2 = L_3 = 0$, etc. A linear relation between the new and Cartesian co-ordinates implies that contours of L_1 are equally placed straight lines parallel to side 2–3 on which $L_1 = 0$, etc.

Indeed it is easy to see that an alternative definition of the co-ordinate L_1 of a point P is by a ratio of the area of the shaded triangle to that of the total triangle.

$$L_1 = \frac{\text{Area } P23}{\text{Area } 123}. \tag{7.24}$$

Hence the name of area co-ordinates.

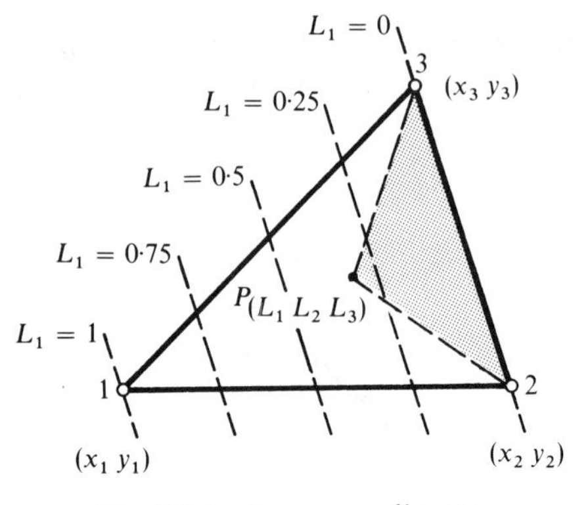

Fig. 7.16 Area co-ordinates

Solving Eq. (7.23) for x and y gives

$$\begin{aligned} L_1 &= (a_1+b_1x+c_1y)/2\Delta \\ L_2 &= (a_2+b_2x+c_2y)/2\Delta \\ L_3 &= (a_3+b_3x+c_3y)/2\Delta \end{aligned} \tag{7.25}$$

in which

$$\Delta = \tfrac{1}{2}\det \begin{vmatrix} 1 & x_1 & y_1 \\ 1 & x_2 & y_2 \\ 1 & x_3 & y_3 \end{vmatrix} = \text{Area } 123 \tag{7.26}$$

and

$$\begin{aligned} a_1 &= x_2y_3-x_3y_2 \\ b_1 &= y_2-y_3 \\ c_1 &= x_3-x_2. \end{aligned}$$

The identity of expressions with those derived in Chapter 4 (Eqs. (4.5*b*), (4.5*c*)) is worth remarking upon.

7.8.2 *Shape functions.* For the first element of the series, Fig. 7.15(*a*), the shape functions are simply the area co-ordinates. Thus

$$N_1 = L_1, \qquad N_2 = L_2, \qquad N_3 = L_3. \tag{7.27}$$

This is obvious as each individually gives unity at one node, zero at others, and varies linearly everywhere.

To derive shape functions for other elements a simple recurrence relation can be derived.[2] However, it is very simple to write an arbitrary triangle of order M in a manner similar to that used for the Lagrangian element of section 7.4.

Denoting a typical node i by three numbers I, J, and K corresponding to the position of co-ordinates L_{1i}, L_{2i}, and L_{3i} we can write the shape function in terms of three Lagrangian interpolations (*vide*, Eq. (7.10))

$$N_i = l_I^I(L_1) l_J^J(L_2) l_K^K(L_3). \tag{7.28}$$

In the above l_I^I etc. are given by expression (7.10), with L_1 taking the place of ξ, etc.

It is easy to verify that the above expression gives

$$N_i = 1 \qquad \text{at} \quad L_1 = L_{1I}, \quad L_2 = L_{2I}, \quad L_3 = L_{3I}$$

and zero at all other nodes.

The highest term occurring in the expansion is

$$L_1^I L_2^J L_3^K$$

and as

$$I + J + K \equiv M$$

for all points the polynomial is of order M.

Expression (7.28) is valid for quite arbitrary distributions of nodes of pattern given in Fig. 7.17 and simplifies if the spacing of the nodal lines is equal (i.e., $1/m$). The formula was first obtained by Argyris *et al.*[11] and formalized in a different manner by others.[7,12]

The reader can verify simply the shape functions for the second and third order elements as given below and indeed derive ones of any higher order easily.

Quadratic triangle (Fig. 7.15(*b*))
For corner nodes

$$N_1 = (2L_1 - 1)L_1, \text{ etc,}$$

mid-side nodes

$$N_4 = 4L_1 L_2, \text{ etc.} \tag{7.29}$$

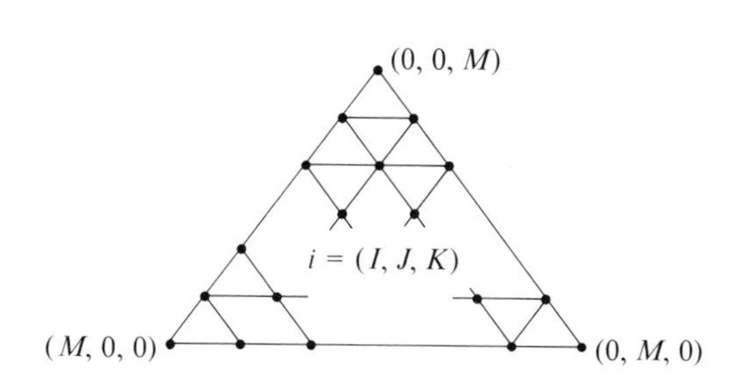

Fig. 7.17 A general triangular element

Cubic triangle (Fig. 7.15(*c*))
For corner nodes

$$N_1 = \tfrac{1}{2}(3L_1-1)(3L_1-2)L_1, \text{ etc,}$$

mid-side nodes

$$N_4 = \tfrac{9}{2}\, L_1 L_2(3L_1-1), \text{ etc,} \tag{7.30}$$

and for the
internal node

$$N_{10} = 27L_1L_2L_3.$$

The last shape again is a function giving zero contribution along boundaries—and will be used in a different context in Chapter 10.

The quadratic triangle was first derived by Veubeke[13] and used in the context of plane stress analysis by Argyris.[14]

When element matrices have to be evaluated it will follow that we are often faced with integration of quantities defined in terms of area co-ordinates over the triangular region. It is useful to note in this context the following integration expression

$$\iint_{\Delta} L_1^a L_2^b L_3^c \, dx\, dy = \frac{a!\, b!\, c!}{(a+b+c+2)!}\, 2\Delta. \tag{7.31}$$

ONE-DIMENSIONAL ELEMENTS

7.9 **Linear Elements**

So far in this book the continuum was considered generally in two or three dimensions. 'One-dimensional' members, being of a kind for which exact solutions are generally available, were treated only as trivial examples in Chapter 3. In many practical two- or three-dimensional problems such elements do in fact appear in conjunction with the more usual continuum elements—and a unified treatment is desirable. In the context of elastic analysis these elements may represent lines of reinforcement (plane and three-dimensional problems) or sheets of thin lining material in axi-symmetric and three-dimensional bodies. In the context of field problems of the type to be discussed in Chapter 17 lines of drains in a porous medium of lesser conductivity can be envisaged.

Once the shape of such a function as displacement is chosen for an element of this kind its properties can be determined, noting, however, that such derived quantities as strain, etc., have to be considered only in one dimension.

Figure 7.18 shows such an element sandwiched between two adjacent cubic-type elements. Clearly for continuity of the function a cubic variation of the unknown with the one variable ξ is all that is required. Thus the shape functions are given directly by the Lagrange polynomial as defined in Eq. (7.10).

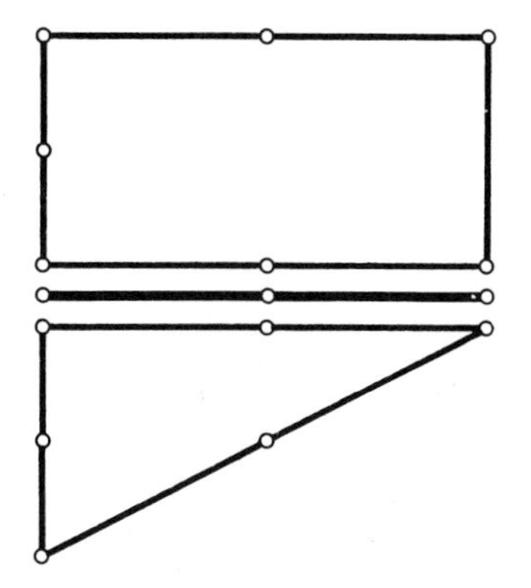

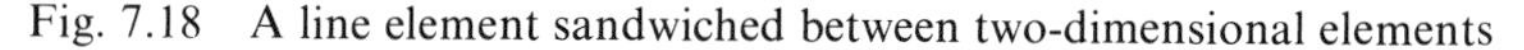

Fig. 7.18 A line element sandwiched between two-dimensional elements

THREE-DIMENSIONAL ELEMENTS

7.10 Rectangular Prisms—'Serendipity' Family[4, 9, 15]

In a precisely analogous way to that given in previous sections equivalent elements of three-dimensional type can be described.

Now, for inter-element continuity the simple rules given previously have to be modified. What is necessary to achieve is that along a whole face of an element the nodal values define a unique variation of the unknown function. With incomplete polynomials, this can be ensured only by inspection.

A family of elements shown in Fig. 7.19 is precisely equivalent to that of Fig. 7.8. Using now three normalized co-ordinates and otherwise following the terminology of section 7.5 we have the following shape functions.

'Linear' element (8 nodes)

$$N_i = \tfrac{1}{8}(1+\xi_0)(1+\eta_0)(1+\zeta_0). \tag{7.32}$$

'Quadratic' element (20 nodes)

Corner nodes

$$N_i = \tfrac{1}{8}(1+\xi_0)(1+\eta_0)(1+\zeta_0)(\xi_0+\eta_0+\zeta_0-2). \tag{7.33}$$

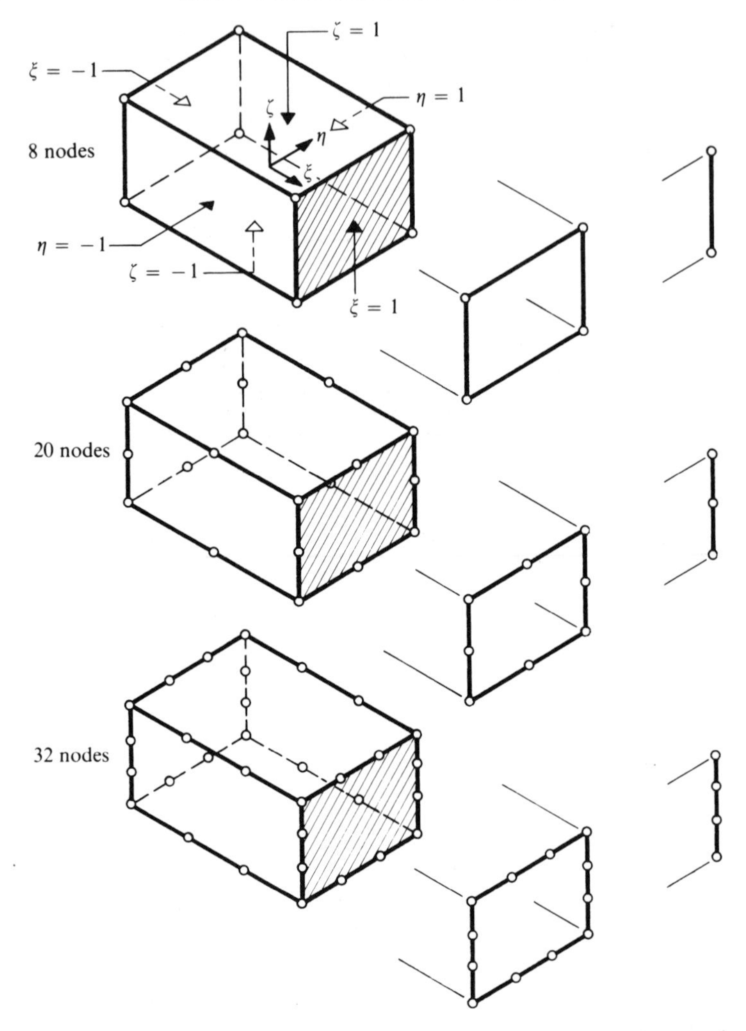

Fig. 7.19 Right prisms of boundary node (Serendipity) family with corresponding sheet and line elements

Typical mid-side node

$$\xi_i = 0, \qquad \eta_i = \pm 1, \qquad \zeta_i = \pm 1$$

$$N_i = \tfrac{1}{4}(1-\xi^2)(1+\eta_0)(1+\zeta_0).$$

Cubic elements (32 nodes)

Corner node

$$N_i = \tfrac{1}{64}(1+\xi_0)(1+\eta_0)(1+\zeta_0)[9(\xi^2+\eta^2+\zeta^2)-19]. \tag{7.34}$$

Typical mid-side node

$$\xi_i = \pm\tfrac{1}{3}, \qquad \eta_i = \pm 1, \qquad \zeta_i = \pm 1$$
$$N_i = \tfrac{9}{64}(1-\xi^2)(1+9\xi_0)(1+\eta_0)(1+\zeta_0).$$

When $\zeta = 1 = \zeta_0$ the above expressions reduce to those of Eqs. (7.14) to (7.16). Indeed such elements of the three-dimensional type can be joined in a compatable manner to sheet or line elements of the appropriate type as shown in Fig. 7.19.

Once again the procedure of generating the shape functions follows that described in Figs 7.9 and 7.10 and once again elements with varying degrees of freedom along the edges can be derived following the same steps.

The equivalent of a Pascal triangle is now a tetrahedron and again we can observe the small number of surplus degrees of freedom—a situation even of greater importance than in two-dimensional analysis.

7.11 Rectangular Prisms—Lagrange Family

Shape function for such elements, illustrated in Fig. 7.20, will be generated by a direct product of three Lagrange polynomials. Extending the notation of Eq. (7.11) we now have

$$N_i \equiv N_{IJK} = l_I^n l_J^m l_K^p \tag{7.35}$$

for n, m, and p subdivisions along each side.

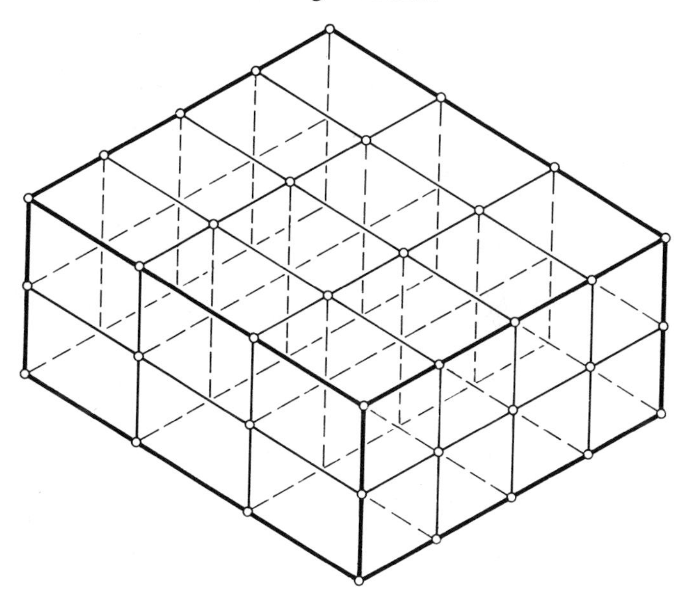

Fig. 7.20 Right prism of Lagrange family

This element again is suggested by Ergatoudis[5] and elaborated upon by Argyris.[6] All the remarks about internal nodes and the limitation of the formulation made in section 7.4 are applicable here and generally the practical application of such elements is inefficient.

7.12 **Tetrahedral Elements**

The tetrahedral family shown in Fig. 7.21 not surprisingly exhibits properties similar to those of the triangle family.

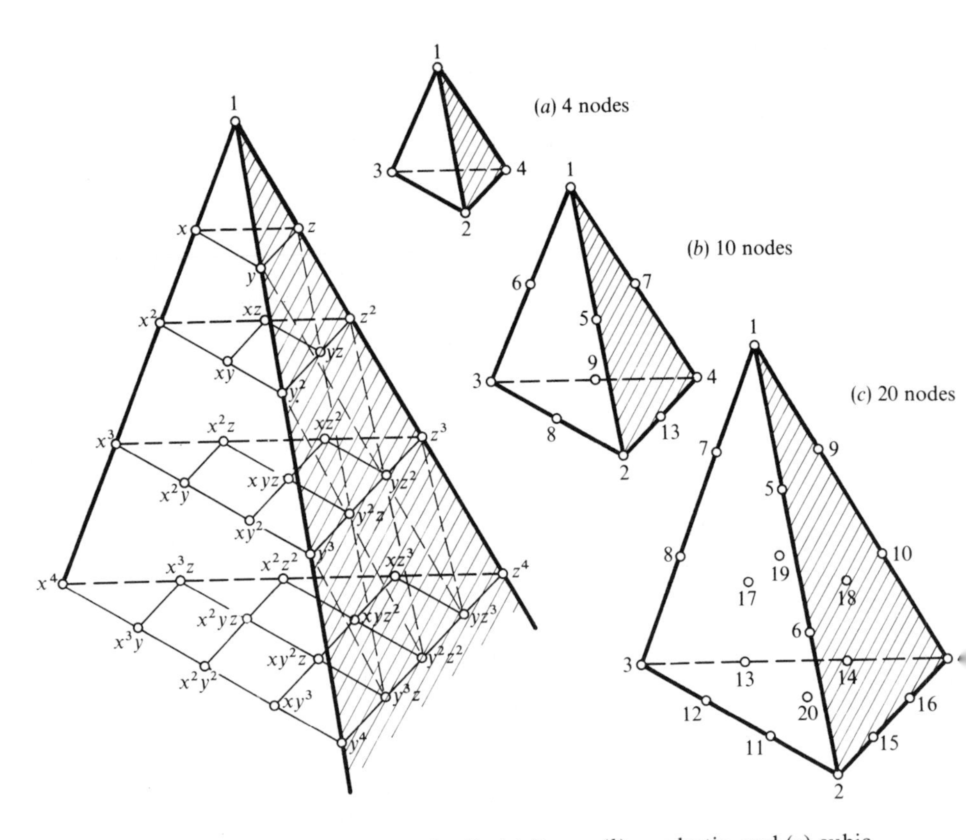

Fig. 7.21 The tetrahedron family (*a*) linear, (*b*) quadratic, and (*c*) cubic

Firstly, once again complete polynomials in three co-ordinates are achieved at each stage. Secondly, as faces are divided in a manner identical with that of the previous triangles, the same order of polynomial in two co-ordinates in the plane of the face is achieved and element compatibility ensured. No surplus terms in the polynomial occur.

7.12.1 *Volume co-ordinates.* Once again special co-ordinates are introduced defined by (Fig. 7.22):

$$
\begin{aligned}
x &= L_1x_1+L_2x_2+L_3x_3+L_4x_4 \\
y &= L_1y_1+L_2y_2+L_3y_3+L_4y_4 \\
z &= L_1z_1+L_2z_2+L_3z_3+L_4z_4 \\
1 &= L_1 \quad +L_2 \quad +L_3 \quad +L_4.
\end{aligned}
\tag{7.36}
$$

Once again the inversion of above leads to expressions of type (7.25) and (7.26) with the constants which can be identified from Chapter 6 (Eq. 6.5). Once again the physical nature of the co-ordinates can be identified as the ratio of volumes of tetrahedra based on an internal point P in the total volume, e.g., as shown in Fig. 7.22

$$L_1 = \frac{\text{Volume } P234}{\text{Volume } 1234} \text{ etc.} \tag{7.37}$$

7.12.2 *Shape function.* As the volume co-ordinates vary linearly with the Cartesian ones from unity at one node to zero at the opposite face then shape functions for the linear element, Fig. 7.21 (*a*), are simply

$$N_1 = L_1, \qquad N_2 = L_2, \text{ etc.} \tag{7.38}$$

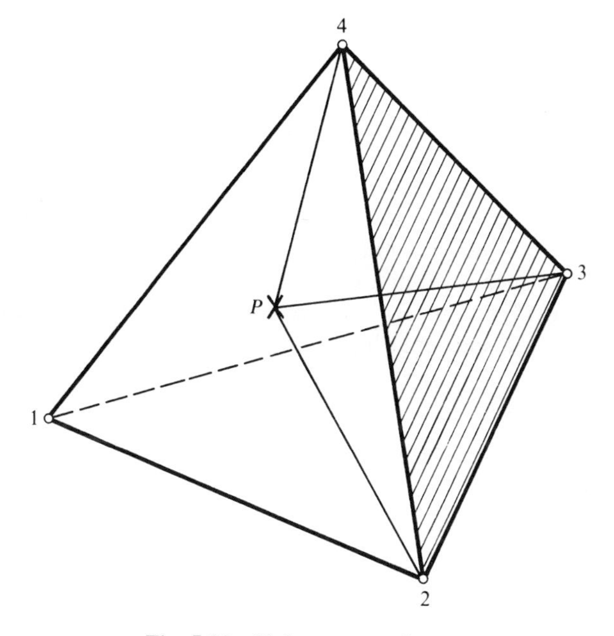

Fig. 7.22 Volume co-ordinates

Formulae for shape functions of higher order tetrahedra are derived in precisely the same manner as for the triangles by establishing an appropriate Lagrange type formulae similar to Eq. (7.28). Leaving this to the reader as a suitable exercise we quote the following.

'*Quadratic*' *tetrahedron* (Fig. 7.21(*b*))

For corner nodes

$$N_1 = (2L_1 - 1)L_1, \text{ etc.} \tag{7.39}$$

For mid-side nodes

$$N_5 = 4L_1L_2, \text{ etc.}$$

'*Cubic*' *tetrahedron*

Corner nodes

$$N_1 = \tfrac{1}{2}(3L_1 - 1)(3L_1 - 2)L_1, \text{ etc.} \tag{7.40}$$

Mid-side nodes

$$N_5 = \tfrac{9}{2}L_1L_2(3L_1 - 1), \text{ etc.}$$

Mid-face nodes

$$N_{17} = 27L_1L_2L_3, \text{ etc.}$$

A useful integration formula again may be here quoted

$$\iiint_{\text{vol.}} L_1^a L_2^b L_3^c L_4^d \, dx\, dy\, dz = \frac{a!\, b!\, c!\, d!}{(a+b+c+d+3)!} 6V. \tag{7.41}$$

7.13 Other Simple Three-dimensional Elements

The possibilities of simple shapes in three dimensions are greater, for obvious reasons, than in two dimensions. A quite useful series of elements can for instance be based on triangular prisms, Fig. 7.23. Here again variants of the product, Lagrange, approach or of the 'Serendipity' type can be distinguished. The first element of both families is identical and indeed the shape functions for it are so obvious as not to need quoting.

For a 'quadratic' element illustrated in Fig. 7.23(*b*) the shape functions are

Corner nodes $L_1 = \xi_1 = 1$

$$N_1 = \tfrac{1}{2}L_1(2L_1 - 1)(1+\zeta) - \tfrac{1}{2}L_1(1-\zeta^2). \tag{7.42}$$

Mid-sides of triangles

$$N_{10} = 2L_1L_2(1+\zeta), \text{ etc.} \tag{7.43}$$

Mid-sides of rectangle

$$N_7 = L_1(1-\zeta^2), \text{ etc.}$$

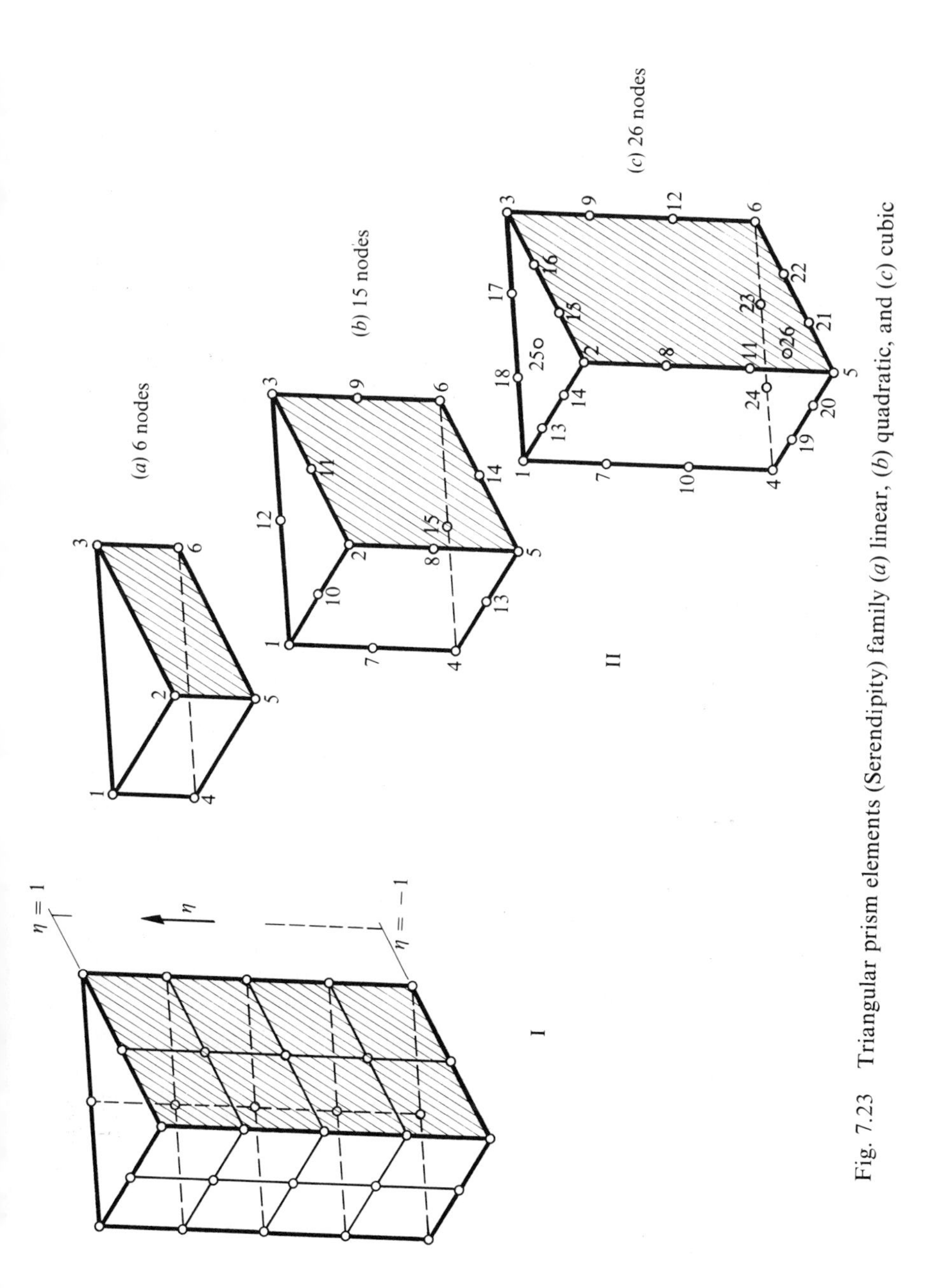

Fig. 7.23 Triangular prism elements (Serendipity) family (*a*) linear, (*b*) quadratic, and (*c*) cubic

Such elements are not purely esoteric but have a practical application as 'fillers' in conjunction with 20 noded parallelepiped elements.

7.14 Concluding Remarks

An unlimited selection of element types has been presented here to the reader—and indeed equally unlimited alternative possibilities exist.[4,9] What of the use of such complex elements in practice? Putting aside the triangle and tetrahedron all the other elements are limited to situations where the real region is of a suitable shape which can be represented as an assembly of right prisms. Such a limitation would be so severe that little practical purpose would have been served by the derivation of such shape functions unless some way could be found of distorting such elements to fit realistic boundaries. In fact, methods for doing this are available and will be described in the next chapter.

References

1. P. C. Dunne, 'Complete polynomial displacement fields for finite element methods', *Trans Roy. Aero. Soc.*, **72**, 245, 1968.
2. B. M. Irons, J. G. Ergatoudis, and O. C. Zienkiewicz, Comment on ref. 1, *Trans. Roy. Aero. Soc.*, **72**, 709–11, 1968.
3. J. G. Ergatoudis, B. M. Irons, and O. C. Zienkiewicz, 'Curved, isoparametric, quadrilateral elements for finite element analysis', *Int. J. Solids Struct.*, **4**, 31–42, 1968.
4. O. C. Zienkiewicz *et al.*, 'Iso-parametric and associate elements families for two and three dimensional analysis', Chapter 13, in *Finite Element Methods in Stress Analysis*, ed. I. Holand and K. Bell, Techn. Univ. of Norway, Tapir Press, Norway, Trondheim, 1969.
5. J. G. Ergatoudis, *Quadrilateral elements in plane analysis: Introduction to solid analysis*, M.Sc. thesis, University of Wales, Swansea, 1966.
6. J. H. Argyris, K. E. Buck, H. M. Hilber, G. Mareczek, and D. W. Scharpf, 'Some new elements for matrix displacement methods', *2nd Conf. on Matrix Methods in Struct. Mech.*, Air Force Inst. of Techn., Wright Patterson Base, Ohio, Oct. 1968.
7. R. L. Taylor, 'On completeness of shape functions for finite element analysis', *Int. J. Num. Meth. Eng.*, **4**, 17–22, 1972.
8. F. C. Scott, 'A quartic, two dimensional isoparametric element', Undergraduate Project, Univ. of Wales, Swansea, 1968.
9. O. C. Zienkiewicz, B. M. Irons, J. Campbell, and F. C. Scott, 'Three dimensional stress analysis', *Int. Un. Th. Appl. Mech. Symposium on High Speed Computing in Elasticity*, Liége, 1970.
10. W. P. Doherty, E. L. Wilson, and R. L. Taylor, *Stress Analysis of Axisymmetric Solids utilizing Higher-Order Quadrilateral Finite Elements*, Report 69–3, Structural Engineering Laboratory, Univ. of California, Berkeley, Jan. 1969.
11. J. H. Argyris, I. Fried, and D. W. Scharpf, 'The TET 20 and the TEA 8 elements for the matrix displacement method', *Aero. J.*, **72**, 618–25, 1968.

12. P. SILVESTER, 'Higher order polynomial triangular finite elements for potential problems', *Int. J. Eng. Sci.*, **7**, 849–61, 1969.
13. B. FRAEIJS DE VEUBEKE, 'Displacement and equilibrium models in the finite element method', Chapter 9 of *Stress Analysis*, ed. O. C. Zienkiewicz and G. S. Holister, J. Wiley & Son, 1965.
14. J. H. ARGYRIS, 'Triangular elements with linearly varying strain for the matrix displacement method', *J. Roy. Aero. Soc. Tech. Note*, **69**, 711–13, Oct. 1965.
15. J. G. ERGATOUDIS, B. M. IRONS, and O. C. ZIENKIEWICZ, 'Three dimensional analysis of arch dams and their foundations', *Symposium on Arch Dams*, Inst. Civ. Eng., London, 1968.

8. Curved, Isoparametric Elements and Numerical Integration

8.1 Introduction

In the previous chapter we have shown how some general families of finite elements can be obtained. A progressively increasing number of nodes and hence improved accuracy characterizes each new member of the family and presumably the number of such elements required to obtain an adequate solution decreases rapidly. To ensure that a small number of elements can represent a relatively complex form of the type which is liable to occur in real, rather than academic, problems, simple rectangles and triangles no longer suffice. This chapter is therefore concerned with the subject of distorting such simple forms into others of more arbitrary shape.

Elements of the basic one-, two- or three-dimensional types will be 'mapped' into distorted forms in the manner indicated in Figs. 8.1 and 8.2.

In these figures it is shown that the ξ, η, ζ, or $L_1 L_2 L_3 L_4$ co-ordinates can be distorted to a new, curvilinear set when plotted in a Cartesian space.

Not only can two-dimensional elements be distorted into others in two dimensions but the mapping of these can be taken into three dimensions as indicated by the flat sheet elements of Fig. 8.2 distorting into a three-dimensional space. This principle applies generally, providing some one-to-one correspondence between Cartesian and curvilinear co-ordinates can be established, i.e., once relations of the type

$$\begin{Bmatrix} x \\ y \\ z \end{Bmatrix} = f \begin{Bmatrix} \xi \\ \eta \\ \zeta \end{Bmatrix} \quad \text{or} \quad f \begin{Bmatrix} L_1 \\ L_2 \\ L_3 \\ L_4 \end{Bmatrix} \tag{8.1}$$

can be established.

Once such co-ordinate relationships are known, shape functions can be specified in local co-ordinates and by suitable transformations the element properties established.

In what follows we shall first discuss the so-called isoparametric form of relationship (8.1) which has found a great deal of practical application.

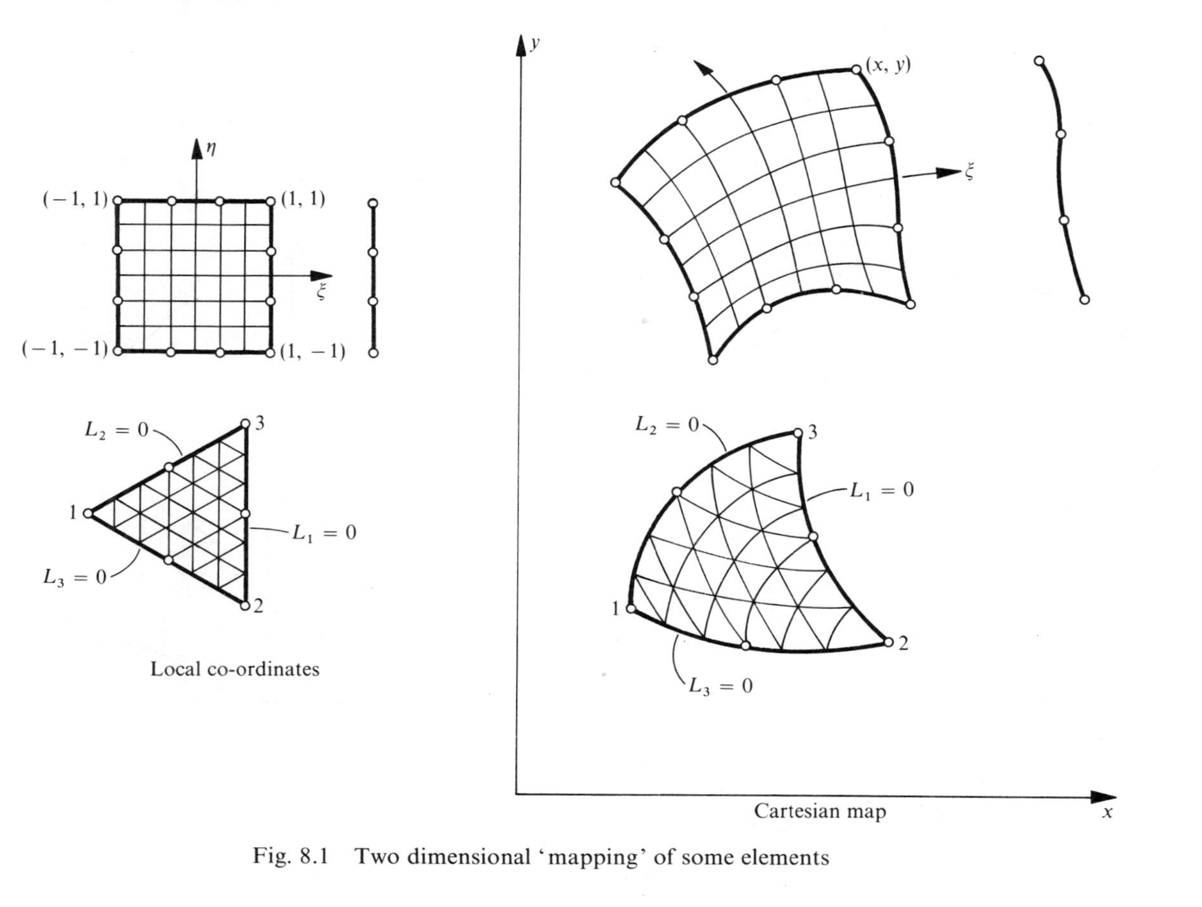

Fig. 8.1 Two dimensional 'mapping' of some elements

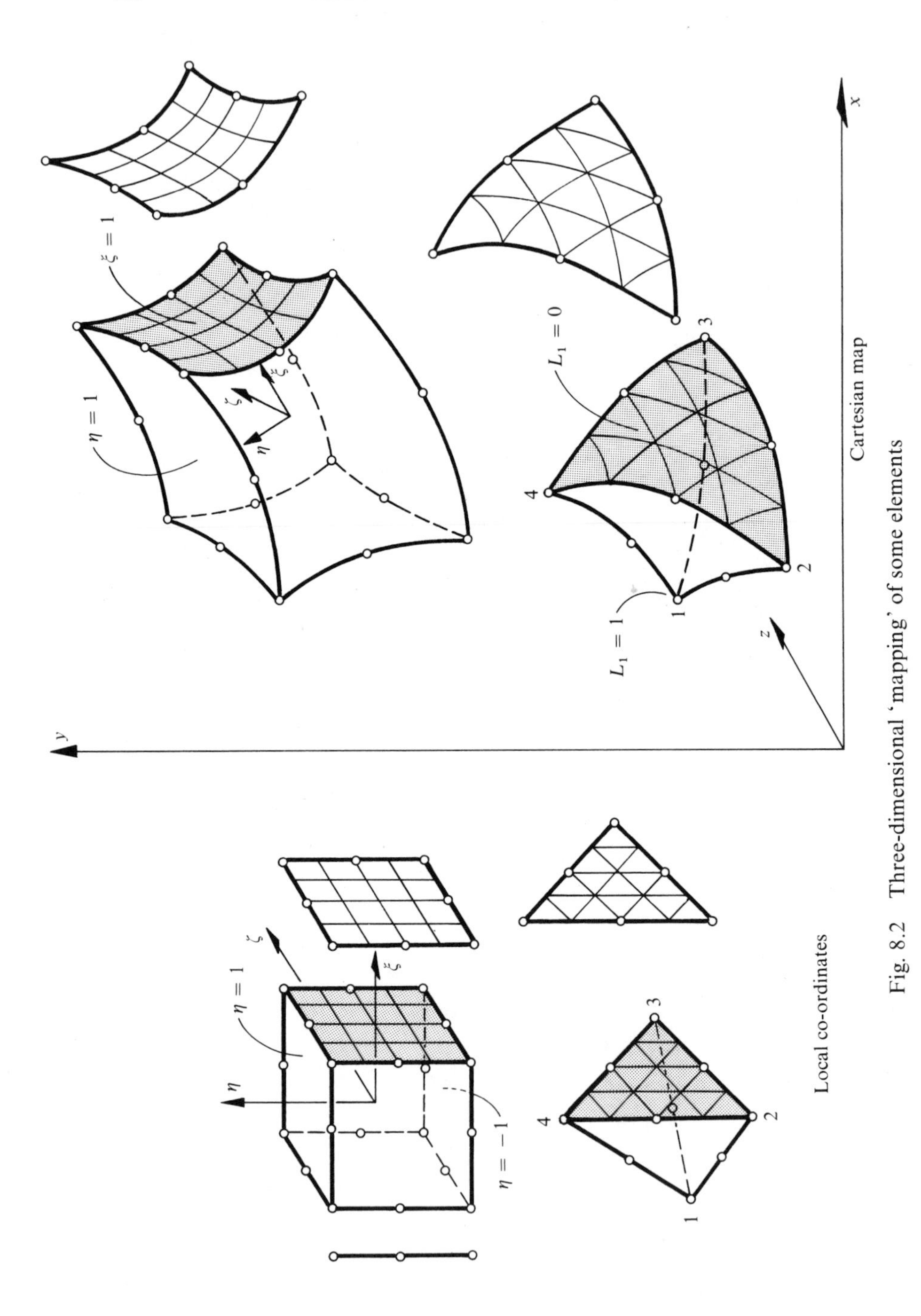

Fig. 8.2 Three-dimensional 'mapping' of some elements

Full details of this formulation will be given, including the establishment of element properties by numerical integration which will be found to be essential.

In the final section we shall show that many other co-ordinate transformations can be used effectively.

PARAMETRIC CURVILINEAR CO-ORDINATES

8.2 Use of 'Shape Functions' in Establishment of Co-ordinate Transformations

A most convenient method of establishing the co-ordinate transformations is to use the shape functions we have already derived to represent the variation of the unknown function.

If we write, for instance, for each element

$$
\begin{aligned}
x &= N'_1 x_1 + N'_2 x_2 + \cdots = \mathbf{N}' \begin{Bmatrix} x_1 \\ x_2 \\ \vdots \end{Bmatrix} = \mathbf{N}'\mathbf{x} \\
y &= N'_1 y_1 + N'_2 y_2 + \cdots = \mathbf{N}' \begin{Bmatrix} y_1 \\ y_2 \\ \vdots \end{Bmatrix} = \mathbf{N}'\mathbf{y} \\
z &= N'_1 z_1 + N'_2 z_2 + \cdots = \mathbf{N}' \begin{Bmatrix} z_1 \\ z_2 \\ \vdots \end{Bmatrix} = \mathbf{N}'\mathbf{z}
\end{aligned}
\tag{8.2}
$$

in which $\mathbf{N}'$ are shape functions given in terms of the local co-ordinates, then immediately a relationship of the required form is available. Further, the points with co-ordinates x_1, y_1, z_1, etc. will lie at appropriate points of the element boundary (as from the general definitions of the shape functions we know that they have a value of unity at the point in question and zero elsewhere).

To each set of local co-ordinates will correspond a set of global Cartesian co-ordinates and in general only one such set. We shall see, however, that a non-uniqueness may arise sometimes with violent distortion.

The concept of using such element shape functions for establishing curvilinear co-ordinates in the context of finite element analysis appears to have been first mentioned by Taig.[1] In his first application basic linear quadrilateral relations were established. Irons [2,3] generalized the idea for other elements.

Quite independently the exercises of devising various practical methods of generating curved surfaces for purposes of engineering design led to the establishment of similar definitions by Coons,[4,5] and indeed today the subjects of surface definitions and analysis are drawing closer together due to this activity.

In Fig. 8.3 an actual distortion of elements based on the cubic and quadratic members of the 'serendipity' family is shown. It is seen here that a

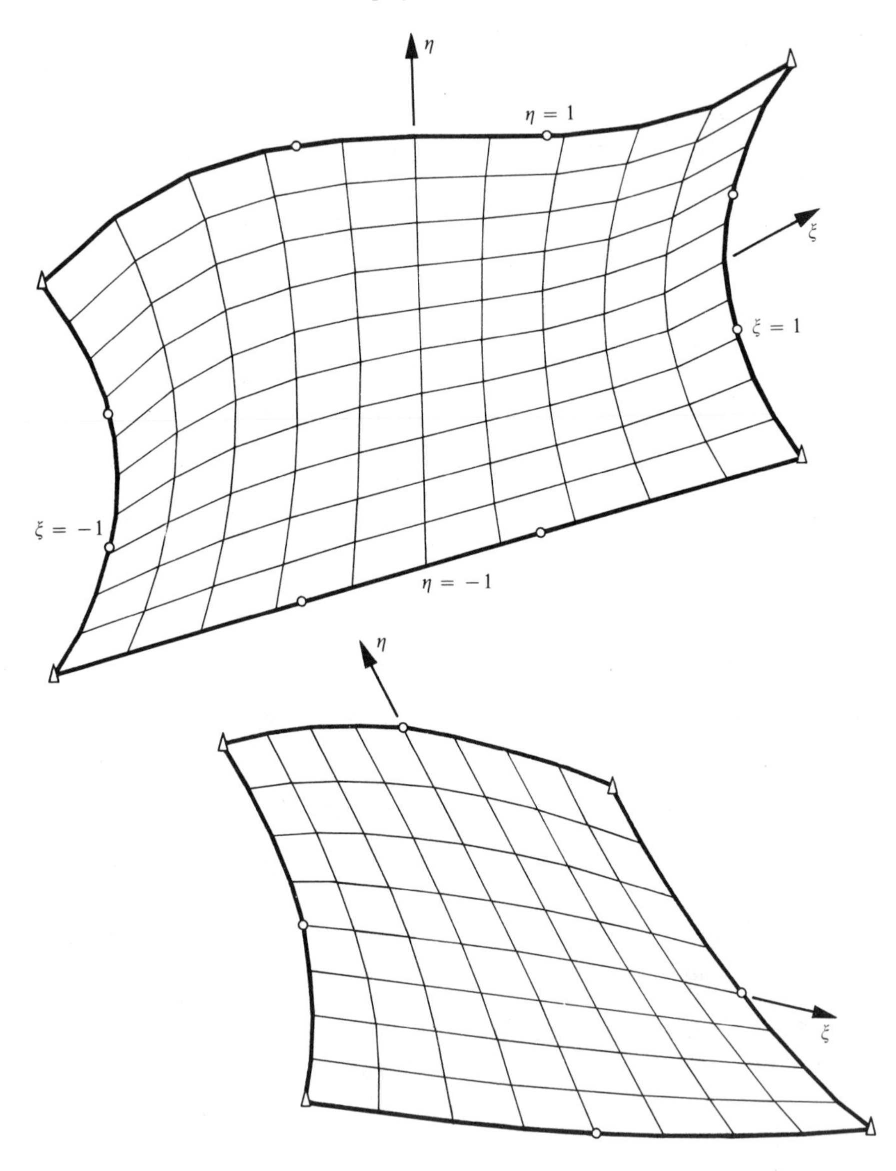

Fig. 8.3 Computer plots of curvilinear co-ordinates for cubic and parabolic elements (reasonable distortion)

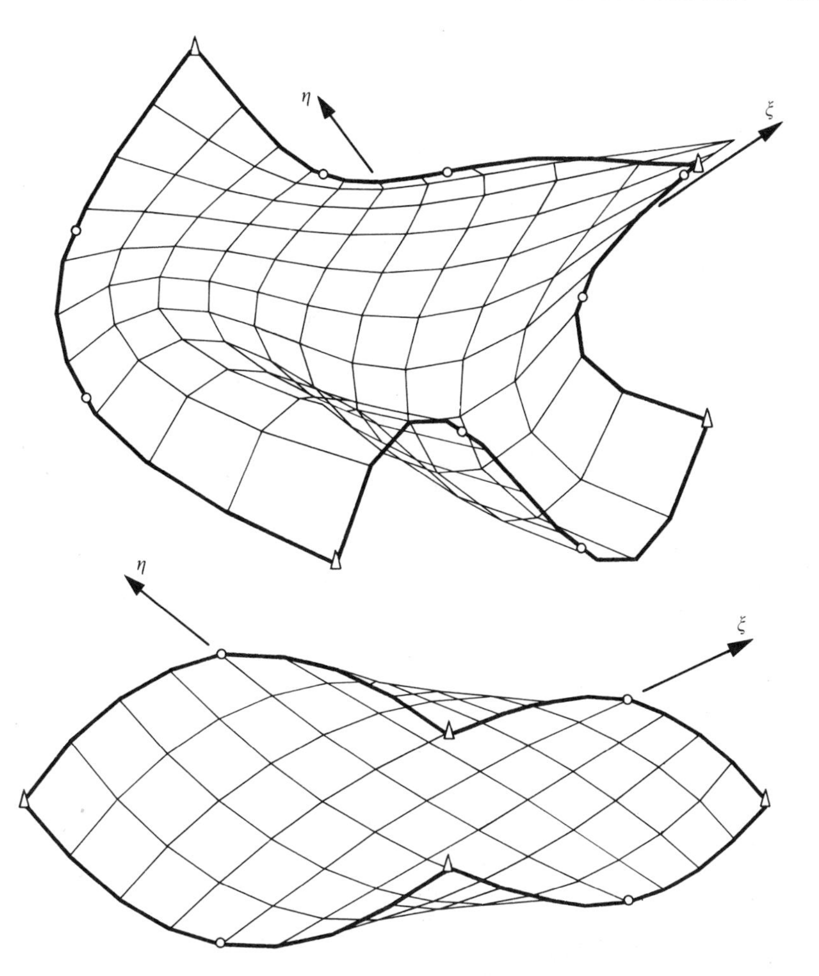

Fig. 8.4 Unreasonable element distortion leading to non-unique mapping, and 'overspill'. Cubic and parabolic elements.

one-to-one relationship exists between the local (ξ, η) and global (x, y) co-ordinates. If the fixed points are such that a violent distortion occurs then a non-uniqueness may occur in the manner indicated for two situations in Fig. 8.4. Here at internal points of the distorted element two sets of local co-ordinates are implied in addition to some internal points being mapped outside the element. Care must be taken in practice to avoid such gross distortion.

Figure 8.5 shows two examples of a two-dimensional (ξ, η) element mapped into a three-dimensional (x, y, z) space.

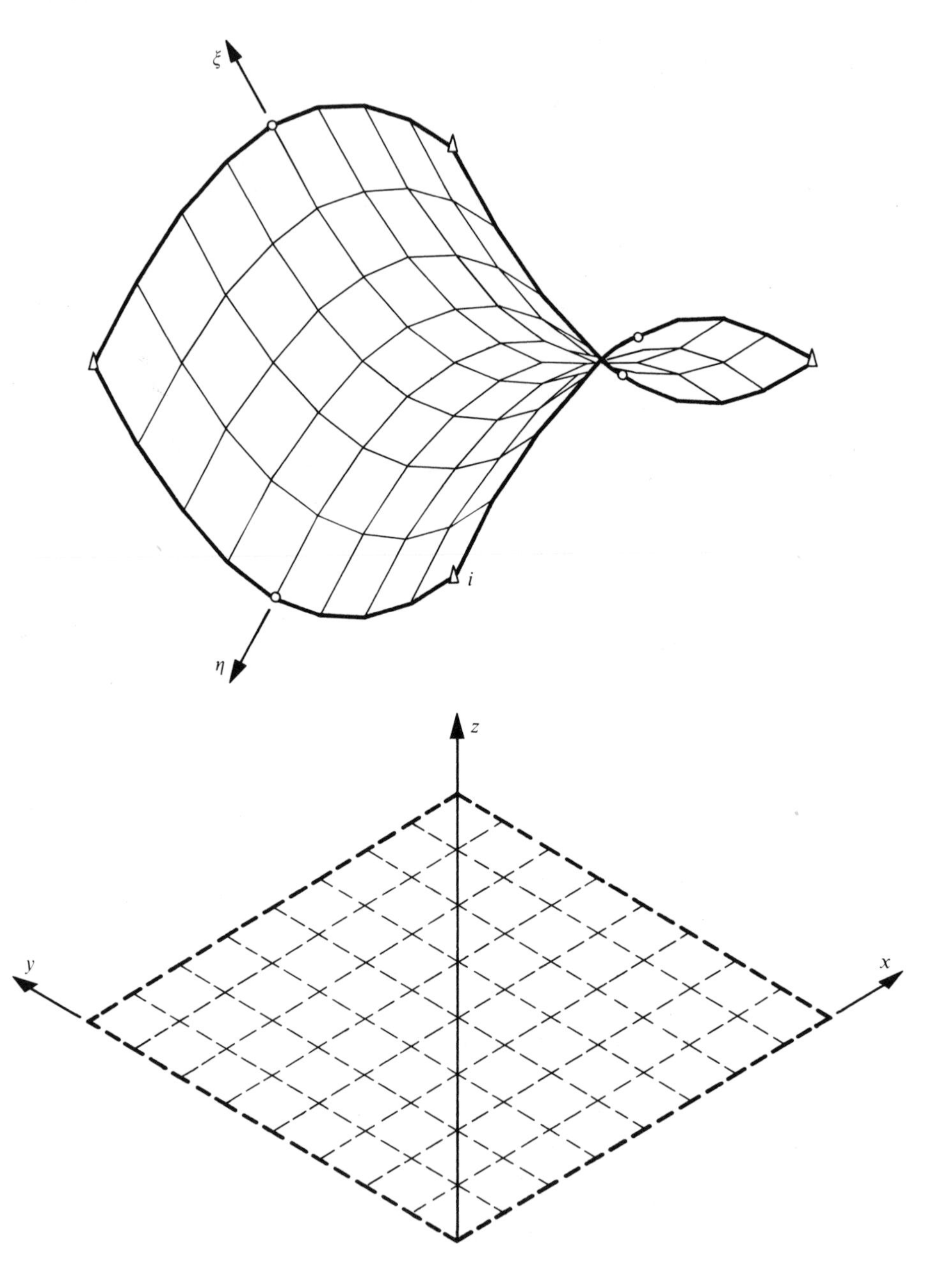

Fig. 8.5 Flat elements (of parabolic type) mapped into three dimensions

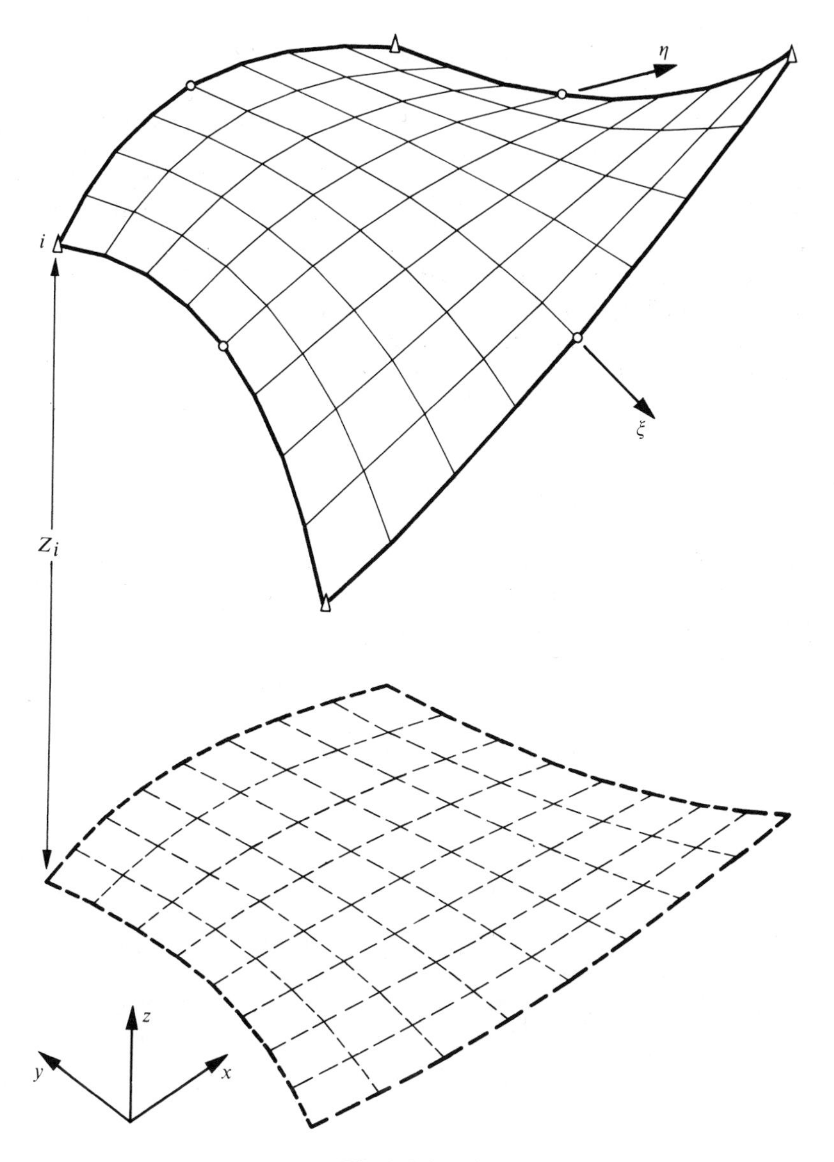

Fig. 8.5 (*cont.*)

In this chapter we shall often refer to the basic element in undistorted, local, co-ordinates as a 'parent' element.

In section 8.5 we shall define a quantity known as the Jacobian determinant. The well-known condition for a *one-to-one* mapping (such as exists in Fig. 8.3 and does not in Fig. 8.4) is that the sign of this quantity should remain unchanged at all the points of the domain mapped.

It can be shown that with a parametric transformation based on linear shape functions, the necessary condition is that no internal angle (such as α in Fig. 8.6(*a*)) be greater than 180°.[6] In transformations based on parabolic type 'Serendipity' functions, it is necessary in addition to this requirement to ensure that the mid-side nodes are in the 'middle third' of the distance between adjacent corners.[7] For cubic functions such general rules are impractical and numerical checks on the sign of the Jacobian determinant are necessary. In practice a parabolic distortion is usually sufficient.

8.3 **Geometrical Conformability of Elements**

While it was shown that by the use of the shape function transformation each parent element maps uniquely a part of the real object, it is important that the subdivision of this into the new, curved, elements should leave no gaps. Possibility of such gaps is indicated in Fig. 8.7.

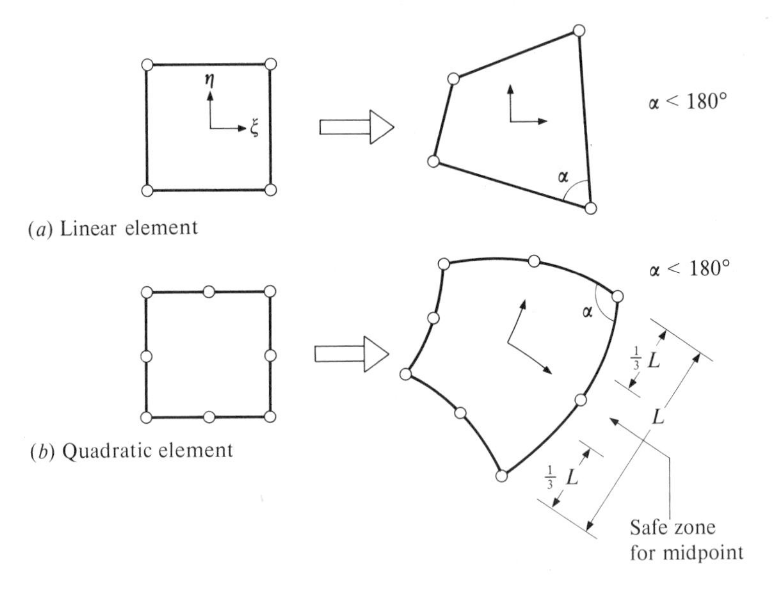

Fig. 8.6 Rules for uniqueness of mapping (*a*) and (*b*)

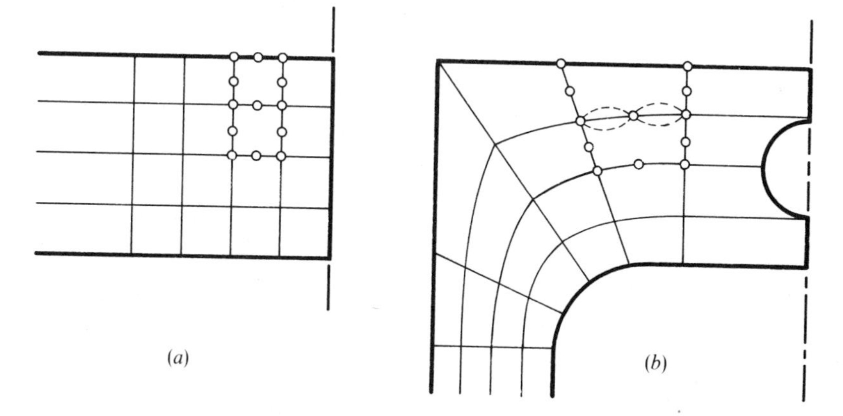

Fig. 8.7 Compatibility requirement in real subdivision of space

THEOREM 1. *If two adjacent elements are generated from 'parents' in which the shape functions satisfy continuity requirements then the distorted elements will be contiguous.*

This theorem is obvious, as in such cases uniqueness of any function ϕ required by continuity is simply replaced by that of uniqueness of the x, y, or z co-ordinate. As adjacent elements are given the same sets of co-ordinates at nodes, continuity is implied.

Nodes of the new distorted elements need not necessarily be placed only at points for which shape functions are specified. Other corresponding sets of nodes can be added on interfaces or boundaries.

8.4 Variation of the Unknown Function within Distorted, Curvilinear, Elements. Continuity Requirements

With the shape of the element now defined by the shape functions $\mathbf{N}'$ the variation of the unknown, ϕ, has to be specified before we can establish element properties. This is most conveniently given in terms of local, curvilinear co-ordinates by the usual expression

$$\phi = \mathbf{N}\mathbf{a}^e \tag{8.3}$$

where $\mathbf{a}^e$ lists the nodal values.

THEOREM 2. *If the shape functions* $\mathbf{N}$ *used in (8.3) are such that continuity of* ϕ *is preserved in the parent co-ordinates—then continuity requirements will be satisfied in distorted elements.*

The proof of this theorem follows the same lines as the previous section.

The nodal values may or may not be associated with the same nodes as used to specify the element geometry. For example in Fig. 8.8 the points marked with a circle are used to define the element geometry. We could use the values of the function defined at nodes marked with a square to define the variation of the unknown.

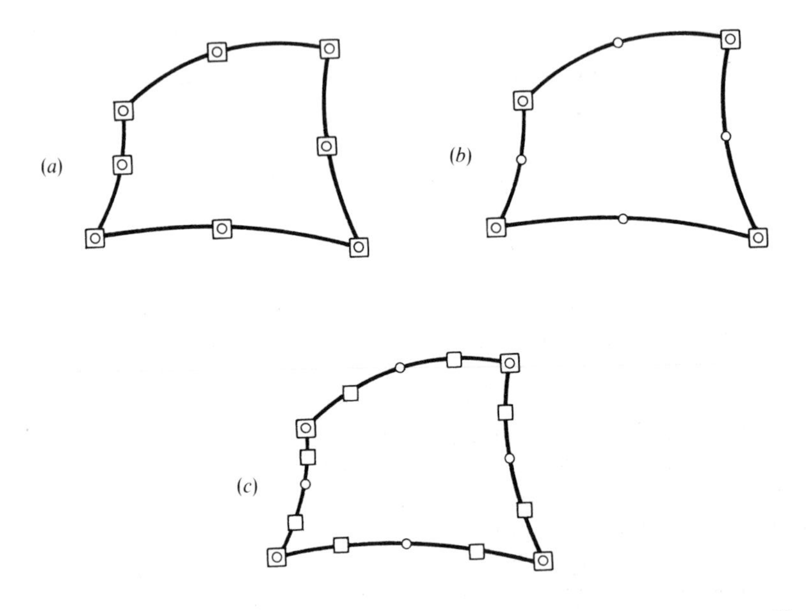

Fig. 8.8 Various element specifications. ○ point at which co-ordinate specified □ points at which function parameter specified. (*a*) Isoparametric, (*b*) Super-parametric, (*c*) Sub-parametric

In Fig. 8.8(*a*) the same points define the geometry and the finite element analysis points. If then

$$\mathbf{N} = \mathbf{N}' \tag{8.4}$$

i.e., the shape functions defining geometry and function are the same, the elements will be called *isoparametric*.

We could, however, use only the four corner points to define the variation of ϕ, Fig. 8.8(*b*). Such an element we shall refer to as *super-parametric*, noting that the variation of geometry is more general than that of the actual unknown.

Similarly if for instance we introduce more nodes to define ϕ than are used to define geometry *sub-parametric* elements will result, Fig. 8.8(*c*). Such elements will be found to be more often of use in practice.

TRANSFORMATIONS

8.5 Evaluation of Element Matrices (Transformation in ξ, η, ζ Co-ordinates)

To perform finite element analysis the matrices defining element properties, e.g., stiffness, etc., have to be found. These will be of the form

$$\int_V \mathbf{G}\, \mathrm{d}V \tag{8.5}$$

in which the matrix **G** depends on **N** or its derivatives with respect to *global co-ordinates.* As an example of this we have the stiffness matrix

$$\int_V \mathbf{B}^{\mathrm{T}}\mathbf{D}\mathbf{B}\, \mathrm{d}V \tag{8.6}$$

and associated load vectors

$$\int_V \mathbf{N}^{\mathrm{T}}\mathbf{b}\, \mathrm{d}V. \tag{8.7}$$

For a particular class of elastic problems the matrices of **B** are given explicitly by their components (*vide* the general form of Eqs. (4.10), (5.6), and (6.11)). Quoting the first of these Eq. (4.10), valid for plane problems we have

$$\mathbf{B}_i = \begin{bmatrix} \dfrac{\partial N_i}{\partial x}, & 0 \\ 0\ , & \dfrac{\partial N_i}{\partial y} \\ \dfrac{\partial N_i}{\partial y}, & \dfrac{\partial N_i}{\partial x} \end{bmatrix}. \tag{8.8}$$

In the elasticity problems the matrix **G** is thus a function of the first derivatives of **N** and this situation will arise in many other classes of problem. In all C_0 continuity is needed and, as we have already noted, this is readily satisfied by the functions of Chapter 7, written in terms of the curvilinear co-ordinates.

To evaluate such matrices we note that two transformations are necessary. In the first place as N_i is defined in terms of local (curvilinear) co-ordinates it is necessary to devise some means of expressing the global derivatives of the type occurring in Eq. (8.8) in terms of local derivatives.

In the second place the element of volume (or surface) over which the integration has to be carried out needs to be expressed in terms of the local co-ordinates with an appropriate change of limits of integration.

Consider for instance the set of local co-ordinates ξ, η, ζ and a corresponding set of global co-ordinates x, y, z. By the usual rules of partial differentiation we can write for instance the ξ derivative as

$$\frac{\partial N_i}{\partial \xi} = \frac{\partial N_i}{\partial x}\frac{\partial x}{\partial \xi} + \frac{\partial N_i}{\partial y}\frac{\partial y}{\partial \xi} + \frac{\partial N_i}{\partial z}\frac{\partial z}{\partial \xi}. \tag{8.9}$$

Performing the same differentiation with respect to the other two co-ordinates and writing in matrix form we have

$$\begin{Bmatrix} \dfrac{\partial N_i}{\partial \xi} \\ \dfrac{\partial N_i}{\partial \eta} \\ \dfrac{\partial N_i}{\partial \zeta} \end{Bmatrix} = \begin{bmatrix} \dfrac{\partial x}{\partial \xi}, & \dfrac{\partial y}{\partial \xi}, & \dfrac{\partial z}{\partial \xi} \\ \dfrac{\partial x}{\partial \eta}, & \dfrac{\partial y}{\partial \eta}, & \dfrac{\partial z}{\partial \eta} \\ \dfrac{\partial x}{\partial \zeta}, & \dfrac{\partial y}{\partial \zeta}, & \dfrac{\partial z}{\partial \zeta} \end{bmatrix} \begin{Bmatrix} \dfrac{\partial N_i}{\partial x} \\ \dfrac{\partial N_i}{\partial y} \\ \dfrac{\partial N_i}{\partial z} \end{Bmatrix} = \mathbf{J} \begin{Bmatrix} \dfrac{\partial N_i}{\partial x} \\ \dfrac{\partial N_i}{\partial y} \\ \dfrac{\partial N_i}{\partial z} \end{Bmatrix}. \tag{8.10}$$

In the above, the left-hand side can be evaluated as the functions N_i are specified in local co-ordinates. Further as x, y, z are explicitly given by the relation defining the curvilinear co-ordinates (Eq. (8.2)), the matrix $\mathbf{J}$, can be found explicitly in terms of the local co-ordinates. This matrix is known as the *Jacobian matrix*.

To find now the global derivatives we invert $\mathbf{J}$ and write

$$\begin{Bmatrix} \dfrac{\partial N_i}{\partial x} \\ \dfrac{\partial N_i}{\partial y} \\ \dfrac{\partial N_i}{\partial z} \end{Bmatrix} = \mathbf{J}^{-1} \begin{Bmatrix} \dfrac{\partial N_i}{\partial \xi} \\ \dfrac{\partial N_i}{\partial \eta} \\ \dfrac{\partial N_i}{\partial \zeta} \end{Bmatrix}. \tag{8.11}$$

In terms of the shape function defining the co-ordinate transformation $\mathbf{N}'$, (which as we have seen are only identical with the shape functions $\mathbf{N}$ when isoparametric formulation is used) we have

$$\mathbf{J} = \begin{bmatrix} \sum \dfrac{\partial N_i'}{\partial \xi} x_i, & \sum \dfrac{\partial N_i'}{\partial \xi} y_i, & \sum \dfrac{\partial N_i'}{\partial \xi} z_i \\ \sum \dfrac{\partial N_i'}{\partial \eta} x_i, & \sum \dfrac{\partial N_i'}{\partial \eta} y_i, & \sum \dfrac{\partial N_i'}{\partial \eta} z_i \\ \sum \dfrac{\partial N_i'}{\partial \zeta} x_i, & \sum \dfrac{\partial N_i'}{\partial \zeta} y_i, & \sum \dfrac{\partial N_i'}{\partial \zeta} z_i \end{bmatrix}$$

$$
= \begin{bmatrix} \dfrac{\partial N'_1}{\partial \xi} & \dfrac{\partial N'_2}{\partial \xi} & \cdots \\ \dfrac{\partial N'_1}{\partial \eta} & \dfrac{\partial N'_2}{\partial \eta} & \cdots \\ \dfrac{\partial N'_1}{\partial \zeta} & \dfrac{\partial N'_2}{\partial \zeta} & \cdots \end{bmatrix} \begin{bmatrix} x_1 & y_1 & z_1 \\ x_2 & y_2 & z_2 \\ \vdots & \vdots & \vdots \end{bmatrix}. \tag{8.12}
$$

To transform the variables and the region with respect to which the integration is made a standard process will be used which involves the determinant of **J**. Thus for instance a volume element becomes

$$
\mathrm{d}x\,\mathrm{d}y\,\mathrm{d}z = \det \mathbf{J}\,\mathrm{d}\xi\,\mathrm{d}\eta\,\mathrm{d}\zeta. \tag{8.13}
$$

This type of transformation is valid irrespective of the number of co-ordinates used. For its justification the reader is referred to standard mathematical texts. A particularly lucid account of this is given by Murnaghan.[8]† (See also Appendix 6)

Assuming that the inverse of **J** can be found we have now reduced the evaluation of the element properties to that of finding integrals of the form of Eq. (8.5).

More explicitly we can write this as

$$
\int_{-1}^{1}\int_{-1}^{1}\int_{-1}^{1} \overline{\mathbf{G}}(\xi, \eta, \zeta,)\,\mathrm{d}\xi\,\mathrm{d}\eta\,\mathrm{d}\zeta \tag{8.14}
$$

if the curvilinear co-ordinates are of the normalized type based on the right prism. Indeed the integration *is carried out within such a prism* and not in the complicated distorted shape, thus accounting for the simple integration limits. One- and two-dimensional problems similarly will result in integrals with respect to one or two co-ordinates within simple limits.

While the limits of the integration are simple in the above case, unfortunately the explicit form of $\overline{\mathbf{G}}$ is not. Excepting the simplest elements, algebraic integration usually defies our mathematical skill, and numerical integration has to be resorted to. This, as will be seen from later sections, is not a severe penalty and has the advantage that algebraic errors are more easily avoided and that general programs, not tied to a particular element, can be written for various classes of problems. Indeed in such numerical calculations the inverses of **J** are never explicitly found.

† The determinant of the Jacobian matrix is known in literature simply as 'the Jacobian' and is often written as

$$
\det \mathbf{J} \equiv \frac{\partial(x, y, z,)}{\partial(\xi, \eta, \zeta)}.
$$

Surface integrals. In elasticity and other applications, surface integrals frequently occur. Typical here the expressions for evaluating the contributions of surface tractions (*vide* Chapter 2, Eq. (2.24b))

$$\mathbf{f} = -\int_A \mathbf{N}^T \bar{\mathbf{t}}\, \mathrm{d}A$$

the element $\mathrm{d}A$ will generally lie on a surface where one of the co-ordinates (say ζ) is constant.

The most convenient process of dealing with the above is to consider $\mathrm{d}A$ as a vector oriented in the direction normal to the surface (see Appendix 6). For three-dimensional problems we form a vector product

$$\mathrm{d}\mathbf{A} = \begin{Bmatrix} \dfrac{\partial x}{\partial \xi} \\ \dfrac{\partial y}{\partial \xi} \\ \dfrac{\partial z}{\partial \xi} \end{Bmatrix} \times \begin{Bmatrix} \dfrac{\partial x}{\partial \eta} \\ \dfrac{\partial y}{\partial \eta} \\ \dfrac{\partial z}{\partial \eta} \end{Bmatrix} \mathrm{d}\xi\, \mathrm{d}\eta$$

and on substitution integrate within a domain $1 \leqslant \xi, \eta \leqslant 1$.

For two dimensions a line length $\mathrm{d}S$ arises and here the magnitude is simply

$$\mathrm{d}S = \begin{Bmatrix} \dfrac{\partial x}{\partial \xi} \\ \dfrac{\partial y}{\partial \xi} \\ \dfrac{\partial z}{\partial \xi} \end{Bmatrix} \mathrm{d}\xi$$

on constant η surfaces.

8.6 Element Matrices. Area and Volume Co-ordinates

The general relationship, Eq. (8.2) for co-ordinate mapping and indeed all the following theorems are equally valid for any set of local co-ordinates and could relate the local $L_1, L_2 \cdots$ co-ordinates, used for triangles and tetrahedra in the previous chapter, to the global Cartesian ones.

Indeed most of the discussion of the previous chapter is valid if we simply rename the local co-ordinates suitably. However, two important differences arise.

The first concerns the fact that the local co-ordinates are not independent and in fact number one more than the Cartesian system. The matrix **J** would apparently therefore become rectangular and would not possess an inverse. The second is simply the difference of integration limits which have to correspond with a triangular or tetrahedral 'parent'.

The simplest, though perhaps not the most elegant, way out of the first difficulty is to consider the last variable as a dependent one. Thus for example we can introduce formally in case of the tetrahedra

$$\begin{aligned} \xi &= L_1 \\ \eta &= L_2 \\ \zeta &= L_3 \\ 1-\xi-\eta-\zeta &= L_4 \end{aligned} \tag{8.15}$$

(by definition of the previous chapter) and thus preserve without change Eq. (8.9) and all the equations up to Eq. (8.14).

As the functions N_i are given in fact in terms of L_1, L_2, etc., we must observe that

$$\frac{\partial N_i}{\partial \xi} = \frac{\partial N_i}{\partial L_1}\frac{\partial L_1}{\partial \xi} + \frac{\partial N_i}{\partial L_2}\frac{\partial L_2}{\partial \xi} + \frac{\partial N_i}{\partial L_3}\frac{\partial L_3}{\partial \xi} + \frac{\partial N_i}{\partial L_4}\frac{\partial L_4}{\partial \xi}. \tag{8.16}$$

On using Eq. (8.15) this becomes simply

$$\frac{\partial N_i}{\partial \xi} = \frac{\partial N_i}{\partial L_1} - \frac{\partial N_i}{\partial L_4}$$

with the other derivatives obtainable by similar expressions.

The integration limits of Eq. (8.14) now change, however, to correspond with the tetrahedron limits. Typically

$$\int_0^1 \int_0^{1-\eta} \int_0^{1-\eta-\zeta} \overline{\mathbf{G}}(\xi, \eta, \zeta)\, d\xi\, d\eta\, d\zeta. \tag{8.17}$$

The same procedure clearly will apply in case of triangular co-ordinates.

It must be noted that once again the expression $\overline{\mathbf{G}}$ will necessitate numerical integration which, however, is carried out over the simple, undistorted, parent region whether this be triangular or tetrahedral.

Finally it should be remarked that any of the elements given in the previous chapter are capable of being mapped. In some, such as the triangular prism, both area and rectangular co-ordinates are used, Fig. 8.9.

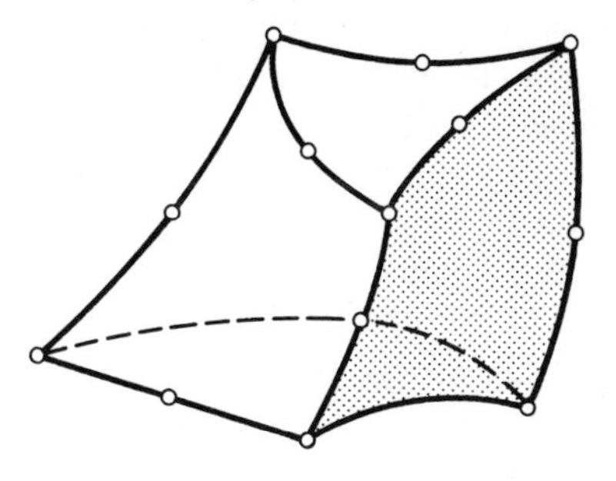

Fig. 8.9 A distorted triangular prism

The remarks regarding the dependence of co-ordinates apply once again with regard to the former but the processes of the present section should make procedures clear.

8.7 **Convergence of Elements in Curvilinear Co-ordinates**

To consider the convergence aspects of the problem posed in curvilinear co-ordinates it is convenient to return to the starting point of the approximation where an energy functional Π, or an equivalent integral form (weak problem statement), was defined by volume integrals essentially similar to those of Eq. (8.5), in which the integrand was a function of ϕ and its first derivatives.

Thus, for instance, the variational principles [*vide* Eq. (3.61)] could be stated for a scalar function ϕ as

$$\Pi = \int_\Omega F\left(\phi, \frac{\partial \phi}{\partial x}, \frac{\partial \phi}{\partial y}, x, y\right) \mathrm{d}\Omega + \int_\Gamma E(\phi, \ldots) \,\mathrm{d}\Gamma. \tag{8.18}$$

The co-ordinate transformation changes the derivatives of any function by the Jacobian relation (8.11). Thus

$$\begin{Bmatrix} \dfrac{\partial \phi}{\partial x} \\ \dfrac{\partial \phi}{\partial y} \end{Bmatrix} = \mathbf{J}^{-1}(\xi, \eta) \begin{Bmatrix} \dfrac{\partial \phi}{\partial \xi} \\ \dfrac{\partial \phi}{\partial \eta} \end{Bmatrix} \tag{8.19}$$

and the functional can be stated simply by a relationship of the form (8.18) with x, y, etc., replaced by ξ, η, etc., with the maximum order of differentiation unchanged.

It follows immediately that if the shape functions are so chosen in the curvilinear co-ordinate space as to observe the usual rules of convergence (continuity and presence of complete first order polynomials), then convergence will occur. Further, all the arguments concerning the order of convergence with the element size h still hold, providing *h is related to the curvilinear co-ordinate system.*

Indeed all that has been said above is applicable to problems involving higher derivatives and to most unique co-ordinate transformations. It should be noted that the patch test as conceived in the $x, y, \ldots,$ co-ordinate system (see Chapters 2 and 11) is no longer generally applicable and in principle should be applied with polynomial fields imposed in the curvilinear co-ordinates. In the case of isoparametric (or subparametric) elements the situation is more advantageous. Here a linear (constant derivative x, y) field is always reproduced by the curvilinear co-ordinate expansion, and thus the lowest order patch test can be applied in the standard manner on such elements.

To prove the point consider a linear field to be obtainable by the expansion.

$$\phi = \sum N_i a_i \equiv \mathbf{N}\mathbf{a}^e = \alpha_1 + \alpha_2 x + \alpha_3 y + \alpha_4 z \qquad \text{with } \mathbf{N} = \mathbf{N}(\xi, \eta, \zeta) \tag{8.20}$$

As at nodes we must have then

$$a_i = \alpha_1 + \alpha_2 x_i + \alpha_3 y_i + \alpha_4 z_i \tag{8.21}$$

the first equality can be rewritten as

$$\begin{aligned} \mathbf{N}\mathbf{a}^e &\equiv \alpha_1 \sum N_i + \alpha_2 \sum N_i x_i + \alpha_3 \sum N_i y_i + \alpha_4 \sum N_i z_i \\ &= \alpha_1 + \alpha_2 x + \alpha_3 y + \alpha_4 z. \end{aligned} \tag{8.22}$$

This will be always satisfied if

$$\begin{aligned} \sum N_i &= 1 \\ \sum N_i x_i &= x \\ \sum N_i y_i &= y \\ \sum N_i z_i &= z. \end{aligned} \tag{8.23}$$

The co-ordinate transformation set out in Eq. (8.2) states that

$$\begin{aligned} \sum N'_i x_i &= x \\ \sum N'_i y_i &= y \\ \sum N'_i z_i &= z \end{aligned} \tag{8.24}$$

and hence:

THEOREM 3. *The constant derivative condition will be satisfied for all isoparametric elements providing* $\sum N_i = 1$.

It can be shown that in fact the same requirement is necessary and the theorem valid for subparametric transformation providing we can express $\mathbf{N}'$ as a linear combination of $\mathbf{N}$, that is,

$$N'_i = \sum C_{ij} N_j. \tag{8.25}$$

NUMERICAL INTEGRATION

8.8 Numerical Integration—One-dimensional

Already in Chapter 5 dealing with a relatively simple problem of axisymmetric stress distribution and simple triangular elements it was noted that exact integration of expressions for element matrices could be troublesome. Now for the more complex distorted elements numerical integration is essential.

Some principles of numerical integration will be summarized here together with tables of convenient numerical coefficients.

To find numerically the integral of a function of one variable we can proceed in one of two basic ways.[9, 10]

Newton–Cotes Quadrature.† In the first, points at which the function is to be found are determined *a priori*—usually at equal intervals—and a polynomial passed through the values of the function at these points and exactly integrated, Fig. 8.10(*a*).

As 'n' values of the function define a polynomial of degree $n-1$, the errors will be of the order $O(\Delta^n)$ where Δ is the point spacing. This leads to the well-known Newton–Cotes 'quadrature' formulae. The integrals can be written as

$$I = \int_{-1}^{1} f(\xi)\,\mathrm{d}\xi = \sum_{1}^{n} H_i f(\xi_i) \tag{8.26}$$

for the range of integration between -1 and $+1$, Fig. 8.10(*a*). For example if $n = 2$, we have the well-known trapezoidal rule,

$$I = f(-1) + f(1) \tag{8.27}$$

For $n = 3$; the Simpson 'one third' rule,

$$I = \tfrac{1}{3}[f(-1) + 4f(0) + f(1)] \tag{8.28}$$

And for $n = 4$;

$$I = \tfrac{1}{4}[f(-1) + 3f(-\tfrac{1}{3}) + 3f(\tfrac{1}{3}) + f(1)] \tag{8.29}$$

Formulae for n up to 21 are given by Kopal.[10]

Gauss quadrature. If in place of specifying the position of sampling points *a priori*, we allow these to be located at points to be determined so as to achieve for best accuracy, for a given number of sampling points an increased accuracy can be obtained. Indeed if we consider again that

$$I = \int_{-1}^{1} f(\xi)\,\mathrm{d}\xi = \sum_{1}^{n} H_i f(\xi_i) \tag{8.30}$$

and assume again a polynomial expression it is easy to see that for n sampling points we have $2n$ unknowns (f_i *and* ξ_i) and hence a polynomial of degree $2n-1$ could be constructed and exactly integrated, Fig. 8.10(*b*). The error thus is of order $O(\Delta^{2n})$.

The simultaneous equations involved are difficult to solve, but some mathematical manipulation[9] will show that the solution can be obtained

† 'Quadrature' is a term used alternatively to 'Numerical Integration'.

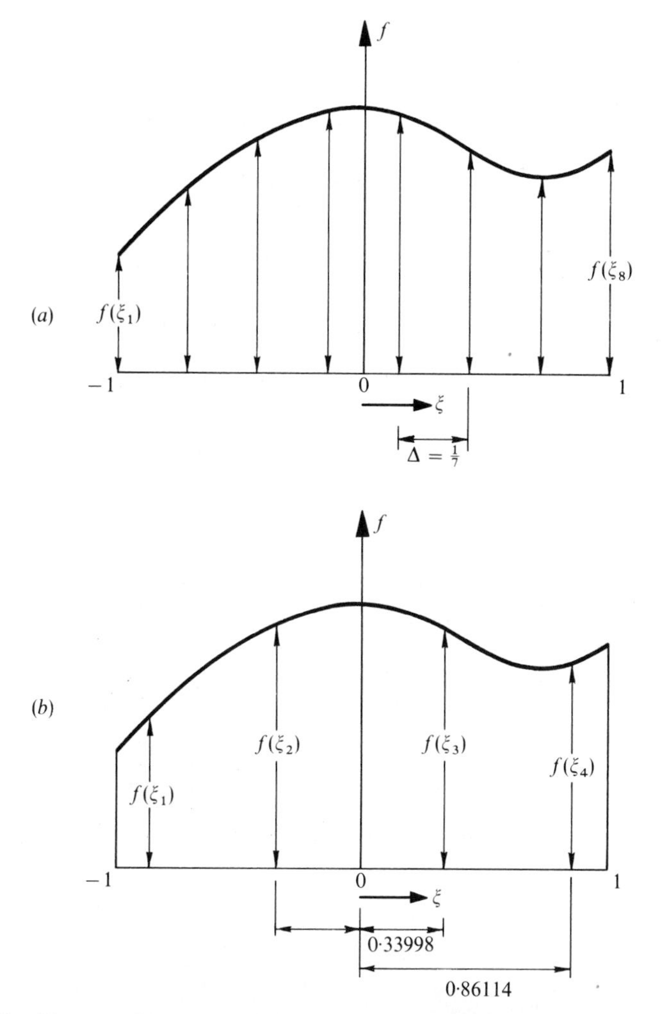

Fig. 8.10 Newton–Cotes (*a*) and Gauss (*b*) Integrations. Each integrates exactly a seventh order polynomial (i.e., error $O(\Delta^8)$)

explicitly in terms of Legendre polynomials. Thus this particular process is frequently known as Gauss–Legendre quadrature.

Table 8.1 shows the positions and weighting coefficients for Gaussian integration.

For purposes of finite element analysis the complex calculations are involved in determining the values of f, the function to be integrated. Thus the Gauss processes, requiring the least number of such evaluations are ideally suited and from now on will be exclusively used.

TABLE 8.1
ABSCISSAE AND WEIGHT COEFFICIENTS OF THE GAUSSIAN QUADRATURE FORMULA

$$\int_{-1}^{1} f(x)\,dx = \sum_{j=1}^{n} H_i f(a_j),$$

$\pm a$		H
	$n = 1$	
0		2·00000 00000 00000
	$n = 2$	
0·57735 02691 89626		1·00000 00000 00000
	$n = 3$	
0·77459 66692 41483		0·55555 55555 55556
0·00000 00000 00000		0·88888 88888 88889
	$n = 4$	
0·86113 63115 94053		0·34785 48451 37454
0·33998 10435 84856		0·65214 51548 62546
	$n = 5$	
0·90617 98459 38664		0·23692 68850 56189
0·53846 93101 05683		0·47862 86704 99366
0·00000 00000 00000		0·56888 88888 88889
	$n = 6$	
0·93246 95142 03152		0·17132 44923 79170
0·66120 93864 66265		0·36076 15730 48139
0·23861 91860 83197		0·46791 39345 72691
	$n = 7$	
0·94910 79123 42759		0·12948 49661 68870
0·74153 11855 99394		0·27970 53914 89277
0·40584 51513 77397		0·38183 00505 05119
0·00000 00000 00000		0·41795 91836 73469
	$n = 8$	
0·96028 98564 97536		0·10122 85362 90376
0·79666 64774 13627		0·22238 10344 53374
0·52553 24099 16329		0·31370 66458 77887
0·18343 46424 95650		0·36268 37833 78362
	$n = 9$	
0·96816 02395 07626		0·08127 43883 61574
0·83603 11073 26636		0·18064 81606 94857
0·61337 14327 00590		0·26061 06964 02935
0·32425 34234 03809		0·31234 70770 40003
0·00000 00000 00000		0·33023 93550 01260
	$n = 10$	
0·97390 65285 17172		0·06667 13443 08688
0·86506 33666 88985		0·14945 13491 50581
0·67940 95682 99024		0·21908 63625 15982
0·43339 53941 29247		0·26926 67193 09996
0·14887 43389 81631		0·29552 42247 14753

Other expressions for integration functions of the type

$$I = \int_{-1}^{1} w(\xi) f(\xi)\, \mathrm{d}\xi = \sum_{1}^{n} H_i f(\xi_i) \tag{8.31}$$

can be derived for prescribed forms of $w(\xi)$ again integrating up to a certain order of accuracy a polynomial expansion of $f(\xi)$.[9]

8.9 Numerical Integration—Rectangular or Right Prism Regions

The most obvious way of obtaining the integral

$$I = \int_{-1}^{1} \int_{-1}^{1} f(\xi, \eta)\, \mathrm{d}\xi\, \mathrm{d}\eta \tag{8.32}$$

is to first evaluate the inner integral keeping η constant, i.e.,

$$\int_{-1}^{1} f(\xi, \eta)\, \mathrm{d}\xi = \sum_{j=1}^{n} H_j f(\xi_j, \eta) = \psi(\eta). \tag{8.33}$$

Evaluating the outer integral in a similar manner, we have

$$\begin{aligned} I = \int_{-1}^{1} \psi(\eta)\, \mathrm{d}\eta &= \sum_{i=1}^{n} H_j \psi(\eta_i) \\ &= \sum_{i=1}^{n} H_i \sum_{j=1}^{n} H_j f(\xi_j, \eta_i) \\ &= \sum_{i=1}^{n} \sum_{j=1}^{n} H_i H_j f(\xi_j, \eta_i). \end{aligned} \tag{8.34}$$

For a right prism we have similarly

$$\begin{aligned} I &= \int_{-1}^{1} \int_{-1}^{1} \int_{-1}^{1} f(\xi, \eta, \zeta)\, \mathrm{d}\xi\, \mathrm{d}\eta\, \mathrm{d}\zeta \\ &= \sum_{m=1}^{n} \sum_{j=1}^{n} \sum_{i=1}^{n} H_i H_j H_m f(\xi_i, \eta_j, \zeta_m). \end{aligned} \tag{8.35}$$

In the above the number of integrating points in each direction was assumed to be the same. Clearly this is not necessary and on occasion it may be of advantage to use different numbers in each direction of integration.

It is of interest to note that in fact the double summation can be readily interpreted as a single one over $(n \times n)$ points for a rectangle (or n^3 points for a cube). Thus in Fig. 8.11 we show the nine sampling points which result in exact integrals of order 5 in each direction.

However, we could approach the problem directly and require an exact integration of a fifth order polynomial in two directions. At any sampling

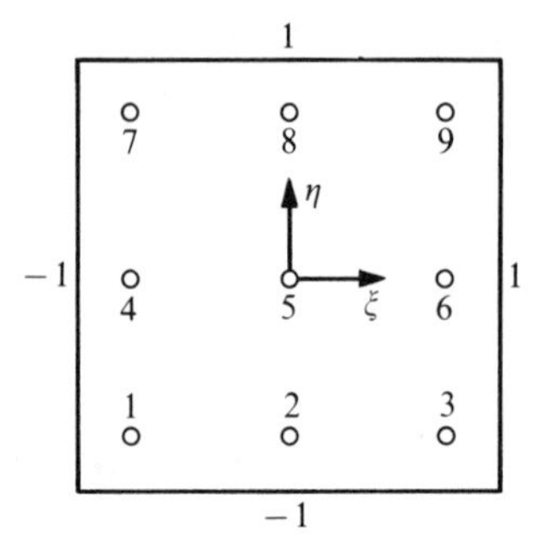

Fig. 8.11 Integrating points for $n = 3$ in a square region. (Exact for polynomial of fifth order in each direction)

point two co-ordinates and a value of f have to be determined in a weighting formula of type

$$I = \int_{-1}^{1} \int_{-1}^{1} f(\xi, \eta) \, d\xi \, d\eta = \sum_{1}^{m} w_i f(\xi_i, \eta_i). \tag{8.36}$$

There it would appear that only seven points would suffice to obtain the same order of accuracy. Some such formulae for three dimensional bricks have been derived by Irons[11] and used successfully.[12]

8.10 Numerical Integration—Triangular or Tetrahedral Regions

For a triangle, in terms of the area co-ordinates the integrals are of the form

$$I = \int_{0}^{1} \int_{0}^{1-L_1} f(L_1 L_2 L_3) \, dL_2 \, dL_1. \tag{8.37}$$

Once again we could use n Gauss points and arrive at a summation expression of the type used in previous section. However, the limits of integration now involve the variable itself and it is convenient to use alternative sampling points for the second integration by use of a special Gauss expression for integrals of type given by Eq. (8.31) in which w is a linear function. These have been devised by Radau,[13] and used successfully in finite element context.[14] It is, however, much more desirable (and aesthetically pleasing) to use special formulae in which no bias is given to any of the natural co-ordinates L_i. Such formulae have been first derived by Hammer *et al.*[15,16] and a series of necessary sampling points and weights is given in Table 8.2.[17] (A more comprehensive list of higher order formulae derived by Cowper is given on p. 184 of reference 6.)

TABLE 8.2
NUMERICAL INTEGRATION FORMULAE FOR TRIANGLES

Order	Fig.	Error	Points	Triangular Co-ordinates	Weights
Linear		$R = O(h^2)$	a	$\frac{1}{3}, \frac{1}{3}, \frac{1}{3}$	1
Quadratic		$R = O(h^3)$	a	$\frac{1}{2}, \frac{1}{2}, 0$	$\frac{1}{3}$
			b	$0, \frac{1}{2}, \frac{1}{2}$	$\frac{1}{3}$
			c	$\frac{1}{2}, 0, \frac{1}{2}$	$\frac{1}{3}$
Cubic		$R = O(h^4)$	a	$\frac{1}{3}, \frac{1}{3}, \frac{1}{3}$	$-\frac{27}{48}$
			b	0·6, 0·2, 0·2	$\frac{25}{48}$ (for b, c, d)
			c	0·2, 0·6, 0·2	
			d	0·2, 0·2, 0·6	
Quintic		$R = O(h^6)$	a	$\frac{1}{3}, \frac{1}{3}, \frac{1}{3}$	0·22500,00000
			b	$\alpha_1, \beta_1, \beta_1$	0·13239,41527 (for b, c, d)
			c	$\beta_1, \alpha_1, \beta_1$	
			d	$\beta_1, \beta_1, \alpha_1$	
			e	$\alpha_2, \beta_2, \beta_2$	0·12593,91805 (for e, f, g)
			f	$\beta_2, \alpha_2, \beta_2$	
			g	$\beta_2, \beta_2, \alpha_2$	

with
$\alpha_1 = 0{\cdot}0597158717$
$\beta_1 = 0{\cdot}4701420641$
$\alpha_2 = 0{\cdot}7974269853$
$\beta_2 = 0{\cdot}1012865073$

Similar extension for tetrahedra could obviously be made. Table 8.3 presents some such formulae based on reference 15.

8.11 **Required Order of Numerical Integration**

With numerical integration used to substitute the exact, an additional error is introduced into the calculation and the first impression is that this

TABLE 8.3

NUMERICAL INTEGRATION FORMULAE FOR TETRAHEDRA

No.	Order	Fig.	Error	Points	Tetrahedral Co-ordinates	Weights
1	Linear		$R = O(h^2)$	a	$\frac{1}{4}, \frac{1}{4}, \frac{1}{4}, \frac{1}{4}$	1
2	Quadratic		$R = O(h^3)$	a	$\alpha, \beta, \beta, \beta$	$\frac{1}{4}$
				b	$\beta, \alpha, \beta, \beta$	$\frac{1}{4}$
				c	$\beta, \beta, \alpha, \beta$	$\frac{1}{4}$
				d	$\beta, \beta, \beta, \alpha$	$\frac{1}{4}$
					$\alpha = 0{\cdot}58541020$	
					$\beta = 0{\cdot}13819660$	
3	Cubic		$R = O(h^4)$	a	$\frac{1}{4}, \frac{1}{4}, \frac{1}{4}, \frac{1}{4}$	$-\frac{4}{5}$
				b	$\frac{1}{2}, \frac{1}{6}, \frac{1}{6}, \frac{1}{6}$	$\frac{9}{20}$
				c	$\frac{1}{6}, \frac{1}{2}, \frac{1}{6}, \frac{1}{6}$	$\frac{9}{20}$
				d	$\frac{1}{6}, \frac{1}{6}, \frac{1}{2}, \frac{1}{6}$	$\frac{9}{20}$
				e	$\frac{1}{6}, \frac{1}{6}, \frac{1}{6}, \frac{1}{2}$	$\frac{9}{20}$

should be reduced as much as possible. Clearly the cost of numerical integration can be quite significant, and indeed in some early programs numerical formulation of element characteristics used a comparable computer time to the subsequent solution of the equations. It is of interest, therefore, to determine (*a*) the minimum integration requirement permitting convergence, and (*b*) the integration requirements necessary to preserve the rate of convergence which would result if exact integration were used.

It will be found later (Chapter 11) that it is in fact often a positive disadvantage to use higher orders of integration than those actually needed under (*b*) as, for very good reasons, a 'cancellation of errors' due to discretization and due to inexact integration occurs.

8.11.1 *Minimum order of integration for convergence.* In problems where the energy functional (or equivalent Galerkin integral statements) define the approximation we have already stated that convergence will occur providing any arbitrary constant value of the mth derivatives can be reproduced. In the present case $m = 1$ and we thus require that in integrals of the form (8.5) a constant value of $\mathbf{G}$ be correctly integrated. Thus the volume of the element $\int_V \mathrm{d}V$ needs to be evaluated correctly for convergence to occur. In curvilinear co-ordinates we thus could argue that $\int_V \det|J| \, \mathrm{d}\zeta \, \mathrm{d}\eta \, \mathrm{d}\xi$ has to be evaluated exactly.[3,6]

Indeed, one could argue that even this condition is too demanding and that, providing $\int_V \mathrm{d}\zeta \, \mathrm{d}\eta \, \mathrm{d}\xi$ is correctly found, convergence will occur. Thus any integration with order of error $O(h)$ suffices. We shall see that such a low integration order is often impractical, although we have in fact used this already in Chapter 4 for axi-symmetric problems.

8.11.2 *Order of integration for no loss of convergence.* In a general problem we have already found that the finite element approximate evaluation of energy (and indeed all the other integrals in a Galerkin type approximation) was exact to the order $2(p-m)$, where p was the order of the complete polynomial present and m the order of differentials occurring in the appropriate expressions.

Providing the integration is exact to the order $2(p-m)$, or shows an error of $O(h^{2(p-m)+1})$, or less, then no loss of convergence order will occur. If in curvilinear co-ordinates we take a curvilinear dimension h of an element, the same rule applies. For C_0 problems (i.e., $m = 1$) the integration formulae should be as follows:

$$
\begin{array}{lll}
p = 1, & \text{linear elements} & O(h) \\
p = 2, & \text{quadratic elements} & O(h^3) \\
p = 3, & \text{cubic elements} & O(h^5)
\end{array}
$$

We shall make use of these results in practice, as will be seen later, but it should be noted that for a linear quadrilateral or triangle a single point integration is adequate. For parabolic quadrilaterals (or bricks) 2×2 (or $2 \times 2 \times 2$), Gauss point integration is adequate and for parabolic triangles (or tetrahedra) three-point (and four-point) formulae of Tables 8.2 and 8.3 are needed.

The basic theorems of this section have been introduced and proved numerically in recent published work.[18,19,20,21]

8.11.3 *Matrix singularity due to numerical integration.* The final outcome of a finite element approximation in linear problems is an equation system

$$
\mathbf{K}\mathbf{a} + \mathbf{f} = 0 \tag{8.38}
$$

in which the boundary conditions have been inserted and which should, on solution for the parameter $\mathbf{a}$, give an approximate solution for the physical situation. If a solution is unique, as is the case with well-posed physical problems, the equation matrix $\mathbf{K}$ should be non-singular. We have *a priori* assumed that this was the case with exact integration and in general have not been disappointed. With numerical integration singularity may arise for low integration orders, and this may make such orders impractical. It is easy to show how, in some circumstances, a singularity of **K** must arise, but it is more difficult to prove that it will not. We shall, therefore, concentrate on the former case.

With numerical integration we replace the integrals by a weighted sum of independent linear relations between the nodal parameters **a**. These linear relations supply the only information from which the matrix **K** is constructed. *If the number of unknowns* **a** *exceeds the number of independent relations supplied at all the integrating points, then the matrix* **K** *must be singular.*

To illustrate this point we shall consider two-dimensional elasticity problems using linear and parabolic quadrilateral elements with one and four point quadratures respectively.

Here at each integrating point *three* independent 'strain relations' are used and the total number of independent relations equals 3 × (number of integration points). The number of unknowns **a** is simply 2 × (number of nodes) less restrained degrees of freedom.

In Fig. 8.12(*a*) and (*b*) we show a single element and an assembly of two elements supported by a minimum number of specified displacements eliminating rigid body motion. The simple calculation shows that only in the assembly of the quadratic elements is elimination of singularity possible, all the other cases remaining strictly singular.

In Fig. 8.12(*c*) a well-supported block of both kinds of elements is considered and here for both element types non-singular matrices may (and will) arise.

The reader may well consider the same assembly but supported again by the minimum restraint of three degrees of freedom. The assembly of linear elements with a single integrating point *will* be singular while the quadratic ones will, in fact, be well behaved.

For the reason just indicated, linear single-point integrated elements are used infrequently while the four-point quadrature is almost universal now for parabolic elements.

The same arguments can be used for assemblies of other two- or three-dimensional elements and we leave the search for such *necessary* singularity as an exercise for the reader.

To conclude, it is of interest to mention that in Chapter 11 we shall in fact *seek* matrix singularity for special purposes by precisely the same arguments.

8.12 Generation of Finite Element Meshes by Mapping

It would have been observed that it is an easy matter to obtain a coarse subdivision of the analysis domain with a small number of isoparametric

× Integrating point (3 independent relations)

○ Nodal point with 2 degrees of freedom

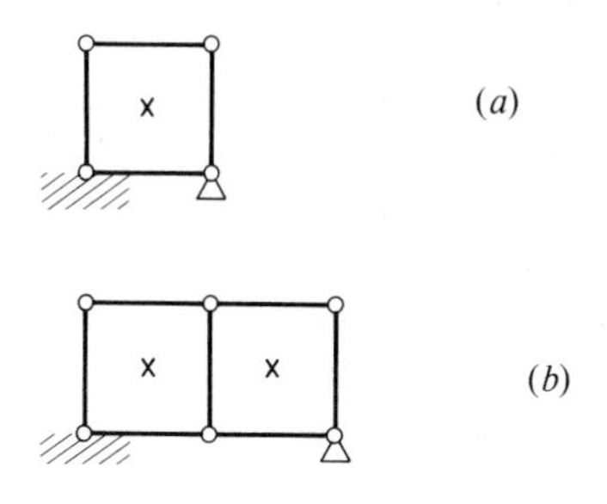

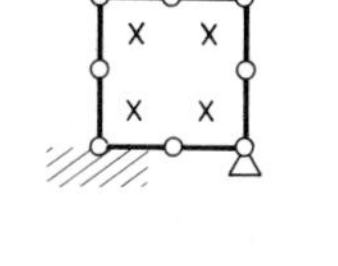

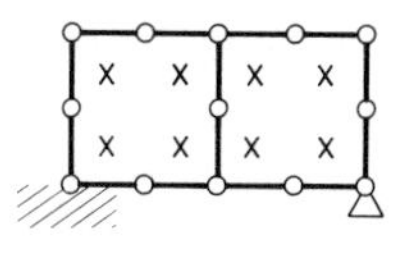

Both d.o.f. suppressed

One d.o.f. suppressed

(c)

	LINEAR		QUADRATIC	
	Degrees of Freedom	Independent Relation	Degrees of Freedom	Independent Relation
(*a*)	$4 \times 2 - 3 = 5$	$> 1 \times 3 = 3$ singular	$2 \times 8 - 3 = 13$	$> 4 \times 3 = 12$ singular
(*b*)	$6 \times 2 - 3 = 9$	$> 2 \times 3 = 6$ singular	$13 \times 2 - 3 = 23$	$< 8 \times 3 = 24$
(*c*)	$25 \times 2 - 18 = 32$	$< 16 \times 3 = 48$	$48 \times 2 = 96$	$< 64 \times 3 = 192$

Fig. 8.12 Check on matrix singularity in two-dimensional elasticity problems (*a*), (*b*), and (*c*)

elements. If second or third order elements are used, the fit of these to quite complex boundaries is reasonable as shown in Fig. 8.13(a) where four parabolic elements specify a sectorial region. This number of elements would be too small for analysis purposes *but a simple subdivision into finer elements* can be done automatically by, say, assigning new positions of nodes of the central points of the curvilinear co-ordinates and deriving thus a larger number of similar elements, as shown in Fig. 8.13(b). Indeed, automatic subdivision could be carried out further to generate a field of triangular elements. The process thus allows, with a small number of original *input data*, to derive a finite element mesh of any refinement desirable. In reference 22 this type of mesh generation is developed for two- and three-dimensional solids and surfaces, and probably presents one of the most efficient means of subdivision.

The main drawback of the mapping and generation suggested is the fact that the originally circular boundaries in Fig. 8.13(a) are approximated by simple parabolae and a geometric error can be developed there. To overcome this difficulty another form of mapping, originally developed for representation of complex motor-car body shapes, can be adopted for this purpose.[23,24] In this mapping blending functions interpolate the unknown ϕ in such a way as to satisfy *exactly* its variations along the edges of a square ξ, η domain. If the co-ordinates x and y are used in a parametric expression of the type given in Eq. (8.2), then any complex can be mapped by a single element. In reference 23, the region of Fig. 8.13

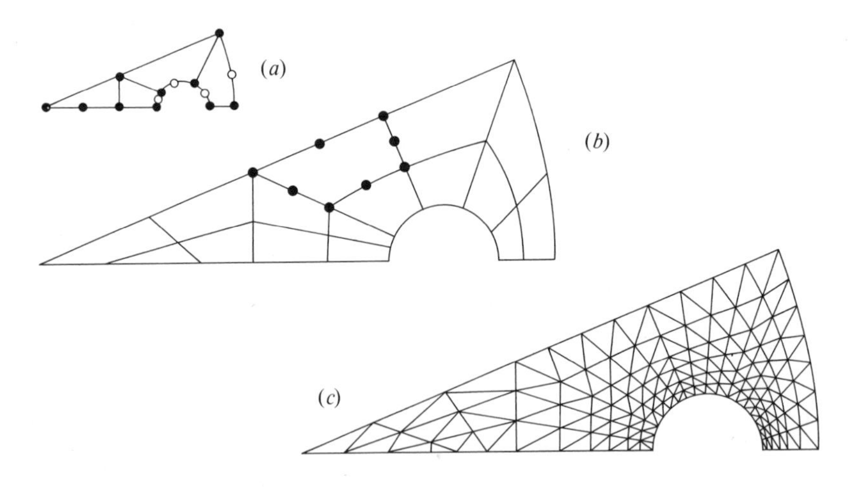

Fig. 8.13 Automatic mesh generation by parabolic isoparametric elements. (a) Specified mesh points. (b) Automatic subdivision into smaller number of isoparametric elements. (c) Automatic subdivision into linear triangles

is in fact so mapped and a mesh subdivision obtained directly without any geometric error on the boundary.

The blending processes are of considerable importance and have been used to construct some interesting element families[25] (which in fact include the standard serendipity elements as a subclass). To explain the process we shall show how a function with prescribed variations along the boundaries can be interpolated.

Consider a region $-1 \leqslant \xi, \eta \leqslant 1$, shown in Fig. 8.14, on the edges of which a function ϕ is specified (i.e., $\phi(-1, \eta)$, $\phi(1, \eta)$, $\phi(\xi, -1)$, $\phi(\xi, 1)$) are given. The problem presented is that of interpolating a function $\phi(\xi, \eta)$ so that a smooth surface reproducing precisely the boundary values is obtained. Writing

$$\begin{aligned} N^1(\xi) &= (1+\xi)/2; \qquad N^2(\xi) = (1-\xi)/2 \\ N^1(\eta) &= (1+\eta)/2; \qquad N^2(\eta) = (1-\eta)/2 \end{aligned} \tag{8.39}$$

for our usual, one-dimensional, linear interpolating functions, we note that

$$P_\eta \phi \equiv N^2(\eta)\phi(\xi, 1) + N^1(\eta)\phi(\xi, -1) \tag{8.40}$$

interpolates linearly between the specified functions in the η direction, as shown in Fig. 8.14(b). Similarly,

$$P_\xi \phi \equiv N^2(\xi)\phi(\eta, 1) - N^1(\xi)\phi(\eta, -1) \tag{8.41}$$

interpolates linearly in the ξ direction (Fig. 8.14(c)). Constructing a third function which is a standard linear, Lagrangian, interpolation of the kind we have already encountered (Fig. 8.14(d)), i.e.,

$$\begin{aligned} P_\xi P_\eta \phi = {} & N^2(\xi)N^2(\eta)\phi(1, 1) + N^2(\xi)N^1(\eta)\phi(1, -1) \\ & + N'(\xi)N\,^2\eta\phi(-1, 1) + N^1(\xi)N^1(\eta)\phi(-1, -1) \end{aligned} \tag{8.42}$$

we note by inspection that

$$\phi = P_\eta \phi + P_\xi \phi + P_\xi P_\eta \phi \tag{8.43}$$

is a smooth surface interpolating exactly the boundary functions.

Extension to functions with higher order blending is almost evident, and immediately the method of mapping the quadrilateral region $-1 \leqslant \xi$, $\eta \leqslant 1$ to any arbitrary shape is obvious.

8.13 Concluding Remarks

In this chapter we have shown how a large number of curvilinear elements can be formulated. The necessity for numerical integration processes led to a description of some of these. For further details the reader is referred to the various texts on Numerical Analysis.

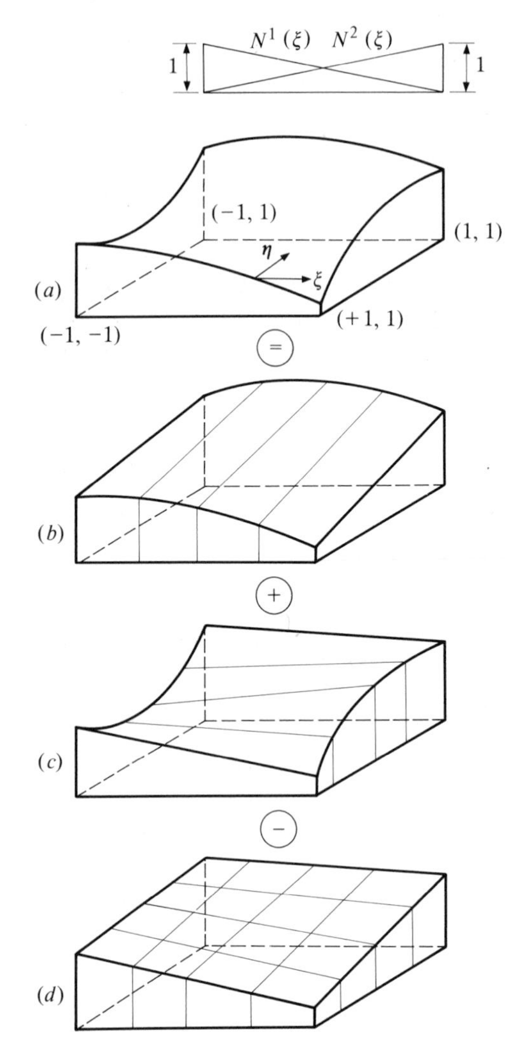

Fig. 8.14 Stages of construction of a blending interpolation (*a*), (*b*), (*c*), and (*d*)

The concepts of this chapter are of fundamental importance in the process of finite element generation and the reader should by now have noted that isoparametric co-ordinates which, at present, are the most widely used finite element form are but one of the numerous co-ordinate possibilities. As an exercise we suggest that the reader formulate explicitly a plane stress element of a sectorial form in which the interpolations are written in terms of radial and angular co-ordinates and where the geometric relation between these constitutes Eq. (8.1). Such elements are occasionally used for special purposes where they suit the geometric requirements of the problem (Fig. 8.15).

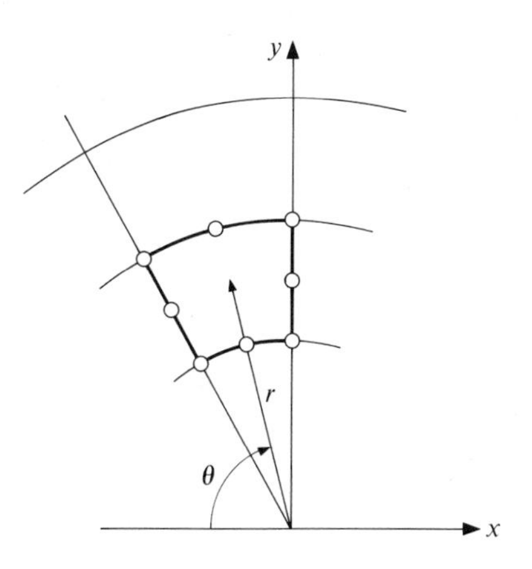

Fig. 8.15 A sectorial element with polar co-ordinates

References

1. I. C. Taig, *Structural analysis by the matrix displacement method*, Engl. Electric Aviation Report No. S017, 1961.
2. B. M. Irons, 'Numerical integration applied to finite element methods', *Conf. Use of Digital Computers in Struct. Eng.* Univ. of Newcastle, 1966.
3. B. M. Irons, 'Engineering application of numerical integration in stiffness method', *J.A.I.A.A.*, **14**, 2035–7, 1966.
4. S. A. Coons, *Surfaces for computer aided design of space form*, M.I.T. Project MAC, MAC-TR-41, 1967.
5. A. R. Forrest, *Curves and surfaces for computer aided design*, Computer Aided Design Group, Cambridge, England, 1968.
6. G. Strang and G. J. Fix, *An Analysis of the Finite Element Method* pp. 156–63, Prentice-Hall, 1973.
7. W. B. Jordan, *The plane isoparametric structural element*, General Elec. Co. Repat KAPL-M-7112, Schenectady, New York, 1970.
8. F. D. Murnaghan, *Finite Deformation of an Elastic Solid*, Wiley, 1951.
9. F. Schied, *Numerical Analysis*, Schaum Series, McGraw-Hill, 1968.
10. Z. Kopal, *Numerical Analysis*, 2nd ed., Chapman & Hall, 1961.
11. B. M. Irons, 'Quadrature rules for brick based finite elements', *Int. J. Num. Meth. Eng.*, **3**, 1971.
12. T. K. Hellen, 'Effective quadrature rules for quadratic solid isoparametric finite elements', *Int. J. Num. Meth. Eng.*, **4**, 597–600, 1972.
13. Radau, *Journ. de Math.*, **3**, 283, 1880.
14. R. G. Anderson, B. M. Irons, and O. C. Zienkiewicz, 'Vibration and stability of plates using finite elements', *Int. J. Solids Struct.*, **4**, 1031–55, 1968.

15. P. C. HAMMER, O. P. MARLOWE, and A. H. STROUD, 'Numerical integration over simplexes and cones', *Math. Tables Aids Comp.*, **10**, 130–7, 1956.
16. C. A. FELIPPA *Refined finite element analysis of linear and non-linear two-dimensional structures*, Structures Materials Research Report No. 66–22, Oct. 1966, Univ. of California, Berkeley.
17. G. R. COWPER, 'Gaussian quadrature formulas for triangles', *Int. J. Num. Meth. Eng.*, **7**, 405–8, 1973.
18. G. J. FIX, 'On the effect of quadrature errors in the finite element method', *Advances in Computational Methods in Structural Mechanics and Design*, pp. 55–68 (eds. J. T. Oden, R. W. Clough, and Y. Yamamoto), Univ. of Alabama Press, 1972. (See also pp. 525–56, *The Mathematical Foundations of the Finite Element Method with Applications to Differential Equations* (ed. A. K. Aziz), Academic Press, 1972.)
19. I. FRIED, 'Accuracy and condition of curved (isoparametric) finite elements', *J. Sound Vibration*, **31**, 345–55, 1973.
20. I. FRIED, 'Numerical integration in the finite element method', *Comp. Struc.*, **4**, 921–32, 1974.
21. M. ZLAMAL, 'Curved elements in the finite element method', *SIAM J. Num. Anal.*, **11**, 347–62, 1974.
22. O. C. ZIENKIEWICZ and D. V. PHILLIPS, 'An automatic mesh generation scheme for plane and curved element domains', *Int. J. Num. Meth. Eng.*, **3**, 519–28, 1971.
23. W. J. GORDON, 'Blending-function methods of bivariate and multivariate interpolation and approximation', *SIAM J. Num. Anal.*, **8**, 158–77, 1971.
24. W. J. GORDON and C. A. HALL, 'Construction of curvilinear co-ordinate systems and application to mesh generation', *Int. J. Num. Meth. Eng.*, **7**, 461–77, 1973.
25. W. J. GORDON and C. A. HALL, 'Transfinite element methods blending-function interpolation over arbitrary curved element domains', *Numer. Math.*, **21**, 109–29, 1973.

9. Some Applications of Isoparametric Elements in Two- and Three-Dimensional Stress Analysis

9.1 Introduction

The high order elements introduced in the previous two chapters require some justification. The additional complexity will require more computer time to be spent in their formulation. The question of economics has therefore to be considered.

Figure 9.1 gives a simple example of a cantilever beam to which various elements are applied. In the first and second pairs of results it will be seen that *a dramatic improvement of accuracy arises with the same number of degrees of freedom when complex elements are used.* This does not necessarily result in a proportional decrease of solution time as with complex elements a larger bandwidth will be encountered—nevertheless a considerable saving occurs for a desired numerical accuracy.

Further, the data preparation is considerably reduced with complex elements. In the examples shown three complex elements replace six and eighteen simple triangles respectively and thus fewer elements have to be specified. Also it is a very simple matter to write into the program a routine which interpolates the positions of mid-side co-ordinates if these sides are straight. Thus the number of co-ordinates needing specification is much smaller.

These points in favour of complex elements can well be countered if efficient automatic mesh generation processes are used—nevertheless the latter will always present more programming difficulty.

On the other side of the picture it will be sometimes seen that the very much reduced number of complex elements may not be adequate to represent all the local geometries of the real problem with the minimum number of elements. In such cases often the balance is in favour of the use of simple formulations.

Probably the most serious economic problem of complex curvilinear elements is the computer time necessary for performing the numerical integrations. Here some economic limit on the accuracy required in this integration must obviously be imposed.

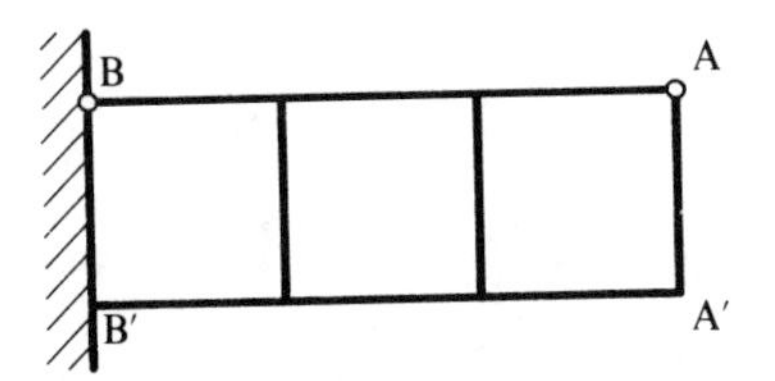

Type of element	Vertical Load of A		Couple at AA′	
	Max. defl. at AA′	Max. stress BB′	Max. defl. at AA′	Max. stress BB′
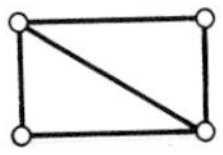	0·26	0·19	0·22	0·22
	0·65	0·56	0·67	0·67
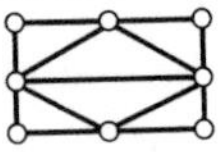	0·53	0·51	0·52	0·55
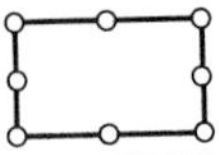	0·99	0·99	1·00	1·00
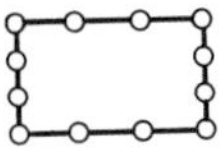	1·00	1·00	1·00	1·00
EXACT	1·00	1·00	1·00	1·00

Fig. 9.1 A cantilever in plane stress analysed by various elements. Accuracy improvement with higher order elements

We have already discussed the problem of 'minimal' integration in the previous chapter. In Chapter 11 we shall show that 'near-minimal' integration is the best for many purposes.

A further economy in computation time can be obtained by an efficient organization of the matrix multiplication involved in stiffness and other element property computation.[1, 2] The algorithms of reference 2 have proved to be most efficient.

9.2 A Computational Advantage of Numerically Integrated Finite Elements[3]

One considerable gain possible in numerically integrated finite elements is the versatility which can be achieved in a single computer program.

It will be observed that for a *given class of problems* the general matrices are always of the same form (*vide* example of Eq. (8.8)) in terms of the shape function and its derivatives.

To proceed to evaluation of the element properties it is necessary first to *specify the shape function* and its derivatives and, second, to *specify the order of integration.*

The computation of element properties thus is composed of three distinct parts as shown in Fig. 9.2. For a *given class of problems* it is only necessary to change the prescription of the shape functions to achieve a variety of possible elements.

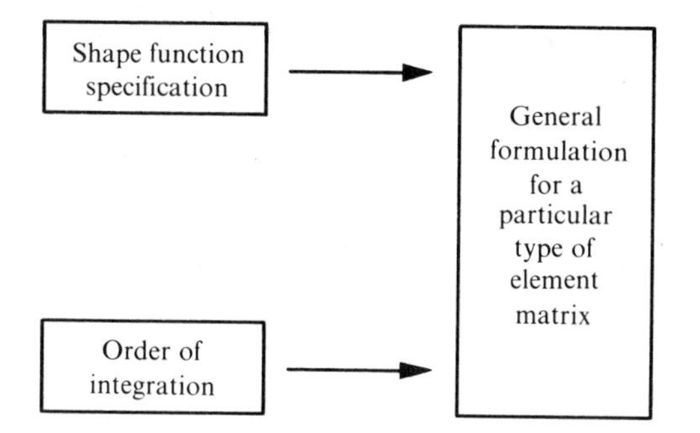

Fig. 9.2 Computation scheme for numerically integrated elements

Conversely the *same shape function* routines can be used in many different classes of problem as is shown in Chapter 24.

Use of different elements, testing the efficiency of a new element in a given context, or extension of programs to deal with new situations can thus be readily achieved, and considerable algebra (with its inherent possibilities of mistakes) avoided.

The computer is thus placed in the position it deserves, i.e., of being the obedient slave capable of saving routine.

The greatest practical advantage of the use of universal shape function routines is that they can be checked decisively for errors by a simple program. Usually it is sufficient to check that the nodal values are correct and that the numbers purporting to be derivatives really are derivatives; this is achieved using simple difference formulae and entering the routine at two closely spaced points. Other tests have been used occasionally. The most interesting is one depending on eigenvalues but its use is somewhat expensive.[4]

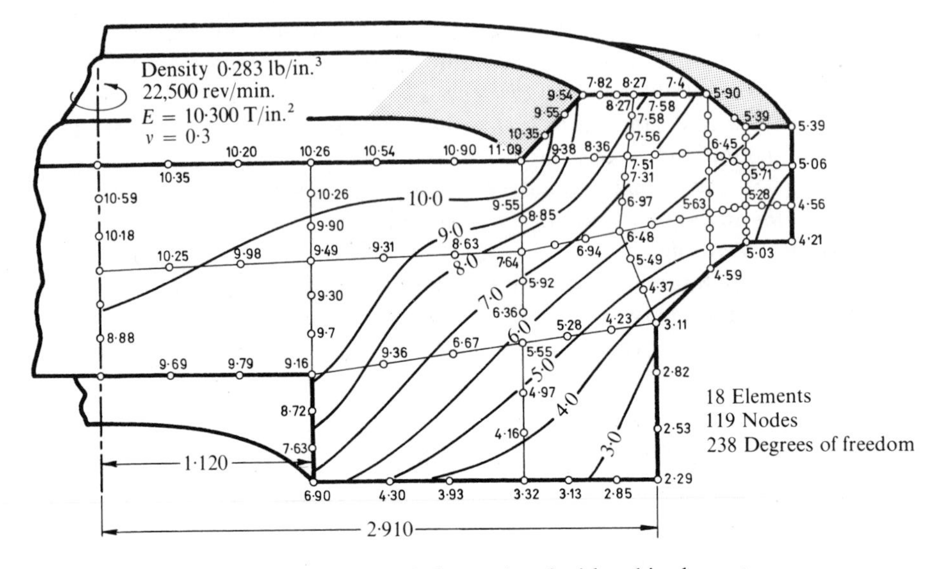

Fig. 9.3 A rotating *disk*—analysed with cubic elements

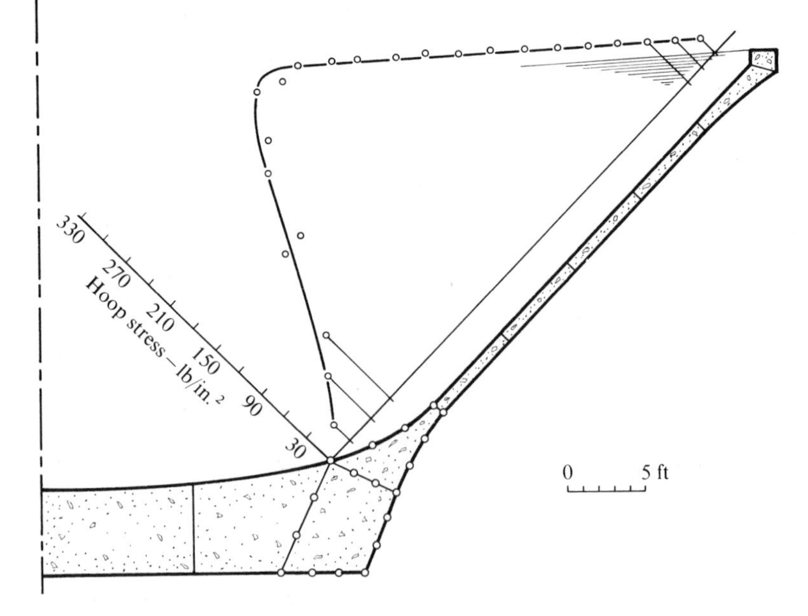

Fig. 9.4 Conical water tank

The incorporation of simple, exactly integrable, elements in such a system is, incidentally, not penalized as time of exact and numerical integration in such cases is almost identical.

9.3 Some Practical Examples of Two-dimensional Stress Analysis[5–11]

Some possibilities of two-dimensional analysis offered by curvilinear elements are illustrated in the following axi-symmetric examples.

Rotating disk (Fig. 9.3). Here eighteen elements only are needed to obtain an adequate solution. It is of interest to observe that all mid-side nodes of the cubic elements are generated within a program and need not be specified.

Conical water tank (Fig. 9.4). In this problem again cubic elements are used. It is worth noting that single element thickness throughout is ade-

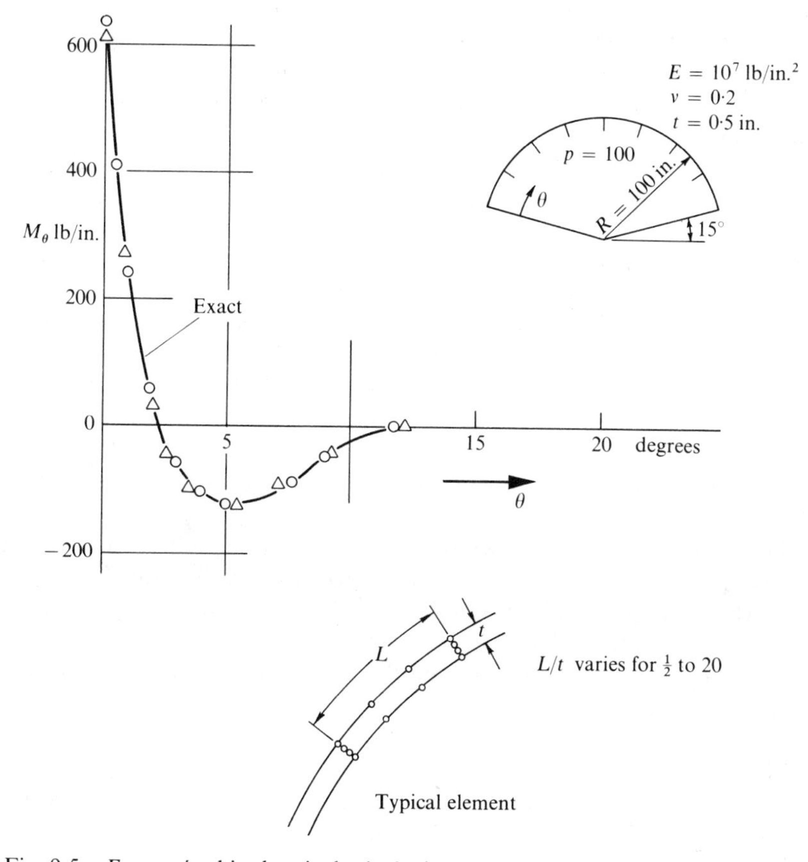

Fig. 9.5 *Encastré*, thin hemispherical shell. Solution with 15 and 24 cubic elements

quate to represent the bending effects in both the thick and thin parts of the container. With simple triangular elements, as we have seen, several layers of elements would have been needed to give an adequate solution.

A hemispherical dome (Fig. 9.5). The possibilities of dealing with shells approached in the previous example are here further exploited to show how a limited number of elements can solve adequately a thin shell problem, with precisely the same program. This type of solution can be further improved upon from the economy viewpoint by making use of the well-known shell assumptions involving a linear variation of displacements across the thickness. Thus the number of degrees of freedom can be reduced. Methods of this kind will be dealt with in detail in Chapter 16.

9.4 **Three-dimensional Stress Analysis**

In three-dimensional analysis, as was already hinted at in Chapter 6, the complex element presents a considerable economic advantage. Some typical examples are shown here in which the quadratic, serendipity type formulation is used almost exclusively. In all problems integration using *three* Gauss points in each direction was used.

Rotating sphere (Fig. 9.6).[6] This example, in which the stresses due to centrifugal action are compared with exact values, is perhaps a test on the efficiency of highly distorted elements. Seven elements are used here and results show reasonable agreement with exact stresses.

Arch dam in rigid valley. This problem, perhaps a little unrealistic from the engineer's viewpoint, was subject of a study carried out by a committee of the Institution of Civil Engineers and provided an excellent test for a convergence study of three-dimensional analysis. In Fig. 9.7 two subdivisions into quadratic and two into cubic elements are shown. In Fig. 9.8 the convergence of displacements in the centre line section is shown, indicating that quite remarkable accuracy can be achieved with even one element.

The comparison of stresses in Fig. 9.9 is again quite remarkable, though showing a greater 'oscillation' with coarse subdivision. The finest subdivision results can be taken as 'exact' from checks by models and alternative methods of analysis.[9]

The above test problems illustrate the general applicability and accuracy. Two further illustrations typical of real situations are included.

Pressure vessel (Fig. 9.10); *An analysis of a biomechanic problem* (Fig. 9.11). Both show subdivisions sufficient to obtain a reasonable engineering accuracy. The pressure vessel, somewhat similar to the one indicated in Chapter 6, Fig. 6.7, shows the very considerable reduction of degrees of freedom possible with the use of more complex elements.

The example of Fig. 9.11 shows a perspective view of the elements used

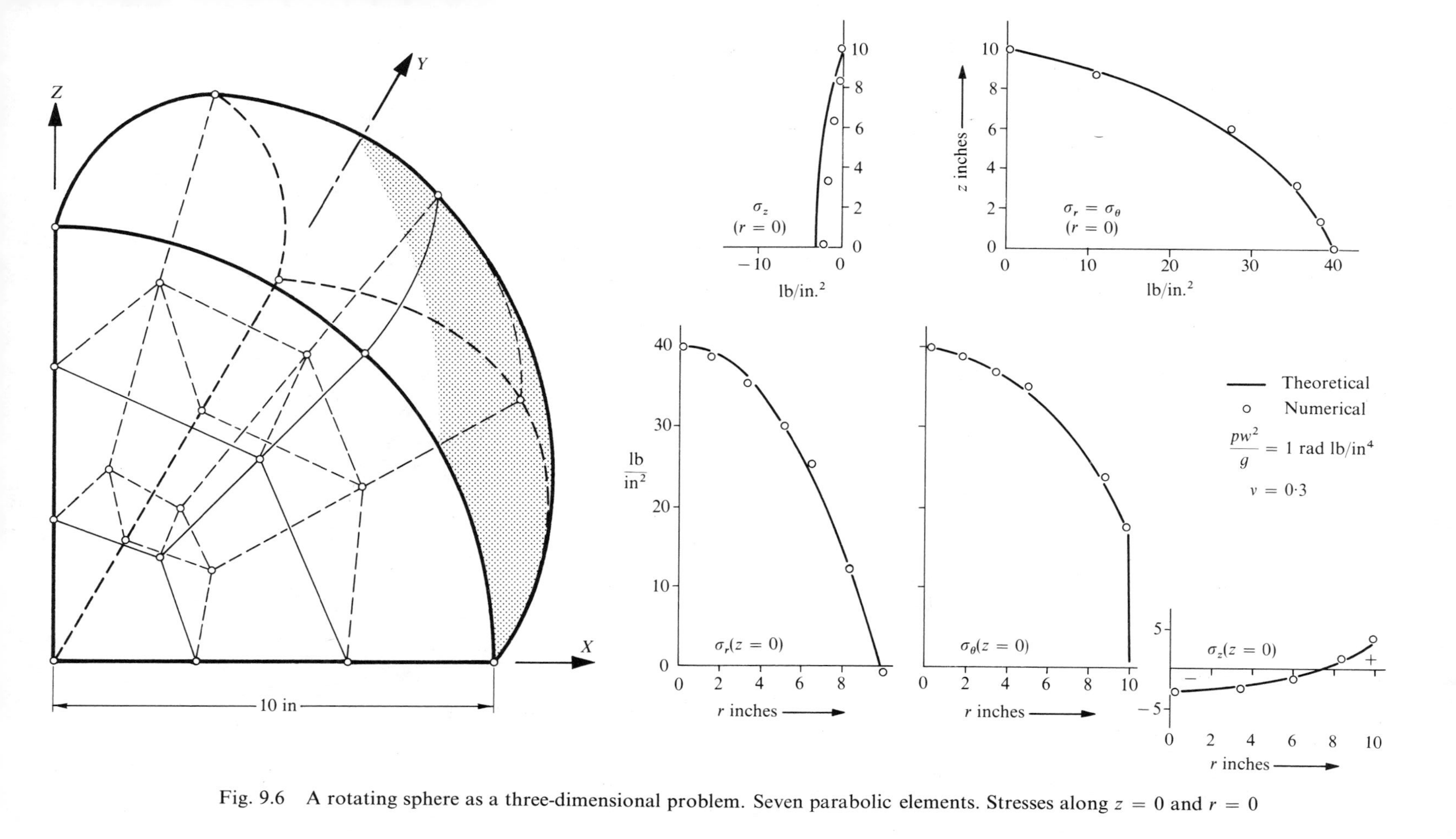

Fig. 9.6 A rotating sphere as a three-dimensional problem. Seven parabolic elements. Stresses along $z = 0$ and $r = 0$

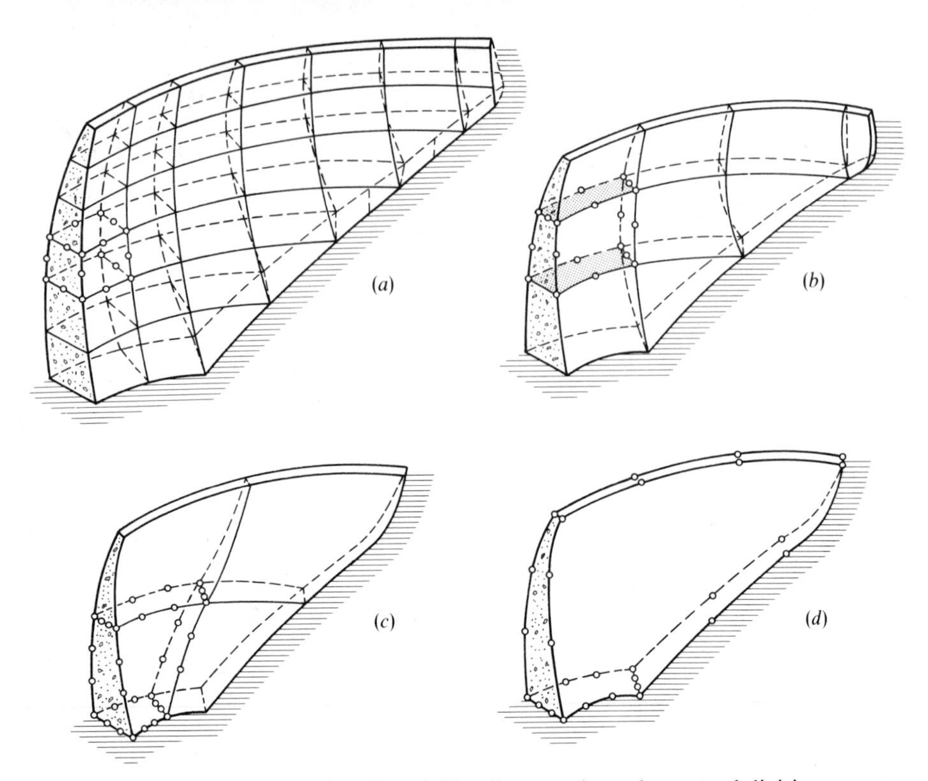

Fig. 9.7 Arch dam in a rigid valley—various element subdivisions

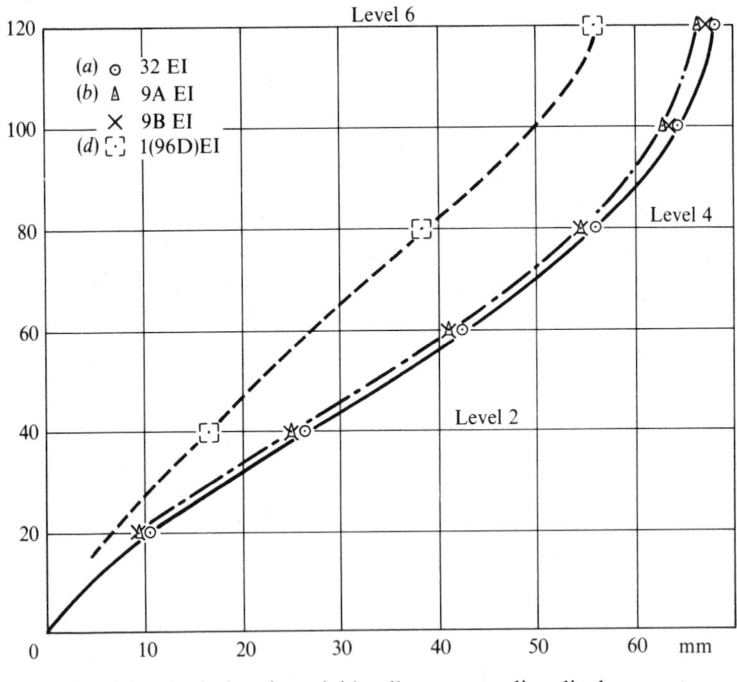

Fig. 9.8 Arch dam in a rigid valley—centre line displacements

obtained directly from the analysis data on an automatic plotter. Such plots are not only helpful in visualization of the problem, they also form an essential part of *data correctness checks* as any gross geometric error can be easily discovered. 'Connectivity' of all specified points is checked automatically.

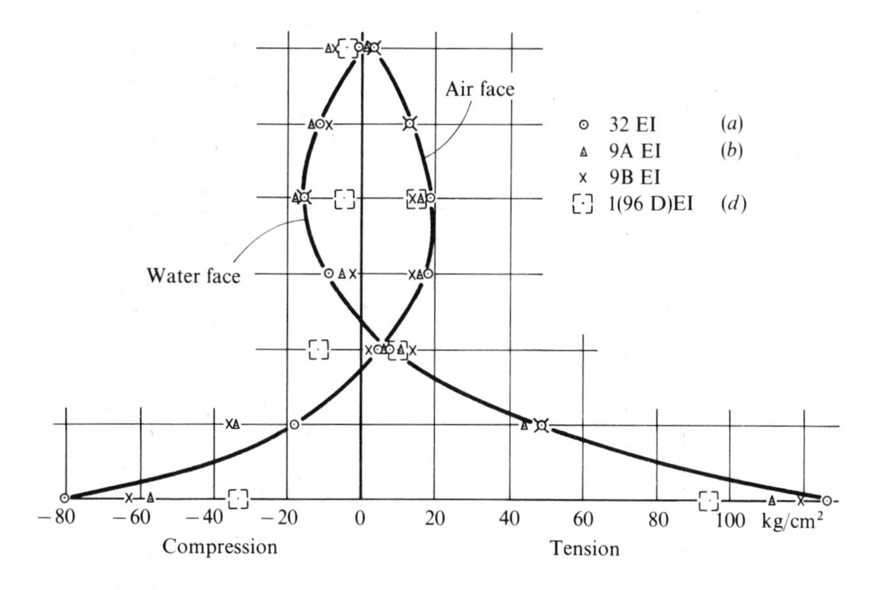

Fig. 9.9 Arch dam in a rigid valley—vertical stresses on centreline

The importance of avoiding data errors in complex three-dimensional problems should be obvious in view of their large usage of computer time. Such, and indeed other,[10] checking methods must form an essential part of any computation system.

9.5 Symmetry and Repeatability

In most of the problems shown, advantage of symmetry in loading and geometry was taken when imposing the boundary conditions, thus reducing the whole problem to manageable proportions. The use of symmetry conditions is so well known to the engineer and physicist that no statement needs to be made about it explicitly. Less known, however, appears to be the use of *repeatability*[12] when an identical structure (and) loading is continuously repeated, as shown in Fig. 9.12 for an infinite blade cascade. Here it is evident that each segment shown shaded behaves identically to the next one, and thus such functions as velocities and

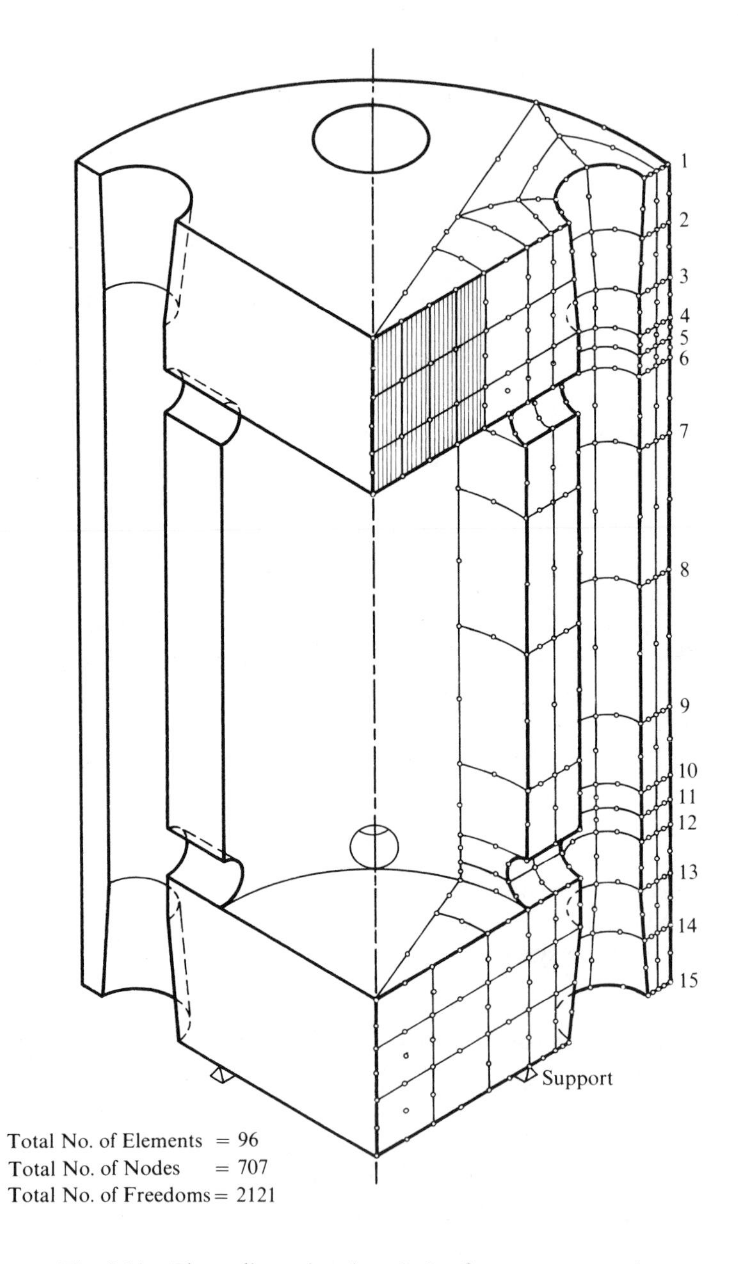

Fig. 9.10 Three-dimensional analysis of a pressure vessel

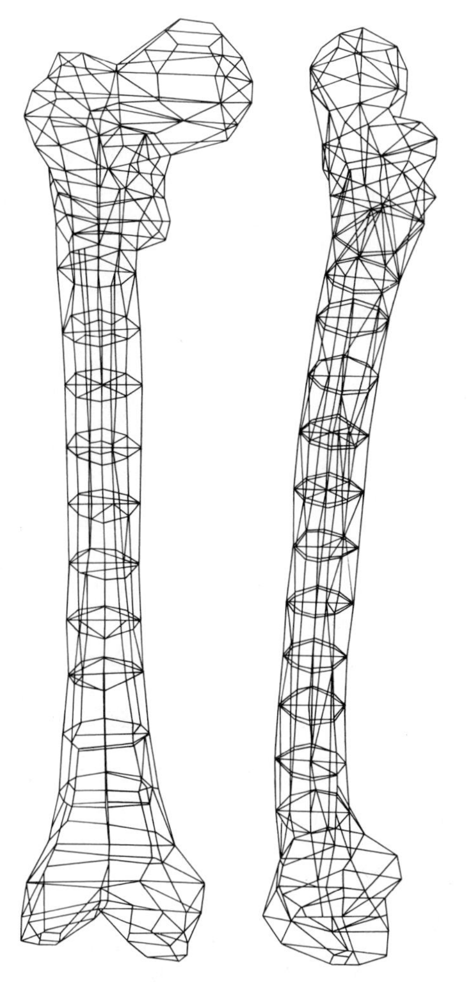

Fig. 9.11 A problem of biomechanics. Plot of linear element form only; curvature of elements omitted. Note degenerate element shapes

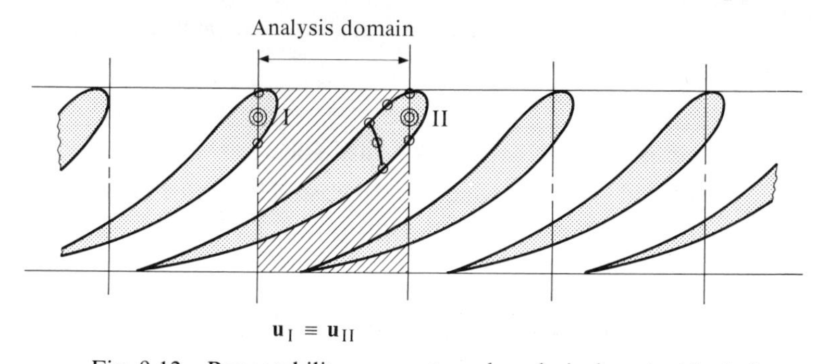

Fig. 9.12 Repeatability segments and analysis domain (shaded)

displacements at corresponding points of AA and BB are simply identified, i.e.,

$$\mathbf{U}_I = \mathbf{U}_{II}.$$

This identification is made directly in a computer program.

Similar repeatability, in radial co-ordinates, occurs very frequently in problems involving turbine or pump impellers. Figure 9.13 shows a typical three-dimensional analysis of such a repeatable segment.

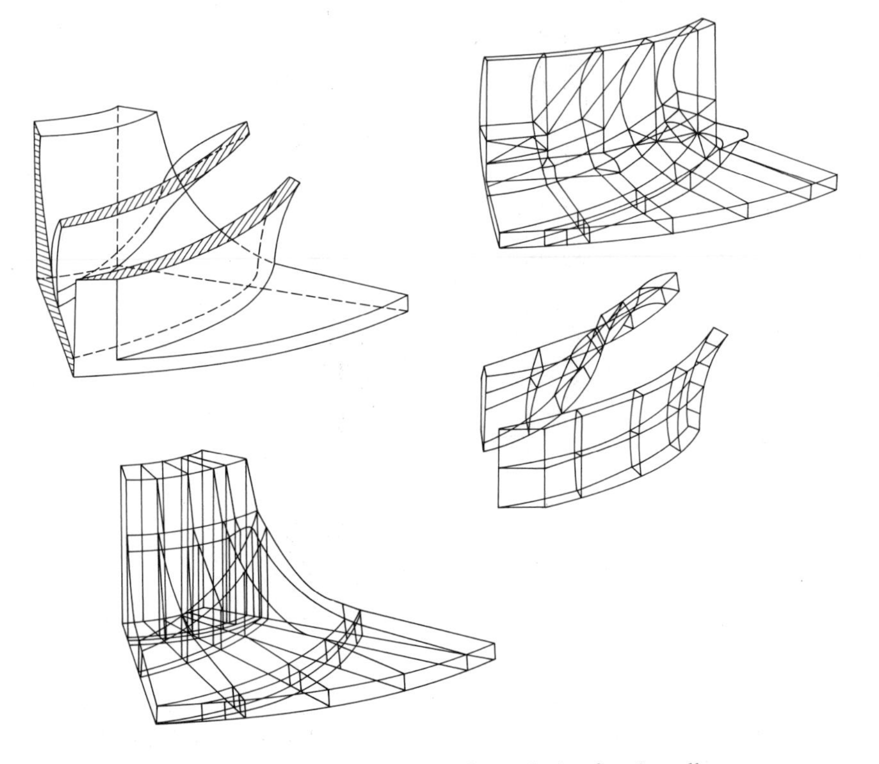

Fig. 9.13 Repeatable sector in analysis of an impeller

9.6 Some General Remarks on Higher Order Elements

With the use of higher order elements, progressively, the departure from an easily conceived physical idealization occurs. This is of little consequence if in fact a better approximation can be achieved but at times this can be embarrassing in practical application. For instance the 'intuitive' allocation of distributed loads is no longer correct.

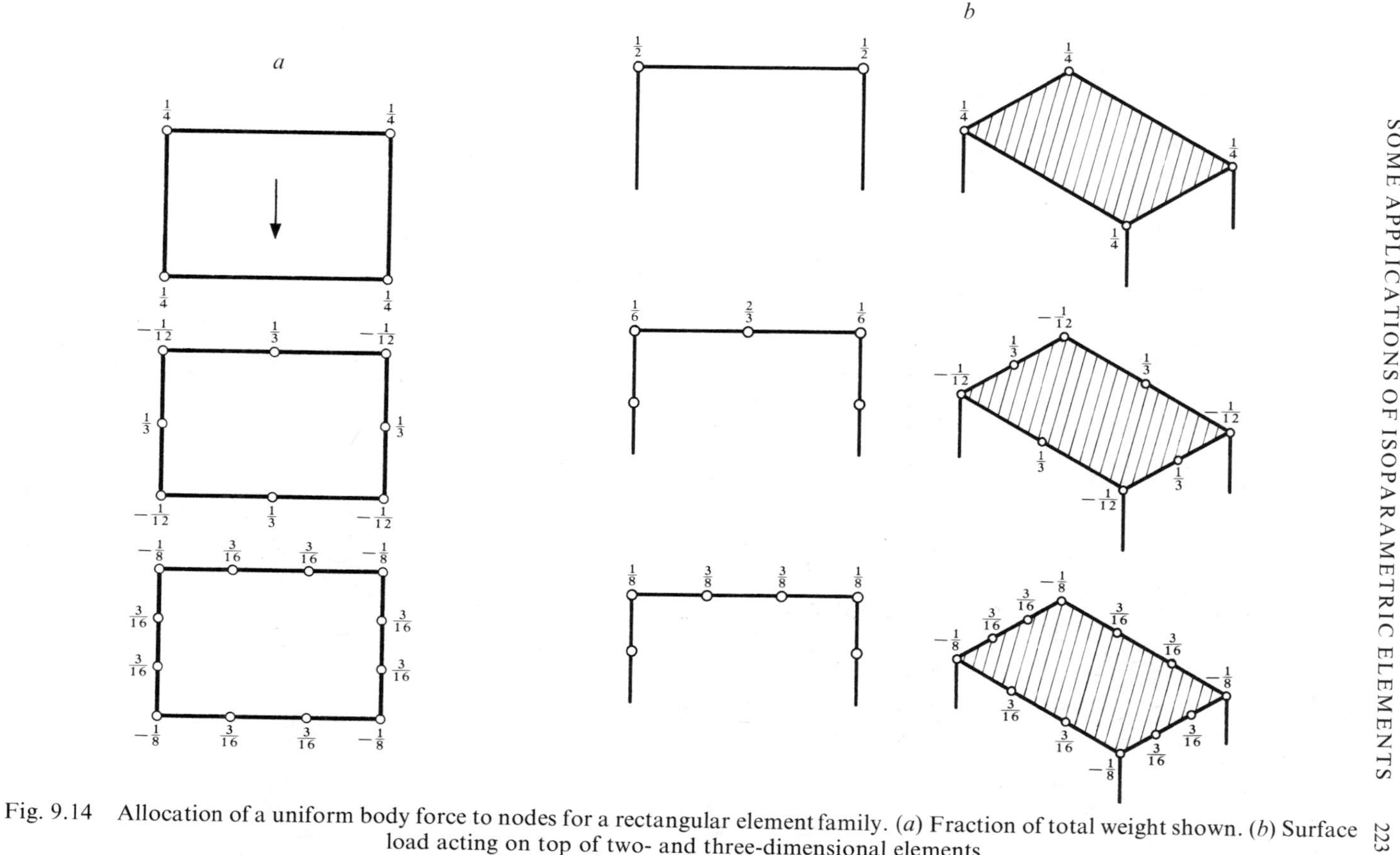

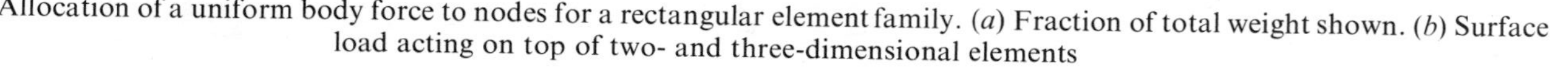

Fig. 9.14 Allocation of a uniform body force to nodes for a rectangular element family. (*a*) Fraction of total weight shown. (*b*) Surface load acting on top of two- and three-dimensional elements

In Chapter 4, for instance, we have shown how properly allocated, consistent nodal forces due to gravity resulted in three equal loads at nodes of a triangular element (section 4.2.7). This result coincides with 'the obvious'. If a similar allocation is carried out for the two-dimensional series of elements of 'serendipity' type (Fig. 7.8, Chapter 7) we get the distribution shown in Fig. 9.14(*a*). Only the first one, for the simplest element of the series, coincides with 'common sense'. In all the others negative allocations at corner nodes exist—a fact not at all 'obvious'.

Indeed, if such elements are curved a yet more complex distribution occurs, and care must be taken to ensure a proper allocation.

The engineer at this point will exclaim that physically the result would still be the same if equal loads were put at each node, in the limit. Certainly this must be the case but in a *finite subdivision* a greater accuracy ensues from the unnatural but consistent distribution.

Surface effects similarly show a non-predictable pattern as given in Fig. 9.14(*b*).

The above considerations affect the *interpretation* of inter-element forces in the usual engineering manner and allowance for this has to be made.

Conversely the representation of stresses in regions close to concentrated load singularities suffers and sometimes unexpected stresses in the vicinity of such loads can be indicated. This indeed is not a sign of decreasing

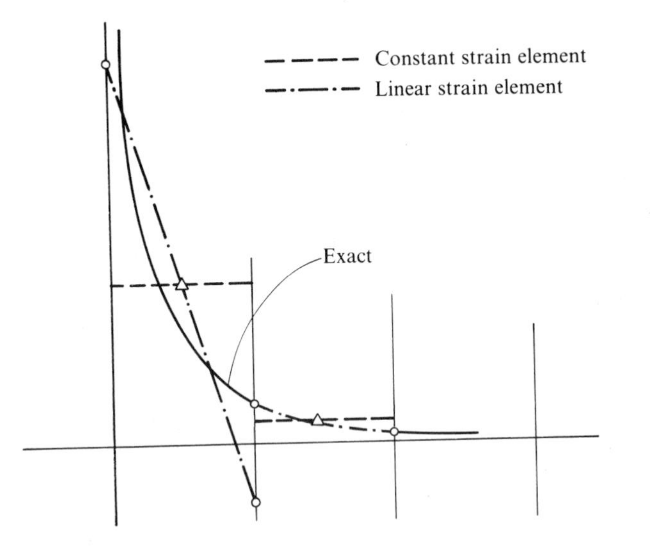

Fig. 9.15 Anomalies which may arise in the vicinity of a concentrated load with complex elements

accuracy but an indication that an element tries on the average, to represent the true effects in an improved way.

In Fig. 9.15 a qualitative comparison between stress representation achieved in constant and linearly varying strain elements is shown near such a singularity. By trying to achieve a closer approximation to the true stress, the more elaborate element gives improved values at the singularity but may result in an unnatural stress reversal close to it which is not indicated by simpler elements. Clearly appropriate smoothing must be resorted to in such cases and proper account taken in interpretation of results. We shall discuss the optimal methods of stress representation in Chapter 11.

References

1. B. M. IRONS, 'Economical computer techniques for numerically integrated finite elements', *Int. J. Num. Meth. Eng.*, **1**, 201–3, 1969
2. A. K. GUPTA and B. MOHRAZ, 'A method of computing numerically integrated stiffness matrices', *Int. J. Num. Meth. Eng.*, **5**, 83–9, 1972.
3. B. M. IRONS, Discussion, p. 328–31, of *Finite Element Techniques in Structural Mechanics*, ed. H. Tottenham and C. Brebbia, Southampton Univ. Press, 1970.
4. B. M. IRONS, 'Testing and assessing finite elements by an eigenvalue technique', *Proc. Conf. on Recent Developments in Stress Analysis, J. Br. Soc. St. An.*, Royal Aero Soc., 1968.
5. O. C. ZIENKIEWICZ, B. M. IRONS, J. G. ERGATOUDIS, S. AHMAD, and F. C. SCOTT, 'Isoparametric and associated element families for two and three dimensional analysis'. *Proc. Course on Finite Element Methods in Stress Analysis*, ed. I. Holand and K. Bell, Trondheim Tech. University, 1969.
6. B. M. IRONS and O. C. ZIENKIEWICZ, 'The isoparametric finite element system —a new concept in finite element analysis', *Proc. Conf. Recent Advances in Stress Analysis*, Royal Aero Soc., 1968.
7. J. G. ERGATOUDIS, B. M. IRONS, and O. C. ZIENKIEWICZ, 'Curved, Isoparametric, "Quadrilateral" elements for finite element analysis', *Int. J. Solids & Struct.*, **4**, 31–42, 1968.
8. J. G. ERGATOUDIS, *Isoparametric elements in two and three dimensional analysis*, Ph.D. Thesis, University of Wales, Swansea, 1968.
9. J. G. ERGATOUDIS, B. M. IRONS, and O. C. ZIENKIEWICZ, 'Three dimensional analysis of arch dams and their foundations', *Symposium on Arch Dams*, Inst. Civ. Eng., London, 1968.
10. O. C. ZIENKIEWICZ, B. M. IRONS, J. CAMPBELL, and F. C. SCOTT, 'Three Dimensional Stress Analysis' *Int. Un. Th. Appl. Mech. Symp. on High Speed Computing in Elasticity*, Liège, 1970.
11. O. C. ZIENKIEWICZ, 'Isoparametric and other numerically integrated elements' in *Numerical and Computer Methods in Structural Mechanics*, pp. 13–41 (ed. S. J. Fenves, N. Perrone, A. R. Robinson and W. C. Schnobrich), Academic Press, 1973.
12. O. C. ZIENKIEWICZ and F. C. SCOTT, 'On the principle of repeatability and its application in analysis of turbine and pump impellers', *Int. J. Num. Meth. Engrs.*, **9**, 445–52, 1972.

10. Bending of Thin Plates C_1-Continuity Problems

10.1 Introduction

In all the problems treated in the earlier chapters the basic stress–strain relationships have been given in their exact form, even though the ultimate solution introduced approximations. In the classical theory of plates[1] certain approximations are introduced initially to simplify the problem to two dimensions. Such assumptions concern the linear variation of strains and stresses on lines normal to the plane of the plate. So-called 'exact' solutions of plate theory are therefore only true if these assumptions are valid. This is so when the plates are thin and the deflection small.

In the solutions presented here the starting point will, once again, be based on the classical plate theory assumptions, and the validity of the approximate, numerical treatment must therefore be tested against plate theory solutions. It will also be subject to precisely the same limitations.

The state of deformation of a plate can be described entirely by one quantity. This is the lateral displacement w of the 'middle plane' of the plate. Continuity conditions between elements have now, however, to be imposed not only on this quantity but on its derivatives. This is to ensure that the plate remains continuous and does not 'kink'.† At each node, therefore, three conditions of equilibrium and continuity will usually be imposed.

Determination of suitable shape functions is now much more complex. Indeed, if complete slope continuity is required on the interfaces between various elements, the mathematical and computational difficulties often rise disproportionately fast. It is, however, relatively simple to obtain shape functions which, while preserving continuity of w, may violate its slope continuity between elements, though naturally not at the node where such continuity is imposed. If such chosen functions satisfy the 'constant strain' criterion and in addition pass the 'patch test' (see Chapters 2 and 11), then convergence will still be found. The first part of this chapter

† If 'kinking' occurs the second derivative or curvature becomes infinite and certain infinite terms occur in the energy expression.

will be concerned with such 'non-conforming' shape functions. In the second part new functions are introduced by which continuity can be restored. The solution with such 'conforming' shape functions will now give bounds to the correct answer but, on many occasions, will yield an inferior accuracy. For practical usage the methods of the first part of the chapter are often recommended.

The simplest type of element shape is now a rectangle and this will be introduced first. Triangular and quadrilateral elements present some difficulties and will be introduced later; for solutions of plates of arbitrary shape or, for that matter, for dealing with shell problems such elements are essential.

The problem of thin plates, where the potential energy functional contains *second derivatives* of the unknown functions, is characteristic of a large class of physical problems associated with *fourth order differential equations*. Thus, although the chapter concentrates on the structural problem the reader with other physical problems in mind will find that the procedures developed will be equally applicable elsewhere.

The difficulty of imposing C_1 continuity on the shape functions has resulted in many alternative approaches to the problems in which this difficulty is side-stepped.

Two basic alternatives are present:

(*a*) Imposition of continuity as a constraint using either Lagrangian multipliers or penalty functions.
(*b*) Complete reformulations of the problem in terms of relationships and energy functionals requiring only C_0 continuity.

Both approaches usually require the introduction of additional degrees of freedom (via new variables). Some of these have proved extremely effective. We shall, however, defer the consideration of such reformulations to Chapters 11 and 12, which will complete the thin plate story.

10.2 **Displacement Formulation of the Plate Problem**

Displacement of a plate, under the usual thin plate theory, is uniquely specified once the deflection, w, is known at all points.

We will write the general form as

$$w = \mathbf{N}\mathbf{a}^e \tag{10.1}$$

in which the shape functions are dependent on Cartesian co-ordinates x, y, and $\mathbf{a}^e$ lists the element (nodal) parameters.

The generalized 'strains' and 'stresses' have now to be specified in such a way that their scalar product gives the internal work in the manner of Chapter 2. Thus we shall define the 'strain' as (Fig. 10.1)

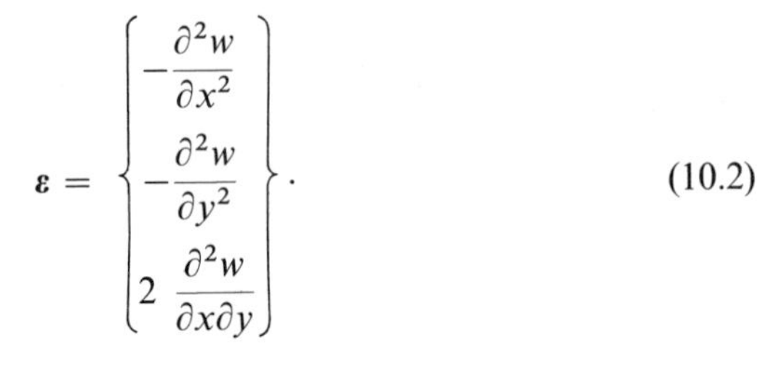

$$\boldsymbol{\varepsilon} = \begin{Bmatrix} -\dfrac{\partial^2 w}{\partial x^2} \\ -\dfrac{\partial^2 w}{\partial y^2} \\ 2\,\dfrac{\partial^2 w}{\partial x \partial y} \end{Bmatrix}. \tag{10.2}$$

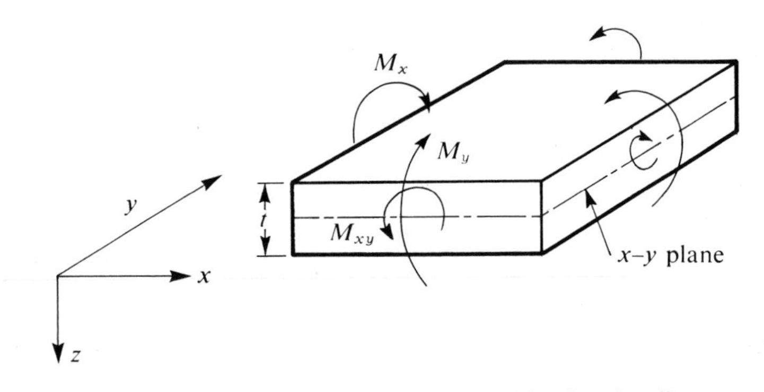

Fig. 10.1 Stress resultants or 'stresses' in plate bending

The corresponding 'stresses' are in fact the usual bending and twisting moments per unit lengths in x and y directions.[1]

$$\boldsymbol{\sigma} = \begin{Bmatrix} M_x \\ M_y \\ M_{xy} \end{Bmatrix}. \tag{10.3}$$

As true strains and stresses vary linearly across the plate thickness[1] these can be found from such expressions as

$$\sigma_x = \frac{12M_x}{t^3}\,z, \text{ etc.}$$

where z is measured from the plate mid-plane and t is the thickness of plate.

The product of the expressions (10.2) and (10.3) will be found to correspond exactly to the internal work requirements.

As the strains are now defined by second derivatives the continuity criterion requires that the shape functions be such that both w and its slope normal to the interface between elements be continuous.

The criterion of constant strain requires that any constant arbitrary value of *second derivative* should be reproducible within the element.

To ensure at least an approximate satisfaction of slope continuity three displacement components are considered as nodal parameters: the first the actual displacement w_n in the z direction, the second a rotation about the x axis $(\theta_x)_n$, and the third a rotation about the y axis $(\theta_y)_n$. Figure 10.2 shows these rotations with their positive directions determined by the right-hand screw rule. Their magnitudes are shown by vectors directed along the axes.

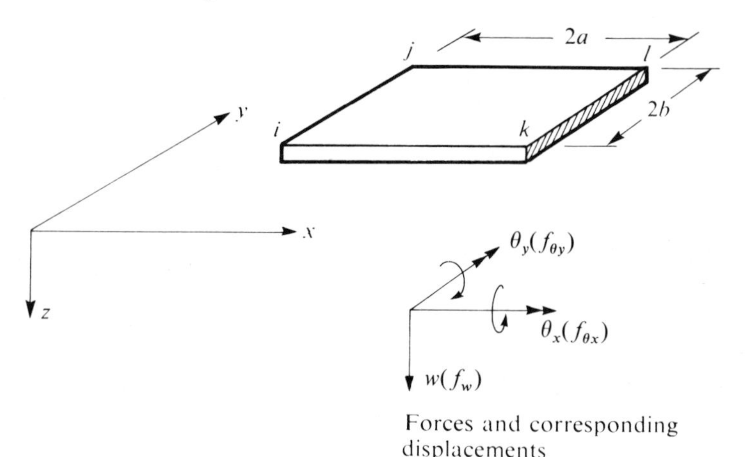

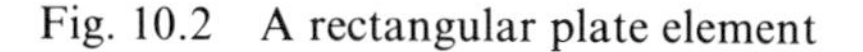
Fig. 10.2 A rectangular plate element

Clearly the slopes of w and the rotation are identical (except for sign) and we can write

$$\mathbf{a}_i = \begin{Bmatrix} w_i \\ \theta_{xi} \\ \theta_{yi} \end{Bmatrix} = \begin{Bmatrix} w_i \\ -\left(\dfrac{\partial w}{\partial y}\right)_i \\ \left(\dfrac{\partial w}{\partial x}\right)_i \end{Bmatrix}. \tag{10.4}$$

The nodal 'forces' corresponding to these displacements can be interpreted as a direct force and two couples

$$\mathbf{f}_i = \begin{Bmatrix} f_{wi} \\ f_{\theta xi} \\ f_{\theta yi} \end{Bmatrix} \tag{10.5}$$

as shown in Fig. 10.2.

The stiffness and other element matrices will be obtained in the usual manner by the expressions of Chapter 2 once the **B** matrix has been determined.

From the definitions, Eqs. (10.1) and (10.2) it follows immediately that

$$\mathbf{B}_i = \begin{Bmatrix} -\dfrac{\partial^2}{\partial x^2}\mathbf{N}_i \\ -\dfrac{\partial^2}{\partial y^2}\mathbf{N}_i \\ 2\dfrac{\partial^2}{\partial x\,\partial y}\mathbf{N}_i \end{Bmatrix}. \tag{10.6}$$

Matrix notation is retained now in the shape functions to denote that this is a quantity dependent on three terms (a 3×1 matrix).

The elasticity matrix $\mathbf{D}$ is involved in the usual definition

$$\boldsymbol{\sigma} \equiv \mathbf{M} = \mathbf{D}(\boldsymbol{\varepsilon}-\boldsymbol{\varepsilon}_0)+\boldsymbol{\sigma}_0. \tag{10.7}$$

For an *isotropic plate* we have (*vide* Timoshenko and Woinowsky-Krieger,[1] p. 81)

$$\mathbf{D} = \frac{Et^3}{12(1-\nu^2)}\begin{bmatrix} 1 & \nu & 0 \\ \nu & 1 & 0 \\ 0 & 0 & (1-\nu)/2 \end{bmatrix}. \tag{10.8}$$

For an *orthotropic slab* with principal directions of orthotropy coinciding with the x and y axes, four constants are needed to define the behaviour i.e.,

$$\mathbf{D} = \begin{bmatrix} D_x & D_1 & 0 \\ D_1 & D_y & 0 \\ 0 & 0 & D_{xy} \end{bmatrix}. \tag{10.9}$$

These can be related to the appropriate elastic constants of the material as shown in Timoshenko and Woinowsky-Krieger[1] but it is more convenient to leave them in the above form as the plate theory is often used to solve grillage problems. In such cases the constants must be related to the properties at the grillage. Clearly, for a most complete case of anisotropy, six constants at most will be needed to define $\mathbf{D}$ since the matrix always has to be symmetric.

We now have all the ingredients for the determination of the stiffness matrices, etc., in the manner indicated in Chapter 2. Indeed, the same form of energy expressions derived there (Eq. (2.29)) is valid providing the boundary work term includes as tractions the bending moment *and* the total edge force.

It is instructive, however, to rederive the whole formulation directly from the equilibrium equations of the plates. If an infinitesimal element

$dx\, dy$, such as shown in Fig. 10.1, is considered it is easy to show that its equilibrium is given by the following equation:

$$\frac{\partial^2 M_x}{\partial x^2}+\frac{\partial^2 M_y}{\partial y^2}-2\frac{\partial^2 M_{xy}}{\partial x\,\partial y}+q=0 \tag{10.10}$$

where q is the lateral distributed load per unit area.

Noting the relationships between the moments and strains implied in Eqs. (10.2), (10.3), and (10.7) we have as a starting point a differential equation of the fourth order

$$\left[\frac{\partial^2}{\partial x^2}, \frac{\partial^2}{\partial y^2}, -2\frac{\partial^2}{\partial x\,\partial y}\right]\mathbf{D}\left[\frac{\partial^2 w}{\partial x^2}, +\frac{\partial^2 w}{\partial y^2}, -2\frac{\partial^2 w}{\partial x\,\partial y}\right]+q=0 \tag{10.11}$$

which, in case of an isotropic plate of constant thickness, reduces to the well-known biharmonic equation

$$\frac{\partial^4 w}{\partial x^4}+2\frac{\partial^4 w}{\partial x^2\,\partial y^2}+\frac{\partial^4 w}{\partial y^4}+q\frac{12(1-\nu^2)}{Et^3}=0. \tag{10.12}$$

If we premultiply Eqs. (10.10) (or (10.11)/(10.12)) by δw and integrate twice by parts in the manner shown in Chapter 3, p. 58, we can rederive all the virtual work (and energy) statements as a purely mathematical exercise. We leave the details of this to the reader, but mention this direct approach as in some mathematical situations the virtual work principles are not directly available.

10.3 Continuity of Requirement for Shape Function (C_1 continuity)

To ensure the continuity of both w and its normal slope across an interface we must have both w and $\partial w/\partial n$ uniquely defined by values along such an interface.

Consider Fig. 10.3 depicting the side 1–2 of a rectangular element. The normal direction n is in fact that of y and we desire w and $\partial w/\partial y$ to be uniquely determined by values of w, $\partial w/\partial x$, $\partial w/\partial y$ at the nodes lying along this line.

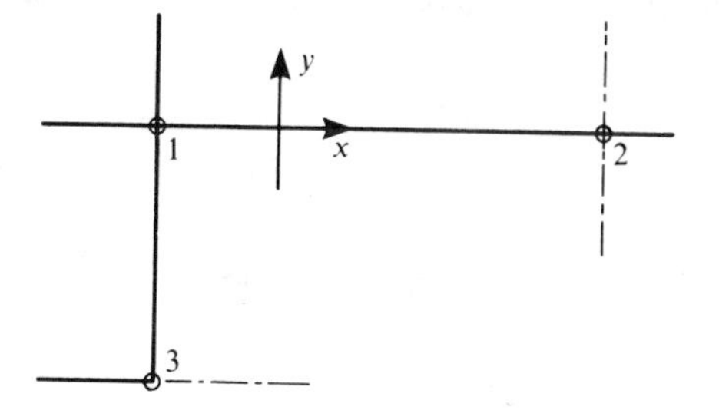

Fig. 10.3 Continuity requirement for normal slopes

Following the principles of Chapter 7 we would write along 1–2

$$w = A_1 + A_2x + A_3x^2 + \cdots \tag{10.13}$$

and

$$\frac{\partial w}{\partial y} = B_1 + B_2x + B_3x^2 + \cdots \tag{10.14}$$

with a number of constants in each expression just sufficient to determine the expressions by nodal parameters associated with the line.

Thus for instance if only two nodes are present a cubic variation of w would be permissible noting that $\partial w/\partial x$ and w are specified at each. Similarly only a linear or, two term, variation of $\partial w/\partial y$ would be permissible.

Note, however, that a similar exercise could be performed along the y direction preserving continuity of $\partial w/\partial x$ along this.

We thus have along (1–2)

$$\frac{\partial w}{\partial y} \quad \text{depending on nodal parameters of line 1–2 only}$$

and along (1–3)

$$\frac{\partial w}{\partial x} \quad \text{depending on nodal parameters of line 1–3 only.}$$

Differentiating the first with respect to x we have on line 1–2

$$\frac{\partial^2 w}{\partial x\,\partial y} \quad \text{depending on nodal parameters of line 1–2 only}$$

and on line 1–3 similarly

$$\frac{\partial^2 w}{\partial y\,\partial x} \quad \text{depending on nodal parameters of line 1–3 only.}$$

At the common point, 1, an inconsistency arises immediately as we cannot automatically have there the necessary identity for continuous functions

$$\frac{\partial^2 w}{\partial x\,\partial y} \equiv \frac{\partial^2 w}{\partial y\,\partial x}$$

for arbitrary values of parameters at nodes 2 and 3.

It is thus impossible to specify simple polynomial expressions for shape functions ensuring full compatibility when only w and its slopes are prescribed at nodes.[2]

If any functions satisfying the compatibility are found with the three nodal variables, they must be such that at corner nodes they are not continuously differentiable and the cross derivative is not unique. Some such functions are discussed in the second part of this chapter.[3–9]

The above proof has been given for a rectangular element. Clearly the arguments can be extended for any two arbitrary directions of interfaces at the corner node 1.

A way out of this difficulty appears to be obvious. We could specify the cross derivative as one of the nodal parameters. This, for an assembly of rectangular elements, is convenient and indeed permissible. Simple functions of that type have been suggested by Bogner *et al.*[10] and used with some success.

Unfortunately the extension to nodes at which a number of element interfaces meet under different angles, Fig. 10.4, is not in general permissible. Here the continuity of cross derivatives in several sets of orthogonal directions implies in fact a specification of *all second derivatives at a node*.

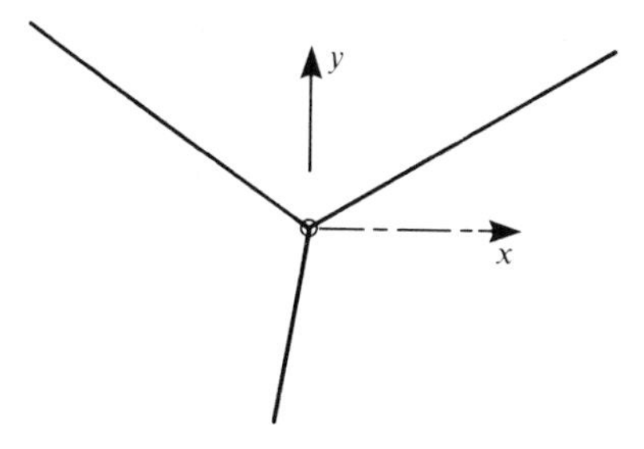

Fig. 10.4 Nodes where elements meet in arbitrary directions

This, however, violates physical requirements if the plate stiffness varies abruptly from element to element as then equality of moments normal to the interfaces cannot be maintained. However, this process has been used with some success in homogeneous plate situations.[11–18]

Indeed Smith, in reference 11 examines the effect of imposing such *excessive continuities* on several orders of higher derivatives.

The difficulties of finding compatible displacement functions here led to several attempts at ignoring the complete slope continuity while still continuing with the other necessary criteria. Proceeding perhaps, from a naïve but intuitive idea that the imposition of slope continuity at nodes only must, in the limit, lead to a complete slope continuity several very successful elements have been developed.[4,19–26]

The convergence of such elements is not obvious but can be proved either by the application of the patch test or by comparison with finite difference algorithms.[27] We have already mentioned the patch test in Chapter 2 but shall return to discuss its importance in Chapter 11. Indeed, some very successful elements have been derived by almost exclusive concentration on this, and these are amongst the most popular plate elements used.

The simplicity and practical use of such elements justifies their special treatment in the following section.

NON-CONFORMING SHAPE FUNCTIONS

10.4 Rectangular Element with Corner Nodes[19, 28–31]

10.4.1 *Shape functions.* Consider a rectangular element of a plate *ijkl* coinciding with the x, y plane as shown on Fig. 10.2. At each node displacements $\mathbf{a}_n$ are introduced. These have three components: the first a displacement in the z direction, w_n, the second a rotation about the x axis, $(\theta_x)_n$, the third a rotation about the y axis $(\theta_y)_n$.

The nodal displacements are defined by Eq. (10.4) while the element displacement will, as usual, be given by the listing of the nodal displacements, now totalling four,

$$\mathbf{a}^e = \begin{Bmatrix} \mathbf{a}_i \\ \mathbf{a}_j \\ \mathbf{a}_l \\ \mathbf{a}_k \end{Bmatrix}. \tag{10.15}$$

A polynomial expression is conveniently used to define the shape functions in terms of the twelve parameters. Certain terms must be omitted from a complete fourth order polynomial. Writing

$$\begin{aligned} w = {} & \alpha_1 + \alpha_2 x + \alpha_3 y + \alpha_4 x^2 + \alpha_5 xy + \alpha_6 y^2 + \alpha_7 x^3 + \alpha_8 x^2 y \\ & + \alpha_9 xy^2 + \alpha_{10} y^3 + \alpha_{11} x^3 y + \alpha_{12} xy^3 \end{aligned} \tag{10.16}$$

has certain advantages. In particular, along any x = constant or y = constant line, the displacement w will vary as a cubic. The element boundaries or interfaces are composed of such lines. As a cubic is uniquely defined by four constants, the two end values of slopes and displacements at the ends of the boundaries will therefore define the displacements along this boundary uniquely. As such end values are common to adjacent elements, continuity of w will be imposed all along any interface.

It will be observed that the gradient of w normal to any of the boundaries also varies along it in a cubic way. (Consider, for instance, $\partial w/\partial x$ along a line on which x is constant.) As on such lines only two values of the normal slope are defined, the cubic is not specified uniquely and, in general, a discontinuity of normal slope will occur. The function is thus 'non-conforming'.

The constants α_1 to α_{12} can be evaluated by writing down the twelve simultaneous equations linking the values of w and its slopes at the nodes when the co-ordinates take up their appropriate values. For instance

$$
\begin{aligned}
w_i &= \alpha_1 + \alpha_2 x_i + \alpha_3 y_i + \text{ etc.} \\
\left(-\frac{\partial w}{\partial y}\right)_i &= \theta_{xi} = \qquad -\alpha_3 \quad + \text{ etc.} \\
\left(\frac{\partial w}{\partial x}\right)_i &= \theta_{yi} = \quad \alpha_2 \qquad + \text{ etc.} \\
&\ldots\ldots\ldots\ldots
\end{aligned}
$$

Listing all twelve equations we can write, in matrix form,

$$\mathbf{a}^e = \mathbf{C}\boldsymbol{\alpha} \tag{10.17}$$

where $\mathbf{C}$ is a twelve by twelve matrix depending on nodal co-ordinates and $\boldsymbol{\alpha}$ a vector of the twelve unknown constants. Inverting we have

$$\boldsymbol{\alpha} = \mathbf{C}^{-1}\mathbf{a}^e. \tag{10.18}$$

This inversion can be carried out by the computer or, if an explicit expression for the stiffnesses, etc., is desired, can be performed algebraically. This was in fact done by Zienkiewicz and Cheung.[19]

It is now possible to write the expression for the displacement within the element in a standard form as

$$\mathbf{u} \equiv w = \mathbf{N}\mathbf{a}^e = \mathbf{P}\mathbf{C}^{-1}\mathbf{a}^e \tag{10.19}$$

where

$$\mathbf{P} = (1, x, y, x^2, xy, y^2, x^3, x^2y, xy^2, y^3, x^3y, xy^3)$$

An explicit form of the above expression was derived by Melosh.[28]

The shape functions can be written simply in terms of normalized co-ordinates of Chapter 7. Thus we can write for any node

$$
\begin{aligned}
\mathbf{N}_i = \tfrac{1}{8}[&(\xi_0+1)(\eta_0+1)(2+\xi_0+\eta_0-\xi^2-\eta^2), \\
&a\xi_i(\xi_0+1)^2(\xi_0-1)(\eta_0+1), \\
&b\eta_i(\xi_0+1)(\eta_0+1)^2(\eta_0-1)]
\end{aligned} \tag{10.20}
$$

with

$$
\begin{aligned}
\xi &= (x-x_c)/a, & \eta &= (y-y_c)/b \\
\xi_0 &= \xi\cdot\xi_i, & \eta_0 &= \eta\cdot\eta_i.
\end{aligned}
$$

The form of **B** is obtained directly from Eq. (10.16) or from Eq. (10.20) using Eq. (10.6). We thus have

$$\boldsymbol{\varepsilon} = \begin{Bmatrix} -2\alpha_4 & -6\alpha_7 x & -2\alpha_8 y & -6\alpha_{11}xy & \\ -2\alpha_6 & -2\alpha_9 x & -6\alpha_{10}y & -6\alpha_{12}xy & \\ 2\alpha_5 & +4\alpha_8 x & +4\alpha_9 y & +6\alpha_{11}x^2 & +6\alpha_{12}y^2 \end{Bmatrix}.$$

We can write

$$\boldsymbol{\varepsilon} = \mathbf{Q}\boldsymbol{\alpha} = \mathbf{Q}\mathbf{C}^{-1}\mathbf{a}^e \quad \text{and thus} \quad \mathbf{B} = \mathbf{Q}\mathbf{C}^{-1} \tag{10.21}$$

in which

$$\mathbf{Q} = \begin{bmatrix} 0 & 0 & 0 & -2 & 0 & 0 & -6x & -2y & 0 & 0 & -6xy & 0 \\ 0 & 0 & 0 & 0 & 0 & -2 & 0 & 0 & -2x & -6y & 0 & -6xy \\ 0 & 0 & 0 & 0 & 2 & 0 & 0 & 4x & 4y & 0 & 6x^2 & 6y^2 \end{bmatrix}. \tag{10.22}$$

It is of interest to remark now that the displacement function chosen does in fact permit a state of constant strain (curvature) to exist† and therefore satisfies one of the criteria of convergence stated in Chapter 2.

10.4.2 *Stiffness and load matrices.* Standard procedure can now be followed, and it is almost superfluous to recount the details.

By Eq. (2.13a) the stiffness matrix relating the nodal *forces* (given by a lateral force and two moments at each node) to the corresponding nodal displacements is

$$\mathbf{K}^e = \int\int_{V^e} \mathbf{B}^{\mathrm{T}}\mathbf{D}\mathbf{B}\,\mathrm{d}x\,\mathrm{d}y \tag{10.23}$$

or substituting Eq. (10.21) and taking t as constant within the element,

$$\mathbf{K}^e = \mathbf{C}^{-1\mathrm{T}}\left(\int\int \mathbf{Q}^{\mathrm{T}}\mathbf{D}\mathbf{Q}\,\mathrm{d}x\,\mathrm{d}y\right)\mathbf{C}^{-1}. \tag{10.24}$$

The terms not containing x and y have now been removed from the operation of integrating. The term within the integration sign can be multiplied out and integrated explicitly without difficulty, if t is constant.

An explicit expression for the stiffness matrix **K** has been evaluated for the case of an orthotropic material and the result is given in Table 10.1.

† If α_7 to α_{12} are zero, then the 'strain' is constant. By Eq. (10.16), corresponding $\mathbf{a}^e$ can be found. As there is a unique correspondence between $\mathbf{a}^e$ and α such a state is therefore unique. All this presumes that $\mathbf{C}^{-1}$ does in fact exist. The algebraic inversion shows that the matrix **C** is never singular.

TABLE 10.1
STIFFNESS MATRIX FOR A RECTANGULAR ELEMENT (FIG. 10.3: ORTHOTROPIC MATERIAL)

Stiffness matrix

$$\mathbf{K} = \frac{1}{60ab}\,\mathbf{L}\{D_x\mathbf{K}_1 + D_y\mathbf{K}_2 + D_1\mathbf{K}_3 + D_{xy}\mathbf{K}_4\}\mathbf{L}$$

with

$$\begin{Bmatrix}\mathbf{f}_i\\ \mathbf{f}_j\\ \mathbf{f}_k\\ \mathbf{f}_l\end{Bmatrix} = \mathbf{K}\begin{Bmatrix}\mathbf{a}_i\\ \mathbf{a}_j\\ \mathbf{a}_k\\ \mathbf{a}_l\end{Bmatrix}$$

$$\mathbf{K}_1 = p^{-2}\begin{bmatrix}
60 & & & & & & & & & & & \\
0 & 0 & & & & & & & & & & \\
30 & 0 & 20 & & & & & & & & & \\
30 & 0 & 15 & 60 & & & & & & & & \\
0 & 0 & 0 & 0 & 0 & & & & & & & \\
15 & 0 & 10 & 30 & 0 & 20 & & & & & & \\
-60 & 0 & -30 & -30 & 0 & -15 & 60 & & & & & \\
0 & 0 & 0 & 0 & 0 & 0 & 0 & 0 & & & & \\
30 & 0 & 10 & 15 & 0 & 5 & -30 & 0 & 20 & & & \\
-30 & 0 & -15 & -60 & 0 & -30 & 30 & 0 & -15 & 60 & & \\
0 & 0 & 0 & 0 & 0 & 0 & 0 & 0 & 0 & 0 & 0 & \\
15 & 0 & 5 & 30 & 0 & 10 & -15 & 0 & 10 & -30 & 0 & 20
\end{bmatrix}\qquad p^{-2} = \frac{b^2}{a^2}\quad \text{Symmetrical}$$

$$\mathbf{K}_2 = p^{2}\begin{bmatrix}
60 & & & & & & & & & & & \\
-30 & 20 & & & & & & & & & & \\
0 & 0 & 0 & & & & & & & & & \\
-60 & 30 & 0 & 60 & & & & & & & & \\
-30 & 10 & 0 & 30 & 20 & & & & & & & \\
0 & 0 & 0 & 0 & 0 & 0 & & & & & & \\
30 & -15 & 0 & -30 & -15 & 0 & 60 & & & & & \\
-15 & 10 & 0 & 15 & 5 & 0 & -30 & 20 & & & & \\
0 & 0 & 0 & 0 & 0 & 0 & 0 & 0 & 0 & & & \\
-30 & 15 & 0 & 30 & 15 & 0 & -60 & 30 & 0 & 60 & & \\
-15 & 5 & 0 & 15 & 10 & 0 & -30 & 10 & 0 & 30 & 20 & \\
0 & 0 & 0 & 0 & 0 & 0 & 0 & 0 & 0 & 0 & 0 & 0
\end{bmatrix}\qquad p^{2} = \frac{a^2}{b^2}\quad \text{Symmetrical}$$

$$\mathbf{K}_3 = \begin{bmatrix}
30 & & & & & & & & & & & \\
-15 & 0 & & & & & & & & & & \\
15 & -15 & 0 & & & & & & & & & \\
-30 & 0 & -15 & 30 & & & & & & & & \\
0 & 0 & 0 & 15 & 0 & & & & & & & \\
-15 & 0 & 0 & 15 & 15 & 0 & & & & & & \\
-30 & 15 & 0 & 30 & 0 & 0 & 30 & & & & & \\
15 & 0 & 0 & 0 & 0 & 0 & -15 & 0 & & & & \\
0 & 0 & 0 & 0 & 0 & 0 & -15 & 15 & 0 & & & \\
30 & 0 & 0 & -30 & -15 & 0 & -30 & 0 & 15 & 30 & & \\
0 & 0 & 0 & -15 & 0 & 0 & 0 & 0 & 0 & 15 & 0 & \\
0 & 0 & 0 & 0 & 0 & 0 & 15 & 0 & 0 & -15 & -15 & 0
\end{bmatrix}\qquad \text{Symmetrical}$$

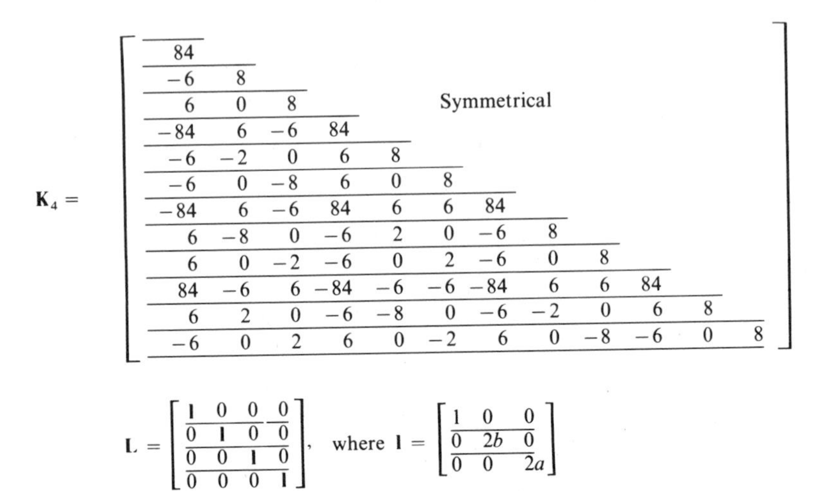

$$\mathbf{K}_4 = \begin{bmatrix} 84 & & & & & & & & & & & \\ -6 & 8 & & & & & & \text{Symmetrical} & & & & \\ 6 & 0 & 8 & & & & & & & & & \\ -84 & 6 & -6 & 84 & & & & & & & & \\ -6 & -2 & 0 & 6 & 8 & & & & & & & \\ -6 & 0 & -8 & 6 & 0 & 8 & & & & & & \\ -84 & 6 & -6 & 84 & 6 & 6 & 84 & & & & & \\ 6 & -8 & 0 & -6 & 2 & 0 & -6 & 8 & & & & \\ 6 & 0 & -2 & -6 & 0 & 2 & -6 & 0 & 8 & & & \\ 84 & -6 & 6 & -84 & -6 & -6 & -84 & 6 & 6 & 84 & & \\ 6 & 2 & 0 & -6 & -8 & 0 & -6 & -2 & 0 & 6 & 8 & \\ -6 & 0 & 2 & 6 & 0 & -2 & 6 & 0 & -8 & -6 & 0 & 8 \end{bmatrix}$$

$$\mathbf{L} = \begin{bmatrix} \mathbf{l} & 0 & 0 & 0 \\ 0 & \mathbf{l} & 0 & 0 \\ 0 & 0 & \mathbf{l} & 0 \\ 0 & 0 & 0 & \mathbf{l} \end{bmatrix}, \quad \text{where } \mathbf{l} = \begin{bmatrix} 1 & 0 & 0 \\ 0 & 2b & 0 \\ 0 & 0 & 2a \end{bmatrix}$$

The corresponding stress matrix for the internal moments of all the nodes is given in Table 10.2.

The external forces at nodes due to distributed loading can be assigned 'by inspection', allocating specific areas as contributing to any node. However, it is more logical and accurate to use once again the standard expression Eq. (2.9) for such an allocation.

If a distributed loading q is acting per unit area of an element in direction of w then, by Eq. (2.11), the contribution of these forces to each of the nodes is

$$\mathbf{f}_i = -\int\int \mathbf{N}^{\mathrm{T}} q \, \mathrm{d}x \, \mathrm{d}y \tag{10.25}$$

or by Eq. (10.19)

$$\mathbf{f}_i = -\mathbf{C}^{-1\mathrm{T}} \int\int \mathbf{P}^{\mathrm{T}} q \, \mathrm{d}x \, \mathrm{d}y. \tag{10.26}$$

The integral is again evaluated simply. It will now be noted that, in general, all three components of external force at any node will have non-zero values. This is a result which the simple allocation of external loads would have missed. Table 10.3 shows the nodal load vector for a uniform loading q.

If initial strains are introduced into the plate the vector of nodal forces due to such initial strains and the initial stresses can be found in a similar way. It is necessary to remark in this connection that initial strain, such as may be due to a temperature rise, is seldom confined in its effects

TABLE 10.2

STRESS MATRIX $\left(p = \dfrac{a}{b}\right)$

RECTANGULAR ELEMENT OF FIG. 10.2 ORTHOTROPIC MATERIAL

$$
\begin{Bmatrix} \mathbf{M}_i \\ \mathbf{M}_j \\ \mathbf{M}_k \\ \mathbf{M}_l \end{Bmatrix} = \frac{1}{4ab}
\left[\begin{array}{cccccccccccc}
\begin{matrix}6p^{-1}D_x\\+6pD_1\end{matrix} & -8aD_1 & 8bD_x & -6pD_1 & -4aD_1 & 0 & -6p^{-1}D_x & 0 & 4bD_x & 0 & 0 & 0 \\
\begin{matrix}6pD_y\\+6p^{-1}D_1\end{matrix} & -8aD_y & 8bD_1 & -6pD_y & -4aD_y & 0 & -6p^{-1}D_1 & 0 & 4bD_1 & 0 & 0 & 0 \\
-2D_{xy} & 4bD_{xy} & -4aD_{xy} & 2D_{xy} & 0 & 4aD_{xy} & 2D_{xy} & -4bD_{xy} & 0 & -2D_{xy} & 0 & 0 \\
-6pD_1 & 4aD_1 & 0 & \begin{matrix}6p^{-1}D_x\\+6pD_1\end{matrix} & 8aD_1 & 8bD_x & 0 & 0 & 0 & -6p^{-1}D_x & 0 & 4bD_x \\
-6pD_y & 4aD_y & 0 & \begin{matrix}6pD_y\\+6p^{-1}D_1\end{matrix} & 8aD_y & 8bD_1 & 0 & 0 & 0 & -6p^{-1}D_1 & 0 & 4bD_1 \\
-2D_{xy} & 0 & -4aD_{xy} & 2D_{xy} & 4bD_{xy} & 4aD_{xy} & 2D_{xy} & 0 & 0 & -2D_{xy} & -4bD_{xy} & 0 \\
-6p^{-1}D_x & 0 & -4bD_x & 0 & 0 & 0 & \begin{matrix}6p^{-1}D_x\\+6pD_1\end{matrix} & -8aD_1 & -8bD_x & -6pD_1 & -4aD_1 & 0 \\
-6p^{-1}D_1 & 0 & -4bD_1 & 0 & 0 & 0 & \begin{matrix}6pD_y\\+6p^{-1}D_1\end{matrix} & -8aD_y & -8bD_1 & -6pD_y & -4aD_y & 0 \\
-2D_{xy} & 4bD_{xy} & 0 & 2D_{xy} & 0 & 0 & 2D_{xy} & -4bD_{xy} & -4aD_{xy} & -2D_y & 0 & 4aD_{xy} \\
0 & 0 & 0 & -6p^{-1}D_x & 0 & -4bD_x & -6pD_1 & 4aD_1 & 0 & \begin{matrix}6p^{-1}\mathrm{D}_x\\+6pD_1\end{matrix} & 8aD_1 & -8bD_x \\
0 & 0 & 0 & -6p^{-1}D_1 & 0 & -4bD_1 & -6pD_y & 4aD_y & 0 & \begin{matrix}6pD_y\\+6p^{-1}D_1\end{matrix} & 8aD_y & -8bD_1 \\
-2D_{xy} & 0 & 0 & 2D_{xy} & 4bD_{xy} & 0 & 2D_{xy} & 0 & -4aD_{xy} & -2D_{xy} & -4bD_{xy} & 4aD_{xy}
\end{array}\right]
\begin{Bmatrix} \mathbf{a}_i \\ \mathbf{a}_j \\ \mathbf{a}_k \\ \mathbf{a}_l \end{Bmatrix}
$$

TABLE 10.3

LOAD MATRIX FOR A RECTANGULAR ELEMENT OF FIG. 10.3 UNDER UNIFORM LOAD q

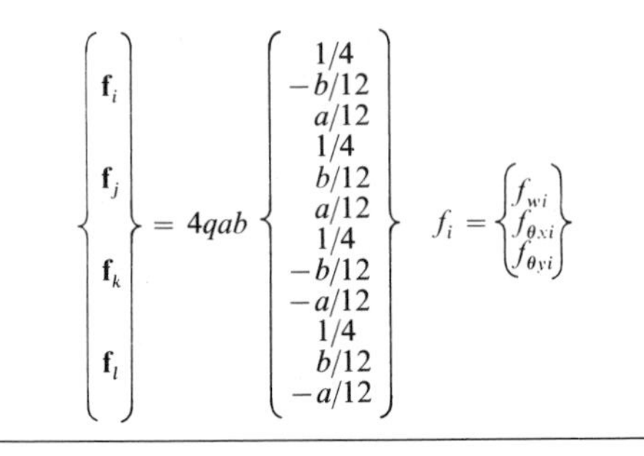

$$\begin{Bmatrix} \mathbf{f}_i \\ \mathbf{f}_j \\ \mathbf{f}_k \\ \mathbf{f}_l \end{Bmatrix} = 4qab \begin{Bmatrix} 1/4 \\ -b/12 \\ a/12 \\ 1/4 \\ b/12 \\ a/12 \\ 1/4 \\ -b/12 \\ -a/12 \\ 1/4 \\ b/12 \\ -a/12 \end{Bmatrix} \qquad f_i = \begin{Bmatrix} f_{wi} \\ f_{\theta xi} \\ f_{\theta yi} \end{Bmatrix}$$

on curvatures. Usually, direct strains in the plate are introduced additionally, and the complete problem can be solved only by consideration of the plane stress problem as well as that of bending.

10.5 **Quadrilateral and Parallelogram Elements**

The rectangular element cannot be easily generalized into quadrilateral shape. Transformation of co-ordinates of the type described in Chapter 8 can be performed but unfortunately now it will be found that the constant curvature criterion is violated. As expected such elements behave badly but by arguments of Chapter 8 (page 194) convergence may still occur providing the patch test is passed in the curvilinear co-ordinates. Henshell *et al.*[32] study the performance of such an element (and also some of a higher order) and conclude that reasonable accuracy is attainable. Their paper gives all the details of transformations required for an isoparametric mapping and the resulting need for numerical integration.

Only for the case of a parallelogram is it possible to achieve states of constant curvature exclusively using functions of ξ and η.

Such an element is suggested in the discussion to reference 19 and the stiffness matrices have been worked out by Dawe.[21]

A somewhat different set of shape functions was suggested by Argyris.[22]

For a parallelogram the local co-ordinates can be related to the global ones by an explicit expression (Fig. 10.5)

$$\begin{aligned} \xi &= (x - y \operatorname{cotan} \alpha)/a \\ \eta &= y \operatorname{cosec} \alpha / b \end{aligned} \tag{10.27}$$

and all expressions can therefore be also derived directly.

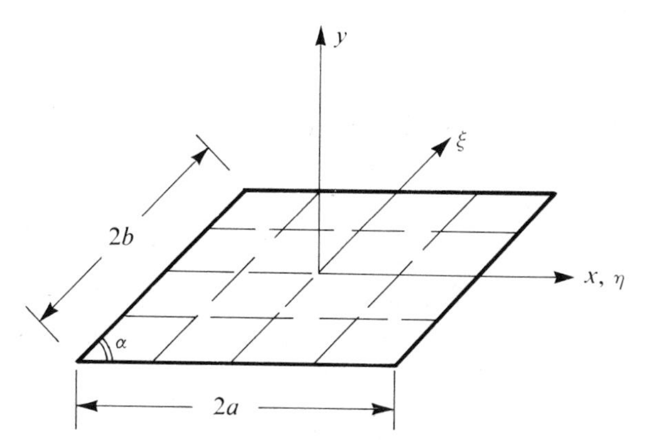

Fig. 10.5 Parallelogram element and skew co-ordinates

10.6 Triangular Element with Corner Nodes

10.6.1 *Shape functions.* At first sight, it would seem that once again a simple polynomial expansion could be used in a manner identical to that of the previous section. As only nine independent movements are imposed, only nine terms of the expansion are permissible. Here, an immediate difficulty arises as the full cubic expansion contains ten terms (Eq. (10.16)) and any omission has to be made rather arbitrarily. To retain a certain symmetry of appearance all ten terms could be retained and two coefficients made equal (e.g., $\alpha_8 = \alpha_9$) to limit the number of unknowns to nine. Several such possibilities have been investigated but a further, much more serious, problem arises. The matrix corresponding to **C** of Eq. (10.17) becomes singular for certain orientations of the triangle sides. This happens, for instance, when two sides of the triangle are parallel to the x and y axes.

An 'obvious' alternative is to add a central node to the formulation and eliminate this by static condensation (see Chapter 7). This would allow a complete cubic to be used but again it was found that an element derived on this basis does not converge.

Difficulties of asymmetry can be avoided by the use of area co-ordinates described in Chapter 8. These are indeed nearly always a natural choice for triangles.

As before we shall use polynomial expansion terms, and it is worth remarking that these are given in area co-ordinates in an unusual form. For instance

$$\alpha_1 L_1 + \alpha_2 L_2 + \alpha_3 L_3$$

gives the three terms of a *complete* linear polynomial and

$$\alpha_1 L_1 L_2 + \alpha_2 L_2 L_3 + \alpha_3 L_3 L_1 + \alpha_4 L_1^2 + \alpha_5 L_2^2 + \alpha_6 L^2$$

gives all the six terms of a quadratic (containing within it the linear terms). The ten terms of a cubic expression are similarly formed by the products of all possible cubic combinations, i.e.,

$$L_1^3,\ L_2^3,\ L_3^3,\ L_1^2 L_2,\ L_2^2 L_3,\ L_3^2 L_1,\ L_1 L_2^2,\ L_2 L_3^2,\ L_3 L_1^2,\ L_1 L_2 L_3.$$

For a nine degree of freedom element any of the above terms can be used in a suitable combination, remembering, however, that only nine independent functions are needed and that constant curvature states have to be obtained. Figure 10.6 shows some functions which are of importance. The first (Fig. 10.6(*a*)) gives one of three functions representing a simple, unstrained, translation of plate. Obviously these modes must be available.

Further, functions of the type $L_1^2 L_2$ of which there are six in the cubic expression will be found to take up a form similar (though not identical) to Fig. 10.6(*b*).

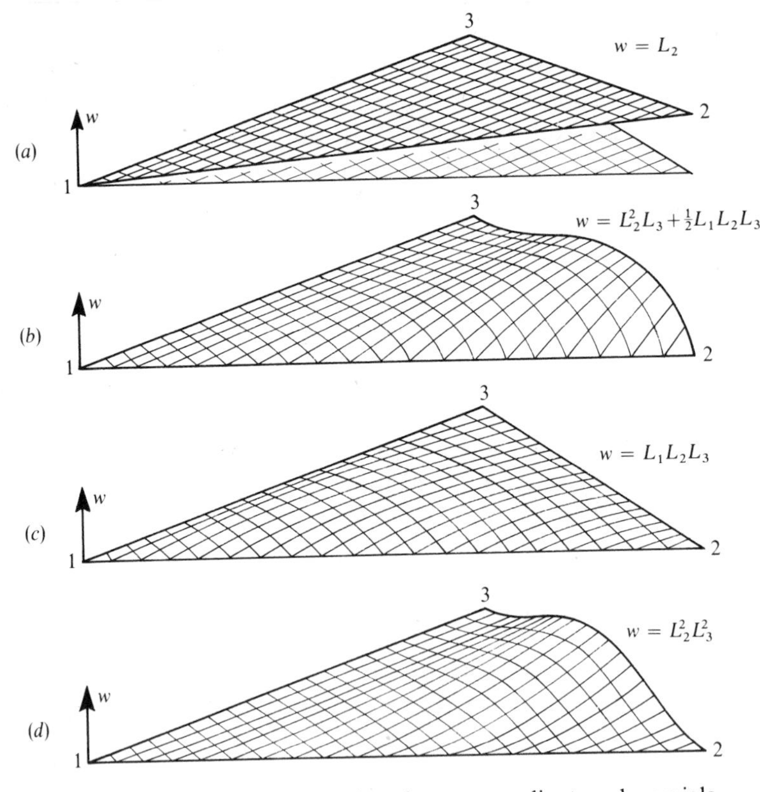

Fig. 10.6 Some basic functions in area coordinate polynomials

Last, a function $L_1L_2L_3$ is shown in Fig. 10.6(*c*) illustrating that this is a purely internal mode with zero values and slopes at all the three corners. This function could thus be useful for a nodeless or internal variable but will not, in isolation, be used as it cannot be prescribed in terms of corner variables. It can, however, be added to any other basic shape in any corner proportion.

The functions of the second kind are therefore of essential interest. They have zero values of w at all corners and indeed always have a zero slope in the direction of one side. A linear combination of two of these (e.g., $L_2^2L_3$ and $L_2^2L_1$) will be capable of providing any desired slopes in the x and y directions at one node while maintaining all other slopes at zero.

We shall, however, consider modes of the type

$$L_2^2L_3+cL_1L_2L_3$$

for an added generality (as the last term does not have a slope contribution at the nodes).

As these modes are the only ones contributing to curvatures it is important to ensure that the general arbitrary state of curvature with zero values of w at nodes is included in a linear combination of six of these functions. In algebraic terms this means that the expression

$$A_1L_1L_2+A_2L_2L_3+\cdots+A_6L_3^2$$

with any set of values of the coefficients A must be achievable by an appropriate combination of

$$B_1(L_2^2L_1+cL_1L_2L_3)+B_2(L_1^2L_2+cL_1L_2L_3)+\cdots$$

with some set of the six constants B. It is possible to show after some algebraic manipulation that this can only be achieved for a value of $c=\frac{1}{2}$. Hence the mode plotted in Fig. 10.6(*b*) is one of the basic ones needed for the generation of shape functions.

We can now describe the displacement of the plate in the form

$$w=\beta_1L_1+\beta_2L_2+\beta_3L_3+\beta_4(L_2^2L_1+\tfrac{1}{2}L_1L_2L_3)+\ldots+\beta_9(L_1^2L_3+\tfrac{1}{2}L_1L_2L_3) \qquad (10.28)$$

and substituting the nodal values of

$$w_i,\ \theta_{xi}=-\left(\frac{\partial w}{\partial y}\right)_i \quad\text{and}\quad \theta_{yi}=\left(\frac{\partial w}{\partial x}\right)_i$$

the constants and hence the shape functions can be determined.

Explicitly the first result can be written in the form given below for a typical shape function using the definitions of Chapter 4, where

$$b_1 = y_2 - y_3 \qquad c_1 = x_3 - x_2, \text{ etc.}$$

$$\mathbf{N}_1^{\mathrm{T}} = \left\{ \begin{array}{c} L_1 + L_1^2 L_2 + L_1^2 L_3 - L_1 L_2^2 - L_1 L_3^2 \\ b_3(L_1^2 L_2 + \tfrac{1}{2} L_1 L_2 L_3) - b_2(L_3 L_1^2 + \tfrac{1}{2} L_1 L_2 L_3) \\ c_3(L_1^2 L_2 + \tfrac{1}{2} L_1 L_2 L_3) - c_2(L_3 L_1^2 + \tfrac{1}{2} L_1 L_2 L_3) \end{array} \right\}. \qquad (10.29)$$

The other two functions for nodes 2 or 3 are written by a cyclic permutation of suffixes $1 \overset{\rightarrow}{\underset{\leftarrow}{-2-}} 3$. The element specified by the above function was first presented in reference 4.

10.6.2 *Stiffness and load matrices.* With the definition of strains of Eq. (10.2) and the general $\mathbf{B}_i$ matrix of Eq. (10.6) we see that second derivatives of $\mathbf{N}$ are necessary.

The only new feature is presented by the fact that differentiation with respect to Cartesian co-ordinates needs to be carried out. This is quite easy noting that

$$\frac{\partial}{\partial x} = \frac{\partial L_1}{\partial x}\frac{\partial}{\partial L_1} + \frac{\partial L_2}{\partial x}\frac{\partial}{\partial L_2} + \frac{\partial L_3}{\partial x}\frac{\partial}{\partial L_3} =$$

$$= \frac{1}{2\Delta}\left(b_1 \frac{\partial}{\partial L_1} + b_2 \frac{\partial}{\partial L_2} + b_3 \frac{\partial}{\partial L_3}\right), \text{ etc.} \qquad (10.30)$$

All expressions remain polynomial in the area co-ordinates and can be simply integrated using the general expression, Eq. (7.31) of Chapter 7. The explicit final form for stiffness and load matrices is somewhat lengthy and the interested reader will find these in reference 30.

However, it is simpler to program using numerical integration as outlined in Chapter 8. As the stiffness matrix involves only quadratic terms a triangle integrating expression using three points only is exact (*vide* Table 8.2 of Chapter 8) and the actual computer times used in such numerical integration are indistinguishable from those involving explicit expressions.

The 'stress' matrix gives moments which vary linearly. However, as the full cubic terms are not allowed in the expansion this leads to a poor approximation and it is usual to evaluate the moments only at centroids and indeed to use nodal averages for further smoothing.

10.7 Convergence of Non-conforming Elements

The two types of elements outlined in the preceding section violate the continuity of slope conditions and therefore only approximate to the principle of minimization of total potential energy. In the next section results will be shown, however, which demonstrate their practical accuracy. The reader may ask whether convergence will always in fact occur to the 'exact' answer with decreasing subdivision. Although this question is somewhat academic it needs an answer.

With regard to rectangular elements Walz *et al.*[27] investigated the algorithm obtained by a set of such elements and, by comparing this with expansions of the governing differential equations for the case of homogeneous plates found that such convergence is then guaranteed. It is not reasonable to extend these conclusions beyond this proven case.

The simple triangle again has been shown by Irons[4] to give an exactly convergent solution when the mesh is generated by three sets of equally spaced parallel lines.

The patch test applied here was simple. If an array of a large number of assembled elements can reproduce an *exact* response to all constant curvature states applied, then, in the limit of subdivision, the plate behaves exactly according to the physical rules applied to an infinitesimal material element. Conversely if such response is *not* available convergence cannot occur.

Indeed, the same test applied to a mesh of type shown in Fig. 10.7 (4 × 4 *B*), in which the triangles are obtained by drawing two diagonals of

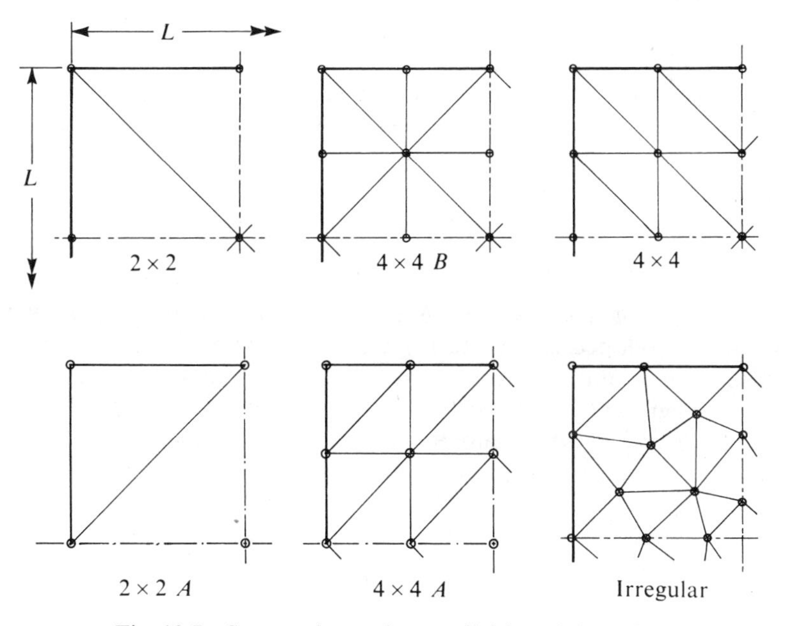

Fig. 10.7 Square plate—element divisions (triangular)

a parallelogram, was shown to give some 1·5 per cent error in displacements. Thus here the non-conforming triangle will not converge to the exact solution but to one in which the errors are of that order.

A similar test was also applied to the non-conforming rectangular element in reference 4 and gave the first convergence proof for this element.

For practical engineering purposes in most cases the accuracy obtained by the non-conforming triangle is adequate. Indeed it gives, at most practical subdivisions, results superior to those attainable with equivalent conforming triangles.[4] This may well be due to the fact that the solution now does not follow the energy bounds given in Chapter 2, and has a greater freedom to take up the best shape.

In the derivation of non-conforming elements given here, it was argued that displacement w should at least be continuous and that the slope condition should be satisfied at all points of interconnection. This always resulted in at least a cubic variation of w. If one relaxes some of these requirements, further interesting possibilities develop. For instance, if a triangle with six nodes is taken and degrees of freedom are six prescribed as w at corners and $\partial w/\partial n$ normal slope at midsides, we find that a *complete quadratic* expansion can be determined. This will result in constant moments and curvatures throughout and would produce the *simplest bending element possible* equivalent to the constant strain triangle.

Such an element has been derived by Morley[23] who shows that despite the apparently rather serious discontinuities the element is a convergent one and gives an approximation comparable with that attainable with the more complex triangle discussed here.

The problem of stiffness derivation for this element is left to the reader as an exercise. Higher order nonconforming elements can be developed and one such element proposed by Narayanaswami has proved to be successful.[24]

10.8 Examples of Solution

10.8.1 *Rectangular elements.* A program based on the displacement functions developed in Eq. (10.16) has been prepared and a few simple test problems computed to illustrate the accuracy and the rate of convergence that can be expected.

Square isotropic plate. Figure 10.8 shows graphically the results obtained by loading, with uniform load, a square plate with clamped edges. Only the results of 2 by 2, 4 by 4, and 6 by 6 division into elements are given, but the accuracy and general convergence are convincing.

The linear distribution of moments tries, as it were, to give the 'best fit' to the exact moment distributions at all stages of the subdivision.

The convergence and accuracy are even more strikingly demonstrated

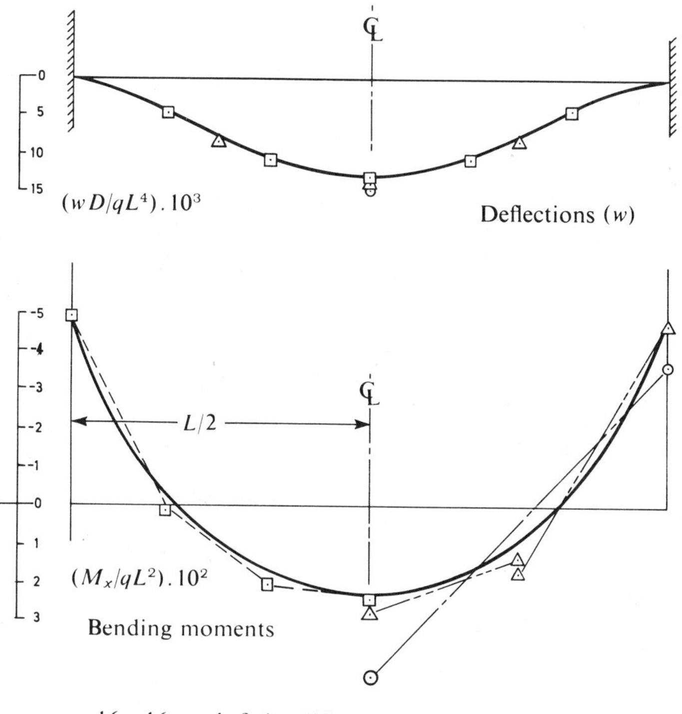

Fig. 10.8 A square plate with clamped edges. Uniform load q. Square elements

in Table 10.4. In this the central deflections are compared for concentrated and distributed loadings with various edge conditions. With an 8 by 8 division into elements the largest error is of the order of 3 per cent. In all cases convergence appears to occur for all subdivisions.

Corner supported plate.[19] A square plate supported by corner columns has been subjected to various experimental and approximate analytical solutions. In Table 10.5 results of a finite element analysis are compared with some other approximate solutions.[33,34] Even in this case, where the corner concentration would tend to cause difficulties, reasonable agreement of both displacements and stresses is apparent.

10.8.2 *Triangular elements—nine-degree-of-freedom elements—square isotropic plate.* Once again a square plate is chosen to illustrate the convergence. This is now divided into different combinations of triangular elements. Some of these are based on a square network, some are completely irregular. Figure 10.7 shows the various types of division into

TABLE 10.4

COMPUTED CENTRAL DEFLECTION OF A SQUARE PLATE FOR SEVERAL MESHES (RECTANGULAR ELEMENTS)

Mesh	Total No. of nodes	Simply Supported Plate		Clamped Plate	
		α (uniform load)	β (concentrated load)	α (uniform load)	β (concentrated (load)
(2 × 2)	9	0·003446	0·013784	0·001480	0·005919
(4 × 4)	25	0·003939	0·012327	0·001403	0·006134
(8 × 8)	81	0·004033	0·011829	0·001304	0·005803
(12 × 12)	169	0·004050	0·011715	0·001283	0·005710
(16 × 16)	289	0·004056	0·011671	0·001275	0·005672
Exact (Timoshenko)		0·004062	0·01160	0·00126	0·00560

$w_{max} = \alpha q L^4/D$ for a uniformly distributed load q;
$w_{max} = \beta P L^2/D$ for a central concentrated load P.
(Based on Tocher, J. L., and Kapur, K. K.).[31]
(Subdivision of whole plate given above)

TABLE 10.5

CORNER SUPPORTED SQUARE PLATE

	Point 1		Point 2	
	w	M_x	w	M_x
Finite element 2 × 2	0·0126	0·139	0·0176	0·095
4 × 4	0·0165	0·149	0·0232	0·108
6 × 6	0·0173	0·150	0·0244	0·109
Marcus[33]	0·0180	0·154	0·0281	0·110
Ballesteros and Lee[34,35]	0·0170	0·140	0·0265	0·109
Multiplier	qL^4/D	qL^2	qL^4/D	qL^2

Point 1, centre of side: point 2, centre of plate

elements while Fig. 10.9 illustrates the displacements obtained for various edge and loading conditions. Again, the accuracy and convergence of displacements are good (though perhaps not quite as good as with rectangular elements).

Figure 10.10 shows the variation of bending moments on typical centre line sections. If the mean values are used then these moments compare well with exact values. No longer can we say, however, that the linear variation of stresses follow the 'best fit' to the actual stress distribution.

For practical problems it is therefore recommended that attention should be focused at stresses (moments) at centroids of the elements.

10.8.3 *Some practical applications.* The range of practical application of the solution program, particularly the one based on triangular elements,

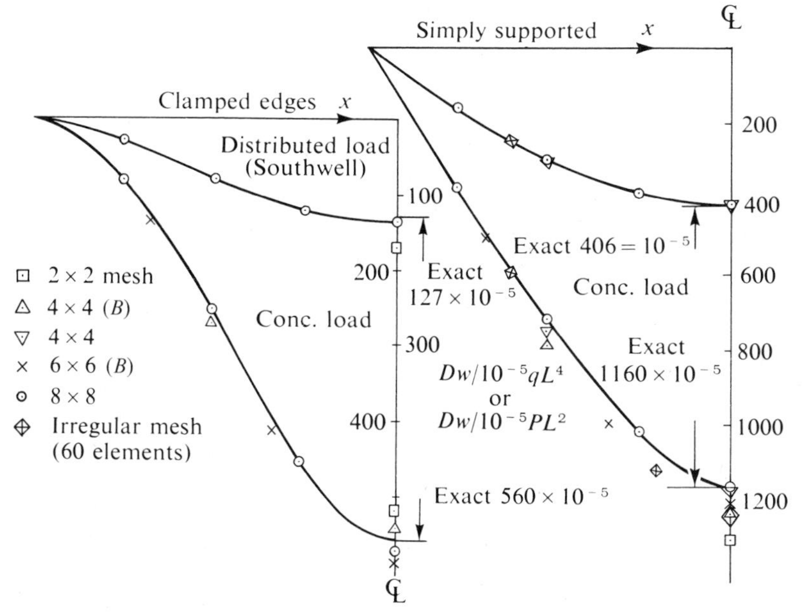

Fig. 10.9 Square plate deflections on centre-line (triangular elements)

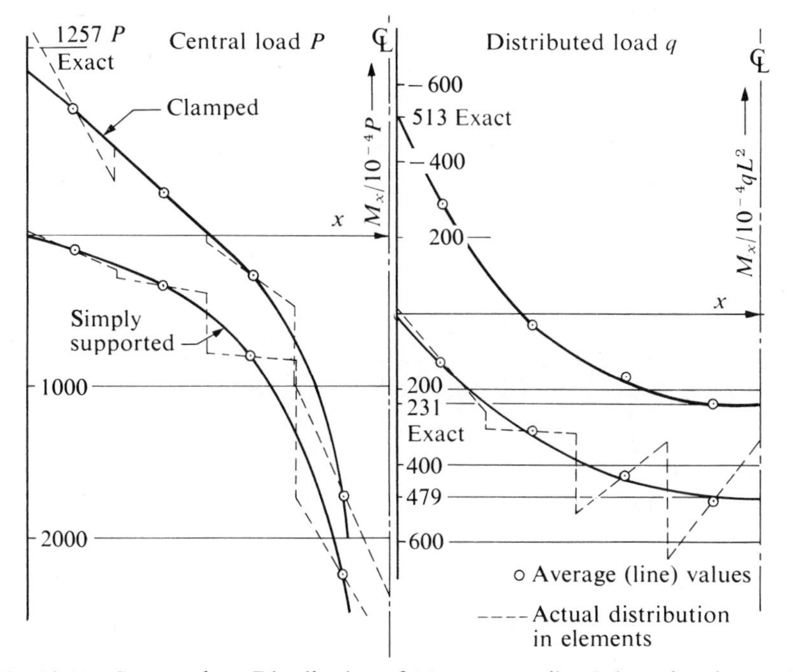

Fig. 10.10 Square plate. Distribution of M_x on centre line (triangular elements)
In distributed load case 'lumped' nodal forces are used

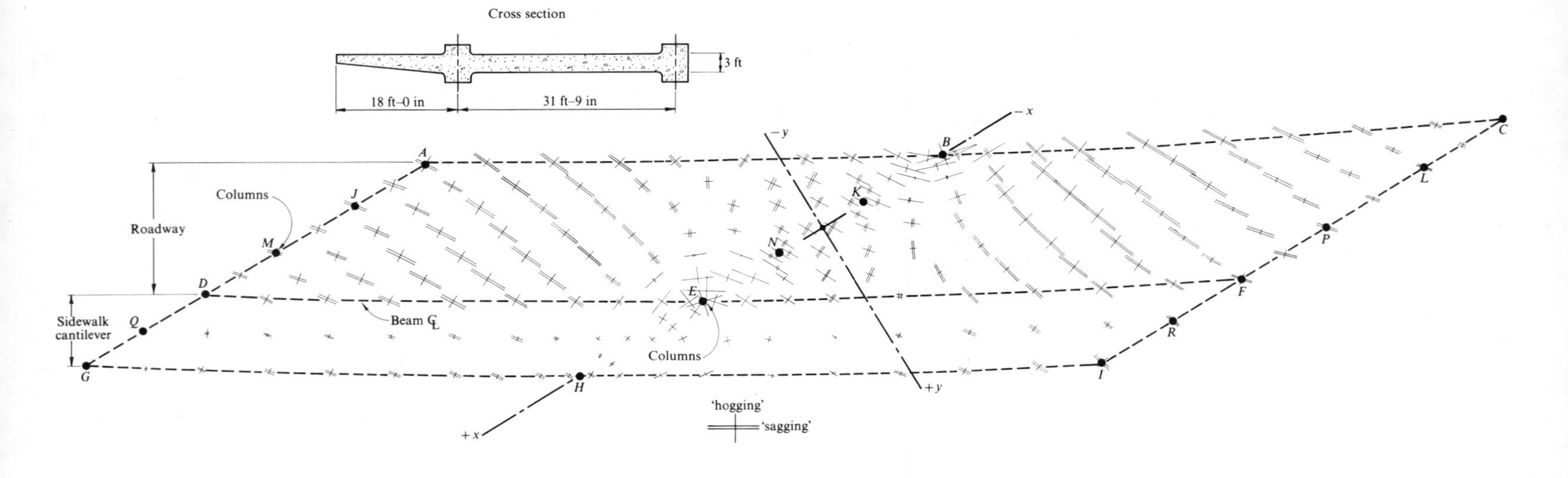

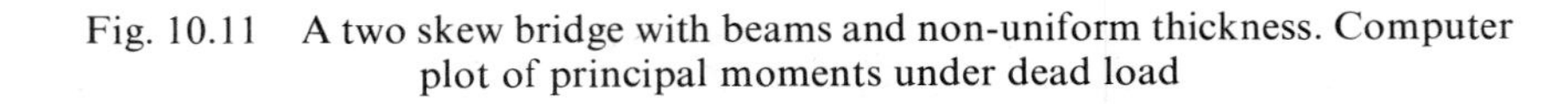

Fig. 10.11 A two skew bridge with beams and non-uniform thickness. Computer plot of principal moments under dead load

is great. Problems of foundation slabs, bridge decks, or ship hulls can be treated with ease.

Indeed the problem of bridge structures is one of extensive practical use and here applications have been very numerous. Figure 10.11 shows an automatic computer plot of prescribed stresses on a multispan bridge.

In Figs. 10.12 and 10.13 a bridge of more complex shape is illustrated. Here component contour plots are given as an alternative presentation. In this example edge beams are present and these have been assumed to have the same neutral axis as the plate. No difficulty exists in coupling such beam elements to the plate structure, assembly proceeding on the usual lines of Chapter 1.

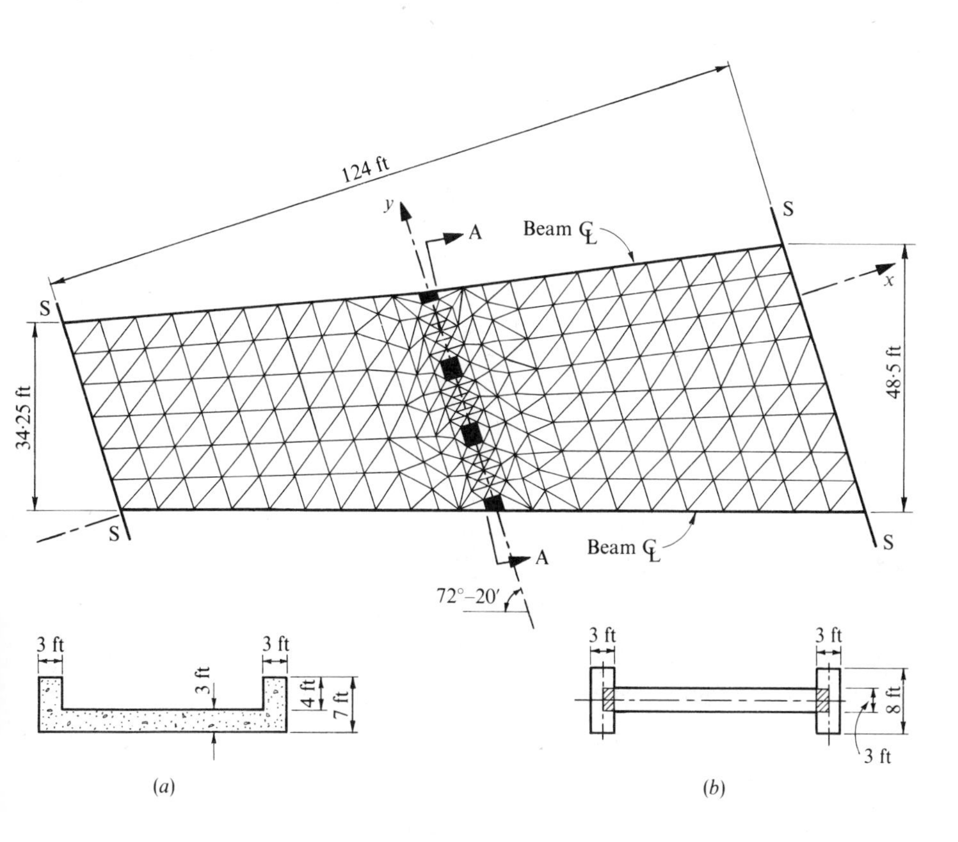

Fig. 10.12 The Castleton bridge. General geometry and detail of finite element subdivisions: (*a*) shows the typical actual sections, while (*b*) shows the nature of idealization involved

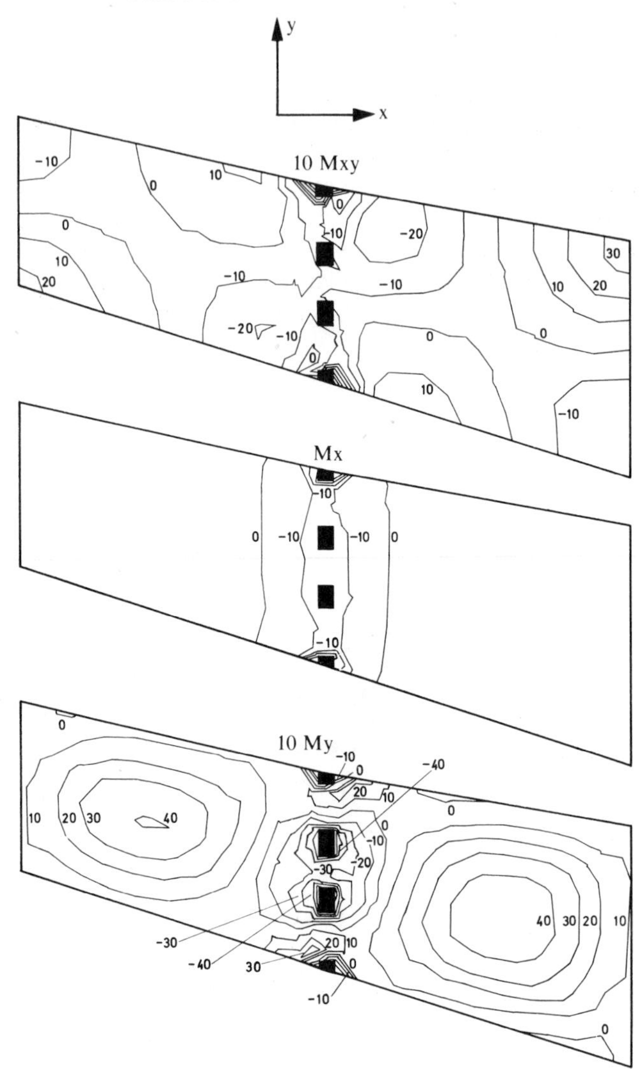

Fig. 10.13 Moment components (ton ft/ft) for bridge of Fig. 10.12 under a uniform load of 150 lb/sq ft. Computer plot of contours. Note that most loads are carried in this example by transverse slab bending

CONFORMING SHAPE FUNCTIONS WITH NODAL SINGULARITIES

10.9 **General Remarks**

It was already seen in section 10.3 that it is impossible to devise a simple polynomial function with only three degrees of nodal freedom which will be able to satisfy slope continuity requirements. The alternative of imposing curvature parameters at nodes has the disadvantage, however, of imposing excessive conditions of continuity. Furthermore it is desirable from many points of view to limit the nodal variables to three quantities only. These, with a simple physical interpretation, allow the generalization of plate elements to shells to be easily interpreted. Also computational advantages arise.

The simple alternative is to provide additional shape functions for which *second order derivatives have non-unique values at nodes*. Providing no infinities occur there, convergence is assured.

Such shape functions will be discussed now in the context of triangular and quadrilateral elements. The simple rectangular shape will be omitted.

10.10 Singular Shape Functions for the Simple Triangular Element

Consider for instance either of the following sets of functions:

$$\varepsilon_{23} = \frac{L_1 L_2^2 L_3^2}{(L_1+L_2)(L_2+L_3)}, \text{ etc.} \tag{10.31}$$

or

$$\varepsilon_{23} = \frac{L_1 L_2^2 L_3^2(1+L_1)}{(L_1+L_2)(L_2+L_3)}, \text{ etc.} \tag{10.32}$$

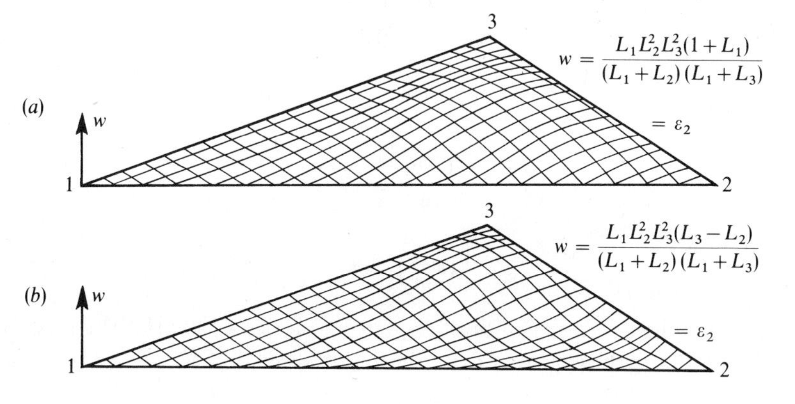

Fig. 10.14 Some singular area co-ordinate functions

Both have the property that along two sides (1–2) and (1–3) of a triangle (Fig. 10.14), their values and the values of their normal slope are zero. On the third side (2–3) their value is zero but a normal slope exists. In both its variation is parabolic. The second function shape is illustrated in Fig. 10.17(*a*).

Now, all the functions used for definition of the non-conforming triangle (*vide* Eq. (10.28)) were cubic and hence permitted also a parabolic variation of the normal slope which is not uniquely defined by the two, end, nodal values (and hence resulted in the non-conformity). However, if we specify as an additional variable the *normal slope of* w at a mid-side point of each side then by combining the new function ε with the other functions previously given a *unique parabolic variation of the normal slope* along inter-element faces is achieved and a compatible element will result.

Apparently, this can be achieved by adding three such additional degrees of freedom to expression (10.28) and proceeding as there described. This will result in an element shown in Fig. 10.15(a) which has six nodes, three corner ones as before and three additional ones at which only normal slope is specified.

Such an element presents some assembly difficulties as different numbers of degrees of freedom are associated with the nodes.

To avoid the above difficulty the mid-side node degree of freedom can now be constrained. For instance we can assume that the normal slope at the centre point of a line is given as the average of the two slopes at the end of that side. This results in a compatible element with exactly the same degrees of freedom as that described in previous sections, Fig. 10.15(b).

The algebra involved in the generation of suitable shape functions on the lines described here is tedious and will not be given. It is developed most simply on the following lines.

First the normal slopes at mid-sides are calculated from the basic element shape functions (Eq. (10.29)) as

$$\left\{\begin{array}{c} \left(\dfrac{\partial w}{\partial n}\right)_4 \\ \left(\dfrac{\partial w}{\partial n}\right)_5 \\ \left(\dfrac{\partial w}{\partial n}\right)_6 \end{array}\right\} = \mathbf{Z}\mathbf{a}^e \tag{10.33}$$

Similarly the average values of the corner slopes normal to the sides are calculated for these points from these functions,

$$\left\{\begin{array}{c} \left(\dfrac{\partial w}{\partial n}\right)_4^a \\ \left(\dfrac{\partial w}{\partial n}\right)_5^a \\ \left(\dfrac{\partial w}{\partial n}\right)_6^a \end{array}\right\} = \mathbf{Y}\mathbf{a}^e \tag{10.34}$$

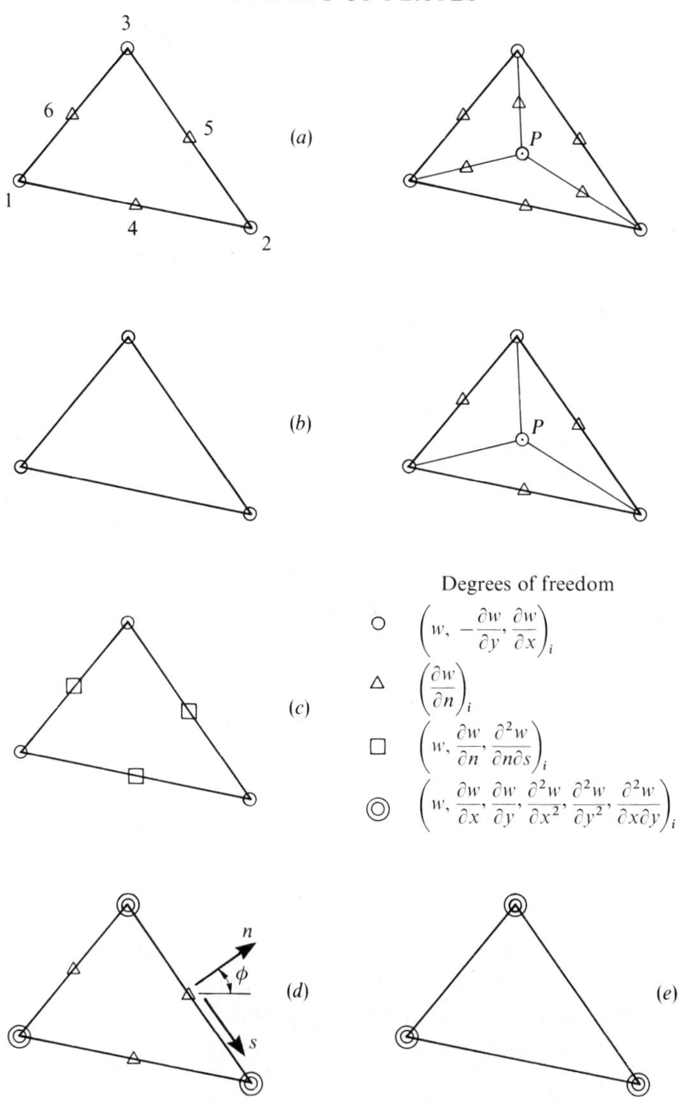

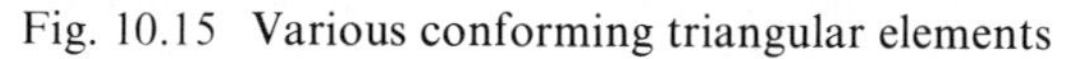
Fig. 10.15 Various conforming triangular elements

Now the contribution of the ε functions to these slopes is added in proportions $\varepsilon_{23} \times \gamma_1$, etc., is simply (as these give unit normal slope)

$$\gamma = \begin{Bmatrix} \gamma_1 \\ \gamma_2 \\ \gamma_3 \end{Bmatrix}. \tag{10.35}$$

On combining Eq. (10.29) and the last three relations we have

$$\mathbf{Y}\mathbf{a}^e = \mathbf{Z}\mathbf{a}^e + \gamma \tag{10.36}$$

from which it immediately follows on finding γ that

$$w = \mathbf{N}^o\mathbf{a}^e + [\varepsilon_{23}, \varepsilon_{31}, \varepsilon_{13}](\mathbf{Y} - \mathbf{Z})\mathbf{a}^e \tag{10.37}$$

in which $\mathbf{N}^o$ are the non-conforming shape functions defined in Eq. (10.29).

Thus the shape functions are now available from Eq. (10.37).

An alternative way of generating compatible triangles was developed by Clough and Tocher.[3] As shown in Fig. 10.15(*a*) each element triangle is first divided into three parts based on an internal point *P*. For each triangle a complete cubic expansion is written involving ten terms. The final expansion is to be expressed in terms of nine conventional degrees of freedom at nodes 1, 2, 3 and normal slopes at nodes 4, 5, 6. As at each corner two triangles have to give the same nodal values, two sets of equations are there provided, i.e., a total of $9 \times 2 + 3 = 21$ equations is thus provided. In addition continuity of displacements and slopes at the centre node *P* provides an additional six equations and continuity of slopes of internal mid-sides a further three.

Thus we have thirty equations and thirty unknowns which suffice in this case to determine the shape functions explicitly and thus achieve an element with twelve degrees of freedom similar to the one previously outlined.

Constraint of normal slopes on exterior sides leads to an element with nine degrees of freedom.

These elements are achieved at the expense of providing two values of second derivatives at the corners. In the previously discussed set, in fact, the shape functions ε provide an infinite number of derivatives depending on the direction in which the corner is approached.

Indeed the derivation of the Clough and Tocher triangles can be approached by defining an alternative set of ε functions as has been shown in reference 4.

As both types of elements lead to almost identical numerical results the preferable one is that leading to simplified computation. If numerical integration is used (as indeed is strongly recommended for such elements) the form of functions continuously defined over the whole triangle as given in Eqs. (10.29) and (10.37) is advantageous, although it can be shown that a very high order of numerical integration is necessary due to the singular nature of the functions.

10.11 An Eighteen-degree-of-freedom Triangular Element with Conforming Shape Functions

An element which presents a considerable improvement over the type illustrated in Fig. 10.15(*a*) is shown in Fig. 10.15(*c*). Here the twelve degrees of freedom are increased to eighteen by considering both the value of w and its cross derivative $\partial^2 w/\partial s\, \partial n$, in addition to the normal slope of $\partial w/\partial n$, at element mid-sides.

Thus an equal number of degrees of freedom is presented at each node giving a computational advantage. Imposition of the continuity of cross derivatives *at mid-sides* does not involve an additional constraint as this indeed must be continuous in physical situations.

The derivation of this element is given by Irons[7] and it will suffice here to say that in addition to the modes already discussed, fourth-order terms of the type illustrated in Fig. 10.6(*d*) and 'twist' functions of Fig. 10.14(*b*) are used. Indeed it can be simply verified that the element contains *all* the fifteen terms of the quartic expansion in addition to the 'singularity' functions.

10.12 Compatible Quadrilateral Elements

Any of the previous triangles can be combined to produce compatible quadrilateral elements with or without internal degrees of freedom. Three such quadrilaterals are illustrated in Fig. 10.16 and in all, no mid-side nodes exist on the external boundaries. This is to avoid the difficulties of assembly already mentioned.

In the first, no internal degrees of freedom are present and indeed no improvement on the comparable triangles is expected. In the following two, 3 and 7 internal degrees of freedom exist respectively. Here normal slope continuity imposed in the last one does not interfere with the assembly, as internal degrees of freedom are in all cases eliminated. Much

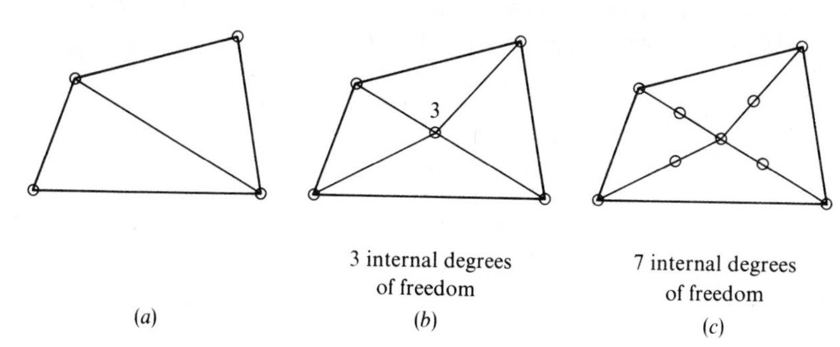

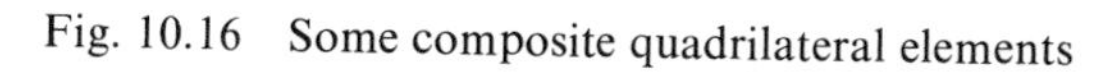
Fig. 10.16 Some composite quadrilateral elements

improved accuracy with these elements has been demonstrated by Clough and Felippa.[8]

An alternative direct derivation of a quadrilateral element was proposed by Sander[5] and Fraeijs de Veubeke.[6, 9] This is along the following lines; within a quadrilateral of Fig. 10.17 a complete cubic with ten constants is taken giving the first component of the displacement which is defined by three functions. Thus

$$w = w^a + w^b + w^c$$

and

$$w^a = \alpha_1 + \alpha_2 x + \cdots + \alpha_{10} y^3. \tag{10.38}$$

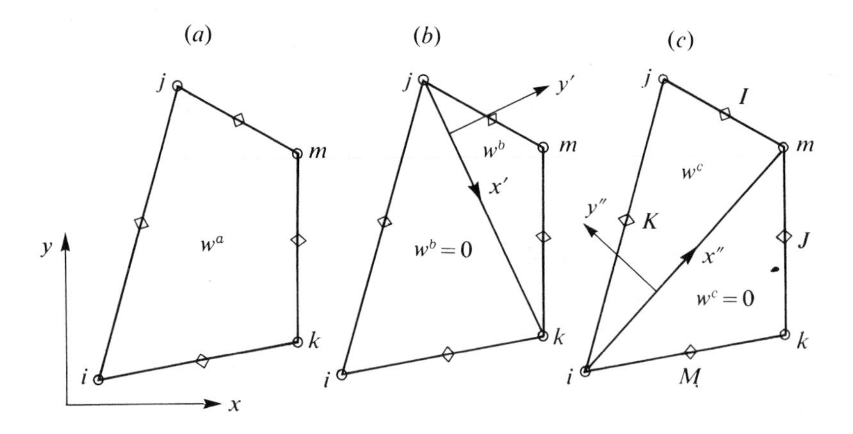

Fig. 10.17 The compatible function suggested by Fraeijs de Veubeke

The second function w^b is defined in a piecewise manner. In the lower triangle of Fig. 10.17(*b*) it is taken as zero; in the upper triangle a cubic expression with three constants merges without slope discontinuity into the field of the lower triangle. Thus in *jkm*

$$w^b = \alpha_{11} y'^2 + \alpha_{12} y'^3 + \alpha_{13} x' y'^2 \tag{10.39}$$

in terms of the locally specified co-ordinates x', y'. Similarly for the third function, Fig. 10.17(*c*), $w^c = 0$ in lower triangle and in *imj*

$$w^c = \alpha_{14} y''^2 + \alpha_{15} y''^3 + \alpha_{16} x'' y''^2. \tag{10.40}$$

The sixteen external degrees of freedom are provided by three usual corner variables and normal mid-side slopes and allow the sixteen constants α_{1-16} to be found by inversion. Compatibility is assured and once again non-unique second derivatives arise at corners.

Again it is possible to constrain the mid-side nodes if desired and thus obtain a twelve-degree-of-freedom element.

The expansion can be found explicitly as shown by Veubeke[9] and a useful element generated.

The element described above cannot be formulated if a corner of the quadrilateral is re-entrant. This is not a serious limitation but needs to be considered on occasion if such an element degenerates to near triangular shape.

10.13 **Which Element? A Numerical Assessment**

A large number of conforming and non-conforming elements have been mentioned and some numerical results for the latter given. In all of these, the simplest nodal degrees of freedom are used, i.e., only displacements and their first derivatives and such elements are admirably suited for an extension to shell problems and indeed to other situations demanding C_1 continuity.

As all the elements converge to the correct answer (or nearly correct answer in the case of the irregular mesh for a nine-degree-of-freedom triangle) all can be used with confidence and the users' choice is governed by convenience and economy. It is not easy to produce a quantitative comparison of solution cost versus accuracy, but the former is roughly dependent on the *total number of unknowns*. If a simple test problem is chosen, then a numerical assessment of various elements is possible. Some such comparisons are available in references 3, 8, and 36. In Fig. 10.18 we show the convergence of some elements already discussed for the case of a simply supported plate under a concentrated, central load.

In this comparison we have included two elements not yet discussed. The first of these is an element based on nine- (and twelve-) degrees-of-freedom conforming elements to which derivative smoothing has been applied[25, 26] (*vide* Chapter 11, p. 278).

The second is an optimized triangular element derived directly from the satisfaction of the patch test condition by Bergan and Hanssen.[37]

There is little doubt that the non-conforming formulations of the nine-degree element are possibly the most convenient approximation in view of the ease of generating arbitrary shapes. Performance for degrees of freedom has so far not been bettered by other elements.

It is at this stage worth while to point out that the conforming triangles give such poor answers that their use as an approximating procedure does not appear justified.

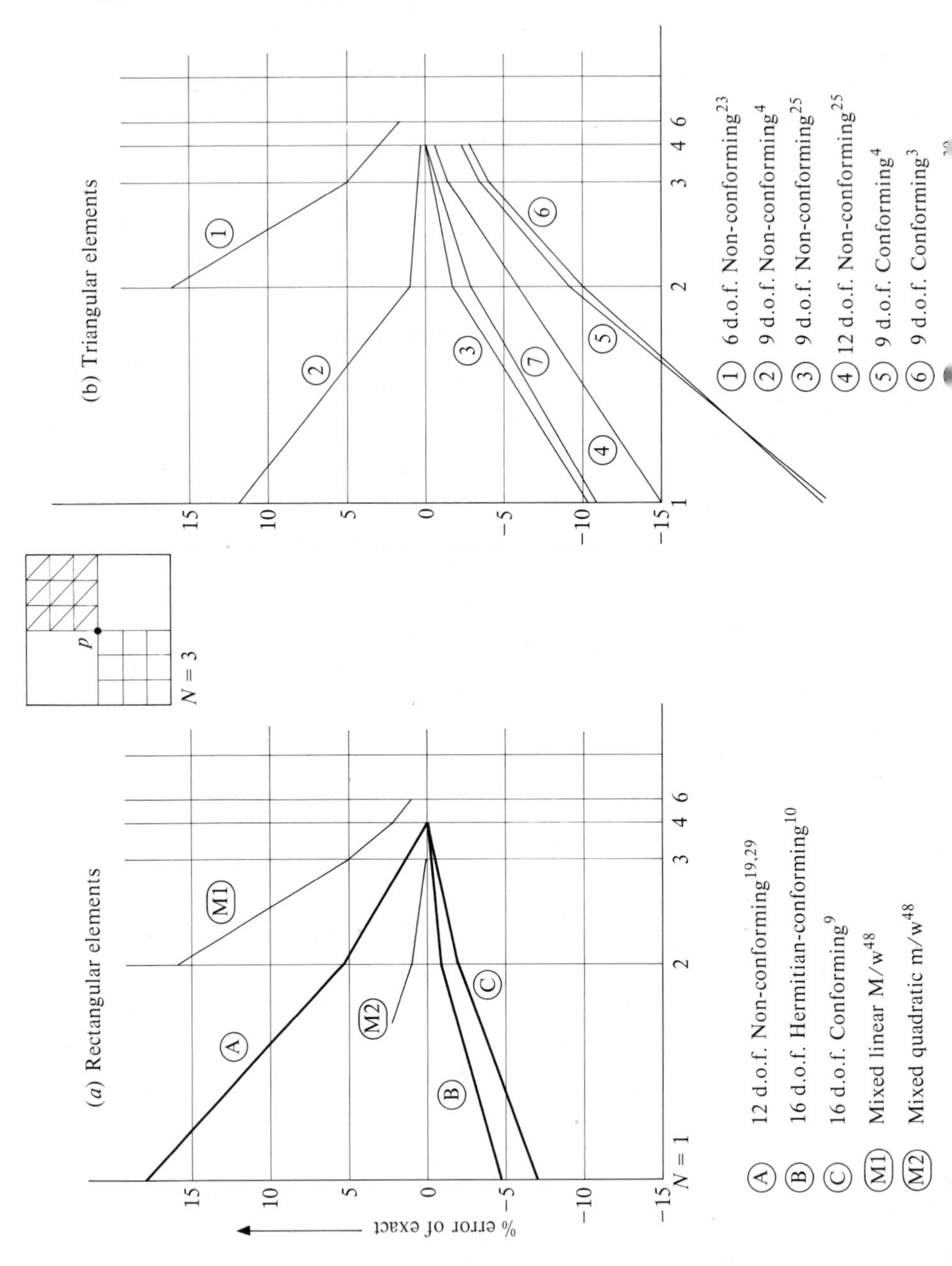

Fig. 10.18 Comparative errors—thin plate, simply supported, central load. (*a*) Rectangular element. (*b*) Triangular element

CONFORMING SHAPE FUNCTION WITH ADDITIONAL DEGREES OF FREEDOM

10.14 Hermitian Rectangle Shape Function

With a rectangular element of Fig. 10.2 the specification of $\partial^2 w/\partial x\, \partial y$ as a nodal parameter is always permissible as it does not involve 'excessive continuity'. It is easy to show that for such an element polynomial shape functions giving compatibility can be easily determined.

A polynomial expansion involving sixteen constants (equal to the number of nodal parameters) could for instance be written retaining terms which do not produce a higher order variation of w or its normal slope than cubic along the sides. Many alternatives will be present here and some may not produce invertible **C** matrices.

An alternative derivation uses Hermitian polynomials which permit the writing down of suitable functions directly. A Hermitian polynomial

$$H^{n}_{mi}(x) \tag{10.41}$$

is a polynomial of order $2n+1$ which gives, when $x = x_i$,

$$\frac{\mathrm{d}^k H}{\mathrm{d}x^k} = 1, \qquad k = m \quad \text{for } m = 0 \text{ to } n$$

and

$$\frac{\mathrm{d}^k H}{\mathrm{d}x^k} = 0, \qquad k \neq m \quad \text{or when } x = x_j.$$

A set of first order Hermitian polynomials is thus a set of cubics giving shape functions for a line element ij at the ends of which slopes and values of the function are used as variables.

Figure 10.19 shows such a set of cubics.

It is easy to verify that the following shape functions

$$\mathbf{N}_i = [H^{(1)}_{0i}(x)H^{(1)}_{0i}(y),\ H^{(1)}_{1i}(x)H^{(1)}_{0i}(y),\ H^{(1)}_{0i}(x)H^{(1)}_{1i}(y),\ H^{(1)}_{1i}(x)H^{(1)}_{1i}(y)] \tag{10.42}$$

correspond to

$$w, \quad \frac{\partial w}{\partial y}, \quad \frac{\partial w}{\partial x}, \quad \frac{\partial^2 w}{\partial x\, \partial y}$$

taking successively unit values at node i and zero elsewhere.

An element based on these shape functions has been developed by Bogner *et al.*[10] and used with some success.

A development of this type of element to include continuity of higher derivatives is simple and is outlined in reference 11.

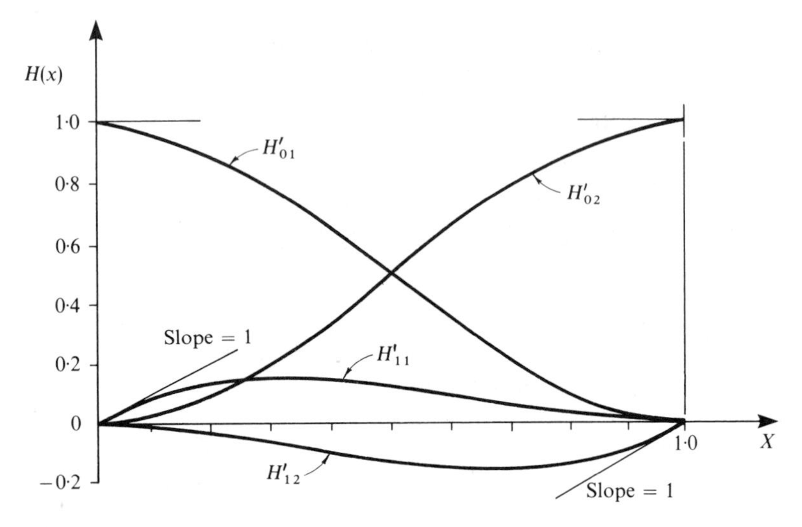

Fig. 10.19 First order Hermitian functions

In its undistorted form the above elements are, as all rectangles, of very limited applicability.

10.15 **Twenty-one- and eighteen-degree-of-freedom Triangles**

If continuity of higher derivatives than the first is accepted at nodes (thus imposing a certain constraint on non-homogeneous situations as explained in section 10.3), the generation of slope and deflection compatible elements presents less difficulty.

Considering as nodal degrees of freedom.

$$w, \quad \frac{\partial w}{\partial x}, \quad \frac{\partial w}{\partial y}, \quad \frac{\partial^2 w}{\partial x^2}, \quad \frac{\partial^2 w}{\partial y^2}, \quad \frac{\partial^2 w}{\partial x\, \partial y}$$

a triangular element will involve at least eighteen degrees of freedom. Now, a complete fifth order polynomial contains twenty-one terms. If therefore we add three normal slopes at mid-side as additional degrees of freedom a sufficient number of equations appear to exist for which the shape function can be found.

Along any edge we have six quantities determining the variation of w (displacement, slopes (and curvature at corner nodes), i.e., specifying a fifth order variation. Thus this is uniquely defined and therefore w is continuous between elements.

Similarly $\partial w/\partial n$ is prescribed by five quantities and varies as a fourth order polynomial. Again this is as required by the deformation and slope continuity between elements.

If we write the complete quintic†

$$w = \alpha_1 + \alpha_2 x + \cdots + \alpha_{21} y^5 \tag{10.43}$$

we can proceed along the lines of the argument used to develop the rectangle in section 10.4 and write

$$\begin{aligned}
w_1 &= \alpha_1 + \alpha_2 x_1 + \qquad\qquad + \alpha_{21} y_1^5 \\
\left(\frac{\partial w}{\partial x}\right)_1 &= \qquad \alpha_2 \qquad\qquad + \alpha_{20} y_1^3 \\
\left(\frac{\partial^2 w}{\partial x^2}\right)_1 &= \qquad\qquad + 2\alpha_4 + \cdots + 2\alpha_{19} y_1^2, \\
\vdots &= \qquad\qquad\qquad\qquad \vdots
\end{aligned}$$

and finally obtain an expression

$$\mathbf{a}^e = \mathbf{C}\boldsymbol{\alpha} \tag{10.44}$$

in which $\mathbf{C}$ is a 21×21 matrix.

The only apparent difficulty in the process which the reader may experience in forming this is the definition of the normal slopes at mid-side nodes. However, if one notes that (Fig. 10.15)

$$\frac{\partial w}{\partial n} = \cos\phi \frac{\partial w}{\partial x} + \sin\phi \frac{\partial w}{\partial y} \tag{10.45}$$

in which ϕ is the angle of a particular side to the x axis, the matter of formulation becomes simple.

Indeed it is not easy to determine an explicit inverse of $\mathbf{C}$ and the stiffness expressions, etc., are evaluated as in Eq. (10.24) by a numerical inversion.

The existence of the mid-side nodes with their single degree of freedom along the sides is an embarrassment. It is possible, however, to constrain these by allowing only a cubic variation of the normal slope along each triangle side. Now, explicitly, the matrix $\mathbf{C}$ and the degrees of freedom can be reduced to eighteen giving an element illustrated in Fig. 10.15(*e*) with three corner nodes and eighteen degrees of freedom. This in fact is the more useful element in practice.

Both these elements were described in several publications appearing during 1968 and obviously quite independently arrived at. This 'simultaneous discovery' fact is one of the curiosities of scientific progress and

† For this derivation use of simple Cartesian co-ordinates is recommended in preference to area co-ordinates. Symmetry is assured as polynomial is complete.

seems to occur in many fields where the stage for a particular development is reached.

Thus, the twenty-one-degree-of-freedom element is described by Argyris,[16] Bell,[12] Bosshard,[15] Irons,[7] and Visser[17] listing the authors alphabetically.

The reduced eighteen-degree-of-freedom version is developed by Argyris,[16] Bell,[12] and Cowper *et al.*[14] An essentially similar, but more complicated, formulation has been developed by Butlin and Ford,[13] and mention of the element shape functions is made earlier by Withum[38] and Fellipa.[39]

It is clear that many more elements of this type could be developed and indeed some are suggested in the above references. A very full study is included in the work of Zenisek.[40] However, it should always be borne in mind that they involve an inconsistency when discontinuous variation of material properties occurs. Further, the existence of higher order derivatives makes it difficult to impose boundary conditions and indeed the simple interpretation of energy derivatives as 'nodal forces' disappears. Thus the engineer may still feel a justified preference for the more intuitive formulation previously described despite the fact that very good accuracy has been demonstrated in the many references quoted for these elements.

10.15 **Concluding Remarks**

A fairly extensive survey of shape functions and the methods of their generation has been included in this chapter. The reason for this is not only the fact that plate bending situations are an important engineering application but that *all the shape functions here presented are applicable to problems in which the functional involves second-order derivatives.* Thus use of these can be made in the context of viscous flow and other physical problems of that type.

Indeed, even two-dimensional stress analysis can, as is well known, be formulated in terms of stress functions and therefore such functionals. As such formulations automatically satisfy the equilibrium conditions an 'upper bound' solution is possible by minimization of 'complementary strain energy'. Such an application was first suggested by Veubeke and Zienkiewicz.[41]

It is for these reasons that many alternative formulations of the plate problem have been here omitted. Some of these are well established,[42–48] and we shall return to such formulations in Chapters 11 and 12.

In the basic formulation of this chapter the classical theory of thin plates has been followed. The shear deformation of plates has thus not been included. This undoubtedly has some importance in very thick plate situations. Some approximate attempts to include shear deformations are described in references 8 and 42. In this book the subject will be dealt with in a different manner in Chapters 11 and 16.

The problem of generating C_1 continuous shape functions—or their close approximation is one of continuing interest. The reader may find it instructive to consult some alternative procedures which were not mentioned in this chapter in references 49, 50 and 51.

References

1. S. Timoshenko and S. Woinowsky-Krieger, *Theory of Plates and Shells*, McGraw-Hill, 2nd ed., 1959.
2. B. M. Irons and J. K. Draper, 'Inadequacy of nodal connections in a stiffness solution for plate bending', *J.A.I.A.A.*, **3**, 5, 1965.
3. R. W. Clough and J. L. Tocher, 'Finite element stiffness matrices for analysis of plates in bending', *Proc. Conf. Matrix Methods in Struct. Mech.*, Air Force Inst. of Tech., Wright Patterson A.F. Base, Ohio, 1965 (October).
4. G. P. Bazeley, Y. K. Cheung, B. M. Irons, and O. C. Zienkiewicz, 'Triangular elements in bending—conforming and non-conforming solutions', *Proc. Conf. Matrix Methods in Struct. Mech.*, Air Force Inst. of Tech., Wright Patterson A.F. Base, Ohio, 1965 (October).
5. G. Sander, 'Bornes supérieures et inférieures dans l'analyse matricielle des plaques en flexion-torsion', *Bull. Soc. Royale des Sc. de Liège*, **33**, 456–94, 1964.
6. B. Fraeijs de Veubeke, 'Bending and Stretching of Plates', *Proc. Conf. Matrix Methods in Struct. Mech.*, Air Force Inst. of Tech., Wright Patterson A.F. Base, Ohio, 1965 (October).
7. B. M. Irons, 'A conforming quartic triangular element for plate bending', *Int. J. Num. Meth. Eng.*, **1**, 29–46, 1969.
8. R. W. Clough and C. A. Felippa, 'A refined quadrilateral element for analysis of plate bending', *Proc. 2nd Conf. Matrix Methods in Struct. Mech.*, Air Force Inst. of Tech., Wright Patterson A.F. Base, Ohio, 1968.
9. B. Fraeijs de Veubeke, 'A conforming finite element for plate bending', *Int. J. Solids Struct.*, **4**, 95–108, 1968.
10. F. K. Bogner, R. L. Fox, and L. A. Schmit, 'The generation of interelement—compatible stiffness and mass matrices by the use of interpolation formulae', *Proc. Conf. Matrix Methods in Struct. Mech.*, Air Force Inst. of Tech., Wright Patterson A.F. Base, Ohio, 1965 (October).
11. I. M. Smith and W. Duncan, 'The effectiveness of nodal continuities in finite element analysis of thin rectangular and skew plates in bending', *Int. J. Num. Mech. Eng.*, **2**, 253–8, 1970.
12. K. Bell, 'A refined triangular plate bending element', *Int. J. Num. Meth. Eng.*, **1**, 101–22, 1969.
13. G. A. Butlin and R. Ford, 'A compatible plate bending element', *Univ. of Leicester Eng. Dept. report*, 68-15, 1968.
14. G. R. Cowper, E. Kosko, G. M. Lindberg, and M. D. Olson, 'Formulation of a new triangular plate bending element', *Trans. Canad. Aero-Space Inst.*, **1**, 86–90, 1968. (See also N.R.C. Aero report LR514, 1968).
15. W. Bosshard, 'Ein neues vollverträgliches endliches Element für Plattenbiegung', *Mt. Assoc. Bridge Struct. Eng. Bulletin*, **28**, 27–40, 1968.

16. J. H. ARGYRIS, I. FRIED, and D. W. SCHARPF, 'The TUBA family of plate elements for the matrix displacement method', *The Aeronautical J. R. Ae. S.*, **72**, 701–9, 1968.
17. W. VISSER, *The finite element method in deformation and heat conduction problems*, Dr. W. Dissertation, T.H., Delft, 1968.
18. B. M. IRONS, Comments on 'Complete polynomial displacement fields for finite element method', by P. C. Dunne, *The Aeronautical J., R. Ae. S.*, **72**, 709, 1968.
19. O. C. ZIENKIEWICZ and Y. K. CHEUNG, 'The finite element method for analysis of elastic isotropic and orthotropic slabs', *Proc. Inst. Civ. Eng.*, **28**, 471–88, 1964.
20. R. W. CLOUGH, 'The finite element method in structural mechanics', chapter 7 of *Stress Analysis*, ed. O. C. Zienkiewicz and G. S. Holister, J. Wiley, 1965.
21. D. J. DAWE, 'Parallelogram element in the solution of rhombic cantilever plate problems', *J. of Strain Analysis*, **3**, 1966.
22. J. H. ARGYRIS, 'Continua and Discontinua', *Proc. Conf. Matrix Methods in Struct. Mech.*, Air Force Inst. of Tech., Wright Patterson A.F. Base, Ohio, 1965 (October).
23. L. S. D. Morley, 'The triangular equilibrium element in the solution of plate bending problems', *Areo Quart.*, **19**, p. 149–69, 1968; 'On the constant moment plate bending element', *J. Strain Analysis*, **6**, 20–4, 1971.
24. NARAYANASWAMI, R., 'New triangular plate-bending element with transverse shear flexibility', *J.A.I.A.A.*, **12**, 1761–3, 1974.
25. A. RAZZAQUE, 'Program for triangular bending element with derivative smoothing', *Int. J. Num. Meth. Eng.*, **5**, 588–9, 1973.
26. B. M. IRONS and A. RAZZAQUE, 'Shape function formulation for elements other than displacement models', *Proc. Conf. Variational Methods in Engineering*, Southampton Univ., 1972.
27. J. E. WALZ, R. E. FULTON, and N. J. CYRUS, 'Accuracy and Convergence of finite element approximation', *Proc. 2nd Conf. Matrix Methods in Struct. Mech.*, Air Force Inst. of Tech., Wright Patterson A.F. Base, Ohio, 1968.
28. R. J. MELOSH, 'Basis of derivation of matrices for the direct stiffness method', *J.A.I.A.A.*, **1**, 1631–7, 1963.
29. A. ADINI and R. W. CLOUGH, *Analysis of plate bending by the finite element method* and Report to Nat. Sci. Found/U.S.A., G.7337, 1961.
30. Y. K. CHEUNG, I. P. KING and O. C. ZIENKIEWICZ, 'Slab bridges with arbitrary shape and support conditions—a general method of analysis based on finite elements', *Proc. Inst. Civ. Eng.*, **40**, 9–36, 1968.
31. J. L. TOCHER and K. K. KAPUR, 'Comment on Basis of derivation of matrices for direct stiffness method', *J.A.I.A.A.*, **3**, 1215–16, 1965.
32. R. D. HENSHELL, D. WALTERS, and G. B. WARBURTON, 'A new family of curvilinear plate bending elements for vibration and stability', *J. Sound and Vibration*, **20**, 327–43, 1972.
33. H. MARCUS, 'Die Theorie elastischer Gewebe und ihre Anwendung auf die Berechnung biegsamer Platten', Springer, Berlin, 1932.
34. P. BALLESTEROS, 'The application of Maclaurin's Series to the analysis of plates in bending', University of Michigan, Ann Arbor, Mich., 59.196, 1958,
35. P. BALLESTEROS and S. L. LEE, 'Uniformly loaded rectangular plate supported at the corners', *Int. J. Mech. Sci.*, **2** (No. 3), 206–11, 1960.
36. A. RAZZAQUE, *Finite element analysis of plates and shells*, Ph.D. Thesis, Univ. of Wales, Civil Engineering Department, Swansea, 1972.

37. P. G. BERGAN and L. HANSSEN, 'A new approach for deriving "good" element stiffness matrices', in *The Mathematics of Finite Elements and Applications*, ed. J. R. Whiteman, Academic Press, 1977.
38. D. WITHUM, *Berechnung von Platten nach dem Ritzsehen verfahren mit Wilfe dreieckförmiger Meshnetze*, Mittl. Inst. Statik Tech. Hochschule, Hanover, 1966.
39. C. A. FELIPPA, *Refined finite element analysis of linear and non-linear two-dimensional structures*, Ph.D., Struct. Eng. Univ. of Calif., Berkeley, 1966.
40. A. ZENISEK, 'Interpolation polynomials on the triangle', *Int. J. Num. Meth. Eng.*, **10**, 283–96, 1976.
41. B. FRAEIJS DE VEUBEKE and O. C. ZIENKIEWICZ, 'Strain Energy Bounds in finite element analysis by slab analogy', *J. Strain Analysis*, **2**, 265–71, 1967.
42. T. H. H. PIAN, 'Derivation of Element Stiffness Matrices by assumed stress distribution', *A.I.A.A. Int.*, **2**, 1332–6, 1964.
43. T. H. H. PIAN and P. TONG, 'Basis of finite element methods for solid continua', *Int. J. Num. Meth. Eng.*, **1**, 3–28, 1969.
44. R. J. ALLWOOD and G. M. M. CORNES, 'A polygonal finite element for plate bending problems using the assumed stress approach', *Int. J. Num. Meth. Eng.*, **1**, 135–50, 1969.
45. R. T. SEVERN and P. R. TAYLOR, 'The finite element method for flexure of slabs where stress distributions are assumed', *Proc. Inst. Civ. Eng.*, **34**, 153–70, 1966.
46. L. R. HERRMANN, 'Finite Element Bending analysis of plates', *Proc. Am. Soc. Eng.*, **93**, EM 5, 1967.
47. B. FRAEIJS DE VEUBEKE, 'An equilibrium model for plate bending', *Int. J. Solids Struct.*, **4**, 447–68, 1968.
48. J. BRON and G. DHATT, 'Mixed quadrilateral elements for bending', *J.A.I.A.A.*, **10**, 1359–61, 1972.
49. J. J. GOËL, 'Construction of basic functions for numerical utilization of Ritz's method', *Numerische Math.*, **12**, 435–47, 1968.
50. G. BIRKHOFF and L. MANSFIELD, 'Compatable triangular finite elements', *J. Math. Analysis and Appl.*, **47**, 531–53, 1974.
51. C. L. LAWSON, 'C'-compatable interpolation over a triangle', *NASA Jet. Prop. Lab.*, T.M., 33–770, 1976.

11. Non-Conforming Elements; Substitute Shape Functions; 'Reduced' Integration and Similar Useful Tricks

11.1 Introduction

In the last chapter dealing with displacement formulation of plates we have, for the first time in this book, introduced non-conforming elements, largely to overcome the difficulties of imposing the slope (C_1) continuity. As these elements violate one of the apparently essential conditions for convergence postulated to enforce integrability in Chapter 2 (p. 33) and Chapter 3 (p. 46), they have appeared to many illegitimate and threatened with excommunication despite the fact that consistently better results have been obtained using those in preference to equivalent conforming elements. Today sufficient conditions for convergence of such elements are established by the patch test to which we have already made reference, and with this the convergence of most of the previously 'illegitimate' elements has now been proved.

As non-conformity appears to improve the element performance, should it be confined only to elements where difficulties arise in obtaining conforming shape functions? Should we not also seek to introduce it deliberately to improve the performance of elements even of C_0 type, where conformity emerges easily as was shown in Chapter 7. It is to such questions that the first part of this chapter is addressed.

The important question is how such non-conformity functions should be generated to pass the patch test and at the same time improve the performance. Physical intuition may well provide the first motive, but a more formal approach is possible utilizing *substitute shape functions* which smooth the appropriate derivatives. We shall show effective use of such substitute shape functions and also indicate that another violation of the true approximations introduced by inexact numerical integration produces the same desirable effects.

Finally we shall show that another beneficial effect of such reduced integration is the introduction of certain singularities necessary to overcome the excessive constraint introduced in some problems by a penalty function formulation. Here we shall first consider problems of incompressibility and then of plate bending in which the slope and displacements are interpolated independently.

Some of the most efficient elements for practical use are introduced by the devices of this chapter which attempts to show that the successes achieved so far are not entirely fortuitous.

11.2 **The Patch Test**

We have already referred to the 'patch test' in section 2.7 of Chapter 2 as a test of acceptability of elements which violate the conditions of conformity in elasticity. Such a test can indeed be derived for any *mathematical problem* approximated in the finite element fashion of Chapter 3.

The original test was introduced in a physical way and could be interpreted as a check which ascertained whether a patch of elements (Fig. 11.1) subject to a constant strain reproduced exactly the constitutive behaviour of the material and resulted in correct stresses[1,2] when it became infinitesimally small. If it did, it could then be argued that the finite element model respresented the real material behaviour and, in the limit, as the size of the elements decreased would therefore reproduce exactly the behaviour of the real structure.

Clearly, although this test would only have to be passed when the size of the element patch became infinitesimal, for most elements in which polynomials are used the patch size did not in fact enter the consideration and the requirement that the patch test be passed for any element size became standard.

Quite obviously a rigid body displacement of the patch would cause no strain, and if the proper constitutive laws were reproduced no stress

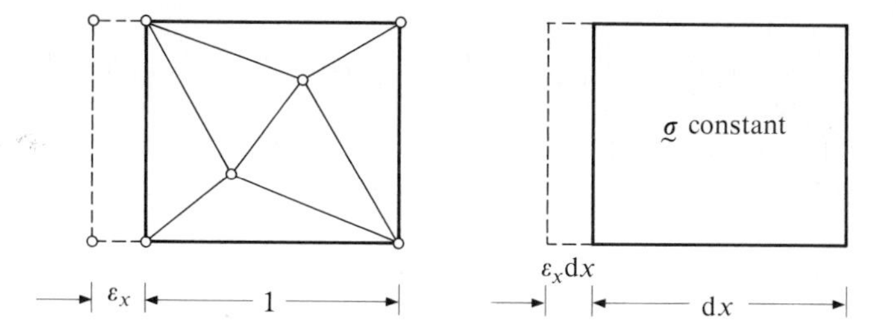

Fig. 11.1 A patch of element and a volume of continuum subject to constant strain ε_x. A physical interpretation of the constant strain or linear displacement field patch test.

changes would result. The patch test thus guarantees that no rigid body motion straining will occur.

When curvilinear co-ordinates are used the patch test still requires to be passed in the limit but generally will not do so for a finite size of the patch. (An exception here is the isoparametric co-ordinate system in problems discussed in Chapter 8.) Thus for many problems such as shells, where local curvilinear co-ordinates are used, this test has to be restricted to infinitesimal patch sizes and, on physical grounds alone, appears to be a *necessary and sufficient condition* for convergence.[3–7]

The original concept of the patch test can be generalized to all finite element approximations and re-derived not only as a sufficient condition for convergence but also as an indicator of the convergence order which is to be expected.[8]

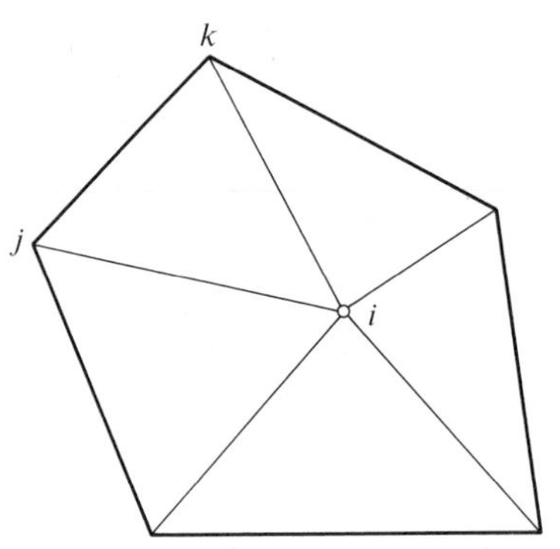

Fig. 11.2 An arbitrary patch of elements with a fully connected node i

Consider an arbitrary element patch such as that shown in Fig. 11.2 in which *at least one fully connected node* (such as i) exists. We know that at such a node

$$\sum_{e=1}^{m} \mathbf{K}_{ij}^{e}\mathbf{a}_j + \mathbf{f}_i^e = 0 \tag{11.1}$$

whether the problem is one of structural origin where $\mathbf{K}_{ij}^{e}$ represent the element stiffness coefficients, or whether the problem arose from a Galerkin finite element approximation to a linear differential equation (*vide* Chapter 3) of type

$$\mathbf{L}.\boldsymbol{\phi}+\mathbf{p} = 0. \tag{11.2}$$

with a trial expansion

$$\hat{\boldsymbol{\phi}} = \sum_{i=1}^{n} \mathbf{N}_i\mathbf{a}_i. \tag{11.3}$$

Now if we consider a local Taylor series expansion of $\boldsymbol{\phi}$, the order of finite element approximations is governed by the neglected terms of this expansion. Thus, if $\boldsymbol{\phi}^p$ represents an exact solution of the differential equation (11.2) by a polynomial of order p (for which the load term takes an appropriate value) then, if the finite element approximation reproduces this solution exactly, the element will be convergent with an order of error of at least

$$O(h^{p+1}).$$

To ascertain the actual convergence order we impose on each node of the element patch shown in Fig. 11.2 nodal displacements corresponding to an arbitrary polynomial expansion of the unknown function $\boldsymbol{\phi}^p$. *If we find that Eq. (11.1) is exactly satisfied, then convergence of the order* $O(h^{p+1})$ *will be assured.*

Naturally in this context the discrete load term has to be calculated using the appropriate loading function $\mathbf{p}^p$.

Clearly we seek solutions also converged in 'energy' and the corresponding strain-like qualities, and the minimum order of p is given by the mth differentiation order defining these (see p. 34). For general convergence, therefore, it is necessary to test the element with polynomial solutions in which $p \geqslant m$. For $p = m$ the test corresponds to the original constant strain condition already discussed. For such a case the loading term $\mathbf{p}^p = 0$ and an exact reproduction of the 'constitutive behaviour' can occur as already mentioned.

If the polynomials imposed on the element patch are quite arbitrary, as also is the shape of the elements, then a single numerical test will suffice as it is extremely unlikely that fortuitous satisfaction of the conditions has been achieved. We shall find, however, that for certain elements the patch test is satisfied only when the elements shape or connectivity possesses a special pattern (e.g., elements are rectangular or parallelograms or triangles connected in a certain regular way, as was indicated in Chapter 10 for a special case).

Given such special patterns or shapes the convergence will be guaranteed if the patch test is passed for those special conditions only.

A slightly different interpretation of the patch test is sometimes used. Here only the edge values of the nodal parameters are imposed on the patch of Fig. 11.2 and Eq. (11.1) is used to determine $\mathbf{a}_i$. If this is found to correspond with the correct value, and the strains or similar derivatives are correctly rendered, the patch test is satisfied.

The difference between the two approaches is minimal—but the second determines additionally that the matrix $\mathbf{K}_{ii}$ is non-singular.

As the patch test provides a test guaranteeing convergence, its application to non-conforming elements will automatically determine whether these can be considered as legitimate.

Much has been written in mathematical literature on the conditions which make non-conforming shape functions suitable to ensure convergence. One of the first such contributions was by Oliveira[9] and this was followed by others.[10] No tests, however, appear to be so easy and simple to apply as that provided by the patch test.

11.3 A Non-conforming C_0 Quadrilateral Element

The linear Lagrangian rectangle and its isoparametric distortion into a general quadrilateral has been discussed in Chapters 7 and 8. This element in its conforming form is illustrated in Fig. 11.3(*a*) together with the basic, linear shape functions. The performance of this element in the context of a plane stress cantilever was illustrated in Fig. 9.1 of Chapter 9 and was found disappointing owing to its inability to follow a simple curved

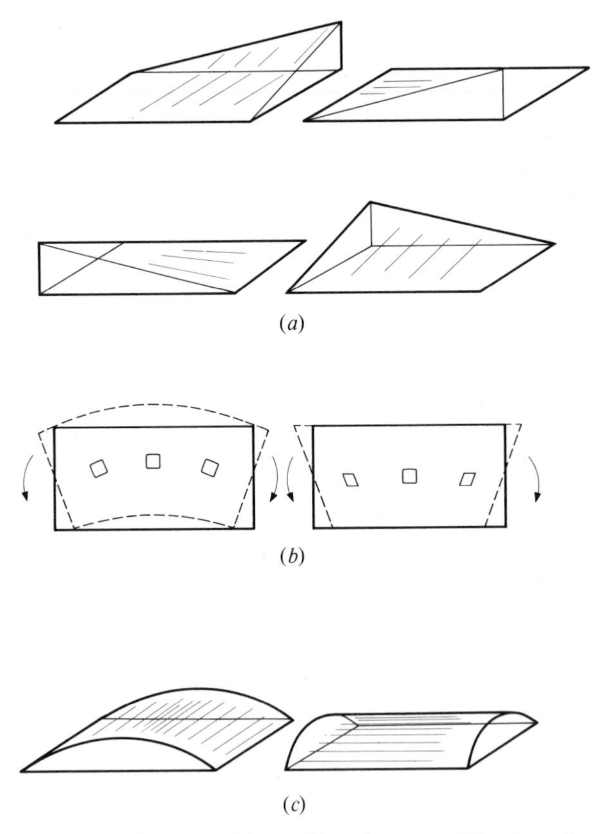

Fig. 11.3 Linear quadrilateral with auxiliary incompatible shape functions (*a*); pure bending and linear displacements causing shear (*b*); auxiliary 'bending' shape functions with internal variables (*c*)

shape induced by bending Fig. 11.3(*b*). A fairly natural suggestion was therefore to introduce two additional displacement modes, illustrated in Fig. 11.3(*c*), to which internal degrees of freedom were attached. Clearly the deformations between elements are now non-conforming but, on the other hand, the beam problem is solved almost exactly. Wilson *et al.*[11] introduce this element and in Fig. 11.4 we show how strikingly the computed deflections and stresses in a cantilever are improved, by comparison with the conforming linear elements.

Does this improvement carry over to other situations and does this new element satisfy the patch test? Results show that it does so only in a rectangular or parallelogram field and it fails (and indeed produces extremely bad results) when distorted to a quadrilateral form, and as a general element is therefore not acceptable.

Rather than reject this element out of hand, Taylor *et al.*[12] suggest an application of a 'medicine' which is *designed entirely to enforce the*

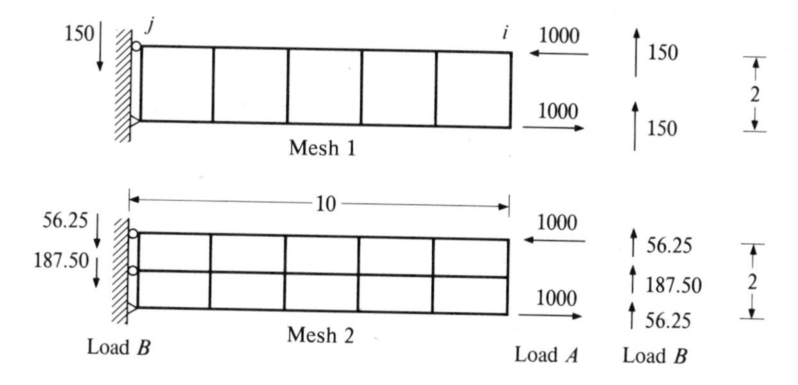

	Displacement at i		Bending stress at j	
	Load A	Load B	Load A	Load B
Beam theory	10.00	103.0	300.0	4050
(*a*) Mesh 1	6.81	70.1	218.2	2945
(*a*) Mesh 2	7.06	72.3	218.8	2954
(*b*) Mesh 1	10.00	101.5	300.0	4050
(*b*) Mesh 2	10.00	101.3	300.0	4050

Fig. 11.4 Performance of the non-conforming quadrilateral in beam bending treated as plane stress. (*a*) conforming linear quadrilateral. (*b*) non-conforming quadrilateral.

satisfaction of the low order (constant strain) patch test and succeed in making this element acceptable and, therefore, adding a simple and effective element of general capability. The procedure of enforcing the patch test is of considerable interest and we shall therefore outline this here.

In the original element the shape functions defining each unknown (in stress analysis, the displacements u and v) are:

for corner nodes (*vide* Eq. (7.14))

$$N_i = (1+\eta_i\eta)(1+\xi_i\xi)/4 \qquad (i = 1,\ldots,4) \tag{11.4a}$$

and for the incompatible modes

$$\begin{aligned} \bar{N}_i &= (1-\xi^2) \qquad (i = 5) \\ \bar{N}_i &= (1-\eta^2) \qquad (i = 6) \end{aligned} \tag{11.4b}$$

The strains can now be written in the usual form

$$\boldsymbol{\varepsilon} = \mathbf{B}^e\mathbf{a}^e = \hat{\mathbf{B}}^e\mathbf{u}^e + \bar{\mathbf{B}}^e\bar{\mathbf{u}}^e \tag{11.5}$$

where parameters $\mathbf{u}^e$ refer to the corner node displacements and $\bar{\mathbf{u}}^e$ to the four internal degrees of freedom associated with the incompatible modes.

The stiffness relations for the element calculated in the normal manner result in nodal forces given as

$$\begin{bmatrix} \mathbf{K}^e_{uu} & \mathbf{K}^e_{u\bar{u}} \\ \mathbf{K}^e_{\bar{u}u} & \mathbf{K}^e_{\bar{u}\bar{u}} \end{bmatrix} \begin{Bmatrix} \mathbf{u}^e \\ \bar{\mathbf{u}}^e \end{Bmatrix} = \begin{Bmatrix} \hat{\mathbf{f}}^e \\ 0 \end{Bmatrix} \tag{11.6}$$

The equations for $\bar{\mathbf{u}}^e$, the internal parameters, are now fully assembled and these parameters can be eliminated at the element level before the assembly. We know that when the compatible modes only are considered, with $\mathbf{u}^e$ taking up at corners values corresponding to a linear displacement field, the nodal forces are such that the patch test is satisfied. *We now want to ensure that the contributions of incompatible modes is zero in such circumstances.* Solving the second of Eq. (11.6) we have

$$\bar{\mathbf{u}}^e = -\mathbf{K}^{e\;-1}_{\bar{u}\bar{u}}\mathbf{K}^e_{\bar{u}u}\mathbf{u}^e \tag{11.7}$$

and we want to make this quantity zero whenever $\mathbf{u}^e$ takes up values corresponding to a linear displacement field

$$\mathbf{u}^e = (\mathbf{u}^e)_l. \tag{11.8}$$

Thus, we require that

$$\mathbf{K}^e_{\bar{u}u}(\mathbf{u}^e)_l \equiv \int_{V^e} \bar{\mathbf{B}}^{e\mathrm{T}}\mathbf{D}\hat{\mathbf{B}}^e(\mathbf{u}^e)_l \, \mathrm{d}V = 0. \tag{11.9}$$

We note that

$$\mathbf{D}\mathbf{B}^e(\mathbf{u}^e)_l \equiv \bar{\boldsymbol{\sigma}}_l \tag{11.10}$$

is the corresponding constant stress state and now, for the patch test to be identically satisfied, we simply require that

$$\int_V \bar{\mathbf{B}}^{e\mathrm{T}}\,\mathrm{d}V \equiv \int_{-1}^{1}\int_{-1}^{1} \bar{\mathbf{B}}^{e\mathrm{T}}|J|\,\mathrm{d}\xi\,\mathrm{d}\eta = 0 \tag{11.11}$$

where $|J|$ is the Jacobian determinant corresponding to the transformation. The various terms of Eq. (11.11) will contain such products as

$$\frac{\partial \bar{N}_i}{\partial \xi}\cdot\frac{\partial x}{\partial \xi}; \qquad \frac{\partial \bar{N}_i}{\partial \eta}\cdot\frac{\partial x}{\partial \xi}, \text{ etc.,}$$

or, due to the quadratic form of the shape functions $\bar{N}_i$, such terms as $\xi(\partial x/\partial \xi)$; $\eta(\partial x/\partial \xi)$, etc. The integral of Eq. (11.11) will be automatically zero if $\partial x/\partial \xi$, $\partial x/\partial \eta$, etc., are all constants. With an isoparametric relation of the co-ordinates to the shape functions giving

$$x = \sum_{i=1}^{4} N_i x_i \qquad y = \sum_{i=1}^{4} N_i y_i \tag{11.12}$$

this will be automatically true if the element is a parallelogram where the Jacobian is a constant quantity, and this accounts for the satisfaction of the patch test in such cases.

Immediately the remedy for the distorted elements is apparent: by setting the values of $\partial x/\partial \xi$, etc., as constants when evaluating the matrices $\mathbf{K}^{e}_{u\bar{u}}$ (or $\mathbf{K}^{e}_{\bar{u}u}$) satisfaction of the patch test is achieved. This can easily be done by evaluating the necessary derivatives at $\xi = \eta = 0$, i.e., the centre of the element. This must only be done for these special matrices.

The new element thus derived is a very successful quadrilateral which generally gives a performance much improved over the compatible linear quadrilaterals. The cost of the new element is only slightly greater as the additional variables are eliminated at the element level. As the element becomes exactly the same as the original incompatible element for parallelogram situations, the virtues demonstrated in Fig. 11.4 are still completely preserved. The arguments associated here with the stiffness elasticity formulations are clearly applicable to other problems in which the functional contains only the first derivatives.

The particular element illustrated here was discussed in detail to show how a derivation based simply on the patch test satisfaction could be used to substantiate it. In a more general context similar approaches have been used by others[13,14] to derive elements which are quite successful, but at the same time a good amount of *intuition* has to be introduced to derive the incompatible modes. We shall show later how a more systematic approach can be derived.

To conclude, it is of interest to note that the incompatible mode element in its rectangular form is identical to one derived intuitively in one of the earliest finite element publications by Turner *et al.*[15] It appears to have been abandoned due to the original 'unscientific' approach used and re-derived again as a special case of a hybrid by Pian[16] (we shall return to this in Chapter 12, p. 320). These identities are brought out clearly in a recent text[17] and will be noted repeatedly with other incompatible elements.

11.4 **Substitute Shape Functions**

The success of the element described in the preceding section, and of the various non-conforming plate bending elements, was based on luck combined with good judgement. A more systematic way of improving the performance of fully compatible elements can surely be made. One way to do this is by the use of *substitute shape functions*, which are so designed that they approach the compatible shape function and its derivatives in a continuous manner as the size of the element is reduced. As convergence is available for the compatible shape functions (and always the patch test is satisfied), any formulation based on such substitute shape functions must also be convergent. (Although the argument here may sound heuristic to a mathematician, it can always be verified *a posteriori* by subjecting a new element so derived to a patch test.)

The poor performance of a conforming element can usually be attributed to terms additional to those of the highest complete polynomial which governs the convergence order—such terms being introduced to satisfy the continuity requirements only. If, thus, we could substitute for each shape function a polynomial of the same order as that of the highest complete polynomial contained in the original shape functions which would approach it in a suitable manner, an equivalent incompatible element, presumably with the same convergence order, would be achieved. An excellent way of doing this is to design the new shape function so that it approximates in a least square sense, the derivatives which enter the (stiffness) computations.

Consider, for instance, the linear quadrilateral of Fig. 11.3(*a*). We note that each shape function illustrated in Fig. 11.5(*a*) is complete only to linear terms and hence the substitute shape functions should be of the form

$$\hat{N}_i = \alpha_1 + \alpha_2 \xi + \alpha_3 \eta \tag{11.13}$$

In the original expansion the term $\xi\eta$ makes the derivatives vary linearly, as shown in Fig. 11.5(b), and their substitutes should be constants approximating to these in the least square sense. Without going into calculations, the substitute derivatives are obviously those shown and the new shape function is available except for the additive constant. This again should be so chosen as to make $\hat{N}_i$ a least square fit to N_i as shown in Fig. 11.5(c).

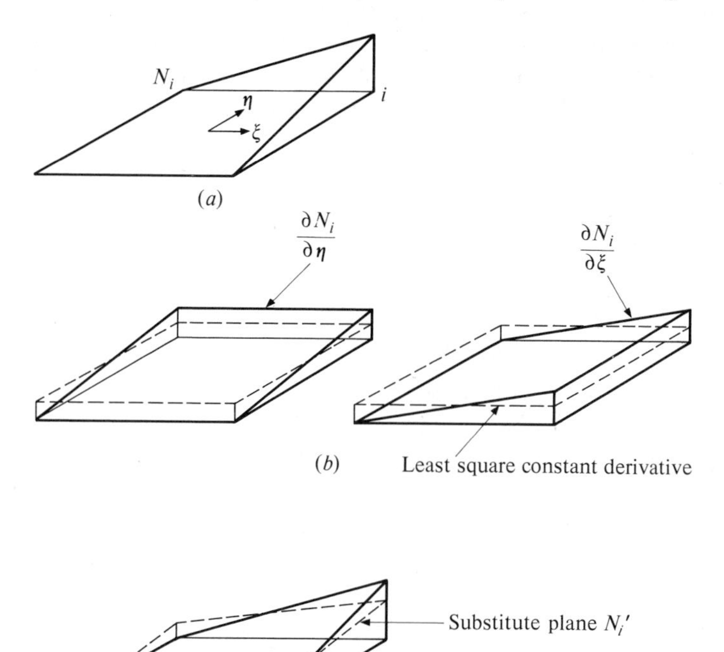

Fig. 11.5 Generation of substitute shape function N'_i for a linear quadrilateral (a), (b), and (c)

The application of the substitute function to the element just described yields satisfaction of the patch test and therefore convergence of the same order as that of the original element, but unfortunately in this case singular stiffness matrices are obtained. A clever way of using such substitute functions which avoids the difficulty of singularity was introduced by Doherty *et al.*[18] by applying these only to the shear components of the matrix which suffer most from the spurious energy terms.

One of the first systematic and successful applications of substitute shape functions was made by Irons and Razzaque[19–21] in the context of triangular plate bending elements. We have seen in Chapter 10 that in a twelve- (or nine-) degree-of-freedom triangle it was necessary to introduce non-polynomial components of the shape function, ε_i, [*vide* Eq. 10.37] in order to satisfy conformity of slope. In the case of the twelve-degree-of-freedom triangle, the other components of shape functions give a complete cubic polynomial. If, thus, a cubic polynomial could be substituted for ε_i in such a manner that its second derivative represented a least square fit of the actual derivatives of ε_i then a possibly successful but certainly convergent, incompatible, approximation could be achieved. In Fig. 11.6 we show such 'smoothed derivatives' replacing the very

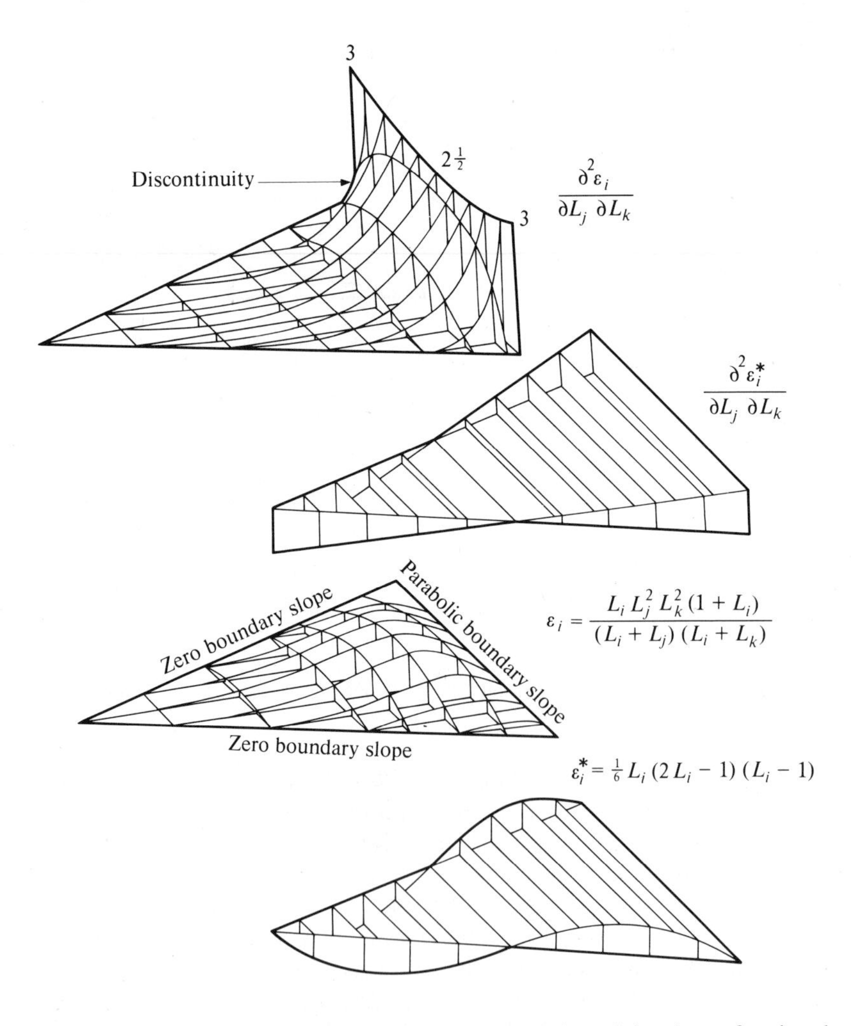

Fig. 11.6 Least square substitute cubic shape function ε_i^* in place of rational function ε_i for plate bending triangles

singular shape of the original derivatives, and the basic substitute function is also shown.

The success of this operation was tested extensively by experiment, and Fig. 10.18 of Chapter 10 includes the results of both the nine- and twelve-degree elements so obtained.

These show performances highly superior to those achieved by normal conforming elements and in the case of non-conforming nine-degree elements similar to that achieved by the alternatively designed non-conforming triangle of section 10.6.

The computations using the substitute shape functions are only carried out at the derivative stage and are quite complex. A completely coded program for the element is, however, given in references 20 and 21 for those interested in the detail. It is worth while remarking that in this case only the derivatives of the shape function were evaluated, as only these enter into the calculations of the stiffness matrices and thus no attempt was made to fit the full shape functions. For consistency of calculating the force matrices, etc., where these enter, the full substitute function should probably be used.

11.5 Why Substitute Functions yield Improved Results. Optimal Sampling and Reduced Integration

11.5.1. The substitute shape functions of the last section gave convergent elements but the reasons for improved performance were rather qualitative. A better understanding, however, is gained from a theorem which states that *minimization of an energy functional* Π *defined as*

$$\Pi = \tfrac{1}{2}\int_{\Omega} (\mathbf{Lu})^{\mathrm{T}}\mathbf{A}(\mathbf{Lu})\, d\Omega + \int_{\Omega} \mathbf{u}^{\mathrm{T}}\mathbf{p}\, d\Omega \tag{11.14}$$

which gives the exact solution $\mathbf{u} = \bar{\mathbf{u}}$ *is equivalent to minimization of another functional* Π^* *defined as*

$$\Pi^* = \tfrac{1}{2}\int_{\Omega} [\mathbf{L}(\mathbf{u}-\bar{\mathbf{u}})]^{\mathrm{T}}\mathbf{A}\mathbf{L}(\mathbf{u}-\bar{\mathbf{u}})\, d\Omega. \tag{11.15}$$

In the above $\mathbf{L}$ is a self-adjoint linear operator and $\mathbf{A}$ and $\mathbf{p}$ are prescribed matrices of position. The above quadratic form (Eq. 11.14) is such as arises in the majority of linear, self adjoint, problems.

This theorem is given in different forms by Herrmann,[22] Moan,[23] and Oden[24] and it shows that the approximate solution for $\mathbf{Lu}$ approaches the exact one $\mathbf{L}\bar{\mathbf{u}}$ as a *weighted least square approximation*.

In the context of elastic analysis, for instance, we can state that the minimization of total potential energy is equivalent to finding a weighted least square fit of the exact strains by those assumed approximately. We shall defer the proof of this theorem to the end of this section in order to draw some conclusions.

With a finite element approximation of order p the $\mathbf{Lu}$ derivatives can at best be approximated by local polynomials of the order $(p-m)$, where m is the differentiation order implied in the operator $\mathbf{L}$. Incomplete polynomials or other terms often add a spurious variation which still, however, in a least square sense approximate the true answer.

As continuity of derivatives is not imposed we can consider the approximation at element level where, in general, the matrix $\mathbf{A}$ tends to be constant. Here the *best fit* available in our finite element approximation will be of a least square fit complete polynomial of order $(p-m)$, Fig. 11.7. We have thus justified the reason for adopting this approximation in the use of substitute shape functions and for their optimal performance, but further important corollaries follow. Before discussing these we shall return to the proof of the theorem stated in Eqs. (11.14) and (11.15).

Proof of the theorem

Variation of Π defined in Eq. (11.14) gives at $\mathbf{u} = \bar{\mathbf{u}}$ (the exact solution)

$$\delta\Pi = \tfrac{1}{2}\int_\Omega (\mathbf{L}\,\delta\mathbf{u})^{\mathrm{T}}\mathbf{A}\mathbf{L}\bar{\mathbf{u}}\,\mathrm{d}\Omega + \tfrac{1}{2}\int_\Omega (\mathbf{L}\bar{\mathbf{u}})^{\mathrm{T}}\mathbf{A}\mathbf{L}\,\delta\mathbf{u}\,\mathrm{d}\Omega + \int_\Omega \delta\mathbf{u}^{\mathrm{T}}\mathbf{p}\,\mathrm{d}\Omega = 0 \tag{11.16}$$

or, if $\mathbf{A}$ is symmetric,

$$\delta\Pi = \int_\Omega (\mathbf{L}\,\delta\mathbf{u})^{\mathrm{T}}\mathbf{A}\mathbf{L}\bar{\mathbf{u}}\,\mathrm{d}\Omega + \int_\Omega \delta\mathbf{u}^{\mathrm{T}}\mathbf{p}\,\mathrm{d}\Omega = 0 \tag{11.17}$$

in which $\delta\mathbf{u}$ is any arbitrary variation. Thus we can write

$$\delta\mathbf{u} = \mathbf{u}$$

and

$$\int_\Omega (\mathbf{L}\mathbf{u})^{\mathrm{T}}\mathbf{A}\mathbf{L}\bar{\mathbf{u}}\,\mathrm{d}\Omega + \int_\Omega \mathbf{u}^{\mathrm{T}}\mathbf{p}\,\mathrm{d}\Omega = 0. \tag{11.18}$$

Subtracting the above from Eq. (11.14), and noting the symmetry of the $\mathbf{A}$ matrix, we can write

$$\Pi = \tfrac{1}{2}\int_\Omega [\mathbf{L}(\mathbf{u}-\bar{\mathbf{u}})]^{\mathrm{T}}\mathbf{A}\mathbf{L}(\mathbf{u}-\bar{\mathbf{u}})\,\mathrm{d}\Omega - \tfrac{1}{2}\int_\Omega (\mathbf{L}\bar{\mathbf{u}})^{\mathrm{T}}\mathbf{A}\mathbf{L}\bar{\mathbf{u}}\,\mathrm{d}\Omega \tag{11.19}$$

where the last term is not subject to variation. Thus

$$\Pi^* = \Pi + \text{const.} \tag{11.20}$$

and its stationality is equivalent to the stationality of Π.

11.5.2 *Optimal sampling points.* In Fig. 11.7 we show a curve which represents an assumed exact variation of the quantity $(\mathbf{L}\bar{\mathbf{u}})$ and a set of piecewise linear least square approximations to it $(\mathbf{Lu})$. It is evident that at some points within each segment the approximate solution *must* equal the exact one. If we knew in advance where such points were located we could always find the exact solution; clearly a dream hardly to be realized.

However, a useful property of numerical integration (Gauss–Legendre) points can help us here. *This property can be stated as follows: If we devise a numerical integration formula with a minimum number of sampling points which just integrates precisely a polynomial of degree* $2M+1$, *then generally at such points a polynomial of order* $M+1$ *is equal to its least square approximations by a polynomial of order* M.

This proposition is exactly true in the case of one-dimensional Gauss point integration and approximately satisfied for other two- and three-dimensional integration expressions.[23]

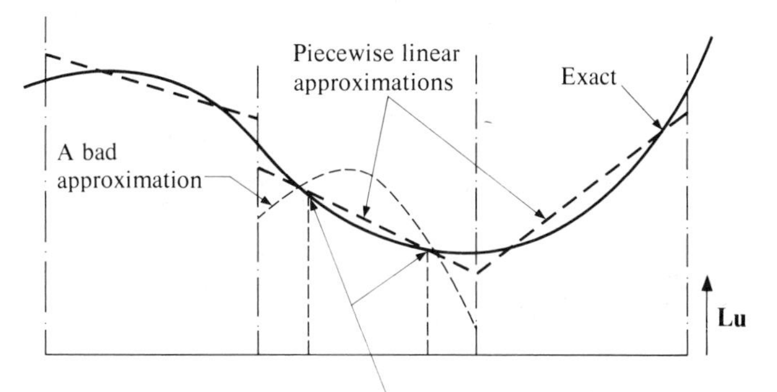

Fig. 11.7 Piecewise linear least square fits to a curve **Lu**

It is immediately obvious in the example shown that if the exact curve were a parabola, then two Gauss points would define uniquely a straight line which is the least square approximation to it. Conversely, if we sampled the approximation **Lu** at these points we would thus obtain an accuracy of one order greater than that available elsewhere by the approximation. Clearly such points are optimal for sampling the quantity **Lu** (or the strains and stresses in an elasticity problem).

We can state quite generally that the approximation to **Lu** is always of the order $O(h^{p-m+1})$, where p is the complete polynomial in the approximating shape function and m the order of the operator **L** (Chapter 3, p. 65). Therefore, *at numerical integration points which just integrate exactly a polynomial of order* $2(p-m)+1$ [i.e., with an error of order $O(h^{2(p-m)+2})$], *the approximation to* **Lu** *will be nearly one order better, i.e.,* $O(h^{p-m+2})$.

Obviously in any finite element computation it pays therefore to sample the strains of such integration points as has been realized by many investigators.[23–26]

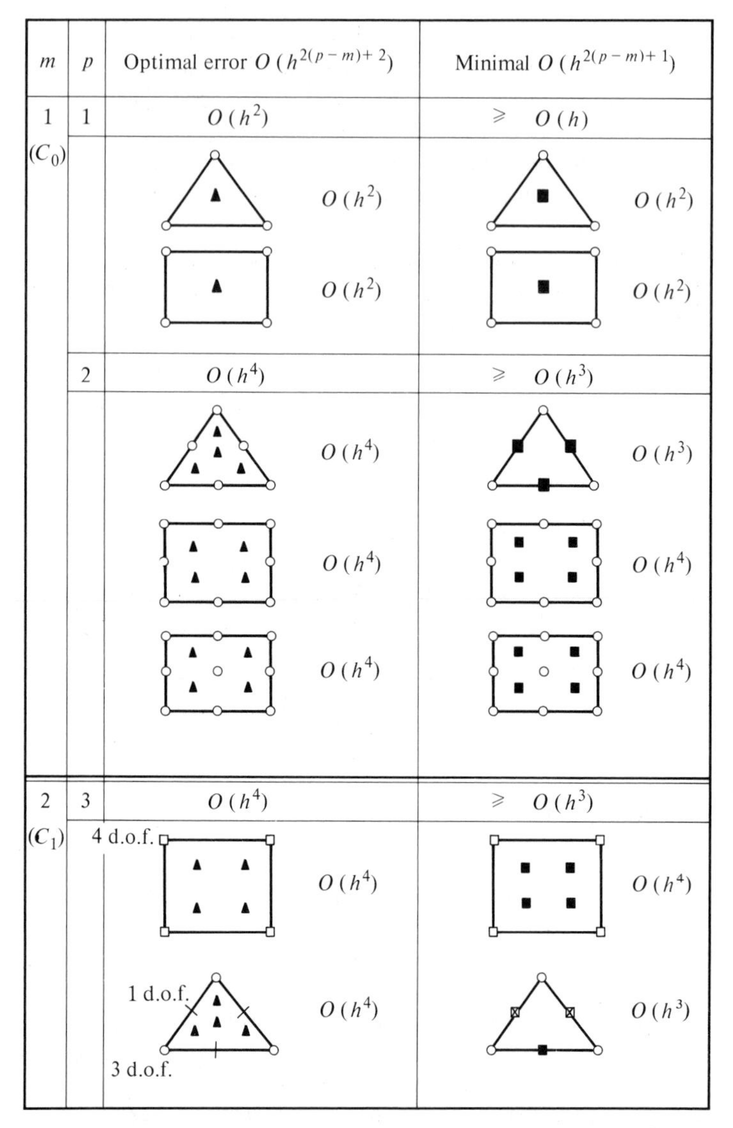

Fig. 11.8 'Optimal' sampling and 'minimum' integration points for some C_0 and C_1 elements

Figure 11.8 shows some such optimal points for sampling for various C_0 type elements ($m = 1$) and some C_1 type ($m = 2$) (plate bending).

The results for linear triangles and quadrilaterals are physically obvious (and we have already remarked in Chapter 4, p. 104, that 'obviously' the stresses would be best represented at centroids). For higher order C_0 elements and for plate bending the results are by no means self-evident—they turn out, however, to be true.

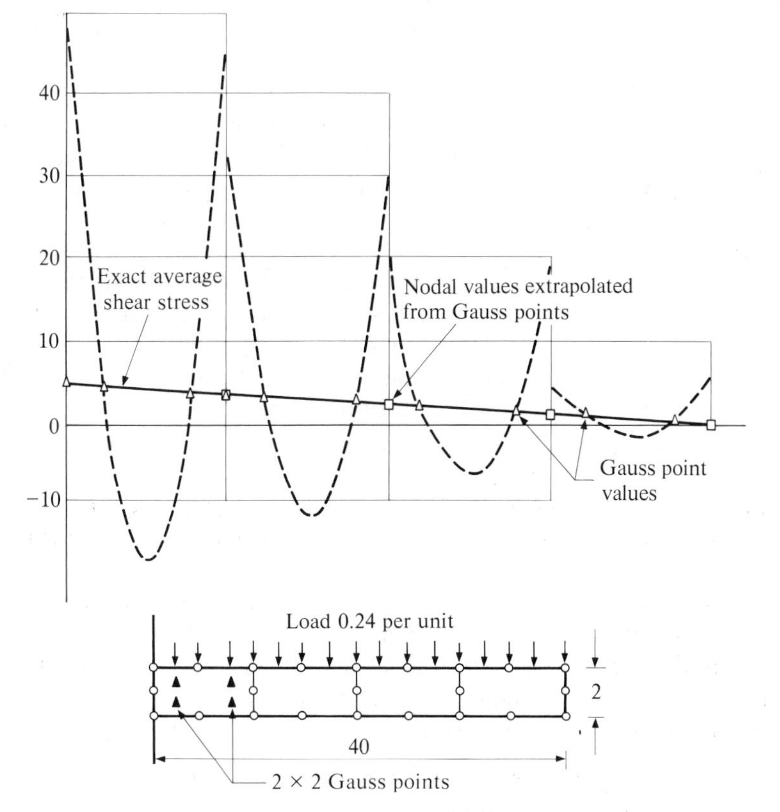

Fig. 11.9 A cantilever beam with four parabolic elements. Stress sampling at Gauss points and linear extrapolation to nodes

In Fig. 11.9 we show, for instance an analysis of a cantilever beam using four quadratic, 'serendipity' type, elements. Whilst the results for deflections and axial stresses are excellent, the shear stresses show a parabolic 'variation' in each element which provides an extremely poor representation of the actual stresses. However, the values sampled at the Gauss points are an excellent representation of the correct mean shear stresses.

Similar improvements can be shown in the context of other elements and problems, although (fortunately) the discrepancies are not always so large.

The example just quoted suggests that in quadratic C_0 elements, whether two- or three-dimensional, the stresses (or similar quantities) *should never be calculated at nodes*. If nodal values are desired, then a simple bilinear extrapolation from Gauss points should be made. Such values are again shown to be excellent in Fig. 11.9. Further examples of such extrapolations are given by Hinton and Campbell.[25] Hinton *et al.*[27] give a very simple extrapolation algorithm for use in such problems.[27]

11.5.3 *Reduced numerical integration.* In Chapter 8 we discussed in detail the procedures of numerical integration orders sufficient to maintain original convergence of the element. This was shown to be such that errors in the integration should be at least of the order $O(h^{2(p-m)+1})$. Such minimum integration points using Gauss–Legendre quadrature formula are shown in Fig. 11.8.

We note that if numerical integration were made at the points corresponding to optimal sampling, convergence would be obtained and we would also expect improved results due to the property of such points discussed in the last section. Further, the use of such points would be generally identical to the use of substitute shape functions, providing the original shape functions contained only polynomial terms of one order higher than those in the complete polynomial.

This is certainly true of all serendipity type elements where the use of 'reduced' integration has proved to lead to a dramatic improvement in results.[28–31]

In Fig. 11.10 we show an example of the type of improvement which can be expected. The problem is one of an elastic hollow sphere subject to internal pressure and is solved as an axi-symmetric situation using both 3×3 and 2×2 Gauss point integration. Different values of the Poisson's ratio are also used. Some interesting points are observed.

(*a*) For *both* types of numerical integration the stresses turn out to be best at the 2×2 Gauss quadrature points.

(*b*) Results are improved if 2×2 integration is used in formulation of matrices (*vide* displacements).

(*c*) The difference between 2×2 and 3×3 integration is most pronounced as Poisson's ratio $\nu \to 0{\cdot}5$, and here the improvement due to reduced integration is quite dramatic.

The first two facts can easily be explained by the arguments of the preceding section. The reason for the improved performance as the behaviour approaches incompressibility is due to a cause which we shall discuss in the next section.

11.6 **Reduced Integration and Penalty Function Formulation**

11.6.1 *Incompressibility and penalty function constraints.* In Chapters 4 to 6 we noted that, even with isotropic elasticity, the displacement formulation becomes inapplicable as the Poisson's ratio approaches 0·5.

Consider for a moment the elasticity matrix **D** corresponding to three-dimensional analysis and given in terms of the elastic modulus E and the

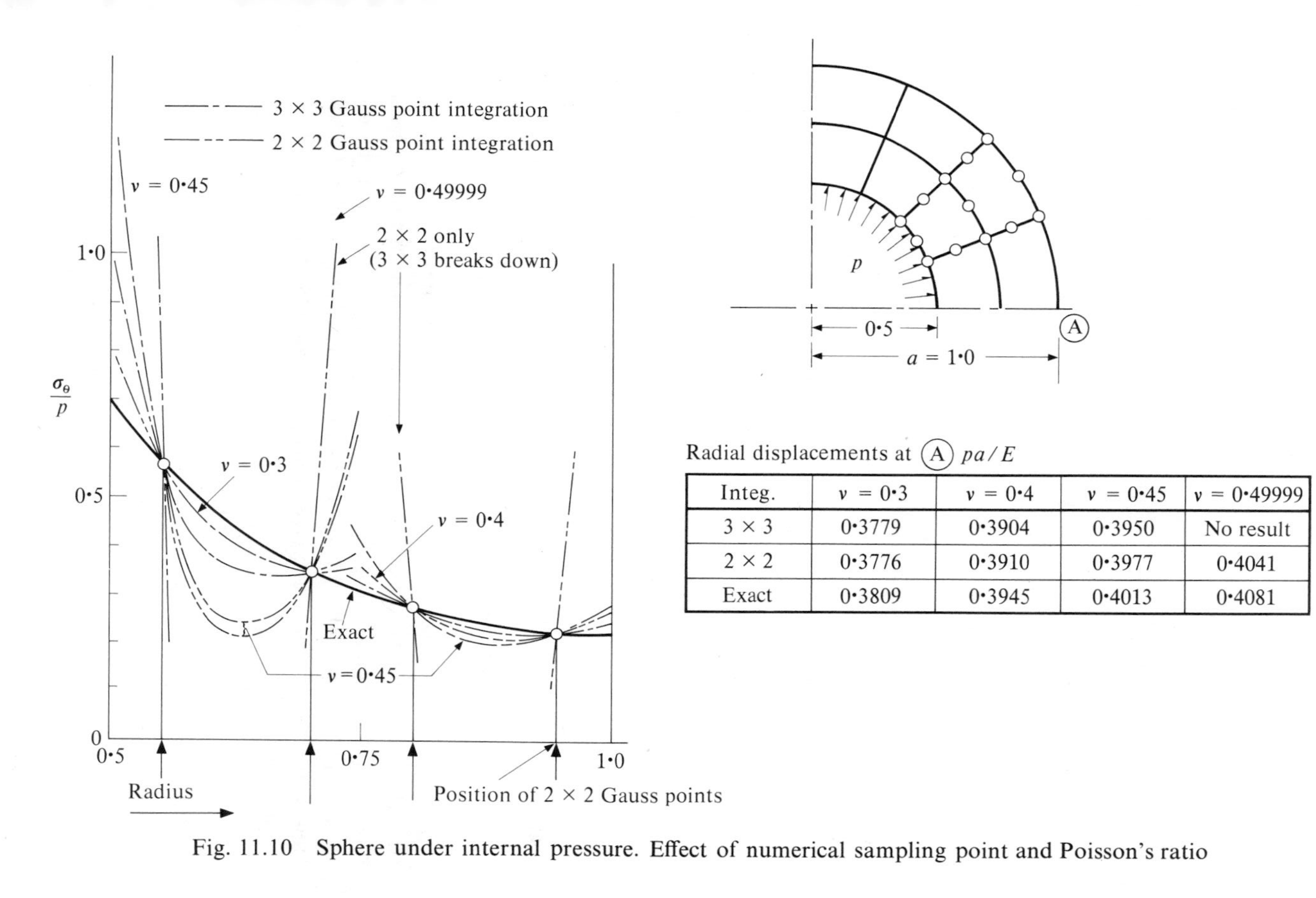

Radial displacements at (A) pa/E

Integ.	ν = 0·3	ν = 0·4	ν = 0·45	ν = 0·49999
3 × 3	0·3779	0·3904	0·3950	No result
2 × 2	0·3776	0·3910	0·3977	0·4041
Exact	0·3809	0·3945	0·4013	0·4081

Fig. 11.10 Sphere under internal pressure. Effect of numerical sampling point and Poisson's ratio

Poisson's ratio ν [Eq. (6.14)]. Clearly, various terms become indeterminate for $\nu = 0{\cdot}5$. Let us rewrite this matrix in terms of two other well-known quantities: the shear modulus G and the bulk modulus K, defined as

$$G = E/2(1+\nu); \qquad K = E/3(1-2\nu). \tag{11.21}$$

We can now express the elasticity matrix as

$$\mathbf{D} = G\begin{bmatrix} 2 & & & & & \\ & 2 & & & \mathbf{0} & \\ & & 2 & & & \\ & & & 1 & & \\ & \mathbf{0} & & & 1 & \\ & & & & & 1 \end{bmatrix} + (K-\tfrac{2}{3}G)\begin{bmatrix} 1 & 1 & 1 & & & \\ 1 & 1 & 1 & & \mathbf{0} & \\ 1 & 1 & 1 & & & \\ & \mathbf{0} & & & \mathbf{0} & \\ & & & & & \end{bmatrix}$$

$$= G\mathbf{D}^s + 2\alpha\mathbf{D}^v \tag{11.22}$$

and we note that as $\nu \to 0{\cdot}5$, G remains finite but $2\alpha \equiv (K-\frac{2}{3}G) \to \infty$.

The strain energy U can now be written as

$$U = \tfrac{1}{2}\int \boldsymbol{\varepsilon}^{\mathrm{T}}(G\mathbf{D}^s)\boldsymbol{\varepsilon}\,\mathrm{d}\Omega + \tfrac{1}{2}\int \boldsymbol{\varepsilon}^{\mathrm{T}}(\mathbf{D}^v\alpha)\boldsymbol{\varepsilon}\,\mathrm{d}\Omega \tag{11.23}$$

or, noting the structure of the $\mathbf{D}^v$ matrix and introducing volumetric strain,

$$\varepsilon_v \equiv \varepsilon_x + \varepsilon_y + \varepsilon_z \equiv \frac{\partial u}{\partial x} + \frac{\partial v}{\partial y} + \frac{\partial w}{\partial z} = \mathbf{m}\boldsymbol{\varepsilon} \tag{11.24}$$

and $$\mathbf{m}^{\mathrm{T}} = [1, 1, 1, 0, 0, 0]$$

$$U = U_s + U_v = \tfrac{1}{2}\int_\Omega \boldsymbol{\varepsilon}^{\mathrm{T}}(G\mathbf{D}^s)\boldsymbol{\varepsilon}\,\mathrm{d}\Omega + \int_\Omega \alpha\varepsilon_v^2\,\mathrm{d}\Omega. \tag{11.25}$$

The two terms in the above can be interpreted as the distortional (shear strain) energy and volumetric strain energy. Further, we note that the notion of incompressible behaviour approached via $\alpha \to \infty$ is identical to formulating the problem in terms of a minimization of distortional energy given in terms of displacements and a simultaneous imposition of a constraint

$$C(\mathbf{u}) \equiv \varepsilon_v = 0. \tag{11.26}$$

If this constraint is introduced by the penalty function procedure[32] (see Chapter 3, p. 83), we would then seek stationality of

$$\Pi^* = U_s + \alpha\int_\Omega \mathbf{C}^{\mathrm{T}}\mathbf{C}\,\mathrm{d}\Omega + W \tag{11.27}$$

where $\mathbf{C}(\mathbf{u})$ is a linear type of constraint and W is the potential energy of the forces. With the usual discretization and functional of the above form will yield

$$\mathbf{u} = \mathbf{N}\mathbf{a}$$

and

$$\mathbf{K}\mathbf{a}+\mathbf{f} \equiv (\mathbf{K}_1+\alpha\mathbf{K}_2)\mathbf{a}+\mathbf{f} = 0. \tag{11.28}$$

In the above the matrices $\mathbf{K}_1$ and $\mathbf{K}_2$ are finite, and we seek a solution with $\alpha \to \infty$.

Clearly as α increases, the solution tends to

$$\alpha\mathbf{K}_2\mathbf{a}+\mathbf{f} = 0$$

and

$$\mathbf{K}_2\mathbf{a} = -\mathbf{f}/\alpha \to 0. \tag{11.29}$$

Thus the constraints take over a dominant role and if the matrix $\mathbf{K}_2$ is non-singular only a trivial result, $\mathbf{a} = 0$, is possible.

Here lies the greatest difficulty of obtaining results for the incompressible problem with a simple finite element formulation and using values of ν tending to, but not equal to, 0·5. With a linear triangle, for instance, very bad results are obtained with values of Poisson's ratio as low as $\nu = 0{\cdot}45$,[33] and we have seen in the example of Fig. 11.10 that the results are equally bad with 3×3 (or exact) integrations for the parabolic elements. This illustrates a problem of over-constraint, which has been discussed extensively in the literature.[34–38]

The remedy for such over-constraint is to impose singularity on the matrix $\mathbf{K}_2$ so that

$$\mathbf{K}_2\mathbf{a} = 0 \qquad \text{but} \qquad \mathbf{a} \neq 0. \tag{11.30}$$

The higher degree of singularity, the bigger the influence of the important distortional energy controlling now the final answer. Such singularity is introduced into the $\mathbf{K}_2$ matrix by the use of low integration orders, as we shall see in the next section, and for problems of such a penalty class the use of reduced integration is particularly recommended.

We shall also note that for 'penalty' problems the introduction of singularity is more important than the use of optimal sampling, and when the two integration orders required to introduce these do not coincide, singularity requirement takes preference.

11.6.2 *Singularity and reduced integration.* In Chapter 8 we have discussed the conditions at which singularity will be introduced into an assembled element matrix by low order of integration. With problems of the penalty-function form given in Eqs. (11.28) we have two contradictory conditions. First we want matrix $\mathbf{K}_2$ to have a high degree of singularity; second we want the matrix $\mathbf{K}$ (at least in its assembled form) to be non-singular. These would seem to indicate that different orders of numerical integration may be necessary for each term to achieve success but usually, though by no means always, it turns out that this is unnecessary and then the general formulation is simple.

The singularity (or its absence) in each of the two matrices depends on the number of independent relations used at each integrating point in these formulations as we have noted in Chapter 8, p. 203.

Consider, for instance, the incompressibility problem cast in two dimensions. When forming the matrix $\mathbf{K}_1$, three independent relations are introduced at each Gauss point, but for the matrix $\mathbf{K}_2$ only one such relation (volumetric strain) is concerned. *If the number of such relations introduced at all the integrating points is less than that of the degrees of freedom available, then singularity must exist.* In Chapter 8 we have made some such calculations for a general field of elements. We shall repeat such calculations here for a field constrained along two edges and note if singularities are introduced by addition of new elements. Thus, if the number of degrees of freedom (nodes introduced $\times$ nodal variables) is equal to or greater than the number of independent relations (integrating points $\times$ independent relations at each) then singularity certainly will exist.

In Fig. 11.11(*a*) we show the results for various types of C_0 elements for the incompressibility problem we have just discussed. We note that singularity of $\mathbf{K}_2$ is introduced with certainty for linear quadrilaterals and single point integration, parabolic triangles with three point integration, and both serendipity and Lagrangian quadratic elements with four point integration. All these elements have indeed proved excellent in the context of incompressible analysis.[30,34,35] For all the elements no singularity of the $\mathbf{K}_1$ matrix apparently develops as noted from the same Fig. 11.11.

The type of calculation outlined shows when singularity must occur, but nearly singular matrices can be introduced by less constrained boundary conditions. Thus for the linear quadrilateral[34,35] four integration points are used to determine the $\mathbf{K}_1$ matrix to avoid singularity of this, despite the fact that these are no longer optimal sampling points. In the serendipity quadratic element this is not necessary, and the optimal four point quadrature do just what is desired and make this element justly popular for many uses.

11.6.3 *Beam and plate bending with independent slope and displacement interpolation.* This section introduces a further useful application of the reduced integration concepts due mainly to the relaxation of the singularity constraints of the $\mathbf{K}_2$ matrix.

In Chapter 10 we have discussed the bending of plates where the slope was given as a derivative of the displacements with the latter representing the only unknowns. Difficulties of introduction of the desired continuity were encountered and many researchers have been tempted to rephrase

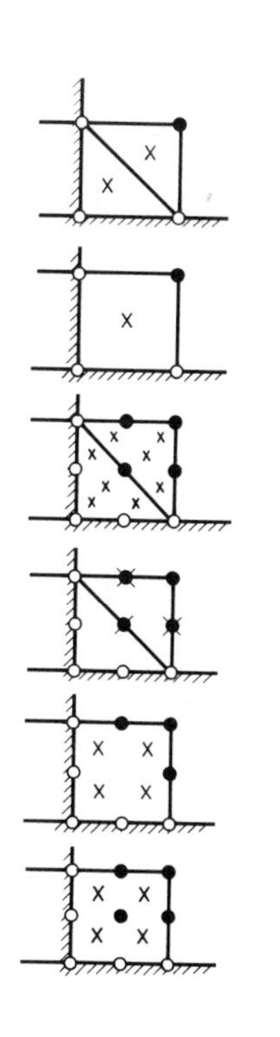

(a) d.o.f.	(a) $\mathbf{K}_1$	(a) $\mathbf{K}_2$	(b) d.o.f.	(b) $\mathbf{K}_1$	(b) $\mathbf{K}_2$
1 × 2 = 2	2 × 3 = 6	2 × 1 = 2	1 × 3 = 3	2 × 3 = 6	2 × 2 = 4
1 × 2 = 2	1 × 3 = 3	1 × 1 = 1 Singular	1 × 3 = 3	1 × 3 = 3	1 × 2 = 2 Singular
4 × 2 = 8	8 × 3 = 24	8 × 1 = 8	4 × 3 = 12	8 × 3 = 24	8 × 2 = 16
4 × 2 = 8	4 × 3 = 12	4 × 1 = 4 Singular	4 × 3 = 12	4 × 3 = 12	4 × 2 = 8 Singular
3 × 2 = 6	4 × 3 = 12	4 × 1 = 4 Singular	3 × 3 = 9	4 × 3 = 12	4 × 2 = 8 Singular
4 × 2 = 8	4 × 3 = 12	4 × 1 = 4 Singular	4 × 3 = 12	4 × 3 = 12	4 × 2 = 8 Singular
d.o.f.	Cons.		d.o.f.	Cons.	

● New node
× New integration points

n - degrees of freedom per node
k - independent relations introduces per integrating points

Fig. 11.11 Singularity of K_1 and K_2 matrices with numerical integration for an *extension* of one element
(*a*) Incompressible elasticity; K_1 = shear strain matrix
K_2 = volumetric strain matrix
(*b*) Plate bending; K_1 = bending strain matrix
K_2 = transverse shear matrix

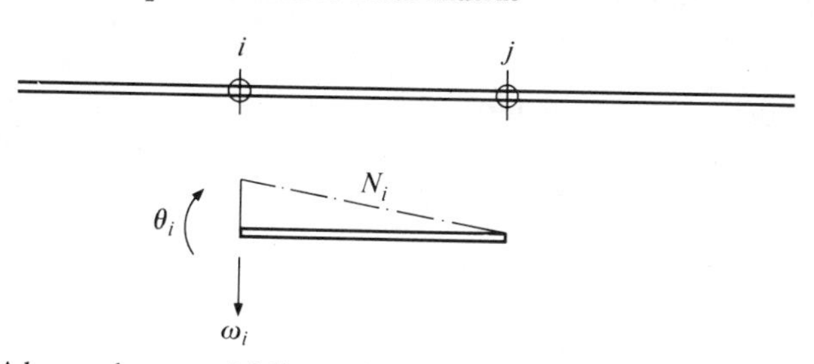

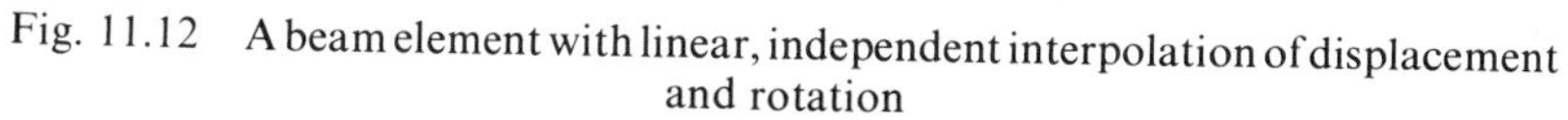

Fig. 11.12 A beam element with linear, independent interpolation of displacement and rotation

the problem using an independent interpolation of slopes and displacements and imposing the relation between these as a constraint.[31,39,40] If this is done we shall find once again that reduced integration, by introducing suitable singularities, produces excellent elements. Again we shall find that the penalty number has a definite physical meaning.

It is simplest to introduce the new concept in terms of beams rather than plate bending. Here 'strain' can be defined in terms of rotation as (see Fig. 11.12)

$$\varepsilon = \frac{d\theta}{dx} \tag{11.31}$$

and the 'stress' is now the bending moment,

$$\boldsymbol{\sigma} \equiv M = D\varepsilon \tag{11.32}$$

where $D = Et^3/12$. The total potential energy can now be written as

$$\Pi = \tfrac{1}{2}\int_0^L \frac{Et^3}{12}\left(\frac{d\theta}{dx}\right)^2 dx - \int_0^L wq\, dx \tag{11.33}$$

where w is the lateral deflection. However, to maintain continuity the energy expression must be minimized with a constraint relating the deflections and slope

$$C(w, \theta) \equiv \frac{dw}{dx} - \theta = 0. \tag{11.34}$$

Introducing this constraint by the penalty method we have thus to minimize

$$\Pi^* = \Pi + \int_0^L \alpha\left(\frac{dw}{dx} - \theta\right)^2 dx, \qquad \alpha \to \infty \tag{11.35}$$

and only C_0 continuity is required now of the variables w and θ [viz. for comparison Chapter 2, p. 37, also Chapter 3, p. 86].

Immediately the physical meaning of the above is apparent (just as it was in the example of incompressible behaviour). If we put

$$\alpha = \kappa G t \tag{11.36}$$

we note that the second term represents the shear strain energy for a beam in which the shear strain is given by (κ accounts for a correction due to non-uniform shear distribution)

$$\varepsilon_\gamma \equiv \frac{dw}{dx} - \theta \neq 0 \tag{11.37}$$

thus the solution with $\alpha \neq \infty$ has now a definite 'physical' meaning, as representing the deflections of the beam with shear distortion.

On minimization and assembly of equations, once again equations of general form (11.28) are obtained, i.e.,

$$(\mathbf{K}_1 + \alpha \mathbf{K}_2)\mathbf{a} + \mathbf{f} = 0. \tag{11.38}$$

(The reader is advised to obtain explicitly the matrices for this exercise which are particularly simple with linear variations of $\mathbf{a}_i^{\mathrm{T}} = [w_i, \theta_i]$.)

With exactly integrated matrices $\mathbf{K}_1$ and $\mathbf{K}_2$ the results are a reasonable approximation for the case of very thick beams (α small) but tend to zero with increasing values of α. With a single point integration (which is, in this case, exact for the bending energy), the matrix $\mathbf{K}_2$ becomes singular and we could expect good answers.

Introducing a non-dimensional parameter, defined as

$$\alpha^1 = \frac{\kappa G}{E} \frac{L^2}{t^2} \tag{11.39}$$

we can compare results directly. In Table 11.1 we show the results for such a linear element applied to the analysis of a cantilever beam carrying an end load.

TABLE 11.1
DEFLECTIONS OF A CANTILEVER BEAM WITH END POINT LOAD[40]
LINEAR INTERPOLATION OF w AND θ. VALUES ARE NORMALIZED AS RATIOS OF APPROXIMATE TO EXACT SOLUTIONS

No. of elements	Thick beam $\alpha^1 = 7{\cdot}2$		Thin beam $\alpha^1 = 7{\cdot}2 \times 10^5$	
	One point int.	Exact int.	One point int.	Exact int.
1	0·752	0·0416	0·750	$0{\cdot}2 \times 10^{-4}$
2	0·940	0·445	0·938	$0{\cdot}3 \times 10^{-4}$
4	0·985	0·762	0·984	$0{\cdot}32 \times 10^{-3}$
8	0·996	0·927	0·996	$0{\cdot}128 \times 10^{-3}$
16	0·999	0·981	0·999	$0{\cdot}512 \times 10^{-3}$

The results are remarkable but easily explained by our previous arguments. In the case of the thick beam ($\alpha^1 = 7{\cdot}2$) both exact and the reduced integration give reasonable answers. But even here there is an improvement with the reduced integration which *can be related to optimal sampling*.

For a very thin beam, $\alpha^1 = 7{\cdot}2 \times 10^5$, the exact integration results tend to zero (as was predicted) but reduced integration by introducing a singular $\mathbf{K}_2$ continues to give excellent results, yielding now, in fact, one of the best simple beam elements available.

Clearly the full imposition of constraints is excessive and its satisfaction at one point is just about right. It is worth remarking that such a constraint could have been introduced directly and not by means of a penalty function. Such approaches are called 'discrete Kirchhoff constraints' and, in the case of the beam, has been discussed in reference 17, pp. 366–7. We shall return to such constraints in the Section 11.7.

All the arguments used here can be extended to the use of reduced integration with higher order beam elements, and all of these turn out correct. However, the plate problem is obviously more interesting as we have already recognized the many difficulties in the standard theory.

Extending the arguments used in deriving the beam bending functional we can formulate the strain energy of an isotropic plate (*vide* Chapter 10) as

$$\Pi^* = \tfrac{1}{2}\int\int_\Omega \boldsymbol{\varepsilon}^{\mathrm{T}}\mathbf{D}\boldsymbol{\varepsilon}\,\mathrm{d}x\,\mathrm{d}y + \int\int_\Omega \alpha_1\left(\frac{\partial w}{\partial x}-\theta_x\right)^2\mathrm{d}x\,\mathrm{d}y +$$

$$\int\int_\Omega \alpha_2\left(\frac{\partial w}{\partial y}-\theta_y\right)^2\mathrm{d}x\,\mathrm{d}y - \int\int_\Omega wq\,\mathrm{d}x\,\mathrm{d}y \qquad (11.40)$$

where $\mathbf{D}$ is the standard plate elasticity matrix and the strains $\boldsymbol{\varepsilon}$ are given in terms of first derivatives of θ_x, θ_y (*vide* Eq. (10.2))

$$\boldsymbol{\varepsilon}^{\mathrm{T}} = \left[-\frac{\partial\theta_x}{\partial x},\ -\frac{\partial\theta_y}{\partial y},\ \frac{\partial\theta_x}{\partial y}+\frac{\partial\theta_y}{\partial x}\right] \qquad (11.41)$$

and the penalty numbers α_1 and α_2 enforce the thin plate constraints

$$\frac{\partial w}{\partial x}-\theta_x = 0$$

$$\frac{\partial w}{\partial y}-\theta_y = 0. \qquad (11.42)$$

Obviously θ_x, θ_y, and w can again be interpolated with C_0 continuity only, and for an isotropic plate we can write, putting non-dimensionally the parameter $\alpha_1 = \alpha_2 = \alpha$, and recognizing its physical meaning

$$(\mathbf{K}_1+\alpha\mathbf{K}_2)\mathbf{a}+\mathbf{f} = 0 \qquad (11.43)$$

where

$$\alpha = \frac{\kappa G}{E}\left(\frac{L}{t}\right)^2 \quad \text{and} \quad \mathbf{a}_i^{\mathrm{T}} = [\theta_{xi}, \theta_{yi}, w_i].$$

Now a whole new series of elements can be derived for plate bending, each one of these based on the well-known C_0 interpolations of Chapter 7 and each one capable of isoparametric distortion. We would expect to get best answers for thick plates with *optimal integration positions* (*see Fig. 11.8(a)*) *and for thin plates with the minimal integration order* (*see Fig. 11.8(b)*) *consistent with a non-singularity of the assembled matrix* (*see Fig.*

11.11(b)): This indeed turns out to be true. In Fig. 11.13 we show a convergence test for some linear and quadratic elements. The linear quadrilateral (A) is sampled for shear stresses alone at the optimal (and minimum) single integrating point[40] and at four integrating points for the bending to ensure non singularity of the total stiffness matrix. The

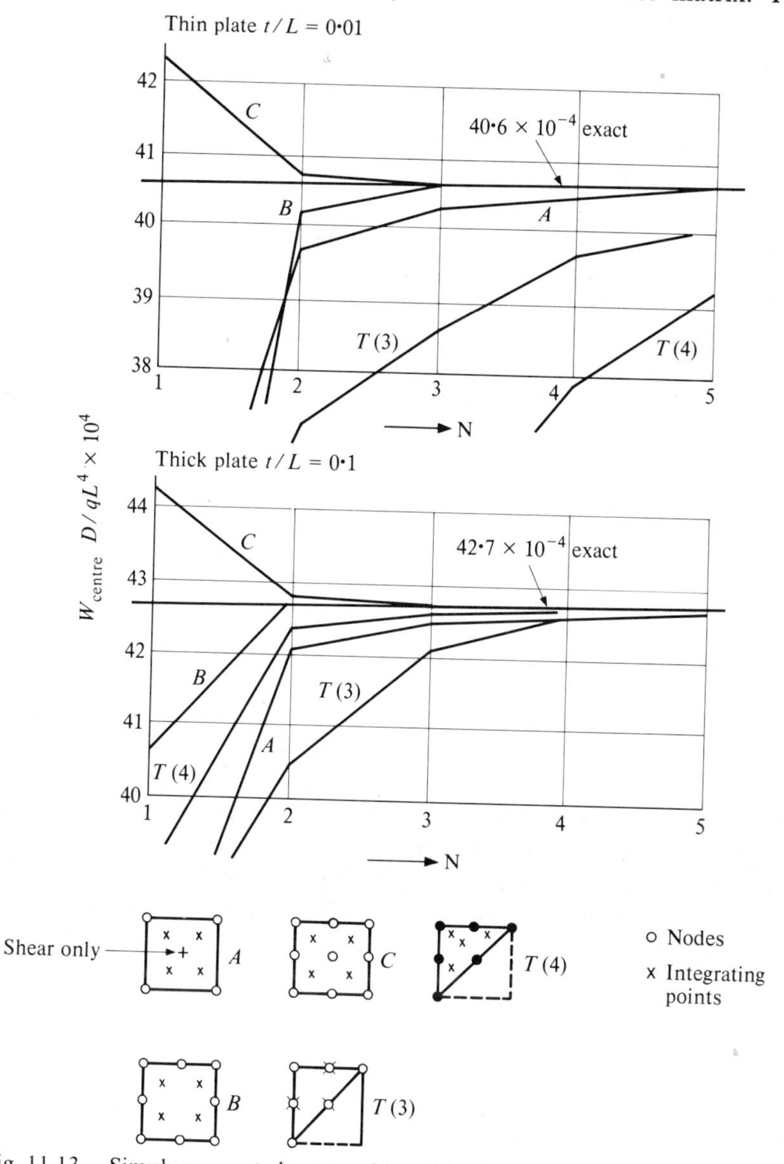

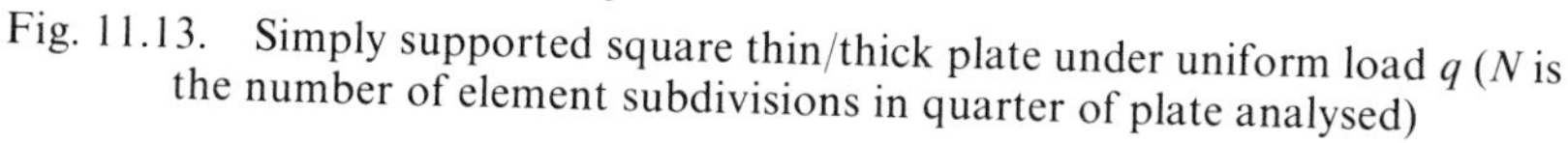

Fig. 11.13. Simply supported square thin/thick plate under uniform load q (N is the number of element subdivisions in quarter of plate analysed)

quadrilaterals (B) and (C) use four Gauss points which, at the same time, are optimal and minimal and provide the necessary singularity of $\mathbf{K}_2$. In triangles the optimum integration is provided by a four-point formula and we would expect this to give the best answers for thick plates, whereas for the introduction of singularity in the $\mathbf{K}_2$ matrix, three-point integration is necessary. The results of T(3) and T(4) bear this out. The same elements with the exact integration yield, in most cases, results which are so far off in the figures that they cannot be shown.

The realization of advantages of reduced integration in the above context came first for serendipity rectangles by physical reasoning[28, 29] and later for other elements reported here.[31, 39, 40, 41, 42]

These elements today compare well in performance with 'conventional' elements of Chapter 10, as shown in the plot of the standard test example in Fig. 11.14 (compare this with Fig. 10.18 of Chapter 10).

11.6.4 *Some further remarks on penalty function forms. Error balancing*, Despite the good behaviour of penalty-function formulation for incompressibility and plate bending, it is quite evident that the limiting solutions cannot be obtained when $\alpha = \infty$ or indeed when it has a finite but very large value. Long before this point, ill-conditioning of equations will occur. We have observed, however, that practical results with which an engineer is well content can be reached when α is sufficiently large. Indeed, it is well worth remarking that in physical situations these are more realistic as the extreme values are never taken up by the coefficients.

We have now to consider how large α should be made (or what limits should be placed in the calculation on the bulk modulus or shear deformability) to obtain acceptable results. Some answers to the above question comes from the realization that $\alpha < \infty$ and $h > 0$ both lead to errors, and it is reasonable to *consider a balancing of these* in such a manner that $\alpha \to \infty$ as $h \to 0$.[37, 43] Limiting convergence is thus still achieved. Indeed, it is reasonable to consider the ratio of such errors. If we look in detail at the plate bending problem, for instance, we note that (*a*) the discretization error is of the order of $O(h/L)^n$, where $n = p+1$, and (*b*) the error of not satisfying the constraints is of the order $O(1/\alpha)$, where

$$\alpha = \frac{\kappa G}{E}\left(\frac{L}{t}\right)^2.$$

The ratio of the errors is thus

$$\gamma = \frac{\kappa G}{E}\left(\frac{L}{t}\right)^2\left(\frac{h}{L}\right)^n$$

or, for $n = 2$,

$$\gamma = \frac{\kappa G}{E}\left(\frac{h}{t}\right)^2. \tag{11.44}$$

Keeping this parameter constant during any series of calculation ensures that in the limit both discretization and constraint errors disappear. Similar reasoning can obviously be applied to the incompressibility problem.

It can be shown that it is the parameter γ which leads to ill-conditioning when it becomes too large, and in computation it is advantageous to

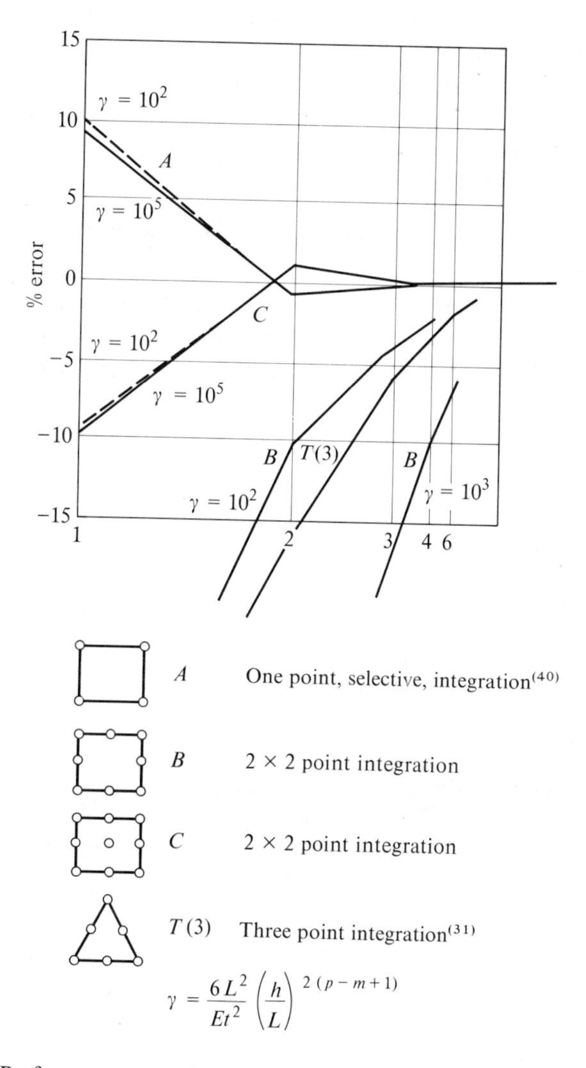

$$\gamma = \frac{6L^2}{Et^2}\left(\frac{h}{L}\right)^{2(p-m+1)}$$

Fig. 11.14 Performance test for plate elements using independent rotation/displacement interpolation and reduced integration (Simply supported square plate—uniform load)

determine the optimal value of γ which would be inserted in the calculation in place of the real physical variables, if exceeded. Indeed this optimum need not necessarily coincide with the maximum value for which ill-conditioning or divergence occurs[39,43] as often, for smaller values, a better cancellation of errors may occur.

In Fig. 11.14 we show the values of γ used for the convergence of the elements concerned.

In concluding this section concerned with plate bending, we must add that the standard patch tests pertaining to thin plates (i.e., imposition of constant curvature tests) can no longer be passed exactly by any of the elements unless the value of the constraint parameter $\alpha = \infty$.

Clearly other types of patch tests need to be considered.

11.7 **Direct application of constraints in the generation of incompatible elements**

In the previous section we have shown how many physical problems can be presented as constrained minimization situations and how such formulations lead (as illustrated by the plate problem) to a relaxation of continuity requirements.

In such cases the constraints have to be applied in an approximate manner so as not to 'lock' the solution to a trivial, zero, answer. Such a relaxation was introduced via use of reduced integration but alternative procedures are available. One of these is the application of the constraints at a limited number of points of the domain only. This introduces a set of additional equations which, if the whole system is treated, lead to intractable non-symmetric matrices. The alternative of applying such

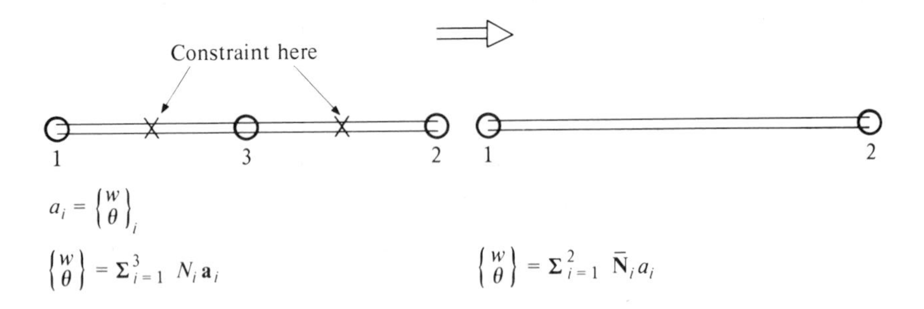

Fig. 11.15 A beam element with independent, Lagrangian, interpolation of w and θ with constraint $\dfrac{\partial w}{\partial x} - \theta = 0$ applied at points X.

constraints at the element level and thus eliminating some of the element variables is clearly possible—but now even the C_0 continuity may be violated. Nevertheless, a number of very successful elements have been so derived and, if the patch test is satisfied, then convergence can be assured.

As an example consider the beam problem given by Eq. 11.33, in which a parabolic interpolation of θ and w is used as shown in Fig. 11.15 and only the bending energy given by Eq. 11.33 is considered.

$$\begin{Bmatrix} w \\ \theta \end{Bmatrix} = \sum_{i=1}^{3} N_i \begin{Bmatrix} w \\ \theta \end{Bmatrix}_i \tag{11.45}$$

Now we shall *enforce* the constraint exactly

$$\frac{\mathrm{d}w}{\mathrm{d}x} - \theta = 0$$

at two points of the beam with co-ordinates x_α and x_β.

We can now write two additional equations which limit the freedom of interpolation given by Eq. 11.45

$$\sum_{i=1}^{3} \dot{N}_i(\alpha) w_i - \sum_{i=1}^{3} N_i(\alpha)\theta_i = 0$$

$$\sum_{i=1}^{3} \dot{N}_i(\beta) w_i - \sum_{i=1}^{3} N_i(\beta)\theta_i = 0$$

$$N(\alpha) \equiv N(x_\alpha) \qquad \dot{N}(\alpha) = \left(\frac{\mathrm{d}N}{\mathrm{d}x}\right)_{x=x_\alpha}, \text{ etc.} \tag{11.46}$$

thus providing an opportunity for elimination of two of the parameters, say, w_2 and θ_2.

Writing above explicitly

$$\begin{bmatrix} \dot{N}_3(\alpha), & -N_3(\alpha) \\ \dot{N}_3(\beta), & -N_3(\beta) \end{bmatrix} \begin{Bmatrix} w_3 \\ \theta_3 \end{Bmatrix} = -\begin{bmatrix} \dot{N}_1(\alpha), & -N_1(\alpha) \\ \dot{N}_1(\beta), & -N_1(\beta) \end{bmatrix} \begin{Bmatrix} w_1 \\ \theta_1 \end{Bmatrix} - \begin{bmatrix} \dot{N}_2(\alpha), & -N_2(\alpha) \\ \dot{N}_2(\beta), & -N_2(\beta) \end{bmatrix} \begin{Bmatrix} w_2 \\ \theta_2 \end{Bmatrix}$$

or

$$\mathbf{A}_3 \begin{Bmatrix} w_3 \\ \theta_3 \end{Bmatrix} = -\mathbf{A}_1 \begin{Bmatrix} w_1 \\ \theta_1 \end{Bmatrix} - \mathbf{A}_2 \begin{Bmatrix} w_2 \\ \theta_2 \end{Bmatrix} \tag{11.47}$$

and substituting into (11.45) shape functions involving only the two end points can be obtained, i.e.,

$$\begin{Bmatrix} w \\ \theta \end{Bmatrix} = \sum_{i=1}^{2} \bar{\mathbf{N}}_i \begin{Bmatrix} w \\ \theta \end{Bmatrix}_i$$

with

$$\begin{aligned} \bar{\mathbf{N}}_i &= N_i\mathbf{I} - N_3\mathbf{A}_3^{-1}\mathbf{A}_i \\ \mathbf{I} &= \begin{bmatrix} 1 & 0 \\ 0 & 1 \end{bmatrix} \end{aligned} \tag{11.48}$$

These shape functions are clearly 'incompatible', but it will be found that using these in the functional of Eq. (11.33) a convergent element is obtained. The reason for this convergence is perhaps self evident as the original problem is correctly posed and the pointwise satisfaction of constraints imposes these uniformly as the element size decreases.

It is of interest to note, that if the points at which the constraint is imposed coincide with the elements two Gauss points then the stiffness matrix obtained is precisely that of the C_1 continuous cubic interpolation described in Chapter 2, pp. 37–38. Clearly the special property of the two Gauss points discussed earlier in this chapter is responsible for this optimum.

It is clear that the process outlined provides now a systematic, but not always successful, means of generating shape functions for other one-dimensional approximations in which C_1 continuity is involved—and indeed for two- (or possibly three-) dimensional problems. The plate problem (and others similar to it) provide an obvious field of application.

In Fig. 11.16, a series of two-dimensional elements for the plate bending problem is shown in which internally applied constraints enforcing the Kirchhoff conditions of Eq. (11.42) are used to eliminate various connecting variables. The 'virgin element' refers to the interpolation used before the application of the constraints and both this and the final elements achieved are indicated. All the elements can clearly be distorted isoparametrically, but not all pass automatically the patch test in such form. The reader should note that in all such formulations the final shape functions are obtained by a process involving a matrix inversion at element level.

The constraint can be applied in a simple manner as in the first element of the series Fig. 11.16(*a*) introduced by Razzaque[21] simply enforcing both conditions of Eq. (11.42) at internal Gauss points. It is found, however, that the element so derived violates continuity sufficiently not to pass the patch test in other than rectangular forms. It appears generally useful to follow the beam example and enforce the condition

$$\gamma_s = \frac{\partial w}{\partial s} - \theta_s = 0 \tag{11.49}$$

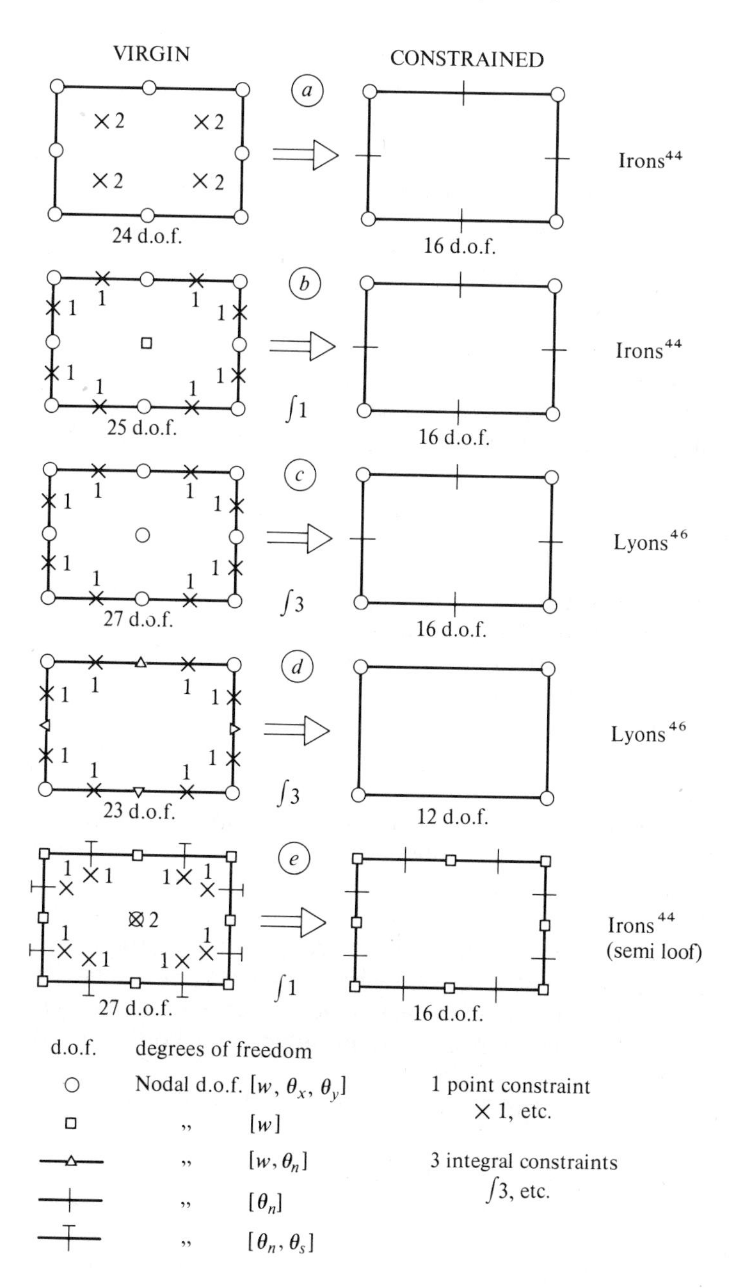

Fig. 11.16 A series of incompatible C^1 (plate bending) elements with constraint at element level. (With the exception of (*b*) all satisfy the patch test in isoparametric distorted forms).

at the Gauss points along the element sides and with s defining the tangential direction. This condition first used by Irons[44] and Baldwin *et al.*[45] appears to lead to elements satisfying the patch test under all configurations (Fig. 11.16(*b*)).

In addition to point constraints, integrals requiring a satisfaction of these 'in the mean' are employed in the above elements. Here for instance

$$\int_{\Gamma^e} \gamma_s \, \mathrm{d}s = 0 \tag{11.50}$$

where Γ^e is the total perimeter is introduced for the element of Fig. 11.16(*b*).

Lyons[46] introduces two further integral constraints requiring that, for an element of area Ω^e

$$\int_{\Omega^e} \gamma_x \, \mathrm{d}\Omega = \int_{\Omega^e} \gamma_y \, \mathrm{d}\Omega = 0 \tag{11.51}$$

within each element. This permits the generation of a series of interesting and useful elements of quadrilateral and triangular shape two of the former being illustrated in Fig. 11.16(*c*) and (*d*).

Again, all such elements are found to pass the patch test in a general isoparametric form and therefore form a useful series.

The most complex of elements of constrained type is undoubtedly that derived by Irons[44] and given the name of SEMILOOF. This element has been successfully used in the context of both plates and shells where it exhibits particularly useful characteristics. In Fig. 11.16(*e*) we show both the 'virgin' and final degrees of freedom and note that here a rather particular interpolation for θ_x and θ_y in the virgin form is adopted. In this two Gauss point nodes on the midsides and a central node are used leading to a form which does not present even a C_0 continuity, but which by the suitable placement of such nodes minimizes discontinuity effects. Such interpolations were initially introduced by Loof[47] are discussed in Ref. 44.

It is to be noted that the discrete imposition of constraints leads to many elements which are efficient from the nodal degree of freedom viewpoint and which perform well. The main drawback is the formulation which requires quite complex inversions and calculations to be carried out at element level—and their economy is not therefore guaranteed.

Clearly many new possibilities are open in both C_1 and C_0 formulations. In the latter the incompressibility problems of section 11.6.1 present several as yet unexplored avenues for deriving elements which could form useful application in fluid mechanics. (*Vide* Chapter 23.)

11.8 Concluding Remarks

We have shown in this chapter how, by judicious judgement, a virtue can be made of the 'variational crimes' of incompatibility and inexact integration. Some of the '*best*' *elements* for two- and three-dimensional analysis and for plate bending problems have been indicated here and we shall show later (Chapters 14 and 16) how such elements can be used for the analysis of shell situations, very effectively.

It has been shown that in some cases of plate problems the use of reduced integration is equivalent to application of 'discrete Kirchhoff constraints'.[17,47]

We shall show in Chapter 12 that the use of substitute shape functions and incompatible modes yields, in many cases, element forms which are identical to those which can be derived by other procedures such as the hybrid formulation. However, the formulation here is more direct and in its own right provides procedures widely applicable in finite element analysis.

In particular we would like to draw attention once again to the identity of the smoothed triangular plate bending elements[20,21] to those derived by hybrid principles by Allman.[49]

References

1. B. M. Irons, 'Numerical integration applied to finite element methods', *Conf. on Use of Digital Computers in Structural Engineering*, Univ. of Newcastle, 1966.
2. G. P. Bazeley, Y. K. Cheung, B. M. Irons and O. C. Zienkiewicz, 'Triangular elements in plate bending. Conforming and nonconforming solutions', *Proc. 1st Conf. Matrix Methods in Structural Mechanics*, pp. 547–76, AFFDL-TR-CC-80, Wright-Patterson A.F. Base, Ohio, 1966.
3. B. M. Irons and A. Razzaque, 'Experience with the patch test for convergence of finite elements method', *Mathematical Foundations of the Finite Element Method*, pp. 557–87 (ed. A. R. Aziz), Academic Press, 1972.
4. G. Strang and G. J. Fix, *Analysis of the Finite Element Method*, Prentice-Hall, 1973.
5. B. Fraeijs de Veubeke, 'Variational principles and the patch test', *Int. J. Num. Meth. Eng.*, **8**, 783–801, 1974.
6. B. M. Irons, 'The patch test for engineers', *Conf. Atlas Computing Center*, March 1974, Harwell, U.K.
7. G. Strang, 'Variational crimes in the finite element method', in 'The Mathematical foundations of the finite element method with application to partial differential equations', pp. 689–710 (ed. A. K. Aziz), Academic Press, 1972.
8. E. R. A. Oliveira, 'Results on the convergence of the finite element method in structural and non-structural cases' in *Finite Element Method in Engineering*, pp. 3–14 (eds. V. A. Pulmano and A. P. Kabaila), Univ. of New South Wales, 1974.
9. E. R. A. Oliveira, 'Theoretical foundations of the finite element method'. *Int. J. Solids Struct.*, **4**, 929–82, 1968.

10. L. R. Herrmann, 'Efficiency of a two dimensional incompatible finite element', *Comp. Struct.*, **3**, 1377–96, 1973.
11. E. L. Wilson, R. L. Taylor, W. P. Doherty and T. Ghabussi, 'Incompatible displacement models' in *Numerical and Computer Methods in Structural Mechanics*, pp. 43–57 (ed. S. T. Fenves *et al.*), Academic Press, 1973.
12. R. L. Taylor, P. J. Beresford and E. L. Wilson, 'A non-conforming element for stress analysis', *Int. J. Num. Meth. Eng.*, **10**, 1211–20, 1976.
13. P. G. Bergan and L. Hanssen, 'A new approach for deriving "good" element stiffness matrices', *The Mathematics of Finite Elements and Applications*, ed. J. R. Whiteman, Academic Press, 1977.
14. G. Sander and P. Beckers, 'Delinquent finite elements for shell idealisation', *Proc. World Congress on Finite Element Methods in Structural Mechanics*, Bournemouth, England, 1975.
15. M. J. Turner, R. W. Clough, H. C. Martin and L. J. Topp, 'Stiffness and deflection analysis of complex structures', *J. Aero. Sci.*, **23**, 805–24, 1956.
16. T. H. H. Pian, 'Derivations of element stiffness matrices by assumed stress distributions', *J.A.I.A.A.*, **2**, 1333–5, 1964.
17. R. H. Gallagher, *Finite Element Analysis Fundamentals*, pp. 275–80, Prentice-Hall, 1975.
18. W. P. Doherty, E. L. Wilson and R. L. Taylor, *Stress analysis of axisymmetric utilizing higher order quadrilateral elements*, Univ. of Calif., Berkeley, Struct. Eng. Lab. Report SESM 61-3, 1969.
19. B. M. Irons and A. Razzaque, 'Shape function formulation for elements other than displacement models', pp. 4/59–71, *Symp. Variational Methods.*, Univ. of Southampton, 1972.
20. A. Razzaque, 'Program for triangular bending elements with derivative smoothing', *Int. J. Num. Meth. Eng.*, **6**, 333–45, 1973.
21. A. Razzaque, *Finite element analysis of plates and shells*, Ph.D. Thesis, Civ. Eng. Dept., Univ. Coll., Swansea, 1972.
22. L. R. Herrmann, 'Interpretation of finite element procedure in stress error minimisation', *Proc. Am. Soc. Civ. Eng.*, **98**, EM5, pp. 1331–6, 1972.
23. (a) T. Moan, 'On the local distribution of errors by the finite element approximation', in *Theory and Practice in Finite Element Standard Analysis* (eds. Y. Yamada and R. H. Gallagher), Univ. of Tokyo Press, 1973.
23. (b) T. Moan, 'Orthogonal polynomials and "best" numerical integration formulas on a triangle', *Z.A.M.M.*, **54**, 501–8, 1974.
24. J. T. Oden, *Finite Elements of Non-linear Continua*, McGraw-Hill, 1971.
25. E. Hinton and J. Campbell, 'Local and global smoothing of discontinuous finite element functions using a least square method', *Int. J. Num. Meth. Eng.*, **8**, 461–80, 1974.
26. J. Barlow, 'Optimal stress locations in finite element models', *Int. J. Num. Meth. Eng.*, **10**, 243–51, 1976.
27. E. Hinton, F. C. Scott and R. E. Ricketts, 'Local least squares stress smoothing for parabolic isoparametric elements', *Int. J. Num. Meth. Eng.*, **9**, 235–56, 1975.
28. O. C. Zienkiewicz, R. L. Taylor and J. M. Too, 'Reduced integration techniques in general analysis of plates and shells', *Int. J. Num. Meth. Eng.*, **3**, 275–90, 1971.
29. S. F. Pawsey and R. W. Clough, 'Improved numerical integration of thick shell finite elements', *Int. J. Num. Meth. Eng.*, **3**, 545–86, 1971.
30. D. J. Naylor, 'Stresses in nearly incompressible materials for finite elements

with application to the calculation of excess pore pressures', *Int. J. Num. Meth. Eng.*, **8**, 443–60, 1974.
31. O. C. ZIENKIEWICZ and E. HINTON, 'Reduced integration, function smoothing and non-conformity in finite element analysis', *J. Franklin Inst.*, **302**, 443–61, 1976.
32. O. C. ZIENKIEWICZ, 'Constrained variational principles and penalty analysis function methods in finite elements', *Conf. on Numerical Solution of Differential Equations*, Dundee, 1973; *Lecture Notes on Mathematics*, Springer, 1973.
33. T. J. R. HUGHES and H. ALLIK, 'Finite element of compressible and incompressible continua', pp. 27–62, *Proc. Symp. Civ. Eng.*, Vanderbilt Univ. (A.S.C.E. publication), 1969.
34. T. J. R. HUGHES, R. L. TAYLOR and J. L. SACKMAN, *Finite element formulations and solutions of contact–impact problems in continua mechanics*, Part III, Univ. of California, Berkeley U.C.S.E.S.M. 75-7, 1975.
35. T. J. R. HUGHES, R. L. TAYLOR and J. F. LEVY, 'A finite element method for incompressible flows', pp. 1–16, *Conf. Finite Element Methods in Flow Problems*, St. Margharita, Italy, 1976.
36. J. C. NAGTEGAAL, D. M. PARKS and J. R. RICE, 'On numerically accurate finite element solutions in the fully plastic range', *Comp. Meth. Appl. Mech. Eng.*, **4**, 153–78, 1974.
37. I. FRIED, 'Finite element analysis of incompressible materials by residual energy balancing', *Int. J. Solids Struct.*, **10**, 993–1002, 1974.
38. J. H. ARGYRIS, P. C. DUNNE, T. ANGELOPOULOS and B. BICHAT, 'Large natural strains and some special difficulties due to non-linearity and incompressibility in finite elements', *Comp. Meth. Appl. Mech. Eng.*, **4**, 219, 1974.
39. E. D. L. PUGH, E. HINTON and O. C. ZIENKIEWICZ, 'A study of quadrilateral plate bending elements with reduced integration', *Int. J. Num. Meth. Eng.* (to be published), 1977–8.
40. T. J. R. HUGHES, R. L. TAYLOR and W. KANOKNUKULCHAI, 'A simple and efficient finite element for plate bending', *Int. J. Num. Meth. Eng.*, **11**, 1529–43, 1977.
41. E. HINTON, A. RAZZAQUE, O. C. ZIENKIEWICZ and J. D. DAVIES, 'Simple finite element solution for plates of homogeneous, sandwich and cellular construction', *Proc. Inst. Civ. Eng.*, Part II, **59**, 43–65, 1975.
42. R. D. COOK, 'More on reduced integration and isoparametric elements', *Int. J. Num. Meth. Eng.*, **3**, 275–90, 1971.
43. I. FRIED, 'Shear in C^0 and C^1 bending finite elements', *Int. J. Solids Struct.*, **9**, 449–60, 1973.
44. B. M. IRONS, 'The Semiloof shell element', Chapter 11, pp. 197–222 of *Finite Elements for thin shells and curved members*, eds. D. G. Ashwell and R. H. Gallagher, Wiley, 1976.
45. J. T. BALDWIN, A. RAZZAQUE and B. M. IRONS, 'Shape function subroutine for an isoparametric thin plate element', *Int. J. Num. Mech. Eng.*, **7**, 431, 1973.
46. L. P. R. LYONS, 'A general finite element system with special reference to the analysis of cellular structures', *Imp. Coll. of Sc. Techn.*, London, 1977.
47. H. W. LOOF, 'The economical computation of stiffness of large structural elements', *Int. Symp. on Use of Comp. in Struct. Eng.*, University of Newcastle upon Tyne, 1966.
48. G. A. WEMPNER, J. T. ODEN and D. A. KROSS, 'Finite element analysis of thin shells', *Proc. Am. Soc. Civ. Eng.*, EM6, 1273–94, 1968.
49. D. J. ALLMAN, 'Triangular finite elements for plate bending with constant and linearly varying bending moments', *IUTAM Symp. on High Speed Computing of Elastic Structures*, Univ. of Liége, 1971.

12. Lagrangian Constraints in Energy Principles of Elasticity. 'Complete Field' and 'Interface Variable' (or Hybrid) Methods

12.1 Introduction

In the preceding chapters we have dealt with elasticity problems using the virtual work equations formed in terms of displacements **u** alone. The alternative formulation which yielded the same results defined the total potential energy

$$\Pi = \Pi(\mathbf{u}) \tag{12.1}$$

and this was minimized again directly with **u** as the dependent variable. The choice of trial functions for **u** was limited only by certain continuity of the field or boundary value satisfaction. Even here, however, we encountered difficulties and at times continuity conditions had to be relaxed. Similar situations exist in most of the mathematically posed problems of Chapter 3 and all of these could be tackled with the direct approximation used.

In some problems, however, the functional is described in terms of variables which themselves have to satisfy quite complex constraints, and these are often conveniently dealt with by use of Lagrangian multipliers defined over the whole domain Ω of the problem (or, on occasions, only on some of its parts). This introduces additional unknowns to discretization. The increase of the unknowns is often compensated for by relaxed continuity requirements (e.g., C_0 continuity in place of C_1 continuity) and thus simpler trial functions can be used.

The Lagrangian multipliers can often be identified physically and in solid mechanics lead to so-called mixed variational principles which we shall introduce in this chapter. Although the concept of Lagrangian multipliers was introduced in Chapter 3 we shall here elaborate on this and indicate areas of fruitful applicability.

Occasionally it is convenient to define the Lagrangian multiplier or some of the other basic variables only on *element interfaces*, with other variables defining the field inside the element. We shall refer to such procedures as *interface variable* or *hybrid* in distinction to *complete field* methods and will show that frequently these approaches can be very useful.

The imposition of constraints by the Lagrangian multipliers is generally approximate, just as is the minimization of the basic functional. Occasionally, however, it is possible to impose the constraints exactly by a discrete number of Lagrangian multipliers, and where this is possible some properties of the original basic functionals can be preserved. We shall try to indicate some such possibilities in structural mechanics but clearly the coverage of this chapter must be restricted by the space available.

12.2 Some General, Complete Field Variational Principles for Elasticity

12.2.1 *Potential and complementary energy.* The virtual work principle given in Chapter 2 (and elaborated in Chapter 3, p. 58) can be stated as a requirement that

$$\int_{\Omega} \delta\boldsymbol{\varepsilon}^{\mathrm{T}}\boldsymbol{\sigma}\,\mathrm{d}\Omega - \int_{\Omega} \delta\mathbf{u}^{\mathrm{T}}\mathbf{b}\,\mathrm{d}\Omega - \int_{\Gamma_t} \delta\mathbf{u}^{\mathrm{T}}\bar{\mathbf{t}}\,\mathrm{d}\Gamma = 0. \tag{12.2}$$

This is true for any system of stresses $\boldsymbol{\sigma}$ with equilibrating tractions $\mathbf{t}$ and body forces $\mathbf{b}$, providing that the arbitrary virtual strains $\delta\boldsymbol{\varepsilon}$ and virtual displacements $\delta\mathbf{u}$ are compatible and observe prescribed boundary conditions, i.e., that

$$\delta\boldsymbol{\varepsilon} = \mathbf{L}\,\delta\mathbf{u} \quad \text{in } \Omega \tag{12.3}$$

where $\mathbf{L}$ is an appropriate strain operator (*vide* Chapter 2, Eq. (2.2)) and

$$\delta\mathbf{u} = 0 \quad \text{on } \Gamma_u \tag{12.4}$$

where boundary displacements are prescribed.

If we express the dependent variables as

$$\boldsymbol{\sigma} = \boldsymbol{\sigma}(\boldsymbol{\varepsilon}) \tag{12.5a}$$

and

$$\boldsymbol{\varepsilon} = \mathbf{L}\mathbf{u} \tag{12.5b}$$

we find that the above leads to a convenient method of discretizing the problem, now expressed in terms of displacements as unknowns. On discretization (and limiting the possible virtual displacements), equilibrium is satisfied approximately but at all stages compatibility is assured. In such solutions we have restricted ourselves to a linear relationship between stress and strain

$$\boldsymbol{\sigma} = \mathbf{D}(\boldsymbol{\varepsilon}-\boldsymbol{\varepsilon}_0)+\boldsymbol{\sigma}_0 \tag{12.6}$$

but the validity of the virtual work statement is universal, as we shall see in Chapter 18 dealing with non-linear materials.

If U, the strain energy, exists such that

$$\delta U = \int_\Omega \delta\boldsymbol{\varepsilon}^{\mathrm{T}}\boldsymbol{\sigma}\,\mathrm{d}\Omega \tag{12.7}$$

the virtual work statement is equivalent to that of minimizing the total potential energy

$$\Pi = U(\boldsymbol{\varepsilon})+V(\mathbf{u}) \tag{12.8}$$

where

$$V = -\int_\Omega \mathbf{u}^{\mathrm{T}}\mathbf{b}\,\mathrm{d}\Omega-\int_{\Gamma_t} \mathbf{u}^{\mathrm{T}}\bar{\mathbf{t}}\,\mathrm{d}\Gamma \tag{12.9}$$

For linear elastic materials the above expression was given explicitly in Chapter 2 (p. 31); the preceding arguments starting from the principle of virtual work can however be paraphrased.

Let $\boldsymbol{\sigma}$ be a state of stress in equilibrium with the body forces $\mathbf{b}$ and prescribed tractions $\bar{\mathbf{t}}$, and the strain $\boldsymbol{\varepsilon}$ be a function of stress obtained by inverting Eq. (12.5a). We now have

$$\boldsymbol{\varepsilon} = \boldsymbol{\varepsilon}(\boldsymbol{\sigma}) \tag{12.10}$$

Further, let $\delta\boldsymbol{\sigma}$ be the virtual stress satisfying equilibrium with both body forces and tractions on Γ_t set to zero.

On equating the sum of internal and external virtual work to zero we assure that the boundary displacements on Γ_u and strains are compatible. Thus the statement

$$\int_\Omega \delta\boldsymbol{\sigma}^{\mathrm{T}}\boldsymbol{\varepsilon}\,\mathrm{d}\Omega-\int_{\Gamma_u} \delta\mathbf{t}^{\mathrm{T}}\bar{\mathbf{u}}\,\mathrm{d}\Gamma = 0 \tag{12.11}$$

where $\delta\mathbf{t} = \mathbf{G}\,\delta\boldsymbol{\sigma}$ ($\mathbf{G}$ defines the boundary tractions in terms of stresses), assures compatible strains and displacements. If we now define a quantity U^*, called complementary energy, such that

$$\delta U^* = \int_\Omega \delta\boldsymbol{\sigma}^{\mathrm{T}}\boldsymbol{\varepsilon}\,\mathrm{d}\Omega; \qquad U^* = U^*(\boldsymbol{\sigma}) \tag{12.12}$$

then the variation of the following functional set to zero

$$\Pi^* = U^*(\boldsymbol{\sigma}) - \int_{\Gamma u} \mathbf{t}^{\mathrm{T}}\bar{\mathbf{u}}\, \mathrm{d}\Gamma = U^* + V^* \tag{12.13}$$

is equivalent to the satisfaction of Eq. (12.11).

For linear stress–strain relations we can write in absence of initial strain or stress

$$U^* = \int_{\Omega} \frac{1}{2} \boldsymbol{\sigma}^{\mathrm{T}}\mathbf{D}^{-1}\boldsymbol{\sigma}\, \mathrm{d}\Omega = U. \tag{12.14}$$

The principle of Eq. (12.13) is known as that of *minimum complementary energy*[1] and is of general validity. It is easy to show that the complementary energy will always exist if potential energy can be defined.

Practical use of the complementary energy principle (or equivalent complementary virtual work) is much more difficult to apply than that of potential energy, as trial approximations for $\boldsymbol{\sigma}$ must now obey the differential equations of equilibrium, i.e.,

$$\mathbf{L}^{\mathrm{T}}\boldsymbol{\sigma} + \mathbf{b} = 0 \quad \text{in } \Omega \tag{12.15}$$

and satisfy the prescribed boundary tractions

$$\mathbf{t} = \bar{\mathbf{t}} \quad \text{on } \Gamma_t. \tag{12.16}$$

Specification of the simple expansion of the stress field in the standard form

$$\boldsymbol{\sigma} = \mathbf{N}\mathbf{a} \tag{12.17}$$

where $\mathbf{a}$ is a set of nodal parameters, is fraught with difficulties as only certain components of $\boldsymbol{\sigma}$ need to be continuous and, therefore, despite its attractiveness, the complementary energy principles (which give an upper bound on the strain energy and can also concentrate on the quantity of main practical interest, i.e., the stress[1,2]) have seldom been programmed directly in terms of the stress variables. This despite many trials of horrifying complexity.

A possible way out is the use of stress functions defining automatically equilibrating stresses. Quite generally stress function (or functions) $\boldsymbol{\phi}$ define stresses as

$$\boldsymbol{\sigma} = \mathbf{S}\boldsymbol{\phi} + \mathbf{s} \tag{12.18}$$

where $\mathbf{S}$ is a linear differential operator and $\mathbf{s}$ a prescribed vector. Stresses defined by such stress functions automatically satisfy the equilibrium equations, i.e.,

$$\mathbf{L}^{\mathrm{T}}(\mathbf{S}\boldsymbol{\phi} + \mathbf{s}) + \mathbf{b} \equiv 0. \tag{12.19}$$

Best known here is the Airy, scalar, stress function ϕ for two-dimensional analysis which gives[3]

$$\boldsymbol{\sigma} = \begin{Bmatrix} \sigma_x \\ \sigma_y \\ \tau_{xy} \end{Bmatrix} = \begin{Bmatrix} \dfrac{\partial^2}{\partial y^2} \\ \dfrac{\partial^2}{\partial x^2} \\ -\dfrac{\partial^2}{\partial x\,\partial y} \end{Bmatrix} \phi + \begin{Bmatrix} \Omega \\ \Omega \\ 0 \end{Bmatrix} \tag{12.20}$$

where Ω is a body force potential giving

$$b_x = -\frac{\partial \Omega}{\partial x}$$

$$b_y = -\frac{\partial \Omega}{\partial y}.$$

The reader can satisfy himself that, with this definition, internal equilibrium is always assured. Further, the boundary values of ϕ can sometimes be presented so that the traction condition is automatically satisfied.[3]

In the context of three-dimensional analysis and of plate bending other stress functions can be produced.[4, 5]

The reader will note that the operator $\mathbf{S}$ associated with the stress function of Eq. (12.20) is of second order and has a great similarity to that defining the strain operators in plate bending. The complementary energy function is now of second order and C_1 continuity is required for the stress functions with all its difficulties discussed in Chapter 10. However, the analogy can be pursued in detail and it can be shown that a plate bending program can be used directly for an equilibrium solution of plane elasticity problems with simple substitution of ϕ for the displacement w and some adjustment of constants. Serious limitations exist, however, if forces do not balance around closed circuits, as multiple values of the function occur.

It can also be shown that the stress functions introduced by Southwell[5] are such as to make it possible to define the plate equilibrium process in terms of quantities analogous to the displacements in plane stress analysis. The recognition of these possibilities of using these stress functions directly in finite element analysis was first put forward by Veubeke and Zienkiewicz,[6] then followed by others.[7, 8] However, considerable difficulty arises in multi-connected problems and a modified approach imposing exact inter-element equilibrium via Lagrangian multiplers identified with displacements proves more attractive.[9, 10] We shall refer to this later.

12.2.2 *Mixed variational principles.* In defining the potential energy functional of Eqs. (12.8) and (12.9) we have assumed that the strains were related to the displacements, i.e., that

$$\boldsymbol{\varepsilon} - \mathbf{L}\mathbf{u} = 0 \quad \text{in } \Omega \tag{12.21a}$$

and that, on the boundary Γ_u,

$$\mathbf{u} - \bar{\mathbf{u}} = 0. \tag{12.21b}$$

If we wish to relax these constraints it is possible to write a new variational principle

$$\Pi_1 = \Pi - \int_\Omega \boldsymbol{\lambda}_1^{\mathrm{T}}(\boldsymbol{\varepsilon} - \mathbf{L}\mathbf{u})\,\mathrm{d}\Omega - \int_{\Gamma_u} \boldsymbol{\lambda}_2^{\mathrm{T}}(\mathbf{u} - \bar{\mathbf{u}})\,\mathrm{d}\Gamma \tag{12.22}$$

by introducing two independent Lagrangian multipliers $\boldsymbol{\lambda}_1$, defined in the domain Ω, and $\boldsymbol{\lambda}_2$, defined only on Γ_u.

Now each quantity can be varied independently. Performing the variation we observe that

$$\begin{aligned}\delta\Pi_1 \equiv 0 = \delta\Pi &- \int_\Omega \delta\boldsymbol{\lambda}_1^{\mathrm{T}}(\boldsymbol{\varepsilon} - \mathbf{L}\mathbf{u})\,\mathrm{d}\Omega - \int_{\Gamma_u} \delta\boldsymbol{\lambda}_2^{\mathrm{T}}(\mathbf{u} - \bar{\mathbf{u}})\,\mathrm{d}\Gamma \\ &- \int_\Omega \boldsymbol{\lambda}_1^{\mathrm{T}}(\delta\boldsymbol{\varepsilon} - \mathbf{L}\,\delta\mathbf{u})\,\mathrm{d}\Omega - \int_{\Gamma_u} \boldsymbol{\lambda}_2^{\mathrm{T}}\delta\mathbf{u}\,\mathrm{d}\Gamma.\end{aligned} \tag{12.23}$$

Noting that (following similar integrations to those of Chapter 3 section 3.6

$$\int_\Omega \boldsymbol{\lambda}_1^{\mathrm{T}}\mathbf{L}\,\delta\mathbf{u}\,\mathrm{d}\Omega \equiv -\int_\Omega \delta\mathbf{u}^{\mathrm{T}}\mathbf{L}^{\mathrm{T}}\boldsymbol{\lambda}_1\,\mathrm{d}\Omega + \int_\Gamma \delta\mathbf{u}^{\mathrm{T}}\mathbf{G}\boldsymbol{\lambda}_1\,\mathrm{d}\Gamma$$

where $\mathbf{G}$ is an operator giving boundary tractions in terms of stresses, and substituting the variation of Π given by Eq. (12.2) we can write Eq. (12.23) explicitly as

$$\begin{aligned}0 \equiv &\int_\Omega \delta\boldsymbol{\varepsilon}^{\mathrm{T}}(\boldsymbol{\sigma}(\boldsymbol{\varepsilon}) - \boldsymbol{\lambda}_1)\,\mathrm{d}\Omega - \int_\Omega \delta\mathbf{u}(\mathbf{L}^{\mathrm{T}}\boldsymbol{\lambda}_1 + \mathbf{b})\,\mathrm{d}\Omega \\ &+ \int_{\Gamma_t} \delta\mathbf{u}^{\mathrm{T}}(\mathbf{G}\boldsymbol{\lambda}_1 - \bar{\mathbf{t}})\,\mathrm{d}\Gamma + \int_{\Gamma_u} \delta\mathbf{u}^{\mathrm{T}}(\mathbf{G}\boldsymbol{\lambda}_1 - \boldsymbol{\lambda}_2)\,\mathrm{d}\Gamma \\ &- \int_\Omega \delta\boldsymbol{\lambda}_1^{\mathrm{T}}(\boldsymbol{\varepsilon} - \mathbf{L}\mathbf{u})\,\mathrm{d}\Omega - \int_{\Gamma_u} \delta\boldsymbol{\lambda}_2^{\mathrm{T}}(\mathbf{u} - \bar{\mathbf{u}})\,\mathrm{d}\Gamma.\end{aligned} \tag{12.24}$$

As the above is true for any variation, we immediately observe that the last two Euler conditions give us the constraint satisfaction and from the others we can see that the Lagrangian multipliers can be identified as follows:

$$\boldsymbol{\lambda}_1 = \boldsymbol{\sigma}$$

which satisfies the equilibrium conditions and boundary tractions and

$$\boldsymbol{\lambda}_2 = \mathbf{t}$$

which gives the boundary tractions. With this identification the variational principle is known as the Hu–Washizu principle, and can be stated as a stationarity requirement for a functional

$$\Pi_1 = \Pi(\boldsymbol{\varepsilon}, \mathbf{u}) - \int_\Omega \boldsymbol{\sigma}^{\mathrm{T}}(\boldsymbol{\varepsilon} - \mathbf{L}\mathbf{u})\, \mathrm{d}\Omega - \int_{\Gamma_u} \mathbf{t}^{\mathrm{T}}(\mathbf{u} - \bar{\mathbf{u}})\, \mathrm{d}\Gamma \qquad (12.25)$$

where Π is the potential energy defined by Eq. (12.8).

In this an independent variation of the following quantities $\boldsymbol{\varepsilon}$, $\mathbf{u}$, and $\boldsymbol{\sigma}$ in the domain Ω and of $\mathbf{t}$ on the boundary is permitted.

This extremely general principle is probably not a very practical one. Let us note, however, that the sum of complementary and strain energies can be written as

$$U + U^* = \int_\Omega \boldsymbol{\sigma}^{\mathrm{T}}\boldsymbol{\varepsilon}\, \mathrm{d}\Omega$$

giving

$$\delta U + \delta U^* = \int_\Omega \delta\boldsymbol{\varepsilon}^{\mathrm{T}}\boldsymbol{\sigma}\, \mathrm{d}\Omega + \int_\Omega \delta\boldsymbol{\sigma}^{\mathrm{T}}\boldsymbol{\varepsilon}\, \mathrm{d}\Omega \qquad (12.26)$$

Immediately another variational principle can be written by replacing U with U^* in the definition of Π given in Eq. (12.8)

$$\Pi_2 = \int_\Omega \boldsymbol{\sigma}^{\mathrm{T}}\boldsymbol{\varepsilon}\, \mathrm{d}\Omega - U^*(\boldsymbol{\sigma}) - \int_\Omega \mathbf{u}^{\mathrm{T}}\mathbf{b}\, \mathrm{d}\Omega - \int_{\Gamma_t} \mathbf{u}^{\mathrm{T}}\bar{\mathbf{t}}\, \mathrm{d}\Gamma$$
$$- \int_\Omega \boldsymbol{\sigma}^{\mathrm{T}}(\boldsymbol{\varepsilon} - \mathbf{L}\mathbf{u})\, \mathrm{d}\Omega - \int_{\Gamma_u} \mathbf{t}^{\mathrm{T}}(\mathbf{u} - \bar{\mathbf{u}})\, \mathrm{d}\Gamma \qquad (12.27)$$

in this again the same variables occur but now in different order.

If we assume further that $\boldsymbol{\varepsilon}$ is not independent and that the two compatibility constraints on it are satisfied (Eq. (12.21)) we can rewrite above as a variational principle known by the name of Reissner and Helinger.[11] This can be stated as

$$\Pi_3 = \int_\Omega \boldsymbol{\sigma}^{\mathrm{T}}\boldsymbol{\varepsilon}\, \mathrm{d}\Omega - U^*(\boldsymbol{\sigma}) - \int_\Omega \mathbf{u}^{\mathrm{T}}\mathbf{b}\, \mathrm{d}\Omega - \int_{\Gamma_t} \mathbf{u}^{\mathrm{T}}\bar{\mathbf{t}}\, \mathrm{d}\Gamma$$
$$(\boldsymbol{\varepsilon} = \mathbf{L}\mathbf{u} \text{ in } \Omega;\ \mathbf{u} - \bar{\mathbf{u}} = 0 \text{ on } \Gamma_u) \qquad (12.28)$$

where now only $\boldsymbol{\sigma}$ and $\mathbf{u}$ are independent variables.

Finally, if we integrate the first term of Eq. (12.28) by parts we observe, after inserting the compatibility conditions, that the above can be simply rewritten as

$$
-\Pi_3 = U^*(\boldsymbol{\sigma}) + \int_\Omega \mathbf{u}^{\mathrm{T}}(\mathbf{L}^{\mathrm{T}}\boldsymbol{\sigma} + \mathbf{b})\,\mathrm{d}\Omega - \int_{\Gamma_t} \mathbf{u}^{\mathrm{T}}(\mathbf{G}\boldsymbol{\sigma} - \bar{\mathbf{t}})\,\mathrm{d}\Gamma
- \int_{\Gamma_u} \bar{\mathbf{u}}^{\mathrm{T}}\mathbf{G}\boldsymbol{\sigma}\,\mathrm{d}\Gamma. \tag{12.29}
$$

Further, if the stresses $\boldsymbol{\sigma}$ are so chosen as to satisfy equilibrium conditions, i.e., Eqs. (12.15) and (12.16), we return to the already proved complementary energy principle:

$$
\Pi_4 = U^*(\boldsymbol{\sigma}) - \int_{\Gamma_u} \bar{\mathbf{u}}^{\mathrm{T}}\mathbf{t}\,\mathrm{d}\Gamma \qquad (\mathbf{t} \equiv \mathbf{G}\boldsymbol{\sigma}). \tag{12.30}
$$

All the above energy principles can be utilized in practice and many more can be derived. However, only a few are practically useful. In particular, the modified form of the Reissner–Helinger principle first introduced by Herrmann[12–14] is to be noted. If we examine the form of the statement of Eq. (12.28) we see that although independent fields of stresses and displacements can be assumed, the latter has still to preserve exactly the same continuity as in the displacement analysis as the strain operator **Lu** remains.

On the other hand, the field of stresses can be completely discontinuous as no differentiation of it is involved. However, integration by parts of the first term

$$
\int_\Omega \boldsymbol{\sigma}^{\mathrm{T}}\boldsymbol{\varepsilon}\,\mathrm{d}\Omega \equiv \int_\Omega \boldsymbol{\sigma}^{\mathrm{T}}\mathbf{L}\mathbf{u}\,\mathrm{d}\Omega \tag{12.31}
$$

can be carried out.

Herrmann modifies thus the variational principle which now imposes some new continuity on the stresses, but relaxes those on the displacements. If the operator **L** is of second order, as it is in plate bending problems, it is easy to see that a single integration by parts will result in first order derivatives only being applied to both fields, thus allowing these to be represented by interpolation of C_0 continuity.

As an example of this we shall consider a beam problem with $\boldsymbol{\varepsilon} \equiv \mathrm{d}^2w/\mathrm{d}x^2$; $\boldsymbol{\sigma} \equiv M = EI(\mathrm{d}^2w/\mathrm{d}x^2)$, and loading $\mathbf{b} \equiv q$. We can now write the Reissner functional of Eq. (12.28), as

$$
\Pi_3 = \int_0^L M\frac{\mathrm{d}^2w}{\mathrm{d}x^2}\,\mathrm{d}x - \int_0^L \frac{1}{2EI}M^2\,\mathrm{d}x - \int_0^L w^{\mathrm{T}}q\,\mathrm{d}x - \frac{\mathrm{d}w}{\mathrm{d}x}\cdot\overline{M}\Big|_0^L
$$

where $\overline{M}$ represents the prescribed end moments and $\mathbf{u} \equiv w$.

Discretization of variables M and w can be performed here, but the latter would have to show continuous slope according to our standard 'integrability' rules.† However, integrating the first term by parts we observe that the functional can be rewritten as

$$\Pi_3 = -\int_0^L \frac{dM}{dx}\cdot\frac{dw}{dx}\,dx - \int_0^L \frac{1}{2EI}M^2\,dx - \int_0^L w\,q\,dx + \frac{dw}{dx}(M-\overline{M})\Big|_0^L$$

which only imposes C_0 continuity on M and w, and both can now be simply interpolated from the nodal parameters M_i and w_i.

The reader is encouraged to formulate this problem in detail for say a linear set of shape functions used in discretizing M and w and also to apply the other variational principles to the same example.

The various mixed variational theorems presented here are but a few of the very large number of possibilities. Some of these are discussed by Oden[17] and others.[16]

Obviously other means of enforcing constraints are possible; the penalty procedure has already been discussed in the Chapter 11 and the least square methods applied directly to the various problems of mechanics can also be used here.[18–20]

In practice, few applications of mixed variational principles have been made with the exception of plate and slab problems. A very full list of examples is given in references 21–33.

In the next sections we shall concern ourselves with hybrid forms which are special cases of mixed formulations with interface variables.

12.3 **Interface Variables and Hybrid Formulations**

12.3.1 *The general concept.* The main difficulties in the use of such functionals as those of potential or complementary energy are those of ensuring the continuity of tractions or displacements across inter-element boundaries (or interfaces).

Some of the mixed variational principles alleviate such difficulties by requiring lower continuity orders, but, however, add the penalty of numerous additional variables. An alternative, wide, range of possibilities

† Strictly speaking an expression such as $\int A(d\phi/dx)\,dx$ does not require a finite value of $d\phi/dx$ for integrability. Such finite values are only required when higher powers of this quantity arise. Clearly, if a jump in ϕ occurs at some point x_i we can write

$$\int_0^L A\frac{d\phi}{dx}\,dx = \int_0^{x_i} A\frac{d\phi}{dx}\,dx + \int_{x_i}^L A\frac{d\phi}{dx} + A(x_i)(\phi_i^+ - \phi_i^-)$$

This type of integration needs the inclusion of jump terms and leads to rather awkward finite element forms such as have been introduced in references 15 and 16.

is to define either compatible or equilibrating fields inside each element without ensuring the inter-element compatibility or equilibrium and then to impose this by Lagrangian multipliers defined on the interfaces alone. If either field is so defined then all or some of its parameters will be associated with one element only and can be thus eliminated locally, reducing the total number of variables in the assembled problem very considerably.

The complete field variational principles of elasticity discussed in the previous section will help us here to identify the meaning of the various Lagrangian variables, but it is more convenient (and more widely applicable) to discuss the principles involved in the context of a general mathematical problem in which we seek the stationarity of some arbitrary functional

$$\Pi = \Pi(\boldsymbol{\phi}) \tag{12.32}$$

with respect to variations of $\boldsymbol{\phi}$.

Quite generally, the inter-element continuity required can be expressed as

$$\mathbf{E}(\boldsymbol{\phi})^1 - \mathbf{E}(\boldsymbol{\phi})^2 = 0 \quad \text{on } \Gamma_I \tag{12.33}$$

where $\mathbf{E}$ is some linear operator and $(\boldsymbol{\phi})^1$ and $(\boldsymbol{\phi})^2$ represents the fields defined in the adjacent elements 1 and 2 joining on the interface Γ_I. For instance, C_0 continuity requires simply that $(\boldsymbol{\phi})^1$ is equal to $(\boldsymbol{\phi})^2$ while C_1 continuity would, in addition, require equality of $\partial\boldsymbol{\phi}/\partial n$, where n is normal to the interface. Other constraints can be included in the general statement of (12.33) in precisely the same way, as we shall see later.

Three possible ways of imposing the constraints (12.33) without excessively multiplying the final set of equations exist. All of these are illustrated in Fig. 12.1.

Class I: Single layer—independent fields (*Fig. 12.1*(*a*))

Here the field of $\boldsymbol{\phi}$ is defined in any element by a set of parameters which are totally independent of those in other elements and satisfy no inter-element compatibility requirements. We write thus

$$\boldsymbol{\phi}^e = \mathbf{N}^e\mathbf{b}^e \tag{12.34}$$

The Lagrangian variable $\boldsymbol{\lambda}$ is now defined on the interfaces only, Γ_I, by a set of nodal variables and suitable interpolation functions

$$\boldsymbol{\lambda}^I = \overline{\mathbf{N}}^I\mathbf{a}. \tag{12.35}$$

The functional imposing the interface conditions (12.33) can now be written as

$$\Pi^* = \Pi + \int_{\Gamma_I} \boldsymbol{\lambda}^{\mathrm{T}}(\mathbf{E}(\boldsymbol{\phi})^1 - \mathbf{E}(\boldsymbol{\phi})^2)\,\mathrm{d}\Gamma \tag{12.36}$$

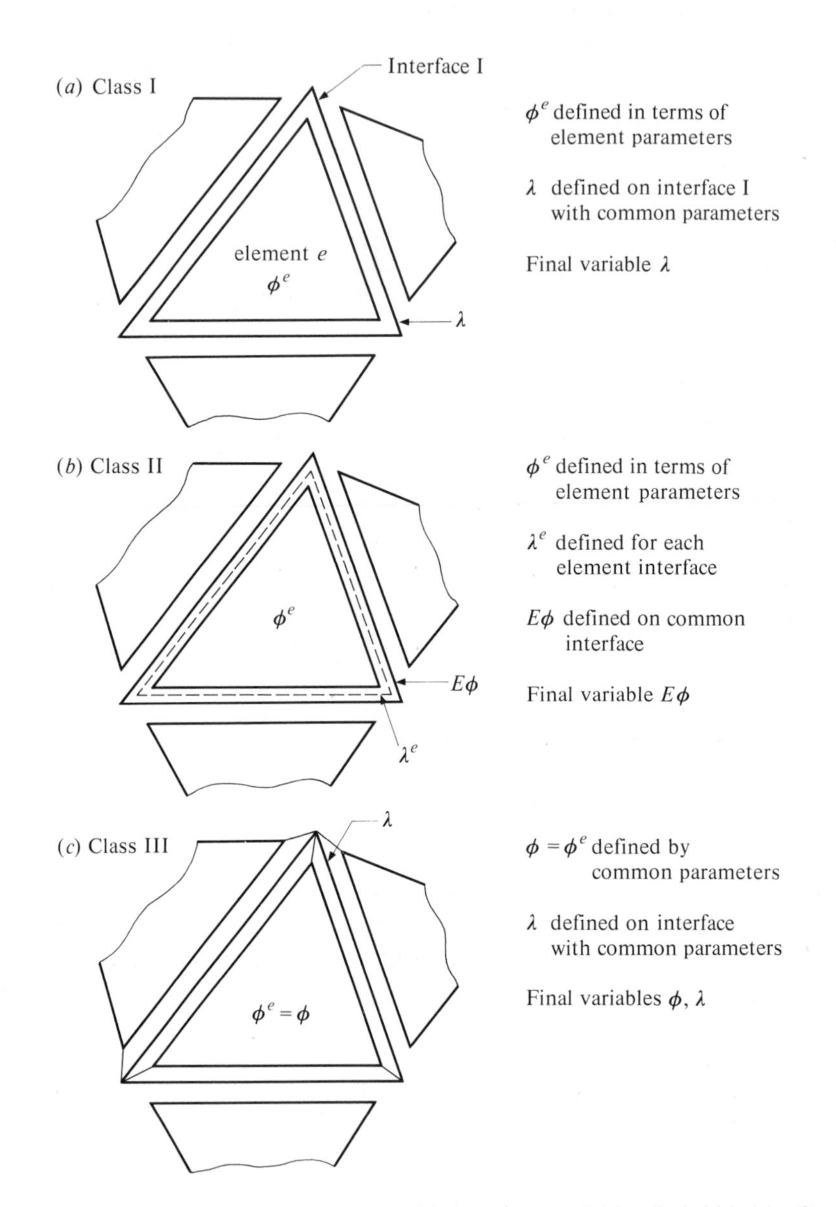

Fig. 12.1 Three classes of elements with interface variables (hybrids) (*a*), (*b*), and (*c*)

where again 1 and 2 stand for any adjoining elements. As parameters $\mathbf{b}^e$ are only defined within a single element, we can write in a general form

$$\frac{\partial \Pi^*}{\partial \mathbf{b}^e} = \frac{\partial \Pi^{*e}}{\partial \mathbf{b}^e} = \mathbf{K}^e\mathbf{b}^e + \mathbf{I}^e\mathbf{a} + \mathbf{f}^e = 0. \tag{12.37}$$

Solving this we can immediately eliminate the internal parameters, writing

$$\mathbf{b}^e = -(\mathbf{K}^e)^{-1}(\mathbf{I}^e\mathbf{a} + \mathbf{f}^e) \tag{12.38}$$

and the final equations of the system become simply

$$\frac{\partial \Pi^*}{\partial \mathbf{a}} = \sum \frac{\partial \Pi^{*e}}{\partial \mathbf{a}} \tag{12.39}$$

These are now totally expressed in terms of the Lagrangian parameters $\mathbf{a}$ by substitution of (12.38). Thus, in the final problem only these parameters play a part.

Class II: 'Frame' methods—Double layer interface

In this class an imaginary frame is thrown around each element and on this frame $\mathbf{E}\boldsymbol{\phi}$ is defined.

The continuity requirements (12.33) are now replaced by the following

$$\mathbf{E}(\boldsymbol{\phi})^1 - \mathbf{E}(\boldsymbol{\phi})^I = 0; \quad \mathbf{E}(\boldsymbol{\phi})^I - \mathbf{E}(\boldsymbol{\phi})^2 = 0 \tag{12.40}$$

(where I stands for the 'frame' values), and thus a functional written as

$$\Pi^{**} = \Pi + \int_{\Gamma_I} \boldsymbol{\lambda}^{\mathrm{T}}(\mathbf{E}(\boldsymbol{\phi})^1 - \mathbf{E}(\boldsymbol{\phi})^I)\, \mathrm{d}\Gamma \tag{12.41}$$

satisfies all continuity requirements.

Now the field $\boldsymbol{\phi}$ inside the element e is written in terms of independent variables $\mathbf{b}^e$ using Eq. (12.34). We can define, in addition, a set of Lagrangian multipliers which although given on the interface are associated with one element at a time. We thus have

$$\boldsymbol{\lambda} = \bar{\bar{\mathbf{N}}}\mathbf{c}^e. \tag{12.42}$$

Now in a single element all the parameters $\mathbf{b}^e$ and $\mathbf{c}^e$ are fully determined and the following equations can be written

$$\frac{\partial \Pi^{**}}{\partial \mathbf{b}^e} \equiv \frac{\partial \Pi^{**e}}{\partial \mathbf{b}^e} = 0 \quad \text{and} \quad \frac{\partial \Pi^{**}}{\partial \mathbf{c}^e} \equiv \frac{\partial \Pi^{**e}}{\partial \mathbf{c}^e} = 0 \tag{12.43}$$

which allows parameters $\mathbf{b}^e$ and $\mathbf{c}^e$ to be eliminated from the final system which is given by

$$\frac{\partial \Pi^{**}}{\partial \mathbf{a}} = \sum \frac{\partial \Pi^{**e}}{\partial \mathbf{a}}. \tag{12.44}$$

Here the parameters $\mathbf{a}$ define the values of $\mathbf{E}(\boldsymbol{\phi})^I$ on the interface. The basic difference of Class II formulation from those of Class I is that the variables on the interface are related to the original quantity $\boldsymbol{\phi}$ rather than being a completely new set of parameters. This, however, is not serious as in many cases a very strict physical meaning on the alternative set is available.

Class III: Single dependent fields

In this procedure the original field is defined as

$$\boldsymbol{\phi} = \mathbf{N}\mathbf{b} \tag{12.45}$$

by a set of parameters which now are not inter-element independent. Continuity is here still imposed as in Class I by specifying λ on interfaces I. This is a simple variant of Class I in which the final equations still contain both sets of the variables $\mathbf{a}$ and $\mathbf{b}$ and no elimination of the total variable number can be made. However, if the originally chosen set of expansions in Eq. (12.45) satisfies a part of the inter-element continuity, then the number of interface Lagrangian variables can be reduced and the final set of equations turns out to be more economical in unknowns than that available from the straight mixed variational principle.

The possibilities of the permutations offered by the interface variable procedures are large, and the concepts are better grasped on specific applications which we shall illustrate in the context of mechanics where the variational principles of the preceding sections will help us to identify the Lagrangian variables.

The most important application of the interface procedures is in Class I using a complementary energy principle supplemented, on the interface, by Lagrangian multipliers which are identified as displacements. Elements of this type are known as 'stress hybrids' and were introduced first by Pian and extensively developed by him and others.[34–47] The advantage of these procedures is that the final finite element type matrix can be made to fit into standard programs and, indeed, mix with ordinary displacement forms so that this class of elements will be discussed in more detail.

Class II approximations have been used less extensively, but examples have been provided by Pin Tong[48] where the potential energy displacement variable provides the basis. Once again the final variables are also displacements now defined on the interfaces and these elements possess all the merits of Class I approximations in the generality and association with straight displacement formulations. In the same category are also elements using a modified variational principle in which the Lagrangian multipliers associated with each element are identified as tractions and directly eliminated by substitution of their expression in terms of the interior displacements. The variational principle of this kind is similar to that discussed in Chapter 3, p. 81, and is the basis of plate bending elements derived in references 49 and 50. This formulation, however, possesses some difficulties recognized by Mang and Gallagher.[51]

In Class III approximations the outstanding example is that provided by Harvey and Kelsey,[52] where a single Lagrangian variable on each side enforces slope continuity for a triangular plate bending element with a complete 10 term cubic polynomial expansion which, as we remarked in Chapter 10, failed to satisfy only the slope compatibility. Despite the inconvenience of mixed variables, this element has been found useful in many applications.

12.3.2 *Consistent choice of parameters; complete constraint enforcement.* With the independent specification of the function $\boldsymbol{\phi}$ in the element and the description of the Lagrangian multipliers $\boldsymbol{\lambda}$ on the interface by another set of parameters, the question of consistency in choosing the correct number of parameters for each expansion arises.

For instance, in Class I hybrids, if the field $\boldsymbol{\phi}$ within the element leads to a polynomial variation of $\mathbf{E}\boldsymbol{\phi}$ along the interface, then the imposition of continuity (Eq. (12.33)) at a certain discrete number of points along a straight part of an interface can ensure complete satisfaction of continuity requirements all along this interface. This is entirely equivalent to limiting the number of parameters being introduced to describe the variation of $\boldsymbol{\lambda}$ in its polynomial expansion along this interface. Clearly there is no point whatsoever in increasing the number of parameters above the number giving the exact satisfaction of constraints. As $\boldsymbol{\lambda}$ in the constrained variational principle does not have to be continuous along the complete interface of any element, the satisfaction of constraints is best enforced by nodes placed along the (straight) sides of each element. Such complete enforcement of inter-element constraints is the basis of fully equilibrating elements obtained as a variant of the stress hybrid method by Veubeke, Sander and Becker.[2,9,53]

Where complete satisfaction of compatibility is impossible, as is the case with some Class I hybrid elements in which the $\boldsymbol{\lambda}$ parameters are specified at the corner nodes, a similar limit still needs to be placed on the number of internal element parameters and an approximation to that limit can easily be found by considering the degree of polynomials present in representation of the constrained quantity and of the Lagrangian parameter along each side. In the context of stress hybrids the question of increasing the number of internal parameters has been discussed and tested extensively by Henshell[39] who shows in fact a deterioration of the results with their increasing number.

In the following sections we shall illustrate by example some elements of Class I in which complete and partial enforcement of constraints is made in the context of assumed stress fields. We hope that this single illustration will provide enough insight to the reader to pursue other applications in more detail. Finally, it should be noted that elements of the hybrid class often reproduce the results of other elements obtained by smoothed incompatible displacement fields described in previous chapters (see Chapter 11).

12.4 Some Examples of Class I Hybrid Formulation

12.4.1 *Equilibrating (complementary energy) elements.* The earliest equilibrium elements for plane stress–strain analysis was described by Veubeke in 1963.[2] This forms a very convenient illustration of the hybrid process which here enforces complete stress equilibrium. In this element a constant stress field is assumed inside each triangular element of Fig. 12.2. This we can write in discrete form simply as

$$\boldsymbol{\sigma} = \begin{Bmatrix} \sigma_x \\ \sigma_y \\ \tau_{xy} \end{Bmatrix} = \mathbf{b}^e. \tag{12.46}$$

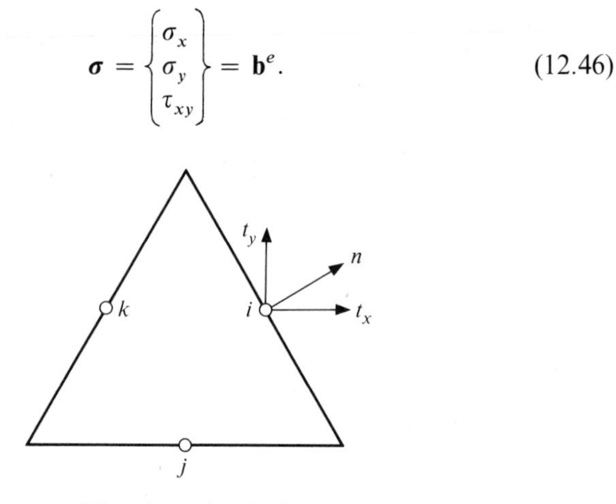

Fig. 12.2 Constant stress equilibrating triangle final variable $\boldsymbol{\lambda}_i = \mathbf{u}_i$ at midside nodes

As this stress automatically satisfies equilibrium conditions in the absence of body forces, the complementary energy can be written for each element as

$$\Pi^e = \frac{1}{2}\int_\Delta \boldsymbol{\sigma}^{\mathrm{T}}\mathbf{C}\boldsymbol{\sigma}\,\mathrm{d}x\,\mathrm{d}y = \frac{1}{2}\mathbf{b}^{e\mathrm{T}}\mathbf{C}\mathbf{b}^e\Delta. \tag{12.47}$$

In the above, Δ stands for the area of the triangle and ($\mathbf{C} = \mathbf{D}^{-1}$) is the material compliance.

For inter-element equilibrium we have to write

$$\mathbf{E}(\boldsymbol{\sigma})^1 - \mathbf{E}(\boldsymbol{\sigma})^2 = \mathbf{t}^1 + \mathbf{t}^2 = 0 \tag{12.48}$$

where $\mathbf{t}$ describes the tractions between adjacent elements. On a typical face we can write such tractions as

$$\mathbf{t} = \begin{Bmatrix} t_x \\ t_y \end{Bmatrix} \equiv \mathbf{G}\boldsymbol{\sigma} \tag{12.49}$$

where

$$\mathbf{G} = \begin{bmatrix} n_x & 0 & n_y \\ 0 & n_y & n_x \end{bmatrix} \tag{12.50}$$

here n_x and n_y are the direction cosines of the normal to the face.

As the stresses (and therefore the tractions) are constant along each side, it suffices to introduce a constant $\boldsymbol{\lambda}$ value. As $\boldsymbol{\lambda}$ can be identified with the displacement $\mathbf{u}$ (via the mixed energy functional of Eq. (12.28)), we write for each element

$$\boldsymbol{\lambda}_i \equiv \mathbf{u}_i \tag{12.51}$$

as a two-component vector associated with, say, the displacements of a centrally placed node. Immediately, for an element we can write

$$\Pi^{*e} = \Delta \frac{1}{2} \mathbf{b}^{e\mathrm{T}} \mathbf{C} \mathbf{b}^e - \sum_{i=1}^{3} \mathbf{u}_i^{\mathrm{T}} l_i \mathbf{G}_i \mathbf{b}^e \tag{12.52}$$

where l_i is the length of the side with node i.

As the parameters $\mathbf{b}^e$ do not enter into any other element, we can write

$$\frac{\partial \Pi^{*e}}{\partial \mathbf{b}^e} = 0 = \Delta \mathbf{C} \mathbf{b}^e - \sum_{i=1}^{3} l_i \mathbf{G}_i^{\mathrm{T}} \mathbf{u}_i \tag{12.53}$$

from which

$$\mathbf{b}^e = \mathbf{C}^{-1} \sum_{i=1}^{3} l_i \mathbf{G}_i^{\mathrm{T}} \mathbf{u}_i / \Delta = \mathbf{D} \sum_{i=1}^{3} l_i \mathbf{G}_i^{\mathrm{T}} \mathbf{u}_i / \Delta. \tag{12.54}$$

For any element we thus have the contribution to the total functional derivative as

$$\frac{\partial \Pi^{*e}}{\partial \mathbf{u}_i} = -l_i \mathbf{G} \mathbf{b}^e = -l_i \mathbf{G}_i \mathbf{D} \sum_{j=1}^{3} l_j \mathbf{G}_j^{\mathrm{T}} \mathbf{u}_j / \Delta \equiv \sum_{j=1}^{3} \mathbf{K}_{ij}^e \mathbf{u}_j \tag{12.55}$$

with

$$\mathbf{K}_{ij}^e = l_i \mathbf{G}_i \mathbf{D} l_j \mathbf{G}_j^{\mathrm{T}} / \Delta.$$

The coefficients $\mathbf{K}_{ij}^e$ are recognized now as the stiffness coefficients and can be assembled for a complete system in the usual manner.†

In an exactly analogous manner higher order equilibrating elements can be derived using a greater number of inter-connecting (Lagrangian) displacements.[9] The choice of an equilibrating field within the element is facilitated by the use of stress functions as already mentioned.

The element just derived occasionally yields a singular stiffness matrix on assembly, and its use therefore has not been extensive.

The procedure described here for plane stress analysis has been used extensively in the context of plate bending[9, 53] and could easily be extended to three-dimensional problems. As an exercise the reader could try to derive an equilibrium triangular or rectangular element with an assumed stress field of the form

$$\begin{aligned} \sigma_x &= b_1^e + b_2^e x + b_3^e y \\ \sigma_y &= b_4^e + b_5^e x - b_6^e y \\ \tau_{xy} &= b_7^e - b_6^e x - b_2^e y \end{aligned} \tag{12.56}$$

which can easily be verified as satisfying the equilibrium conditions. As a linear variation of traction occurs, at most two sets of Lagrangian multipliers (displacements) are necessary on each side, as shown on Fig. 12.3.

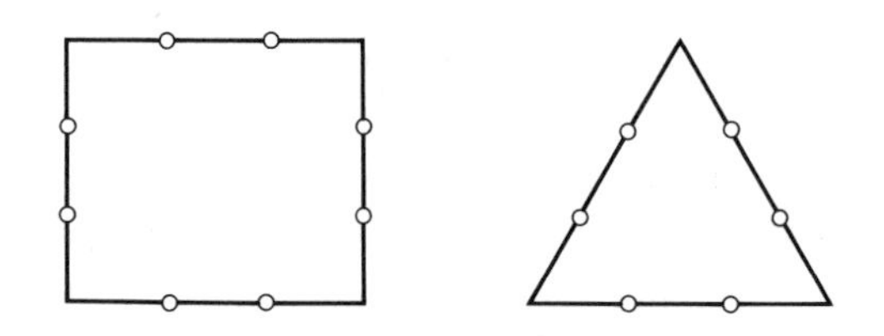

Fig. 12.3 Fully equilibrating hybrids with linear stress distributions

12.4.2 *Hybrid 'stress' elements.* The derivation of hybrid stress elements follows precisely the procedures of the previous section, but now the number of Lagrangian constraints is not necessarily sufficient to satisfy precisely the interface constraints of traction continuity. One of the first elements derived here was the plane–strain analysis rectangle shown in Fig. 12.4, and we shall describe it in some detail.

† It is interesting to note that precisely the same result would have been obtained in this case by assuming a linear, incompatible, displacement field given in terms of nodes placed at the element mid-sites. The proof requires some tedious algebra which we leave to the reader but, the point made about hybrid and incompatible elements being identified at times is here shown again by example.

Let the stress field in the element be described as

$$\boldsymbol{\sigma} = \begin{Bmatrix} \sigma_x \\ \sigma_y \\ \tau_{xy} \end{Bmatrix} \begin{bmatrix} 1, y, 0, 0, 0 \\ 0, 0, 1, x, 0 \\ 0, 0, 0, 0, 1 \end{bmatrix} \begin{Bmatrix} b_1^e \\ \vdots \\ b_5^e \end{Bmatrix} \equiv \mathbf{N}\mathbf{b}^e \qquad (12.57)$$

which again is easily verified as satisfying the equilibrium equations.

On the interfaces the Lagrangian displacement parameter is defined in terms of shape functions

$$\boldsymbol{\lambda} \equiv \mathbf{u} = \overline{\mathbf{N}}\mathbf{a}_i \qquad (12.58)$$

with $\mathbf{a}_i$ representing the values of displacements at the four corner nodes.

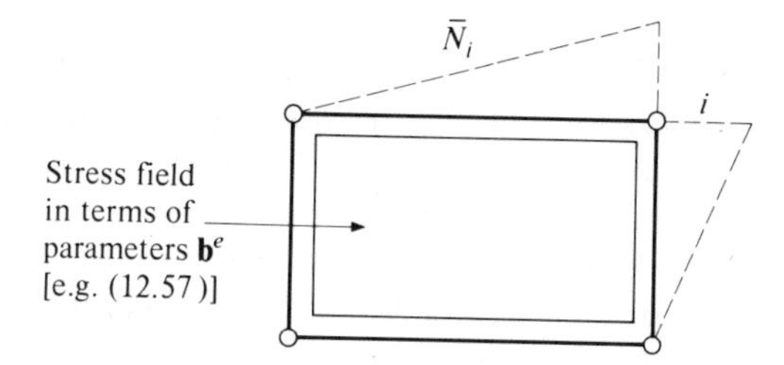

Fig. 12.4 Plane stress/strain hybrid with corner nodes, and linear λ interpolation

With the tractions defined as in Eq. (12.49), we note that n_x and n_y are either 0 or 1 on the faces of the rectangle, making the analysis easy.

Writing the modified functional for the element we have

$$\Pi^{*e} = \frac{1}{2}\mathbf{b}^{e\mathrm{T}}\left(\int_{\Omega^e} \mathbf{N}^{\mathrm{T}}\mathbf{C}\mathbf{N}\,\mathrm{d}x\,\mathrm{d}y\right)\mathbf{b}^e - \int_{\Gamma_I} (\overline{\mathbf{N}}\mathbf{a})^{\mathrm{T}}\mathbf{G}\mathbf{N}\mathbf{b}^e\,\mathrm{d}\Gamma. \qquad (12.59)$$

The above can be written in the form

$$\Pi^{*e} = \tfrac{1}{2}\mathbf{b}^{e\mathrm{T}}\mathbf{H}\mathbf{b}^e - \mathbf{a}^{e\mathrm{T}}\mathbf{P}\mathbf{b}^e \qquad (12.60)$$

with

$$\mathbf{H} = \int_{\Omega^e} \mathbf{N}^{\mathrm{T}}\mathbf{C}\mathbf{N}\,\mathrm{d}x\,\mathrm{d}y$$

$$\mathbf{P} = \int_{\Gamma_I} \overline{\mathbf{N}}^{\mathrm{T}}\mathbf{G}\mathbf{N}\,\mathrm{d}\Gamma.$$

Assuming linear interpolation functions $\overline{\mathbf{N}}$ and the form of $\mathbf{N}$ defined in Eq. (12.57) the reader is invited to work out the above expressions in detail.

Writing

$$\frac{\partial \Pi^{*e}}{\partial \mathbf{b}^e} = 0 = \mathbf{H}\mathbf{b}^e - \mathbf{P}^{\mathrm{T}}\mathbf{a}^e \tag{12.61}$$

determines

$$\mathbf{b}^e = \mathbf{H}^{-1}\mathbf{P}^{\mathrm{T}}\mathbf{a}^e$$

and the 'stiffness' of the element becomes

$$\frac{\partial \Pi^{*e}}{\partial \mathbf{a}^e} = \mathbf{P}\mathbf{b}^e = \mathbf{P}\mathbf{H}^{-1}\mathbf{P}^{\mathrm{T}}\mathbf{a}^e \tag{12.62}$$

with

$$\mathbf{K}^e = -\mathbf{P}\mathbf{H}^{-1}\mathbf{P}^{\mathrm{T}}$$

A little tedious algebra will identify the matrix of this element to be identical to that of the rectangular element with two internal incompatible parabolic modes (described in Chapter 11, p. 272).

Many other variants of the rectangle are possible using an equilibrating stress form defined by a larger number of internal parameters. However, none of these improves the element performance.

In the context of two-dimensional analysis just shown, the hybrid procedure is of rather academic interest, but applied to plate bending the formulation has produced a series of useful elements. These range from a constant moment triangle,[8] which turns out to be an equilibrating element, to more complex forms.[33,35,40,41]

It is interesting to remark that once again many of these are in fact identical with the alternatives derived by incompatible shape function assumptions. The equilibrating triangle of reference 8 is a case in point, discussed in Chapter 10, p. 246, and the triangles with nine or twelve degrees of freedom derived in reference 33 turn out to be identical to those obtained by 'derivative smoothing' in Chapter 11, p. 278.

12.4.3 *Generation of equilibrium stress fields and isoparametric forms.* In the examples we have assumed several stress fields [Eqs. (12.56) and (12.57)] which satisfy the equilibrium relations.

In general, any continuous Airy stress functions will generate such fields in two dimensions (and equivalent Finzi functions do so in three dimensions). For instance, a complete cubic stress function expansion imposed by polynomial form could generate automatically a stress field of the type given by Eq. (12.56). Automatically, other expansions can be obtained. As the first three linear terms of the stress function polynomial generate a zero stress, only 7 constants arise although the cubic itself is defined by 10. Obviously the stress function itself could equally well be used as a direct variable in determining the interior stress field.

The advantage of stress functions is that they can be defined easily in isoparametric or indeed any other curvilinear co-ordinate system and thus generate equilibriating interior stress fields.

Attempts to derive complete equilibrium elements of curvilinear forms have so far failed, but some quite useful hybrids have been derived in reference 54.

12.5 Variational Principles for Incompressibility (or Near Incompressibility)

In section 11.6.1 of Chapter 11 we discussed problems of incompressibility in elastic analysis and the difficulties caused by the fact that the elasticity matrix is indeterminate when the bulk modulus tends to infinity.

The problem was identified as one of minimization of potential energy defined in terms of distortional strain energy and the application of incompressibility constraints. For near incompressibility and incompressibility as the bulk modulus k becomes very large, we can state the problem as the one of stationarity of the distortional potential energy $\Pi(\mathbf{u})$ with the constraints

$$\varepsilon_v \equiv \frac{\partial u}{\partial x}+\frac{\partial v}{\partial y}+\frac{\partial w}{\partial z} = \frac{p}{k}$$

$$p = \frac{\sigma_x+\sigma_y+\sigma_z}{3} \tag{12.63}$$

The new variational principle can be easily written as

$$\Pi^* = \Pi(u)+\int_\Omega \lambda\left(\frac{\partial u}{\partial v}+\frac{\partial v}{\partial y}+\frac{\partial w}{\partial z}-\frac{p}{k}\right)\mathrm{d}\Omega \tag{12.64}$$

and after some standard manipulations the Lagrangian multiplier is identified as

$$\lambda \equiv p$$

This constrained variational principle was identified first by Herrmann[55] and is used extensively for the study of incompressible materials. Its modified form is given in reference 56.

Applications of the principle yield very similar results to those of Chapter 11, and again over-constraint need to be guarded against.

As p does not have to be continuous, independent element approximations can be made for it, although in general practice standard C_0 interpolations are used for both the displacement and p variables (usually with a lower order interpolation for the latter, thus reducing the constraint conditions and proving advantageous).

The variational principle of (12.64) is a variant of the Reissner theorem, though a question of interest is whether alternative approaches to incompressibility are possible. Clearly the full equilibrating complementary energy approach is one in which the matrix $\mathbf{C} = \mathbf{D}^{-1}$ exists and is well-defined, and such complementary energy principles with stress variables are useful. The stress hybrid or similar equilibrating elements need, however, again an inversion of the $\mathbf{C}$ matrix and difficulties arise. A way of overcoming this is mentioned in reference 57.

An interesting alternative has not yet been explored. By introducing, for instance, an equivalent 'stream function' it is easy to define displacement fields which have automatic incompressibility, e.g., in two dimensions we can write for such fields

$$u = -\frac{\partial \phi}{\partial y}; \qquad v = \frac{\partial \phi}{\partial x}$$

where ϕ is any continuous function.

If such an incompressible displacement field is defined within an element of Class II, it is possible to formulate a hybrid with standard nodal displacement variables which preserves automatically approximate incompressibility without introducing Lagrangian parameters. This possibility has yet to be fully explored.

12.6 **Concluding Remarks**

Only a rather brief description could be given here of the great variety of mixed and hybrid procedures, but it is hoped that the generalization of the latter would have indicated to the reader the many possibilities. While all of these provide excellent academic exercises and lead to somewhat esoteric calculations, a few words are necessary about the practical value of the processes described.

The first comments concern hybrid types of Class I and II in which the final result leads to similar element forms to those already achieved by direct displacement approaches. On several examples we have shown that with a limited number of parameters describing the stress field the *stiffness matrices become identical to those available from non-conforming displacement analysis* (which, if convergent, have always yielded better results than fully conforming ones). As the number of parameters describing the equilibrating stress field increases within the frame of the element, the displacements within become progressively more compatible until, with an *infinite number*, *a fully conforming displacement element is reached* (and as already remarked many times, worse results obtained).

If the simpler non-conforming formulation gives the same results as the hybrid form, perhaps the only advantage of the latter is the assurance of convergence without a patch test.[37,58]

The use of Class I principles to derive fully equilibrating elements has, however, led to new formulations of elements—usually with nodes placed only along the sides of the elements; the possibility of these not being obvious in standard displacement methods. Here again probably a non-conforming displacement pattern could be devised which would lead to identical results. Nevertheless, these procedures have introduced families of elements of considerable importance.

The mixed variational principles—including those used for incompressible materials and Class III hybrids—are gaining some popularity, but as all these introduce Lagrangian variables into the final equation this always results in semi-definite matrices of the type discussed in Chapter 3, p. 80.

Here the final equation system are for two, dimensionally different, sets of parameters **a** and **b** and are always of the form

$$\begin{bmatrix} \mathbf{K}_{11} & \mathbf{K}_{12} \\ \mathbf{K}_{12}^{\mathrm{T}} & \mathbf{0} \end{bmatrix} \begin{Bmatrix} \mathbf{a} \\ \mathbf{b} \end{Bmatrix} + \begin{Bmatrix} \mathbf{f}_1 \\ \mathbf{f}_2 \end{Bmatrix} = 0$$

Such equations may be troublesome in the final solution as zero or negative pivots are possible. Experience has shown, however, that with a proper order of elimination standard programs can still generally be used with no chance of ill-conditioning, providing a unique solution exists. Undoubtedly for some problems such elements can prove superior to straightforward displacement approaches, and indeed in a later chapter dealing with fluid mechanics we shall show that such formulations arise naturally.

References

1. K. Washizu, *Variational Methods in Elasticity and Plasticity*, 2nd ed., Pergammon Press, 1975.
2. B. Fraeijs de Veubeke, 'Displacement and equilibrium models in the finite element method', Chapter 9 of *Stress Analysis* (eds. O. C. Zienkiewicz and G. S. Holister), Wiley, 1965.
3. S. Timoshenko and J. N. Goodier, *Theory of Elasticity*, 2nd ed., McGraw-Hill, 1951.
4. B. Finzi, 'Integrazione delle equatzione indefinite della mechanica dei systemi continui', *Comp. Rend. Lincei*, **19**, 1934.
5. R. V. Southwell, 'On the analogues relating flexure and displacement of flat plates', *Quart. J. Mech. Appl. Math.*, **3**, 257–70, 1950.
6. B. Fraeijs de Veubeke and O. C. Zienkiewicz, 'Strain energy bounds in finite element analysis by slab analogy', *J. Strain Analysis* **2**, 265–7, 1967.
7. Z. M. Elias, 'Duality in Finite Element methods'. *Proc. Am. Soc. Civ. Eng.*, **94**, EM4, 931–46, 1968.
8. L. S. D. Morley, 'A triangular equilibrium element with linearly varying bending moments for plate bending problems', *J. Roy. Aero. Soc.*, **71**, 715–21, 1967.

9. G. SANDER, 'Applications of the dual analysis principle', *Proc. IUTAM Conf. on High Speed Computing of Elastic Structures*, Univ. of Liége, 1970.
10. W. PRAGER, 'Variational principles of linear elastostatics for discontinuous displacements, strains and stresses', *The F. Odqvist Volume*, J. Wiley & Son, 1967.
11. E. REISSNER, 'On a variational theorem in elasticity', *J. Math. Phys.*, **29**, 90–5, 1950.
12. L. R. HERRMANN, 'A bending analysis of plates', *Proc. 1st Conf. Matrix Methods in Structural Mechanics*, Wright-Patterson, A. F. Base, AFFDL-TR-80, Ohio, 1965.
13. L. R. HERRMANN, 'Finite element bending analysis of plates'. *Proc. Am. Soc. Civ. Eng.*, **94**, EM5, 13–25, 1968.
14. L. R. HERRMANN and D. M. CAMPBELL, 'A finite element analysis for thin shells', *J.A.I.A.A.*, **6**, 1842–7, 1968.
15. E. L. WACHSPRESS and J. M. BECKER, 'Variational synthesis with discontinuous trial functions', *Proc. Conf. on Appl. of Computor Methods to Reactor Problems*, AEC Report ANL 7105, p. 191, 1965.
16. S. NEMAT NASSER and K. N. LEE, 'Finite element formulations for elastic plates by general variational statements with discontinuous fields', *Comp. Meth. Appl. Mech. Eng.*, **2**, 33–41, 1973.
17. J. T. ODEN, 'Some contributions to the mathematical theory of mixed finite element approximations' in *Theory and Practice in Finite Element Structural Analysis*, pp. 3–24 (eds. Y. Yamada and R. Gallagher), Univ. of Tokyo Press, 1973.
18. O. C. ZIENKIEWICZ, D. R. J. OWEN and K. N. LEE, 'Least square finite element for elasto-static problems; use of "reduced" integration', *Int. J. Num. Meth. Eng.*, **8**, 341–58, 1974.
19. P. P. LYNN and S. K. ARYA, 'Use of least square criterion in finite element formulation', *Int. J. Num. Meth. Eng.*, **6**, 75–88, 1973.
20. P. P. LYNN, 'Least square finite element analysis of laminar boundary layer flows', *Int. J. Num. Meth. Eng.*, **8**, 865, 1974.
21. R. S. DUNHAM and K. PISTER, 'A finite element application of the Hellinger–Reissner variational theorem', *Proc. 2nd Conf. on Matrix Methods*, AFFDL-TR-150, pp. 471–487.
22. W. VISSER, 'A refined mixed type plate bending element', *J.A.I.A.A.*, **7**, 1801–3, 1969.
23. W. VISSER, *The application of a curved, mixed-type shell element*, Report SM-38, Div. of Engng., Harvard, 1970.
24. J. CONNOR, 'Mixed models for plates', in *Finite Element Techniques* (eds. H. Tottenham and C. Brebbia), Stress Analysis Publ., Southampton, 1971.
25. J. CONNOR, 'Mixed models for shells', in *Finite Element Techniques* (eds. H. Tottenham and C. Brebbia), Stress Analysis Publ., Southampton, 1971.
26. C. PRATO, 'Shell finite element via Reissner's principle', *Int. J. Solids Struct.*, **5**, 1119–33, 1969.
27. J. CONNOR and G. WILL, 'A mixed finite element method shallow shell formulation', in *Advances in Matrix Methods of Structural Analysis and Design* (ed. R. Gallagher), Univ. of Alabama Press, 1971.
28. W. PRAGER, 'Variational principles for elastic plates with relaxed continuity requirements', *Int. J. Solids Struct.*, **4**, 837–44, 1968.
29. Z. M. ELIAS, 'A mixed finite element method for axisymmetric shells', *Int. J. Num. Meth. Eng.*, **4**, 261–78, 1972.

30. K. HELLAN, 'On the unity of constant stress–constant moment finite elements', *Int. J. Num. Meth. Eng.*, **6**, 2, 191–209, 1973.
31. A. CHATTERJEE and A. V. SETLUR, 'A mixed finite element formulation for plate problems', *Int. J. Num. Meth. Eng.*, **4**, 67–84, 1972.
32. J. BRON and G. DHATT, 'Mixed quadrilateral elements for bending', *J.A.I.A.A.*, **10** (No. 10), 1359–61, Oct. 1972.
33. D. J. ALLMAN, 'Finite element analysis of plate buckling using a mixed variational principle', *Proc. 3rd Air Force Conf. on Matrix Methods in Structural Mechanics*, Dayton, Ohio, 1971.
34. T. H. H. PIAN, 'Derivation of element stiffness matrices by assumed stress distributions', *J.A.I.A.A.*, **2**, 1333–5, 1964.
35. T. H. H. PIAN, 'Element stiffness matrices for boundary compatibility and for prescribed boundary stresses', *Proc. Conf. on Matrix Methods in Structural Mechanics*, AFFDL-TR-66-80, pp. 457–78.
36. R. D. COOK and J. AT-ABDULLA, 'Some plane quadrilateral "Hybrid" finite elements', *J.A.I.A.A.*, **7**, 1969.
37. T. H. H. PIAN and P. TONG, 'Basis of finite element methods for solid continua', *Int. J. Num. Meth. Eng.*, **1**, 3–28, 1969.
38. S. ATLURI, 'A new assumed stress hybrid finite element model for solid continua', *J.A.I.A.A..*, **9**, 1647–9, 1971.
39. R. D. HENSHELL, 'On hybrid finite elements' in *The Mathematics of Finite Elements and Applications*, pp. 299–312 (ed. J. R. Whiteman), Academic Press, 1973.
40. R. DUNGAR and R. T. SEVERN, 'Triangular finite elements of variable thickness', *J. Strain Analysis*, **4**, 10–21, 1969.
41. R. J. ALLWOOD and G. M. M. CORNES, 'A polygonal finite element for plate bending problems using the assumed stress approach', *Int. J. Num. Meth. Eng.*, **1**, 135–49, 1969.
42. T. H. H. PIAN, 'Hybrid models' in *Numerical and Computer Methods in Applied Mechanics* (eds. S. J. Fenves *et al.*), Academic Press, 1971.
43. R. ALI, S. GOPALACHARYULU and P. W. SHARMAN, 'The development of a series of hybrid-stress finite elements', *Proc. World Congress on Finite Element Methods in Structural Mechanics*, **2**, 13.1–13.27.
44. Y. YOSHIDA, 'A hybrid stress element for thin shell analysis' in *Finite Element Methods in Engineering*, pp. 271–286 (eds. V. Pulmano and A. Kabaila), Univ. of New South Wales, Australia, 1974.
45. R. D. COOK and S. G. LADKANY, 'Observations regarding assumed-stress hybrid plate elements', *Int. J. Num. Meth. Eng.*, **8** (No. 3), 513–20, 1974.
46. J. P. WOLF, 'Generalized hybrid stress finite element models', *J.A.I.A.A.*, **11**, 1973.
47. P. L. GOULD and S. K. SEN, 'Refined mixed method finite elements for shells of revolution', *Proc. 3rd Air Force Conf. on Matrix Methods in Structural Mechanics*, Wright-Patterson A.F. Base, Ohio, 1971.
48. P. TONG, 'New displacement hybrid finite element models for solid continua', *Int. J. Num. Meth. Eng.*, **2**, 73–83, 1970.
49. F. KIKUCHI and Y. ANDO, 'A new variational functional for the finite element method and its application to plate and shell problems', *Nucl. Eng. Des.*, **21**, 95–113, 1972.
50. F. KIKUCHI and Y. ANDO, 'Some finite element solutions for plate bending problems by simplified hybrid displacement method', *Nucl. Eng. Des.*, **23**, 155–78, 1972.

51. H. A. Mang and R. H. Gallagher, 'A critical assessment of the simplified hybrid displacement method', *Int. J. Num. Meth. Eng.*, **11**, 145–68, 1977.
52. J. W. Harvey and S. Kelsey, 'Triangular plate bending elements with enforced comutivity', *J.A.I.A.A.*, **9**, 1023–6, 1971.
53. B. Fraeijs de Veubeke, G. Sander and P. Beckers, 'Dual analysis by finite elements: linear and nonlinear applications', *Proc. 3rd Air Force Conf. on Matrix Methods in Structural Mechanics*, AFFDL-TR-72–93, Wright-Patterson A.F. Base, Ohio, 1972.
54. J. Robinson, *Integrated Theory of Finite Element Methods*, Wiley, 1973.
55. L. R. Herrmann, 'Elasticity equations for incompressible and nearly incompressible materials by a variation theorem', *J. A. I. A. A.*, **3**, 1896–1900, 1965.
56. S. W. Key, 'A variational principle for incompressible and nearly incompressible anisotropic elasticity', *Int. J. Solids Struct.*, 1970.
57. P. Tong, 'An assumed stress hybrid finite element method for an incompressible and near-incompressible material', *Int. J. Solids Struct.*, **5**, 455–61, 1969.
58. I. Babuska, J. T. Oden and J. K. Lee, *Mixed-hybrid finite element approximations of second order elliptic boundary value problems*, Texas Inst. of Comp. Mech., Report 75–7, Austin, 1975.

13. Shells as an Assembly of Flat Elements

13.1 Introduction

A shell is, in essence, a structure which can be derived from a thin plate by initially forming the middle plane to a singly (or doubly) curved surface. Although the same assumptions regarding the transverse distribution of strains and stresses are again valid, the way in which the shell supports external loads is quite different from that of a flat plate. The stress resultants acting parallel to the middle plane of the shell now have components normal to the surface and carry a major part of the load, a fact which explains the economy of shells as load-carrying structures and their well-deserved popularity.

The derivation of detailed governing equations for a curved-shell problem presents many difficulties and, in fact, leads to many alternative formulations, each depending on the approximations introduced. For details of classical shell treatment the reader is referred to standard texts on the subject, e.g., the well-known treatise by Flügge.[1]

In the finite element treatment of shell problems to be described in this chapter the difficulties referred to above are eliminated, at the expense of introducing a further approximation. This approximation is of a physical, rather than mathematical, nature. In this it is assumed that the behaviour of a continuously curved surface can be adequately represented by the behaviour of a surface built up of small, flat, elements.

Intuitively, as the size of the subdivision decreases it would seem that convergence must occur, and indeed experience indicates such a convergence.

It will be argued by many shell experts that when we compare the *exact* solution of a shell approximated by flat facets to the exact solution of a truly curved shell considerable differences in the distribution of bending moments, etc., occur. This is undoubtedly true, *but for simple elements* the discretization error is approximately of the same order and excellent results can be obtained with the flat shell element approximation. The mathematics of this problem is discussed in detail by Ciarlet.[2]

In a shell, the element will be subject, generally, both to bending and 'in-plane' forces. For a flat element these cause independent deformations, provided the local deformations are small, and therefore the ingredients for obtaining the necessary stiffness matrices are available in the material already covered in this book.

In the division of an arbitrary shell into flat elements only triangular elements can be used. Although the concept of the use of such elements in the analysis has been suggested as early as 1961 by Greene *et al.*,[3] the success of such analysis was hampered by the lack of a good stiffness matrix for triangular plate elements in bending.[4–7] The developments described in Chapter 10 open the way to adequate models for representing the behaviour of shells with such a division.

Some shells, for example those with general cylindrical shapes, can be well represented by flat elements of rectangular or quadrilateral shape. With good stiffness matrices available for such elements the progress here has been more satisfactory. Practical problems of arch dam design, and others for cylindrical shape roofs, have been solved earlier with such subdivisions.[8,9]

Clearly, the possibilities of analysis of shell structures by the finite element method are enormous. Problems presented by openings, variation of thickness or anisotropy are no longer of consequence once general programs are written.

A special case is presented by axi-symmetrical shells. Although it is obviously possible to deal with these in the way described in this chapter, a simpler approach can be used. This will be presented in Chapter 14.

As an alternative to the type of analysis described here, curved shell elements could be used. Here curvilinear co-ordinates are essential and general procedures of Chapter 8 can be extended to define these. The physical approximation involved in flat elements is now avoided at the expense of re-introducing an arbitrariness of various shell theories. Several approaches using a direct displacement approach are given in references 10–29 and 'mixed' variation principles in references 30–33.

A very simple and effective way of deriving curved shell elements is to use the so called 'shallow' shell theory approach.[19,20]

Here the displacement components, w, u, v, define the *normal and tangential* components of displacement to the curved surface and if all the elements are assumed tangent to each other, no need arises to transfer those from local to global values.

The element is assumed to be 'shallow' with respect to a local co-ordinate system representing its projection on to a plane defined by nodal points and its strain energy is defined by appropriate equations which include derivatives with respect to *co-ordinates in the plane of projection*. Thus, precisely the same shape functions can be used as in flat elements discussed in this chapter and all integrations are in fact carried out in the plane as before.

Such shallow shell elements, by coupling the effects of membrane and bending strain in the energy expressions, are slightly more efficient than flat ones where such coupling occurs on boundary only. For simple, small elements the gains are marginal but with few complex large elements advantages show up. A particularly good discussion of such a formulation is given in reference 21.

However, for many practical purposes the flat element approximation gives very adequate answers and indeed permits an easy coupling with edge beam and rib members, a facility sometimes not present in curved element formulation. Indeed in many practical problems the structure is in fact composed of flat surfaces at least in part and these can be simply reproduced. For these reasons curved general thin shell forms will not be discussed here and instead a general formulation of thick curved shells (based directly on three-dimensional behaviour and avoiding the shell equation ambiguities) will be presented in Chapter 15.

In the context of axi-symmetric shells given in the next chapter both straight and curved elements will be considered.

In most arbitrary-shaped, curved shell elements so far derived the co-ordinates used are such that complete smoothness of the surface between elements is not guaranteed. The shape discontinuity occurring there, and, indeed, on any shell where 'branching' occurs, is precisely of the same type as that encountered in this chapter and therefore the methodology of assembly discussed here is perfectly general.

13.2 **Stiffness of a Plane Element in Local Co-ordinates**

Consider a typical polygonal flat element subject simultaneously to 'in plane' and bending actions (Fig. 13.1).

Taking first the *in plane* (plane stress) action, we know from Chapter 4 that the state of strain is uniquely described in terms of the u and v displacement of each typical node i. The minimization of the total energy potential led to the stiffness matrices described there and gives 'nodal' forces due to displacement parameters $\mathbf{a}^p$ as

$$\mathbf{f}^{ep} \equiv \mathbf{K}^{ep}\mathbf{a}^p \qquad \text{with} \qquad \mathbf{a}_i^p = \begin{Bmatrix} u_i \\ v_i \end{Bmatrix} \qquad \mathbf{f}_i^p = \begin{Bmatrix} U_i \\ V_i \end{Bmatrix} \tag{13.1}$$

Similarly, when bending was considered, the state of strain was given uniquely by the nodal displacement in the z direction (w) and the two rotations θ_x and θ_y. This resulted in stiffness matrices of the type

$$\mathbf{f}^{eb} = \mathbf{K}^{eb}\mathbf{a}^{b} \qquad \text{with} \qquad \mathbf{a}_i^b = \begin{Bmatrix} w_i \\ \theta_{xi} \\ \theta_{yi} \end{Bmatrix} \qquad \mathbf{f}_i^b = \begin{Bmatrix} W_i \\ M_{xi} \\ M_{yi} \end{Bmatrix} \tag{13.2}$$

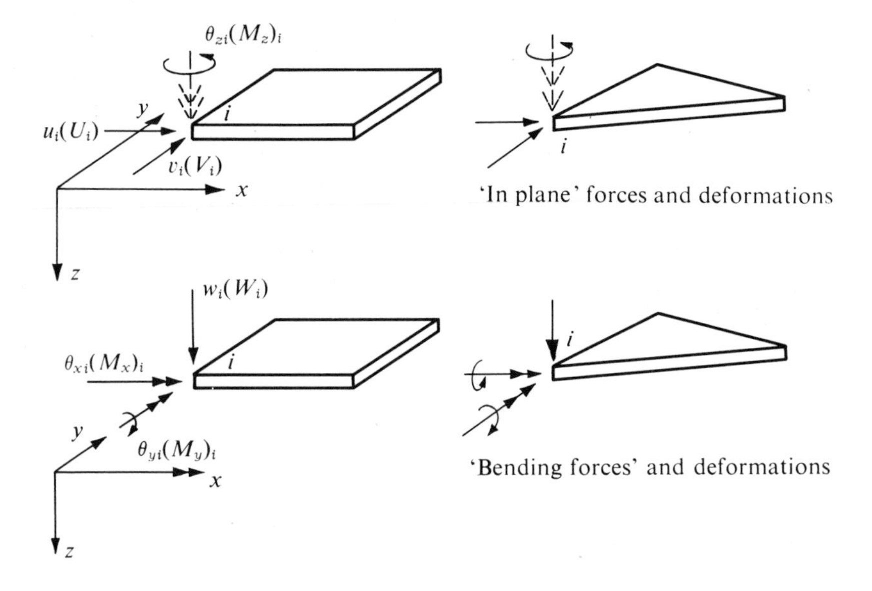

Fig. 13.1 A flat element subject to 'in plane' and bending actions

Before combining these stiffnesses it is important to note two facts. The first, that the displacements prescribed for 'in plane' forces do not affect the bending deformations and vice versa. The second, that rotation θ_z does not enter as a parameter into definition of deformations in either mode. While one could neglect this entirely at the present stage it is convenient, for reasons which will be apparent later when assembly is considered, to take this rotation into account now, and associate with it a fictitious couple M_z. The fact that it does not enter into the minimization procedure can be accounted for simply by inserting an appropriate number of zeros into the stiffness matrix.

Redefining now the combined nodal displacements as

$$\mathbf{a}_i = \begin{Bmatrix} u_i \\ v_i \\ w_i \\ \theta_{xi} \\ \theta_{yi} \\ \theta_{zi} \end{Bmatrix} \tag{13.3}$$

and the appropriate 'forces' as

$$\mathbf{f}_i^e = \begin{Bmatrix} U_i \\ V_i \\ W_i \\ M_{xi} \\ M_{yi} \\ M_{zi} \end{Bmatrix} \tag{13.4}$$

we can write

$$\mathbf{f}^e = \mathbf{K}^e \mathbf{a}. \tag{13.5}$$

The stiffness matrix is now made up from the following submatrices

$$\mathbf{K}_{rs} = \left[\begin{array}{cc|ccc|c} \multicolumn{2}{c|}{\multirow{2}{*}{$\mathbf{K}_{rs}^p$}} & 0 & 0 & 0 & 0 \\ & & 0 & 0 & 0 & 0 \\ \hline 0 & 0 & & & & 0 \\ 0 & 0 & & \mathbf{K}_{rs}^b & & 0 \\ 0 & 0 & & & & 0 \\ \hline 0 & 0 & 0 & 0 & 0 & 0 \end{array}\right] \tag{13.6}$$

if we note that

$$\mathbf{a}_i = \begin{Bmatrix} \mathbf{a}_i^p \\ \mathbf{a}_i^b \\ \theta_{zi} \end{Bmatrix} \tag{13.7}$$

The above formulation is valid for any shape of polygonal element and, in particular, for the two important cases illustrated in Fig. 13.1.

13.3 **Transformation to Global Co-ordinates and Assembly of the Elements**

The stiffness matrix derived in the previous section used a system of local co-ordinates as the 'in plane', and bending components are originally derived for this system.

Transformation of co-ordinates to a common global system (which now will be denoted by xyz, and the local system by $x'y'z'$) will be necessary to assemble the elements and to write the appropriate equilibrium equations.

In addition it will be initially more convenient to specify the element nodes by their global co-ordinates and to establish from these the local co-ordinates, thus requiring an inverse transformation. Fortunately, all the transformations are accomplished by a simple process.

The two systems of co-ordinates are shown in Fig. 13.2. The forces and displacements of a node transform from the global to the local system by a matrix L giving

$$\mathbf{a}_i' = \mathbf{L}\mathbf{a}_i \qquad \mathbf{f}_i' = \mathbf{L}\mathbf{f}_i \tag{13.8}$$

in which

$$\mathbf{L} = \begin{bmatrix} \boldsymbol{\lambda} & 0 \\ 0 & \boldsymbol{\lambda} \end{bmatrix} \tag{13.9}$$

with $\boldsymbol{\lambda}$ being a three by three matrix of direction cosines of angles formed between the two sets of axes, i.e.,

$$\boldsymbol{\lambda} = \begin{bmatrix} \lambda_{x'x} & \lambda_{x'y} & \lambda_{x'z} \\ \lambda_{y'x} & \lambda_{y'y} & \lambda_{y'z} \\ \lambda_{z'x} & \lambda_{z'y} & \lambda_{z'z} \end{bmatrix} \tag{13.10}$$

in which $\lambda_{xx'}$ = cosine of angle between x and x' axes, etc.

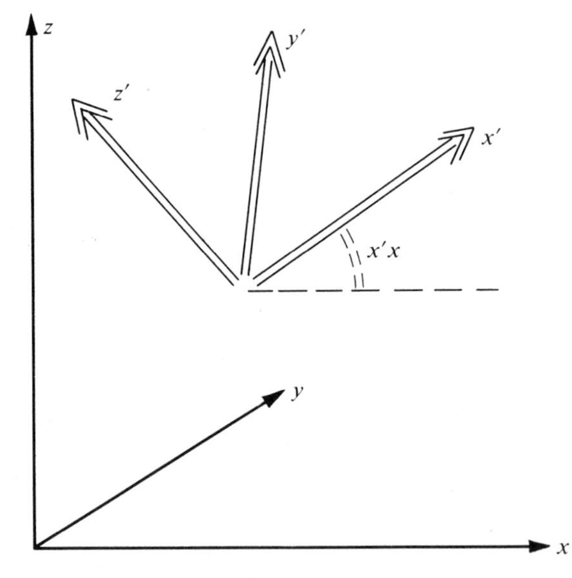

Fig. 13.2 Local and global co-ordinates

For the whole set of forces acting on the nodes of an element we can therefore write

$$\mathbf{a}'^e = \mathbf{T}\mathbf{a}^e, \quad \text{etc.} \tag{13.11}$$

By the rules of orthogonal transformation (*see* section 1.8, pp. 17–18) the stiffness matrix of an element in the global co-ordinates becomes

$$\mathbf{K}^e = \mathbf{T}^{\mathrm{T}}\mathbf{K}'^e\mathbf{T} \tag{13.12}$$

In both of the above equations $\mathbf{T}$ is given by

$$\mathbf{T} = \begin{bmatrix} \mathbf{L} & 0 & 0 & \cdots \\ 0 & \mathbf{L} & 0 & \\ 0 & 0 & \mathbf{L} & \\ \vdots & & & \end{bmatrix} \tag{13.13}$$

a diagonal matrix built up of $\mathbf{L}$ matrices in a number equal to that of the nodes in the element.

It is simple to show that the typical stiffness submatrix now becomes

$$\mathbf{K}^e_{rs} = \mathbf{L}^{\mathrm{T}}\mathbf{K}'^e_{rs}\mathbf{L} \tag{13.14}$$

in which $\mathbf{K}'^e_{rs}$ is determined by Eq. (13.6) in the local co-ordinates.

The determination of local co-ordinates follows a similar pattern. If the origins of both local and global systems are identical then

$$\begin{Bmatrix} x' \\ y' \\ z' \end{Bmatrix} = \boldsymbol{\lambda} \begin{Bmatrix} x \\ y \\ z \end{Bmatrix} \tag{13.15}$$

As in the computation of stiffness matrices the position of the origin is immaterial, this transformation will always suffice for determination of the local co-ordinates in the plane (or a plane parallel to the element).

Once the stiffness matrices of all the elements have been determined in the common, global, co-ordinate system the assembly of the elements and the final solution follow the standard pattern. The resulting displacements calculated are referred to the global system, and before the stresses can be computed it is necessary to change these for each element of the local system. The usual stress matrices for 'in plane' bending components can then be used.

13.4 A Fictitious Rotational Stiffness—Six-degree-of-freedom Assembly

In the formulation just described a difficulty arises if all the elements meeting at a node are co-planar. This is due to the assignment of a zero stiffness in the θ_{zi} direction of Fig. 13.1.

If the set of assembled equilibrium equations *in local co-ordinates* is considered at such a point we have six equations of which the last (corresponding to θ_z direction) is simply

$$0 = 0. \tag{13.16}$$

As such, an equation of this type, presents no special difficulties (although in some computation programs it would lead to an error message). However, if the global co-ordinate directions differ from the local ones and a transformation is accomplished, six apparently correct equations are achieved. These, being derived by a suitable addition of the above multiplied suitably by real numbers, are singular.†

Thus two alternatives are present

(*a*) to assemble the equations at points where elements are co-planar in local co-ordinates (and to delete the $0 = 0$ equation)

or

(*b*) to insert an arbitrary stiffness coefficient $k'_{\theta z}$ at such points only.

This leads in the local co-ordinates to replacing Eq. (13.16) by

$$k'_{\theta z}\,\theta_{zi} = 0 \tag{13.17}$$

which, on transformation, leads to a perfectly well-behaved set of equations from which, by usual processes, all displacements, now including θ_{zi}, are obtained. As θ_{zi} does not affect the stresses and indeed is uncoupled from all equilibrium equations any arbitrary value of $k'_{\theta z}$ can be inserted as an external stiffness without affecting the result.

Both alternatives suggested above lead to certain programing difficulty (although the second one is in fact simpler) and some work has proceeded to determine the real stiffness coefficient for rotations of the type described by considering these as an additional degree of freedom in plane analysis.[20]

In a program used by the author[7] a fictitious set of rotation stiffness coefficients was simply used in all elements whether co-planar or not. For a triangular element these were defined by a matrix such that in local co-ordinates equilibrium is not disturbed, i.e.,

$$\begin{Bmatrix} M_{zi} \\ M_{zj} \\ M_{zk} \end{Bmatrix} = \alpha E t \Delta \begin{bmatrix} 1, & -0{\cdot}5, & -0{\cdot}5 \\ & 1 & -0{\cdot}5 \\ \text{sym.} & & 1 \end{bmatrix} \begin{Bmatrix} \theta_{zi} \\ \theta_{zj} \\ \theta_{zk} \end{Bmatrix} \tag{13.18}$$

where α is some coefficient yet to be specified.

† The reader will recall the apparently logical practice based on multiplying such an equality and deriving $2 = 4$, etc.

Now this additional stiffness does in fact affect the results because it occurs also at nodes which are not co-planar and indeed the device represents an approximation. However, the effects of varying α between very wide limits are quite small. For instance in Table 13.1 given below, a set of displacements of an arch dam analysed in reference 3 is given for various values of α.

TABLE 13.1

NODAL ROTATION COEFFICIENT IN DAM ANALYSIS[3]

$\alpha =$	1·00	0·50	0·10	0·03	0·00
radial displacement (mm)	61·13	63·35	64·52	64·78	65·28

The displacements for $\alpha = 0$ are nearly exact. For practical purposes extremely small values of α are possible providing a large computer[34] precision is available.

13.5 **Elements with mid-side slope connections only**

Many of the difficulties encountered with the nodal assembly in global co-ordinates disappear if the element is so constructed as to require only the continuity of displacements u, v and w at the corner nodes with continuity of the normal slope being imposed along the element sides. Clearly, the corner assembly is now simple and the introduction of the sixth nodal variable is unnecessary. As the normal slope rotation along the sides is the same both in local and global co-ordinates its transformation there is necessary.

Elements of this type arise naturally in hybrid forms (*vide* Chapter 12) and we have already referred to a plate bending element of a suitable type in Chapter 10. This element of the simplest possible kind has been used in shell problems by Dawe[25] with some success. A considerably more sophisticated and complex element of such a type is derived by Irons[26] with a suggested curious name of 'Semiloof'. This element is briefly mentioned in Chapter 11, and although its derivation is far from simple it performs well in many situations.

13.6 **Local Direction Cosines**

Once the direction cosine matrix $\boldsymbol{\lambda}$ has been determined for each element the problem presents no difficulties, and the solution follows the usual lines. The determination of the direction cosine matrix gives rise to some algebraic difficulties, and indeed, is not unique since the direction of one of the axes is arbitrary, provided it lies in the plane of the element.

We shall first deal with the assembly of rectangular elements in which this problem is particularly simple.

13.6.1 *Rectangular elements.* Such elements are limited in use to representing a cylindrical or box type of surface and it is convenient to take one side of the elements and the corresponding co-ordinate x' parallel to the global, x, axis. For a typical element $ijkm$, illustrated in Fig. 13.3, it is now easy to calculate all the relevant direction cosines.

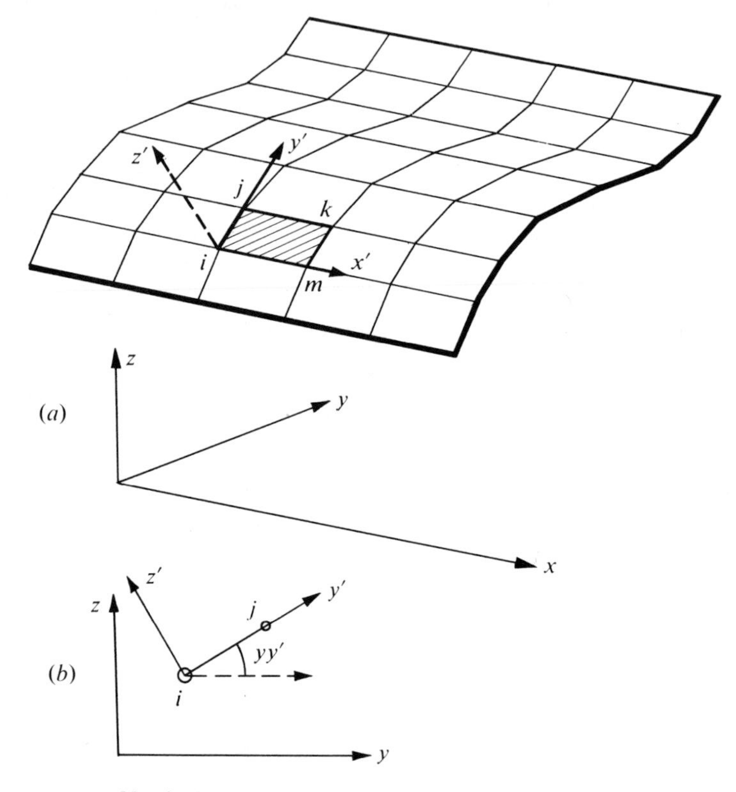

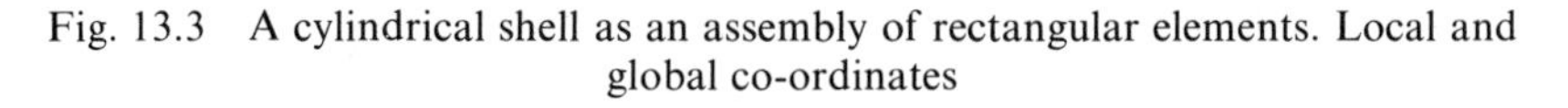
Fig. 13.3 A cylindrical shell as an assembly of rectangular elements. Local and global co-ordinates

Direction cosines of x' are, obviously

$$
\begin{aligned}
\lambda_{x'x} &= 1 \\
\lambda_{x'y} &= 0 \\
\lambda_{x'z} &= 0.
\end{aligned}
\tag{13.19}
$$

The direction cosines of the y' axis have to be obtained by consideration of the co-ordinates of the various nodal points. Thus

$$\begin{aligned} \lambda_{y'x} &= 0 \\ \lambda_{y'y} &= +\frac{y_j - y_i}{\sqrt{\{(z_j - z_i)^2 + (y_j - y_i)^2\}}} \\ \lambda_{y'z} &= +\frac{z_j - z_i}{\sqrt{\{(z_j - z_i)^2 + (y_j - y_i)^2\}}} \end{aligned} \tag{13.20}$$

simple geometrical relations which can be obtained by consideration of the sectional plane passing vertically through ij.

Similarly, from the same section we have for the z' axis

$$\begin{aligned} \lambda_{z'x} &= 0 \\ \lambda_{z'y} &= -\frac{z_j - z_i}{\sqrt{\{(z_j - z_i)^2 + (y_j - y_i)^2\}}} \\ \lambda_{z'z} &= +\frac{y_j - y_i}{\sqrt{\{(z_j - z_i)^2 + (y_j - y_i)^2\}}} \end{aligned} \tag{13.21}$$

Clearly, the numbering of points in a consistent fashion is important to preserve the correct signs of the expression.

13.6.2 *Triangular elements arbitrarily oriented in space.* An arbitrary shell divided into triangular elements is shown in Fig. 13.4(a). Each element is in an orientation in which the angles with the co-ordinate planes are arbitrary. The problems of defining local axes and their direction cosines are therefore considerably more complex than in the previous simple example. The most convenient way of dealing with the problem is to use some features of geometrical vector algebra and for readers who may have forgotten some of this background a brief resumé of its essentials is included in Appendix 5.

One choice of local axis direction is arbitrary and a decision on this has to be made *a priori*. We shall specify this, x', axis, to be directed along the side ij of the triangle as shown in Fig. 13.4(b).

The vector $\mathbf{V}_{ij}$ defines this side and in terms of global co-ordinates we have

$$\mathbf{V}_{ij} = \begin{Bmatrix} x_j - x_i \\ y_j - y_i \\ z_j - z_i \end{Bmatrix}. \tag{13.22}$$

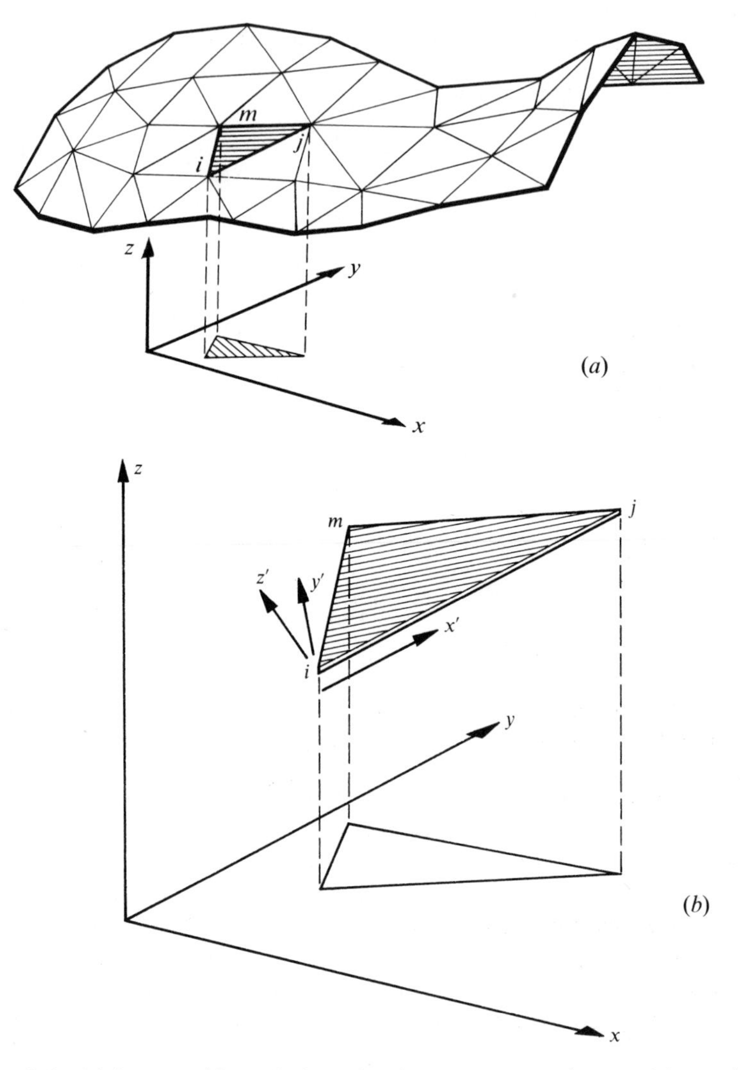

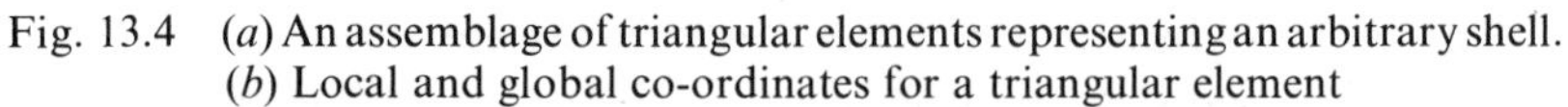
Fig. 13.4 (*a*) An assemblage of triangular elements representing an arbitrary shell. (*b*) Local and global co-ordinates for a triangular element

The direction cosines are given by dividing the components of this vector by its length, i.e., defining a vector of unit length

$$\mathbf{v}_{x'} = \begin{Bmatrix} \lambda_{x'x} \\ \lambda_{x'y} \\ \lambda_{x'z} \end{Bmatrix} = \frac{1}{l_{ij}} \begin{Bmatrix} x_{ji} \\ y_{ji} \\ z_{ji} \end{Bmatrix} \tag{13.23}$$

with

$$l_{ij} = \sqrt{x_{ji}^2 + y_{ji}^2 + z_{ji}^2}$$

in which $x_{ji} = x_j - x_i$, etc., for brevity.

Now the z' direction, which must be normal to the plane of the triangle, needs to be established. By properties of the cross product of two vectors we can obtain this direction from a 'vector' (cross) product of two sides of the triangle. Thus

$$\mathbf{V}_{z'} = \mathbf{V}_{ij} \times \mathbf{V}_{im} = \begin{Bmatrix} y_{ji}z_{mi} - z_{ji}y_{mi} \\ \cdot \quad \cdot \quad \cdot \quad \cdot \quad \cdot \\ \cdot \quad \cdot \quad \cdot \quad \cdot \quad \cdot \end{Bmatrix} \tag{13.24}$$

represents a vector normal to the plane of the triangle whose length, by definition (see Appendix 5), is equal to twice the area of the triangle. Thus

$$l_{z'} = \sqrt{(y_{ji}z_{mi} - z_{ji}y_{mi})^2 + (\ldots)^2 + (\ldots)^2} = 2\Delta.$$

The direction cosines of the z' axis are available simply as the direction cosines of $\mathbf{V}_{z'}$ and we have a unit vector

$$\mathbf{v}_{z'} = \begin{Bmatrix} \lambda_{z'x} \\ \lambda_{z'y} \\ \lambda_{z'z} \end{Bmatrix} = \frac{1}{2\Delta} \begin{Bmatrix} y_{ji}z_{mi} - z_{ji}y_{mi} \\ \cdot \quad \cdot \quad \cdot \quad \cdot \\ \cdot \quad \cdot \quad \cdot \quad \cdot \end{Bmatrix} \tag{13.25}$$

Finally the direction cosines of the y' axis are established in a similar manner as the direction cosines of a vector normal to both the x' and z' directions. If vectors of unit length are taken in each of these directions as in fact defined by Eqs. (13.23) and (13.25) we have simply

$$\mathbf{v}_{y'} = \begin{Bmatrix} \lambda_{y'x} \\ \lambda_{y'y} \\ \lambda_{y'z} \end{Bmatrix} = \mathbf{v}_{z'} \times \mathbf{v}_{x'} = \begin{Bmatrix} \lambda_{z'y}\lambda_{x'z} - \lambda_{z'z}\lambda_{x'y} \\ \cdot \quad \cdot \quad \cdot \quad \cdot \quad \cdot \quad \cdot \\ \cdot \quad \cdot \quad \cdot \quad \cdot \quad \cdot \quad \cdot \end{Bmatrix} \tag{13.26}$$

without having to divide by the length of the vector which is now simply unity.

Indeed the vector operations involved can be written as a special computer routine in which vector products, normalizing (i.e., division by length), etc. are automatically carried out[35] and there is no need to specify in detail the various operations given above.

In the preceding outline the direction of the x' axis was taken as lying along one side of the element. A useful alternative is to specify this by the section of the triangle plane with a plane parallel to one of the co-ordinate planes. Thus for instance if we should desire to erect the x' axis along a horizontal contour of the triangle (i.e., a section parallel to the xy plane) we can proceed as follows.

First the normal direction cosines $\mathbf{v}_{z'}$ are defined as in Eq. (13.25).

Now, the matrix of direction cosines of x' has to have a zero component in the z direction. Thus we have

$$\mathbf{v}_{x'} = \begin{Bmatrix} \lambda_{x'x} \\ \lambda_{x'y} \\ 0 \end{Bmatrix}. \tag{13.27}$$

As the length of the vector is unity

$$\lambda_{x'x}^2 + \lambda_{x'y}^2 = 1 \tag{13.28}$$

and as further the *scalar* product of the $\mathbf{v}_{x'}$ and $\mathbf{v}_{z'}$ must be zero, we can write

$$\lambda_{x'x} . \lambda_{z'x} + \lambda_{x'y} . \lambda_{z'y} = 0 \tag{13.29}$$

and from these two equations $\mathbf{v}_{y'}$ can be uniquely determined. Finally, as before

$$\mathbf{v}_{y'} = -\mathbf{v}_{x'} \times \mathbf{v}_{z'} . \tag{13.30}$$

Yet another alternative of a unique specification of the x' axis is given in Chapter 15.

13.7 Choice of Element

Numerous 'in plane' and bending element formulations are now available, and, in both, conformity was achievable in flat assemblies. Clearly if the plates are not coplanar conformity will, in general, be violated (except in the limit as smooth shell conditions are reached). As we have shown, non-conforming elements are usually superior in performance and we shall, therefore, use these in our illustrative examples.

It would appear consistent to use expansions of similar accuracy in both the membrane and bending approximations, but much depends on which action is predominant. The simplest triangular element would thus appear to be one with linear in plane displacement field and a quadratic bending displacement—thus approximating the stresses as constants in plane and in bending. Such an element is used by Dawe[25] but gives rather poor (though convergent) results.

In the examples shown we use the following elements which give adequate performance:

Element A: Non-conforming, in plane, rectangle with four corner nodes (see Chapter 11, p. 272) combined with the non-conforming bending rectangle with four corner nodes (Chapter 10, p. 234). This was first used in references 8 and 9.

Element B: Constant strain triangle with three nodes (basic element of Chapter 2) combined with the incompatible bending triangle with nine degrees of freedom (Chapter 10, p. 241). Use of this in shell context is given in references 7 and 36.

Element C: In this a more consistent linear strain triangle with six nodes is combined with a twelve-degree-of-freedom bending triangle using shape function smoothing. This element has been introduced by Razzaque.[37]

13.8 Some Practical Examples

The first example given here is that of the solution of an arch dam shell. A simple geometrical configuration, shown in Fig. 13.5, was taken for this particular problem as results of model experiments and alternative numerical approaches were available.

A division based on rectangular elements (type A) was used as the simple cylindrical shape permitted this, although a rather crude approximation to the fixed foundation line had to be used.

Two sizes of division into elements were used, and the results given in Figs. 13.6 and 13.7 for both deflections and stresses on the centre-line section show that little refinement was achieved by the use of the

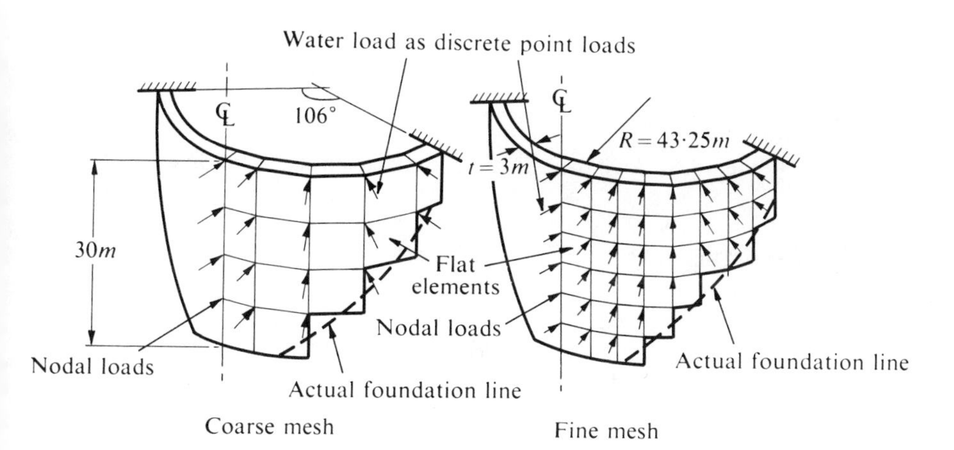

Fig. 13.5 An arch dam as an assembly of rectangular element

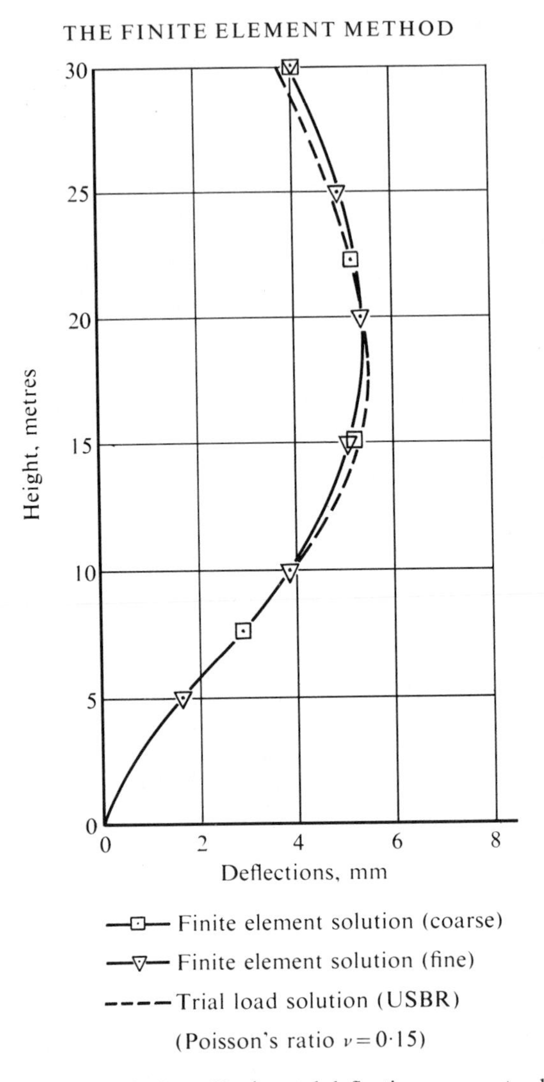

Fig. 13.6 Arch dam. Horizontal deflections on centre-line

finer mesh. This indicates that the convergence of both the physical approximation to the true shape by flat elements, and of the mathematical approximation involved in the finite element formulation, is excellent. For comparison, stresses and deflection obtained by another, approximate, method of calculation are shown.

A doubly curved arch dam was similarly analysed using the triangular flat element (Type B) representation. The results show an even better approximation.[7]

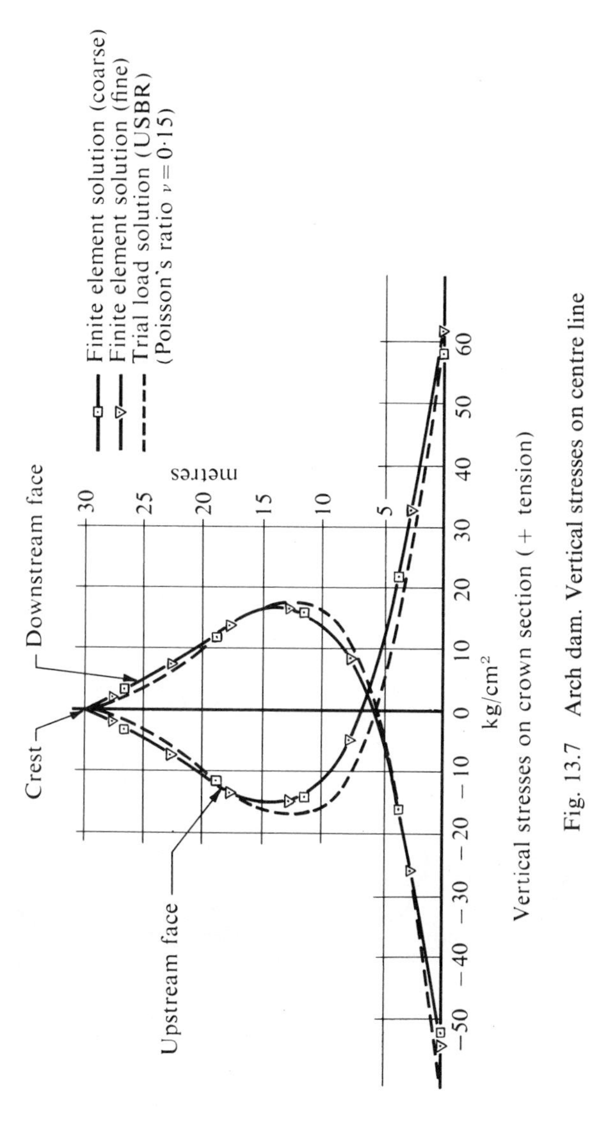

Fig. 13.7 Arch dam. Vertical stresses on centre line

A large number of examples have been computed by Parekh[36] using the triangular, non-conforming element (B) and indeed show for equal division its general superiority over the conforming triangular version presented by Clough and Johnson.[6] Some examples of such analyses are shown here.

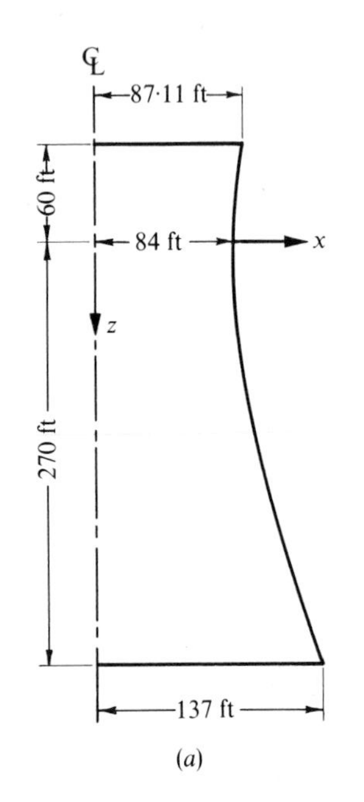

(a)

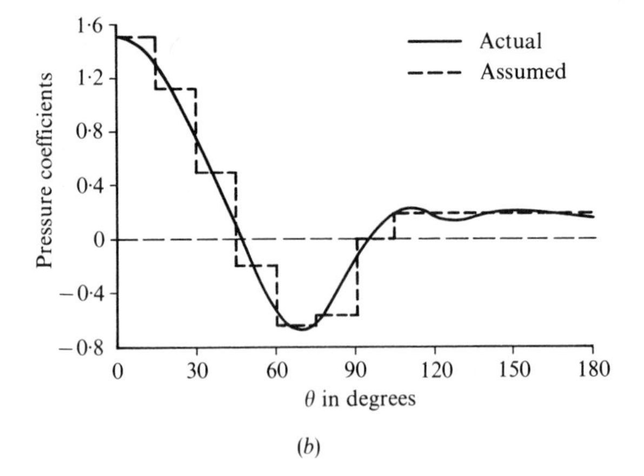

(b)

Fig. 13.8 Cooling tower. Geometry and pressure load variation about circumference

Cooling tower. This problem of a general axi-symmetric shape could, obviously, be more efficiently dealt with by the processes of Chapters 15 or 16. However, here this example is used as a general illustration of the accuracy attainable. The answers against which the numerical solution is compared have been derived by Albasiny and Martin.[38] Figures 13.9 to 13.10 show the geometry mesh used and some results. Unsymmetric wind loading is used here.

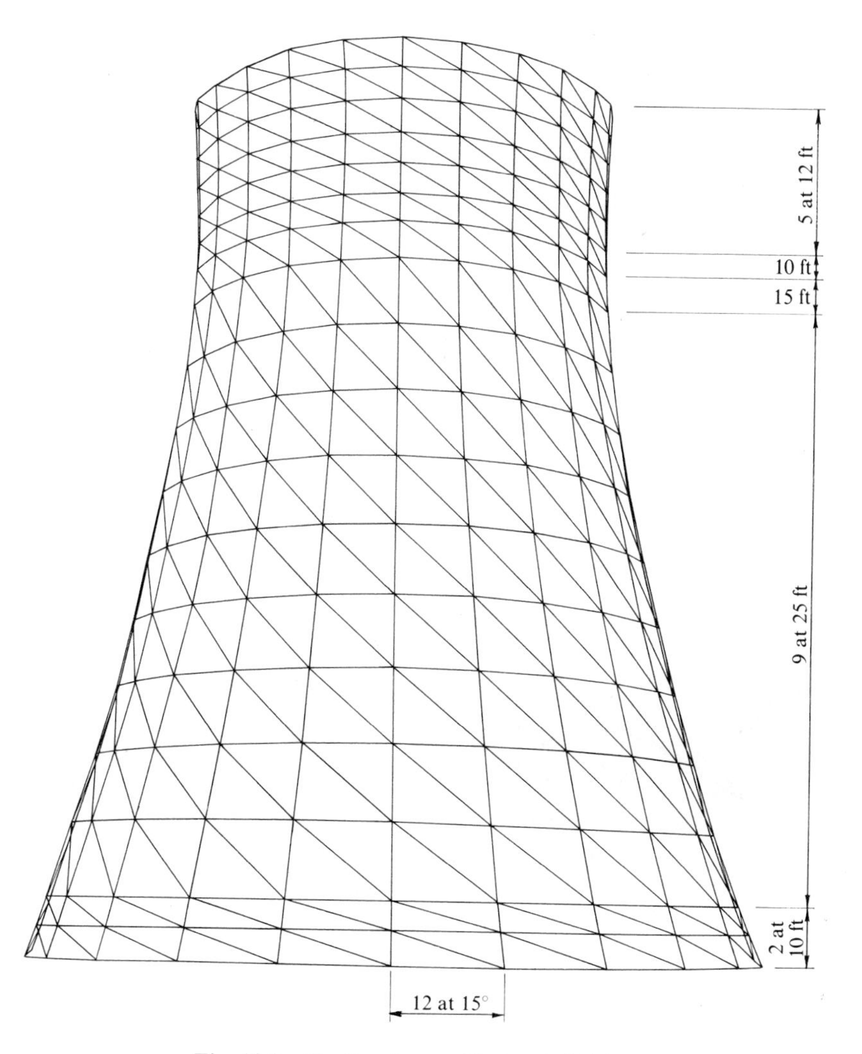

Fig. 13.9 Cooling tower. Mesh subdivisions

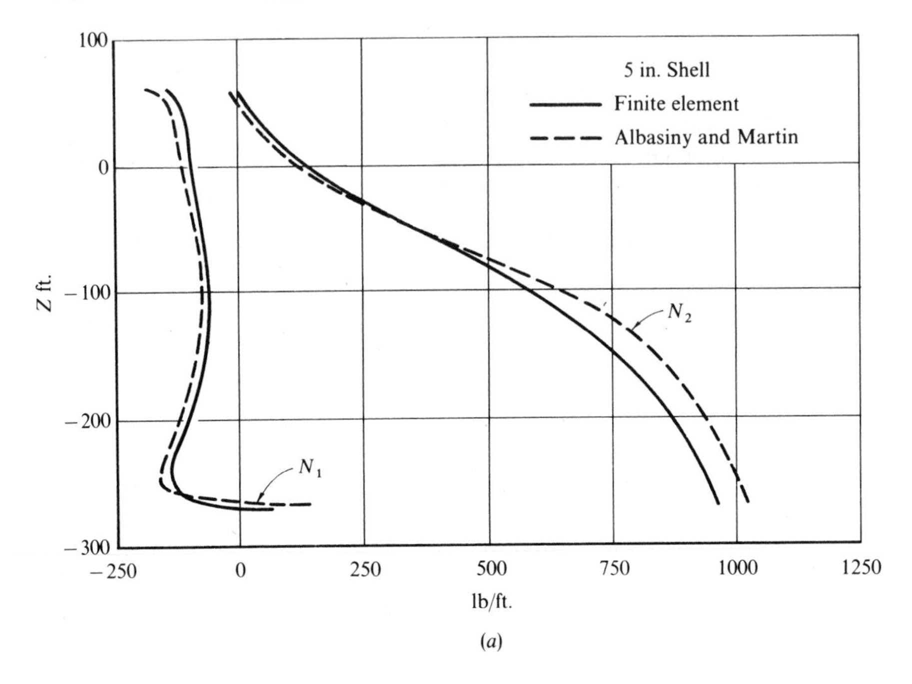

(*a*)

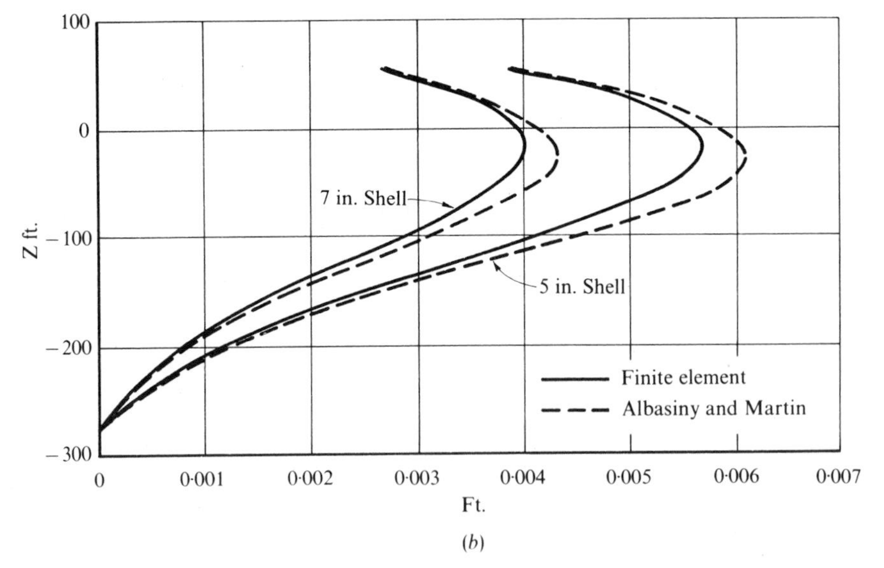

(*b*)

Fig. 13.10 Cooling tower of Fig. 13.9 (*a*) membrane forces at $\theta = 0°$, N_1 = tangential forces, N_2 = meridional force. (*b*) Radial displacements at $\theta = 0°$. (*c*) Bending moments at $\theta = 0°$, M_1 = tangential moment, M_2 = Meridional moment

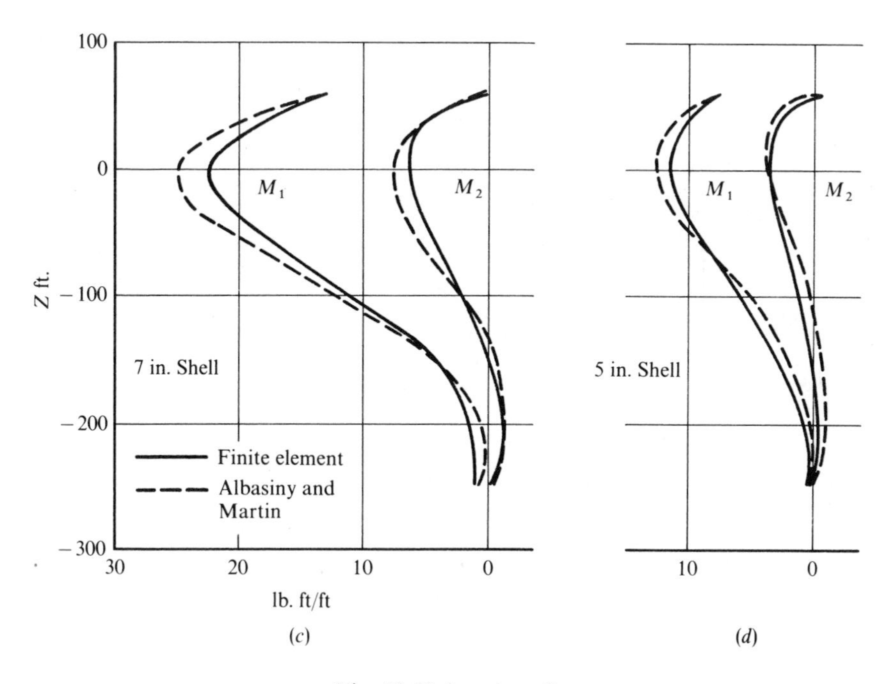

Fig. 13.10 (continued)

Barrel vault. This typical shell used in civil engineering is analysed using conventional processes by Scordelis and Lo.[39] The barrel is supported on rigid diaphragms and is loaded by its own weight. Figures 13.11 and 13.12 show some comparative answers, obtained by elements of Type B or C of the previous section. The latter are obviously more accurate involving more degrees of freedom and with a mesh of 6×6 elements the results are almost indistinguishable from exact ones. This problem has become a classic one on which various shell elements are compared and we shall return to it in Chapter 15. It is worth while remarking that only a few, second order, curved elements give superior results to those presented here with a flat element approximation.

Folded plate structure. As no exact solution to this problem is known comparison is made with a set of experimental results obtained by Mark and Riesa.[40]

This example demonstrates a problem in which actual flat finite element representation is physically exact. Also a frame stiffness is included in analysis by suitable beam elements.

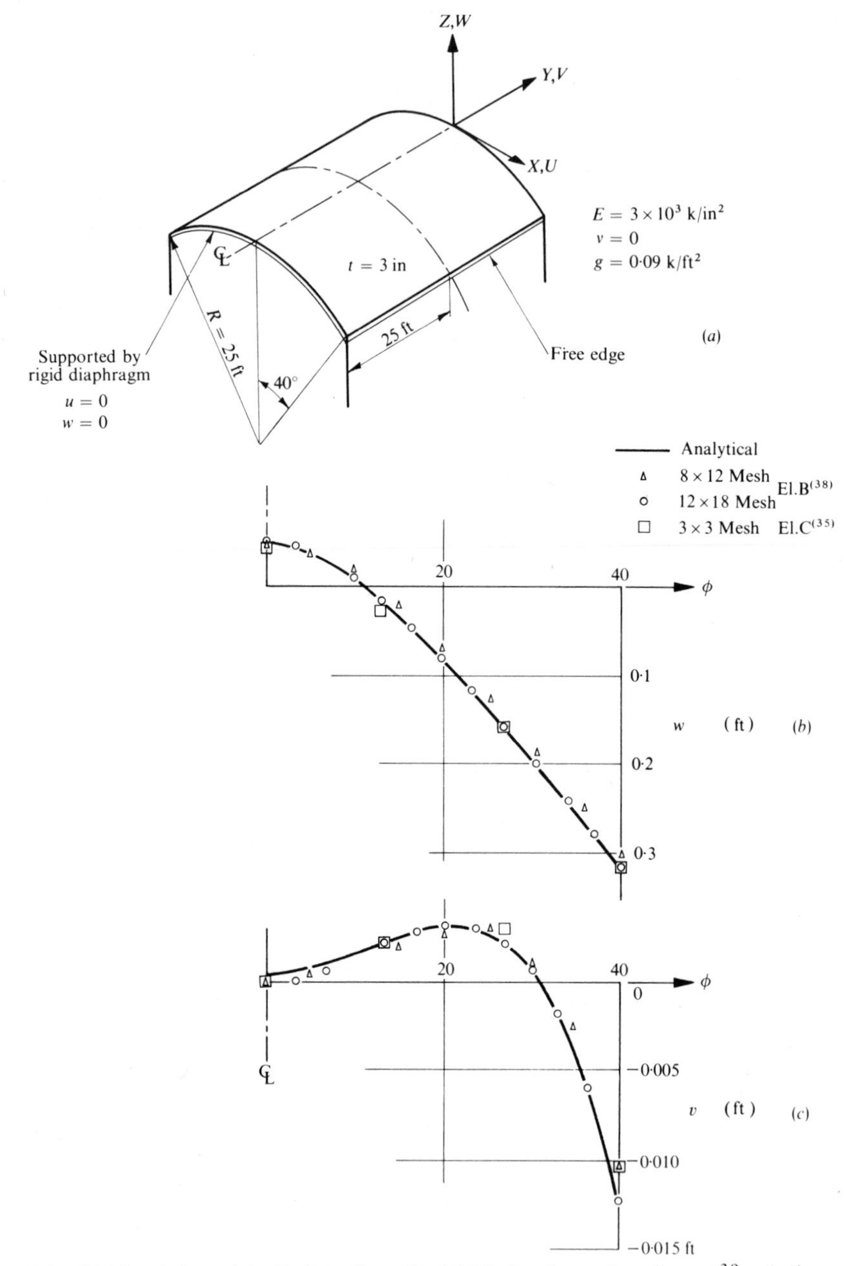

Fig. 13.11 A barrel (cylindrical) vault. (*a*) Finite element and exact[39] solutions under dead loads. (*b*) Vertical displacement of central sections. (*c*) Longitudinal displacement of support
$E = 3 \times 10^6$ lb/in² $v = 0$ weight of shell = 90 lb/ft²

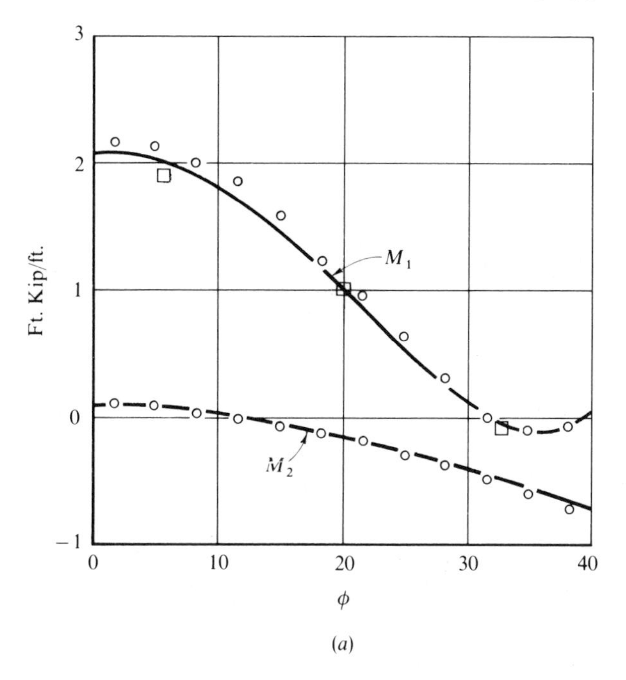

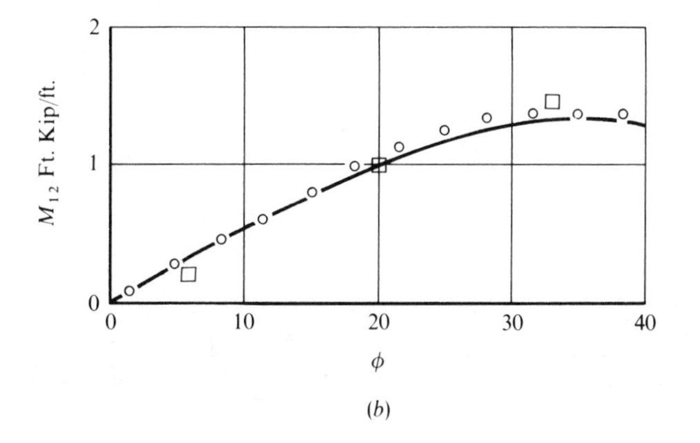

Fig. 13.12 Barrel vault of Fig. 13.11. (*a*) M_1 = transverse moments, M_2 = Longitudinal moments at central section. (*b*) M_{12} = twisting moment at support

Figures 13.13 and 13.14 show the results. Similar applications are of considerable importance in the analysis of box type structures, etc.

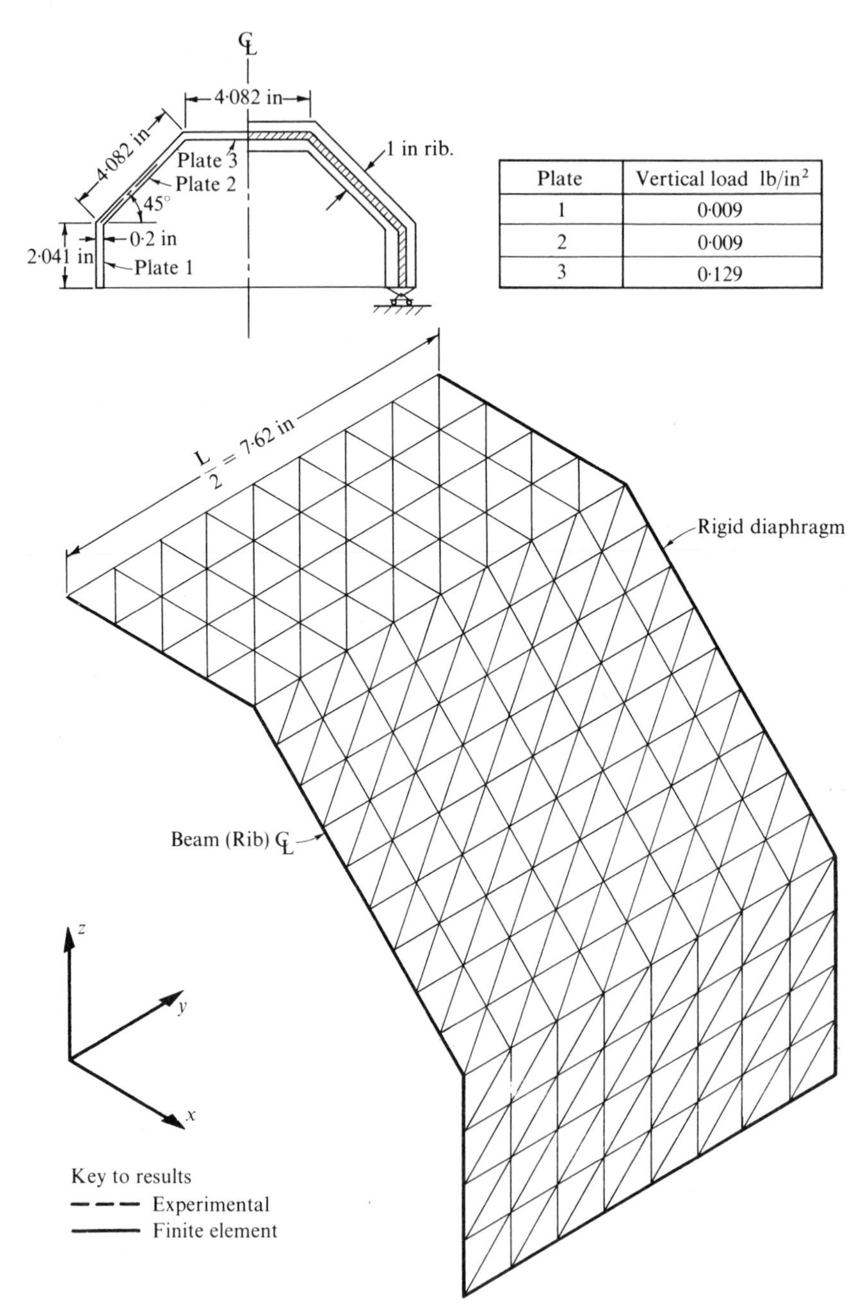

Plate	Vertical load lb/in²
1	0·009
2	0·009
3	0·129

Fig. 13.13 A folded plate structure.[40] Model geometry, loading and mesh
$E = 3560\ \text{lb/in}^2 \quad v = 0{\cdot}43$

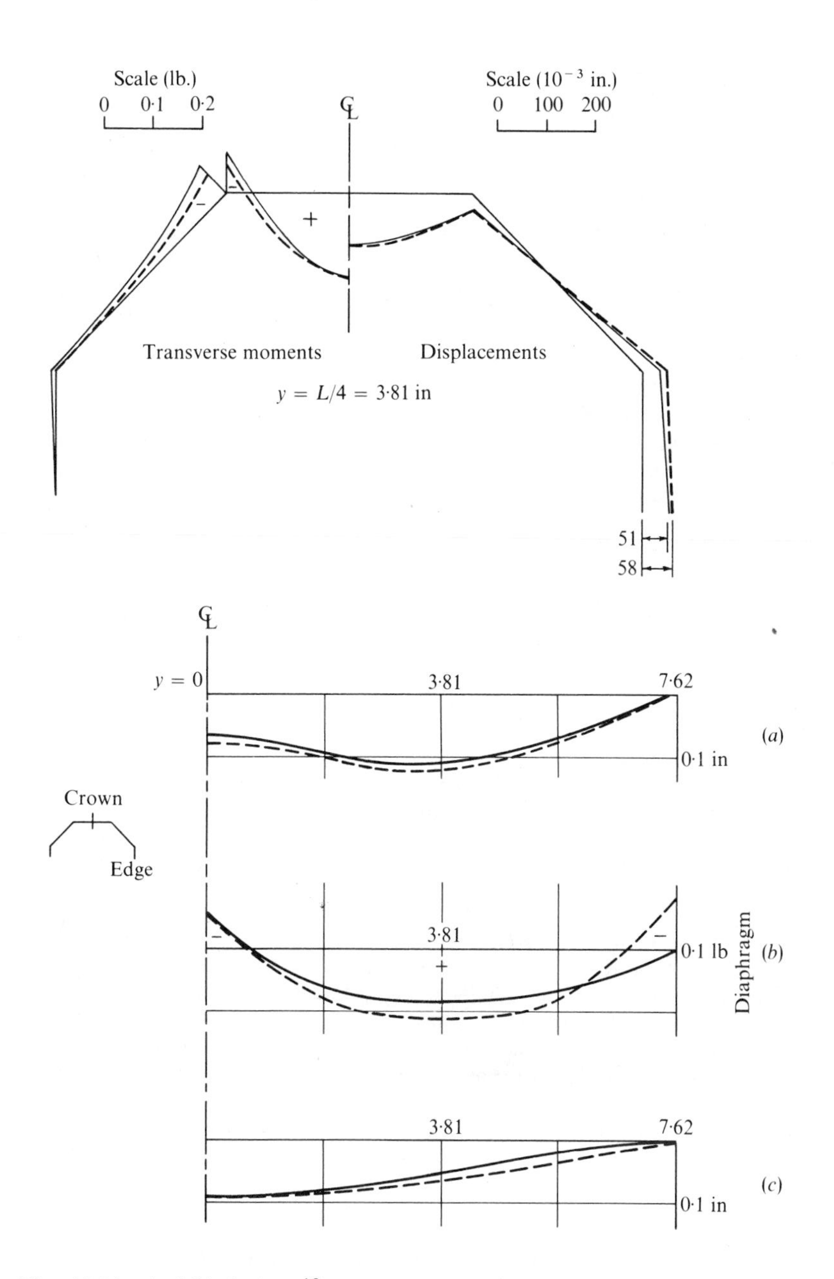

Fig. 13.14 A folded plate.[40] Moments and displacements on centre section. (*a*) Vertical displacements along the crown; (*b*) longitudinal moments along the crown; (*c*) horizontal displacements along edge

References

1. W. FLÜGGE, *Stresses in Shells*, Springer-Verlag, 1960.
2. P. G. CIARLET, 'Conforming finite element method for shell problem', in *The Mathematics of Finite Elements and Applications II* (ed. J. Whiteman), Academic Press, 1977.
3. B. E. GREENE, D. R. STROME, and R. C. WEIKEL, 'Application of the stiffness method to the analysis of shell structures', *Proc. Aviation Conf. Amer. Soc. Mech. Eng.*, Los Angeles, March 1961.
4. R. W. CLOUGH and J. L. TOCHER, 'Analysis of thin arch dams by the finite element method', *Proc. Symp. on Theory of Arch Dams*, Southampton Univ., 1964 (Pergamon Press, 1965).
5. J. H. ARGYRIS, 'Matrix displacement analysis of anisotropic shells by triangular elements', *J. Roy. Aero. Soc.*, **69**, 801–5, 1965.
6. R. W. CLOUGH and C. P. JOHNSON, 'A finite element approximation for the analysis of thin shells', *J. Solids Struct.*, **4**, 43–60, 1968.
7. O. C. ZIENKIEWICZ, C. J. PAREKH, and I. P. KING, 'Arch dams analysed by a linear finite element shell solution program', *Proc. Symp. Arch Dams, Inst. Civ. Eng.*, London, 1968.
8. O. C. ZIENKIEWICZ and Y. K. CHEUNG, 'Finite element procedures in the solution of plate and shell problems', Chapter 8 of *Stress Analysis* (eds. O. C. Zienkiewicz and G. S. Holister), Wiley, 1965.
9. O. C. ZIENKIEWICZ and Y. K. CHEUNG. 'Finite element method of analysis for arch dam shells and comparison with finite difference procedures', *Proc. Symp. Theory of Arch Dams*, Southampton Univ., 1964 (Pergamon Press, 1965).
10. R. H. GALLAGHER, 'Shell elements', *World Conf. on Finite Element Methods in Structural Mechanics*, Bournemouth, 1975.
11. D. J. DAWE, 'Rigid-body motions and strain-displacement equations of curved shell finite elements', *Int. J. Mech Sci.*, **14**, 569–78, 1972.
12. G. CANTIN, 'Strain–Development relationships for cylindrical shells', *J.A.I.A.A.*, **6** (No. 9), 1787–88, 1968.
13. D. G. ASHWELL, 'Strain elements with applications to arches, rings, and cylindrical shells' in *Finite Element Thin Shell Analysis* (eds. D. Ashwell and R. H. Gallagher), Wiley, 1976.
14. F. K. BOGNER, R. L. FOX, and L. A. SCHMIT, 'A cylindrical shell element', *J.A.I.A.A.*, **5**, 745–50, 1967.
15. G. CANTIN and R. W. CLOUGH, 'A refined, curved, cylindrical shell element', *A.I.A.A. Conf.*, Paper 68–176, New York, 1968.
16. G. BONNES, G. DHATT, Y. M. GIROUX, and L. P. A. ROBICHAUD, 'Curved triangular elements for analysis of shells, *Proc. 2nd Conf. Matrix Methods in Structural Mechanics*, Air Force Inst. Tech., Wright-Patterson A.F. Base, Ohio, 1968.
17. G. E. STRICKLAND and W. A. LODEN, 'A doubly curved triangular shell element', *Proc. 2nd Conf. Matrix Methods in Structural Mechanics*, Air Force Inst. Tech., Wright-Patterson A.F. Base, Ohio, 1968.
18. B. E. GREENE, R. E. JONES, and D. R. STROME, 'Dynamic analysis of shells using doubly curved finite elements', *Proc. 2nd Conf. Matrix Methods in Structural Mechanics*, Air Force Inst. Tech., Wright-Patterson A.F. Base, Ohio, 1968.
19. J. CONNOR and C. BREBBIA, 'Stiffness matrix for shallow rectangular shell element', *Proc. Am. Soc. Civ. Eng.*, **93**, EM 43–65, 1967.

20. A. J. CARR, *A refined element analysis of thin shell structures including dynamic loading*, SEL Report No. 67–9, Univ. of California, Berkeley, 1967.
21. G. R. COWPER, G. M. LINDBERG, and M. D. OLSON, 'A shallow shell finite element of triangular shape', *Int. J. Solids Struct.*, **6**, 1133–56, 1970.
22. S. UTKU, 'Stiffness matrices for thin triangular elements of non-zero Gaussian curvature', *J.A.I.A.A.*, **5**, 1659–67, 1967.
23. S. AHMAD, *Curved finite elements in the analysis of solid shell and plate structures*, Ph.D. Thesis, Univ. of Wales, Swansea, 1969.
24. S. W. KEY and Z. E. BEISINGER, 'The analysis of thin shells by the finite element method', in *High Speed Computing of Elastic Structures*, Tome 1, pp. 209–52, Univ. of Liége Press, 1971.
25. D. J. DAWE, *The analysis of thin shells using a facet element*, CEGB Report No. RD/B/N2038, Berkeley Nuclear Lab., England, 1971.
26. B. M. IRONS, 'The Semiloof Shell Element', chapter 11, pp. 197–222, of *Finite Elements for thin shells and curved members*, ed. D. G. Ashwell and R. H. Gallagher. Wiley, 1976.
27. D. G. ASHWELL and A. SABIR, 'A new Cylindrical shell finite element based on simple independent strain functions', *Int. J. Mech. Sci.*, **4**, 37–47, 1973.
28. G. R. THOMAS and R. H. GALLAGHER, *A Triangular thin shell finite element: linear analysis*, NASA CR-2582, 1975.
29. G. DUPUIS and J. J. GOËL, 'A curved finite element for thin elastic shells', *Int. J. Solids Struct.*, **6**, 987–96, 1970.
30. C. PRATO, 'Shell finite element via Reissner's principle', *Int. J. Solids Struct.*, **5**, 1119–33, 1969.
31. J. CONNOR and G. WILL, 'A mixed finite element shallow shell formulation', *Advances in Matrix Methods of Structural Analysis and Design*, pp. 105–37 (eds. R. Gallagher *et al.*), Univ. of Alabama Press, 1969.
32. L. R. HERRMANN and W. E. MASON, 'Mixed formulations for finite element shell analysis', *Conf. on Computer-Oriented Analysis of Shell Structures*, AFFDL-TR-71-79, June 1971.
33. G. EDWARDS and J. J. WEBSTER, 'Hybrid cylindrical shell elements', *Finite Element Thin Shell Analysis* (eds. D. Ashwell and R. Gallagher), Wiley, 1976.
34. R. W. CLOUGH and E. L. WILSON, 'Dynamic finite element analysis of arbitary thin shells', *Computers & Struct.*, **1**, 35, 1971.
35. S. AHMAD, B. M. IRONS, and O. C. ZIENKIEWICZ, 'A simple matrix-vector handling scheme for three-dimensional and shell analysis', *Int. J. Num. Meth. Eng.*, **2**, 509–22, 1970.
36. C. J. PAREKH, *Finite element solution system*, Ph.D. Thesis, Univ. of Wales, Swansea, 1969.
37. A. RAZZAQUE, *Finite element analysis of plates and shells*, Ph.D. Thesis, Univ. College of Swansea, 1972.
38. E. L. ALBASINY and D. W. MARTIN, 'Bending and membrane equilibrium in cooling towers', *Proc. Am. Soc. Civ. Eng.*, **93**, EM3, 1–17, 1967.
39. A. C. SCORDELIS and K. S. LO, 'Computer analysis of cylindrical shells', *J. Am. Concr. Inst.*, **61**, May 1964.
40. R. MARK and J. D. RIESA, 'Photoelastic analysis of folded plate structures', *Proc. Am. Soc. Civ. Eng.*, **93**, EM4, 79–83, 1967.

14. Axi-Symmetric Shells

14.1 Introduction

The problem of axi-symmetric shells is of sufficient practical importance to include in this chapter special methods of dealing with their solution.

While the general method described in the previous chapter is obviously applicable here, it will be found that considerable simplification can be achieved if account is taken of axial symmetry of the structure. In particular, if both the shell and the loading are axi-symmetric it will be found that the elements become 'one dimensional'. This is the simplest type of element, to which little attention was given in earlier chapters.

The first approach to the finite element solution of axi-symmetric shells was presented by Grafton and Strome.[1] In this, the elements are simple conical frustra and a direct approach via displacement functions is used. Refinements in the derivation of element stiffnesses are presented in Popov *et al.*[2] and Jones and Strome,[3] and an extension to the case of unsymmetrical loads, which was suggested in Grafton and Strome,[1] is elaborated in Percy *et al.*,[4] Klein,[5] and others.[6,7]

More recently much work has been accomplished to extend the processes to curved elements and indeed to refine the approximations involved. The literature on the subject is considerable, no doubt promoted by the interest in missile behaviour, and a complete bibliography is here impracticable. References 8 to 16 show how curvilinear co-ordinates of various kinds can be introduced to the analysis while 11, and 13 discuss the use of additional nodeless degrees of freedom in improving the accuracy. 'Mixed' formulation (Chapter 12) have found here some use.[17] The subject is reviewed comprehensively by Gallagher[18] and others[19] where very comprehensive bibliographies can be found.

In axi-symmetric shells, in common with all other shells, both bending and 'in plane' or 'membrane' forces will occur. These will be specified uniquely in terms of the generalized 'strains', which now involve extensions and curvatures of the middle surface. If the displacement of each point of the middle surface is specified, such 'strains' and the internal stress resultants, or simply 'stresses', can be determined by formulae available in standard texts dealing with shell theory.

For example, in an axi-symmetric shell under axi-symmetric loading such as is shown in Fig. 14.1, the displacement of a point on the middle surface is uniquely determined by two components u and w in the tangential and normal directions respectively.

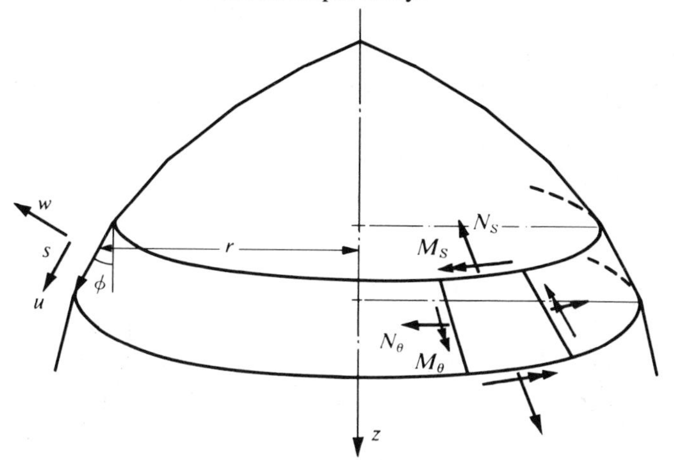

Fig. 14.1 Axi-symmetric shell and loading. Displacements and stress resultants. Shell represented as a stress of conical frustra

The four strain components are given by the following expression, using Kirchhoff–Love assumption, provided the angle ϕ does not vary (i.e., elements are straight)[20–22]

$$\{\varepsilon\} = \begin{Bmatrix} \varepsilon_s \\ \varepsilon_\theta \\ \chi_s \\ \chi_\theta \end{Bmatrix} = \begin{Bmatrix} \mathrm{d}u/\mathrm{d}s \\ (w \cos\phi + u \sin\phi)/r \\ -\mathrm{d}^2 w/\mathrm{d}s^2 \\ -\dfrac{\sin\phi}{r}\dfrac{\mathrm{d}w}{\mathrm{d}s} \end{Bmatrix}. \tag{14.1}$$

This results in four internal stress resultants, shown in Fig. 14.1, and related to the strains by an elasticity matrix $\mathbf{D}$

$$\boldsymbol{\sigma} = \begin{Bmatrix} N_s \\ N_\theta \\ M_s \\ M_\theta \end{Bmatrix} = \mathbf{D}\boldsymbol{\varepsilon}. \tag{14.2}$$

For an isotropic shell the matrix $\mathbf{D}$ becomes

$$\mathbf{D} = \frac{Et}{(1-\nu^2)} \begin{bmatrix} 1 & \nu & 0 & 0 \\ \nu & 1 & 0 & 0 \\ 0 & 0 & t^2/12 & \nu t^2/12 \\ 0 & 0 & \nu t^2/12 & t^2/12 \end{bmatrix} \tag{14.3}$$

the upper part being a plane stress and the lower a bending stiffness matrix, the shear terms being omitted in both.

14.2 Element Characteristics—Axi-Symmetrical Loads—Straight Elements

Let the shell be divided by nodal surfaces into a series of conical frustra, as shown in Fig. 14.2. The nodal displacements at points such as i and j will have to define uniquely the deformations of the element via prescribed shape functions.

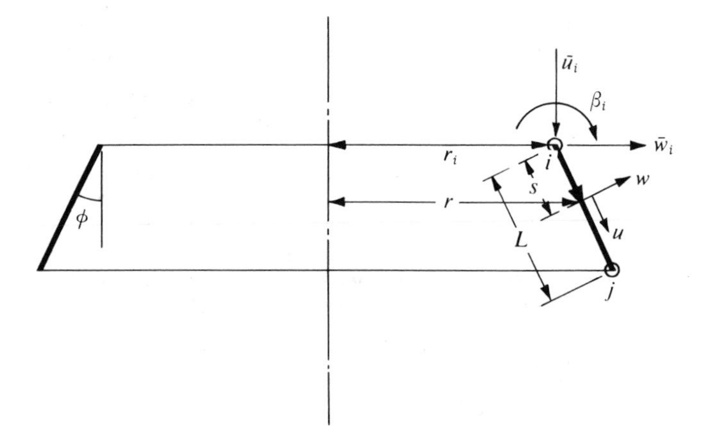

Fig. 14.2 An element of an axi-symmetric shell

At each node the axial and radial movements and a rotation will be prescribed. All three components are necessary as the shell can carry bending moments. The displacement of a node i can thus be defined by three components, the first two being in global directions,

$$\mathbf{a}_i = \begin{Bmatrix} \bar{u}_i \\ \bar{w}_i \\ \beta_i \end{Bmatrix}. \tag{14.4}$$

The element with two nodes, ij, thus possesses six degrees of freedom, determined by the element displacements

$$\mathbf{a}^e = \begin{Bmatrix} \mathbf{a}_i \\ \mathbf{a}_j \end{Bmatrix}. \tag{14.5}$$

The displacements within the element have to be uniquely determined by the nodal displacements $\mathbf{a}^e$ and the position s, and maintain slope and displacement continuity.

Thus in local co-ordinates we have

$$\mathbf{u} = \begin{Bmatrix} u \\ w \end{Bmatrix} = \mathbf{N}\mathbf{a}^e. \tag{14.6}$$

If u is taken as varying linearly with s and w as a cubic in s we shall have six undetermined constants, which can be determined from the nodal values of $\bar{u}$, $\bar{w}$, and β.

At the node i

$$\begin{Bmatrix} u_i \\ w_i \\ (\mathrm{d}w/\mathrm{d}s)_i \end{Bmatrix} = \begin{bmatrix} \cos\phi & +\sin\phi & 0 \\ -\sin\phi & \cos\phi & 0 \\ 0 & 0 & 1 \end{bmatrix} \begin{Bmatrix} \bar{u}_i \\ \bar{w}_i \\ \beta_i \end{Bmatrix} = \boldsymbol{\lambda}\mathbf{a}_i. \tag{14.7}$$

Writing

$$\begin{aligned} u &= \alpha_1 + \alpha_2 s \\ w &= \alpha_3 + \alpha_4 s + \alpha_5 s^2 + \alpha_6 s^3 \end{aligned} \tag{14.8}$$

it is an easy matter to state the six end conditions and arrive at†

$$\begin{Bmatrix} u \\ w \end{Bmatrix} = \left[\begin{array}{c|c|c|c|c|c} 1-s' & 0 & 0 & s' & 0 & 0 \\ 0 & 1-3s'^2+2s'^3 & L(s'-2s'^2+s'^3) & 0 & 3s'^2-2s'^3 & (-s'^2+s'^3)L \end{array}\right] \begin{Bmatrix} u_i \\ w_i \\ (\mathrm{d}w/\mathrm{d}s)_i \\ u_j \\ w_j \\ (\mathrm{d}w/\mathrm{d}s)_j \end{Bmatrix} \tag{14.9}$$

in which

$$s' = s/L.$$

Calling the above two by six matrix $\mathbf{N}'$ we can now write

$$\begin{Bmatrix} u \\ w \end{Bmatrix} = \mathbf{N}' \begin{bmatrix} \boldsymbol{\lambda} & 0 \\ 0 & \boldsymbol{\lambda} \end{bmatrix} \mathbf{a}^e = [\mathbf{N}_i'\boldsymbol{\lambda}, \mathbf{N}_j'\boldsymbol{\lambda}]\, \mathbf{a}^e = \mathbf{N}\mathbf{a}^e. \tag{14.10}$$

From Eq. (14.10) it is a simple matter to obtain the strain matrix $\mathbf{B}$ by the use of the definition Eq. (14.1). This gives

$$\boldsymbol{\varepsilon} \equiv \mathbf{B}\mathbf{a}^e = [\mathbf{B}_i'\boldsymbol{\lambda}, \mathbf{B}_j'\boldsymbol{\lambda}]\mathbf{a}^e \tag{14.11}$$

† The functions which occur there are, in fact, Hermitian polynomials of order 0 and 1, see Chapter 10, section 10.14.

in which

$$
\mathbf{B}'_i = \begin{bmatrix} -1/L & 0 & 0 \\ (1-s')\sin\phi/r & (1-3s'^2+2s'^3)\cos\phi/r & L(s'-2s'^2+s'^3)\cos\phi/r \\ 0 & (6-12s')/L^2 & (4-6s')/L \\ 0 & (6s'-6s'^2)\sin\phi/rL & (-1+4s'-3s'^2)\sin\phi/r \end{bmatrix}
$$

$$
\mathbf{B}'_j = \begin{bmatrix} 1/L & 0 & 0 \\ s'\sin\phi/r & (3s'^2-2s'^3)\cos\phi/r & L(-s'^2+s'^3)\cos\phi/r \\ 0 & (-6+12s')/L^2 & (2-6s')/L \\ 0 & (-6s'+6s'^2)\sin\phi/rL & (2s'-3s'^2)\sin\phi/r \end{bmatrix} \tag{14.12}
$$

Now all the 'ingredients' required for computing the stiffness matrix (or load, stress, and initial stress matrices) by standard formulae of Chapter 2 are known. The integrations required are carried out over the area, A, of the element, i.e., with

$$
\mathrm{d}A = 2\pi r \,\mathrm{d}s = 2\pi r L \,\mathrm{d}s' \tag{14.13}
$$

with s' varying from 0 to 1.

Thus, the stiffness matrix $\mathbf{K}$ becomes by Eq. (2.13a)

$$
\mathbf{K} = \int_0^1 \mathbf{B}'^{\mathrm{T}}\mathbf{D}\mathbf{B}' 2\pi r L \,\mathrm{d}s'. \tag{14.14}
$$

On substitution, the element $\mathbf{K}_{rs}$ of this matrix is given by

$$
\mathbf{K}_{rs} = \boldsymbol{\lambda}^{\mathrm{T}}\left(\int_0^1 \mathbf{B}_r'^{\mathrm{T}}\mathbf{D}\mathbf{B}_s' r \,\mathrm{d}s'\right)\boldsymbol{\lambda} 2\pi L. \tag{14.15}
$$

The radius r has to be expressed as a function of s before such integrations are carried.

Once again it is convenient to use numerical integration. Grafton and Strome[1] give an explicit formula for the stiffness matrix based on a single average value of the integrand and using a $\mathbf{D}$ matrix corresponding to an orthotropic material. Even with this crude approximation extremely good results can be obtained, provided small elements are used.

Percy *et al.*[4] and Klein[5] carry out a 7-point numerical integration and a slightly improved matrix is obtained.

It should be remembered that if any external line loads or moments are present, their full circumferential value must be used in the analysis just as was the case with axi-symmetric solids discussed in Chapter 5.

14.3 **Examples and Accuracy**

In the treatment of axi-symmetric shells described here continuity is satisfied at all times. For a polygonal shape of shell, therefore, convergence will always occur.

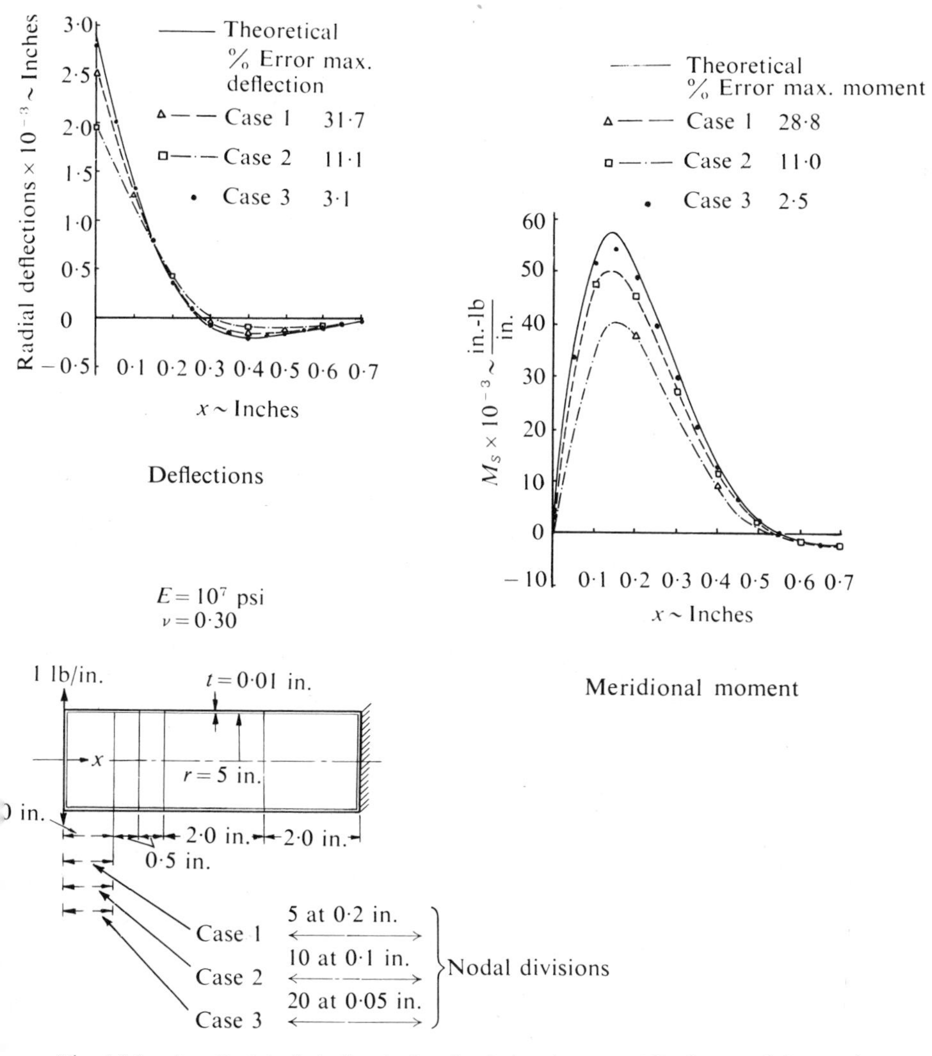

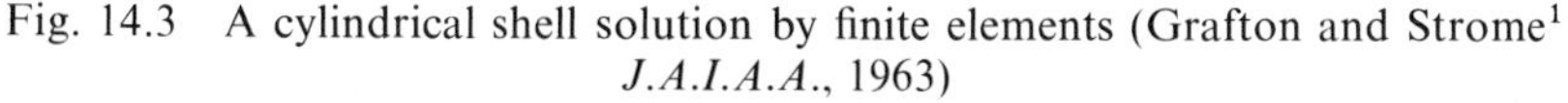

Fig. 14.3 A cylindrical shell solution by finite elements (Grafton and Strome[1] *J.A.I.A.A.*, 1963)

The problem of the physical approximation to a curved shell by a polygonal shape is similar to the one discussed in Chapter 13. Intuitively,

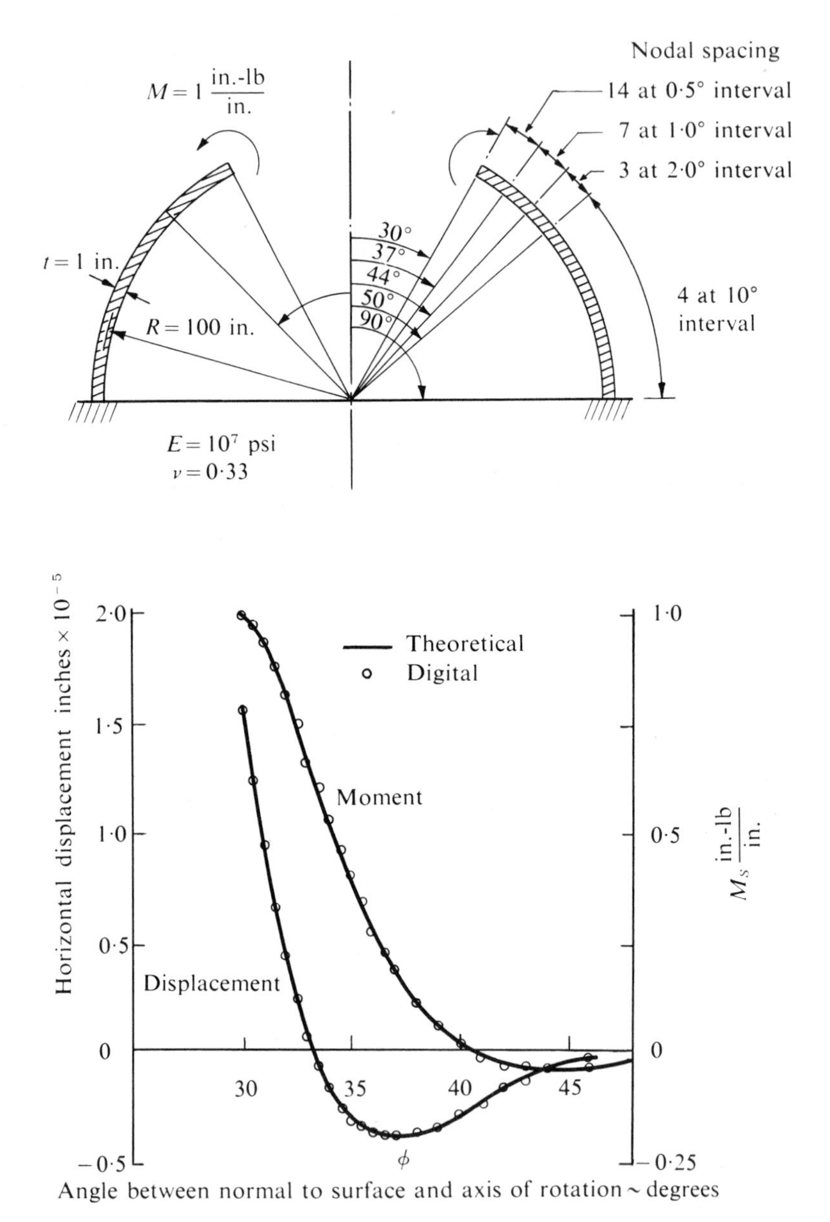

Fig. 14.4 A hemispherical shell solution by finite elements (Grafton and Strome[1] *J.A.I.A.A.*, 1963)

convergence can be expected, and indeed numerous examples indicate this.

When the loading is such as to cause predominantly membrane stresses, discrepancies in bending moment values have been found to exist even with reasonably fine subdivision. Again, however, these disappear as the size of the subdivision decreases particularly if correct (consistent) sampling (Chapter 11) is used. This is necessary to eliminate the physical approximation involved in representing the shell as a series of conical frustra.

Figures 14.3 and 14.4 illustrate some typical examples taken from Grafton and Strome,[1] which show quite remarkable accuracy.

14.4 Curved Elements and their Shape Functions

Use of curved elements has already been described in Chapter 8, in the context of analysis which involved, in the definition of strain, only first derivatives. Here second derivatives exist (*vide* Eq. (14.1)) and some of the theorems of Chapter 8 are no longer applicable.

It was previously mentioned that many possible definitions of curved elements have been proposed and used in the context of axisymmetric shells.[9–12] The derivation used here is one due to Delpak[11] and, to use the nomenclature of Chapter 8, is of the sub-parametric type.

The basis of curved element definition is one which gives a common tangent between adjacent elements (or alternatively a specified tangent

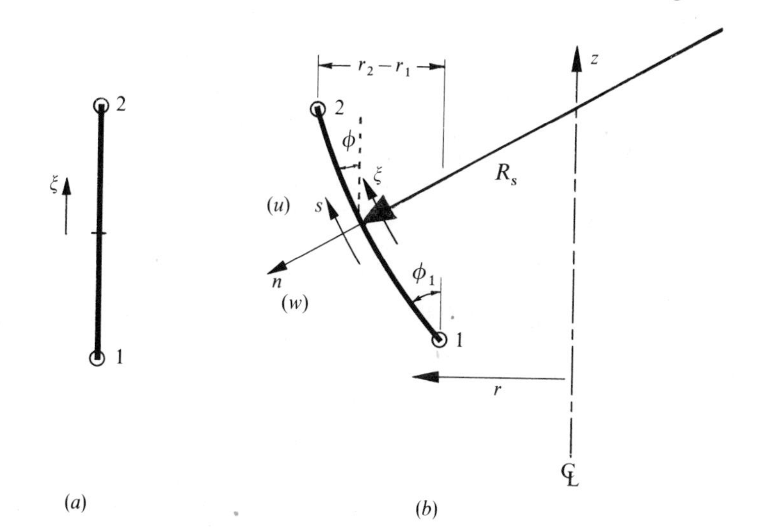

Fig. 14.5 Curved, isoparametric, shell element for axi-symmetric problems. (*a*) Parent element. (*b*) Curvilinear co-ordinates

direction). This is physically necessary to avoid 'kinks' in the description of what in practice is possibly a smooth shell.

If a general curved form of a shell of revolution is considered as shown in Fig. 14.5, the expressions for strain quoted in Eq. (14.1) have to be modified to take into account the curvature of the shell in the meridianal plane.[20–22] These now become

$$\boldsymbol{\varepsilon} = \begin{Bmatrix} \varepsilon_s \\ \varepsilon_\theta \\ \chi_s \\ \chi_\theta \end{Bmatrix} = \begin{Bmatrix} \mathrm{d}u/\mathrm{d}s + w/R_s \\ (w \cos\phi + u \sin\phi)/r \\ -\mathrm{d}^2 w/\mathrm{d}s^2 + \mathrm{d}(u/R_s)/\mathrm{d}s \\ -(\sin\phi/r)\times(\mathrm{d}w/\mathrm{d}s - u/R_s) \end{Bmatrix} \tag{14.16}$$

In the above the angle ϕ is a function of s, i.e.,

$$\mathrm{d}r/\mathrm{d}s = \sin\phi.$$

R_s is the principal radius in the meridianal plane and the second principal curvature radius R_θ is given by

$$R_\theta = r/\cos\phi.$$

The reader can verify that for $R_s = \infty$ this coincides with expression (14.1).

We shall now consider a curved element 1–2 shown in Fig. 14.5(*b*), where the co-ordinate in its 'parent' form $-1 \leqslant \xi \leqslant 1$ is shown in Fig. 14.5(*a*). The co-ordinates and the unknowns are 'mapped' in the manner of Chapter 8. As we wish to interpolate a quantity ψ with slope continuity we can write

$$\psi = \sum_{i=1}^{2} \left(N_i' \psi_i + N_i'' \left(\frac{\mathrm{d}\psi}{\mathrm{d}\xi}\right)_i \right) = \mathbf{N}\boldsymbol{\Psi}^e. \tag{14.17}$$

In this N' and N'' are scalar shape functions and for simplest representation will be cubics (similar to those used in Eq. (14.9) for the variation of w).

Explicitly we can write these cubic functions as

$$\begin{aligned} N_i' &= \tfrac{1}{4}\{\xi_0 \xi^2 - 3\xi_0 + 2\} \\ N_i'' &= \tfrac{1}{4}(1-\xi_0)^2(1+\xi_0) \quad \text{with } \xi_0 = \xi_i \xi. \end{aligned} \tag{14.18}$$

Now we can simultaneously use the above functions to describe the variation of the global displacements $\bar{u}$ and $\bar{w}$† *and* of the co-ordinates r and z which define the shell (mid-surface).

† One immediate difference will be observed from that of the previous formulation. Now both displacement components vary in at least a cubic manner along an element while previously a linear variation of the tangental displacement was permitted. This additional degree of freedom does not, however, introduce any excessive continuities in this case providing the shell is itself continuous in thickness.

Indeed if the thickness of the element is also variable the same interpolation could be applied to it.

Such an element would then be isoparametric. (See Chapter 8.)

Thus we can define the geometry as

$$r = \sum_{1}^{2} \left(N_i' r_i + N_i'' \left(\frac{\mathrm{d}r}{\mathrm{d}\xi} \right)_i \right)$$

and (14.19)

$$z = \sum_{1}^{2} \left(N_i' z_i + N_i'' \left(\frac{\mathrm{d}z}{\mathrm{d}\xi} \right)_i \right)$$

and providing the nodal values in the above can be specified, a one-to-one relation between ξ and the position on the curved element surface is defined, Fig. 14.5(*b*).

While specification of r_i and z_i is obvious, at the ends only the slope

$$(\tan \phi)_i = \left(\frac{\mathrm{d}r}{\mathrm{d}z} \right)_i \tag{14.20}$$

is defined. What specification is to be adopted with regard to the derivatives occurring in Eq. (14.19) depends on the *scaling* of ξ along the tangent length s.

Only the ratio

$$\left(\frac{\mathrm{d}r}{\mathrm{d}z} \right)_i = \left(\frac{\mathrm{d}r}{\mathrm{d}\xi} \right)_i \Big/ \left(\frac{\mathrm{d}z}{\mathrm{d}\xi} \right)_i \tag{14.21}$$

is unambiguously specified. $(\mathrm{d}r/\mathrm{d}\xi)_i$ (or $\mathrm{d}z/\mathrm{d}\xi$) can be given an arbitrary value. Here, however, practical considerations intervene as with the wrong choice of value a very uneven relationship between s and ξ will be achieved. Indeed with an unsuitable choice the shape of the curve can depart from the smooth one illustrated and loop between the end values.

To achieve a reasonably uniform spacing it suffices for well-behaved surfaces to approximate

$$\frac{\mathrm{d}r}{\mathrm{d}\xi} = \frac{\Delta r}{\Delta \xi} = \frac{r_2 - r_1}{2} \tag{14.22}$$

noting that the whole range of ξ is 2 between the nodal points.

14.5 Strain Expressions and Properties of Curved Elements

The variation of global displacements has been specified while, by Eq. (14.16), the strains are determinate in terms of the derivatives of locally directed displacements with respect to the tangent, s. Some transformations are therefore necessary before the strains can be determined.

If thus we take the global displacement variation to be defined by the shape function, Eq. (14.17) as

$$\begin{aligned} \bar{u} &= \sum_{i=1}^{2} \left(N_i' \bar{u}_i + N_i'' \left(\frac{d\bar{u}}{d\xi} \right)_i \right) \\ \bar{w} &= \sum_{i=1}^{2} \left(N_i' \bar{w}_i + N_i'' \left(\frac{d\bar{w}}{d\xi} \right)_i \right) \end{aligned} \tag{14.23}$$

we can find the locally directed displacements u, v from the transformation implied in Eq. (14.7), i.e.,

$$\begin{Bmatrix} u \\ w \end{Bmatrix} = \begin{bmatrix} \cos\phi & \sin\phi \\ -\sin\phi & \cos\phi \end{bmatrix} \begin{Bmatrix} \bar{u} \\ \bar{w} \end{Bmatrix} = \mathbf{L} \begin{Bmatrix} \bar{u} \\ \bar{w} \end{Bmatrix}. \tag{14.24}$$

where ϕ is the angle of the tangent to the curve and z axis (Fig. 14.5). However, before we can proceed further it is necessary to express this transformation in terms of the ξ co-ordinate. We have

$$\tan\phi = \left(\frac{dr}{d\xi} \right) \Big/ \left(\frac{dz}{d\xi} \right) \tag{14.25}$$

and hence this can now be accomplished by (14.19).

Before proceeding further we must consider whether continuity can be imposed at nodes on the parameters of Eq. (14.23). Clearly the global displacements must be continuous. However, on previous occasions we have specified a continuity of *rotation* of the tangent only. Here we shall allow usually the continuity of both the s derivatives in displacements. Thus the parameters

$$\frac{d\bar{u}}{ds} \quad \text{and} \quad \frac{d\bar{w}}{ds}$$

will be given common values at nodes.

As

$$\frac{d\bar{u}}{ds} = \frac{d\bar{u}}{d\xi} \Big/ \frac{ds}{d\xi}, \qquad \frac{d\bar{w}}{ds} = \frac{d\bar{w}}{d\xi} \Big/ \frac{ds}{d\xi}$$

and

$$\frac{ds}{d\xi} = \sqrt{\left(\frac{dr}{d\xi} \right)^2 + \left(\frac{dz}{d\xi} \right)^2} \tag{14.26}$$

no difficulty exists in substituting these new variables in Eqs. (14.23) and (14.24) which now take the form

$$\begin{Bmatrix} u \\ w \end{Bmatrix} = [N(\xi)]\mathbf{a}^e \quad \text{with } \mathbf{a}_i \begin{Bmatrix} \bar{u}_i \\ \bar{w}_i \\ (d\bar{u}/ds)_i \\ (d\bar{w}/ds)_i \end{Bmatrix}. \tag{14.27}$$

The form of the (2×4) submatrices is complicated but can be explicitly determined.[11]

We note that the curvature radius R_s can be explicitly calculated from the mapped, parametric, form of the element.

We thus write

$$R_s = \frac{\left[\left(\frac{\mathrm{d}r}{\mathrm{d}\xi}\right)^2 + \left(\frac{\mathrm{d}z}{\mathrm{d}\xi}\right)^2\right]^{3/2}}{\left[\frac{\mathrm{d}r}{\mathrm{d}\xi}\frac{\mathrm{d}^2 z}{\mathrm{d}\xi^2} - \frac{\mathrm{d}z}{\mathrm{d}\xi}\frac{\mathrm{d}^2 r}{\mathrm{d}\xi^2}\right]} \tag{14.28}$$

in which all the derivatives are directly evolved from expression (14.19).

If shells which branch or in which abrupt thickness changes occur are to be treated, the nodal parameters specified in Eq. (14.27) are not satisfactory. It is better then to rewrite these as

$$\mathbf{a}_i = \begin{Bmatrix} \bar{u}_i \\ \bar{w} \\ \beta_i \\ (\mathrm{d}\bar{u}/\mathrm{d}s)_i \end{Bmatrix} \tag{14.29}$$

where $\beta_i = \mathrm{d}w/\mathrm{d}s$ is the nodal rotation, and to connect only the first three parameters. The fourth is now an unconnected element parameter with respect to which, however, the usual minimization is still carried out.

Transformations needed in the above are implied in Eq. (14.24).

In the derivation of the **B** matrix expressions which define the strains, both first and second derivatives with respect to s occur as seen in the definition of Eq. (14.16).

If we observe that the derivatives can be obtained by the simple rules already implied in Eq. (14.26) for any function F we can write

$$\frac{\mathrm{d}F}{\mathrm{d}s} = \frac{\mathrm{d}F}{\mathrm{d}\xi} \Big/ \frac{\mathrm{d}s}{\mathrm{d}\xi}$$

and (14.30)

$$\frac{\mathrm{d}^2F}{\mathrm{d}s^2} = \frac{\mathrm{d}^2F}{\mathrm{d}\xi^2} \Big/ \left(\frac{\mathrm{d}s}{\mathrm{d}\xi}\right)^2 - \frac{\mathrm{d}F}{\mathrm{d}\xi}\left(\frac{\mathrm{d}^2 s}{\mathrm{d}\xi^2}\right) \Big/ \left(\frac{\mathrm{d}s}{\mathrm{d}\xi}\right)^3$$

and all the expressions of **B** can be found.

Finally the stiffness matrix is obtained in a similar way as in Eq. (14.14) changing the variable

$$\mathrm{d}s = \frac{\mathrm{d}s}{\mathrm{d}\xi}\,\mathrm{d}\xi \tag{14.31}$$

and integrating within limits -1 and $+1$.

Once again the quantities contained in the integral expressions prohibit explicit integration and numerical integration must be used. As this is carried out in one co-ordinate only it is not very time-consuming and an adequate number of Gauss points can be used to determine the stiffness very accurately.

Stress and other matrices are similarly obtained.

The particular isoparametric formulation presented in outline here differs somewhat from the alternatives of references 8, 9, 10, and 12 and has the advantage that due to its *isoparametric* form rigid body displacement modes and indeed the states of constant first derivative are available. Proof of this is similar to that contained in section 8.5 of Chapter 8. The fact that the forms given in the alternative formulations strain under rigid body displacements may not be serious in some applications as discussed by Haisler and Stricklin.[23] However, in some modes of non-axi-symmetric loads (*vide* Chapter 15) this incompleteness may be a serious drawback and may indeed lead to very wrong results.

Constant states of curvature cannot be obtained for a *finite* element of any kind described here and indeed are not physically possible. When the size of the element decreases it will be found that such arbitrary constant curvature states are available in the limit.

14.6 **Additional Nodeless Variables**

Addition of nodeless variables in the analysis of axi-symmetric shells is particularly valuable as large curved elements are capable of reproducing with good accuracy the geometric shapes.

Thus an addition of a set of internal element variable

$$\sum_{j=1}^{n} N_j''' a_j \tag{14.32}$$

to the definition of the normal displacement defined in Eq. (14.6 or 14.23), in which a_j is a set of internal element parameters and N_j''' is a set of functions having zero values and zero first derivatives at the nodal points, allows considerable improvement in representation of the displacements to be achieved without violating any of the convergence requirements (*vide* Chapter 2).

For tangental displacements the requirements of zero first derivative of nodes can be omitted.

Webster[13] uses such additional functions in the context of straight elements.

Whether the element is in fact straight or curved does not matter and indeed we can supplement the definitions of displacements contained in Eq. (14.23) by Eq. (14.32) for each of the components. If this is done only in the displacement definition and *not* in the co-ordinate definition (Eq. (14.19)) the element becomes now of the category of sub-parametric.† As proved in Chapter 8 the same advantages are retained as in isoparametric forms.

The question as to the expression to be used for the additional, internal shape functions is of some importance though the choice is wide. While it is no longer necessary to use polynomial representation, Delpak[11] does so and uses a special form of Légendre polynomials. The general shapes are shown in Fig. 14.6.

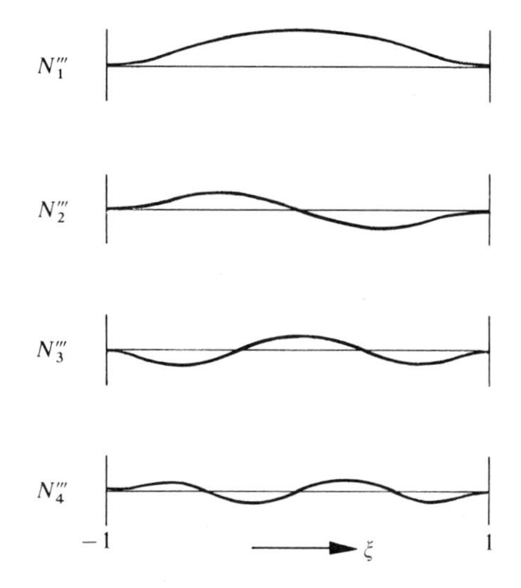

Fig. 14.6 Internal shape functions for a linear element

A series of examples shown in Figs. 14.7, 14.8, and 14.9 illustrate the applications of the isoparametric curvilinear element of the previous section with additional internal parameters.

In Fig. 14.7 a spherical dome with clamped edges is analysed and compared with analytical results of reference 21. Figures 14.8 and 14.9 show, respectively, more complex examples. In the first a torus analysis is made and compared with alternative finite element results.[12, 15, 24, 25] The second case is one where branching occurs, and here analytical alternative results are given by Kraus.[26]

† While it would obviously be possible to include the new shape function in the element shape definition little practical advantage would be gained as a cubic represents the realistic shapes adequately.

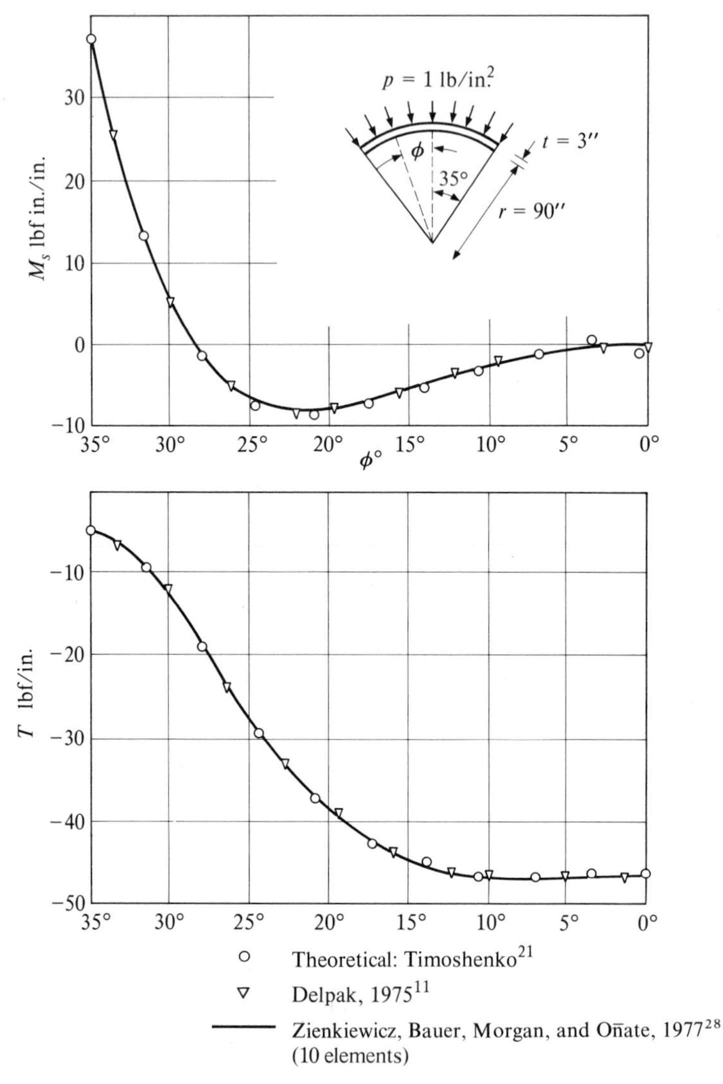

Fig. 14.7 Spherical dome under uniform pressure

14.7 Independent Slope–Displacement Interpolation with Penalty Functions (Thick or Thin Shell Formulations)

In Chapter 11 we discussed, in the context of beams and plates (section 11.6.3), the possibility of an independent slope and displacement interpolation, with the continuity conditions being imposed by use of a penalty function. The latter was, indeed, identified as the shear rigidity and had a real significance in the context of thick beams or plate theory.

The success of such procedures was entirely dependent on the use of 'reduced integration' to ensure the singularity of the appropriate constraint matrix.

Identical methods are obviously applicable in the context of axi-symmetric shells and we shall here illustrate the general formulation. Further, we shall develop in detail the simplest possible element of this class. This is a direct descendant of the linear beam element[27] discussed on p. 290. Here a linear interpolation is used with a single integrating point for the transverse shear/penalty number terms.[28]

Consider the strain expressions of Eq. (14.1) for a straight element. When using these the need for C_1 continuity was implied by the second derivative of w existing there. If now we replace

$$\frac{\mathrm{d}w}{\mathrm{d}s} = -\beta \tag{14.33}$$

the strain expression becomes

$$\boldsymbol{\varepsilon} = \begin{Bmatrix} \varepsilon_s \\ \varepsilon_\theta \\ \chi_s \\ \chi_\theta \end{Bmatrix} = \begin{Bmatrix} \mathrm{d}u/\mathrm{d}s \\ (u \sin\phi + w\cos\phi)/r \\ \mathrm{d}\beta/\mathrm{d}s \\ (\beta \sin\phi)/r \end{Bmatrix} \tag{14.34}$$

As β can vary independently a constraint has to be imposed:

$$C(w, \beta) \equiv \frac{\mathrm{d}w}{\mathrm{d}s} + \beta = 0. \tag{14.35}$$

This can be done using the energy functional with a penalty multiplier α. We can thus write

$$\Pi = \frac{1}{2}\int \boldsymbol{\varepsilon}^{\mathrm{T}}\mathbf{D}\boldsymbol{\varepsilon}\; 2\pi r\,\mathrm{d}s + \frac{1}{2}\int \alpha\left(\frac{\mathrm{d}w}{\mathrm{d}s} + \beta\right)^2 2\pi r\,\mathrm{d}s + \mathrm{l.t.} \tag{14.36}$$

where l.t. stands for loading terms and $\boldsymbol{\varepsilon}$ and $\mathbf{D}$ are defined as previously. Immediately α can be identified as the shear rigidity

$$\alpha = \kappa G t \qquad \kappa = \frac{5}{6} \tag{14.37}$$

The penalty functional (14.36) can, indeed, be identified on purely physical grounds. Washizu[22] quotes this on pp. 199–201, and the general theory indeed follows that earlier suggested by Naghdi[29] for shells with shear deformation.

With first derivatives only occurring in the energy expression C_0, continuity is now required only in the interpolation for u, w, β, and in

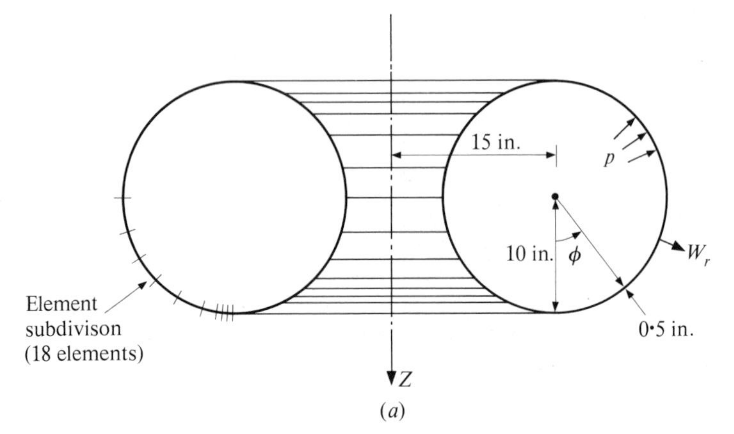

(a)

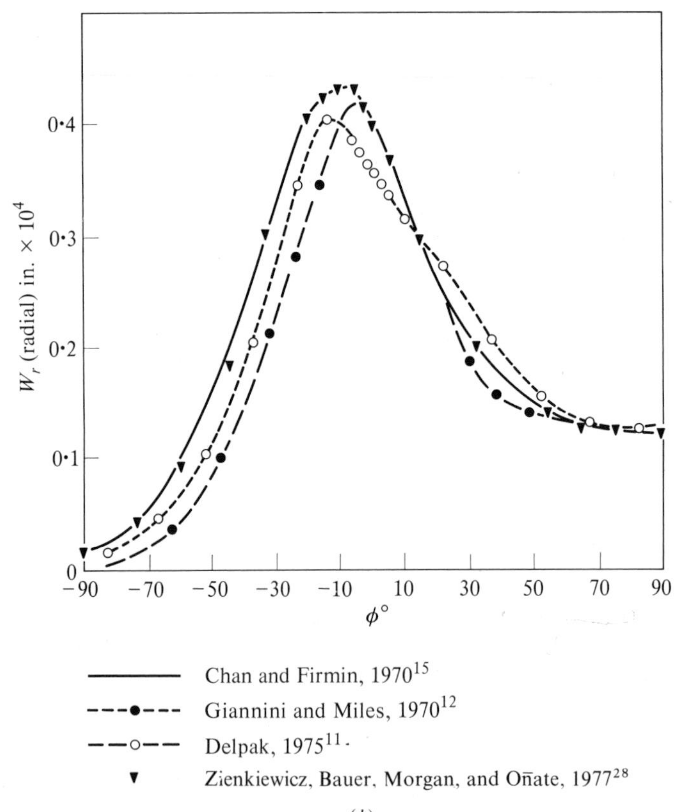

(b)

Fig. 14.8 Toroidal shell under internal pressure. (a) element subdivision. (b) radial displacements

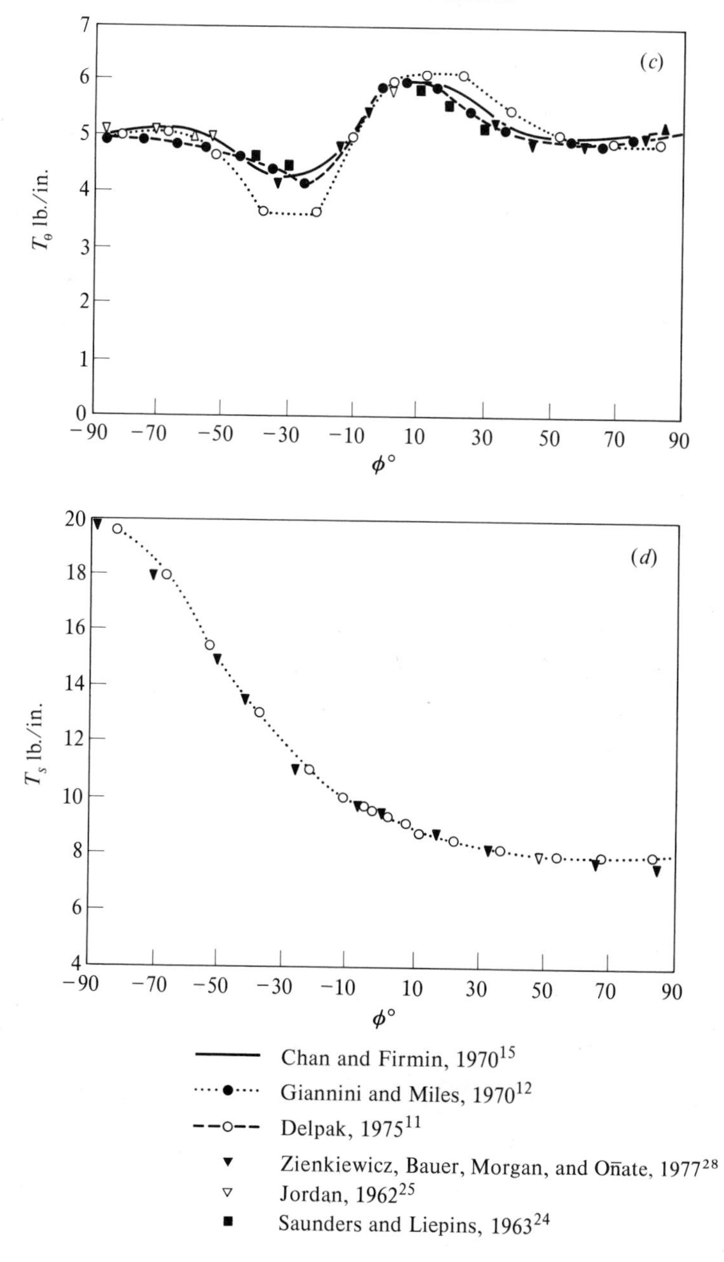

Fig. 14.8 (continued). (*c*)/(*d*) In plane stress resultants

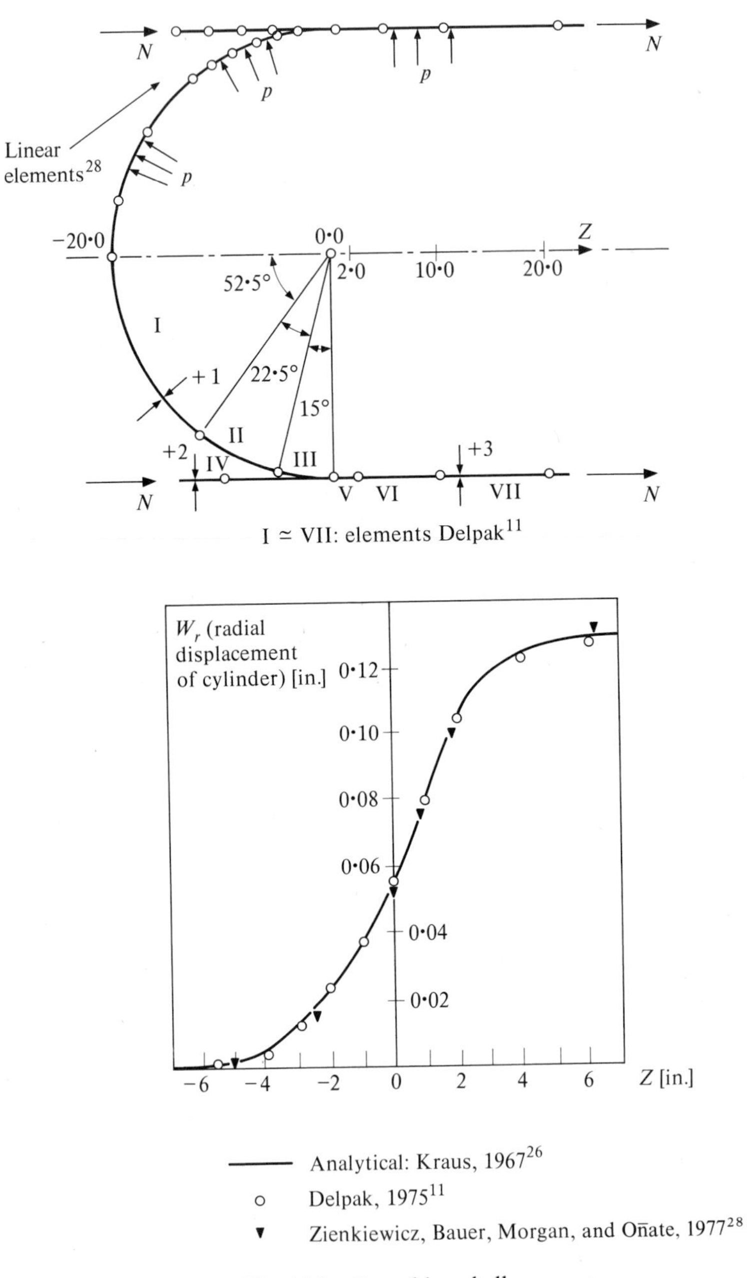

Fig. 14.9 Branching shell

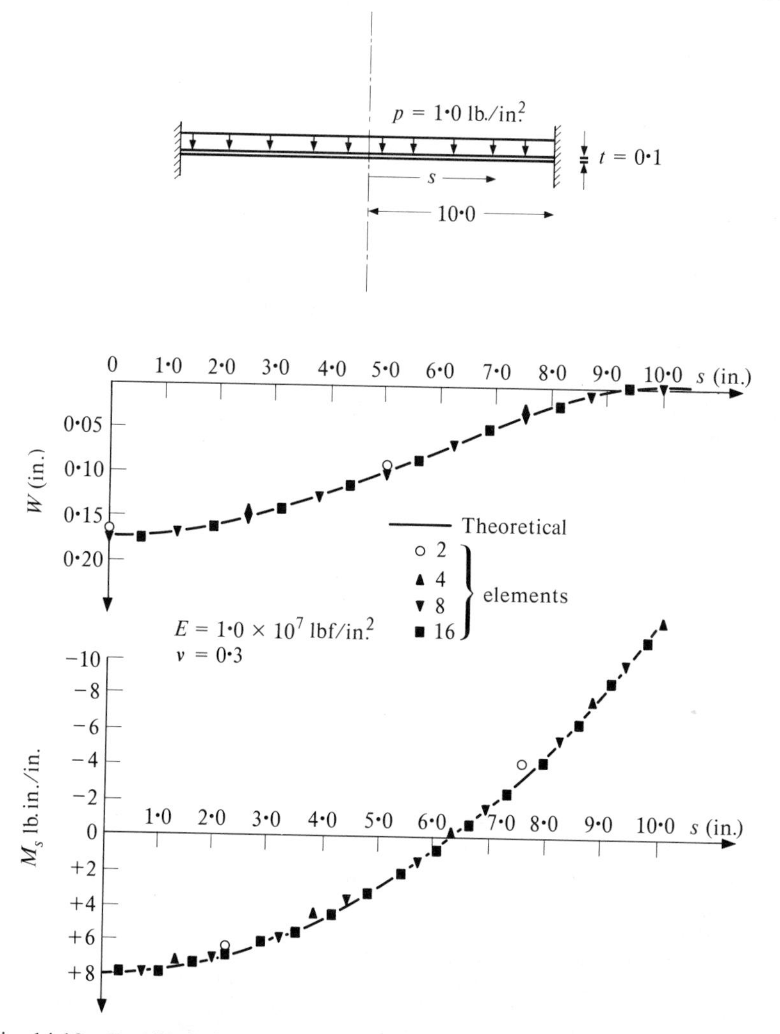

Fig. 14.10 Bending of circular plate under uniform load. Convergence study

place of Eqs. (14.6) to (14.10) we can write directly

$$\mathbf{u} = \begin{Bmatrix} u \\ w \\ \beta \end{Bmatrix} = N\boldsymbol{\lambda}\mathbf{a}^e \quad \text{where } N = N(\xi) \tag{14.38}$$

$$\mathbf{a}_i^{\mathrm{T}} = [\bar{u}, \bar{w}, \beta]_i.$$

Here for $N(\xi)$ we can use any of the one-dimensional, C_0 interpolations of Chapter 7. Once again isoparametric transformation could be used for curvilinear elements with strains defined now by Eq. (14.16) and indeed a formulation which we shall discuss in Chapter 16 is but an alternative to this process. If linear elements are used, we can write the expression without consequent use of isoparametric transformation. With the symbols used in Eq. (14.8) we can now simply write

$$\begin{aligned} \bar{u} &= \bar{u}_i(1-s')+\bar{u}_j s' \\ \bar{w} &= \bar{w}_i(1-s')+\bar{w}_j s' \\ \beta &= \beta_i(1-s')+\beta_j s' \end{aligned} \tag{14.39}$$

and evaluate the integrals arising from expression (14.36) at one Gauss point which is sufficient to maintain convergence and yet here does not give a singularity.

This extremely simple form will give very poor results with exact integration even for thick shells, but now with reduced integration shows a surprisingly good performance.

On Figures 14.7 to 14.9 we superpose results obtained with this simple, straight element; the results speak for themselves.

For other examples the reader can consult reference 28, but in Fig. 14.10 we show a very simple example of a bending of a circular plate using different numbers of equal elements. This purely bending problem shows the type of results and convergence attainable.

References

1. P. E. Grafton and D. R. Strome, 'Analysis of axi-symmetric shells by the direct stiffness method', *J.A.I.A.A.*, **1**, 2342–7, 1963.
2. E. P. Popov, J. Penzien, and Z. A. Lu, 'Finite element solution for axi-symmetric shells', *Proc. Am. Soc. Civ. Eng.*, EM, 119–45, 1964.
3. R. E. Jones and D. R. Strome, 'Direct stiffness method of analysis of shells of revolution utilising curved elements', *J.A.I.A.A.*, **4**, 1519–25, 1966.
4. J. H. Percy, T. H. H. Pian, S. Klein, and D. R. Navaratna, 'Application of matrix displacement method to linear elastic analysis of shells of revolution', *J.A.I.A.A.*, **3**, 2138–45, Jan. 1965.
5. S. Klein, 'A study of the matrix displacement method as applied to shells of revolution', *Proc. Conf. on Matrix Methods in Structural Mechanics*, Air Force Inst. Tech., Wright-Patterson A.F. Base, Ohio, Oct. 1965.
6. R. E. Jones and D. R. Strome, 'A survey of analysis of shells by the displacement method', *Proc. Conf. on Matrix Methods in Structural Mechanics*, Air Force Inst. Tech., Wright-Patterson A.F. Base, Ohio, Oct. 1965.
7. O. E. Hansteen, 'A conical element for displacement analysis of axi-symmetric shells', *Finite Element Methods*, TAPIR, Trondheim, 1969.
8. P. L. Gould and S. K. Sen, 'Refined mixed method finite elements for shells of revolution', *3rd Conf. on Matrix Methods in Structural Mechanics*, Wright-Patterson A.F. Base, Ohio, 1971.
9. J. A. Stricklin, D. R. Navaratna, and T. H. H. Pian, 'Improvements in the analysis of shells of revolution by matrix displacement method (curved elements)', *A.I.A.A. Int.*, **4**, 2069–72, 1966.
10. M. Khojasteh-Bakht, *Analysis of elastic-plastic shells of revolution under axi-symmetric loading by the finite element method*, Dept. Civ. Eng., Univ. of California, SESA 67–8, 1967.

11. R. Delpak, *Role of the curved parametric element in linear analysis of thin rotational shells*, Ph.D. Thesis, Department of Civil Engineering and Building, The Polytechnic of Wales, 1975.
12. M. Giannini and G. A. Miles, 'A curved element approximation in the analysis of axi-symmetric thin shells', *Int. J. Num. Meth. Eng.*, **2**, 459–76, 1970.
13. J. J. Webster, 'Free vibration of shells of revolution using ring elements', *Int. J. Mech. Sci.*, **9**, 559, 1967.
14. S. Ahmad, B. M. Irons, and O. C. Zienkiewicz, 'Curved thick shell and membrane elements with particular reference to axi-symmetric problems', *Proc. 2nd Conf. on Matrix Methods in Structural Mechanics*, Wright-Patterson A.F. Base, Ohio, AFFDL-TR-68-150, 1968.
15. A. S. L. Chan and A. Firmin, 'The analysis of cooling towers by the matrix finite element method', *Aeronaut. J.*, **74**, 826–35, 1970.
16. E. A. Witmer and J. J. Kotanchik, 'Progress report on discrete element elastic and elastic-plastic analysis of shells of revolution subjected to axi-symmetric and asymmetric loading'. *Proc. 2nd Conf. on Matrix Methods in Structural Mechanics*, Wright-Patterson A.F. Base, Ohio, AFFDL-TR-68-150, 1968.
17. Z. M. Elias, 'Mixed finite element method for axisymmetric shells', *Int. J. Num. Meth. Eng.*, **4**, 261–77, 1972.
18. (a) R. H. Gallagher, 'Analysis of plate and shell structures' in *Applications of Finite Element Method in Engineering*, pp. 155–205, Vanderbilt Univ., ASCE, 1969.
18. (b) R. H. Gallagher, 'Shell element', *World Conf. on Finite Element Methods in Structural Mechanics*, E1–E35, Bournemouth, Dorset, England, Oct. 1975.
19. J. A. Stricklin, 'Geometrically nonlinear static and dynamic analysis of shell of revolution', *High Speed Computing of Elastic Structures*, 383–411, Univ. of Liége, 1976.
20. V. V. Novozhilov, *Theory of Thin Shells* (translation), P. Noordhoff, 1959.
21. S. Timoshenko and S. Woinowsky-Krieger, *Theory of Plates and Shells*, 2nd ed., pp. 533–5, 1959.
22. K. Washizu, *Variational Methods in Elasticity and Plasticity*, 2nd ed., pp. 189–99, Pergamon Press, 1975.
23. W. E. Haisler and J. A. Stricklin, 'Rigid body displacements of curved elements in the analysis of shells by the matrix displacement method'. *J.A.I.A.A.*, **5**, 1525–7, 1967.
24. J. L. Sanders, Jr. and A. Liepins, 'Toroidal membrane under internal pressure', *J.A.I.A.A.*, **1**, 2105–10 (1963).
25. F. F. Jordan, 'Stresses and deformations of the thin-walled pressurized torus', *J. Aero. Sci.*, **29**, 213–25, 1962.
26. H. Kraus, *Thin Elastic Shells*, pp. 168–78, Wiley, 1967.
27. T. J. R. Hughes, R. L. Taylor, and W. Kanoknukulchai, 'A simple and efficient finite element for plate bending' (to be published in *Int. J. Num. Meth. Eng.*).
28. O. C. Zienkiewicz, J. Bauer, K. Morgan and E. Oñate, 'A simple element for axi-symmetric shells with shear deformation (to be published in *Int. J. Num. Meth. Eng.*).
29. P. M. Naghdi, 'Foundations of elastic shell theory', *Progress in Solid Mechanics*, Vol. IV, Chapter 1 (eds. I. N. Sneddon and R. Hill), North-Holland, 1963.

15. Semi-Analytical Finite Element Processes— Use of Orthogonal Functions

15.1 **Introduction**

The standard finite element methods have been shown to be capable, in principle, of dealing with any two- or three- (or even four)† dimensional situations. Nevertheless the cost of solutions increases greatly with each dimension added and indeed, on occasion, overtaxes the capabilities of available machines. It is therefore always desirable to search for alternatives which may reduce the computational labour. One such class of processes of quite a wide applicability will be illustrated here.

In many physical problems the situation is such that the *geometry* and *material properties* do not vary along one co-ordinate direction. However, the 'load' terms may still exhibit a variation in that direction preventing the use of such simplifying assumptions as those which, for instance, permitted a two-dimensional, plane strain, analysis to be substituted for a full three-dimensional treatment. In such cases it is possible still to consider a 'substitute' problem, not involving the particular co-ordinate (along which the properties do not vary), and to synthesize the true answer from a series of such simplified solutions.

The method to be described is of quite general use and, obviously, is not limited to structural situations. It will be convenient, however, to use the nomenclature of structural mechanics and to use the potential energy minimization as an example.

We shall confine our attention to problems of minimizing a quadratic functional such as described in Chapters 2 and 3. The interpretation of the processes involved as the application of partial discretization of Chapter 3, p. 60, followed by the use of a Fourier series expansion should be noted.

Let (x, y, z) be the co-ordinates describing the domain (in this context

† *Vide*: finite elements in the time domain Chapter 21.

these do not necessarily have to be the Cartesian co-ordinates). The last one of these, z, is the co-ordinate along which the geometry and material properties do not change and which is limited to lie between two values

$$0 \leqslant z \leqslant a.$$

The boundary values are thus specified at $z = 0$ and $z = a$.

We shall assume that the shape functions defining the variation of displacements $\mathbf{u}$ (Eq. (2.1)) can be written in a product form as

$$\begin{aligned} \mathbf{u} &= \mathbf{N}(x, y, z)\, \mathbf{a}^e \\ &= \sum_{l=1}^{L} \left\{ \overline{\mathbf{N}}(x, y) \cos \frac{l\pi z}{a} + \overline{\overline{\mathbf{N}}}(x, y) \sin \frac{l\pi z}{a} \right\} \mathbf{a}_l^e. \end{aligned} \tag{15.1}$$

In this type of representation completeness is preserved in view of the capability of the Fourier series to represent any continuous function within a given region (naturally assuming that the shape functions $\overline{\mathbf{N}}$ and $\overline{\overline{\mathbf{N}}}$ in the domain x, y satisfy the same requirements).

The loading terms will similarly be given a form

$$\mathbf{b} = \sum_{l=1}^{L} \left(\overline{\mathbf{b}}_l \cos \frac{l\pi z}{a} + \overline{\overline{\mathbf{b}}}_l \sin \frac{l\pi z}{a} \right) \tag{15.2}$$

with similar form for concentrated loads and boundary tractions (see Chapter 2).

Indeed initial strains and stresses, if present, would be expanded again in the above form.

Applying the standard processes of Chapter 2 to the determination of the element contribution to the equation minimizing the potential energy, and limiting our attention to the contribution of forces $\mathbf{b}$ only we can write

$$\frac{\partial \Pi}{\partial \mathbf{a}^e} = \mathbf{K}^e \begin{Bmatrix} \mathbf{a}_1^e \\ \vdots \\ \mathbf{a}_L^e \end{Bmatrix} + \begin{Bmatrix} \mathbf{f}_1^e \\ \vdots \\ \mathbf{f}_L^e \end{Bmatrix}. \tag{15.3}$$

In the above, to avoid summation signs, the vectors $\mathbf{a}^e$, etc., are expanded listing the contribution of each value of l separately.

Now a typical submatrix of $\mathbf{K}^e$ is

$$(\mathbf{K}^{lm})^e = \iiint_V \mathbf{B}^{l\mathrm{T}} \mathbf{D} \mathbf{B}^m \, dx\, dy\, dz \tag{15.4}$$

and a typical term of the 'force' vector becomes

$$(\mathbf{f}^l)^e = \iiint_V \mathbf{N}^{\mathrm{T}} \mathbf{b} \, dx\, dy\, dz. \tag{15.5}$$

Without going into details it is obvious that the matrix given by Eq. (15.4) will contain the following integrals as products of various submatrices and the following integrals

$$I_1 = \int_0^a \sin\frac{l\pi z}{a}\cos\frac{m\pi z}{a}\,dz$$
$$I_2 = \int_0^a \sin\frac{l\pi z}{a}\sin\frac{m\pi z}{a}\,dz. \qquad (15.6)$$
$$I_3 = \int_0^a \cos\frac{l\pi z}{a}\cos\frac{m\pi z}{a}\,dz$$

These integrals arise from products of the derivatives contained in the definition of **B** and, due to the well-known orthogonality property, give

$$I_2 = I_3 = 0 \quad \text{for } l \neq m \qquad (15.7)$$

when $l = 1, 2, \ldots$ and $m = 1, 2, \ldots$.'

I_1 is only zero when l and m are both even or odd numbers. The term involving I_1, however, vanishes in most applications.

This means that the matrix $\mathbf{K}^e$ becomes a diagonal one and that the assembled final equations of the system have the form

$$\begin{bmatrix} \mathbf{K}^{11} & & & \\ & \mathbf{K}^{22} & & \\ & & \ddots & \\ & & & \mathbf{K}^{LL} \end{bmatrix} \begin{Bmatrix} \mathbf{a}_1 \\ \vdots \\ \mathbf{a}_L \end{Bmatrix} + \begin{Bmatrix} \mathbf{f}_1 \\ \vdots \\ \mathbf{f}_L \end{Bmatrix} = 0 \qquad (15.8)$$

and the large system of equations splits into L separate problems

$$\mathbf{K}^{ll}\mathbf{a}_l + \mathbf{f}^l = 0 \qquad (15.9)$$

in which

$$\mathbf{K}_{ij}^{ll} = \iiint_V \mathbf{B}_i^{lT}\mathbf{D}\mathbf{B}_j^l \,dx\,dy\,dz. \qquad (15.10)$$

Further, from Eqs. (15.5) and (15.2) we observe that due to the orthogonal property of the integrals given by Eqs. (15.6), thr typical load term becomes simply

$$\mathbf{f}_i^l = \iiint_V \mathbf{N}_i^{lT}\mathbf{b}^l\,dx\,dy\,dz. \qquad (15.11)$$

This means that the force term of the lth harmonic only affects the lth system of Eq. (15.9) and contributes nothing to the other equations. This extremely important property is of considerable practical significance for, *if the expansion of the loading factors involves only one term, only one set of equations need be solved.* The solution of this will tend to the exact one with increasing subdivision in the x–y domain only. Thus, what was originally a three-dimensional problem, has now been reduced to a two-dimensional one with the consequent reduction of computational effort.

The preceding derivation was illustrated on a three-dimensional, elastic situation. Clearly the arguments could be equally well applied for reduction of two-dimensional problems to one-dimensional ones, etc., and the arguments are not restricted to problems of elasticity. Any physical problem governed by a minimization of a quadratic functional (Chapter 3), or by linear differential equations, is amenable to the same treatment, which under various guises has been used since time immemorial in applied mechanics.

A word of warning should be added regarding the boundary conditions imposed on $\mathbf{u}$. For a complete decoupling to be possible these must be satisfied separately by each and every term of the expansion given by Eq. (15.1). Insertion of a zero displacement in the final reduced problem implies in fact a zero displacement fixed through the z direction by definition. Care must be taken not to treat the final matrix therefore as a simple reduced problem. Indeed this is one of the limitations of the process described.

When the loading is complex and many Fourier components need to be considered the advantages of the approach outlined here reduce and the full solution sometimes becomes superior in economy.

Other permutations of the basic definitions of the type given by Eq. (15.1) are obviously possible. For instance two independent sets of parameters $\mathbf{a}^e$ may be specified with each of the trigonometric terms.

Indeed on occasion use of other orthogonal functions may be possible.

As trigonometric functions will arise frequently it is convenient to remind the reader of the following integrals

$$\begin{gathered}\int_0^a \sin\frac{l\pi z}{a}\cos\frac{l\pi z}{a}\,\mathrm{d}z = 0 \quad \text{when } l = 0, 1, \ldots \\ \int_0^a \sin^2\frac{l\pi z}{a}\,\mathrm{d}z = \int_0^a \cos^2\frac{l\pi z}{a}\,\mathrm{d}z = \frac{a}{2} \quad \text{when } l = 1, 2, \ldots .\end{gathered} \tag{15.12}$$

15.2 Prismatic Bar

Consider a prismatic bar illustrated in Fig. 15.1 which is assumed to be held at $z = 0$ and $z = a$ in a manner preventing all displacements in the x–y plane but permitting unrestricted motion in the z direction (traction $t_z = 0$).

The problem is fully three dimensional and three components of displacement u, v, and w have to be considered.

Subdividing into finite elements in the xy plane we can prescribe the lth displacement components in the x direction as

$$u^l = [N_1, N_2, \ldots] \sin\frac{l\pi z}{a}\,\mathbf{u}^l \tag{15.13}$$

with similar expressions for the v^l and w^l but with a cosine term in the last.

In this N, etc., are simply the (scalar) shape functions appropriate to the element used. If, as shown in Fig. 15.1, simple triangles are used then the shape functions are given by Eq. (4.8) of Chapter 4—but any of the more elaborate elements described in Chapter 7 (with or without the transformation of Chapter 8) would be equally suitable.

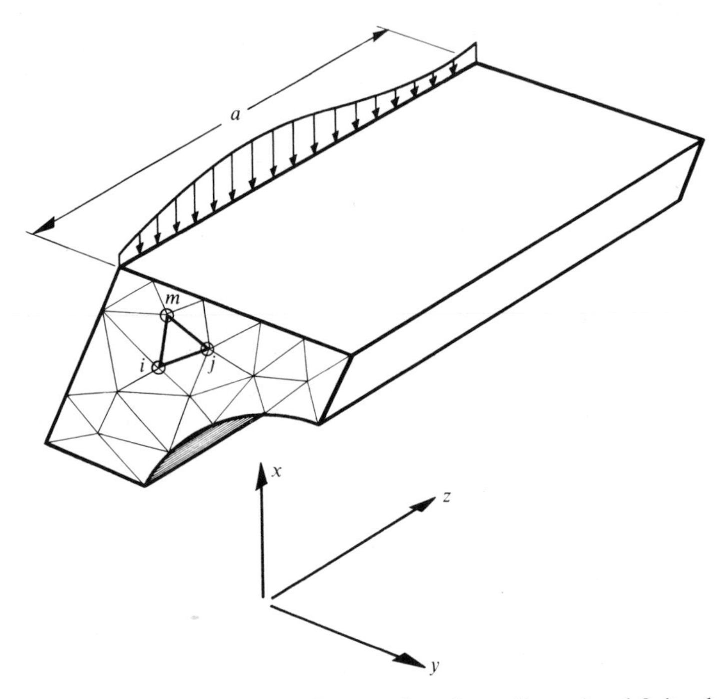

Fig. 15.1 A prismatic bar reduced to a series of two-dimensional finite element solutions

The expansion ensures zero u and w displacements and zero axial stresses at the ends.

The load terms can still be expressed in terms of a similar Fourier series, giving, for components in x–y plane

$$\mathbf{b}^l = \bar{\bar{\mathbf{b}}}^l \sin\frac{l\pi z}{a}; \; \bar{\bar{\mathbf{b}}}^e = \bar{\bar{\mathbf{b}}}^e(x, y) \tag{15.14}$$

The problem being fully three dimensional, the appropriate expression for strain involving all six components needs to be considered. This expression is given in Chapter 6 by Eqs. (6.9)–(6.11). On substitution of the shape function given by Eq. (15.13) for a typical term of the $\mathbf{B}$ matrix we have.

$$\mathbf{B}_i^l = \begin{bmatrix} \frac{\partial N_i}{\partial x}\sin\gamma & 0 & 0 \\ 0 & \frac{\partial N_i}{\partial y}\sin\gamma & 0 \\ 0 & 0 & -N_i\frac{l\pi}{a}\sin\gamma \\ \frac{\partial N_i}{\partial y}\sin\gamma & \frac{\partial N_i}{\partial x}\sin\gamma & 0 \\ 0 & N_i\frac{l\pi}{a}\cos\gamma & \frac{\partial N_i}{\partial y}\cos\gamma \\ N_i\frac{l\pi}{a}\cos\gamma & 0 & \frac{\partial N_i}{\partial x}\cos\gamma \end{bmatrix} \tag{15.15}$$

with $\gamma = l\pi z/a$. It is convenient to separate the above as

$$\mathbf{B}_i^l = \bar{\mathbf{B}}_i^l \sin\frac{\pi lz}{a} + \bar{\bar{\mathbf{B}}}_i^l \cos\frac{\pi lz}{a}. \tag{15.16}$$

In all of the above it is assumed that the parameters are listed in usual order

$$\mathbf{a}_i^l = \begin{Bmatrix} u_i^l \\ v_i^l \\ w_i^l \end{Bmatrix} \tag{15.17}$$

and that the axes are as shown in Fig. 15.1.

The stiffness matrix can be computed in the usual manner noting that

$$(\mathbf{K}_{ij}^{ll})^e = \iiint_{V^e} \mathbf{B}_i^{l\mathrm{T}}\mathbf{D}\mathbf{B}_j^l \,\mathrm{d}x\,\mathrm{d}y\,\mathrm{d}z. \tag{15.18}$$

On substitution of Eq. (15.16), multiplying out, and noting the value of the integrals from Eq. (15.12), this reduces to

$$(\mathbf{K}_{ij}^{ll})^e = \frac{a}{2}\iint_{A^e} \{\bar{\mathbf{B}}_i^{l\mathrm{T}}\mathbf{D}\bar{\mathbf{B}}_j^l + \bar{\bar{\mathbf{B}}}_{ij}^{l\mathrm{T}}\mathbf{D}\bar{\bar{\mathbf{B}}}_{j}^l\}\,\mathrm{d}x\,\mathrm{d}y \tag{15.19}$$

when $l = 1, 2 \ldots$.

The integration is now simply carried out over the element *area*.†

Similarly the contributions due to distributed loads, initial stresses, etc., are found as the loading terms. Concentrated line loads for instance

† It should be noted that now, even for a simple triangle, the integration is not trivial as some linear terms will remain in $\bar{\bar{\mathbf{B}}}$.

would be expressed directly as nodal forces

$$\mathbf{f}_i^l = \int_0^a \sin\frac{\pi l z}{a} \begin{Bmatrix} \bar{\bar{\mathbf{f}}}_{xi}^l \\ \bar{\bar{\mathbf{f}}}_{yi}^l \\ \bar{\bar{\mathbf{f}}}_{zi}^l \end{Bmatrix} \sin\frac{\pi l z}{a} \mathrm{d}z = \bar{\bar{\mathbf{f}}}_i^l \frac{a}{2} \tag{15.20}$$

in which $\bar{\bar{\mathbf{f}}}_i^e$ are intensities per unit length.

The boundary conditions used here have been of a type ensuring *simply supported* conditions for the prism. Other conditions can be inserted by suitable expansions.

The method of analysis outlined here can be applied to a range of practical problems—one of these being a popular type of concrete bridge illustrated in Fig. 15.2. Here a particularly convenient type of element is the distorted, 'serendipity', quadratic or cubic of Chapters 7 and 8.[1]

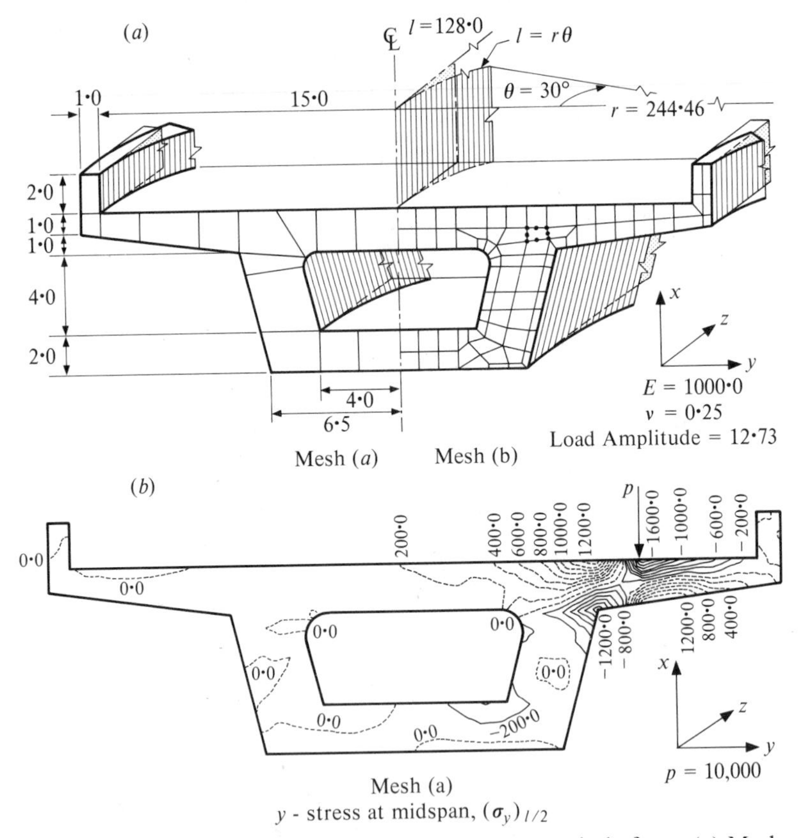

Fig. 15.2 A thick box bridge prism of straight or curved platform. (*a*) Mesh of isoparametric elements. (*b*) Distribution of σ_y stress on midpsan; computer stress plot. Point load on cantilevered span.

Finally it should be mentioned that some restrictions placed on the general shapes defined by Eqs. (15.1) or (15.13) can be raised by doubling the number of parameters and writing expansions in the form of two sums

$$\mathbf{u} = \sum_{l=1}^{L} \overline{\mathbf{N}}(x, y) \cos\frac{l\pi z}{a}\, \mathbf{a}^{Al} + \sum_{l=1}^{L} \overline{\overline{\mathbf{N}}}(x, y) \sin\frac{l\pi z}{a}\, \mathbf{a}^{Bl}. \qquad (15.21)$$

Parameters $\mathbf{a}^{Al}$ and $\mathbf{a}^{Bl}$ are independent and for every component of displacement two values have to be found and two equations formed.

An alternative to the above process is to write the expansion as

$$\mathbf{u} = \sum [\mathbf{N}(x, y)\, e^{i(l\pi z/a)}]\, \mathbf{a}^e$$

and to observe that both **N** and **a** are complex quantities.

Complex algebra is now available on standard computers and the identity of the above expression with Eq. (15.21) will be observed noting that

$$e^{i\theta} = \cos\theta + i\sin\theta.$$

15.3 Thin Membrane Box Structures

In the previous section a three-dimensional problem was reduced to that of two dimensions. Here we shall see how a somewhat similar problem can be reduced to one-dimensional elements. (Fig. 15.3)

A box-type structure is made up of thin sheet components capable of sustaining stresses only in its own plane.

Now, just as in the previous case, three displacements have to be considered at every point and indeed similar variation can be prescribed for these. However, a typical element *ij* is 'one dimensional' in the sense that integrations have to be carried out only along the line *ij* and stresses in that

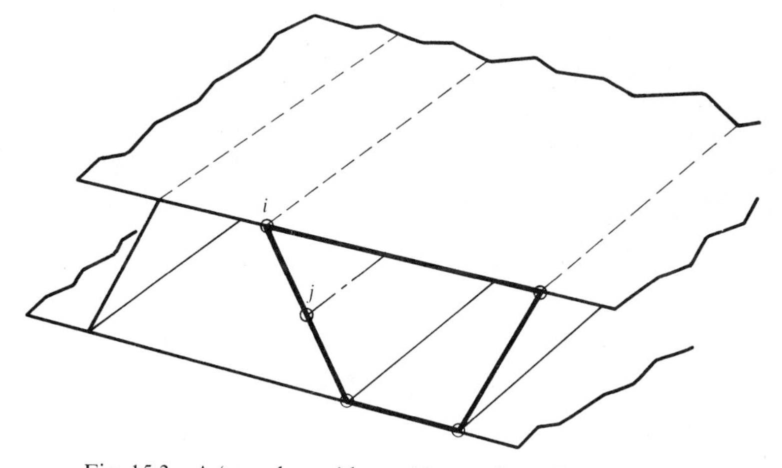

Fig. 15.3 A 'membrane' box with one-dimensional elements

direction only considered. Indeed it will be found that the situation and the solution are similar to that of a pin-jointed framework.

15.4 Plates and Boxes with Flexure

Consider now a rectangular plate simply supported at the ends and in which all strain energy is contained in flexure. Now, only one displacement, w, is needed to specify fully the state of strain (see Chapter 10).

For consistency of notation, the direction in which geometry and material properties do not change has been taken as y (see Fig. 15.4). To preserve slope continuity the functions need to include now the 'rotation' parameter θ_i.

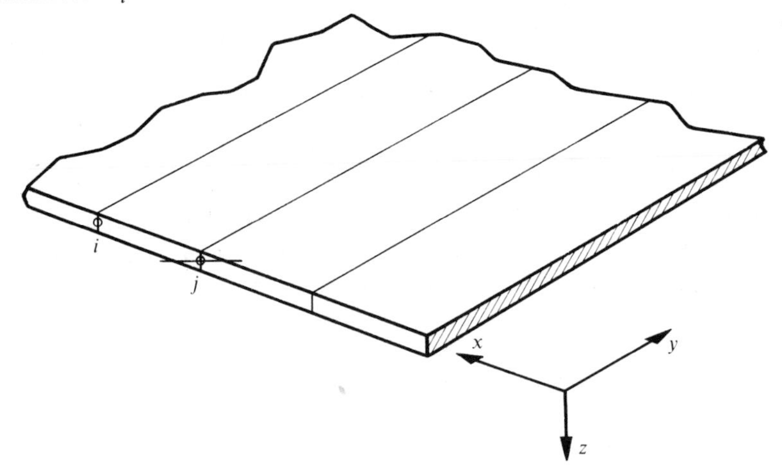

Fig. 15.4 The 'strip' method in slabs

Use of simple beam functions is easy and for a typical element ij we can write

$$w^l = \overline{\mathbf{N}}(x) \sin\frac{l\pi y}{a} (\mathbf{a}^l)^e \tag{15.22}$$

ensuring *simply supported* end conditions.
In this, the typical nodal parameters are

$$\mathbf{a}_i^l = \begin{Bmatrix} w_i \\ \theta_i \end{Bmatrix}. \tag{15.23}$$

The shape functions of the cubic type are easy to write and are in fact identical to those used for the axi-symmetric shell problem (Chapter 14).

Using all the definitions of Chapter 10 the strains (curvatures) are found and the **B** matrices determined, now with C_1 continuity satisfied in a trvial manner, the problem of a two-dimensional kind has here been reduced to that of one dimension.

This application has been developed by Cheung,[2-16] named by him the 'finite strip' method, and used to solve many rectangular plate problems, box girders, shells and various folded plates.

It is illuminating to quote an example from the above papers here. This refers to a square, uniformly loaded plate with three sides simply supported and one free. Ten strips or elements in the x direction were used in the solution and Table 15.1 gives the results corresponding to the first three harmonics.

TABLE 15.1

SQUARE PLATE, UNIFORM LOAD q
THREE SIDES SIMPLY SUPPORTED, ONE CLAMPED

$\nu = 0{\cdot}3$	Central deflection	Central M_x	Max. negative M
$l = 1$	0·002832	0·0409	−0·0858
$= 2$	−0·000050	−0·0016	0·0041
$= 3$	0·000004	0·0003	−0·0007
Σ	0·002786	0·0396	−0·0824
Exact	0·0028	0·039	−0·084
Multiplier	qa^4/D	qa^2	

Not only is an accurate solution of each l term a simple one involving only some nine unknowns but the importance of higher terms in the series is seen to decrease rapidly.

Extension of the process to box structures in which both *membrane and bending effects* are present is almost obvious when this example is considered together with the ones of the previous section.

In another paper Cheung[5] shows how functions other than trigonometric ones can be used to advantage although only partial decoupling then occurs.

In the examples just quoted a thin plate theory using the single displacement variable w and enforcing C_1 compatibility in the x direction was employed. Obviously any of the independently interpolated slope and displacement elements of Chapter 11 could be used here again employing reduced integration. Parabolic type elements are thus employed in references 6 and 7, and the linear interpolation with a simple integration point is shown to be effective in reference 8.

Other applications for plate and box type structures abound and additional information is given in the text of reference 17.

15.5 Axi-symmetric Solids with Non-symmetrical Load

One of the most natural and indeed earliest applications of the one-way Fourier expansion occurs in axi-symmetric bodies subject to non-axi-symmetric loads.

Now, not only the radial (u) and axial (v) displacements (as in Chapter 5) will have to be considered but also a tangential component (w) associated with angular direction θ (Fig. 15.5). It is in this direction that the geometric and material properties do not vary and hence here the elimination will be applied.

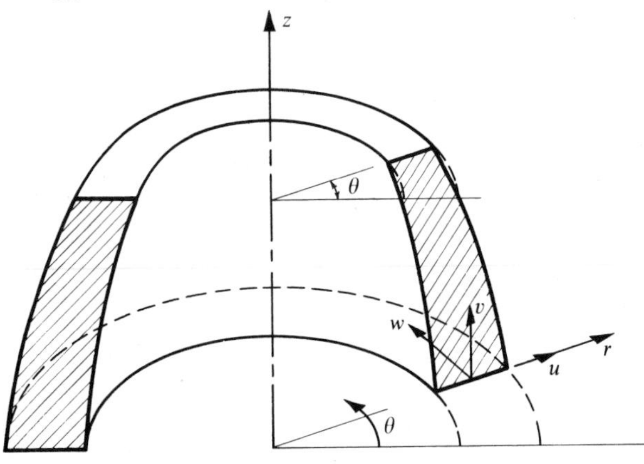

Fig. 15.5 An axi-symmetric solid. Co-ordinates and displacement components in an axi-symmetric body

To simplify matters we shall consider first components of load which are symmetric about the $\theta = 0$ axis and later those which are anti-symmetric. Describing now only the nodal loads (with similar expansions holding for body forces, boundary conditions, initial strains, etc.) we specify forces per unit length of circumference as

$$
\begin{aligned}
R &= \sum_{1}^{L} \bar{R}^{l} \cos l\theta \\
Z &= \sum_{1}^{L} \bar{Z}^{l} \cos l\theta \qquad (15.24) \\
T &= \sum_{1}^{L} \bar{T}^{l} \sin l\theta
\end{aligned}
$$

in direction of the various co-ordinates for symmetric loads, Fig. 15.6(a). The apparently non-symmetric sine expansion is used for T, as to achieve symmetry the direction of T has to change for $\theta > \pi$.

The displacement components are described again in terms of the two-dimensional (r, z) shape functions appropriate to the element subdivision and observing symmetry we write as in Eq. (15.13)

$$
\begin{aligned}
u^l &= [N_1, N_2, \ldots] \cos l\theta \mathbf{u}^{le} \\
v^l &= [N_1, N_2, \ldots] \cos l\theta \mathbf{v}^{le} \\
w^l &= [N_1, N_2, \ldots] \sin l\theta \mathbf{w}^{le}.
\end{aligned} \tag{15.25}
$$

To proceed further it is necessary to specify the general, three-dimensional expression for strains in cylindrical co-ordinates. This is (*vide* Love[18])

$$
\boldsymbol{\varepsilon} = \begin{Bmatrix} \varepsilon_r \\ \varepsilon_z \\ \varepsilon_\theta \\ \gamma_{rz} \\ \gamma_{r\theta} \\ \gamma_{z\theta} \end{Bmatrix} = \begin{Bmatrix} \dfrac{\partial u}{\partial r} \\ \dfrac{\partial v}{\partial z} \\ \dfrac{u}{r} + \dfrac{1}{r}\dfrac{\partial w}{\partial \theta} \\ \dfrac{\partial u}{\partial z} + \dfrac{\partial v}{\partial r} \\ \dfrac{1}{r}\dfrac{\partial u}{\partial \theta} + \dfrac{\partial w}{\partial r} - \dfrac{w}{r} \\ \dfrac{1}{r}\dfrac{\partial v}{\partial \theta} + \dfrac{\partial w}{\partial z} \end{Bmatrix}. \tag{15.26}
$$

As before, uncoupling will occur between the modes and we can proceed to evaluation of the stiffness matrices, etc., in each harmonic. Typically, we have on substitution of Eq. (15.25) into Eq. (15.26), and grouping the variables as in Eq. (15.17):

$$
\mathbf{B}_i = \begin{bmatrix} \dfrac{\partial N_i}{\partial r} \cos l\theta & 0 & 0 \\ 0 & \dfrac{\partial N_i}{\partial z} \cos l\theta & 0 \\ \dfrac{N_i}{r} \cos l\theta & 0 & \dfrac{lN_i}{r} \cos l\theta \\ \dfrac{\partial N_i}{\partial z} \cos l\theta & \dfrac{\partial N_i}{\partial r} \cos l\theta & 0 \\ -\dfrac{lN_i}{r} \sin l\theta & 0 & \left(\dfrac{\partial N_i}{\partial r} - \dfrac{N_i}{r}\right) \sin l\theta \\ 0 & -\dfrac{lN_i}{r} \sin l\theta & \dfrac{\partial N_i}{\partial z} \sin l\theta \end{bmatrix} \tag{15.27}
$$

The remaining steps of the formulation follow precisely the previous derivations and could be repeated by the reader as an exercise.

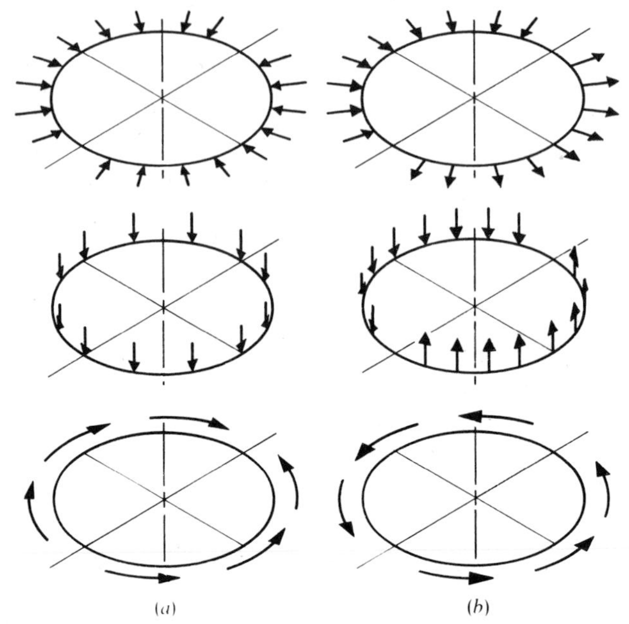

Fig. 15.6 Symmetric (*a*) and antisymmetric (*b*). Load and displacement components in an axi-symmetric body

For the antisymmetric loading of Figs. 15.6(*b*) we shall simply replace the sine by cosine and vice versa in Eqs. (15.24) and (15.25).

The load terms in each harmonic will be obtained by virtual work as

$$\mathbf{f}_i^l = \int_0^{2\pi} \begin{Bmatrix} \bar{R}^l \cos^2 l\theta \\ \bar{Z}^l \cos^2 l\theta \\ \bar{T}^l \sin^2 l\theta \end{Bmatrix} \mathrm{d}\theta = \pi \begin{Bmatrix} \bar{R}^l \\ \bar{Z}^l \\ \bar{T}^l \end{Bmatrix} \quad \text{when } l = 1, 2, \ldots$$

$$= 2\pi \begin{Bmatrix} \bar{R}^l \\ \bar{Z}^l \\ 0 \end{Bmatrix} \quad \text{when } l = 0 \tag{15.28}$$

for the symmetric case. Similarly for the antisymmetric case

$$\mathbf{f}_i^l = \pi \begin{Bmatrix} \bar{R}^l \\ \bar{Z}^l \\ \bar{T}^l \end{Bmatrix} \quad \text{when } l = 1, 2, \ldots = \begin{Bmatrix} 0 \\ 0 \\ \bar{T}^l \end{Bmatrix} \quad \text{when } l = 0. \tag{15.29}$$

We see from this and from the expansion of $\mathbf{K}^e$ that, as expected, for $l = 0$ the problem reduces to only two variables and the axi-symmetric case is retrieved when symmetric terms only are involved.

Similarly, when $l = 0$ only one set of equations remains in the variable w for the antisymmetric case. This corresponds to constant tangential

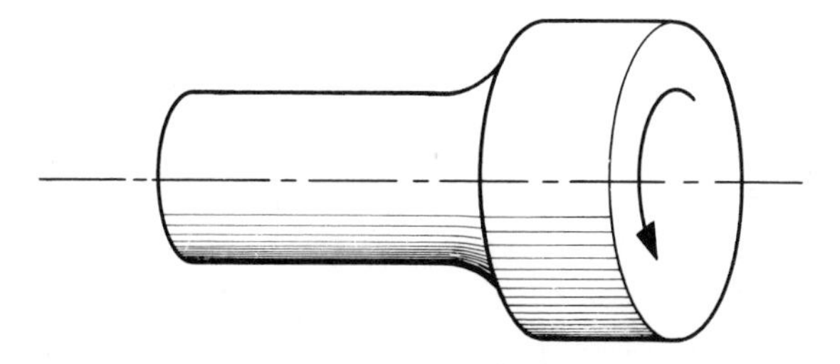

Fig. 15.7 Torsion of a variable section bar

traction and solves simply the torsion problems of shafts subject to known torques, Fig. 15.7. This problem is classically treated by the use of a stress function[19] and indeed has been solved using a finite element formulation.[20] Here an alternative, more physical approach is available.

The first application of the above concepts to the analysis of axi-symmetric solids was made by Wilson.[21]

A simple example illustrating the effects of various harmonics is shown in Fig 15.8(a) and (b).

15.6 **Axi-symmetric Shells with Non-symmetric Loading**

The extension of analysis of axi-symmetric shells as described in Chapter 14 to the case of non-axi-symmetric loads is simple and will again follow the standard pattern.

It is, however, necessary to extend the definition of strains and to include now all three displacements and force components, Fig. 15.9. Three membrane and three bending effects are now present and extending Eq. (14.1) involving straight generators we now define strains as[22]†

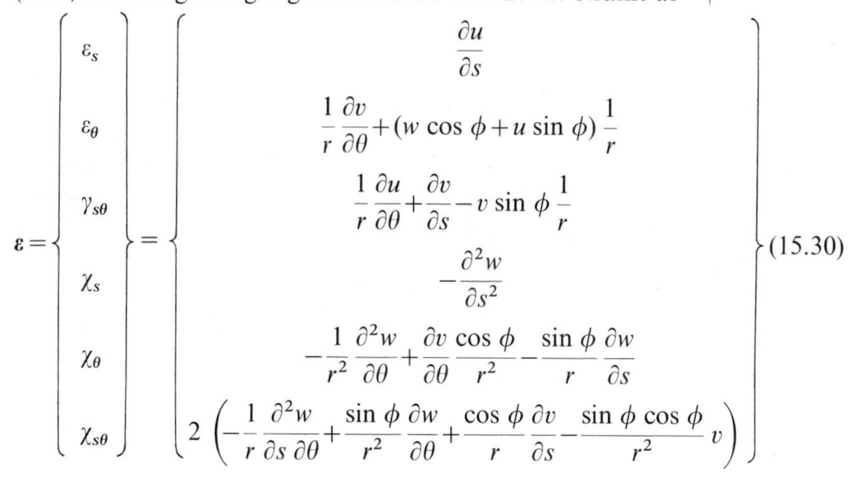

$$\boldsymbol{\varepsilon}=\begin{Bmatrix}\varepsilon_s\\ \varepsilon_\theta\\ \gamma_{s\theta}\\ \chi_s\\ \chi_\theta\\ \chi_{s\theta}\end{Bmatrix}=\begin{Bmatrix}\dfrac{\partial u}{\partial s}\\ \dfrac{1}{r}\dfrac{\partial v}{\partial \theta}+(w\cos\phi+u\sin\phi)\dfrac{1}{r}\\ \dfrac{1}{r}\dfrac{\partial u}{\partial \theta}+\dfrac{\partial v}{\partial s}-v\sin\phi\,\dfrac{1}{r}\\ -\dfrac{\partial^2 w}{\partial s^2}\\ -\dfrac{1}{r^2}\dfrac{\partial^2 w}{\partial \theta}+\dfrac{\partial v}{\partial \theta}\dfrac{\cos\phi}{r^2}-\dfrac{\sin\phi}{r}\dfrac{\partial w}{\partial s}\\ 2\left(-\dfrac{1}{r}\dfrac{\partial^2 w}{\partial s\,\partial \theta}+\dfrac{\sin\phi}{r^2}\dfrac{\partial w}{\partial \theta}+\dfrac{\cos\phi}{r}\dfrac{\partial v}{\partial s}-\dfrac{\sin\phi\cos\phi}{r^2}v\right)\end{Bmatrix} \tag{15.30}$$

† Various alternatives are here present due to the multiplicity of shell theories. This one is fairly generally accepted.

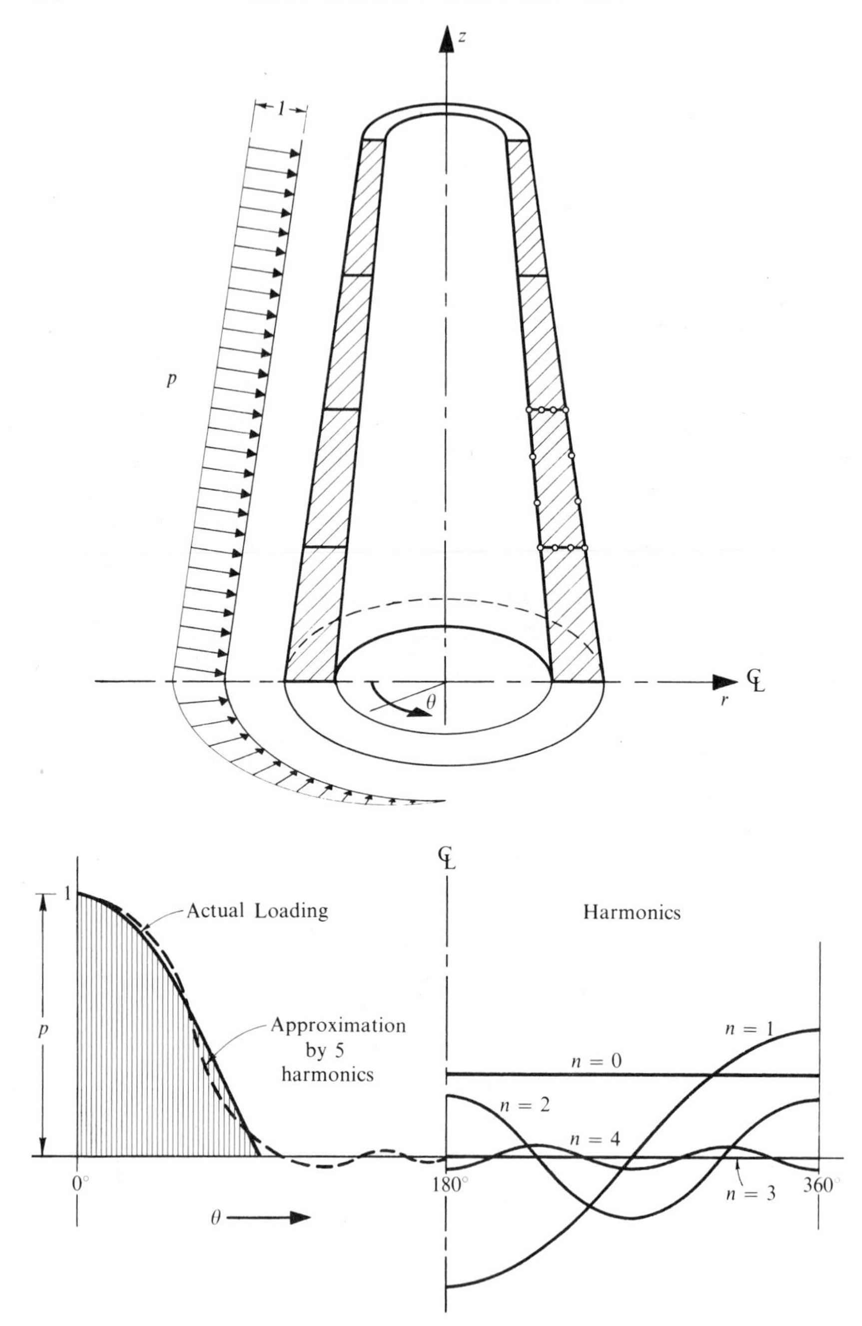

Fig. 15.8(*a*) An axi-symmetric tower under non-symmetric load. Four cubic elements used in solution. Harmonics of load expansion used in analysis are shown.

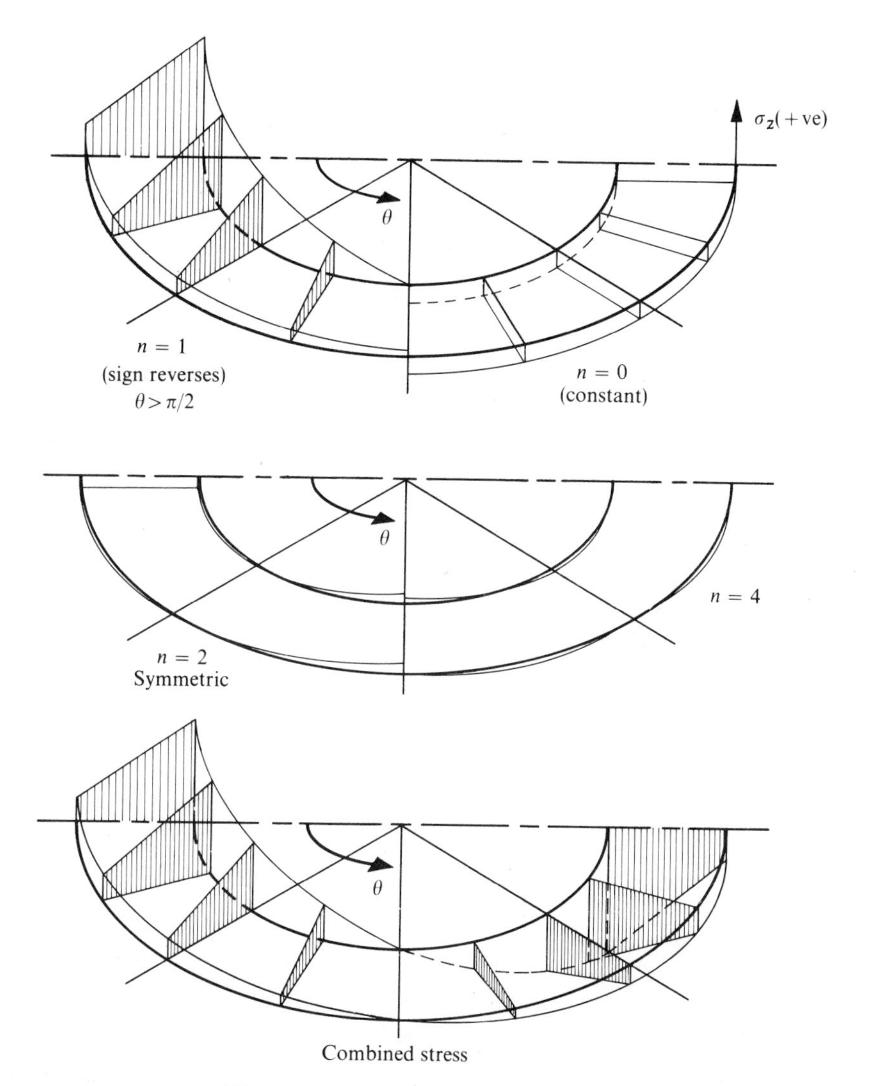

Fig. 15.8(*b*) Distribution of σ_z—the vertical stress on base due to various harmonics and their combination (third harmonic identically zero). First two harmonics give practically complete answer

The corresponding stress matrix is

$$\boldsymbol{\sigma} = \begin{Bmatrix} N_s \\ N_\theta \\ N_{s\theta} \\ M_s \\ M_\theta \\ M_{s\theta} \end{Bmatrix} \tag{15.31}$$

with the three membrane and bending 'stresses' defined as in Fig. 15.9.

Once again symmetric and antisymmetric variation of loads and displacements can be assumed as in the previous section.

As the processes involved in executing this extension of the application are now obvious no further elaboration is needed here but note again should be made of the more elaborate form of equations necessary when curved elements are involved (*vide* Chapter 14, Eq. (14.16)).

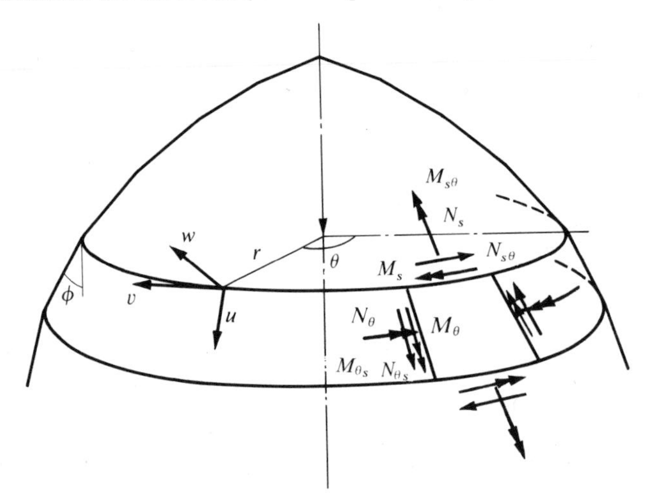

Fig. 15.9 Axi-symmetric shell with non-symmetric load. Displacements and stress resultants

The reader is referred to the original paper by Grafton and Strome[23] in which this problem is first treated and to the many later papers on the subject listed in Chapter 14.

Some examples illustrating the process in the context of thick shell analysis are given in Chapter 16.

15.7 **Concluding Remarks**

A fairly general process combining some of the advantages of finite element analysis with the economy of expansion in terms of orthogonal functions has been illustrated in several applications. Certainly these only touch on the possibilities offered but it should be borne in mind that the economy is only achieved in certain geometrically constrained situations and those to which the number of terms requiring solution is limited.

Similarly other 'prismatic' situations can be dealt with in which only a segment of a body of revolution is developed. (Fig. 15.10.) Clearly, the expansion must now be taken in terms of the angle $l\pi\theta/\alpha$ but otherwise the approach is identical to that described previously.[1]

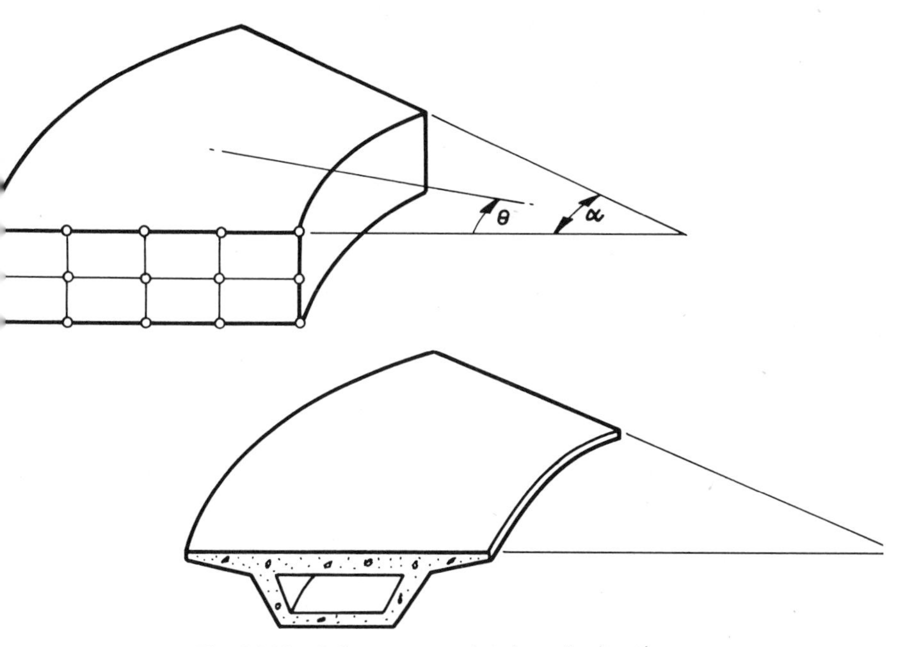

Fig. 15.10 Other segmental, prismatic situations

In the methods of this chapter it was assumed that material properties remain invariant with one co-ordinate direction. This restriction can on occasion be lifted with the same general process maintained. An interesting example of this type is outlined by Stricklin and De Andrade.[24]

In Chapter 3 dealing with the general formulation of the finite element discretization we have referred to semi-discretization (see section 3.7). In this, one of the problem variables (say z) is retained and the problem is reduced to an ordinary differential equation in terms of the nodal parameters $\mathbf{a}$ and its derivatives with respect to z.

We shall have occasion to make use of such partial discretization in Chapters 20 and 21 where the variable z is the time domain in which the problem is 'prismatic'. However, all the problems we have described in this chapter could be derived in terms of such semi-discretization. We would thus *first* semi-discretize, describing the problem in terms of a differential equation of the form

$$\mathbf{K}_1\frac{\mathrm{d}^2\mathbf{a}}{\mathrm{d}z^2}+\mathbf{K}_2\frac{\mathrm{d}\mathbf{a}}{\mathrm{d}z}+\mathbf{K}_3\mathbf{a}+\mathbf{f} = 0.$$

Second, the above equation system would be solved in the domain $0<z<a$ using orthogonal functions which *naturally* enter the problem as solutions at ordinary differential equations *with constant coefficients*. This second solution is most easily found using a diagonalization process described in dynamic applications (Chapter 20).

Clearly the final result of such computations would turn out to be identical with the procedures here described—but on occasion the above formulation is more self-evident.

The use of analytical processes to extend the power of the finite element method is not limited to the processes described here. We shall return to it in a wider context in Chapter 23.

References

1. O. C. Zienkiewicz and J. J. M. Too, 'The finite prism in analysis of thick simply supported bridge boxes', *Proc. Inst. Civ. Eng.*, **53**, 147–72, 1972.
2. Y. K. Cheung, 'The finite strip method in the analysis of elastic plates with two opposite simply supported ends', *Proc. Inst. Civ. Eng.*, **40**, 1–7, 1968.
3. Y. K. Cheung, 'Finite strip method of analysis of elastic slabs', *Proc. Am. Soc. Civ. Eng.*, **94**, EM6, 1365–78, 1968.
4. Y. K. Cheung, 'Folded plate structures by the finite strip method', *Proc. Am. Soc. Civ. Eng.*, **95** ST, 2963–79, 1969.
5. Y. K. Cheung, 'The analysis of cylindrical orthotropic curved bridge decks', *Publ. Int. Ass. Struct. Eng.*, **29**-II, 41–52, 1969.
6. A. S. Mawenya and J. D. Davies, 'Finite strip analysis of plate bending including transverse shear effects', *Building Science*, **9**, 175–80, 1974.
7. P. R. Benson and E. Hinton, 'A thick finite strip solution for static, free vibration and stability problems', *Int. J. Num. Meth. Eng.*, **10**, 665–78, 1976.
8. E. Hinton and O. C. Zienkiewicz, 'A note on a simple thick finite strip', *Int. J. Num. Meth. Eng.*, **11**, 905–9, 1977.
9. Y. K. Cheung, M. S. Cheung, and A. Ghali, 'Analysis of slab and girder bridges by the finite strip method', *Building Science*, **5**, 95–104, 1970.
10. Y. C. Loo and A. R. Cusens, 'Development of the finite strip method in the analysis of cellular bridge decks', *Conf. on Developments in Bridge Design and Construction* (ed. Rockey *et al.*), Crosby Lockwood, 1971.
11. Y. K. Cheung and M. S. Cheung, 'Static and dynamic behaviour of rectangular plates using higher order finite strips', *Building Science*, **7**, 151–8, 1972.
12. T. G. Brown and A. Ghali, 'Semi-analytic solution of skew plates in bending', *Proc. Inst. Civ. Eng.* **57**-II, 165–75, 1974.
13. G. S. Tadros and A. Ghali, 'Convergence of semi-analytical solution of plates', *Proc. Am. Soc. Civ. Eng.*, **99**, EM5, 1023–35, 1973.

14. A. R. CUSENS and Y. C. LOO, 'Application of the finite strip method in the analysis of concrete box bridges', *Proc. Inst. Civ. Eng.*, **57**-II, 251–73, 1974.
15. Y. K. CHEUNG, 'Folded plate structures by the finite strip method', *Proc. Am. Soc. Civ. Eng.*, No. ST12, 2 & 63–79, 1969.
16. Y. K. CHEUNG, 'The analysis of cylindrical orthotropic curves bridge decks', *Int. Assoc. for Bridges and Structural Engineering*, **29**-II, 41–51, 1969.
17. Y. K. CHEUNG, *Finite strip method in structural analysis*, Pergamon Press, 1976.
18. A. E. H. LOVE, *The Mathematical Theory of Elasticity*, 4th ed., Cambridge Univ. Press, 1927, p. 56.
19. S. TIMOSHENKO and J. N. GOODIER, *Theory of Elasticity*, 2nd ed., McGraw-Hill, 1951.
20. O. C. ZIENKIEWICZ and Y. K. CHEUNG, 'Stresses in shafts', *The Engineer*, 24 Nov. 1967.
21. E. L. WILSON, 'Structural analysis of axi-symmetric solids', *J.A.I.A.A.*, **3**, 2269–74, 1965.
22. V. V. NOVOZHILOV, *Theory of Thin Shells* (Translation), P. Noordhoff, 1959.
23. P. E. GRAFTON and D. R. STROME, 'Analysis of axi-symmetric shells by the direct stiffness method', *J.A.I.A.A.*, **1**, 2342–7, 1963.
24. J. A. STRICKLIN and J. C. DE ANDRADE, 'Linear and non linear analysis of shells of revolution with asymmetrical stiffness properties', *Proc. 2nd Conf. Matrix Methods in Struct. Mech.*, Air Force Inst. Techn., Wright-Patterson A.F. Base, Ohio, 1968.

16. Shells as a Special Case of Three-Dimensional Analysis

16.1 Introduction

In Chapters 8 and 9 the formulation and use of complex, curved, two- and three-dimensional elements was illustrated. It seems obvious that use of such elements could be made directly in the analysis of curved shells simply by reducing their dimension in the shell thickness direction as shown in Fig. 16.1. Indeed in an axi-symmetric situation such an application has been illustrated in the example of Fig. 9.6 in Chapter 9.

With a straightforward use of the three-dimensional concept, however, certain difficulties will be encountered.

In the first place the retention of three degrees of freedom at each node leads to large stiffness coefficients for relative displacements along an edge corresponding to the shell thickness. This presents numerical problems and may lead to ill-conditioned equations when shell thicknesses become small compared with the other dimensions in the element.

The second factor is that of economy. The use of several nodes across the shell thickness ignores the well-known fact that even for thick shells the 'normals' to the middle surface remain practically straight after deformation. Thus an unnecessarily high number of degrees of freedom has to be carried, involving penalties of computer time.

Here, specialized formulation is presented overcoming both these difficulties.[1,2,3] The constraint of straight 'normals' is introduced to improve economy and the strain energy corresponding to stresses perpendicular to the middle surface is ignored to improve numerical conditioning. With these modifications an efficient tool for analysing curved thick shells becomes available. Its accuracy and wide range of applicability is demonstrated in several examples.

The reader will note that the two constraints introduced correspond only to a part of the usual assumptions in shell theory. Thus, the statement that after deformation the normals remain normal to the deformed middle surface has been deliberately omitted. This omission permits the shell to

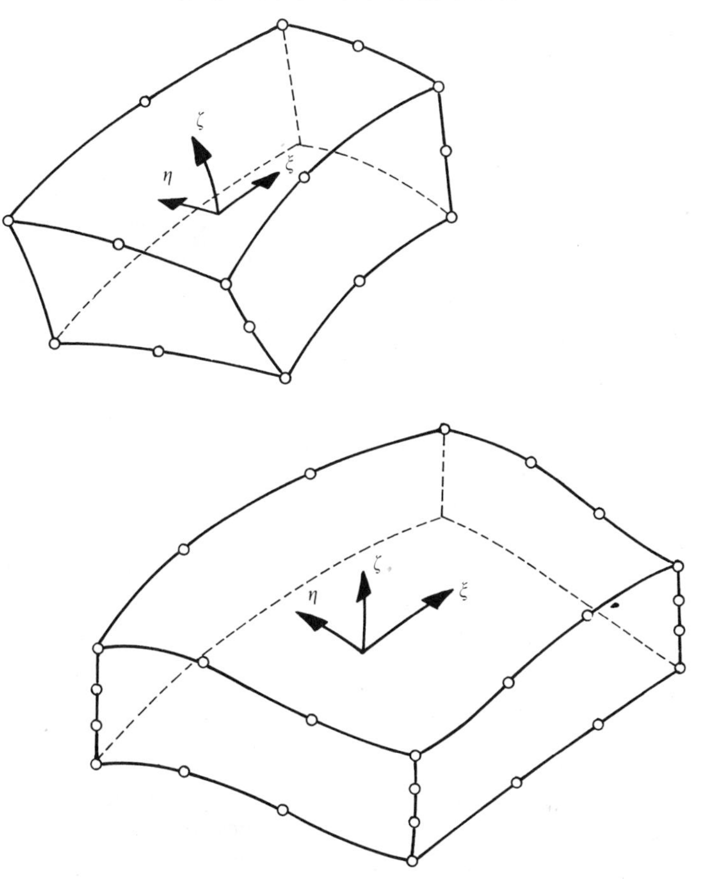

Fig. 16.1 Curved, isoparametric hexahedra in a direct approximation to a curved shell

experience shear deformations—an important feature in thick shell situations.

The formulation presented here leads to some more complication than that of straightforward use of a three-dimensional element—and indeed the reader may be tempted to a direct use of three-dimensional formulation, especially as it appears that the use of an element with only a linear variation of displacements across the thickness would be permissible. Only six degrees of freedom corresponding to a specified mid-plane point would be necessary *vis-à-vis* five which will arise with the formulation of this chapter, and this would appear to be a small penalty if the ill-conditioning due to large stiffness ratios could be overcome. This is indeed feasible, as shown by Wood[4,5] and Wilson[6] who use as variables the *differences* of *displacements* at the two surfaces and a high precision computer. However, another difficulty now becomes apparent if linear interpolation in the direction of the shell normal is used. When the Poisson's ratio is not zero the results converge to a solution which is in error by a factor $\frac{(1-\nu)^2}{1-2\nu}$. The reason for this is easily explained. With pure bending a zero strain is obtained in the direction normal to the middle plane and, consequently, stresses develop in that direction if $\nu \neq 0$ restraining the *in-plane* strains. To overcome this effect *either* artificial anisotropic properties of the material have to be assumed or a full parabolic displacement used which will make the computation uneconomic.

The elements developed here are in essence an alternative formulation of the processes discussed in Chapters 11 and 14 for which an independent interpolation of slopes and displacement was used with a penalty function imposition of the continuity requirements. The use of reduced integration is thus once again imperative if thin shells are to be dealt with—and it was, indeed, in this context that this procedure was first discovered.[7–10]

16.2 Geometric Definition of the Element

Consider a typical shell element of Fig. 16.2. The external faces of the element are curved, while the sections across the thickness are generated by straight lines. Pairs of points, i_{top} and i_{bottom}, each with given Cartesian co-ordinates, prescribe the shape of the element.

Let ξ, η be the two curvilinear co-ordinates in the middle plane of the shell and ζ a linear co-ordinate in the thickness direction. If further we assume that ξ, η, ζ vary between -1 and 1 on the respective faces of the element we can write a relationship between the Cartesian co-ordinates of any point of the shell and the curvilinear co-ordinates in the form

$$\begin{Bmatrix} x \\ y \\ z \end{Bmatrix} = \sum N_i(\xi, \eta) \frac{(1+\zeta)}{2} \begin{Bmatrix} x_i \\ y_i \\ z_i \end{Bmatrix}_{\text{top}} + \sum N_i(\xi, \eta) \frac{(1-\zeta)}{2} \begin{Bmatrix} x_i \\ y_i \\ z_i \end{Bmatrix}_{\text{bottom}} \quad (16.1)$$

Here $N_i'(\xi, \eta)$ is a shape function taking a value of unity at the nodes i and zero of all other nodes (Chapter 8). If the basic functions N_i are derived as 'shape functions' of a 'parent', two-dimensional element, square or even triangular† in plan and are so 'designed' that compatibility is achieved at interfaces, then the curved space elements will fit into each other. Arbitrary curved shapes of the element can be achieved by using shape functions of different orders. Only parabolic and cubic types are shown in Fig. 16.2. By placing a larger number of nodes on the surfaces of the element more

† Area co-ordinates would be used in this case in place of ξ and η as in Chapter 7.

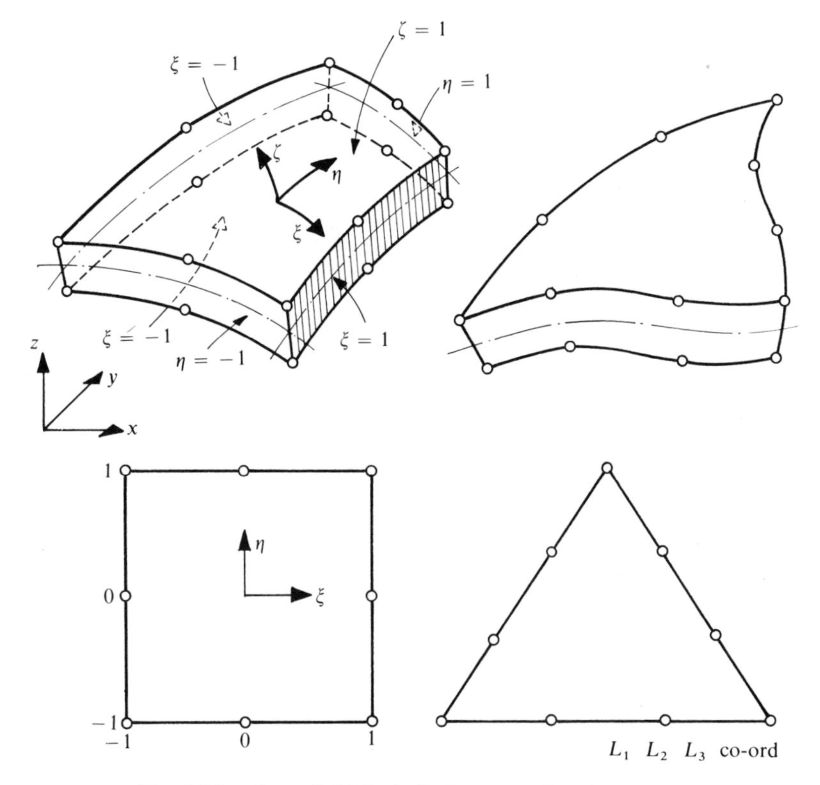

Fig. 16.2 Curved thick shell elements of various types

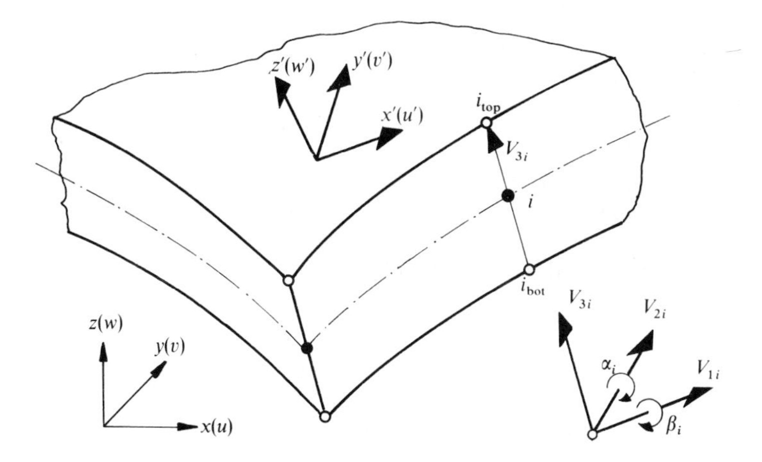

Fig. 16.3 Local and global co-ordinates

elaborate shapes can be achieved if so desired. Indeed any of the two-dimensional shape functions of Chapter 7 can be used here.

The relation between the Cartesian and curvilinear co-ordinates is now established and it will be found desirable to operate with the curvilinear co-ordinates as the basis.

It should be noted that the co-ordinate direction ζ is *only approximately normal* to the middle surface.

It is convenient to rewrite the relationship, Eq. (16.1) in a form specified by the 'vector' connecting the upper and lower points (i.e., a vector of length equal to the shell thickness t) and the mid-surface co-ordinates. Thus† we can rewrite Eq. 16.1 as (Fig. 16.3)

$$\begin{Bmatrix} x \\ y \\ z \end{Bmatrix} = \sum N_i \begin{Bmatrix} x_i \\ y_i \\ z_i \end{Bmatrix}_{\text{mid.}} + \sum N_i \frac{\zeta}{2} \mathbf{V}_{3i}$$

with (16.2)

$$\mathbf{V}_{3i} = \begin{Bmatrix} x_i \\ y_i \\ z_i \end{Bmatrix}_{\text{top}} - \begin{Bmatrix} x_i \\ y_i \\ z_i \end{Bmatrix}_{\text{bottom}} .$$

defining a vector whose length is the shell thickness.

16.3 Displacement Field

The displacement field has now to be specified for the element. As the strains in the direction normal to the mid-surface will be assumed to be negligible, the displacement throughout the element will be taken to be uniquely defined by the *three Cartesian components* of the mid-surface node displacement and two rotations of the nodal vector $\mathbf{V}_{3i}$ about orthogonal directions normal to it. If two such orthogonal directions are given by vectors $\mathbf{v}_{2i}$ and $\mathbf{v}_{1i}$ of unit magnitude, with corresponding (scalar) rotations α_i and β_i we can write, similarly to Eq. (16.2) but now dropping the suffix 'mid' for simplicity

$$\begin{Bmatrix} u \\ v \\ w \end{Bmatrix} = \sum N_i \begin{Bmatrix} u_i \\ v_i \\ w_i \end{Bmatrix} + \sum N_i \zeta \frac{t_i}{2} [\mathbf{v}_{1i}, -\mathbf{v}_{2i}] \begin{Bmatrix} \alpha_i \\ \beta_i \end{Bmatrix} \tag{16.3}$$

from which the usual form is readily obtained as

$$\begin{Bmatrix} u \\ v \\ w \end{Bmatrix} = \mathbf{N} \begin{Bmatrix} \mathbf{a}_i^e \\ \vdots \\ \mathbf{a}_j^e \end{Bmatrix} \quad \text{with} \quad \mathbf{a}_i^e = \begin{Bmatrix} u_i \\ v_i \\ w_i \\ \alpha_i \\ \beta_i \end{Bmatrix}$$

† For details of vector algebra see Appendix 5.

where u, v, and w are displacements in the directions of the global, x, y, and z axes.

As an infinity of vector directions normal to a given direction can be generated, a particular scheme has to be devised to ensure a *unique* definition.

Some such schemes were discussed in Chapter 13. Here a simpler unique alternative will be given,[2] but other possibilities are open.[10]

Thus if $\mathbf{V}_{3i}$ is the vector to which a normal direction is to be constructed we form the first normal axis in a direction perpendicular to the plane defined by this vector and the x axis.†

A vector $\mathbf{V}_{1i}$ of this description is given by the cross product

$$\mathbf{V}_{1i} = \mathbf{i} \times \mathbf{V}_{3i}. \tag{16.4}$$

In this

$$\mathbf{i} = \begin{Bmatrix} 1 \\ 0 \\ 0 \end{Bmatrix}$$

is a unit vector in direction of the x axis. Dividing this by its length we can write the unit vector $\mathbf{v}_{1i}$.

The last vector normal to the other two is simply

$$\mathbf{V}_{2i} = \mathbf{V}_{1i} \times \mathbf{V}_{3i} \tag{16.5}$$

and all the direction cosines of the local axes can be determined by normalizing this to $\mathbf{v}_{2i}$. We have thus three local, orthogonal axes defined by unit vectors

$$\mathbf{v}_{1i}, \quad \mathbf{v}_{2i}, \quad \text{and} \quad \mathbf{v}_{3i}. \tag{16.6}$$

Once again if N_i are compatible functions then displacement compatibility is maintained between adjacent elements.

The element co-ordinate definition is now given by the relation, Eq. (16.1), which has more degrees of freedom than the definition of the displacements. The element is therefore of the super-parametric kind (*vide* Chapter 8, section 8.3) and the constant strain criteria are not automatically satisfied.

Nevertheless, it will be seen from the definition of strain components involved that both rigid body motions and constant strain conditions are available.

Physically, it has been assumed in the definition of Eq. (16.3) that no strains occur in the 'thickness' direction ζ. While this direction is not

† This process fails if $\mathbf{V}_{3i}$ corresponds in direction with the x axis. A program checking this possibility is easily written and in such a case the local directions are obtained using the y axis.

exactly normal to the middle surface it still represents to a good approximation one of the usual shell assumptions.

At each mid-surface node i of Fig. 16.3 we have now the five basic degrees of freedom and the connection of elements will follow precisely the patterns described in Chapter 13 (sections 13.3 and 13.4).

16.4 **Definition of Strains and Stresses**

To derive the properties of a finite element the essential strains and stresses have to be defined. The components in directions of *orthogonal axes* related to the surface ζ = constant are essential if account is to be taken of the basic shell assumptions. Thus if at any point in this surface we erect a normal z' with two other orthogonal axes x' and y' tangent to it (Fig. 16.3) the strain components of interest are given simply by the three-dimensional relationships of Chapter 6.

$$\boldsymbol{\varepsilon}' = \begin{Bmatrix} \varepsilon_{x'} \\ \varepsilon_{y'} \\ \gamma_{x'y'} \\ \gamma_{x'z'} \\ \gamma_{y'z'} \end{Bmatrix} = \begin{Bmatrix} \dfrac{\partial u'}{\partial x'} \\ \dfrac{\partial v'}{\partial y'} \\ \dfrac{\partial u'}{\partial y'} + \dfrac{\partial v'}{\partial x'} \\ \dfrac{\partial w'}{\partial x'} + \dfrac{\partial u'}{\partial z'} \\ \dfrac{\partial w'}{\partial y'} + \dfrac{\partial v'}{\partial z'} \end{Bmatrix} \tag{16.7}$$

with the strain in direction z' neglected so as to be consistent with the usual shell assumptions. It must be noted that in general none of these directions coincide with those of the curvilinear co-ordinates ξ, η, ζ, although x', y' are in the ξ–η plane (ζ = constant).†

The stresses corresponding to these strains are defined by a matrix $\boldsymbol{\sigma}'$ and are related by the usual elasticity matrix $\mathbf{D}'$. Thus

$$\boldsymbol{\sigma}' = \begin{Bmatrix} \sigma_{x'} \\ \sigma_{y'} \\ \tau_{x'y'} \\ \tau_{x'z'} \\ \tau_{y'z'} \end{Bmatrix} = \mathbf{D}'(\boldsymbol{\varepsilon}' - \boldsymbol{\varepsilon}'_0) + \boldsymbol{\sigma}'_0 \tag{16.8}$$

where $\boldsymbol{\varepsilon}'_0$ in $\boldsymbol{\sigma}'_0$ may represent any 'initial' strains or stresses.

† Indeed these directions will only approximately agree with the nodal direction $\mathbf{v}_{1i}$, etc., previously derived as in general the vector $\mathbf{v}_{3i}$ is only approximately normal to the mid-surfaces.

The 5×5 matrix $\mathbf{D}'$ can now include any anisotropic properties and indeed may be prescribed as a function of ζ if sandwich (layered) construction is used. For the present moment we shall define it only for an isotropic material. Here

$$\mathbf{D}' = \frac{E}{1-\nu^2}\begin{bmatrix} 1 & \nu & 0 & 0 & 0 \\ & 1 & 0 & 0 & 0 \\ & & \dfrac{1-\nu}{2} & 0 & 0 \\ & & & \dfrac{1-\nu}{2k} & 0 \\ \text{sym.} & & & & \dfrac{1-\nu}{2k} \end{bmatrix} \tag{16.9}$$

in which E and ν are Young's modulus and Poisson's ratio respectively. The factor k included in the last two shear terms is taken as 1·2 and its purpose is to improve the shear displacement approximation. From the displacement definition it will be seen that the shear distribution is approximately constant through the thickness, whereas in reality the shear distribution is approximately parabolic. The value $k = 1{\cdot}2$ is the ratio of relevant strain energies.

It is important to note that this matrix is *not* derived simply by deleting appropriate terms from the equivalent three-dimensional stress matrix of Chapter 6 (Eq. (6.14)). It must be derived by substituting $\sigma'_z = 0$ into Eq. (6.13) and a suitable elimination so that this important shell assumption is satisfied.

16.5 **Element Properties and Necessary Transformations**

The stiffness matrix—and indeed all other 'element' property matrices—involve integrals over the volume of the element, which are quite generally of the form

$$\int_{V^e} \mathbf{S}\, \mathrm{d}x\, \mathrm{d}y\, \mathrm{d}z \tag{16.10}$$

where the matrix $\mathbf{S}$ is a function of the co-ordinates.

In the stiffness matrix

$$\mathbf{S} = \mathbf{B}^{\mathrm{T}}\mathbf{D}\mathbf{B} \tag{16.11}$$

for instance, with the usual definition of Chapter 2

$$\boldsymbol{\varepsilon} = \mathbf{B}\mathbf{a}^e \tag{16.12}$$

we have **B** defined in terms of the displacement derivatives with respect to the local Cartesian co-ordinates $x'y'z'$ by Eq. (16.7).

Now, therefore, *two sets of transformations* are necessary before the element can be integrated with respect to the curvilinear co-ordinates ξ, η, ζ.

First, by identically the same process as we used in Chapter 8, the derivatives with respect to the x, y, z directions are obtained.

As Eq. (16.3) relates the global displacements u, v, w to the curvilinear co-ordinates, the derivatives of these displacements with respect to the global x, y, z co-ordinates are given by a matrix relation.

$$\begin{bmatrix} \frac{\partial u}{\partial x} & \frac{\partial v}{\partial x} & \frac{\partial w}{\partial x} \\ \frac{\partial u}{\partial y} & \frac{\partial v}{\partial y} & \frac{\partial w}{\partial y} \\ \frac{\partial u}{\partial z} & \frac{\partial v}{\partial z} & \frac{\partial w}{\partial z} \end{bmatrix} = \mathbf{J}^{-1} \begin{bmatrix} \frac{\partial u}{\partial \xi} & \frac{\partial v}{\partial \xi} & \frac{\partial w}{\partial \xi} \\ \frac{\partial u}{\partial \eta} & \frac{\partial v}{\partial \eta} & \frac{\partial w}{\partial \eta} \\ \frac{\partial u}{\partial \zeta} & \frac{\partial v}{\partial \zeta} & \frac{\partial w}{\partial \zeta} \end{bmatrix}. \tag{16.13}$$

In this, the Jacobian matrix is defined as before

$$\mathbf{J} = \begin{bmatrix} \frac{\partial x}{\partial \xi} & \frac{\partial y}{\partial \xi} & \frac{\partial z}{\partial \xi} \\ \frac{\partial x}{\partial \eta} & \frac{\partial y}{\partial \eta} & \frac{\partial z}{\partial \eta} \\ \frac{\partial x}{\partial \zeta} & \frac{\partial y}{\partial \zeta} & \frac{\partial z}{\partial \zeta} \end{bmatrix} \tag{16.14}$$

and is calculated from the co-ordinate definitions of Eq. (16.2).

Now, for every set of curvilinear co-ordinates the global displacement derivatives can be obtained numerically. A further transformation to local displacement directions x', y', z' will allow the strains, and hence the **B** matrix, to be evaluated.

Second, the directions of the local axes have to be established. A vector normal to the surface ζ-constant can be found as a vector product of any two vectors tangent to the surface. Thus

$$\mathbf{V}_3 = \begin{Bmatrix} \frac{\partial x}{\partial \xi} \\ \frac{\partial y}{\partial \xi} \\ \frac{\partial z}{\partial \xi} \end{Bmatrix} \times \begin{Bmatrix} \frac{\partial x}{\partial \eta} \\ \frac{\partial y}{\partial \eta} \\ \frac{\partial z}{\partial \eta} \end{Bmatrix} = \begin{Bmatrix} \frac{\partial y}{\partial \xi}\frac{\partial z}{\partial \eta} - \frac{\partial y}{\partial \eta}\frac{\partial z}{\partial \xi} \\ \frac{\partial x}{\partial \eta}\frac{\partial z}{\partial \xi} - \frac{\partial x}{\partial \xi}\frac{\partial z}{\partial \eta} \\ \frac{\partial x}{\partial \xi}\frac{\partial y}{\partial \eta} - \frac{\partial x}{\partial \eta}\frac{\partial y}{\partial \xi} \end{Bmatrix}. \tag{16.15}$$

Following the process which defines uniquely two perpendicular vectors, given previously, and reducing to unit magnitudes, we construct a matrix of unit vectors in x', y', z' directions (which is in fact the direction cosine matrix)

$$\boldsymbol{\theta} = [\mathbf{v}_1, \mathbf{v}_2, \mathbf{v}_3]. \tag{16.16}$$

The global derivatives of displacements u, v, and w are now transformed to the local derivatives of the local orthogonal displacements by a standard operation

$$\begin{bmatrix} \dfrac{\partial u'}{\partial x'} & \dfrac{\partial v'}{\partial x'} & \dfrac{\partial w'}{\partial x'} \\ \dfrac{\partial u'}{\partial y'} & \dfrac{\partial v'}{\partial y'} & \dfrac{\partial w'}{\partial y'} \\ \dfrac{\partial u'}{\partial z'} & \dfrac{\partial v'}{\partial z'} & \dfrac{\partial w'}{\partial z'} \end{bmatrix} = \boldsymbol{\theta}^{\mathrm{T}} \begin{bmatrix} \dfrac{\partial u}{\partial x} & \dfrac{\partial v}{\partial x} & \dfrac{\partial w}{\partial x} \\ \dfrac{\partial u}{\partial y} & \dfrac{\partial v}{\partial y} & \dfrac{\partial w}{\partial y} \\ \dfrac{\partial u}{\partial z} & \dfrac{\partial v}{\partial z} & \dfrac{\partial w}{\partial z} \end{bmatrix} \boldsymbol{\theta}. \tag{16.17}$$

From this the components of the $\mathbf{B}'$ matrix can now be found explicitly, noting that five degrees of freedom exist at each node

$$\boldsymbol{\varepsilon}' = \mathbf{B}' \begin{Bmatrix} \mathbf{a}_i^e \\ \vdots \\ \mathbf{a}_j^e \end{Bmatrix}; \quad \mathbf{a}_i^e = \begin{Bmatrix} u_i \\ v_i \\ w_i \\ \alpha_i \\ \beta_i \end{Bmatrix}. \tag{16.18}$$

The infinitesimal volume is given in terms of the curvilinear co-ordinates as

$$\mathrm{d}x\,\mathrm{d}y\,\mathrm{d}z = \det|\mathbf{J}|\,\mathrm{d}\xi\,\mathrm{d}\eta\,\mathrm{d}\zeta \tag{16.19}$$

and this standard expression completes the basic formulation.

Numerical integration within the appropriate -1, $+1$ limits is carried out in exactly the same way as for three-dimensional elements discussed in Chapter 8.

Identical processes serve to define all the other relevant element matrices.

As the variation of the strain quantities in the thickness, ζ direction, is linear, only two Gauss points in that direction are required while three or four in the ξ, η directions are used for parabolic and cubic shape functions respectively.

It should be remarked here that, in fact, the integration with respect to ζ can be performed exactly if desired, thus saving computation time.[7]

16.6 Some Remarks on Stress Representation

The element properties are now defined, and the assembly and solution are standard processes.

It remains to discuss the presentation of the stresses, and this problem is of some consequence. The strains being defined in local directions, $\boldsymbol{\sigma}'$ is readily available. Such components are indeed directly of interest but as the directions of local axes are not easily visualized it is sometimes convenient to transfer the components to the global system using the following expression

$$\begin{bmatrix} \sigma_x & \tau_{xy} & \tau_{xz} \\ \tau_{xy} & \sigma_y & \tau_{yz} \\ \tau_{xz} & \tau_{yz} & \sigma_z \end{bmatrix} = \boldsymbol{\theta} \begin{bmatrix} \sigma_{x'} & \tau_{x'y'} & \tau_{x'z'} \\ \tau_{x'y'} & \sigma_{y'} & \tau_{y'z'} \\ \tau_{x'z'} & \tau_{y'z'} & 0 \end{bmatrix} \boldsymbol{\theta}^{\mathrm{T}}. \tag{16.20}$$

If the stresses are calculated at a nodal point where several elements meet then they are averaged.

In a general shell structure, the stresses in a global system do not, however, give a clear picture of shell surface stresses. It is thus convenient always to compute the principal stresses by a suitable transformation.

However, regarding the shell surface stresses more rationally, one may note that the shear components $\tau_{x'z'}$ and $\tau_{y'z'}$ are in fact zero there, and can indeed be made zero at the stage before converting to global components. The values directly obtained for these shear components are the average values across the section. The maximum transverse shear value occurs on the neutral axis and is equal to 1·5 times the average value.

16.7 Special Case of Axi-symmetric, Curved, Thick Shells

For axi-symmetric shells the formulation is, obviously, simplified.[1] Now the element mid-surface is defined by only two co-ordinates ξ, η and a considerable saving in computer effort is obtained.

The element now is derived in a similar manner but starting from a two-dimensional definition of Fig. 16.4.

Equations (16.1) and (16.2) are now replaced by their two-dimensional equivalents defining the relation between the co-ordinates as

$$\begin{aligned} \begin{Bmatrix} r \\ z \end{Bmatrix} &= \sum N_i(\xi)\frac{(1+\eta)}{2}\begin{Bmatrix} r_i \\ z_i \end{Bmatrix}_{\text{top}} + \sum N_i(\xi)\frac{(1-\eta)}{2}\begin{Bmatrix} r_i \\ z_i \end{Bmatrix}_{\text{bottom}} \\ &= \sum N_i(\xi)\begin{Bmatrix} r_i \\ z_i \end{Bmatrix}_{\text{mid}} + \sum N_i(\xi)\frac{\eta}{2}\,\mathbf{V}_{3i} \end{aligned} \tag{16.21}$$

with

$$\mathbf{V}_{3i} = t_i \begin{Bmatrix} \cos\phi_i \\ \sin\phi_i \end{Bmatrix}$$

in which ϕ_i is the angle defined in Fig. 16.4(*b*) and t_i the shell thickness.

Similarly the displacement definition is specified by following the lines of Eq. (16.3).

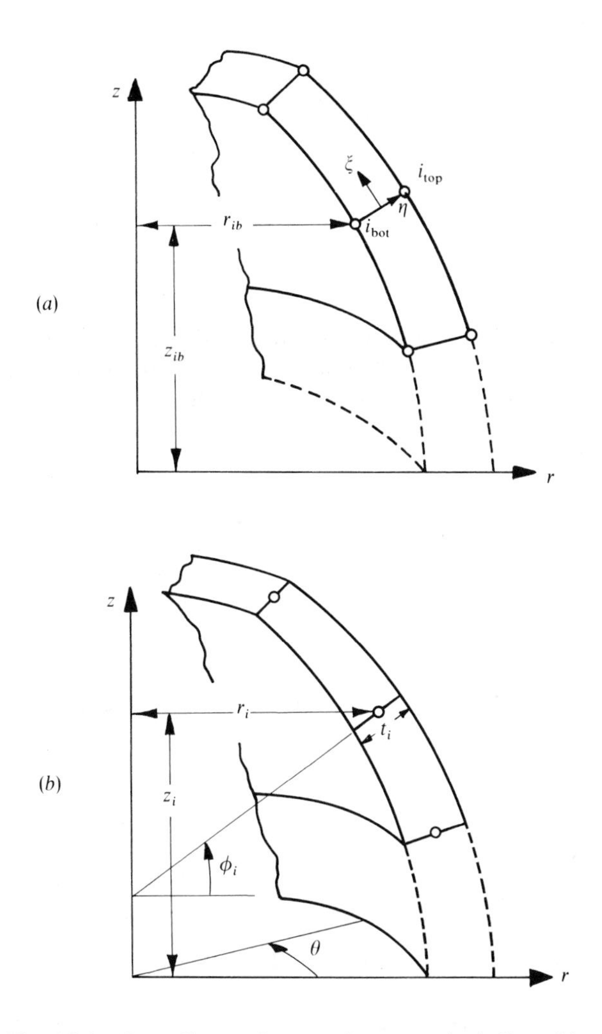

Fig. 16.4 Co-ordinates for an axi-symmetric shell problem

For generality we shall consider the case of non-symmetric loading only noting the terms which can be eliminated *a priori* for the simple case of symmetry. Indeed the decomposition into trigonometric components will be tacitly assumed as it follows precisely the lines described in Chapter 14.

Thus generally we specify the three displacement components of the nth harmonic as

$$\begin{Bmatrix} u^n \\ v^n \\ w^n \end{Bmatrix} = \begin{bmatrix} \cos n\theta & 0 & 0 \\ 0 & \cos n\theta & 0 \\ 0 & 0 & \sin n\theta \end{bmatrix}$$

$$\times \left(\sum N_i \begin{Bmatrix} u_i^n \\ v_i^n \\ w_i^n \end{Bmatrix} + \sum N_i \eta \frac{t_i}{2} \begin{bmatrix} -\sin\phi_i & 0 \\ \cos\phi_i & 0 \\ 0 & 1 \end{bmatrix} \begin{Bmatrix} \alpha_i^n \\ \beta_i^n \end{Bmatrix} \right) \quad (16.22)$$

In this α_i stands for the rotation illustrated in Fig. 16.5, u_i, etc., for the displacement of the middle surface node and β_i is the rotation about the vector tangential (approximately) to the middle surface.

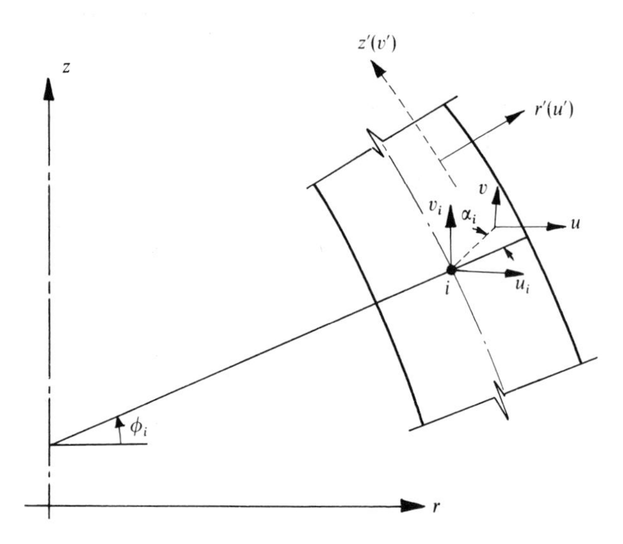

Fig. 16.5 Global displacements in an axi-symmetric shell

For the purely axi-symmetric case a further simplification arises by omitting the w terms, the first matrix of trigonometric constants and the rotation β_i.

Local strains are now more conveniently defined by the relationship Eq. (16.7) written in global cylindrical co-ordinates.

$$\boldsymbol{\varepsilon} = \begin{Bmatrix} \varepsilon_r \\ \varepsilon_z \\ \varepsilon_\theta \\ \gamma_{rz} \\ \gamma_{r\theta} \\ \gamma_{z\theta} \end{Bmatrix} = \begin{Bmatrix} \dfrac{\partial u}{\partial r} \\ \dfrac{\partial v}{\partial z} \\ \dfrac{u}{r} + \dfrac{1}{r}\dfrac{\partial w}{\partial \theta} \\ \dfrac{\partial u}{\partial z} + \dfrac{\partial v}{\partial r} \\ \dfrac{1}{r}\dfrac{\partial u}{\partial \theta} + \dfrac{\partial w}{\partial r} - \dfrac{w}{r} \\ \dfrac{1}{r}\dfrac{\partial v}{\partial \theta} + \dfrac{\partial w}{\partial z} \end{Bmatrix}. \tag{16.23}$$

These strains are transformed to the local co-ordinates and component normal to η = constant neglected.

The $\mathbf{D}'$ matrix takes, however, a form identical to that defined by Eq. (16.9). For the axi-symmetric case once again the appropriate terms are simply deleted.

All the transformations follow the pattern described in previous sections and need not be further commented upon except perhaps to remark that they are now only carried out between sets of directions ξ, η; r, z; and r', z' involving only two variables.

Similarly the integration of element properties is carried out numerically with respect to ξ and η only, noting, however, that the volume element is

$$\mathrm{d}x\,\mathrm{d}y\,\mathrm{d}z = \det|\mathbf{J}|\,\mathrm{d}\xi\,\mathrm{d}\eta\,r\,\mathrm{d}\theta. \tag{16.24}$$

By suitable choice of shape functions $N_i(\xi)$, straight parabolic or cubic shapes of variable thickness elements can be used as shown in Fig. 16.6.

16.8 Special Case of Thick Plates

The transformations necessary in this chapter are somewhat involved and indeed the programming needed is sophisticated. However, the application of the principle involved is available for thick plates and the reader is advised to test his comprehension on such a simple problem.

Here the following obvious simplifications arise

(1) $\zeta = z$ and unit vectors $\mathbf{v}_{1i}, \mathbf{v}_{2i}, \mathbf{v}_{3i}$ can be taken in directions of x, y, and z axes respectively.

(2) α_i and β_i are simply the rotations θ_y and θ_x (*vide* Chapter 10).

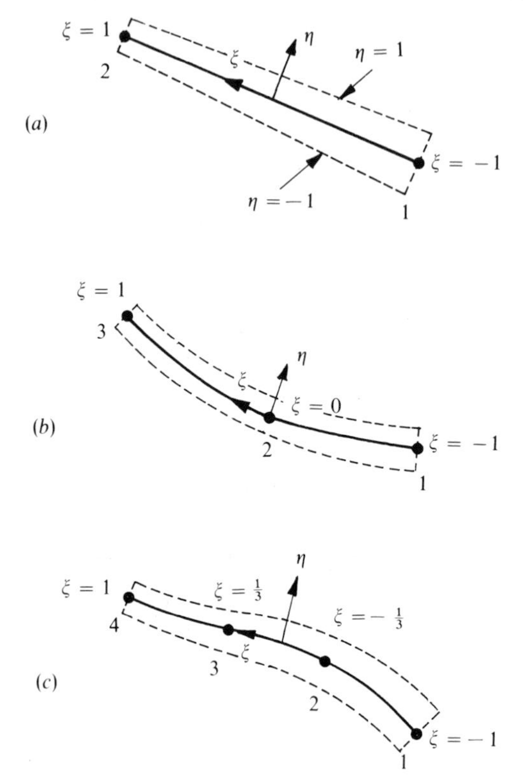

Fig. 16.6 Linear (*a*), parabolic (*b*), and cubic (*c*), axi-symmetric shell elements

(3) it is no longer necessary to transform stress and strain components to a local system of axis x' y' z' and global definitions can be used throughout. For elements of this type, numerical integration can be avoided and as an exercise the reader is encouraged to derive stiffnesses, etc., for say linear, rectangular, elements. He will then find a form identical to that derived in Chapter 11 with an independent displacement and rotation interpolation and using shear constraints. This shows the essential identity of the alternative procedures.

16.9 **Convergence**

While in three-dimensional analysis it is possible to talk about absolute convergence to the true exact solution of the elasticity problem, in equivalent plate and shell problems such a convergence cannot happen. The so-called convergent solution of a plate bending problem converges, as the element size decreases, only to the exact solution of the approximate model implied in the formulation. Thus, here again convergence of the above formulation will only occur to the exact solution constrained by the requirement that plane sections remain plane during deformation.

In elements of finite size it will be found that pure bending deformation modes are accompanied always by some shear stresses which in fact do not exist in the conventional, thin plate or shell bending theory. Thus large elements deforming mainly under bending action (as would be the case of the shell element degenerated to a flat plate) tend to be appreciably too stiff. In such cases certain limits of the ratio of side of element to its thickness have to be imposed.

However, it will be found that such restrictions can be relaxed by a simple expedient of *reducing the integration order*.[7]

Figure 16.7 shows, for instance, the application of the parabolic element to a square plate situation. Here results for integration with 3×3 and 2×2 Gauss points are given and results plotted for different thickness to span ratios. For reasonably thick situations, the results are similar and both give the additional shear deformation not available in thin plate theory,

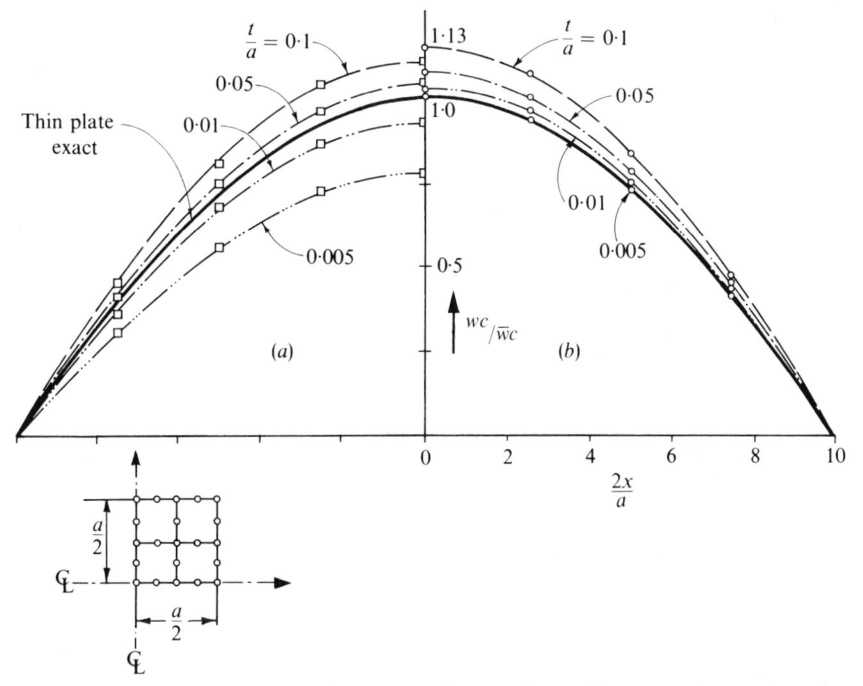

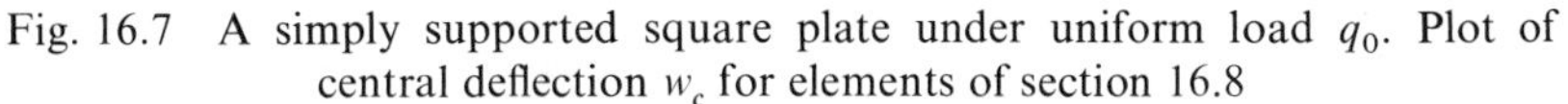

Fig. 16.7 A simply supported square plate under uniform load q_0. Plot of central deflection w_c for elements of section 16.8

(*a*) with 3×3 Gauss point integration
(*b*) ,, 2×2 (reduced), Gauss point integration

$\bar{w}_c$ central deflection for thin plate theory

but for thin plates the results with the more exact integration tend to diverge rapidly from the now correct thin plate results whereas the reduced integration still gives excellent results. The reasons for this improved performance are fully discussed in Chapter 11.

The reader is referred to Chapter 11 for further plate examples using different types of shape functions.

16.10 Some shell examples

A limited number of examples which show the accuracy and range of application of the shell formulation just described will be given. For a fuller selection the reader is referred to references 1, 2, 3, 7, and 10.

Spherical dome under uniform pressure. The 'exact' solution of shell theory is known for this axi-symmetrical problem illustrated in Fig. 16.8. Twenty-four cubic-type elements were used here. These were of graded size more closely spaced towards the abutments.

The solution appears to be more accurate than the 'exact' one distinguishing between the application of pressure on the inner and outer surfaces.

Edge loaded cylinder. A further axi-symmetric example is shown in Fig. 16.9 to study the effect of subdivision. Two, six, or fourteen elements of unequal length are used and the results for both the latter subdivisions are almost coincident with the exact solution. Even the two-element solution gives reasonable results and departs only in the vicinity of the loaded edge.

Once again the solutions are basically identical to those derived with independent slope and displacement interpolation in the manner presented in Chapter 14 (section 14.7).

Cylindrical vault. This is a test example of application of the full process to a shell in which bending action is severe, due to supports restraining deflection at the ends (see also pages 349–51).

In Fig. 16.10 the geometry, physical details of the problem, and subdivision are given while in Fig. 16.11 the comparison of the effects of 3×3 and 2×2 integration using parabolic elements is shown on the displacements calculated. Both integrations result as expected in convergence. For the more exact integration, this is rather slow while, with reduced integration order, very accurate results are obtained even with one element. This example illustrates most dramatically the advantages of this simple expedient and is described more fully in references 7 and 9. The 'exact' solution for this problem is one derived on more conventional lines by Scordelis and Lo.[11]

The improved convergence of displacements is incidentally matched by the convergence of stress components.

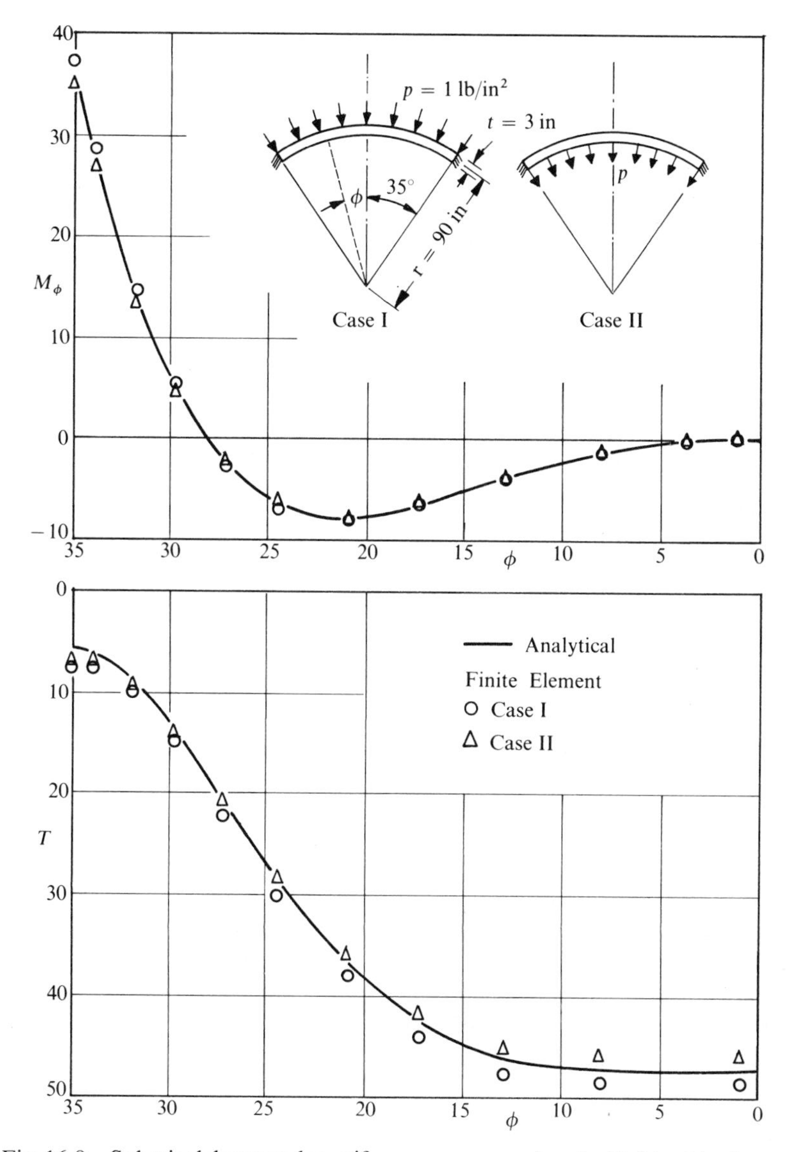

Fig. 16.8 Spherical dome under uniform pressure analysed with 24 cubic elements (first elements subtends an angle of 0·1° from fixed end, others in arithmetic progression)

M_ϕ = meridional bending moment in lb/in
T = hoop force lb/in
$\nu = \frac{1}{6}$

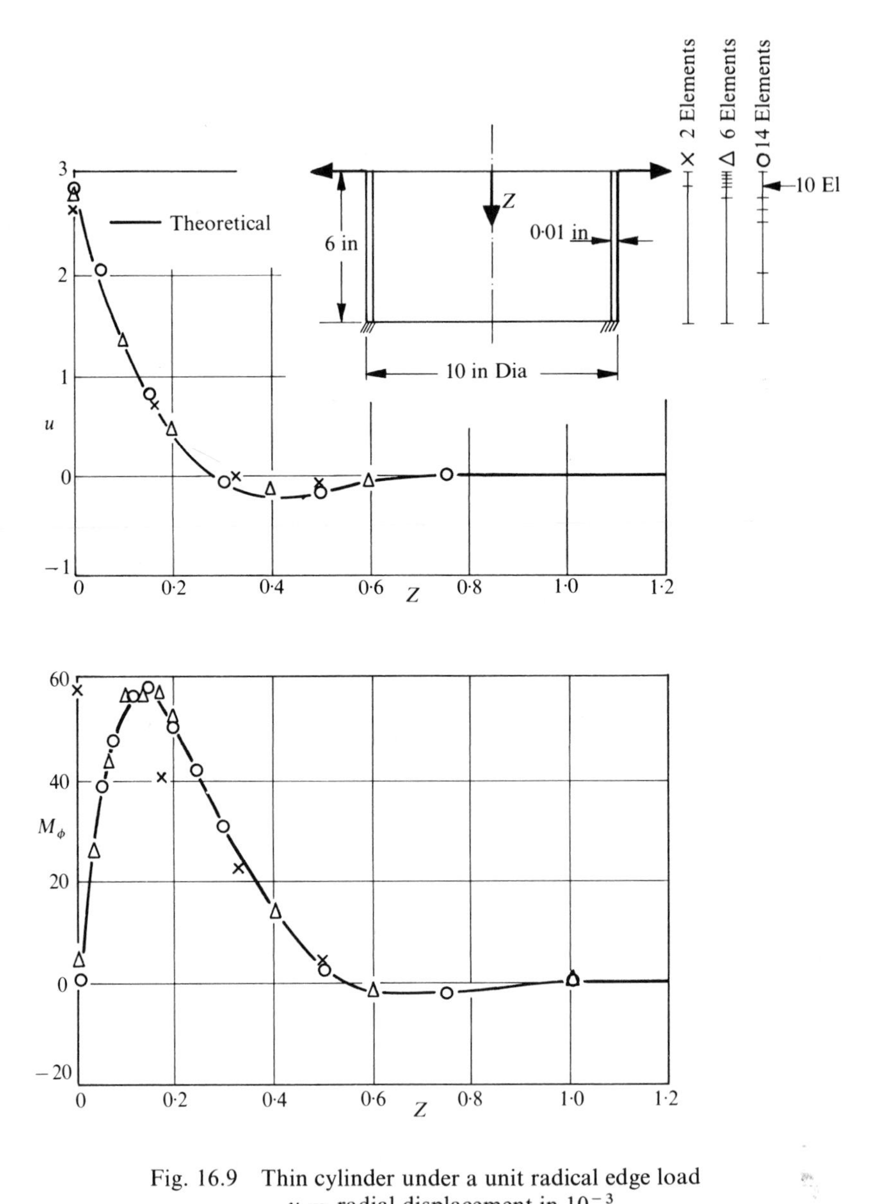

Fig. 16.9 Thin cylinder under a unit radical edge load
u = radial displacement in 10^{-3}
M_ϕ = meridional moment in lb/in
$E = 10^7$ lb/in^2
$\nu = 0{\cdot}3$

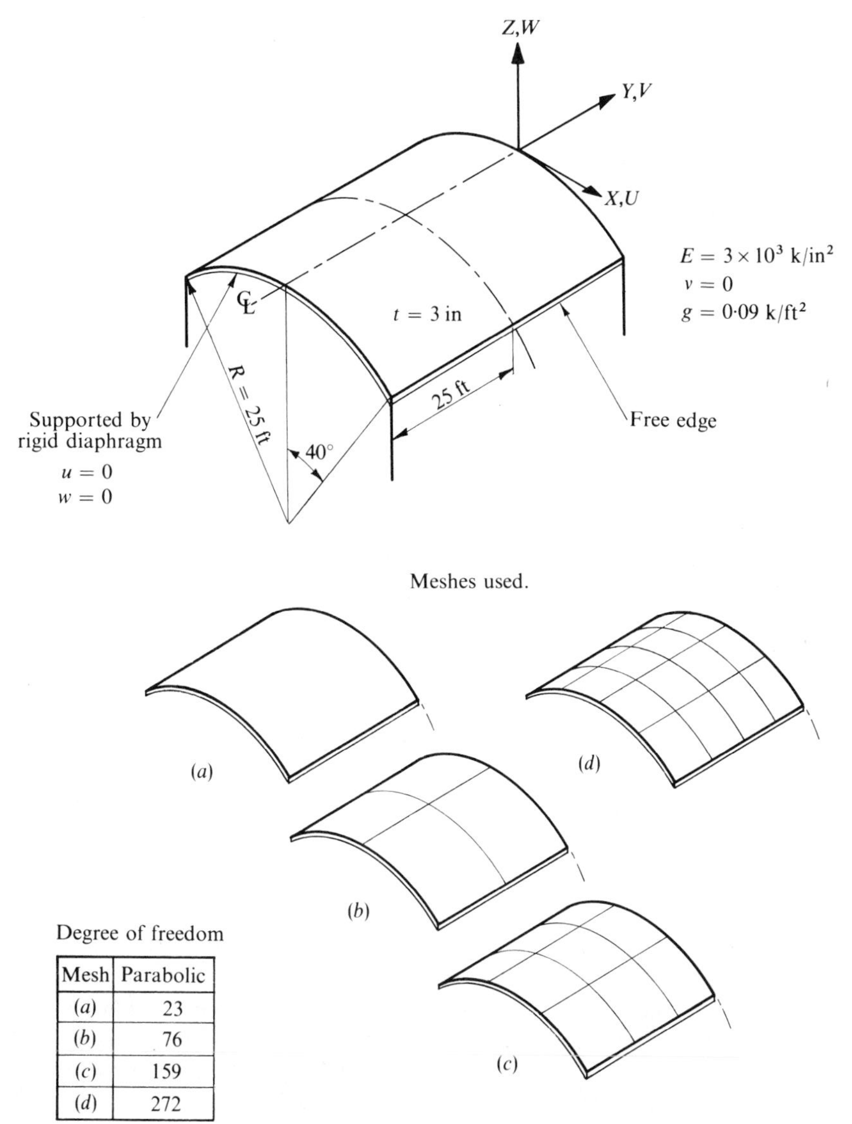

Mesh	Parabolic
(*a*)	23
(*b*)	76
(*c*)	159
(*d*)	272

Fig. 16.10 Cylindrical shell example, self-weight behaviour

Cooling tower. The cooling tower already referred to in Chapter 13 (Figs. 13.9–13.10) has been again analysed dividing the axi-symmetric shell into fifteen elements of cubic type. Using ten harmonics the unsymmetric (wind) loading is adequately represented and the results coincide

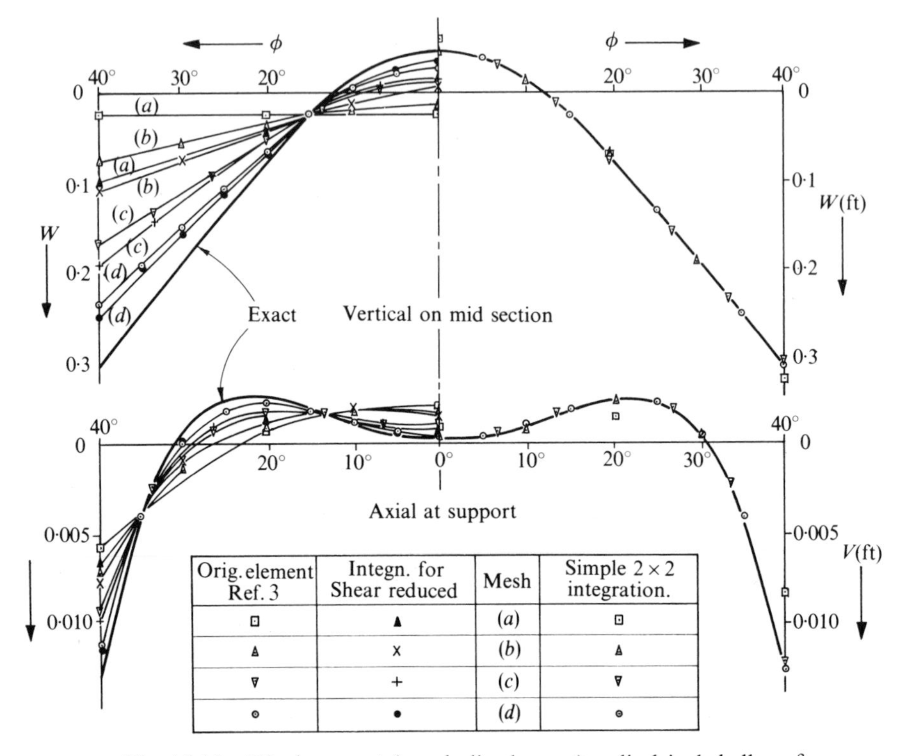

Fig. 16.11 Displacement (parabolic element), cylindrical shell roof

with those of the test analysis against which the results of Chapter 13 are compared so that additional plots are not necessary.

Curved dams. All the previous examples were rather thin shells and indeed demonstrated the applicability of the process to these situations. At the other end of the scale this formulation has been applied to the doubly curved dams illustrated in Chapter 9 (Fig. 9.8). Indeed exactly the same subdivision was again used and *results reproduced almost exactly those of the three-dimensional solution.*[3] This remarkable result was achieved at a very considerable saving in both degrees of freedom of solution and computer time.

Clearly the range of application of this type of element is very wide.

Pipe penetration[12] *and spherical cap.*[10] The last two examples of Fig. 16.12/13 and 16.14 illustrate applications in which irregular shape of elements are used. Both illustrate practical problems of some interest and show that with reduced integration a useful and very general shell element is available even when the elements are quite distorted.

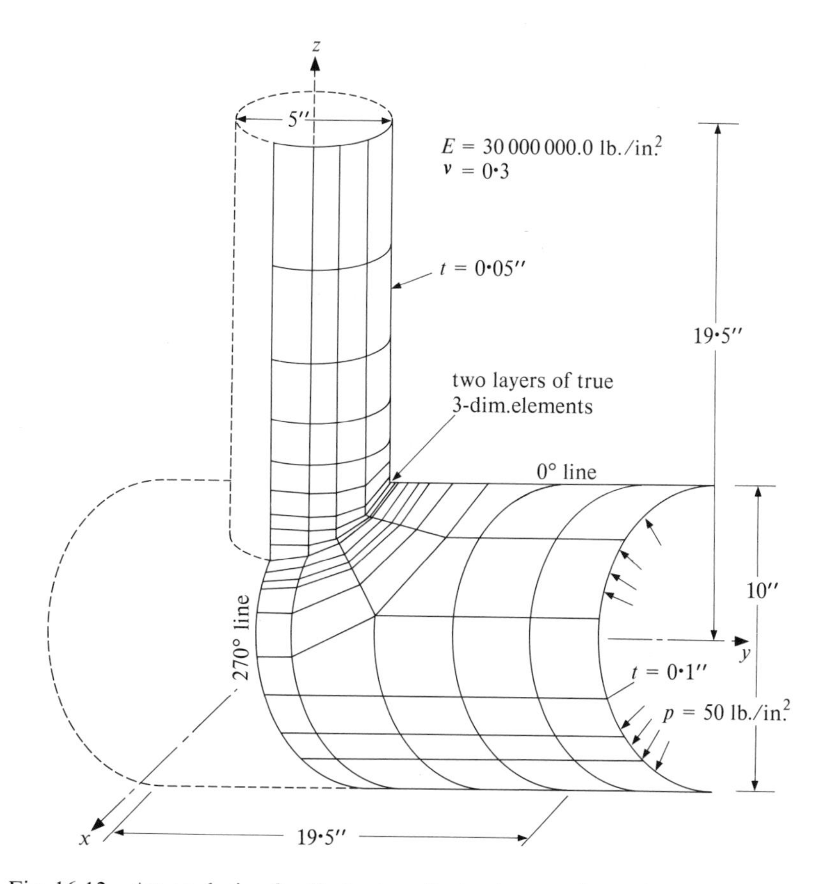

Fig. 16.12 An analysis of cylinder-to-cylinder intersection using reduced integration shell type elements.[12]

16.11 **Concluding Remarks**

The elements described in this chapter using a degeneration of solid elements have been noted in plate and axi-symmetric problems to be identical with those in which an independent slope and displacement interpolation was directly used in the middle plane in Chapters 11 and 14. For the general curved shell the analogy is less obvious and the approach used here is relatively simple. However, the possibilities of alternative approach are now being investigated.

We should note from Chapter 11 that on many occasions bilinear

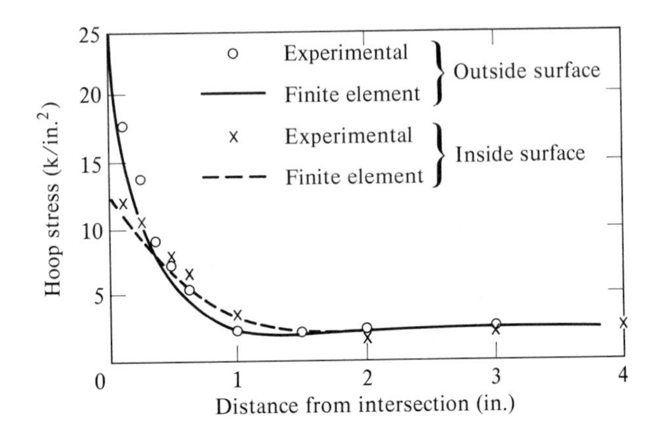

(*a*)

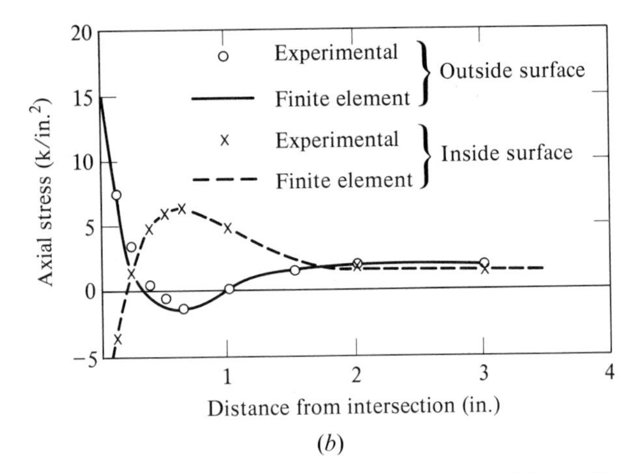

(*b*)

Fig. 16.13 Cylinder-to-cylinder intersections of Fig. 16.12. (*a*) Hoop stresses near 0° line. (*b*) axial stresses near 0° line

interpolation and biparabolic (9 node) interpolation prove better than those using the 8-node 'serendipity' elements.

Cook[13, 14] showed that in the plate context the use of interior degrees of freedom in an 8-node element is advantageous—and this appears to be true in the shell context.

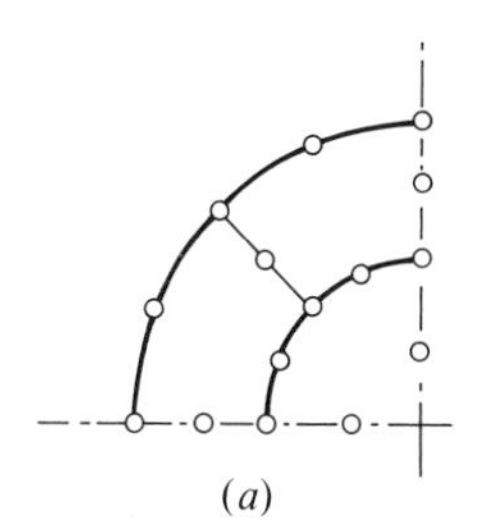

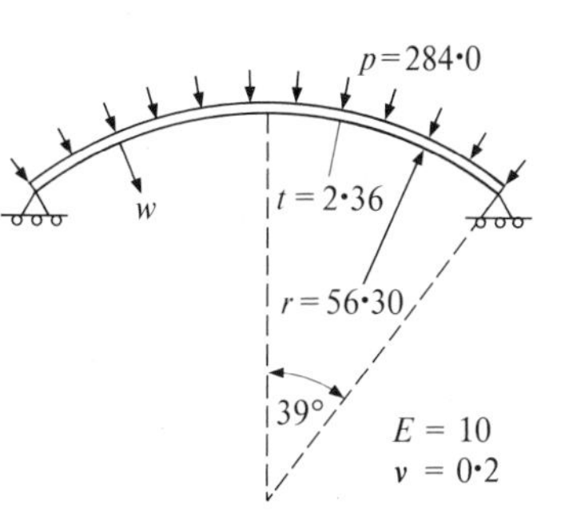

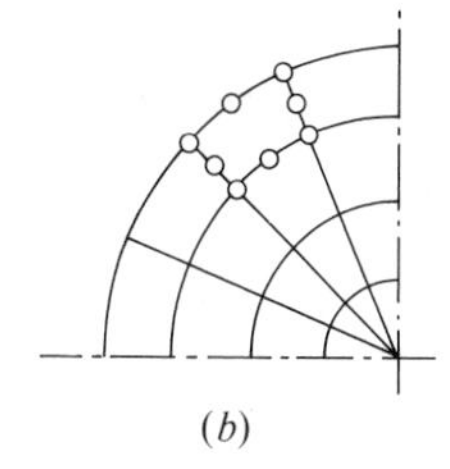

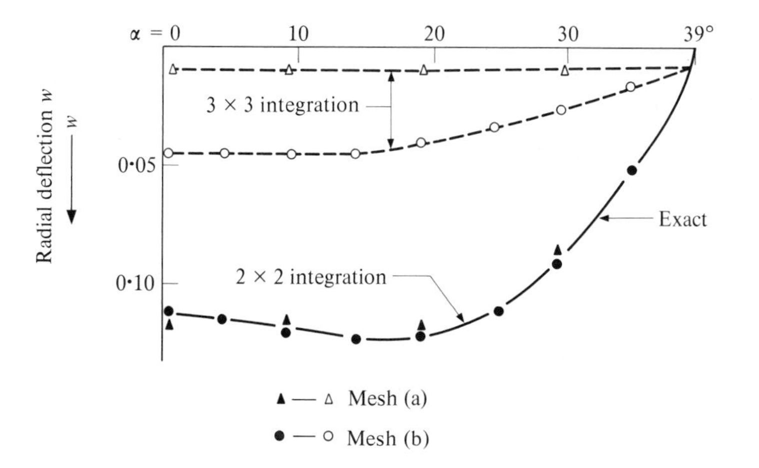

Fig. 16.14 A spherical cap analysis with irregular isoparametric shell elements using full 3×3 and reduced 2×2 integration

References

1. S. Ahmad, B. M. Irons, and O. C. Zienkiewicz, 'Curved thick shell and membrane elements with particular reference to axi-symmetric problems', *Proc. 2nd Conf. Matrix Methods in Structural Mechanics*, Wright-Patterson A.F. Base, Ohio, 1968.
2. S. Ahmad, *Curved finite elements in the analysis of solid, shell and plate structures*, Ph.D. Thesis, Univ. of Wales, Swansea, 1969.
3. S. Ahmad, B. M. Irons and O. C. Zienkiewicz, 'Analysis of thick and thin shell structures by curved elements', *Int. J. Num. Meth. Eng.*, **2**, 419–51, 1970.
4. R. D. Wood, *The application of finite element methods to geometrically non-linear analysis*, Ph.D. Thesis, Univ. of Wales, Swansea, 1973.
5. R. D. Wood and O. C. Zienkiewicz, 'Geometrically non-linear finite element analysis of beams, frames, arches and axisymmetric shells' (to be published).
6. E. L. Wilson, 'Finite elements for foundations, joints and fluids', *Proc. Conf. on Numerical Methods in Soil and Rock Mechanics*, Univ. of Karlsruhe, to be published, Wiley, 1977.
7. O. C. Zienkiewicz, J. Too and R. L. Taylor, 'Reduced integration technique in general analysis of plates and shells', *Int. J. Num. Meth. Eng.*, **3**, 275–90, 1971.
8. S. F. Pawsey and R. W. Clough, 'Improved numerical integration of thick slab finite elements', *Int. J. Num. Meth. Eng.*, **3**, 575–86, 1971.
9. S. F. Pawsey, Dept. of Structural Mechanics, Ph.D. Thesis, Univ. of California, Berkeley, 1970.
10. J. J. M. Too, *Two dimensional, plate, shell and finite prism isoparametric elements and their applications*, Ph.D. Thesis, Univ. of Wales, Swansea, 1971.
11. A. C. Scordelis and K. S. Lo, 'Computer analysis of cylindrical shells', *J. Am. Concr. Inst.*, **61**, 539–61, 1969.
12. S. A. Bakhrebah and W. C. Schnobrich, *Finite element analysis of intersecting cylinders*, Univ. of Illinois, Civil Eng. Studies, UILU-ENG-73-2018 1973.
13. R. D. Cook, 'More on reduced integration and isoparametric elements', *Int. J. Num. Meth. Eng.*, **5**, 141–2, 1972.
14. R. D. Cook, *Concepts and Applications Finite Element Analysis*, Wiley, 1974.

17. Steady-State Field Problems—Heat Conduction, Electric and Magnetic Potential, Fluid Flow, etc.

17.1 Introduction

While, in detail, most of the previous chapters dealt with problems of an elastic continuum the general procedures can be applied to a variety of physical problems. Indeed some such possibilities have been indicated in Chapter 3 and here more detailed attention will be given to a particular but wide class of such situations.

Primarily we shall deal with situations governed by the general 'quasi-harmonic' equation, the particular cases of which are the well-known Laplace and Poisson equations.[1–6] The range of physical problems falling into this category is large. To list but a few frequently encountered in engineering practice we have:

- Heat conduction
- Seepage through porous media
- Irrotational flow of ideal fluids
- Distribution of electrical (or magnetic) potential
- Torsion of prismatic shafts
- Bending of prismatic beams, etc.
- Lubrication of pad bearings.

The formulation developed in this chapter is applicable equally to all, and hence little reference will be made to the actual physical quantities. Isotropic or anisotropic regions can be treated with equal ease.

Two-dimensional problems are discussed in the first part of the chapter. A generalization to three dimensions follows. It will be observed that the same, C_0, 'shape functions' as those used previously in two- or three-dimensional formulation of elasticity problems will again be encountered. The main difference will be that now only one unknown scalar quantity (the unknown function) is associated with each point in space. Previously, several unknown quantities, represented by the displacement vector, were sought.

In Chapter 3 we indicated both the 'weak form' and a variational principle applicable to the Poisson and Laplace equations (*vide* sections 3.3 and 3.10.1). In the following sections we shall generalize these approaches to a general, quasi-harmonic equation and indicate the ranges of applicability of a *single, unified, approach* by which one computer program can solve a large variety of physical problems.

17.2 The General Quasi-harmonic Equation

17.2.1 *The general statement.* In many physical situations we are concerned with the *diffusion* or flow of some quantity such as heat, mass or a chemical, etc. In such problems the rate of transfer per unit area, $\mathbf{q}$ can be written in terms of its Cartesian components as

$$\mathbf{q}^{\mathrm{T}} = [q_x, q_y, q_z]. \tag{17.1}$$

If the rate at which the relevant quantity is generated (or removed) per unit volume is Q, then for steady-state flow the balance or continuity requirement gives

$$\frac{\partial q_x}{\partial x}+\frac{\partial q_y}{\partial y}+\frac{\partial q_z}{\partial z} = Q. \tag{17.2}$$

Introducing the gradient operator

$$\boldsymbol{\nabla} = \begin{Bmatrix} \dfrac{\partial}{\partial x} \\ \dfrac{\partial}{\partial y} \\ \dfrac{\partial}{\partial z} \end{Bmatrix} \tag{17.3}$$

we can write the above as

$$\boldsymbol{\nabla}^{\mathrm{T}}\mathbf{q}-Q = 0. \tag{17.4}$$

Generally the rates of flow will be related to *gradients* of some potential quantity ϕ. This may be temperature in the case of heat flow, etc. A very general relationship will be of a form

$$\mathbf{q} = \begin{Bmatrix} q_x \\ q_y \\ q_z \end{Bmatrix} = -\mathbf{k}\begin{Bmatrix} \dfrac{\partial \phi}{\partial x} \\ \dfrac{\partial \phi}{\partial y} \\ \dfrac{\partial \phi}{\partial z} \end{Bmatrix} = -\mathbf{k}\,\boldsymbol{\nabla}\phi \tag{17.5}$$

where **k** is a three by three matrix. This is generally of a symmetric form due to energy arguments.

The final governing equation for the 'potential' ϕ is obtained by substitution of Eq. (17.5) into (17.4), leading to

$$\boldsymbol{\nabla}^{\mathrm{T}}\mathbf{k}\boldsymbol{\nabla}\phi + Q = 0 \tag{17.6}$$

which has to be solved in the domain Ω. On the boundaries of such a domain we shall usually encounter one or other of the following conditions:

(*a*) on Γ_ϕ

$$\phi = \bar{\phi} \tag{17.7a}$$

i.e., the potential is specified, or

(*b*) on Γ_q the normal component of flow, q_n, is given as

$$q_n = \bar{q} + \alpha\phi \tag{17.7b}$$

where α is a transfer or radiation coefficient.

As

$$q_n = \mathbf{q}^{\mathrm{T}}\mathbf{n}; \qquad \mathbf{n}^{\mathrm{T}} = [n_x, n_y, n_z]$$

where **n** is a vector of direction cosines of the normal to the surface, this condition can immediately be rewritten as

$$-(\mathbf{k}\ \boldsymbol{\nabla}\phi)^{\mathrm{T}}\mathbf{n} - \bar{q} - \alpha\phi = 0 \tag{17.7c}$$

in which $\bar{q}$ and α are given.

17.2.2 *Particular forms.* If we consider the general statement of Eq. (17.5) as being determined for an arbitrary set of co-ordinate axes x, y, z we shall find that it is always possible to determine locally another set of axes x', y', z' with respect to which the matrix $\mathbf{k}'$ becomes diagonal. With respect to such axes we have

$$\mathbf{k}' = \begin{bmatrix} k_{x'} & 0 & 0 \\ 0 & k_{y'} & 0 \\ 0 & 0 & k_{z'} \end{bmatrix} \tag{17.8}$$

and the governing equation (Eq. (17.6)) can be written (now dropping the prime)

$$\frac{\partial}{\partial x}\left(k_x \frac{\partial \phi}{\partial x}\right) + \frac{\partial}{\partial y}\left(k_y \frac{\partial \phi}{\partial y}\right) + \frac{\partial}{\partial z}\left(k_z \frac{\partial \phi}{\partial z}\right) + Q = 0 \tag{17.9}$$

with a suitable change of boundary conditions.

Lastly, for an isotropic material we can write

$$\mathbf{k} = kI \tag{17.10}$$

where I is an identity matrix. This leads to simple form of Eq. (3.10) which was discussed in detail in Chapter 3.

17.2.3 *Weak form of general quasi harmonic equation* [*Eq.* (*17.6*)]. Following the principles of Chapter 3, section 3.2, we can obtain the weak form of Eq. (17.6) by writing that

$$\int_{\Omega} v[\nabla^{T}\mathbf{k}\nabla\phi + Q]\,d\Omega - \int_{\Gamma_q} v[(\mathbf{k}\nabla\phi)^{T}\mathbf{n} - \bar{q} - \alpha\phi]\,d\Gamma = 0 \tag{17.11}$$

for all functions v which are zero on Γ_{ϕ}.

Integration by parts (see Appendix 3) will result in the following weak statement which is equivalent to satisfying the governing equations and the *natural* boundary conditions (17.7b).

$$\int_{\Omega} \nabla^{T} v\mathbf{k}\nabla\phi\,d\Omega - \int_{\Omega} vQ\,d\Omega - \int_{\Gamma_q} v(\alpha\phi + \bar{q})\,d\Gamma = 0. \tag{17.12}$$

The *forced* boundary condition (17.7a) still needs to be imposed.

17.2.4 *The variational principle*. We shall leave as an exercise to the reader the verification that the functional

$$\Pi = \tfrac{1}{2}\int_{\Omega} (\nabla\phi)^{T}\mathbf{k}\nabla\phi\,d\Omega - \int_{\Omega} Q\phi\,d\Omega - \tfrac{1}{2}\int_{\Gamma_q} \alpha\phi^{2}\,d\Gamma - \int_{\Gamma_q} \bar{q}\phi\,d\Gamma \tag{17.13}$$

gives on minimization (subject to constraint of Eq. (17.7a)) the satisfaction of the original problem set in Eqs. (17.6) and (17.7).

The algebraic manipulations required to verify the above principle follow precisely the lines of section 3.10 of Chapter 3 and can be carried out as an exercise.

17.3 **Finite Element Discretization**

This can now proceed on the assumption of a trial function expansion

$$\phi = \sum N_i a_i = \mathbf{N}\mathbf{a} \tag{17.14}$$

using either the weak formulation of Eq. (17.12) or the variational statement of Eq. (17.13). If, in the first, we take

$$v = N_i \tag{17.15}$$

according to the Galerkin principle, an identical form will arise with that obtained from the minimization of the variational principle.

Substituting thus Eq. (17.15) into (17.12) we have a typical statement giving

$$\left[\int_{\Omega} \nabla^{\mathrm{T}} N_i \mathbf{k}\, \nabla \mathbf{N}\, \mathrm{d}\Omega - \int_{\Gamma_q} N_i \alpha \mathbf{N}\, \mathrm{d}\Gamma\right] \mathbf{a} -$$

$$-\int_{\Omega} N_i Q\, \mathrm{d}\Omega - \int_{\Gamma_q} N_i \bar{q}\, \mathrm{d}\Gamma = 0 \qquad (i = 1, \ldots, n) \qquad (17.16)$$

or a set of standard discrete equations of the form

$$\mathbf{Ha} + \mathbf{f} = 0 \qquad (17.17)$$

with

$$H_{ij} = \int_{\Omega} \nabla^{\mathrm{T}} N_i \mathbf{k}\, \nabla N_j\, \mathrm{d}\Omega - \int_{\Gamma_q} N_i \alpha N_j\, \mathrm{d}\Gamma$$

$$f_i = -\int_{\Omega} N_i Q\, \mathrm{d}\Omega - \int_{\Gamma_q} N_i \bar{q}\, \mathrm{d}\Gamma$$

on which prescribed values of $\bar{\phi}$ have to be imposed on boundaries Γ_{ϕ}.

We note now that an additional 'stiffness' is contributed on boundaries for which a radiation constant α is specified but that otherwise a complete analogy with the elastic-structural problem exists.

Indeed in a computer program the same standard operations will be followed even including an evaluation of quantities analogous to the stresses. These, obviously, are the flow rates

$$q \equiv -\mathbf{k}\, \nabla\phi = -(\mathbf{k}\, \nabla N)\mathbf{a} \qquad (17.18)$$

and, in accordance with indications of Chapter 11, these should be sampled at the optimal (integration) points according to the expansion order used.

Any of the C_0 expansions, isoparametric transformations, etc. given in Chapters 7 and 8 can again be used.

17.4 Some Economic Specializations

17.4.1 *Anisotropic and non-homogeneous media.* Clearly material properties defined by the **k** matrix can vary from element to element in a discontinuous manner. This is implied in both the weak and variational statements of the problem.

The material properties are usually known only with respect to the principal (or symmetry) axes, and if these directions are constant within the element it is convenient to use in the formulation local axes specified within each element, as shown in Fig. 17.1.

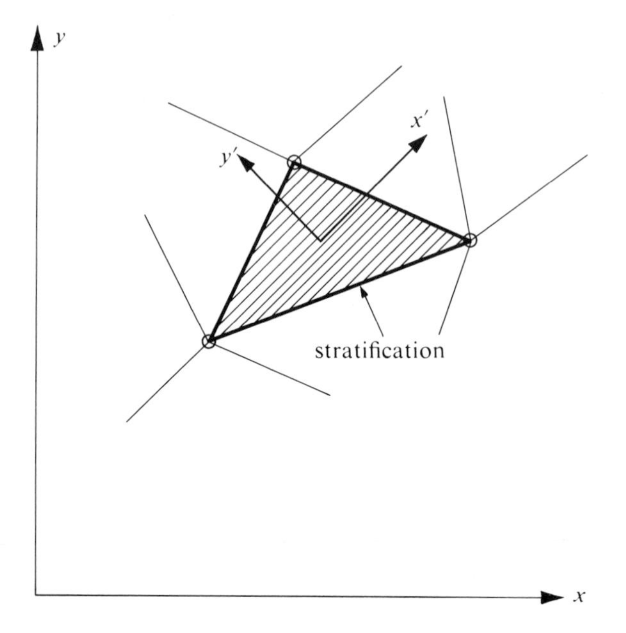

Fig. 17.1 Anisotropic material. Local co-ordinates coincide with the principal directions of stratification

With respect to such axes only three coefficients k_x, k_y, and k_z need be specified and indeed considerable computational economy is achieved as only a multiplication by a diagonal matrix is needed in formulating the coefficients of the matrix **H** [Eq. (17.17)].

It is important to note that as the parameters **a** *correspond to scalar values, no transformation of matrices computed in local co-ordinates is necessary before assembly of the global matrices.*

Thus, in most computer programs only a diagonal specification of the **k** matrix is used.

17.4.2 *Two-dimensional problem.* The general governing Eq. (17.9) specialized to local co-ordinates becomes, in two dimensions,

$$\frac{\partial}{\partial x}\left(k_x \frac{\partial \phi}{\partial x}\right) + \frac{\partial}{\partial y}\left(k_y \frac{\partial \phi}{\partial y}\right) + Q = 0 \tag{17.19}$$

On discretization by Eq. (17.17) a slightly simplified form of the matrices will now be found. Dropping the terms with α and $\bar{q}$ we can write

$$H^e_{ij} = \int_{V^e} \left(k_x \frac{\partial N_i}{\partial x}\frac{\partial N_j}{\partial x} + k_y \frac{\partial N_i}{\partial y}\frac{\partial N_j}{\partial y}\right) \mathrm{d}x\,\mathrm{d}y. \tag{17.20}$$

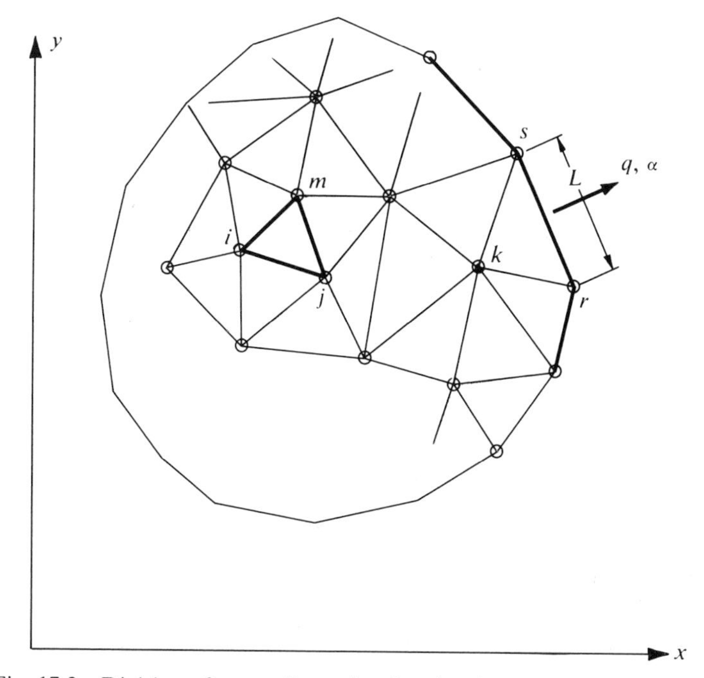

Fig. 17.2 Division of a two-dimensional region into triangular elements

No further discussion at this point appears necessary. However, it may be worth while to particularize here to the most simple yet very useful triangular element, Fig. 17.2.

With

$$N_i = (a_i + b_i x + c_i y)/2\Delta$$

as in Eq. (4.8) of Chapter 4, we can write down the element 'stiffness' matrix as

$$\mathbf{H}^e = \frac{k_x}{4\Delta}\begin{bmatrix} b_i b_i & b_i b_j & b_i b_m \\ & b_j b_j & b_j b_m \\ \text{sym.} & & b_m b_m \end{bmatrix} + \frac{k_y}{4\Delta}\begin{bmatrix} c_i c_i & c_i c_j & c_i c_m \\ & c_j c_j & c_j c_m \\ \text{sym.} & & c_m c_m \end{bmatrix}. \tag{17.21}$$

The load matrices follow a similar simple pattern and thus for instance the reader can show that due to Q we have

$$\mathbf{f}^e = -\frac{Q\Delta}{3}\begin{Bmatrix} 1 \\ 1 \\ 1 \end{Bmatrix} \tag{17.22}$$

a very simple (almost 'obvious') result.

Alternatively the equation may be specialized to cylindrical co-ordinates and used for solution of axi-symmetric situations. Now the differential equation is

$$\frac{\partial}{\partial r}\left(k_r r \frac{\partial \phi}{\partial r}\right)+\frac{\partial}{\partial z}\left(k_z r \frac{\partial \phi}{\partial z}\right)+Q = 0. \tag{17.23}$$

The variational principle could now be again suitably transformed but it is simpler to substitute the values $(k_r r)$ and $(k_z r)$ as modified 'conductivities' and use the previous expressions directly. Integration now will be best carried out numerically as in equivalent problems of Chapter 5.

17.5 **Examples—An Assessment of Accuracy**

It is very easy to show that by assembling explicitly worked out 'stiffnesses' of triangular element for 'regular' meshes shown in Fig. 17.3(*a*), the discretized equations are *identical* with those which can be derived by well-known finite difference methods.[7]

Obviously the solutions obtained by the two methods will be identical, and so will also be the orders of approximation.†

If an 'irregular' mesh based on a square arrangement of nodes is used a difference between the two approaches will be evident (Fig. 17.3(*b*)). This is confined to the 'load' vector $\mathbf{f}^e$. The assembled equations will show 'loads' which differ by small amounts from node to node, but the sum of which is still the same as that due to the finite difference expressions. The solutions therefore differ only locally and will represent the same averages.

In Fig. 17.4 a test comparing the results obtained on an 'irregular' mesh with a relaxation solution of the lowest order finite difference approximation is shown. Both give results of the similar accuracy, as indeed would be anticipated. However, it can be shown that in one-dimensional problems the finite element algorithm gives *exact* answers of nodes, while the finite difference method generally does not. In general, therefore, superior accuracy is available with the finite element discretization. Further advantages of the finite element process are:

(*a*) It can deal simply with non-homogeneous and anisotropic situations (particularly when the direction of anisotropy is variable).

(*b*) The elements can be graded in shape and size to follow arbitrary boundaries and to allow for regions of rapid variation of the function sought.

† This is only true in the case where the boundary values $\bar{\phi}$ only are prescribed.

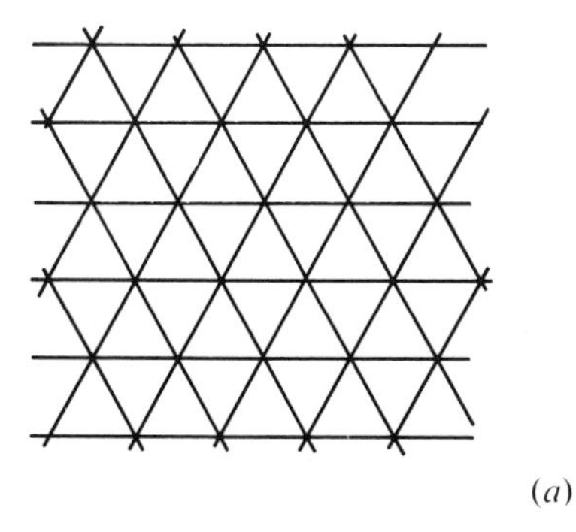

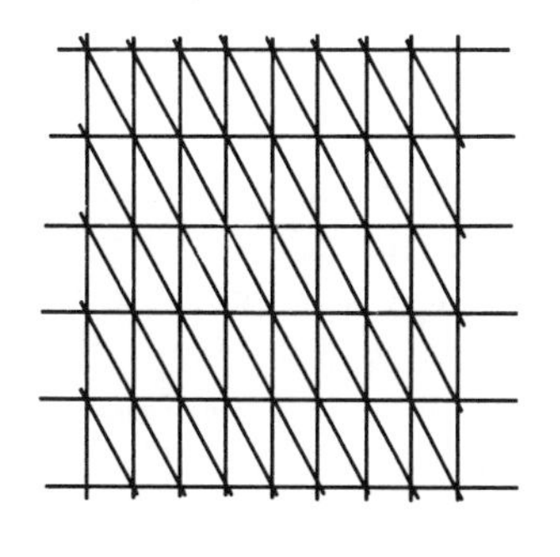

(a)

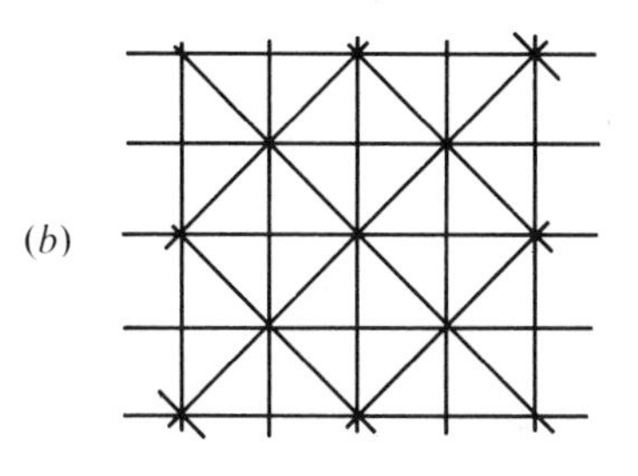

(b)

Fig. 17.3 'Regular' and 'irregular' subdivision patterns

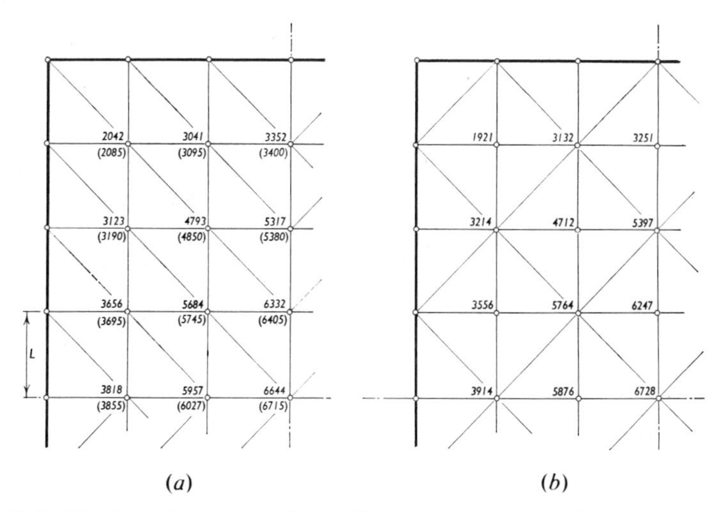

(a) (b)

Fig. 17.4 Torsion of a rectangular shaft. Numbers in parentheses show a more accurate solution due to Southwell using a 12 × 16 mesh (values of $\phi/G\theta L^2$)

(*c*) Specified gradient or 'radiation' boundary conditions are introduced naturally and with a better accuracy than in standard finite difference procedures.

(*d*) Higher order elements can be readily used to improve accuracy without complicating boundary conditions—a difficulty always arising with finite difference approximations of a higher order.

(*e*) Finally, but of considerable importance in the computer age: standard (structural) programs may be used for assembly and solution.

Two more sophisticated examples are given at this stage to illustrate the accuracy attainable in practice. The first is the problem of pure torsion of a non-homogeneous shaft illustrated in Fig. 17.5. The basic differential equation here is

$$\frac{\partial}{\partial x}\left(\frac{1}{G}\frac{\partial \phi}{\partial x}\right)+\frac{\partial}{\partial y}\left(\frac{1}{G}\frac{\partial \phi}{\partial y}\right)+2\theta=0 \tag{17.24}$$

in which ϕ is the stress function, G is the shear modulus, and θ the angle of twist per unit length of the shaft.

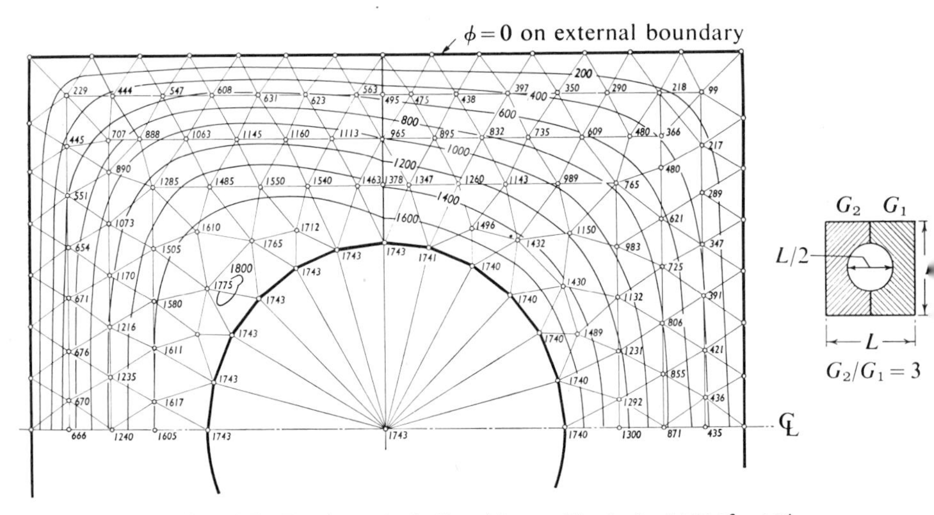

Fig. 17.5 Torsion of a hollow bi-metallic shaft. $\phi/G\theta L^2 \times 10^4$

In the finite element solution presented, the hollow section was represented by a material for which G has a value of the order of 10^{-3} compared with the other materials.† The results compare well with the contours derived from an accurate finite difference solution.[8]

† This was done to avoid difficulties due to the 'multiple connection' of the region and to permit the use of a standard program.

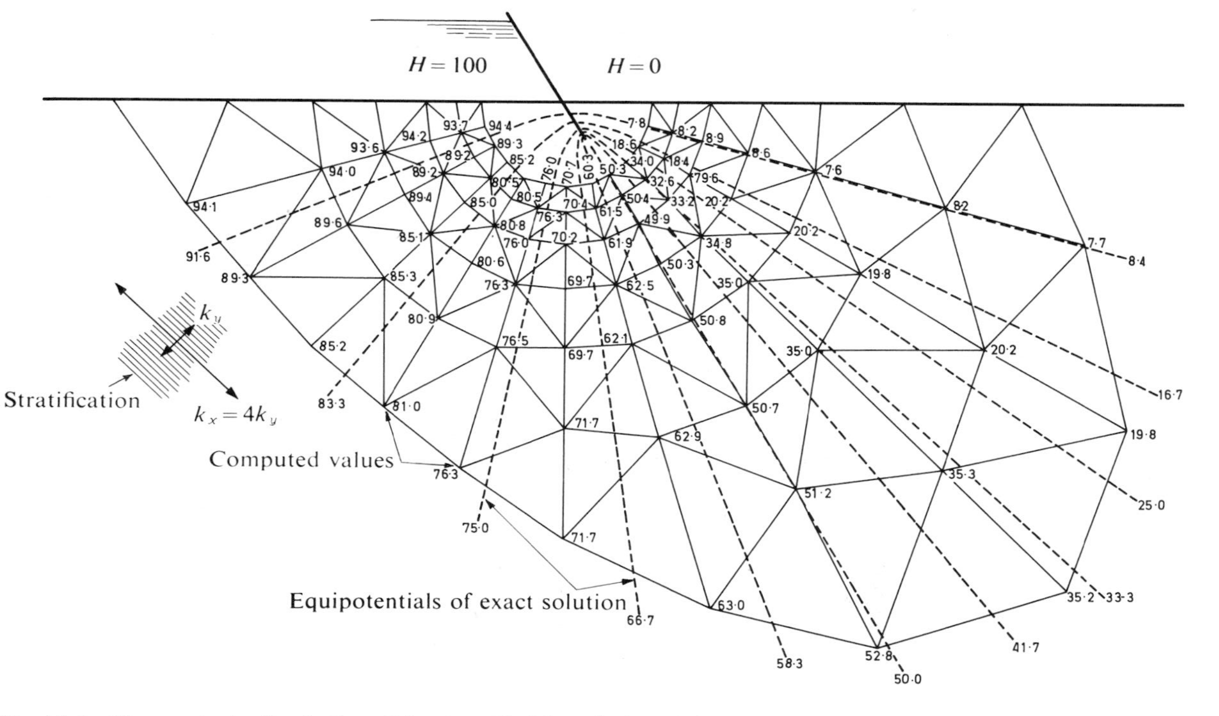

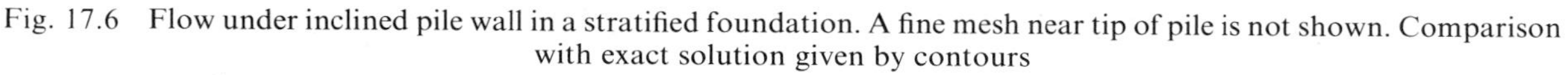
Fig. 17.6 Flow under inclined pile wall in a stratified foundation. A fine mesh near tip of pile is not shown. Comparison with exact solution given by contours

An example concerning flow through an anisotropic porous foundation is shown in Fig. 17.6.

Here the governing equation is

$$\frac{\partial}{\partial x}\left(k_x \frac{\partial H}{\partial x}\right) + \frac{\partial}{\partial y}\left(k_y \frac{\partial H}{\partial y}\right) = 0 \tag{17.25}$$

in which k_x and k_y represent the permeability coefficients in direction of the (inclined) principal axes. The answers are here compared against contours derived by an exact solution. The possibilities of the use of a graded size of subdivision are evident in this example.

17.6 Some Practical Applications

17.6.1 *Anisotropic seepage.* The first of the problems is concerned with the flow through highly non-homogeneous, anisotropic and contorted strata. The basic governing equation is still Eq. (17.25). However, a special feature has to be incorporated in the computer program to allow for changes of x' and y' principal directions from element to element.

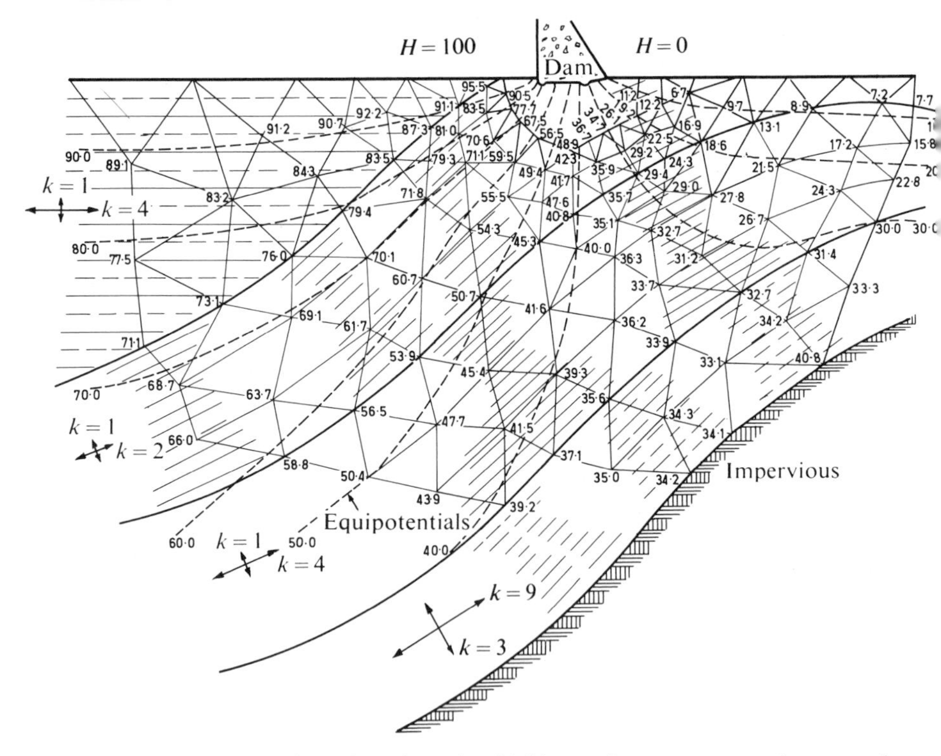

Fig. 17.7 Flow under a dam through a highly non-homogeneous and contorted foundation

No difficulties are encountered in computation, and the problem together with its solution is given in Fig. 17.7.[3]

17.6.2 *Axi-symmetric heat flow*. The axi-symmetric heat flow equation can be written in the standard form as

$$\frac{\partial}{\partial r}\left(rk\frac{\partial T}{\partial r}\right)+\frac{\partial}{\partial z}\left(rk\frac{\partial T}{\partial z}\right)=0 \qquad (17.26)$$

if no heat generation occurs. In the above, T is the temperature and k conductivity. The co-ordinates x and y are now replaced by r and z, the radial and axial distances.

In Fig. 17.8 the temperature distribution in a nuclear reactor pressure vessel[1] is shown for a steady-state heat conduction when a uniform temperature increase is applied on the inside.

17.6.3 *Hydrodynamic pressures on moving surfaces*. If a submerged surface moves in a fluid with prescribed accelerations and a small amplitude

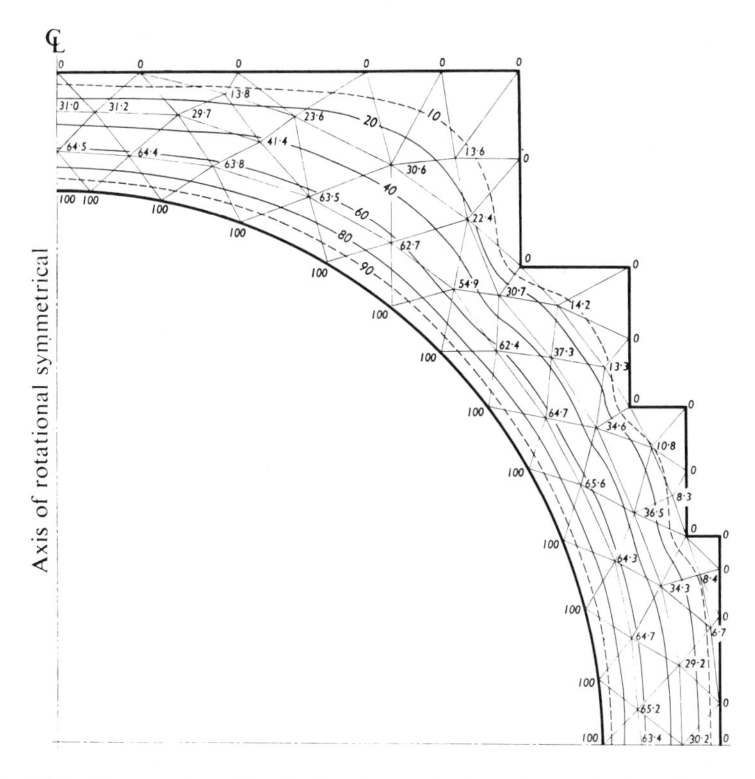

Fig. 17.8 Temperature distribution in a steady-state conduction for an axi-symmetrical pressure vessel

of movement, then it can be shown[9] that if compressibility is ignored the excess pressures, developed obey the Laplace equation (see Chapter 20, section 20.3).

$$\nabla^2 p = 0.$$

On moving (or stationary) boundaries the boundary condition is of type (*b*) (*vide* Eq. (17.7b) and is given by

$$\frac{\partial p}{\partial n} = -\rho a_n \qquad (17.27)$$

in which ρ is the density of the fluid and a_n is the normal component of acceleration of the boundary.

On free surfaces the boundary condition is (if surface waves are ignored) simply

$$p = 0. \qquad (17.28)$$

The problems clearly therefore come into the category of those discussed in this chapter.

As an example, let us consider the case of a vertical wall in a reservoir, shown in Fig. 17.9, and determine the pressure distribution at points along the surface of the wall and at the bottom of the reservoir for any prescribed motion of the boundary points 1 to 7.

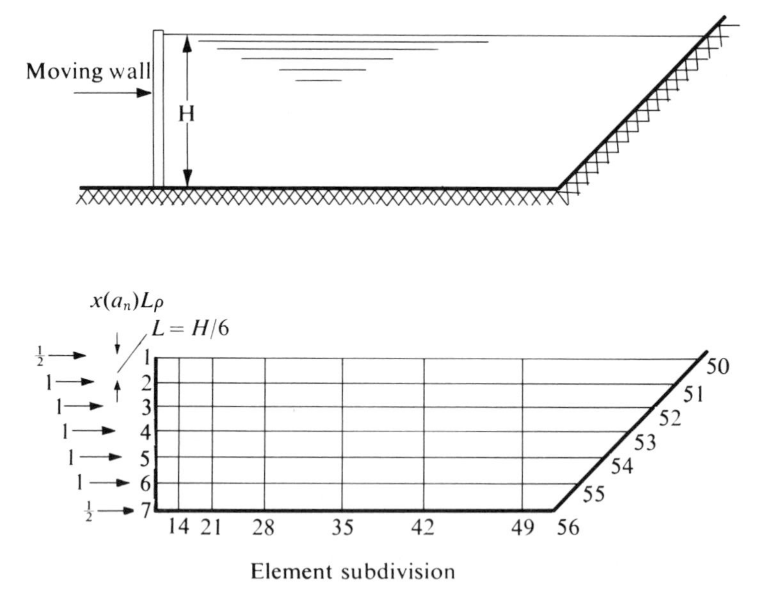

Fig. 17.9 Problem of a wall moving horizontally in a reservoir

The division of the region into elements (42 in number) is shown. Now elements of quadrilateral shape are used. So that results can be made valid for *any* acceleration system, seven separate problems are solved. In each, in turn, the portion of the boundary adjacent to the point in question is given a unit acceleration, resulting in 'loads' $\rho\frac{1}{2}L$, $\rho L, \ldots, \rho L$, $\rho\frac{1}{2}L$ being applied, in turn, to points 1 to 7. For any arbitrary distribution of acceleration the pressures developed at points 1 to 56 can be listed as a matrix dependent on acceleration of the points 1 to 7. This becomes

$$\begin{Bmatrix} p_1 \\ \vdots \\ p_7 \\ p_{14} \\ p_{21} \\ p_{28} \\ p_{35} \\ p_{42} \\ p_{49} \\ p_{56} \end{Bmatrix} = \mathbf{M} \begin{Bmatrix} a_1 \\ \vdots \\ a_7 \end{Bmatrix} \qquad (17.29)$$

in which the matrix $\mathbf{M}$ is given in Table 17.1.

TABLE 17.1

$\mathbf{M} = \rho \dfrac{H}{6}$ ×

1	0	0	0	0	0	0	0
2	0	0·7249	0·3685	0·2466	0·1963	0·1743	0·0840
3	0	0·3685	0·9715	0·5648	0·4210	0·3644	0·1744
4	0	0·2466	0·5648	1·1459	0·7329	0·5954	0·2804
5	0	0·1963	0·4210	0·7329	1·3203	0·9292	0·4210
6	0	0·1744	0·3644	0·5954	0·9292	1·5669	0·6489
7	0	0·1680	0·3488	0·5607	0·8420	1·2977	1·1459
14	0	0·1617	0·3332	0·5260	0·7548	1·0285	0·6429
21	0	0·1365	0·2754	0·4171	0·5573	0·6793	0·3710
28	0	0·0879	0·1731	0·2519	0·3187	0·3657	0·1918
35	0	0·0431	0·0838	0·1195	0·1478	0·1661	0·0863
42	0	0·0186	0·0359	0·0150	0·0626	0·0699	0·0362
49	0	0·0078	0·0150	0·0213	0·0261	0·0291	0·0151
56	0	0·0069	0·0134	0·0190	0·0232	0·0259	0·0134

$(L = H/6)$

Now for any distribution of accelerations the pressures can be found. For example, if acceleration a is uniform the pressures can be computed taking

$$\begin{Bmatrix} a_1 \\ \vdots \\ a_7 \end{Bmatrix} = \bar{a} \begin{Bmatrix} 1 \\ \vdots \\ 1 \end{Bmatrix}. \qquad (17.30)$$

The resulting pressure distribution on the wall and the bottom of the reservoir is shown in Fig. 17.10. The results for the pressures on the wall agree to within 1 per cent with the well-known, exact solution derived by Westergaard.[10]

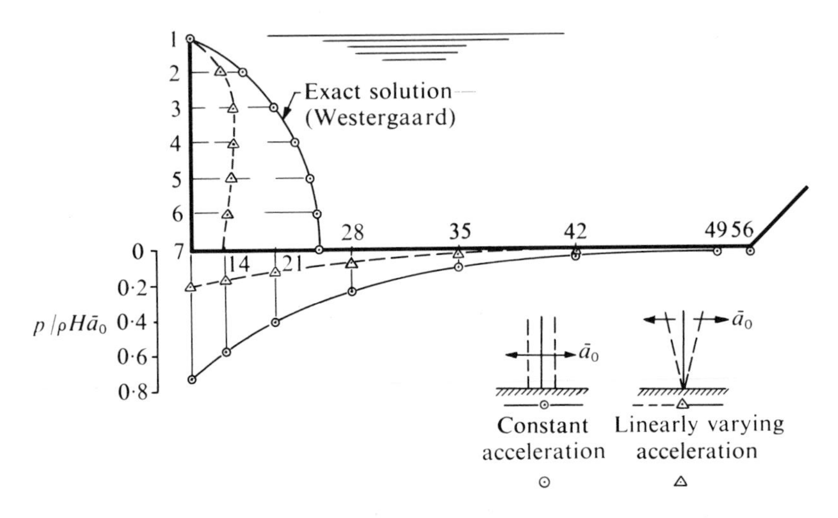

Fig. 17.10 Pressure distribution on moving wall and reservoir bottom

For any other motion the pressures can be similarly derived. If, for instance, the wall is hinged at the base and oscillates around this point with the top (point no. 1) accelerating by $\bar{a}$, then

$$\begin{Bmatrix} a_1 \\ \vdots \\ a_7 \end{Bmatrix} = \bar{a} \begin{Bmatrix} 1 \\ \frac{5}{6} \\ \frac{4}{6} \\ \vdots \\ 0 \end{Bmatrix}. \tag{17.31}$$

Again, the pressure distribution obtained is given by expression, and the results are plotted in Fig. 17.10.

The importance of deriving such an 'influence matrix' is relevant to vibration problems. If the 'wall' oscillates, then in general its acceleration is unknown. From Eq. (17.29) we can write for pressures at points 1 to 7, taking the upper part of matrix $\mathbf{M}$, say $\mathbf{M}_0$,

$$\begin{Bmatrix} p_1 \\ \vdots \\ p_7 \end{Bmatrix} = \mathbf{M}_0 \begin{Bmatrix} a_1 \\ \vdots \\ a_7 \end{Bmatrix} = \mathbf{M}_0\mathbf{a}. \tag{17.32}$$

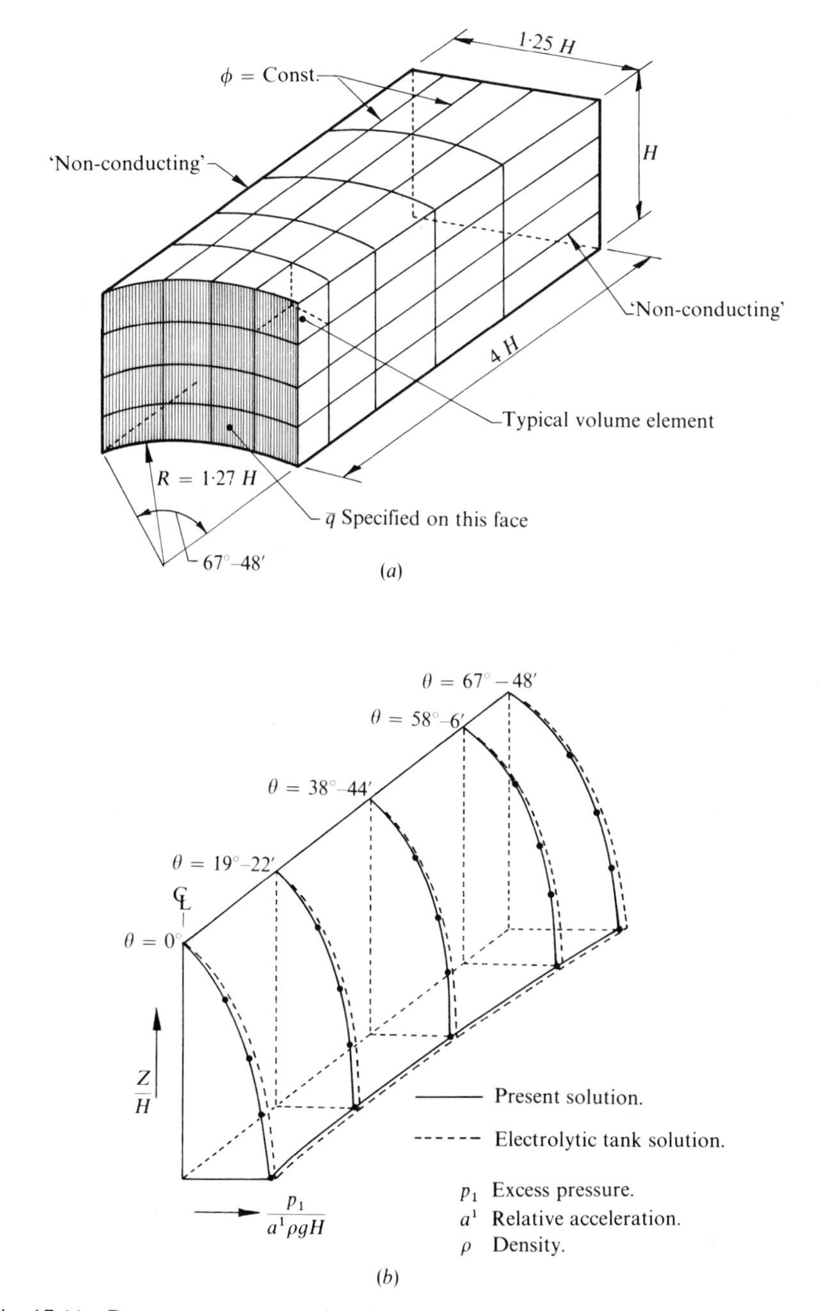

Fig. 17.11 Pressures on an accelerating surface of dam in an incompressible fluid

These pressures result in nodal forces

$$\mathbf{f} = \begin{Bmatrix} \mathbf{f}_1 \\ \vdots \\ \mathbf{f}_7 \end{Bmatrix} = \mathbf{A}\mathbf{M}_0 \begin{Bmatrix} \mathbf{a}_1 \\ \vdots \\ \mathbf{a}_7 \end{Bmatrix} \tag{17.33}$$

in which $\mathbf{A}$ is a suitable load assignment matrix and $\mathbf{a}$ represents the acceleration of nodal points on the wall. This can be coupled to the dynamic equations of the wall. This and related problems will be discussed in more detail in Chapter 20.

In Fig. 17.11 the solution of a similar problem in three dimensions is shown.[4] Here simple tetrahedral elements were used and very good accuracy obtained.

In many practical problems the computation of such simplified 'added' masses is sufficient—and the process described here has become widely used in this context.[11–13]

17.6.4 *Electrostatic and magnetostatic problems.* In this area of activity frequent need arises to determine appropriate field strengths and the governing equations are usually of the standard quasi-harmonic type discussed here. Thus the formulations are directly transferable. One of the first applications made here was to fully three-dimensional electrostatic field distributions governed by simple Laplace equations as early as 1967[4] (Fig. 17.12).

In Fig. 17.13 a similar use of triangular elements was made in the context of magnetic two-dimensional fields by Winslow[6] in 1966. These early works stimulated a considerable activity in this area and much work has now been published.[14–17]

The magnetic problem is of particular interest as its formulation usually involves the introduction of a *vector potential* with three components which leads to a formulation different from those discussed in this chapter. It is, therefore, worth while to introduce here a recent variant which allows the standard programs of this section to be utilized for this problem.[18]

In electro-magnetic theory for steady-state fields the problem is governed by Maxwell equations which are

$$\begin{aligned} \nabla^{\mathrm{T}} \times \mathbf{H} &= -\mathbf{J} \\ \mathbf{B} &= \mu \mathbf{H} \\ \nabla^{\mathrm{T}} \mathbf{B} &= 0 \end{aligned} \tag{17.34}$$

with the boundary condition specified at an infinite distance from the disturbance requiring $\mathbf{H}$ and $\mathbf{B}$ to tend to zero there. In the above $\mathbf{J}$ is a prescribed electric current density confined to conductors, $\mathbf{H}$ and $\mathbf{B}$ are vector quantities with three components denoting the magnetic field

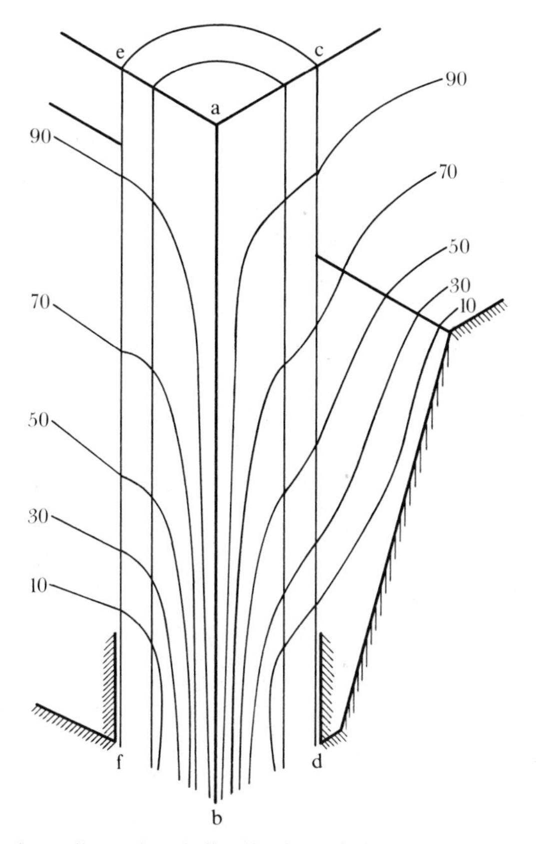

Fig. 17.12 A three-dimensional distribution of electrostatic potential around a porcelain insulator in an earthed trough[9]

strength and flux density respectively, μ is the magnetic permeability which varies (in an absolute set of units) from unity *in vacuo* to several thousand in magnetizing materials, and $\times$ denotes a vector product, defined in Appendix 6.

The formulation presented here depends on the fact that it is a relatively simple matter to determine a field $\mathbf{H}_s$ which exactly solves Eq. (17.34) when $\mu \equiv 1$ everywhere. This is given at any point defined by a vector co-ordinate $\mathbf{r}$ by an integral.

$$\mathbf{H}_s = \tfrac{1}{4}\pi \int_{\Omega} \frac{\mathbf{J} \times (\mathbf{r}-\mathbf{r}')}{(\mathbf{r}-\mathbf{r}')^2}\, \mathrm{d}\Omega. \tag{17.35}$$

In the above, $\mathbf{r}'$ refers to the co-ordinates of $\mathrm{d}\Omega$ and obviously the integration domain only involves the electric conductors where $\mathbf{J} \neq 0$.

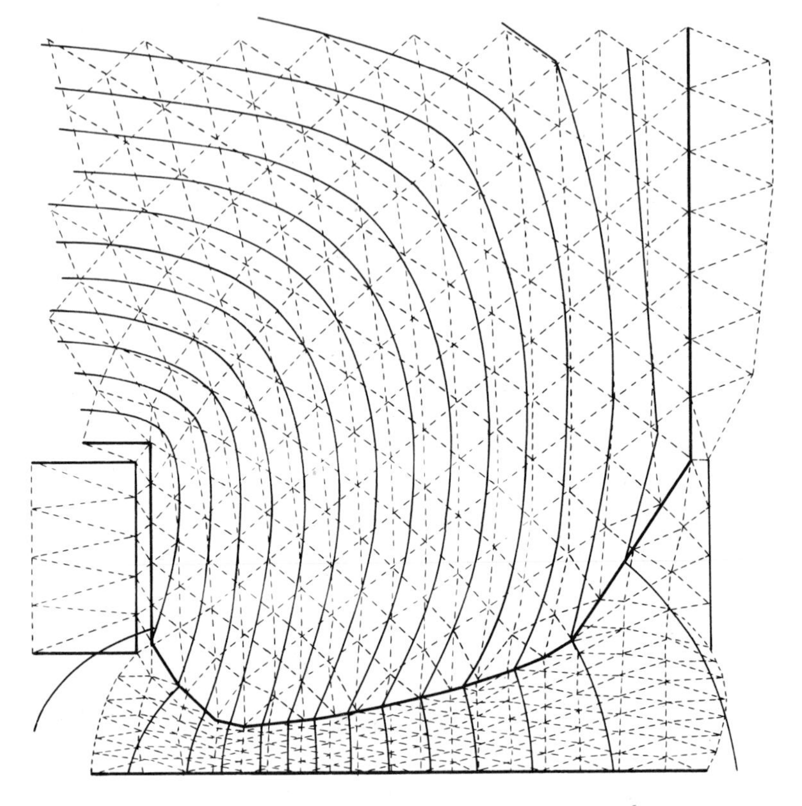

Fig. 17.13 Field near a magnet (after Winslow[6])

With $\mathbf{H}_s$ known we can write

$$\mathbf{H} = \mathbf{H}_s + \mathbf{H}_m$$

and, on substitution into Eq. (17.34), we have a system

$$\begin{aligned} \boldsymbol{\nabla}^{\mathrm{T}} \times \mathbf{H}_m &= 0 \\ \mathbf{B} &= \mu(\mathbf{H}_s + \mathbf{H}_m) \\ \boldsymbol{\nabla}^{\mathrm{T}} \mathbf{B} &= 0 \end{aligned} \tag{17.36}$$

If we now introduce a *scalar* potential ϕ, defining $\mathbf{H}_m$ as

$$\mathbf{H}_m \equiv \boldsymbol{\nabla} \phi \tag{17.37}$$

we find the first of Eqs. (17.36) automatically satisfied and, on eliminating $\mathbf{B}$ in the other two, the governing equation becomes

$$\boldsymbol{\nabla}^{\mathrm{T}} \mu \boldsymbol{\nabla} \phi + \boldsymbol{\nabla}^{\mathrm{T}} \mu \mathbf{H}_s = 0 \tag{17.38}$$

with $\phi \to 0$ at infinity. This is precisely of the standard form discussed in this chapter [Eq. (17.6)] with the second term, which is now specified, replacing Q.

An apparent difficulty exists, however, if μ varies in a discontinuous manner, as indeed one would expect it to do on the interfaces of two materials.

Here the term Q is now undefined and, in the standard discretization of Eq. (17.16) or (17.17) the term

$$\int_{\Omega} N_i Q \, d\Omega \equiv \int_{\Omega} N_i \, \nabla^{T} \mu \mathbf{H}_s \, d\Omega \tag{17.39}$$

apparently has no meaning.

Integration by parts comes once again to the rescue and we note that

$$\int_{\Omega} N_i \, \nabla^{T} \mu \mathbf{H}_s \, d\Omega \equiv -\int_{\Omega} \nabla^{T} N_i \mu \mathbf{H}_s + \int_{\Gamma} N_i \mu \mathbf{H}_s \mathbf{n} \, d\Gamma. \tag{17.40}$$

As in regions of constant μ, $\nabla^{T}\mathbf{H}_s \equiv 0$; the only contribution to the forcing terms comes as a line integral of the second term at discontinuity interfaces.

Introduction of the sealer potential makes both two- and three-dimensional magnetostatic problems solvable by a standard program used for all the problems in this section. Fig. 17.14 shows a typical three-dimensional solution for a transformer. Here isoparametric brick elements were used.[18]

In the typical magnetostatic problems a high non-linearity exists with

$$\mu = \mu(|\mathbf{H}|) \qquad \text{where} \quad \mathbf{H} = \sqrt{H_x^2 + H_y^2 + H_z^2}. \tag{17.41}$$

Treatment of such non-linearity will be discussed in Chapter 18.

Considerable economy in this and other problems of infinite extent can be achieved by the use of special *infinite* elements to be discussed in Chapter 23.

17.6.5 *Lubrication problems*. There once again a standard Poisson type of equation is encountered in the two-dimensional domain of a bearing pad. In the simplest case of constant lubricant density and viscosity the equation to be solved is (Reynolds equation)[19]

$$\frac{\partial}{\partial x}\left(h^3 \frac{\partial p}{\partial x}\right) + \frac{\partial}{\partial y}\left(h^3 \frac{\partial p}{\partial y}\right) = 6\mu V \frac{\partial h}{\partial x} \tag{17.42}$$

where h is the film thickness, p pressure developed, μ viscosity and V the velocity of the pad in the x direction.

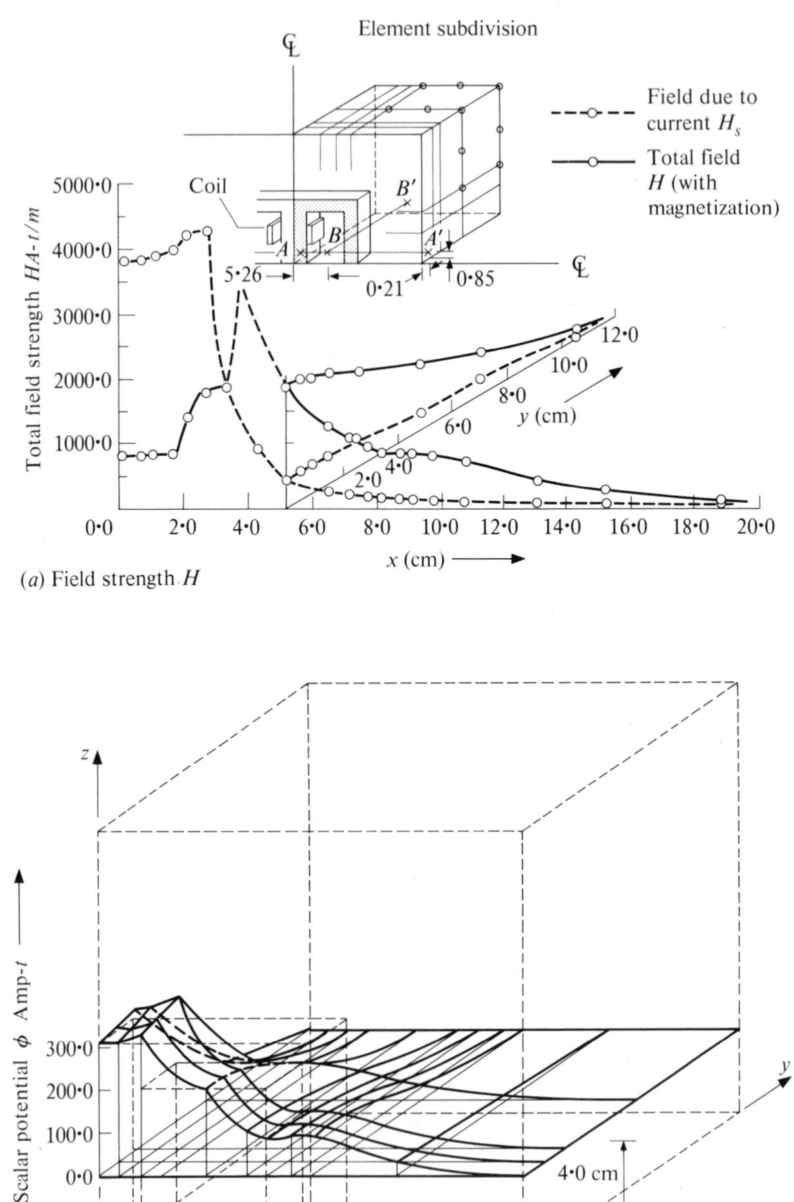

(*a*) Field strength H

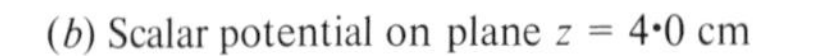
(*b*) Scalar potential on plane z = 4·0 cm

Fig. 17.14 Three-dimensional transformer

Figure 17.15 shows the pressure distribution in a typical case of a stepped pad.[20] The boundary condition is simply that of zero pressure and it is of interest to note that the step causes an equivalent of a 'line load' on integration of the right hand side of Eq. (17.42), just as in the case of magnetic discontinuity mentioned above.

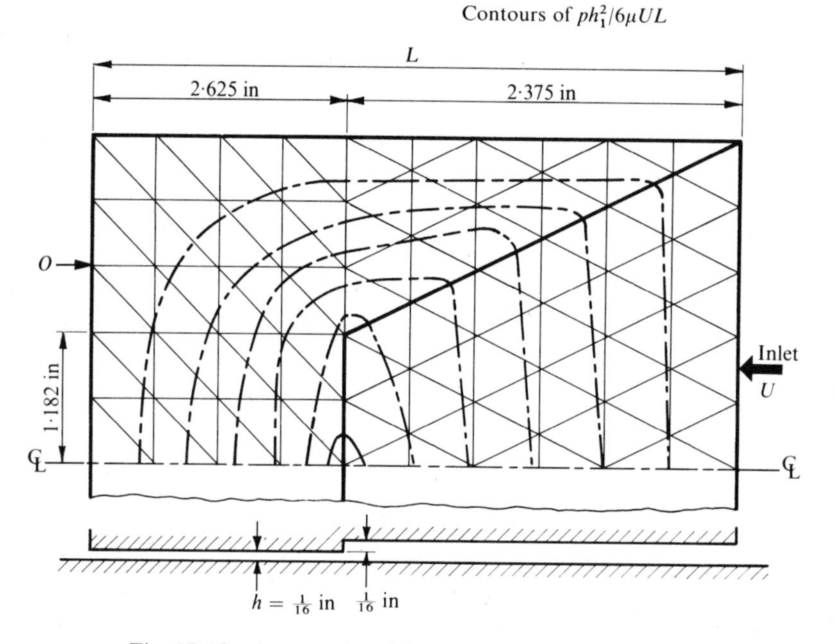

Fig. 17.15 A stepped pad bearing. Pressure distribution

More general cases of lubrication problems, including vertical pad movements (squeeze films) and compressibility, can obviously be dealt with, and much work has recently been done here.[21–28]

Irrotational and free surface flows. The basic Laplace equation which governs the flow of viscous fluid in seepage problems is also applicable in the problem of irrotational fluid flow outside the boundary layer created by viscous effects. The examples already given are adequate to illustrate the general applicability in this context. Further examples are quoted by Martin[29] and others.[30–35]

If no viscous effects exist, then it can be shown that for a fluid starting at rest the motion must be irrotational, i.e.,

$$\omega_z \equiv \frac{\partial u}{\partial y} - \frac{\partial v}{\partial x} = 0, \text{ etc.} \tag{17.43}$$

where u and v are appropriate velocity components.

This implies the existence of a velocity potential, giving

$$u = -\frac{\partial \psi}{\partial x} \qquad v = -\frac{\partial \psi}{\partial y}$$

$$(\text{or } \mathbf{u} = -\boldsymbol{\nabla}\phi) \tag{17.44}$$

If, further, the flow is incompressible the continuity equation [*vide* Eq. (17.2)] has to be satisfied, i.e.,

$$\boldsymbol{\nabla}^{\mathrm{T}} u = 0 \tag{17.45}$$

and therefore

$$\boldsymbol{\nabla}^{\mathrm{T}}.\boldsymbol{\nabla}\phi = 0. \tag{17.46}$$

Alternatively, for two-dimensional flow a stream function may be introduced defining the velocities as

$$u = -\frac{\partial \psi}{\partial y} \qquad v = \frac{\partial \psi}{\partial x} \tag{17.47}$$

and this identically satisfies the continuity equation. The irrotationality condition now must ensure that

$$\boldsymbol{\nabla}^{\mathrm{T}}.\boldsymbol{\nabla}\psi = 0 \tag{17.48}$$

and thus problems of ideal fluid flow can be posed in one form or the other. As the standard formulation is again applicable, there is little more that needs to be added, and for examples the reader can well consult the literature quoted.

The similarity with problems of seepage flow, which has already been discussed, is obvious.[36, 37]

A particular class of fluid flow deserves mention. This is the case when free surface limits the extent of the flow and this surface is not known *a priori*.

The class of problem is typified by two examples—that of a freely overflowing jet, Fig. 17.16(*a*) and that of a flow through an earth dam, Fig. 17.16(*b*). In both, the free surface represents a streamline and in both the position of the free surface is unknown *a priori* but has to be determined so that an *additional condition* on this surface is satisfied. For instance in the second problem, if formulated in terms of potential H, Eq. (17.25) governs the problem.

The free surface, being a streamline, imposes the condition

$$\frac{\partial H}{\partial n} = 0 \tag{17.49}$$

to be satisfied there.

In addition, however, the pressure must be zero on the surface as this is exposed to atmosphere. As

$$H = p/\gamma + y \tag{17.50}$$

where γ is the fluid specific weight, p is the fluid pressure, and y elevation above some (horizontal) datum, we must have on the surface

$$H = y. \tag{17.51}$$

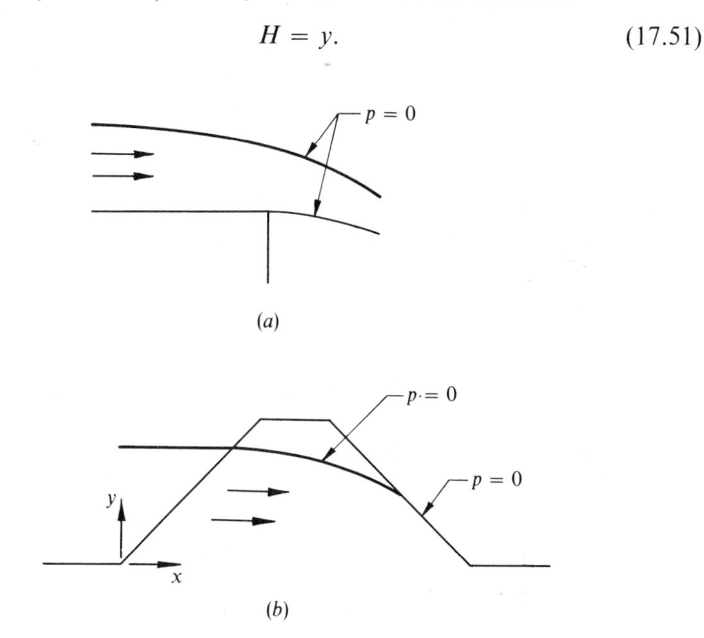

Fig. 17.16 Typical free surface problems with a streamline also satisfying an additional condition of pressure = 0. (*a*) Jet overflow. (*b*) Seepage through an earth dam

The solution may be approached iteratively. Starting with a prescribed free surface streamline the standard problem is solved. A check is carried out to see if Eq. (17.51) is satisfied and if not, an adjustment of the surface is carried out to make new y equal to H just found. A few iterations of this kind show that convergence is reasonably rapid. Taylor and Brown[38] show such a process. In Chapter 21 an alternative method is described. Special variational principles for dealing with this problem have been devised.[39, 40]

17.7 Concluding Remarks

We have shown how a general formulation for the solution of a steady-state quasi-harmonic problem can be written—and how a single program of such a form can be applied to a wide variety of physical situations. Indeed, the selection of problems dealt with is by no means exhaustive and many other examples of application are of practical interest. The reader will doubtless find appropriate analogies for his own problems.

References

1. O. C. Zienkiewicz and Y. K. Cheung, 'Finite elements in the solution of field problems', *The Engineer*, 507–10, Sept. 1965.
2. W. Visser, 'A finite element method for the determination of non-stationary temperature distribution and thermal deformations', *Proc. Conf. on Matrix Methods in Structural Mechanics*, Air Force Inst. Tech., Wright-Patterson A.F. Base, Ohio, 1965.
3. O. C. Zienkiewicz, P. Mayer and Y. K. Cheung, 'Solution of anisotropic seepage problems by finite elements', *Proc. Am. Soc. Civ. Eng.*, **92**, EM1, 111–20, 1966.
4. O. C. Zienkiewicz, P. L. Arlett and A. K. Bahrani, 'Solution of three-dimensional field problems by the finite element method', *The Engineer*, 27 October 1967.
5. L. R. Herrmann, 'Elastic torsion analysis of irregular shapes', *Proc. Am. Soc. Civ. Eng.*, **91**, EM6, 11–19, 1965.
6. A. M. Winslow, 'Numerical solution of the quasi-linear Poisson equation in a non-uniform triangle "mesh"', *J. Comp. Phys.*, **1**, 149–72, 1966.
7. D. N. de G. Allen, *Relaxation Methods*, p. 199, McGraw-Hill, 1955.
8. J. F. Ely and O. C. Zienkiewicz, 'Torsion of compound bars—a relaxation solution', *Int. J. Mech. Sci.*, **1**, 356–65, 1960.
9. O. C. Zienkiewicz and B. Nath, 'Earthquake hydrodynamic pressures on arch dams—an electric analogue solution', *Proc. Inst. Civ. Eng.*, **25**, 165–76, 1963.
10. H. M. Westergaard, 'Water pressure on dams during earthquakes', *Trans. Am. Soc. Civ. Eng.*, **98**, 418–33, 1933.
11. O. C. Zienkiewicz and R. E. Newton, 'Coupled vibrations of a structure submerged in a compressible fluid', *Proc. Symp. on Finite Element Techniques*, pp. 359–71, Stuttgart, 1969.
12. R. E. Newton, 'Finite element analysis of two-dimensional added mass and damping', *Finite Elements in Fluid*, Vol. I, pp. 219–32 (eds. R. H. Gallagher, J. T. Oden, C. Taylor and O. C. Zienkiewicz), Wiley, 1975.
13. P. A. A. Back, A. C. Cassell, R. Dungar and R. T. Severn, 'The seismic study of a double curvature dam', *Proc. Inst. Civ. Eng.*, **43**, 217–48, 1969.
14. P. Silvester and M. V. K. Chari, 'Non-linear magnetic field analysis of D.C. machines', *Trans. I.E.E.E.*, No. 7, 5–89, 1970.
15. P. Silvester and M. S. Hsieh, 'Finite element solution of two dimensional exterior field problems', *Proc. I.E.E.E.*, **118**, 1971.
16. B. H. McDonald and A. Wexler, 'Finite element solution of unbounded field problems', *Proc. I.E.E.E.*, MTT-20, No. 12, 1972.
17. E. Munro, 'Computer design of electron lenses by the finite element method', *Image Processing and Computer Aided Design in Electron Optics*, p. 284, Academic Press, 1973.
18. O. C. Zienkiewicz, J. F. Lyness and D. R. J. Owen, 'Three dimensional magnetic field determination using a scalar potential. A finite element solution' (to be published *Magn. Trans, I.E.E.E..*, 1977).

19. W. A. GROSS, *Gas Film Lubrication*, Wiley, 1962.
20. D. V. TANESA and I. C. RAO, *Student project report on lubrication*, Royal Naval College, Dartmouth, 1966.
21. M. M. REDDI, 'Finite element solution of the incompressible lubrication problem', *Trans. Am. Soc. Mech. Eng.*, **91** (Ser. F), 524, 1969.
22. M. M. REDDI and T. Y. CHU, 'Finite element solution of the steady state compressible lubrication problem', *Trans. Am. Soc. Mech. Eng.*, **92** (Ser. F), 495, 1970.
23. J. H. ARGYRIS and D. W. SCHARPF, 'The incompressible lubrication problem', *J. Roy. Aero. Soc.*, **73**, 1044–6, 1969.
24. J. F. BOOKER and K. H. HUEBNER, 'Application of finite element methods to lubrication; an engineering approach', *J. Lubr. Techn., Trans. Am. Soc. Mech. Eng.*, **14** (Ser. F), 313, 1972.
25. K. H. HUEBNER, 'Application of finite element methods to thermo-hydrodynamic lubrication', *Int. J. Num. Meth. Eng.*, **8**, 139–68, 1974.
26. S. M. ROHDE and K. P. OH, 'Higher order finite element methods for the solution of compressible porous bearing problems', *Int. J. Num. Meth. Eng.*, **9**, 903–12, 1975.
27. A. K. TIEU, 'Oil film temperature distributions in an infinitely wide glider bearing: an application of the finite element method', *J. Mech. Eng. Sci.*, **15**, 311, 1973.
28. K. H. HUEBNER, 'Finite element analysis of fluid film lubrication—a survey', *Finite Elements in Fluids*, Vol. II, pp. 225–54 (eds. R. H. Gallagher, J. T. Oden, C. Taylor and O. C. Zienkiewicz), Wiley, 1975.
29. H. C. MARTIN, 'Finite element analysis of fluid flows', *Proc. 2nd Conf. on Matrix Methods in Structural Mechanics*, Air Force Inst. Tech., Wright-Patterson A.F. Base, Ohio, 1968.
30. G. DE VRIES and D. H. NORRIE, *Application of the finite element technique to potential flow problems*, Reports 7 and 8, Dept. Mech. Eng., Univ. of Calgary, Alberta, Canada, 1969.
31. J. H. ARGYRIS, G. MARECZEK and D. W. SCHARPF, 'Two and three dimensional flow using finite elements', *J. Roy. Aero. Soc.*, **73**, 961–64, 1969.
32. L. J. DOCTORS, 'An application of finite element technique to boundary value problems of potential flow', *Int. J. Num. Meth. Eng.*, **2**, 243–52, 1970.
33. G. DE VRIES and D. H. NORRIE, 'The application of the finite element technique to potential flow problems', *J. Appl. Mech., Am. Soc. Mech. Eng.*, **38**, 978–802, 1971.
34. S. T. K. CHAN, B. E. LAROCK and L. R. HERRMANN, 'Free surface ideal fluid flows by finite elements', *Proc. Am. J. Civ. Eng.*, **99**, HY6, 1973.
35. B. E. LAROCK, 'Jets from two dimensional symmetric nozzles of arbitrary shape', *J. Fluid Mech.*, **37**, 479–83, 1969.
36. C. S. DESAI, 'Finite element methods for flow in porous media', *Finite Elements in Fluids*, Vol. 1, pp. 157–82 (ed. R. H. Gallagher), 1975.
37. I. JAVANDEL and P. A. WITHERSPOON, 'Applications of the finite element method to transient flow in porous media', *Trans. Soc. Petrol. Eng.*, **243**, 241–51, 1968.
38. R. L. TAYLOR and C. B. BROWN, 'Darcy flow solutions with a free surface', *Proc. Am. Soc. Civ. Eng.*, **93**, HY2, 25–33, 1967.
39. J. C. LUKE, 'A variational principle for a fluid with a free surface', *J. Fluid Mech.*, **27**, 395–7, 1957.
40. K. WASHIZU, *Variational Methods in Elasticity and Plasticity*, 2nd ed., Pergamon Press, 1975.

18. Non-Linear Material Problems. Plasticity, Creep (Visco-plasticity), Non-Linear Field Problems, etc.

18.1 **Introduction**

In all the problems discussed so far the differential equations governing the situations were linear—leading to the standard quadratic form of the functional. In elastic solid mechanics this was implied in

(*a*) a linear form of strain–displacement relationships (*vide* Eq. (2.2), Chapter 2)

and

(*b*) a linear form of stress–strain relationships (*vide* Eq. (2.5), Chapter 2).

In various field problems similar linearity was implied by such 'constants' as the permeability k remaining independent of the variation of the unknown potential ϕ (*vide* Eq. (17.6), Chapter 17).

Many problems of practical consequence exist in which such linearity is not preserved and it is of interest to extend the numerical processes described to cover these. In this context we have a whole range of *solid mechanics* situations in which such phenomena as plasticity, creep, or other *complex constitutive relations* supersede the simple linear elasticity assumptions.

Similarly in flow-type situations the dependence of viscosity on velocity distribution, the inapplicability of Darcy's seepage laws in a porous medium due to onset of turbulence, or dependence of magnetic permeability on flux densities give non-linearity with respect to material properties.

These classes of problems can often be simply dealt with without reformulation of the complete problem (i.e., without recourse to rewriting of the basic variational postulates). If, a solution to the 'linear' problem can be arrived at by some 'trial and error' process in which, at the final stage, the material constants are so adjusted that the appropriate new constitutive law is satisfied, then a solution is achieved.

However, if the strain-displacement relationship is non-linear, then a more fundamental reorganization of the formulation is necessary. For this reason alone such problems have been removed from this chapter and will be dealt with separately in Chapter 19. It will nevertheless be found that the basic iteration processes remain unchanged and indeed combination of both types of non-linearities may easily be achieved.

One important point needs, however, to be mentioned. While in linear problems the solution was always unique this no longer is the case in many non-linear situations. Thus, if *a solution* is achieved it may not necessarily be *the solution* sought. Physical insight into the nature of the problem and, on occasion, small-step, incremental approaches are essential to obtain the significant answers.

To find a successful solution, the technique to be applied to a particular problem is often arrived at by intuition and physical reasoning. Indeed, many procedures now used widely were introduced in this manner and later recognized as classical techniques of non-linear numerical analysis. In this chapter we shall reverse the process. First, a description of *general processes* applicable to discrete non-linear analysis will be given. This will be followed by application to

(*a*) *time independent material non-linearity in solids*
(*b*) *time dependent* (*creep*) *phenomena in solids*, and
(*c*) *non-linear field problems*.

18.2 General Procedures for Solutions of Non-linear Discrete Problems

18.2.1 *Introduction.* The discretized non-linear system can generally be written as a set of algebraic equations in one of the following forms

$$\boldsymbol{\Psi}(\mathbf{a}) \equiv \mathbf{P}(\mathbf{a})+\mathbf{f} \equiv \mathbf{K}(\mathbf{a}).\mathbf{a}+\mathbf{f} = 0. \tag{18.1}$$

The explicit form that is most conveniently derived depends usually on the problem in question and the method of discretization (weighted residual, variational principles, etc.). In the above the parameters $\mathbf{a}$ describe, as usual, the approximations to the unknown function or functions.

While the solution of a linear equation system

$$\mathbf{Ka}+\mathbf{f} = 0 \tag{18.2}$$

can be accomplished without difficulty in a direct manner, this is not possible for non-linear systems, but the various techniques to be described rely on our ability to solve such linear systems repeatedly until convergence is obtained.[1–4]

18.2.2 *Direct iteration.* The most obvious and direct solution procedure is one of iteration, which starts from the form

$$\mathbf{K}\mathbf{a}+\mathbf{f} = 0 \tag{18.3}$$

in which

$$\mathbf{K} = \mathbf{K}(\mathbf{a}).$$

If, initially, some value of $\mathbf{a} = \mathbf{a}^{\circ}$ is assumed, an improved approximation is obtained as

$$\mathbf{a}^{1} = -(\mathbf{K}^{\circ})^{-1}\mathbf{f} \tag{18.4}$$

where

$$\mathbf{K}^{\circ} = \mathbf{K}(\mathbf{a}^{\circ}).$$

Repetition of the process can be written as

$$\mathbf{a}^{n} = -(\mathbf{K}^{n-1})^{-1}\mathbf{f} \tag{18.5}$$

and this is terminated when the 'error', i.e.,

$$\mathbf{e} = \mathbf{a}^{n}-\mathbf{a}^{n-1} \tag{18.6}$$

becomes sufficiently small. Usually some *norm* of the error is determined and iteration continues until this is sufficiently small. Various norms and convergence criteria can be used, e.g.,

$$|\mathbf{e}| = \max \mathbf{e}_i \text{ (maximum value of any component)}$$

or

$$|\mathbf{e}| = \sqrt{\mathbf{e}^{\mathrm{T}}\mathbf{e}} \tag{18.7}$$

and convergence can be considered achieved when

$$|\mathbf{e}| \leqslant \alpha\,|\mathbf{a}|. \tag{18.8}$$

where α is some, specific, fraction.

Direct iteration is useful in some field problems as here the formulation leads directly to the form of Eq. (18.3) but often results in a lack of convergence. Figure 18.1 illustrates the convergent and divergent possibilities of such an iteration in a problem of one variable. In this process a full system of different linear equations has to be solved at each step of the computation.

18.2.3 *Newton–Raphson method.* Here, if an approximate solution $\mathbf{a} = \mathbf{a}^{n}$ is reached to Eq. (18.1) we can write an improved solution using a

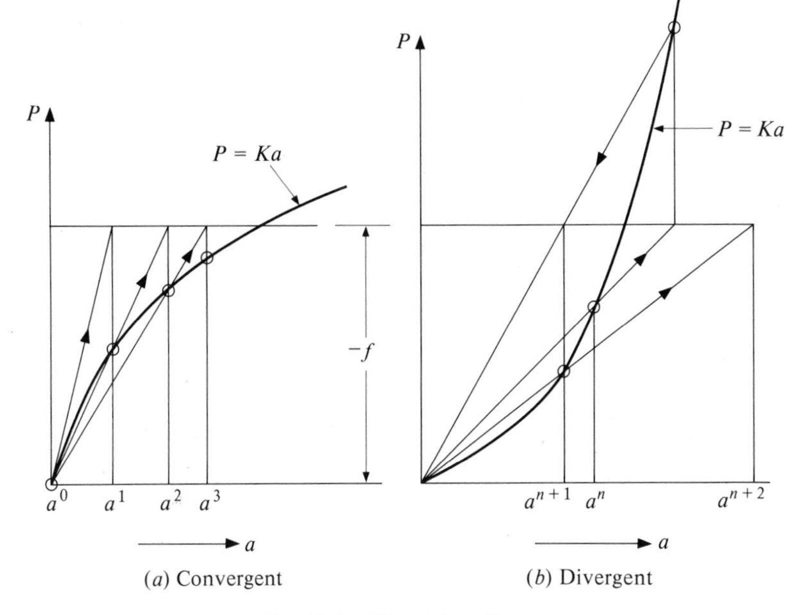

(*a*) Convergent (*b*) Divergent

Fig. 18.1 Direct iteration

curtailed Taylor expression as

$$\boldsymbol{\Psi}(\mathbf{a}^{n+1}) \equiv \boldsymbol{\Psi}(\mathbf{a}^n)+\left(\frac{\mathrm{d}\boldsymbol{\Psi}}{\mathrm{d}\mathbf{a}}\right)_n \Delta\mathbf{a}^n = 0 \tag{18.9}$$

with

$$\mathbf{a}^{n+1} = \mathbf{a}^n + \Delta\mathbf{a}^n.$$

In the above

$$\frac{\mathrm{d}\boldsymbol{\Psi}}{\mathrm{d}\mathbf{a}} \equiv \frac{\mathrm{d}\mathbf{P}}{\mathrm{d}\mathbf{a}} \equiv \mathbf{K}_{\mathrm{T}}(\mathbf{a})$$

represents a *tangential* matrix. The improved value of $\mathbf{a}^{n+1}$ can then be obtained by computing

$$\begin{aligned}\Delta\mathbf{a}^n &= -(\mathbf{K}_{\mathrm{T}}^n)^{-1}\,\boldsymbol{\Psi}^n \\ &= -(\mathbf{K}_{\mathrm{T}}^n)^{-1}\,(\mathbf{P}^n+\mathbf{f}).\end{aligned} \tag{18.10}$$

The process is illustrated in Fig. 18.2, and we note that once again at every step of the computation a new set of linearized equations has to be solved for $\Delta\mathbf{a}^n$.

The process is usually convergent in the vicinity of the solution, but if the initial 'guess' is not close again a divergence can occur.

At this stage it is of interest to remark that the tangential matrix $\mathbf{K}_T$ will *always be symmetric* if the original discretized equation arose via a variational principle (see Eq. 3.67, Chapter 3). This symmetry does not necessarily arise if the direct iteration process is used.

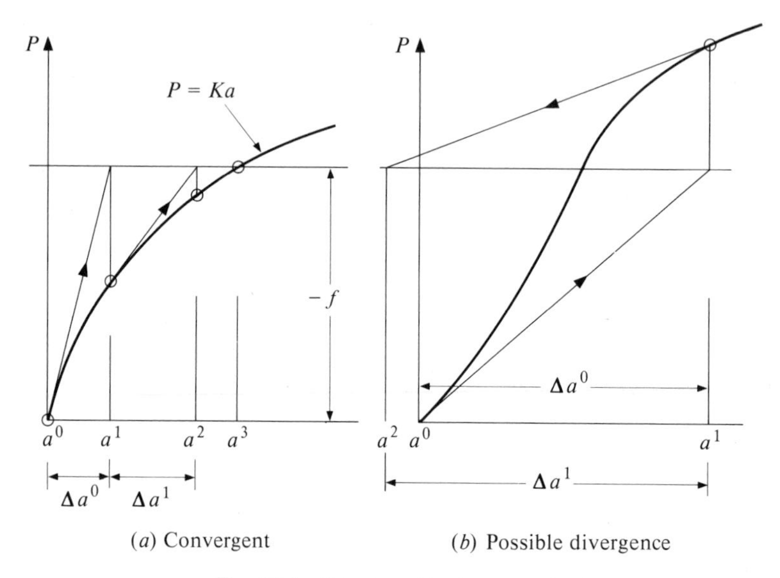

(*a*) Convergent (*b*) Possible divergence

Fig. 18.2 Newton–Raphson method

18.2.4 *Modified Newton–Raphson matrix.* To overcome the difficulty of having to solve a completely new system of equations at each iteration stage, frequently an approximation is made by writing

$$\mathbf{K}_T^n = \mathbf{K}_T^\circ \tag{18.11}$$

This modifies the algorithm of Eqs. (18.9) and (18.10) to

$$\Delta\mathbf{a}^n = -(\mathbf{K}_T^0)^{-1}(\mathbf{P}^n + \mathbf{f}) \tag{18.12}$$

and a simple *resolution of the same equation system* is repeatedly used. This is clearly more economical at each step but the convergence is now slower. In many cases, however, this process has an overall economy. It is illustrated in Fig. 18.3. A variation of this process would allow an updating of $\mathbf{K}_T$ after some iterations to, say, $\mathbf{K}_T^m$ and occasionally this may be useful.

18.2.5 *Incremental methods.* None of the processes so far described is guaranteed to converge in all circumstances, although in specific problems such convergence can be proved. An alternative set of procedures makes use of the fact that often the solution for **a** is known when the 'load' term **f** of Eq. (18.1) is zero. Indeed, if **f** is an actual force and **a** stands for structural displacement, both generally are zero at the reference start of the problem. In such circumstances it is convenient to study the behaviour of **a** as the vector **f** is incremented.

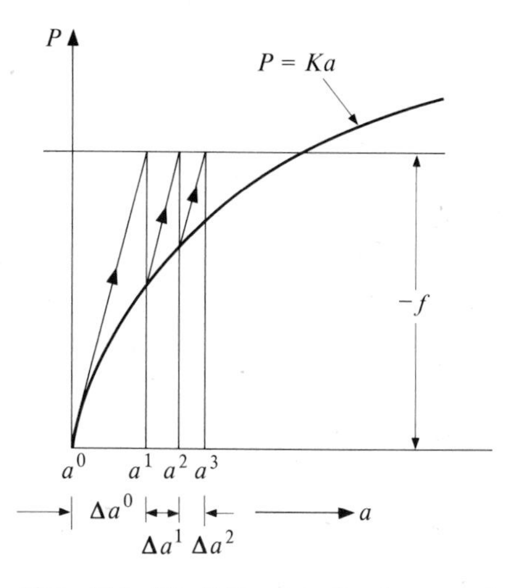

Fig. 18.3 'Modified' Newton–Raphson method

With such methods choosing a suitably small increment of **f** convergence can be guaranteed and, indeed, results of a reasonable kind will generally be available. Further, the intermediate results allow useful information on the 'loading' process to be achieved.

To explain the method it is convenient to rewrite Eq. (18.1) as

$$\mathbf{P}(\mathbf{a})+\lambda\mathbf{f}_0 = 0. \tag{18.13}$$

On differentiation with respect to λ, this results in

$$\frac{\mathrm{d}\mathbf{P}}{\mathrm{d}\mathbf{a}}\frac{\mathrm{d}\mathbf{a}}{\mathrm{d}\lambda}+\mathbf{f}_0 \equiv \mathbf{K}_\mathrm{T}\frac{\mathrm{d}\mathbf{a}}{\mathrm{d}\lambda}+\mathbf{f}_0 = 0 \tag{18.14a}$$

or

$$\frac{\mathrm{d}\mathbf{a}}{\mathrm{d}\lambda} = -\mathbf{K}_\mathrm{T}(\mathbf{a})^{-1}\,\mathbf{f}_0 \tag{18.14b}$$

where $\mathbf{K}_\mathrm{T}$ is the tangential matrix already described.

The problem posed in Eq. (18.14b) is a classical one of numerical analysis for which many solution (integration) techniques are available.

The simplest one (the Euler method) states

$$\mathbf{a}_{m+1}-\mathbf{a}_m = -\mathbf{K}_\mathrm{T}(\mathbf{a}_m)^{-1}\,\mathbf{f}_0\,\Delta\lambda_m = -(\mathbf{K}_\mathrm{T})_m^{-1}\,\Delta\mathbf{f}_m \tag{18.15}$$

where the subscript refers to the increments of λ (or $\mathbf{f}$), i.e.,

$$\lambda_{m+1} = \lambda_m+\Delta\lambda_m$$

or

$$\mathbf{f}_{m+1} = \mathbf{f}_m+\Delta\mathbf{f}_m \tag{18.16}$$

Improved integration schemes, such as the various predictor-corrector of Runge–Kutta formulae, improve accuracy (albeit by additional effort). A 'corrected' Euler process (equivalent to the second order Runge–Kutta process) is particularly useful. Here after calculation of $\mathbf{a}_{m+1}^0$ by expression (18.15), we re-compute an improved value of $\mathbf{a}_{m+1}$ as

$$\mathbf{a}_{m+1}-\mathbf{a}_m = -(\mathbf{K}_\mathrm{T})_{m+\theta}^{-1}\,\Delta\mathbf{f}_m \tag{18.17}$$

where

$$(\mathbf{K}_\mathrm{T})_{m+\theta} \equiv \mathbf{K}_\mathrm{T}(\mathbf{a}_{m+\theta})$$

$$\mathbf{a}_{m+\theta} = (1-\theta)\,\mathbf{a}_m+\theta\,\mathbf{a}_{m+1}^\circ\,;\, 0<\theta<1$$

thus obtaining an improved estimate $\mathbf{a}_{m+1}$.

In the straightforward procedures of numerical integration no use has been made of the actual Eq. (18.13) but only of its incremental form (18.14), and occasionally a drift of results may occur. To overcome this a slight re-interpretation of the process is convenient. In this we write Eq. (18.13) for the $m+1$ increment of λ and its $n+1$ iterate using the Newton–Raphson process [cf. Eqs. (18.9) and (18.10)]

$$\mathbf{\Psi}_{m+1}^{n+1} \equiv \mathbf{P}(\mathbf{a}_{m+1}^{n+1})+\lambda_{m+1}\mathbf{f}_0 = \mathbf{P}(\mathbf{a}_{m+1}^n)+\lambda_{m+1}\mathbf{f}_0+(\mathbf{K}_\mathrm{T})_{m+1}^n\,\Delta\mathbf{a}_{m+1}^n = 0 \tag{18.18}$$

with

$$\mathbf{a}_{m+1}^{n+1} = \mathbf{a}_{m+1}^n+\Delta\mathbf{a}_{m+1}^n$$

where $(\mathbf{K}_\mathrm{T})_{m+1}^n$ is the successively improved tangent matrix of λ_{m+1} and *start* the iteration using $\mathbf{a}_{m+1}^0 = \mathbf{a}_m$. Successive corrections lead to an exact satisfaction of Eq. (18.13) at the stage $m+1$.

If a single iteration is used

$$\Delta\mathbf{a}_{m+1} = \Delta\mathbf{a}_{m+1}^0 = -(\mathbf{K}_\mathrm{T})_m^{-1}\,(\mathbf{P}_m+\lambda_{m+1}\mathbf{f}_0)$$

and if the governing equation is satisfied at the previous step

i.e., $$\mathbf{P}_m+\lambda_m\mathbf{f}_0 = 0 \qquad \Delta\mathbf{a}_{m+1} = -(\mathbf{K}_\mathrm{T})_m^{-1}\,\mathbf{f}_0\Delta\lambda_m \tag{18.19}$$

and we have once again the Euler algorithm of Eq. (18.15).

The alternative iteration scheme allows an exact satisfaction of the governing equation to be achieved—but for its success it is important that the function **P(a)** be explicitly known.

18.2.6 *Acceleration of convergence.* An addition to the processes previously described involves various acceleration techniques. All the processes of iteration start from *some initial guessed* value. If the process of convergence of the error in the iteration procedure can be observed, it is often possible to extrapolate and start from an improved value. Alternatively, an 'over-relaxation' can be used in, say, the Newton–Raphson or the modified Newton–Raphson procedure in which the correction $\Delta \mathbf{a}^n$ is multiplied by a constant, often in the vicinity of 2, to improve the convergence rate.[5–9]

Most of such procedures have to be determined experimentally for a particular case but their use can be very advantageous.

18.2.7 *Final remarks.* The many alternatives for a numerical solution of a non-linear discretized equation system present such a wide choice to the user that the question of 'what is best' must inevitably be raised. The answer, unfortunately, cannot be given unequivocably as processes most economical in one context may be divergent in another. However, if only one process were to be implemented in a general program, the author would choose that of the incremental procedure outlined in section 18.2.5. Here, as we have already mentioned, convergence can always be obtained using sufficiently small increments. Further, in many cases we shall find that the explicit form of Eq. (18.1) cannot readily be written for some problems (e.g., those of plasticity). Here only an incremental $\mathbf{K}_T$ matrix is reasonably well defined, and clearly such procedures will allow an integration to be performed.

At this stage it is necessary to mention a further possible variant of the non-linear solution which, so far, has found little appreciation but which promises to be computationally advantageous. When solving a system of *linear* algebraic equations it is generally found that direct solution algorithms are faster than numerical iteration and are particularly economic if multiple 'load vectors' need to be considered. In non-linear situations the latter point is of little consequence as superposition of the effects is not valid, and it is possible to combine the iteration of the solution with an iteration necessary to achieve non-linearity. Perhaps the main reason that such processes are seldom used is the existence of powerful *linear computer programs* which can be adapted to non-linear solutions.

A potentially useful method of obtaining non-linear solutions iteratively is mentioned briefly in Chapter 21, p. 603, where a possible technique is suggested. In that approach a *dynamic* problem formulation is used and the static solution is approached asymptotically as oscillations die out.

18.3 Non-linear Constitutive Problems in Solid Mechanics. Non-linear Elasticity[10]

Returning to the basic problem of solid mechanics formulation in terms of displacement parameters **a**, we note that the equilibrium equations state that [Chapter 2, Eq. (2.11)]

$$\int_V \mathbf{B}^T \boldsymbol{\sigma} \, dV + \mathbf{f} = 0 \tag{18.20}$$

where the displacements and strains are defined respectively [Eqs. (2.1) and (2.2)] as

$$\mathbf{u} = \mathbf{N}\mathbf{a}; \qquad \boldsymbol{\varepsilon} = \mathbf{L}\mathbf{u} \equiv \mathbf{B}\mathbf{a}. \tag{18.21}$$

This derivation, based on the virtual work (and not energy) principles, is valid for any material behaviour. If now, for instance, we assume a *non-linear elastic behaviour* (replacing (Eq. 2.5) of linear elasticity) by

$$\boldsymbol{\sigma} = \boldsymbol{\sigma}(\boldsymbol{\varepsilon}) \tag{18.22}$$

then, relations (18.20)–(18.22) define completely the form (18.1) given as

$$\mathbf{P}(\mathbf{a}) + \mathbf{f} = 0 \tag{18.23}$$

and any of the techniques of solution discussed in section 18.2 can be used. As relationship (18.22) is unique, i.e., for any given strain a unique stress is given (albeit, on occasion, not explicitly), **P(a)** is also uniquely defined.

The direct iteration process of section 18.2.2 is, however, seldom convenient as in general it is not an easy matter to write **P(a)** in the form **Ka**; thus, the techniques of sections 18.2.3 to 18.2.5 are preferable. We note that the tangential matrix $\mathbf{K}_T$ is now given by

$$\mathbf{K}_T \equiv \frac{d\mathbf{P}}{d\mathbf{a}} = \int_V \mathbf{B}^T \frac{d\boldsymbol{\sigma}}{d\boldsymbol{\varepsilon}} \frac{d\boldsymbol{\varepsilon}}{d\mathbf{a}} \, dV = \int_V \mathbf{B}^T \mathbf{D}_T \mathbf{B} \, dV \tag{18.24}$$

where

$$\mathbf{D}_T = \frac{d\boldsymbol{\sigma}}{d\boldsymbol{\varepsilon}} \tag{18.25}$$

is known as the *tangential elasticity matrix*. The above form is particularly convenient as it is precisely the same as that pertaining to elasticity, and if $\mathbf{D}_T$ is a symmetric matrix, the same computational routines as those for the solution of elasticity problems can once again be used.

It is useful to interpret the residual vector $\boldsymbol{\Psi}(\mathbf{a}^n)$ of Eqs. (18.9) and (18.10) as a *residual* or *unbalanced* force vector. This provides a convenient physical measure of the error by which the equations are not satisfied.

As, frequently, departures from linearity occur only at higher values of stress or strain, it is convenient to compare relationship (18.22) with the linear one [Eq. (2.5)] i.e.

$$\boldsymbol{\sigma} = \mathbf{D}(\boldsymbol{\varepsilon} - \boldsymbol{\varepsilon}_0) + \boldsymbol{\sigma}_0. \tag{18.26}$$

Clearly, it is possible to make this equivalent to the general statement (18.22) by expressing $\boldsymbol{\varepsilon}_0$ or $\boldsymbol{\sigma}_0$ as functions of the strain level, where these functions have zero values at small strain levels and, indeed, can be applied as a *correction* to the linear process. We note also that

$$\text{when } \mathbf{a} = 0, \qquad \boldsymbol{\varepsilon} = 0 \quad \text{and} \quad \mathbf{D}_{\mathrm{T}} = \mathbf{D}. \tag{18.27}$$

If all the non-linearity is expressed in the initial stress term $\boldsymbol{\sigma}_0 = \boldsymbol{\sigma}_0(\boldsymbol{\varepsilon})$, a purely linear elastic solution with a constant matrix $\mathbf{D}$ will result in Eq. (18.20) being in error by a force term given by

$$\int_V \mathbf{B}^{\mathrm{T}} \boldsymbol{\sigma}_0 \, \mathrm{d}V \tag{18.28}$$

and this *unbalanced force vector* has to be corrected.

The correction can be carried out by a subsequent elastic solution using either the tangent or the original elastic modulus, and the reader will recognize that this is precisely the application of the Newton–Raphson or modified Newton–Raphson processes discussed in the previous section in which we test the error by comparing the stress which has been computed as a linear elastic one with that determined from the non-linear relationship.

Such techniques are, therefore, known as the *stress transfer*[6] or *initial stress methods* and are essentially identical to those described in the algorithms (18.10) or (18.12) in which

$$\mathbf{P}^n = \int \mathbf{B}^{\mathrm{T}} \boldsymbol{\sigma}^n \, \mathrm{d}V \tag{18.29}$$

and

$$\boldsymbol{\sigma}^n = \boldsymbol{\sigma}^n(\boldsymbol{\varepsilon}) = \boldsymbol{\sigma}^n(\mathbf{a}) \tag{18.30}$$

Little computational advantage comes from this interpretation but it was this physical insight which in fact led to most of the successful procedures now widely used.

A difficulty arises in occasional problems where the explicit determination of $\boldsymbol{\sigma}$ in terms of $\boldsymbol{\varepsilon}$ is not possible, but in which the reverse, i.e.,

$$\boldsymbol{\varepsilon} = \boldsymbol{\varepsilon}(\boldsymbol{\sigma}) \tag{18.31}$$

is well determined.

Here it is convenient to use an 'initial strain' process where at every stage of iteration for which initial values of $\boldsymbol{\sigma}$ and $\boldsymbol{\varepsilon}$ are known, $\boldsymbol{\varepsilon}_0$ is computed by comparing Eqs. (18.26) and (18.31). Iteration is continued until no changes in $\boldsymbol{\varepsilon}_0$ occur. Such processes are known as those of *initial strain* and are sometimes used in practice for 'locking materials' in which the strain has a limited value irrespective of the stress level.

18.4 **Plasticity**

18.4.1 *General theory.* 'Plastic' behaviour of solids is characterized by a non-unique stress–strain relationship—as opposed to that of non-linear elasticity discussed previously. Indeed, one definition of plasticity may be the presence of irrecoverable strains on load removal.

If uniaxial behaviour of a material is considered, as shown in Fig. 18.4(*a*), a non-linear relationship on loading alone does not determine whether non-linear elastic or plastic behaviour is exhibited. Unloading will immediately discover the difference with the elastic material following the same path and the plastic material showing a *history dependent*, different, path.

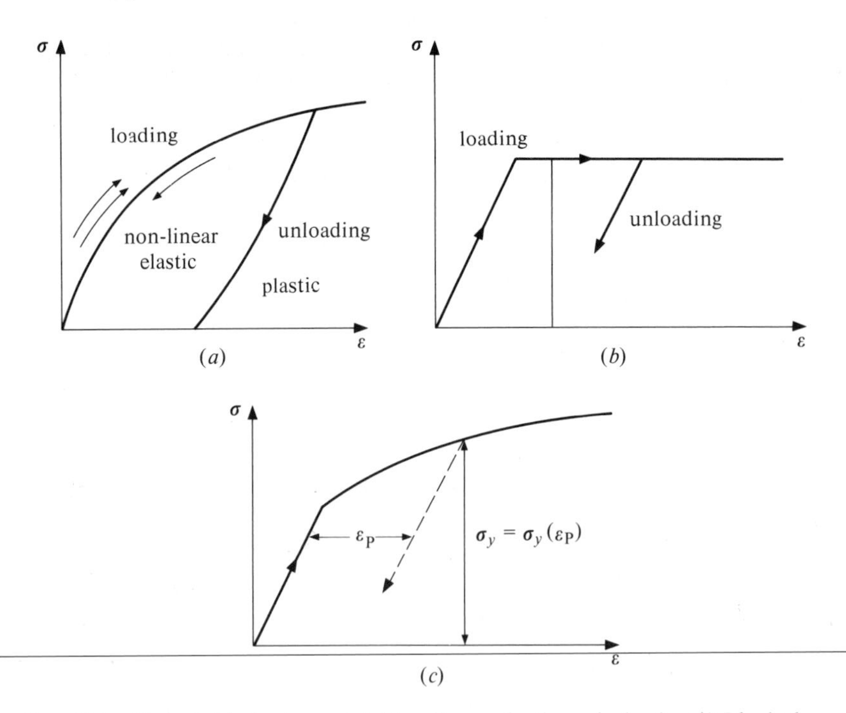

Fig. 18.4 Uniaxial behaviour: (*a*) Non-linear elastic and plastic. (*b*) Ideal plasticity. (*c*) Strain hardening plasticity

Many materials show an *ideal plastic* behaviour in which a limiting yield stress, σ_y, exists at which the strains are indeterminate. For all stresses below such yield a linear (or non-linear) elasticity relationship is assumed. Figure 18.4(*b*) illustrates this.

A further refinement of this model is one of *hardening/softening plastic material* in which the yield stress depends on some parameter κ (such as plastic strain $\boldsymbol{\varepsilon}_p$), Fig. 18.4(*c*).

It is with the latter two kinds of plasticity that this section is concerned and for which much theory has been developed.[11–17]

In a general state of stress $\boldsymbol{\sigma}$ the theory needs some expansion and the concepts of yield stresses need to be generalized.

Yield surface. It is quite generally postulated, as an experimental fact, that yielding can occur only if the stresses $\boldsymbol{\sigma}$ satisfy the general yield criterion

$$F(\boldsymbol{\sigma}, \kappa) = 0. \tag{18.32}$$

where κ is a 'hardening' parameter. This yield condition can be visualized as a surface in n-dimensional space of stress with the position of the surface dependent on the instantaneous value of the parameter κ (Fig. 18.5).

Flow rule (Normality principle). Von Mises[11] first suggested the basic constitutive relation defining the plastic strain increments in relation to the yield surface. Heuristic arguments for the validity of the relationship proposed have been given by various workers in the field[12,13] and at the present time the following hypothesis appears to be generally accepted: If $\mathrm{d}\boldsymbol{\varepsilon}_p$ denotes the increment of plastic strain then

$$\mathrm{d}\boldsymbol{\varepsilon}_p = \lambda\frac{\partial F}{\partial \boldsymbol{\sigma}} \tag{18.33}$$

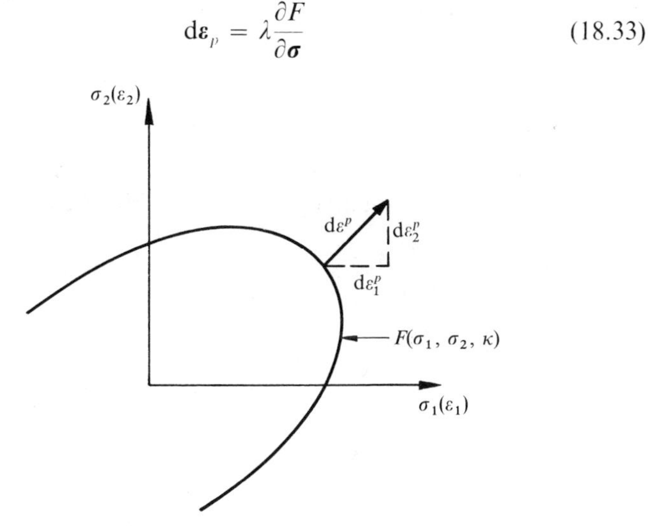

Fig. 18.5 Yield surface and normality criterion in two-dimensional stress space

or for any component n

$$d\varepsilon_{n,p} = \lambda \frac{\partial F}{\partial \sigma_n}.$$

In this λ is a proportionality constant, as yet undetermined. The rule is known as the *normality* principle because relation (18.33) can be interpreted as requiring the normality of the plastic strain increment 'vector' to the yield surface in the space of n stress dimensions.

A reduction of the restriction of the above rule can be obtained by specifying a *plastic potential*

$$Q = Q(\boldsymbol{\sigma}, \kappa) \tag{18.34}$$

which defines the plastic strain increment as similar to Eq. (18.33), i.e., giving this as

$$d\boldsymbol{\varepsilon}_p = \lambda \frac{\partial Q}{\partial \boldsymbol{\sigma}}. \tag{18.35}$$

The particular case of $Q = F$ is known as *associated plasticity*. When this relation is not satisfied the plasticity is *non-associated*. In what follows the more general form will be considered.

Total stress-strain relations. During an infinitesimal increment of stress, changes of strain are assumed to be divisible into elastic and plastic parts. Thus

$$d\boldsymbol{\varepsilon} = d\boldsymbol{\varepsilon}_e + d\boldsymbol{\varepsilon}_p. \tag{18.36}$$

The elastic strain increments are related to stress increments by a symmetric matrix of constants $\mathbf{D}$ as usual. We can thus write Eq. (18.36) incorporating the plastic relation (18.35) as

$$d\boldsymbol{\varepsilon} = \mathbf{D}^{-1}\, d\boldsymbol{\sigma} + \frac{\partial Q}{\partial \boldsymbol{\sigma}}\, \lambda. \tag{18.37}$$

When plastic yield is occurring the stresses are on the yield surface given by Eq. (18.32). Differentiating this we can write therefore

$$dF = \frac{\partial F}{\partial \sigma_1}\, d\sigma_1 + \frac{\partial F}{\partial \sigma_2}\, d\sigma_2 + \cdots + \frac{\partial F}{\partial \kappa}\, d\kappa = 0$$

or (18.38)

$$\left\{\frac{\partial F}{\partial \boldsymbol{\sigma}}\right\}^{\mathrm{T}} d\boldsymbol{\sigma} - A\lambda = 0$$

in which we make the substitution

$$A = -\frac{\partial F}{\partial \kappa}\, d\kappa\, \frac{1}{\lambda}. \tag{18.39}$$

Equations (18.37) and (18.38) can be written in a single symmetric matrix form as†

$$\begin{Bmatrix} d\boldsymbol{\varepsilon} \\ 0 \end{Bmatrix} = \begin{bmatrix} \mathbf{D}^{-1} & \dfrac{\partial Q}{\partial \boldsymbol{\sigma}} \\ \left(\dfrac{\partial F}{\partial \boldsymbol{\sigma}}\right)^{\mathrm{T}} & -A \end{bmatrix} \begin{Bmatrix} d\boldsymbol{\sigma} \\ \lambda \end{Bmatrix} \tag{18.40}$$

The indeterminate constant λ can be eliminated (taking care not to multiply or divide by A which may be zero in general). This results in an explicit expansion which determines the *stress changes* in terms of imposed *strain changes* with[17]

$$d\boldsymbol{\sigma} = \mathbf{D}^*_{ep}\, d\boldsymbol{\varepsilon} \tag{18.41}$$

$$\mathbf{D}^*_{ep} = \mathbf{D} - \mathbf{D}\left\{\frac{\partial Q}{\partial \boldsymbol{\sigma}}\right\}\left\{\frac{\partial F}{\partial \boldsymbol{\sigma}}\right\}^{\mathrm{T}} \mathbf{D}\left[A + \left\{\frac{\partial F}{\partial \boldsymbol{\sigma}}\right\}^{\mathrm{T}} \mathbf{D}\left\{\frac{\partial Q}{\partial \boldsymbol{\sigma}}\right\}\right]^{-1} \tag{18.42}$$

The elasto-plastic matrix $\mathbf{D}^*_{ep}$ takes the place of the elasticity matrix $\mathbf{D}_{\mathrm{T}}$ in incremental analysis.

This matrix is symmetric only when the plasticity is associated. The non-associated material will present special difficulties if tangent modulus procedures other than the modified Newton–Raphson method are used.

The matrix is defined even for ideal plasticity when $A = 0$. Explicit formulation of plasticity in this form was first introduced by Yamada *et al.*[18] and Zienkiewicz *et al.*[19]

Significance of parameter 'A'. Clearly for ideal plasticity with no hardening, A is simply zero. If hardening is considered, attention must be given to the nature of the parameter (or parameters) κ on which the shifts of the yield surface depend.

With a 'work hardening' material κ is taken to be represented by the amount of plastic work done during plastic deformation. Thus

$$d\kappa = \sigma_1\, d\varepsilon_1^p + \sigma_2\, d\varepsilon_2^p + \cdots = \boldsymbol{\sigma}^{\mathrm{T}}\, d\boldsymbol{\varepsilon}_p. \tag{18.43}$$

† To accomplish the elimination multiply the first set of Eqs. (18.40) by $(\partial F/\partial \boldsymbol{\sigma})^{\mathrm{T}}\mathbf{D}$, giving

$$\left(\frac{\partial F}{\partial \boldsymbol{\sigma}}\right)^{\mathrm{T}} d\boldsymbol{\sigma} = \left(\frac{\partial F}{\partial \boldsymbol{\sigma}}\right)^{\mathrm{T}} \mathbf{D}\, d\boldsymbol{\varepsilon} - \left(\frac{\partial F}{\partial \boldsymbol{\sigma}}\right)^{\mathrm{T}} \mathbf{D} \frac{\partial Q}{\partial \boldsymbol{\sigma}} \lambda$$

substituting into the second set yields

$$\left(\frac{\partial F}{\partial \boldsymbol{\sigma}}\right)^{\mathrm{T}} \mathbf{D}\, d\boldsymbol{\varepsilon} - \left[\left(\frac{\partial F}{\partial \boldsymbol{\sigma}}\right)^{\mathrm{T}} \mathbf{D} \frac{\partial Q}{\partial \boldsymbol{\sigma}} + A\right]\lambda = 0.$$

Elimination of λ from the first equation now gives Eqs. (18.41) and (18.42).

Substituting the flow rule Eq. (18.33) we have simply

$$d\kappa = \lambda \boldsymbol{\sigma}^{\mathrm{T}} \frac{\partial Q}{\partial \boldsymbol{\sigma}}. \tag{18.44}$$

By Eq. (18.39) we now see that λ disappears and we can write

$$A = -\frac{\partial F}{\partial \kappa} \boldsymbol{\sigma}^{\mathrm{T}} \frac{\partial Q}{\partial \boldsymbol{\sigma}} \tag{18.45}$$

a strictly determinate form if explicit relationship between F and κ is known.

Prandtl–Reuss relations. To illustrate some of the concepts consider the particular case of the well-known Huber–von Mises yield surface with an associated flow rule. This is given by

$$F = [\tfrac{1}{2}(\sigma_1 - \sigma_2)^2 + \tfrac{1}{2}(\sigma_2 - \sigma_3)^2 + \tfrac{1}{2}(\sigma_3 - \sigma_1)^2 + 3\sigma_4^2 + 3\sigma_5^2 + 3\sigma_6^2]^{1/2} - \sigma_y \equiv \bar{\sigma} - \sigma_y \tag{18.46}$$

in which suffixes 1, 2, 3 refer to the normal stress components and 4, 5, 6 to shear stress components in a general three-dimensional stress state.

On differentiation it will be found that

$$\frac{\partial F}{\partial \sigma_1} = \frac{3\sigma_1'}{2\bar{\sigma}}, \quad \frac{\partial F}{\partial \sigma_2} = \frac{3\sigma_2'}{2\bar{\sigma}}, \quad \frac{\partial F}{\partial \sigma_3} = \frac{3\sigma_3'}{2\bar{\sigma}}$$
$$\frac{\partial F}{\partial \sigma_4} = \frac{3\sigma_4}{\bar{\sigma}}, \quad \frac{\partial F}{\partial \sigma_5} = \frac{3\sigma_5}{\bar{\sigma}}, \quad \frac{\partial F}{\partial \sigma_6} = \frac{3\sigma_6}{\bar{\sigma}} \tag{18.47}$$

in which the dashes stand for so called deviatoric stresses, i.e.,

$$\sigma_1' = \sigma_1 - \frac{(\sigma_1 + \sigma_2 + \sigma_3)}{3} \text{ etc.} \tag{18.48}$$

The quantity $\sigma_y = \sigma_y(\kappa)$ is the uniaxial stress at yield. If a plot of the uniaxial test giving $\bar{\sigma}$ versus the *plastic* uniaxial strain ε_{up} is available and if simple work hardening is assumed, then

$$d\kappa = \sigma_y \, d\varepsilon_{up}$$

and

$$-\frac{\partial F}{\partial \kappa} = \frac{\partial \sigma_y}{\partial \kappa} = \frac{\partial \sigma_y}{\partial \varepsilon_{up}} \cdot \frac{1}{\sigma_y} = \frac{H'}{\sigma_y} \tag{18.49}$$

in which H' is the slope of the plot at the particular value of $\bar{\sigma}$.

On substituting into Eq. (18.45) we obtain, after some transformation, simply

$$A = H'. \tag{18.50}$$

This re-establishes the well-known Prandtl–Reuss stress–strain relations.

For a generalization of the concepts to a yield surface possessing 'corners' the reader is referred to the work of Koiter.[13]

Other yield surfaces. Clearly the general procedures outlined allow the determination of the tangent matrices for almost any yield surface applicable in practice. If the yield surface (and the material) is isotropic it is convenient to express it in terms of the three stress invariants. A particularly useful form of these is given below[20]

$$\sigma_m = \frac{J_1}{3} = \frac{(\sigma_x+\sigma_y+\sigma_z)}{3}$$

$$\bar{\sigma} = J_2{}^{1/2} = [\tfrac{1}{2}(s_x^2+s_y^2+s_z^2)+\tau_{xy}^2+\tau_{yz}^2+\tau_{zx}^2]^{1/2} \tag{18.51}$$

$$\theta = \frac{1}{3}\sin^{-1}\left[-\frac{3\sqrt{3}}{2}\frac{J_3}{\bar{\sigma}^3}\right] \quad \text{with } -\frac{\pi}{6}<\theta<\frac{\pi}{6}$$

where

$$J_3 = s_x s_y s_z + 2\tau_{xy}\tau_{yz}\tau_{zx} - s_x\tau_{yz}^2 - s_y\tau_{xz}^2 - s_z\tau_{xy}^2$$

and

$$s_x = \sigma_x - \sigma_m, \quad s_y = \sigma_y - \sigma_m, \quad s_z = \sigma_z - \sigma_m$$

It is shown in reference 20 that the yield surface for several classical yield conditions can be given as:

1. Tresca:

$$F = 2\bar{\sigma}\cos\theta - Y(\kappa) = 0 \tag{18.52}$$

where $Y(\kappa)$ is the yield stress from uniaxial tests.

2. Huber–von Mises:

$$\mathrm{F} = \sqrt{3}\bar{\sigma} - Y(\kappa) = 0. \tag{18.53}$$

Both 1 and 2 are well verified in metal plasticity. For soils, concrete, and other 'frictional' materials the Mohr–Coulomb law and its approximation given by Drucker and Prager are frequently used.[21]

3. Mohr–Coulomb:

$$F = \sigma_m \sin\phi + \bar{\sigma}\cos\theta - \frac{\bar{\sigma}}{\sqrt{3}}\sin\phi\sin\theta - c\cos\phi = 0 \tag{18.54}$$

where $c(\kappa)$ and $\phi(\kappa)$ are the cohesion and angle of friction, respectively, which could depend on some strain hardening parameter κ.

4. Drucker–Prager:[21]

$$F = 3\alpha'\sigma_m + \bar{\sigma} - K = 0 \tag{18.55}$$

where

$$\alpha' = \frac{2 \sin \phi}{\sqrt{3}\,(3 - \sin \phi)}, \qquad K = \frac{6c \cos \phi}{\sqrt{3}\,(3 - \sin \phi)}$$

and again c and ϕ can depend on a strain hardening parameter.

These forms are simpler than the alternatives used previously, and lead to a very convenient definition of the gradient vectors $\partial F/\partial \boldsymbol{\sigma}$ or $\partial Q/\partial \boldsymbol{\sigma}$, irrespective of whether the surface is used as a yield condition or potential. Thus we can always write:

$$\frac{\partial F}{\partial \boldsymbol{\sigma}} = \frac{\partial F}{\partial \sigma_m}\frac{\partial \sigma_m}{\partial \boldsymbol{\sigma}} + \frac{\partial F}{\partial J_2}\frac{\partial J_2}{\partial \boldsymbol{\sigma}} + \frac{\partial F}{\partial J_3}\frac{\partial J_3}{\partial \boldsymbol{\sigma}} \tag{18.56}$$

Noting

$$\frac{\partial F}{\partial J_3} = \frac{\partial F}{\partial \theta}\frac{\partial \theta}{\partial J_3} \tag{18.57}$$

and using Eq. (18.51) one can write the gradient vector as

$$\frac{\partial F}{\partial \boldsymbol{\sigma}} = \left(\frac{\partial F}{\partial \sigma_m}\mathbf{M}^0 + \frac{\partial F}{\partial J_2}\mathbf{M}^{\mathrm{I}} + \frac{\partial F}{\partial J_3}\mathbf{M}^{\mathrm{II}}\right)\boldsymbol{\sigma} \tag{18.58}$$

where the form of the square matrices $\mathbf{M}^0$, $\mathbf{M}^{\mathrm{I}}$, and $\mathbf{M}^{\mathrm{II}}$ is given in Table 18.1.

The values of the three derivatives with respect to the invariants are shown in Table 18.2 for the various yield surfaces mentioned. The reader can verify that the Prandtl–Reuss relations (18.47) are herein contained.

TABLE 18.1

MATRICES **M** OF EQ. (18.58)

$$\mathbf{M}^0 = \frac{1}{9\sigma_m}\begin{bmatrix} 1 & 1 & 1 & 0 & 0 & 0 \\ & 1 & 1 & 0 & 0 & 0 \\ & & 1 & 0 & 0 & 0 \\ & & & 0 & 0 & 0 \\ \text{Sym.} & & & & 0 & 0 \\ & & & & & 0 \end{bmatrix} \qquad \mathbf{M}^{\mathrm{I}} = \begin{bmatrix} \frac{2}{3} & -\frac{1}{3} & -\frac{1}{3} & 0 & 0 & 0 \\ & \frac{2}{3} & -\frac{1}{3} & 0 & 0 & 0 \\ & & \frac{2}{3} & 0 & 0 & 0 \\ & & & 2 & 0 & 0 \\ \text{Sym.} & & & & 2 & 0 \\ & & & & & 2 \end{bmatrix}$$

$$\mathbf{M}^{\mathrm{II}} = \begin{bmatrix} \frac{1}{3}\sigma_x & \frac{1}{3}\sigma_z & \frac{1}{3}\sigma_y & -\frac{2}{3}\tau_{yz} & \frac{1}{3}\tau_{zx} & \frac{1}{3}\tau_{xy} \\ & \frac{1}{3}\sigma_y & \frac{1}{3}\sigma_x & +\frac{1}{3}\tau_{yz} & -\frac{2}{3}\tau_{zx} & \frac{1}{3}\tau_{xy} \\ & & \frac{1}{3}\sigma_z & +\frac{1}{3}\tau_{yz} & \frac{1}{3}\tau_{zx} & -\frac{2}{3}\tau_{xy} \\ & & & -\sigma_x & \tau_{xy} & \tau_{zx} \\ \text{Sym.} & & & & -\sigma_y & \tau_{yz} \\ & & & & & -\sigma_z \end{bmatrix} + \sigma_m \begin{bmatrix} -\frac{1}{3} & -\frac{1}{3} & -\frac{1}{3} & 0 & 0 & 0 \\ & -\frac{1}{3} & -\frac{1}{3} & 0 & 0 & 0 \\ & & -\frac{1}{3} & 0 & 0 & 0 \\ & & & 1 & 0 & 0 \\ \text{Sym.} & & & & 1 & 0 \\ & & & & & 1 \end{bmatrix}$$

TABLE 18.2

INVARIANT DERIVATIVES FOR VARIOUS YIELD CONDITIONS

Yield conditions	$\frac{\partial F}{\partial \sigma_m}$	$\sqrt{J_2}\,\frac{\partial F}{\partial J_2}$	$J_2\,\frac{\partial F}{\partial J_3}$
Tresca	0	$2\cos\theta(1+\tan\theta\tan 3\theta)$	$\frac{\sqrt{3}\sin\theta}{\cos 3\theta}$
Huber–von Mises	0	$\sqrt{3}$	0
Mohr–Coulomb	$\sin\phi$	$\frac{\cos\theta}{2}\left[(1+\tan\theta\sin 3\theta) + \sin\phi(\tan 3\theta-\tan\theta)/\sqrt{3}\right]$	$\frac{\sqrt{3}\sin\theta+\sin\phi\cos\theta}{2\cos 3\theta}$
Drucker–Prager	$3\alpha'$	1·0	0

The form of the various yield surfaces given above is shown in principal stress space in Fig. 18.6.

Generalized hardening/softening rules. We have so far assumed that the parameter κ is associated with the plastic work done and this, being a scalar quantity, will obviously change the yield surface by a simple expansion or contraction (isotropic hardening). Such models are found to be deficient in reproducing the true behaviour of some materials and kinematic hardening theories have been introduced in which dependence on the direction of plastic straining is noted.[15,22,23]

An alternative, utilizing the finite element procedure directly, is that of modelling the material by an '*overlay*' technique.[24,25]†

18.4.2 *Computation procedures for plasticity problems.* The main difference from the non-linear elasticity problems for which the numerical procedure follows the classical processes of section 18.2 lies in the fact that an explicit relationship of the form of Eq. (18.22) is no longer available.

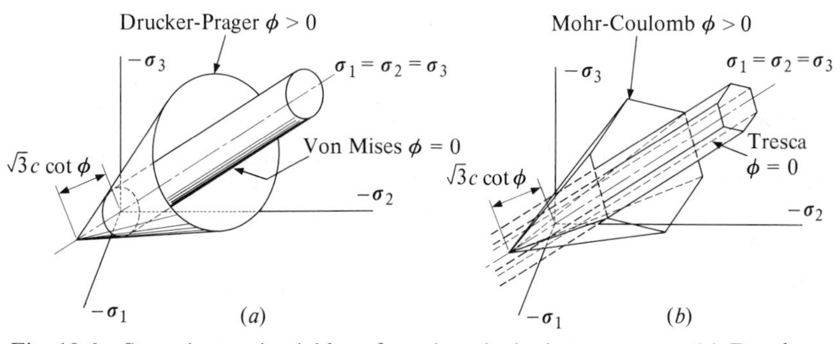

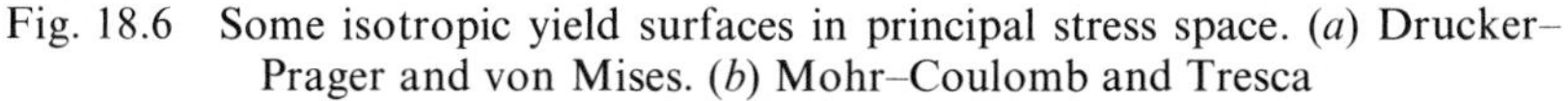

Fig. 18.6 Some isotropic yield surfaces in principal stress space. (*a*) Drucker–Prager and von Mises. (*b*) Mohr–Coulomb and Tresca

† Here the finite element method demonstrates yet another of the facets permitting each material component to be modelled by an element with simple behaviour, yet achieving a complex response in superposition.

Although the stresses for any level of strain have to lie on or within the current yield surface, the exact value of each component cannot be determined. The solution has thus to be determined utilizing the fact that (*a*) the tangential matrix

$$\mathbf{D}_{\mathrm{T}} = \mathbf{D}^{*}_{ep} \tag{18.59}$$

is known for a specified stress value and loading direction, and (*b*) the stresses can be *integrated* using the fact that

$$\mathrm{d}\boldsymbol{\sigma} = \mathbf{D}^{*}_{\mathrm{T}}\,\mathrm{d}\boldsymbol{\varepsilon}. \tag{18.60}$$

Obviously, *incremental* procedures of the type described in section 18.2.5 must now be used. The algorithm of Eq. (18.18) is applicable, but an indirect procedure of evaluating

$$\mathbf{P}^{n}_{m+1} = \int_V \mathbf{B}^{\mathrm{T}}\boldsymbol{\sigma}^{n}_{m+1}\,\mathrm{d}V \tag{18.61}$$

needs to be used. Here we can write

$$\boldsymbol{\sigma}^{n}_{m+1} = \boldsymbol{\sigma}_m + \Delta\boldsymbol{\sigma}^{n-1}_{m} \tag{18.62}$$

in which

$$\Delta\boldsymbol{\sigma}^{n-1}_{m} = (\mathbf{D}_{\mathrm{T}})^{n-1}_{m+\theta}\,\Delta\boldsymbol{\varepsilon}^{n-1}_{m} \tag{18.63}$$

where $(\mathbf{D}_{\mathrm{T}})^{n-1}_{m+\theta}$ is the tangent matrix computed for a stress intermediate between σ_m and σ^{n-1}_{m+1}.

An additional control requires that σ_{m+1} does not exceed the yield condition and here, often, a proportional scaling is employed (as the yield surface is undoubtedly a better known quantity than the flow rule which determines $\mathbf{D}_{\mathrm{T}}$).

In the various iterative steps given by Eq. (18.18) the matrix $\mathbf{K}_{\mathrm{T}}$ can be replaced by the constant elastic matrix $\mathbf{K}$, thus leading again to the modified Newton–Raphson or *initial stress* method. Algorithms of this type are described in detail in reference 17.

Clearly plasticity computation, due to its incremental nature, is complex and we shall see that more straightforward algorithms are available through a viscoplastic formulation (see section 18.6).

In the earliest finite element solutions both initial strain[26–29] and initial stress[17,19] approaches were used, employing in all iterations a simple resolution of the constant linear elastic problem. While this is in general economical for some problems, the incremental approach using the tangent matrix and a limited number of iterations within each load increment is preferred, and such techniques are commonly used today.[30–37]

The finite element discretization technique in plasticity problems follows precisely the same procedure as that of corresponding elasticity problems.

Any of the elements already discussed can be used and again we find that generally higher order elements with reduced integration show an improved performance with the *best* sampling point of the same kind as those described in Chapter 11.

The use of such reduced integration is important in metal plasticity as the von Mises' flow rules do not permit any volume changes. As the extent of plasticity spreads at the collapse load the deformation becomes nearly incompressible, and with conventional exactly integrated elements the system *locks* and a true collapse load cannot be obtained.[38]

While the elasto-plastic matrix deduced previously is valid for a general three-dimensional continuum, in two-dimensional plasticity it has to be reduced to special forms. In *plane stress*, for instance, this reduction is obvious by a simple deletion of the appropriate columns in Eq. (18.40) to which zero stress components are assigned; in a *plane strain* situation, all stresses exist but appropriate strain components have to be made zero. Appropriate elimination has now to be carried out and explicit expressions will be found in reference 19. It is of interest to note that in such cases the diagonal term corresponding to A is no longer zero even in the case of ideal plasticity.

Finally, we should remark that the possibility of solving plastic problems is not limited to a displacement formulation alone. Equilibrium fields and, indeed, most of the formulations described in Chapter 12 form a suitable vehicle,[39,40,41] but owing to their convenience and easy interpretation displacement forms are most commonly used.

18.4.3 *Some examples. Perforated plate with and without strain hardening.*[19,31,40] Figure 18.7 shows the configuration and the division into simple triangular elements. In this example plane stress conditions are assumed and solution is obtained for both ideal plasticity and strain hardening. The von Mises criterion is used and, in the case of strain hardening, a constant slope of the uniaxial hardening curve, H' [Eq. (18.50)] was taken. The spread of plastic zones at various load levels is shown in Fig. 18.7(*b*) and (*c*).

Although the plasticity relation is only incremental, if the loads are applied in a single large step the initial stress process will still yield an equilibrating solution and one which does not exceed the yield stresses. Such a single-step solution for a very large load increment is shown in Fig. 18.7(*d*). It is of interest to note that even now despite the violation of the incremental strain laws very similar results for plastic zones are achieved.

It is even more significant to note that the maximum strains reached at the point of first yield are almost identical with those achieved incrementally, Fig, 18.8.

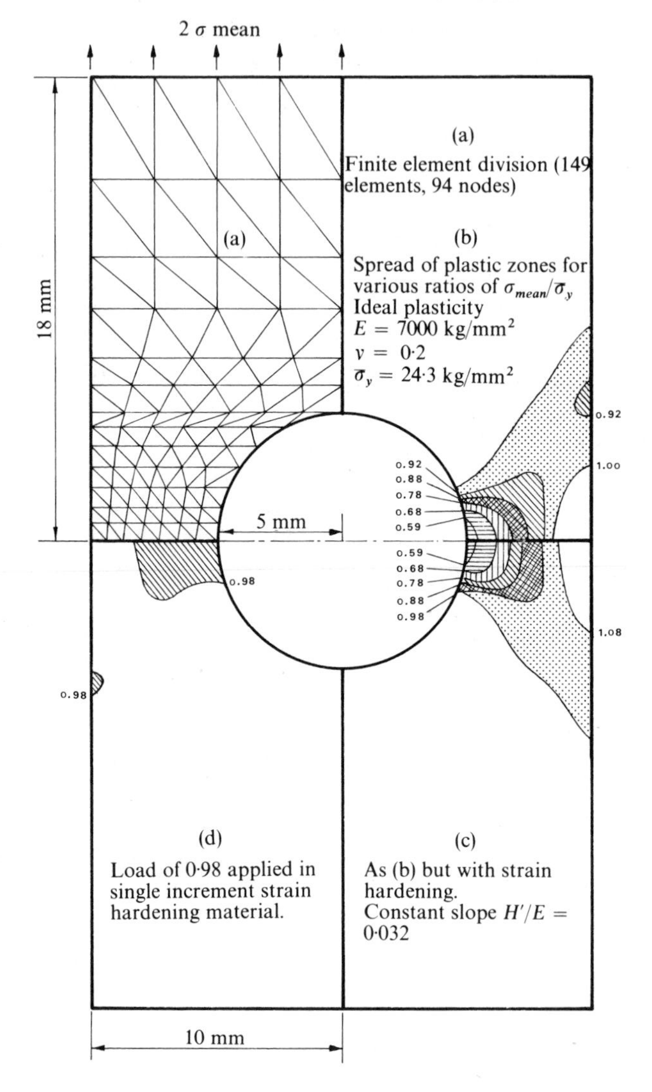

Fig. 18.7 Perforated tension strip (plane stress)

A notched specimen (Fig. 18.9). In the previous example simple triangular elements were used; now a comparative study of such elements with higher order isoparametric ones is shown.[17,42] The much more consistent spread of plasticity zones with such elements and the more rapid convergence of the results will be noted.

Steel pressure vessel. This final example, for which test results obtained

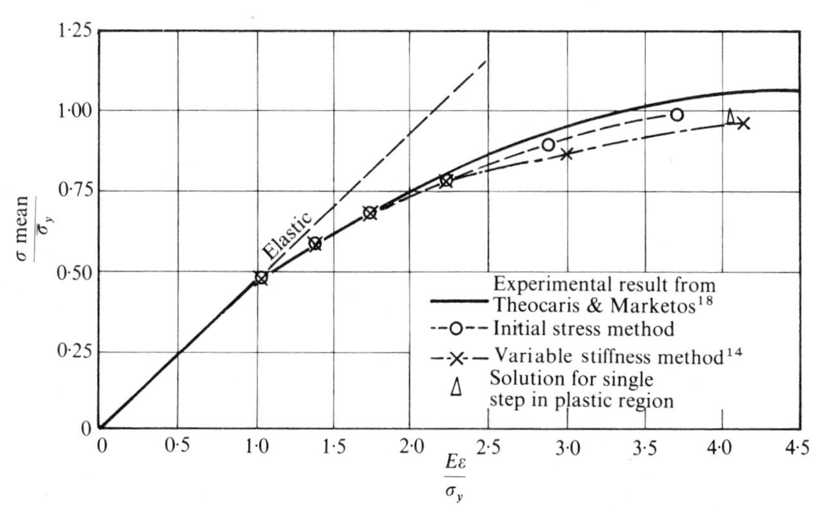

Fig. 18.8 Perforated plate—strain hardening material. Development of maximum strain at point of first yield. $H'/E = 0{\cdot}032$. Load increment $= 0{\cdot}2 \times$ first yield load

by Dinno and Gill[43] were available, illustrates a practical application and its objectives are twofold.

Firstly, to show that this problem which can really be described as a thin shell can be adequately represented by a limited number (53) of isoparametric elements. Indeed this model simulates the over-all behaviour as well as the local stress concentration [Figure 18.10(*a*)].

Secondly, it was decided to push the solution almost to the failure point while incrementing the pressure rather than displacement.

Comparison of calculated and measured deflections in Fig. 18.10(*b*) show how well the objectives are achieved.

TIME-DEPENDENT PROBLEMS IN SOLID MECHANICS (CREEP; VISCOPLASTICITY AND VISCOELASTICITY)

18.5 The Basic Formulation of Creep Problems

The phenomenon of 'creep' is manifested by a time-dependent deformation under a constant strain. Thus, in addition to an instantaneous strain, the material develops creep strains $\boldsymbol{\varepsilon}_c$ which generally increase with duration of loading. The constitutive law of creep will usually be of a form in which the *rate of creep strain* is defined as some function of stresses and total creep strains, i.e.,

$$\dot{\boldsymbol{\varepsilon}}_c \equiv \frac{d\boldsymbol{\varepsilon}_c}{dt} = \boldsymbol{\beta}(\boldsymbol{\sigma}, \boldsymbol{\varepsilon}_c). \qquad (18.64)$$

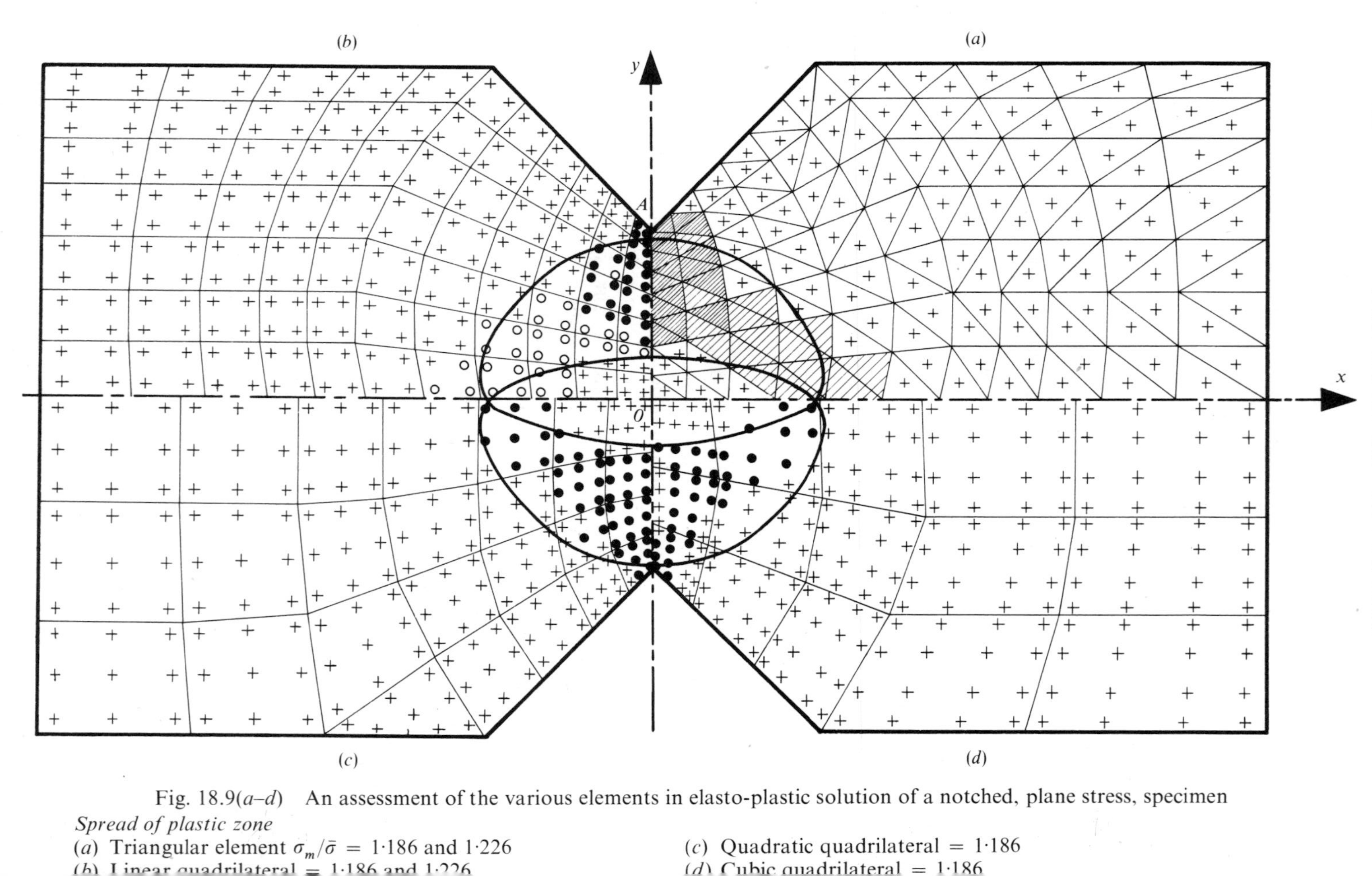

Fig. 18.9(*a–d*) An assessment of the various elements in elasto-plastic solution of a notched, plane stress, specimen
Spread of plastic zone
(*a*) Triangular element $\sigma_m/\bar{\sigma}$ = 1·186 and 1·226
(*b*) Linear quadrilateral = 1·186 and 1·226
(*c*) Quadratic quadrilateral = 1·186
(*d*) Cubic quadrilateral = 1·186

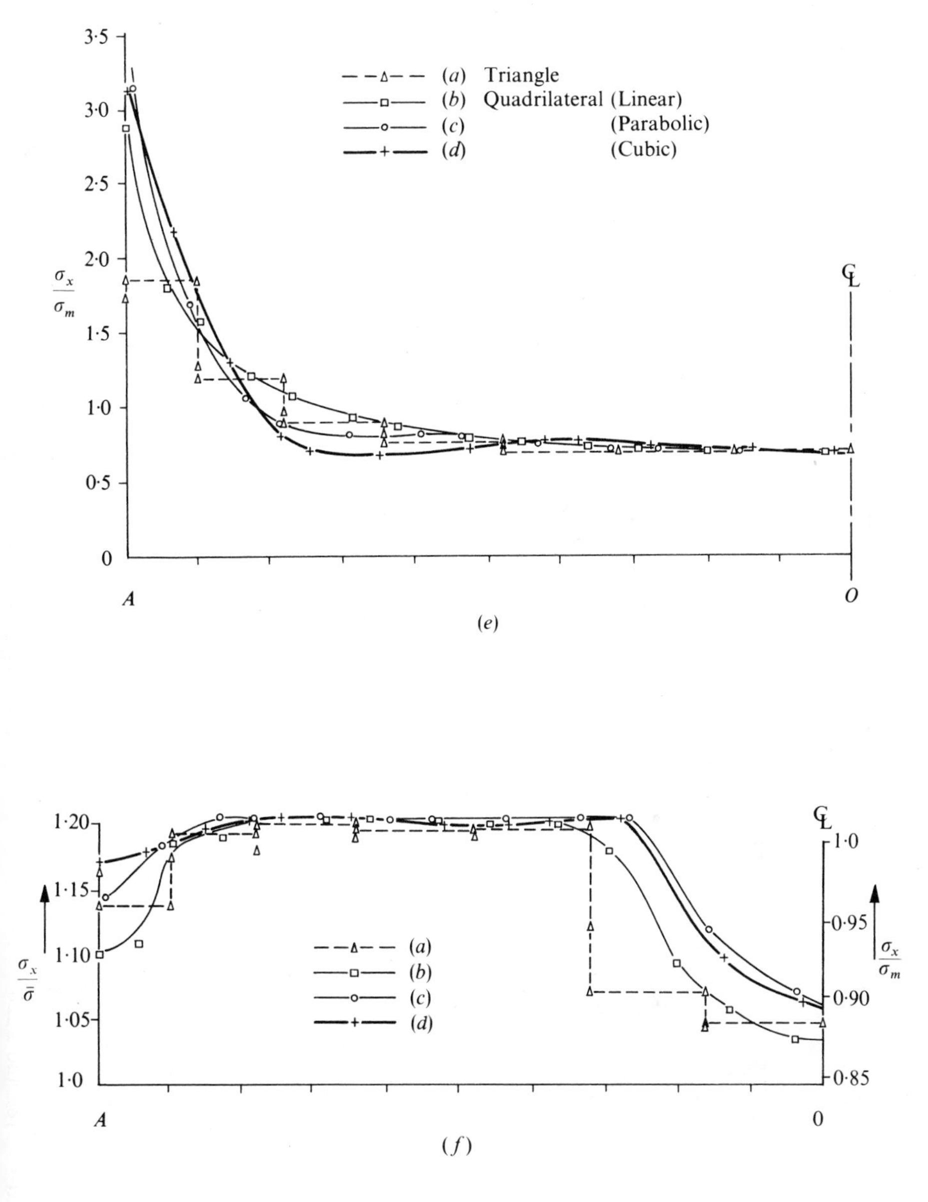

Fig. 18.9 *cont.*
Distribution of stresses in notch section
(*e*) Elastic
(*f*) Elasto plastic for $\sigma_m/\bar{\sigma} = 1{\cdot}186$
Number of degrees of freedom approximately equal of 172–178 in all the four solutions

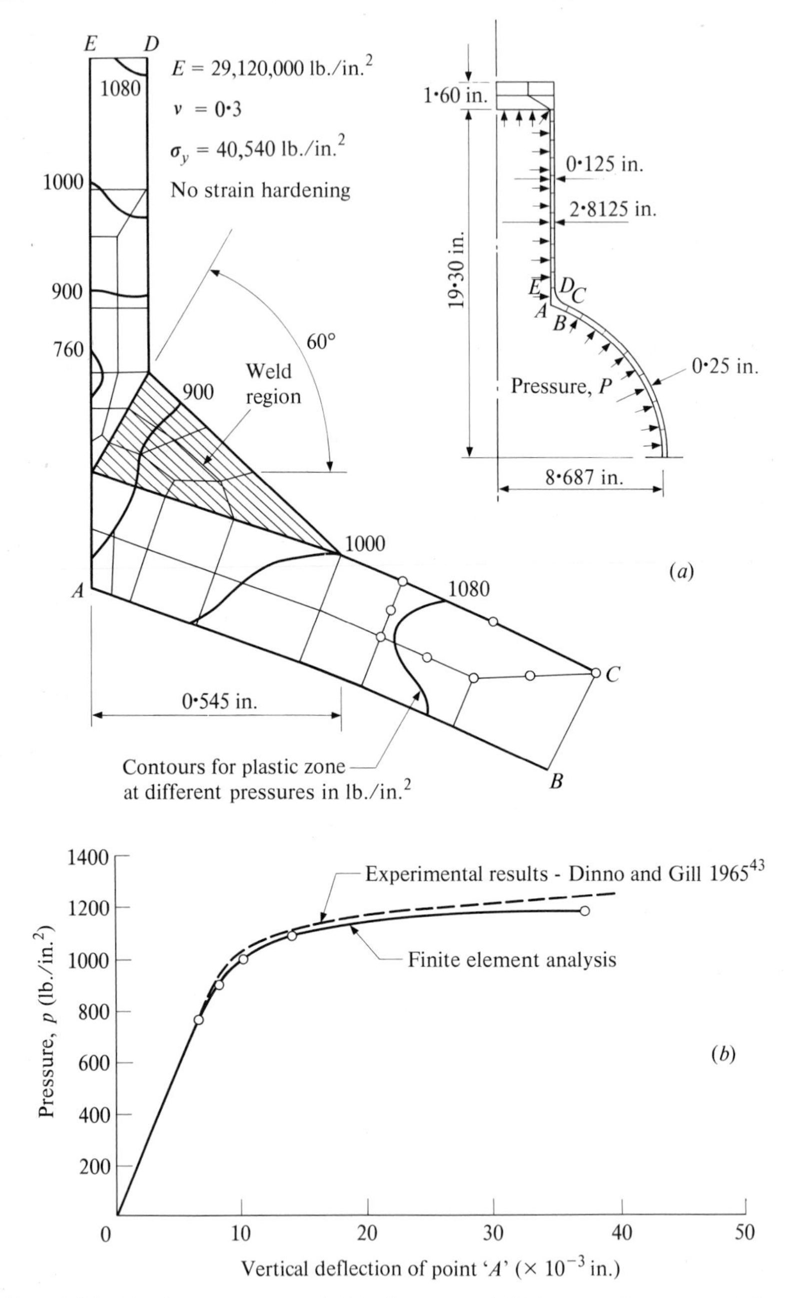

Fig. 18.10 Steel pressure vessel. (*a*) Element subdivision and spread of plastic zones[43] (von Mises yield and ideal plasticity). (*b*) Vertical deflection at point A with increasing pressure

If we consider that the instantaneous strains are elastic, the total strain can be written as

$$\boldsymbol{\varepsilon} = \boldsymbol{\varepsilon}_e + \boldsymbol{\varepsilon}_c \tag{18.65}$$

with

$$\boldsymbol{\varepsilon}_e = \mathbf{D}^{-1}\boldsymbol{\sigma} \tag{18.66}$$

neglecting any initial (thermal) strains or initial (residual) stresses. As usual the equilibrium conditions

$$\int_V \mathbf{B}^{\mathrm{T}}\boldsymbol{\sigma}\,\mathrm{d}V + \mathbf{f} = 0 \tag{18.67}$$

hold at all times and, if the initial conditions of the system are known, the system of Eqs. (18.64)–(18.67) gives a solvable first order system of ordinary differential equations with non-linear coefficients. In Chapter 21 we shall discuss in detail the solution of such equations but, as the most important non-linearity concerns here the material behaviour, we shall anticipate some of the general procedures.

In particular, if we consider an interval of time Δt_m at the beginning of which the state characterized by a set of nodal displacement parameters $\mathbf{a}_m$, stress $\boldsymbol{\sigma}_m$, and forces $\mathbf{f}_m$ is known, we can write a set of non-linear algebraic equations linking the final conditions with time. Thus we have

$$\boldsymbol{\Psi}_{m+1} \equiv \int_V \mathbf{B}^{\mathrm{T}}\boldsymbol{\sigma}_{m+1}\,\mathrm{d}V + \mathbf{f}_{m+1} = 0 \tag{18.68}$$

as the equilibrium condition.

From Eq. (18.65) and (18.66) we can write

$$\begin{aligned} \boldsymbol{\sigma}_{m+1} - \boldsymbol{\sigma}_m &= \mathbf{D}(\boldsymbol{\varepsilon}_{m+1} - \boldsymbol{\varepsilon}_m) - \mathbf{D}(\boldsymbol{\varepsilon}_{c,\,m+1} - \boldsymbol{\varepsilon}_{c,\,m}) \\ &= \mathbf{D}\mathbf{B}(\mathbf{a}_{m+1} - \mathbf{a}_m) - \mathbf{D}(\boldsymbol{\varepsilon}_{c,\,m+1} - \boldsymbol{\varepsilon}_{c,\,m}). \end{aligned} \tag{18.69}$$

From the rate equation [Eq. (18.64)]—from which we drop for simplicity the dependence on $\boldsymbol{\varepsilon}_c$—we can write approximately

$$\boldsymbol{\varepsilon}_{c,\,m+1} - \boldsymbol{\varepsilon}_{c,\,m} = \Delta t_m\,\boldsymbol{\beta}_{m+\theta} \tag{18.70}$$

where

$$\boldsymbol{\sigma}_{m+\theta} = (1-\theta)\boldsymbol{\sigma}_m + \theta\boldsymbol{\sigma}_{m+1} \quad (0 \leqslant \theta \leqslant 1)$$

and

$$\boldsymbol{\beta}_{m+\theta} = \boldsymbol{\beta}(\boldsymbol{\sigma}_{m+\theta}).$$

Equations (18.69) and (18.70) can be combined to give a set of non-linear equation

$$\bar{\boldsymbol{\Psi}}_{m+1} = \boldsymbol{\sigma}_{m+1} - \boldsymbol{\sigma}_m - \mathbf{D}\mathbf{B}(\mathbf{a}_{m+1} - \mathbf{a}_m) + \mathbf{D}\,\Delta t_m\boldsymbol{\beta}_{m+\theta} = 0. \tag{18.71}$$

The system of Eqs. (18.68) and (18.71) is a non-linear one from which $\boldsymbol{\sigma}_{m+1}$ and $\mathbf{a}_{m+1}$ need to be determined.

The approach to the solution via the Newton–Raphson procedure can now be made (and the reader will note at once the similarity with that used for the incremental process of section 18.2.5).

Taking the first 'guess' of $\boldsymbol{\sigma}_{m+1}^0$ and $\mathbf{a}_{m+1}^0$ as

$$\boldsymbol{\sigma}_{m+1}^0 = \boldsymbol{\sigma}_m \quad \text{and} \quad \mathbf{a}_{m+1}^0 = \mathbf{a}_m$$

we can write for Eq. (18.71) a set of successive iterations

$$\bar{\boldsymbol{\Psi}}_{m+1}^{n+1} = 0 = \boldsymbol{\Psi}_{m+1}^n + \Delta\boldsymbol{\sigma}_{m+1}^n - \mathbf{DB}\,\Delta\mathbf{a}_{m+1}^n + \mathbf{D}\,\Delta t\theta \mathbf{S}^n\,\Delta\boldsymbol{\sigma}_{m+1}^n \tag{18.72}$$

where $\mathbf{S} = (\partial\boldsymbol{\beta}/\partial\boldsymbol{\sigma})_{m+\theta}$. Similarly for Eq. (18.68)

$$\boldsymbol{\Psi}_{m+1}^{n+1} = 0 = \boldsymbol{\Psi}_{m+1}^n + \int_V \mathbf{B}^{\mathrm{T}}\,\Delta\boldsymbol{\sigma}_{m+1}^n\,\mathrm{d}V. \tag{18.73}$$

From Eqs. (18.72) and (18.73) successive corrections, $\Delta\mathbf{a}_{m+1}^n$ and $\Delta\boldsymbol{\sigma}_{m+1}^n$, can be determined giving

$$\begin{aligned} \mathbf{a}_{m+1}^{n+1} &= \mathbf{a}_{m+1}^n + \Delta\mathbf{a}_{m+1}^n \\ \boldsymbol{\sigma}_{m+1}^{n+1} &= \boldsymbol{\sigma}_{m+1}^n + \Delta\boldsymbol{\sigma}_{m+1}^n \end{aligned} \tag{18.74}$$

It is convenient to rewrite Eq. (18.72) as

$$\Delta\boldsymbol{\sigma}_{m+1}^n = -(\bar{\mathbf{D}}^n\mathbf{D}^{-1})\,\bar{\boldsymbol{\Psi}}_{m+1}^n + \bar{\mathbf{D}}^n\mathbf{B}\,\Delta\mathbf{a}_{m+1}^n \tag{18.75}$$

where

$$\bar{\mathbf{D}}^n \equiv [\mathbf{D}^{-1} + \Delta t\theta\,\mathbf{S}^n]^{-1} \tag{18.76}$$

and to eliminate $\Delta\boldsymbol{\sigma}_{m+1}^n$ from Eq. (18.73), thus obtaining an explicit algorithm for $\Delta\mathbf{a}_{m+1}^n$.

As an approximation is already involved at the stage of writing Eq. (18.70), the iterative solution is generally curtailed after one or two cycles. Further, a considerable simplification (as well as loss of accuracy) occurs if $\theta = 0$ is taken. The various possibilities include some well-known incremental processes.

Method 1—Euler: $\theta = 0;\quad n = 1$

From Eq. (18.71) and the initial conditions we have

$$\bar{\boldsymbol{\Psi}}_{m+1}^0 = \mathbf{D}\,\Delta t\boldsymbol{\beta}_m$$

and from Eq. (18.76)

$$\bar{\mathbf{D}}^n = \mathbf{D}.$$

From Eq. (18.75) we have

$$\Delta\boldsymbol{\sigma}_{m+1}^{0} = -\overline{\boldsymbol{\Psi}}_{m+1}^{0} + \mathbf{DB}\,\Delta\mathbf{a}_{m+1}^{0} = \mathbf{D}(\mathbf{B}\,\Delta\mathbf{a}_{m+1}^{0} - \Delta t\boldsymbol{\beta}_{m}) \qquad (18.77)$$

where the last term represents simply the creep strain increments.

Substituting this into Eqs. (18.73) and (18.68) gives

$$\left(\int_{V} \mathbf{B}^{\mathrm{T}}\mathbf{DB}\,\mathrm{d}V\right)\Delta\mathbf{a}_{m+1}^{0} + \Delta\mathbf{f}_{m} - \int_{V} \mathbf{B}^{\mathrm{T}}\mathbf{D}\,\Delta t\boldsymbol{\beta}_{m}\,\mathrm{d}V = 0 \qquad (18.78)$$

from which $\Delta\mathbf{a}_{m+1}^{0}$ and hence $\mathbf{a}_{m+1}^{1} = \mathbf{a}_{m}^{0} + \Delta\mathbf{a}_{m+1}^{0}$ are obtained.

The process is thus simply one of *initial strain* which is estimated from the rate $\dot{\boldsymbol{\varepsilon}}_{c} = \boldsymbol{\beta}_{m}$ computed at the start of the interval, and a simple elastic solution with the well-known constant stiffness matrix

$$\mathbf{K} = \int_{V} \mathbf{B}^{\mathrm{T}}\mathbf{DB}\,\mathrm{d}V$$

is involved at each time step.

This process is deservedly popular[44–47] as the computation at each time step is simply one of resolution, but it is obviously less accurate than other alternatives. Further, if the time interval is too large, unstable results may be obtained (*vide* Chapter 21). Thus it is necessary for

$$\Delta t \leqslant \Delta t_{\mathrm{crit}} \qquad (18.79)$$

where Δt_{crit} is determined in a suitable manner.

A rule of thumb which proves quite effective in practice is that the increment of creep strain should not exceed one-half of the total elastic strain, in that[48]

$$\Delta t_{\mathrm{crit}}\boldsymbol{\beta}_{m} \leqslant \tfrac{1}{2}\boldsymbol{\varepsilon}_{e} = \tfrac{1}{2}\mathbf{D}^{-1}\boldsymbol{\sigma}_{m}. \qquad (18.80)$$

Method 2—Tangential: $\theta \neq 0$; $n = 1$

Again from Eq. (18.71) and the initial conditions we have

$$\boldsymbol{\Psi}_{m+1}^{0} = \mathbf{D}\,\Delta t\boldsymbol{\beta}_{m} \quad \text{as } \boldsymbol{\beta}_{m+\theta}^{0} \equiv \boldsymbol{\beta}_{m}$$

but now from Eq. (18.76)

$$\overline{\mathbf{D}}^{0} = [\mathbf{D}^{-1} + \Delta t_{m}\theta\mathbf{S}^{0}]$$

in which $\mathbf{S}^{0}$ is computed using stresses at the start of the interval.

With the same sequence as in Method 1 from Eq. (18.75)

$$\Delta\boldsymbol{\sigma}_{m+1}^{0} = \overline{\mathbf{D}}^{0}(\mathbf{B}\,\Delta\mathbf{a}_{m+1}^{0} - \Delta t\boldsymbol{\beta}_{m}) \qquad (18.81)$$

and using again Eq. (18.68) and Eq. (18.73)

$$\left(\int_{V} \mathbf{B}^{\mathrm{T}}\overline{\mathbf{D}}^{0}\mathbf{B}\,\mathrm{d}V\right)\Delta\mathbf{a}_{m+1}^{0} + \Delta\mathbf{f}_{m} - \int_{V} \mathbf{B}^{\mathrm{T}}\overline{\mathbf{D}}^{0}\,\Delta t\boldsymbol{\beta}_{m}\,\mathrm{d}V = 0. \qquad (18.82)$$

from which $\Delta\mathbf{a}_{m+1}^{0}$ and hence $\Delta\boldsymbol{\sigma}_{m+1}^{0}$ are obtained.

This process using $\theta = \frac{1}{2}$ was introduced by Cyr and Teter[49] and Zienkiewicz.[50] Other values of $\theta > \frac{1}{2}$ have proved even more reliable[51] and it can be shown that with any $\theta \geqslant \frac{1}{2}$ the scheme is unconditionally stable and the only limit to the time step is that of accuracy.[52]

In either method successive iterations can be used to improve the results.

The penalty for the improved performance and stability of Method 2 is the fact that at each time step a new matrix needs to be formed and inverted; further. on occasion, such matrices are non-symmetric. The method becomes more complex if $\boldsymbol{\varepsilon}_c$ enters expression (18.64). No such difficulty is encountered, however, with the first procedure.

18.6 **Viscoplasticity**

18.6.1 *General.* The purely plastic behaviour of solids postulated earlier is probably a fiction as the maximum stress which can be carried is invariably associated with the rate at which this is applied. A purely elasto-plastic behaviour in a uniaxial loading is described in a model of Fig. 18.11(*a*) in which the plastic strain rate is zero for stresses below yield, i.e.,

$$\dot{\varepsilon}_p = 0 \quad \text{if } \sigma - \sigma_y < 0$$

and $\dot{\boldsymbol{\varepsilon}}_p$ is indeterminable when $\sigma - \sigma_y = 0$.

The elasto-viscoplastic material, on the other hand, can be modelled as shown in Fig. 18.11(*b*), where a dashpot is placed in parallel with the plastic element. Now stresses can exceed σ_y for strain rates other than zero.

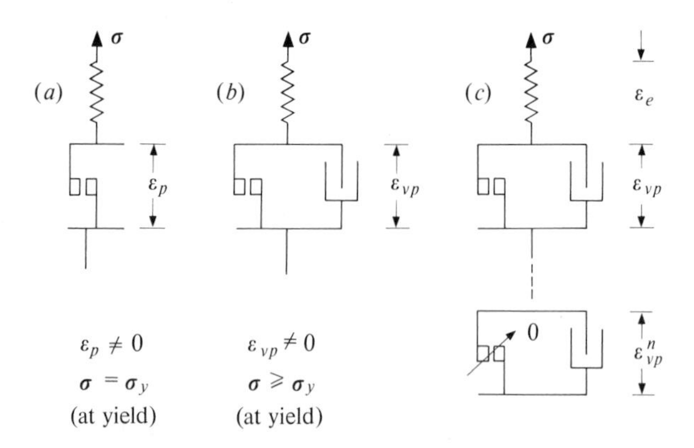

Fig. 18.11 (*a*) Elasto-plastic. (*b*) Elasto-viscoplastic. (*c*) A series of elasto-viscoplastic models

The viscoplastic (or creep) strain rate is now given by a general expression

$$\dot{\varepsilon}_{vp} = \gamma\langle\phi(\sigma-\sigma_y)\rangle$$

where the arbitrary function ϕ is such that

$$\langle\phi(\sigma-\sigma_y)\rangle = 0 \quad \text{if } \sigma-\sigma_y \leqslant 0$$

$$\langle\phi(\sigma-\sigma_y)\rangle = \phi(\sigma-\sigma_y) \quad \text{if } \sigma-\sigma_y > 0$$

The model suggested is, in fact, of a creep type category described in the previous sections and this appears to be more realistic than that of pure plasticity.

A generalization of a viscoplastic model to a general stress state follows precisely the arguments of the plasticity section.

First, we note that the strain rate will be a function of the yield condition

$$F(\boldsymbol{\sigma})$$

defined in Eq. (18.28). If this is less than zero, no plastic flow will occur.

Second, we shall postulate a viscoplastic potential $Q(\boldsymbol{\sigma})$, the normality to this governing the ratio of the various strain rate components [*vide* Eq. (18.35)].

Thus, quite generally we can write

$$\dot{\boldsymbol{\varepsilon}}_{vp} = \gamma\langle\phi(F)\rangle\frac{\partial Q}{\partial \boldsymbol{\sigma}} \equiv \boldsymbol{\beta}(\boldsymbol{\sigma}) \tag{18.83}$$

where

$$\langle\phi(F)\rangle = 0 \quad \text{if } F \leqslant 0$$

$$\langle\phi(F)\rangle = \phi(F) \quad \text{if } F > 0$$

Once again *associated* or *non-associated* flows can be invoked, depending on whether $Q = F$ or not. Further, any of the yield surfaces described in section 18.4.1 can be used to define the appropriate flow in detail.

The concept of viscoplasticity in one of its earliest versions was introduced by Bingham in 1922[53] and a complete survey of such modelling is given in reference 54.

The computational procedure of using the viscoplastic model can obviously follow any of the general procedures described in the previous section. Most commonly the straightforward *Euler* method of the previous section is used.[55–58] The detailed stability requirements for this have been considered for several types of yield condition by Cormeau.[57] The *tangential* process can occasionally be economically used, but unless the viscoplastic flow is associated (i.e., $Q = F$), non-symmetric systems of equations have to be solved at each step.[51,52]

The viscoplastic law can easily be generalized to include a series of components, as shown in Fig. 18.11(*c*).

Now we write

$$\dot{\boldsymbol{\varepsilon}}_{vp} = \dot{\boldsymbol{\varepsilon}}_{vp}^{1} + \dot{\boldsymbol{\varepsilon}}_{vp}^{2} + \cdots = \boldsymbol{\beta}(\boldsymbol{\sigma}) \tag{18.84}$$

and again the standard formulation suffices.

If, as shown in the last element of Fig. 18.11(*c*), the plastic element is put as zero, a 'pure' creep situation arises in which flow occurs at all stress levels.

The viscoplastic model can also be used conveniently for the solution of purely plastic problems, indeed providing here a simpler algorithm than those previously discussed. The process is simply one of imposing a constant load and performing the time integration *until all strain rates become zero*. At that stage a static plastic solution is clearly reached (as all the 'dashpots' become inactive).

Again, for such solutions all of the previously described processes have been used.[55–58] Recent work indicates that, as the exact nature of the time variation is not important and indeed plays a purely fictional role, an iterative process of the previous section with one very large time step using a 'backward difference', i.e., $\theta = 1$, may be economical.[52]

18.6.2 *Creep of metals.* If an associated form of viscoplasticity using the von Mises yield criterion of Eq. (18.53) is considered, the viscoplastic strain rate can be written as

$$\dot{\boldsymbol{\varepsilon}}_{vp} = \gamma \langle \phi(\bar{\sigma} - \sigma_y) \rangle \frac{\partial \bar{\sigma}}{\partial \boldsymbol{\sigma}}. \tag{18.85}$$

If σ_y, the yield stress, is put to zero, ϕ is made an exponential form, and we make use of the expressions of Table 18.1, the above can be written as

$$\dot{\boldsymbol{\varepsilon}}_{vp} = \dot{\boldsymbol{\varepsilon}}_{c} = \gamma \bar{\sigma}^{m} \mathbf{M}^{I} \boldsymbol{\sigma} = \boldsymbol{\beta}(\boldsymbol{\sigma}) \tag{18.86}$$

and we obtain well-known Norton–Soderberg creep law. In this, generally, the parameter γ is a function of time, temperature, and the total creep strain. For a survey of such laws the reader should consult specialized references.[59–61]

An example initially solved using a large number of triangular elements[47] is presented in Fig. 18.12, where a much smaller number of isoparametric quadrilaterals are used in a general viscoplastic program.[56d]

18.6.3 *Plasticity solutions by viscoplastic algorithm—soil mechanics.* As we have already mentioned, the viscoplastic model provides a simple and powerful tool for the solution of all plasticity problems. In reference 56d many classical problems are thus resolved and the reader is directed

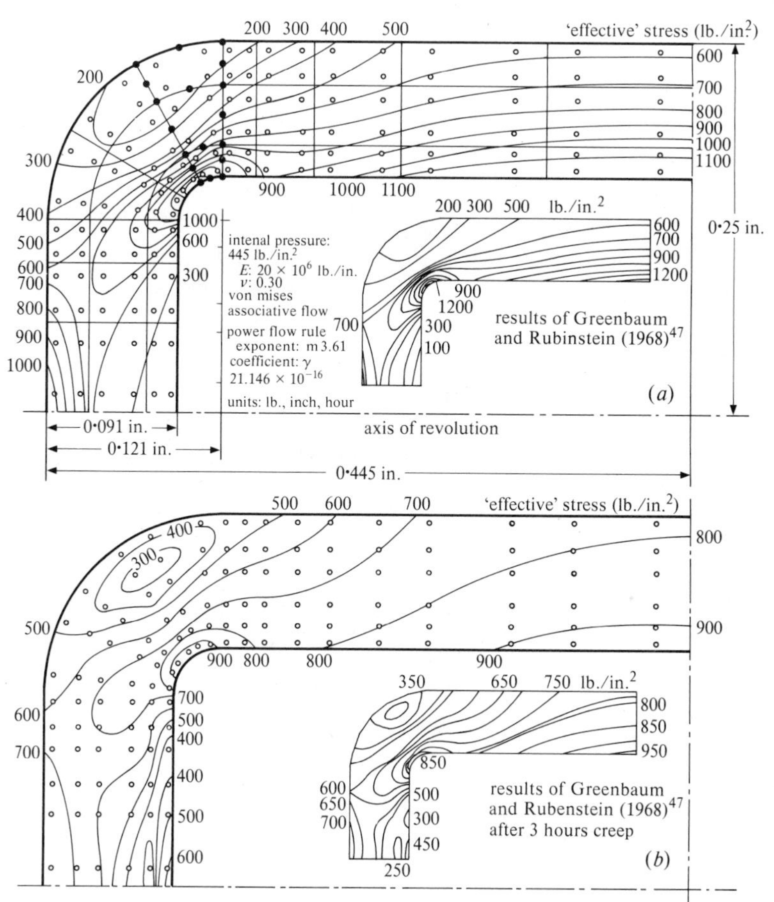

Fig. 18.12 Creep in a pressure vessel (*a*) and (*b*)

therein for details. In this section some problems of soil mechanics are discussed in which the facility of the process for solving non-associated behaviour is demonstrated. The whole subject of the behaviour of soils and similar porous media is one in which much has yet to be done to formulate good constitutive models. For a fuller discussion the reader is referred to several recent texts, conferences, and papers on the subject.[58,62–64]

One particular controversy centres on associated versus non-associated nature of soil behaviour. In the example of Fig. 18.13, dealing with an axi-symmetric sample, the effect of these different assumptions is investigated.[57] Here a Mohr–Coulomb law is used to describe the yield

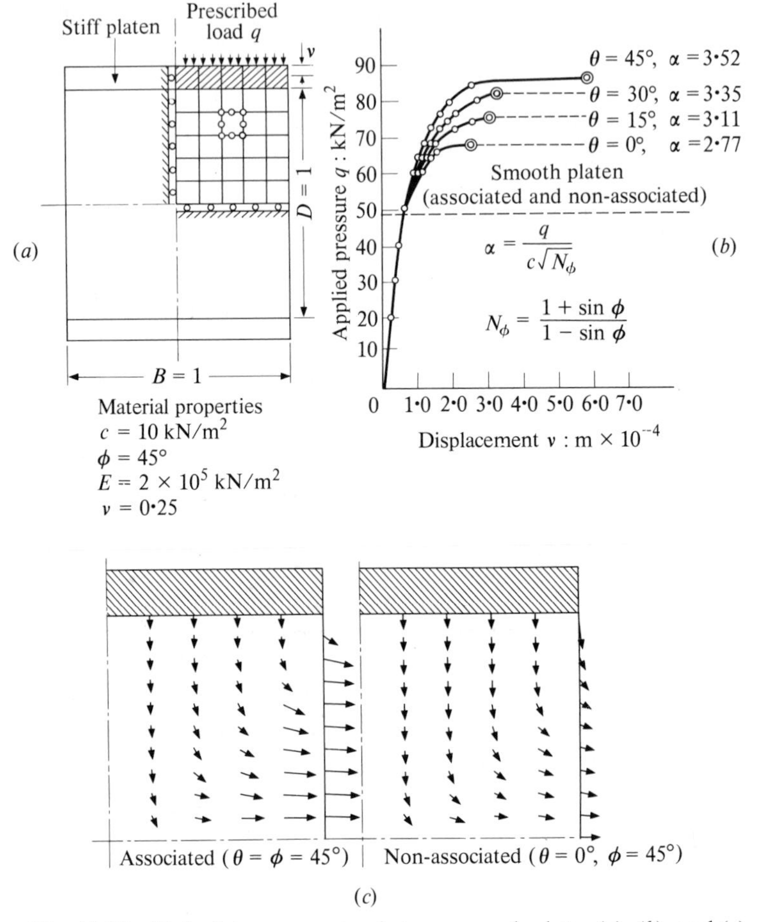

Fig. 18.13 Uniaxial compression between rough plates (*a*), (*b*), and (*c*)

surface and a similar form, but with a different friction angle $\phi = \theta$, is used as the plastic potential ($\theta = 0$ reducing the plastic potential to the Tresca form of Fig. 18.6 and suppressing volumetric strain changes). As can be seen from the results, only moderate changes in the collapse load occur although very appreciable differences in plastic flow patterns exist.

Figure 18.14 shows a similar study carried out for an embankment. Here, despite quite different flow patterns, a prediction of collapse load was almost unaffected by the flow rate law assumed.[58]

18.7 Visco-elasticity

History dependence of creep. Visco-elastic phenomena are characterized by the fact that the rate at which creep strains develop depends not only

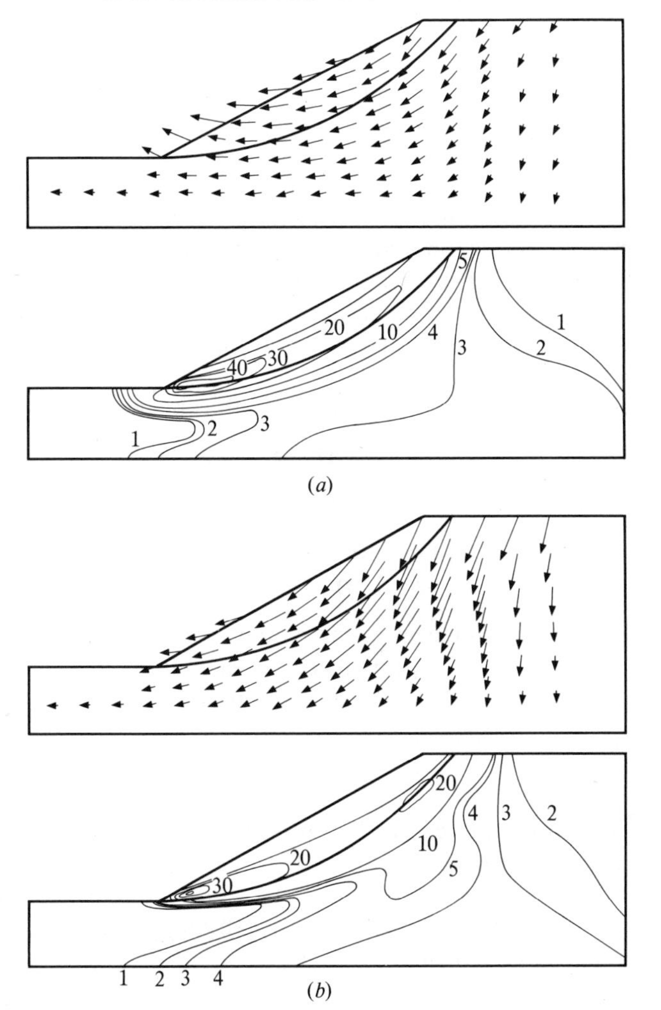

Fig. 18.14 Embankment—relative plastic velocities at collapse and effective shear strain rate contours at collapse (*a*) and (*b*)

on the current state of stress and strain but, in general, on the *full history* of their development. Thus to determine the increment of strain $\Delta\boldsymbol{\varepsilon}_c$ at a particular time interval it is necessary to know the state of stress and strains at all *preceding time intervals*. While in the computation process these can in fact be obtained, *in principle* the problem presents little difficulty. Practical limitations, however, appear immediately. Even with the largest computers available it is not practicable to store the full history on core and the repeated use of backing storage is too slow and therefore too costly to be contemplated.

A method of overcoming this difficulty was described by Zienkiewicz *et al.*[45] in the context of *linear visco-elastic* analysis and presents possibilities for suitably formulated *non-linear* visco-elastic materials.

In linear visco-elasticity it is always possible to write the stress-strain relationship in a form similar to that of elasticity, with the various terms of the **D** matrix representing now in place of elastic constants, suitable differential or integral operators.[65] Thus in an isotropic continuum a pair of operators corresponding to an appropriate pair of elastic constants will appear—while for anisotropic materials up to 21 separate operators may be necessary.

Typically the creep part of the strain may thus be described by

$$\boldsymbol{\varepsilon}_c = \overline{\mathbf{D}}^{-1}\boldsymbol{\sigma}$$

where each term of the 'visco-elastic matrix', $\overline{\mathbf{D}}^{-1}$, may take up a form

$$\bar{d}_{rs} = \frac{a_0 + a_1(\mathrm{d}/\mathrm{d}t) + a_2(\mathrm{d}^2/\mathrm{d}t^2) + \cdots}{b_0 + b_1(\mathrm{d}/\mathrm{d}t) + b_2(\mathrm{d}^2/\mathrm{d}t^2) + \cdots} \tag{18.87}$$

if the operators are written in a differential form. If this expansion is finite, then separating any instantaneous elastic effects one can usually rewrite Eq. (18.87) in terms of partial fractions as

$$\bar{d}_{rs} = \frac{A_1}{\mathrm{d}/\mathrm{d}t + B_1} + \frac{A_2}{\mathrm{d}/\mathrm{d}t + B_2} + \cdots \tag{18.88}$$

This, as is well known, can be interpreted as a response of a series of 'Kelvin' elements illustrated in Fig. 18.15 (even though physically no significance need be attached to such models) with each term representing one Kelvin unit. A typical contribution to a strain component is thus an addition of terms of the form

$$\varepsilon^n = \frac{A_n}{\mathrm{d}/\mathrm{d}t + B_n}\sigma_s \tag{18.89}$$

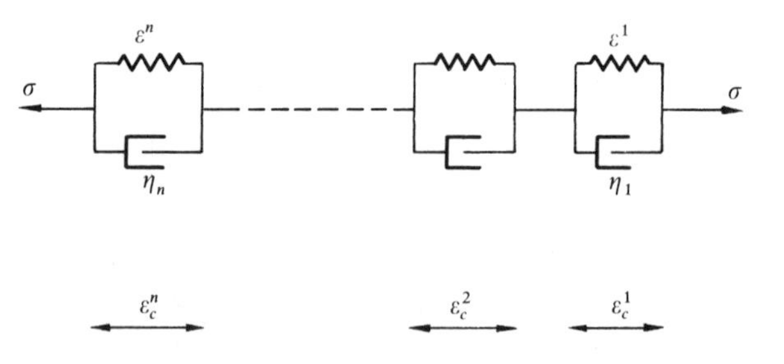

Fig. 18.15 A series of Kelvin elements

or

$$\frac{\mathrm{d}}{\mathrm{d}t}\boldsymbol{\varepsilon}^n = A_n\sigma_s - B_n\varepsilon_n. \qquad (18.90)$$

and the full expression for $\dot{\boldsymbol{\varepsilon}}_c$ is obtained once again in the form of Eq. (18.64) for which the standard methods described in section 18.5 are applicable. The Euler procedure is in fact used in reference 45, but the tangential procedures lead to a more rapid and accurate solution if we observe that the matrix **S** is constant in linear visco-elastic forms.

In practice only a limited number of Kelvin elements are needed to represent the material behaviour and, additionally, only a few 'visco-elastic' operators exist. For instance, in an isotropic incompressible material only one operator serves to define the $\overline{\mathbf{D}}^{-1}$ matrix. With two elements of expansion, Eq. (18.88), defining this operator only two quantities will have to be stored during the computation process.[45]

The values of A_n and B_n for each Kelvin model can be age and temperature dependent without introduction of any complexity of calculation—thus dealing with problems of thermo-visco-elasticity, such as occur in the creep of concrete or plastics.

The visco-elastic computations can, on occasion, be simplified by the use of special devices for expanding the history-dependent creep strain. A particularly efficient method is suggested by Taylor *et al.*[66]

The labour of step-by-step solutions for linear visco-elastic media can, on occasion, be substantially reduced. In the case of a homogeneous structure with linear isotropic visco-elasticity and constant Poisson ratio operator, the Alfrey-McHenry analogies allow single-step elastic solutions to be used to obtain stresses and displacements at a given time by the use of *equivalent loads*, *displacements*, and *temperatures*.[67]

Some extensions of these analogies have been proposed by Hilton.[68]

Further, when creep deformation is of the type tending to a constant value at infinite time it is possible to determine the final stress distribution even in cases where the above analogies are not applicable. Thus, for instance, where the visco-elastic properties are temperature dependent and the structure is subject to a system of loads and temperatures which remain constant with time, long term 'equivalent elastic constants' can be found and the problem solved as a single, non-homogeneous elastic one of a linear kind.[69]

18.8 Some Special Problems of Rock, Concrete, etc.

18.8.1 *The no-tension material.* A hypothetical material capable of sustaining only compressive stresses and straining without resistance in tension is in many respects similar to an ideal plastic material. While in practice probably such an ideal material does not exist, it gives a close approximation to the behaviour of randomly jointed rock and other granular materials.

While an explicit stress-strain relation cannot be generally written it suffices to carry out the analysis elastically and wherever tensile stresses develop to reduce these to zero. The initial stress process is here natural and indeed was developed in this context.[16]

The steps of calculation are obvious but it is important to remember that the *principal tensile stresses* have to be eliminated.

The 'constitutive law' as stated above can at best approximate to the true situation, no account being taken of closing of fissures on re-application of compressive stresses. However, these results certainly give a clearer insight into the behaviour of real rock structures.

An underground power station. Figure 18.16(*a*) and (*b*) show an application of this model to a practical problem. In Fig. 18.11(*a*) an elastic solution is shown for stresses in the vicinity of an underground power station with cable prestressing applied in the vicinity of the opening. The zones in which tension exists are indicated. In Fig. 18.16(*b*) a *no tension* solution is given for the same problem indicating the rather small general stress redistribution and the zones where 'cracking' has occurred.

Reinforced concrete. A variant on this type of material may be one in which a finite tensile strength exists but when this is once exceeded the strength drops to zero (on fissuring). Such an analysis was used by Valliappan and Nath[70] in the study of the behaviour of reinforced concrete beams. Very good correlation with experimental results for over-reinforced beams (in which development of compressive yield is not important) have been obtained. The beam is one for which tests have been carried out by Krahl *et al.*[71] Figure 18.17 shows some relevant results.

Much development work on the behaviour of reinforced concrete has taken place with various plasticity forms being introduced to allow for compressive failure and procedures which take into account the crack-closing history. References 72–80 list some of the more important papers on this subject.

18.8.2 *'Laminar' material and joint elements.* Another idealized material model is one which is assumed to be built up of a large number of isotropic and elastic laminae. When under compression these can transmit shear stress parallel to their direction providing this does not exceed the frictional resistance. No tensile stresses can, however, be transmitted in the normal direction to the laminae.

This idealized material has obvious uses in the study of rock masses with parallel joints but as will be seen later has a much wider applicability.

Figure 18.18 shows a two-dimensional situation involving such a material. With a local co-ordinate axis x' oriented in the direction of the

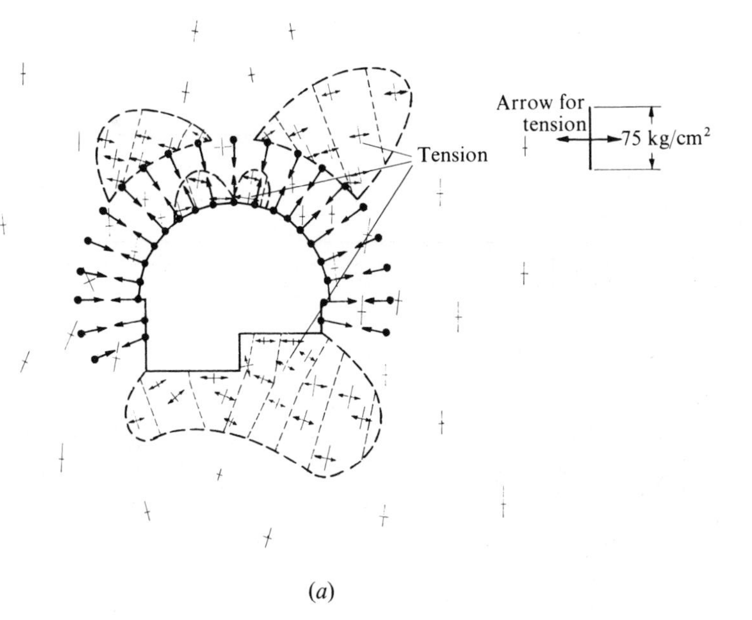

(a)

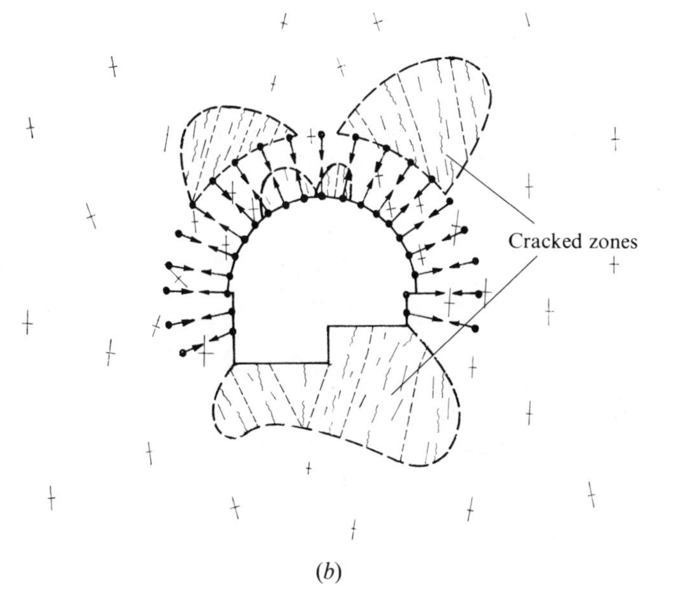

(b)

Fig. 18.16 Underground power station gravity and prestressing (a) elastic stresses; (b) 'no-tension' stresses

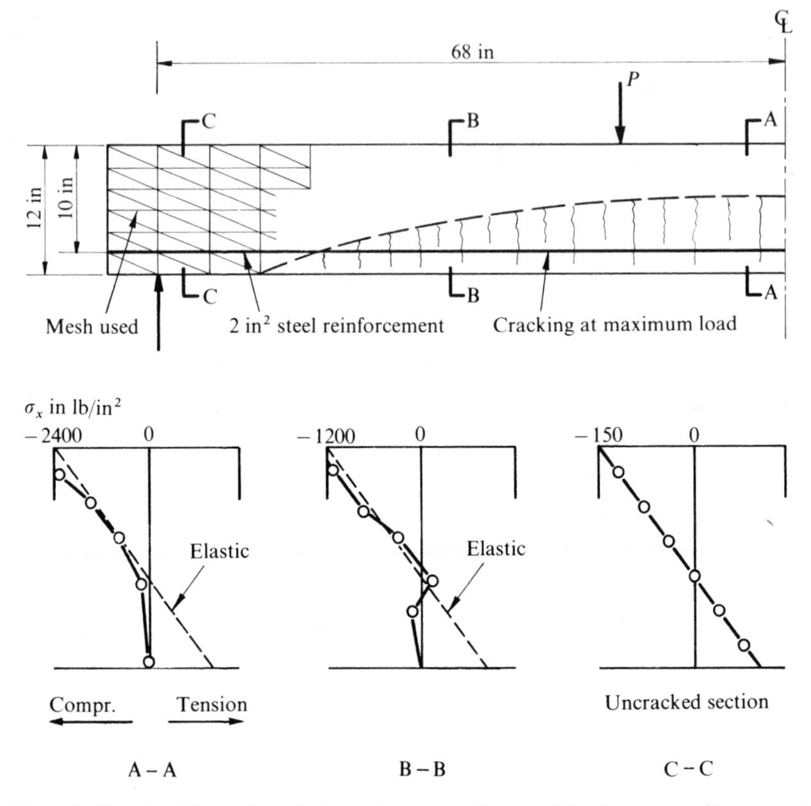

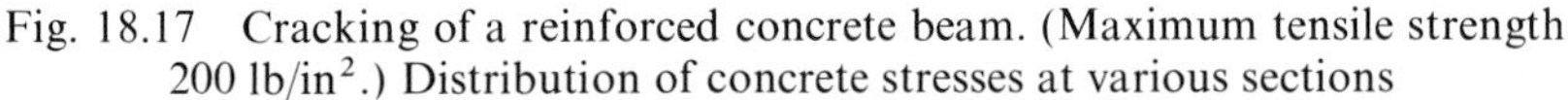

Fig. 18.17 Cracking of a reinforced concrete beam. (Maximum tensile strength 200 lb/in^2.) Distribution of concrete stresses at various sections

laminae we can write

$$|\tau_{x'y'}| \leqslant \mu\sigma_y \tag{18.91a}$$

and

$$\sigma_{y'} \leqslant 0 \tag{18.91b}$$

for stresses at which purely elastic behaviour occurs. In the above, μ is the friction coefficient applicable between the laminae.

If elastic stresses exceed the limits imposed the stresses have to be reduced to the limiting values given above.

The application of the initial stress process in this context is again self evident, and the problem is very similar to that implied in the *no tension* material of previous section. At each step of elastic calculation first the existence of tensile stresses $\sigma_{y'}$ is checked and if these develop, a corrective initial stress reducing these and the shearing stresses to zero is applicable.

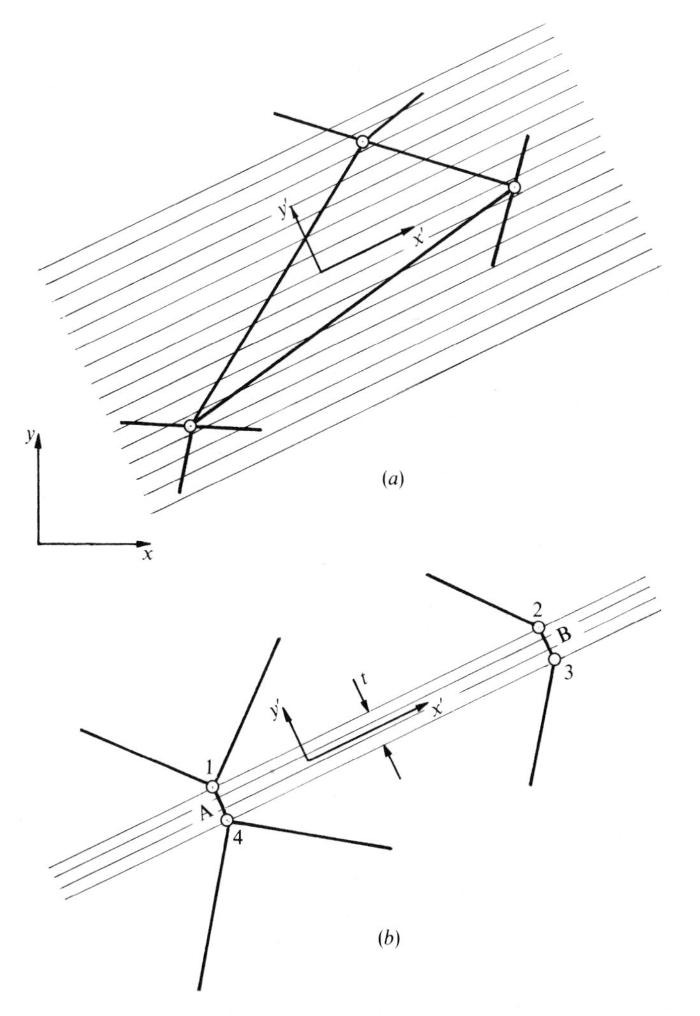

Fig. 18.18 'Laminar' material (*a*) General. (*b*) In a narrow joint

If $\sigma_{y'}$ = stresses are compressive, the absolute magnitude of the shearing stresses $\tau_{x'y'}$ are checked and again, if these exceed the value given by Eq. (18.91a) they are reduced to their proper limit.

However, such a procedure poses the question of the manner in which the stresses are reduced, as two components have to be considered. It is, therefore, preferable to use the statements of Eqs. (18.91a)–(18.91b) as definitions of plastic yield surfaces (F). The assumption of an additional plastic potential (Q) will now define the position, and we note that associated behaviour with Eq. (18.91), used as potential, will imply a simultaneous separation and sliding of the laminar (as the corresponding strain rates $d\gamma_{x'y'}$ and $d\varepsilon_{y'}$ are finite). Non-associated plasticity (as visco-plasticity) techniques have therefore to be used.

Once again, if stress reversal is possible it is necessary to note the opening of the laminae, i.e., the yield surface is made strain dependent.

In some instances the laminar behaviour is confined to a narrow joint between relatively homogeneous elastic masses. This may well be of a nature of a geological fault or a major crushed rock zone. In such cases it is convenient to use narrow, generally rectangular, elements whose geometry may be specified by mean co-ordinates of two ends A and B (Fig. 18.18b) and the thickness. The element still has, however, separate points of continuity (1–4) with the adjacent masses.[81]

Such joint elements can be simple rectangles as shown here, but equally well can take up more complex shapes if isoparametrically specified (see Chapter 8).

Goodman *et al.*[82] describe somewhat similar joint elements used in the context of stability of joint rock masses.

Laminations may not be confined to one direction only—and indeed the material itself may possess a plastic limit. The use of such multi-laminate models in the context of rock mechanics has been developed recently.[83]

18.9 **Concluding Remarks—Solid Mechanics**

In the preceding sections the general processes of dealing with complex, non-linear constitutive relations have been examined and some particular applications were discussed. Clearly, the subject is so large and of so great a practical importance that presentation in a single chapter is impracticable. For different materials different forms of constitutive relations can be proposed and experimentally verified. *Once such constitutive relations are available the standard processes of this chapter can be adapted;* indeed it is now possible to build standard computing systems applicable to a wide variety of material properties in which new specifications of behaviour are simply inserted as an appropriate 'black box'.

What must be once more re-stated is that in non-linear problems

(*a*) non-uniqueness of solution may arise
(*b*) convergence can never be, *a priori*, guaranteed
(*c*) cost of solution is invariably greater than in linear solutions.

In this section we have limited ourselves to small strain behaviour. Extension to large strain is a subject beyond the scope of this book, but readers who are interested can consult other appropriate references.[84,85]

In the next chapter we shall deal mainly with large deformations—

small strain behaviour, but the basic techniques described here will again be found to be applicable.

18.10 Non-linear Quasi-harmonic Field Problems

Non-linearity may arise in many problems beyond those of solid mechanics—but the techniques described at the start of this chapter are still universally applicable. In subsequent chapters we shall touch upon such non-linearities in the context of transient problems (Chapter 21) and various fluid mechanics situations (Chapter 22). Here we shall look again at one class of these problems which is governed by the field equations of Chapter 17.

With Eq. (17.9)—written for simplicity for an isotropic material and two dimensions of space—we have

$$\nabla^{\mathrm{T}} k \nabla \phi + Q = \frac{\partial}{\partial x} k \frac{\partial \phi}{\partial x} + \frac{\partial}{\partial y} k \frac{\partial \phi}{\partial y} + Q = 0 \tag{18.92}$$

with suitable boundary conditions.

The discretization based on Galerkin procedures is still valid if k and/or Q (and indeed the boundary conditions) are dependent on ϕ or its derivatives. Thus the discretized form given by Eq. (17.17), i.e.,

$$\mathbf{\Psi}(\mathbf{a}) \equiv \mathbf{Ha} + \mathbf{f} = 0; \qquad \mathbf{H} = \mathbf{H}(\mathbf{a}); \qquad \mathbf{f} = \mathbf{f}(\mathbf{a}) \tag{18.93}$$

is such that the integrand of each term depends in a scalar fashion on ϕ (or its derivative).

Equations (18.93) are a particular case of the general non-linear problem given by Eq. (18.1) for which *direct iteration techniques are possible*. However, as these occasionally fail to converge it is necessary to determine the tangential matrix $\mathrm{d}\mathbf{\Psi}/\mathrm{d}\mathbf{a}$ and use one of the other techniques.

Let us examine this in detail, making use of Eq. (17.17) which defines the appropriate matrices. Using these we can write

$$\frac{\mathrm{d}\mathbf{\Psi}}{\mathrm{d}\mathbf{a}} \mathrm{d}\mathbf{a} \equiv \mathbf{H}\,\mathrm{d}\mathbf{a} + \mathrm{d}\mathbf{H}\mathbf{a} + \mathrm{d}\mathbf{f}. \tag{18.94}$$

If k and Q are direct functions of ϕ, we can write†

$$\mathrm{d}\mathbf{H}\mathbf{a} = \mathbf{A}\,\mathrm{d}\mathbf{a} \quad \text{with } A_{ij} = \int_{\Omega} \nabla N_i^{\mathrm{T}} (\nabla \mathbf{N}\mathbf{a}) N_j k' \,\mathrm{d}\Omega \tag{18.95}$$

† To derive the matrix terms of $\mathrm{d}\mathbf{Ha}$ it is convenient to consider the jth row of $\mathrm{d}\mathbf{\Psi}_i$ only, i.e.,

$$\mathrm{d}\left(\int (\nabla N_i^T)\, k \nabla \mathbf{N}\, \mathrm{d}\Omega\right) \mathbf{a} =$$

$$= \int \nabla N_i^{\mathrm{T}} (\nabla N_1 k' (N_1\, \mathrm{d}a_1 + N_2\, \mathrm{d}a_2 + \cdots) a_1 + \nabla N_2 k' (N_1\, \mathrm{d}a_1 + \cdots) + \cdots$$

The coefficient of $\mathrm{d}a_j$ immediately gives $\mathbf{A}_{ij}$.

and

$$\mathrm{d}\mathbf{f} = \left(\int_\Omega \mathbf{N}^\mathrm{T} Q' \mathbf{N}\, \mathrm{d}\Omega\right) \mathrm{d}\mathbf{a} = \mathbf{C}\, \mathrm{d}\mathbf{a}$$

$$\text{with} \quad C_{ij} = \int_\Omega N_i^\mathrm{T} Q' N_j\, \mathrm{d}\Omega \tag{18.96}$$

where

$$k' = \mathrm{d}k/\mathrm{d}\phi \qquad Q' = \mathrm{d}Q/\mathrm{d}\phi. \tag{18.97}$$

The tangential matrix thus becomes

$$\frac{\mathrm{d}\boldsymbol{\Psi}}{\mathrm{d}\mathbf{a}} = \mathbf{H} + \mathbf{A} + \mathbf{B} \tag{18.98}$$

in which the second term is non-symmetric. Newton–Raphson procedures in such a case are inconvenient and modifications of the formulation are sometimes made.

Indeed it is easy to show that a variational principle in such cases is not the same as that given in Chapter 17. Special forms of a variational principle can be devised and will lead to symmetry.[86]

In many physical problems, however, the values of k depend on the absolute value of the gradient of $\nabla\phi$, i.e.,

$$V = \sqrt{(\nabla\phi)^\mathrm{T}(\nabla\phi)} = \sqrt{\left(\frac{\partial\phi}{\partial x}\right)^2 + \left(\frac{\partial\phi}{\partial y}\right)^2}$$
$$\bar{k}' = \frac{\mathrm{d}k}{\mathrm{d}V}. \tag{18.99}$$

In such cases we can fortunately write

$$\mathrm{d}\mathbf{H}\mathbf{a} = \bar{\mathbf{A}}\, \mathrm{d}\mathbf{a} \tag{18.100}$$

where

$$\bar{A}_{ij} = \int_\Omega (\nabla N_i)^\mathrm{T} (\nabla\mathbf{N}\mathbf{a})^\mathrm{T} \bar{k}' (\nabla\mathbf{N}\mathbf{a})\, \nabla N_j\, \mathrm{d}\Omega \tag{18.101}$$

and symmetry is preserved.

Situations of this kind arise in seepage flow where the permeability is dependent on the absolute value of the flow velocity,[87–89] in magnetic fields[90–92] where magnetic formulations in a function of the absolute field strength, in slightly compressible yield flow, and indeed in many other physical situations.[93,94]

Figure 18.19 from reference 90 illustrates a typical non-linear magnetic field solution.

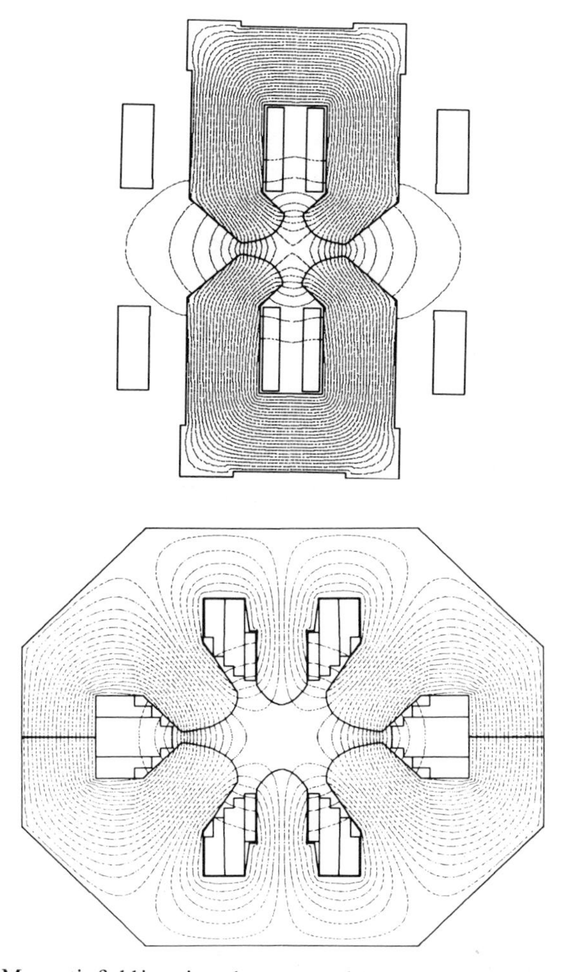

Fig. 18.19 Magnetic field in a six pole magnet with non-linearity due to saturation[90]

Whilst many more interesting problems could be quoted we conclude with one in which the only non-linearity is that due to the heat generation term Q in Eq. (18.92). This particular problem of spontaneous ignition,[95] in which Q depends exponentially on the temperature, serves to illustrate the point about the possibility of multiple solutions and indeed the non-existence of any solution in certain non-linear cases.

Taking $k = 1$ and $Q = \delta e^{\phi}$, an eliptic domain is examined in Fig. 18.20. Using various values of δ, a Newton–Raphson iteration is used to obtain a solution and we find that no convergence (and indeed *no solution*) exists when $\delta > \delta_{crit}$ exists. Above the critical value of δ_{crit} the temperature rises indefinitely and *spontaneous ignition* of the material occurs. For values

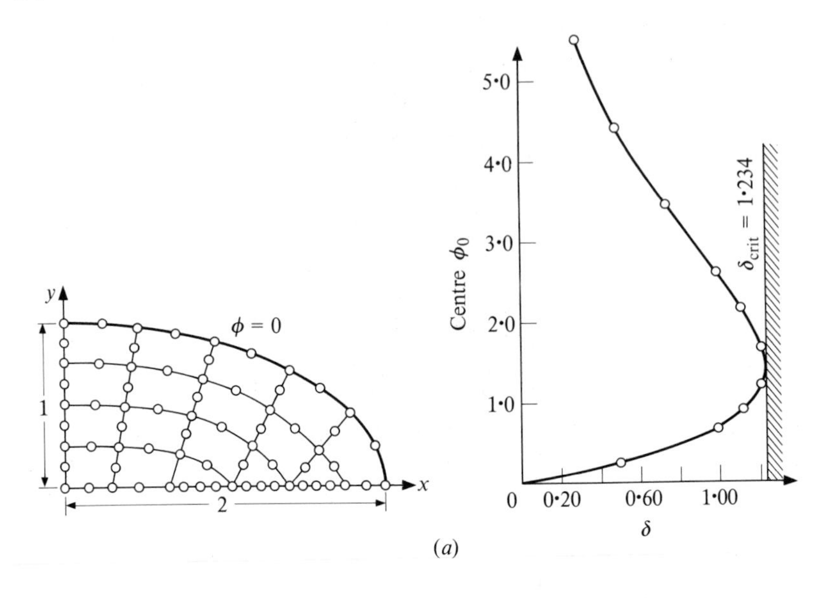

(a)

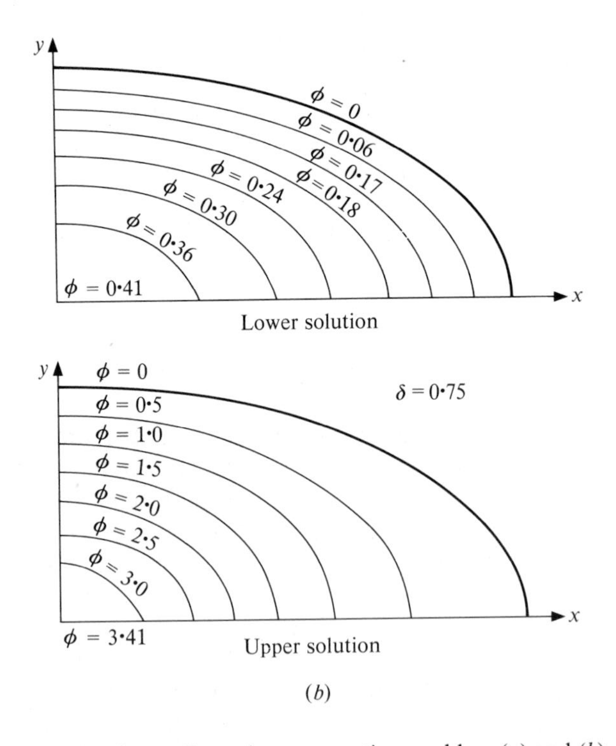

(b)

Fig. 18.20 A non-linear heat generation problem (a) and (b)

below this *two* solutions are possible and the starting point of the iteration determines which one is in fact obtained.

This last point illustrates that an insight into the problem is, in non-linear solutions, even more important than elsewhere.

References

1. R. Beckett and J. Hurt, *Numerical Calculations and Algorithms*, McGraw-Hill, 1967.
2. A. Ralston, *A first course in numerical analysis*, McGraw-Hill, 1965.
3. L. Fox, *An introduction to numerical linear algebra*, Oxford University Press, 1965.
4. L. B. Rall, 'Computational solution of non-linear operator equations', J. Wiley & Son, 1969.
5. B. M. Irons and R. C. Tuck, 'A version of the Aitken accelerator for computer iteration', *Int. J. Num. Meth. Eng.*, **1**, 275–8, 1969.
6. O. C. Zienkiewicz and B. M. Irons, 'Matrix iteration and acceleration processes in finite element problems of structural mechanics', Chapter 9 of *Numerical Methods for Non-linear Algebraic Equations* (ed. P. Rabinowitz), Gordon and Breach, 1970.
7. G. C. Nayak and O. C. Zienkiewicz, 'Note on the "alpha"-constant stiffness method for analysis of non-linear problems', *Int. J. Num. Meth. Eng.*, **4**, 579–82, 1972.
8. I. M. Smith and R. Hobbs, 'Finite element analysis of centrifuged and build-up slopes', *Geotechnique*, **24** (No. 4), 531–59, 1974.
9. O. C. Zienkiewicz, B. Best, C. Dullage, and K. G. Stagg, 'Analysis of non-linear problems in rock mechanics with particular reference to jointed rock systems', *Proc. 2nd Int. Cong. on Rock Mechanics*, Belgrade, 1970.
10. J. T. Oden, 'Numerical formulation of non-linear elasticity problems', *Proc. Am. Soc. Civ. Eng.*, **93**, ST3, 235–55, 1967.
11. R. von Mises, 'Mechanik der plastischen Formanderung der Kristallen', *Z. angew. Math. Mech.*, **8**, 161–85, 1928.
12. D. C. Drucker, 'A more fundamental approach to plastic stress–strain solutions', *Proc. 1st U.S. Natn. Cong. Appl. Mech.*, 487–91, 1951.
13. W. T. Koiter, 'Stress–strain relations, uniqueness and variational theorems for elastic plastic materials with a singular yield surface', *Q. J. Appl. Math.*, **11**, 350–4, 1953.
14. R. Hill, *The Mathematical Theory of Plasticity*, Clarendon Press, 1950.
15. (a) W. Johnson and P. W. Mellor, *Plasticity for Mechanical Engineers*, Van Nostrand, 1962.
15. (b) W. Prager, *An Introduction to Plasticity*, Addison Wesley, 1959.
15. (c) A. Mendelson, *Plasticity—Theory and Application*, Macmillan, 1968.
16. O. C. Zienkiewicz, S. Valliappan, and I. P. King, 'Stress analysis of rock as a "no-tension" material', *Geotechnique*, **18**, 56–66, 1968.
17. G. C. Nayak and O. C. Zienkiewicz, 'Elasto-plastic stress analysis. Generalization for various constitutive relations including strain softening', *Int. J. Num. Meth. Eng.*, **5**, 113–35, 1972.
18. Y. Yamada, N. Yishimura, and T. Sakurai, 'Plastic stress–strain matrix and its application for the solution of elastic–plastic problems by the finite element method', *Int. J. Mech. Sci.*, **10**, 343–54, 1968.
19. O. C. Zienkiewicz, S. Valliappan, and I. P. King, 'Elasto-plastic solutions

of engineering problems. Initial-stress, finite element approach', *Int. J. Num. Meth. Eng.*, **1**, 75–100, 1969.
20. G. C. NAYAK and O. C. ZIENKIEWICZ, 'Convenient forms of stress invariants for plasticity', *Proc. Am. Soc. Civ. Eng.*, **98**, 949–54, 1972.
21. D. C. DRUCKER and W. PRAGER, 'Soil mechanics and plastic analysis or limit design', *Q. J. Appl. Math.*, **10**, 157–65, 1952.
22. J. F. BESSELING, 'A theory of elastic, plastic and creep deformations of an initially isotropic material', *J. Appl. Mech.*, **25**, 529–36, 1958.
23. Z. MROZ, 'An attempt to describe the behaviour of metals under cyclic loads using more general work hardening model', *Act. Mech.*, **7**, 199, 1969.
24. O. C. ZIENKIEWICZ, G. C. NAYAK, and D. R. J. OWEN, 'Composite and "overlay" models in numerical analysis of elasto-plastic continua', *Foundations of Plasticity*, pp. 107–22 (ed. A. Sawczuk), Noordhoff Press, 1972.
25. D. R. J. OWEN, A. PRAKASH, and O. C. ZIENKIEWICZ, 'Finite element analysis of non-linear composite materials by use of overlay systems', *Comp. Struct.*, **4**, 1251–67, 1974.
26. R. H. GALLAGHER, J. PADLOG, and P. P. BIJLAARD, 'Stress analysis of heated complex shapes', *J. Am. Rocket Soc.*, **32**, 700–7, 1962.
27. J. H. ARGYRIS, 'Elasto-plastic matrix displacement analysis of three-dimensional continua', *J. Roy. Aero. Soc.*, **69**, 633–5, 1965.
28. G. G. POPE, *A discrete element method for analysis of plane elasto-plastic strain problems*, R.A.E. Farnborough, T.R. 65028, 1965.
29. J. L. SWEDLOW, M. L. WILLIAMS, and W. M. YANG, *Elasto-plastic stresses in cracked plates*, Calcit Report SM 65–19, California Institute of Technology, 1965.
30. J. L. SWEDLOW, 'Elastic plastic cracked plates in plane strain', *Int. J. Fract. Mech.*, **5**, 33–44, 1969.
31. P. V. MARÇAL and I. P. KING, 'Elastic-plastic analysis of two-dimensional stress systems by the finite element method', *Int. J. Mech. Sci.*, **9**, 143–55, 1967.
32. S. F. REYES and D. U. DEERE, 'Elasto-plastic analysis of underground openings by the finite element method', *Proc. 1st Int. Cong. Rock Mechanics*, **11**, 477–86, Lisbon, 1966.
33. E. P. POPOV, M. KHOJASTEH-BAKHT, and S. YAGHMAI, 'Bending of circular plates of hardening material', *Int. J. Solids Struct.*, **3**, 975–88, 1967.
34. J. H. ARGYRIS and D. W. SCHARPF, 'Methods of elasto-plastic analysis', *Symp. on Finite Element Techniques*, Stuttgart, June, 1969.
35. J. L. SWEDLOW, 'A procedure for solving problems of elasto-plastic flow', *Int. J. Comp. Struct.*, **3**, 878–98, 1973.
36. J. F. BESSELING, 'Non-linear analysis of structures by the finite element method as a supplement to linear analysis', *Comp. Meth. Appl. Mech. Eng.*, **3**, 173–94, 1974.
37. R. M. MCMEEKING and J. R. RICE, 'Finite element formulations for problems of large elastic plastic deformation', *Int. J. Solids Struct.*, **11**, 601–16, 1975.
38. J. C. NAGTEGAAL, D. M. PARKS, and J. R. RICE, 'On numerically accurate finite element solutions in the fully plastic range', *Comp. Meth. Appl. Mech. Eng.*, **4**, 153–78, 1974.
39. R. H. GALLAGHER and A. K. DHALLA, 'Direct flexibility finite element elasto-plastic analysis', 6 Part, *First SMIRT Conf.*, Berlin, 1971.
40. E. F. RYBICKI and L. A. SCHMIT, 'An incremental complementary energy method of non-linear stress analysis', *J.A.I.A.A.*, 1105–12, 1970.

41. J. A. STRICKLIN, W. E. HEISLER, and W. VON RUSMAN, 'Elvauation of solution procedures for material and/or geometrically non-linear structural analysis', *J.A.I.A.A.*, **11**, 292–9, 1973.
42. D. R. J. OWEN, G. C. NAYAK, A. P. KFOURI, and J. R. GRIFFITHS, 'Stresses in a partly yielded notched bar', *Int. J. Num. Meth. Eng.*, **6**, 63–72, 1973.
43. K. S. DINNO and S. S. GILL, 'An experimental investigation into the plastic behaviour of flush nozzles in spherical pressure vessels', *Int. J. Mech. Sci.*, **7**, 817, 1965.
44. A. MENDELSON, M. H. HISCHBERG, and S. S. MANSON, 'A general approach to the practical solution of creep problems', *J. Basic Eng., Trans. Am. Soc. Mech. Eng.*, **81** (Ser. D), 585–98, 1959.
45. O. C. ZIENKIEWICZ, M. WATSON, and I. P. KING, 'A numerical method of visco-elastic stress analysis', *Int. J. Mech. Sci.*, **10**, 807–27, 1968.
46. O. C. ZIENKIEWICZ, *The Finite Element Method in Structural and Continuum Mechanics*, 1st ed., McGraw-Hill, 1967.
47. G. A. GREENBAUM and M. F. RUBINSTEIN, 'Creep analysis of axi-symmetric bodies using finite elements', *Nucl. Eng. Des.*, **7**, 379–97, 1968.
48. P. V. MARÇAL, Private communication, 1972.
49. N. A. CYR and R. D. TETER, 'Finite element elastic plastic creep analysis of two-dimensional continuum with temperature dependent material properties', *Comp. Struct.*, **3**, 849–63, 1973.
50. O. C. ZIENKIEWICZ, 'Visco-plasticity, plasticity, creep and visco-plastic flow. (Problems of small, large and continuing deformation'. *Computational Mechanics*, Texas Inst. Comp. Mech. Lect. Notes on Math. 461, Springer-Verlag, 1975.
51. M. B. KANCHI, D. R. J. OWEN, and O. C. ZIENKIEWICZ, *An implicit scheme for finite element solution of problems of visco plasticity and creep*, C/R/252/75, Univ. Coll. of Swansea, 1975.
52. T. J. R. HUGHES and R. L. TAYLOR, 'Unconditionally stable algorithms for quasi-static elasto/visco-plastic finite element analysis', *Computers and Structures*, **8**, 169–173, 1978.
53. E. C. BINGHAM, *Fluidity and Plasticity*, Chapter VIII, pp. 215–18, McGraw-Hill, 1922.
54. P. PERZYNA, 'Fundamental problems in viscoplasticity', *Adv. Appl. Mech.*, **9**, 243–377, 1966.
55. O. C. ZIENKIEWICZ and I. C. CORMEAU, 'Viscoplasticity solution by finite element process', *Arch. Mech.*, **24**, 5–6, 873–88, 1972.
56. (a) O. C. ZIENKIEWICZ and I. C. CORMEAU, 'Visco-plasticity and plasticity. An alternative for finite element solution of material non-linearities', *Proc. Colloque Methodes Calcul. Sci. Tech.*, 171–99, IRIA, Paris, 1973.
56. (b) J. ZARKA, 'Généralisation de la théorie du potential multiple en visco-plasticité', *J. Mech. Phys. Solids*, **20**, 179–95, 1972.
56. (c) Q. A. NGUYEN and J. ZARKA, 'Quelques méthodes de resolution numérique en elastoplasticité classique et en elasto-viscoplasticité', *Seminaire Plasticité et Viscoplasticité*, École Polytechnique, Paris, 1972; also *Sciences et technique de l'armement*, **47**, 407–36, 1973.
56. (d) O. C. ZIENKIEWICZ and I. C. CORMEAU, 'Visco-plasticity, plasticity, and creep in elastic solids—a unified numerical solution approach', *Int. J. Num. Meth. Eng.*, **8**, 821–45, 1974.
57. I. C. CORMEAU, 'Numerical stability in quasi-static elasto-visco-plasticity', *Int. J. Num. Meth. Eng.*, **9**, 109–28, 1975.

58. O. C. Zienkiewicz, C. Humpheson, and R. W. Lewis, 'Associated and non-associated visco-plasticity and plasticity in soil mechanics', *Geotechnique*, **25**, 671–89, 1975.
59. F. A. Leckie and J. B. Martin, 'Deformation bounds for bodies in a state of creep', *J. Appl. Mech., Am. Soc. Mech. Eng.*, 411–17, June 1967.
60. I. Finnie and W. R. Heller, *Creep of Engineering Materials*, McGraw-Hill, 1959.
61. A. E. Johnson, 'Complex stress creep', *Met. Rev.*, **5**, 447, 1960.
62. C. S. Desai (ed.), *Numerical Methods in Geomechanics*, Vol. I-III, *Am. Soc. Civil Eng.* Special Publication, 1976.
63. G. von Gudehus (ed.), *Finite Elements in Geomechanics*, J. Wiley & Son, 1977.
64. O. C. Zienkiewicz, R. W. Lewis, V. A. Norris, and C. Humpheson, *Numerical analysis for foundations of offshore structures with special reference to progressive deformation*, Soc. Petrol. Eng., *Am. Soc. Mech. Eng.*, Paper No. SPE 5760, 1976.
65. E. H. Lee, 'Visco-elasticity' in *Handbook of Engineering Mechanics* (ed. W. Flügge), McGraw-Hill, 1962.
66. R. L. Taylor, K. Pister, and G. Goudreau, 'Thermo-mechanical analysis of visco-elastic solids', *Int. J. Num. Meth. Eng.*, **2**, 45–60, 1970.
67. (a) O. C. Zienkiewicz, 'Analysis of visco-elastic behaviour of concrete structures with particular reference to thermal stresses', *Proc. Am. Concr. Inst.*, **58**, 383–94, 1961.
67. (b) T. Alfrey, *Mechanical Behaviour of High Polymers*, Interscience Pub., N.Y., 1948.
67. (c) D. McHenry, 'A new aspect of creep in concrete and its application to design', *Proc. Am. Soc. Test. Mat.*, **43**, 1064, 1943.
68. H. H. Hilton and H. G. Russell, 'An extension of Alfrey's analogy to thermal stress problems in temperature dependent linear visco-elastic media', *J. Mech. Phys. Solids*, **9**, 152–64, 1961.
69. O. C. Zienkiewicz, M. Watson, and Y. K. Cheung, 'Stress analysis by the finite element method—thermal effects', *Proc. Conf. on Prestressed Concrete Pressure Vessels*, Inst. Civ. Eng., London, 1967.
70. S. Valliappan and P. Nath, 'Tensile crack propagation in reinforced concrete beams by finite element techniques', *Int. Conf. on Shear Torsion and Bond in Reinforced Concrete*, Coimbatore, India, January 1969.
71. N. W. Krahl, W. Khachaturian, and C. P. Seiss, 'Stability of tensile cracks in concrete beams', *Proc. Am. Soc. Civ. Eng.*, **93**, ST1, 235–54, 1967.
72. K. Muto, N. Ohmori, T. Sugano, T. Miyashita, and H. Shimizu, 'Non-linear analysis of reinforced concrete buildings', *Proc. 1973, Tokyo Seminar on Finite Element Analysis*, Univ. of Tokyo Press, 1973.
73. A. Scanlon and D. W. Murray, 'An analysis to determine the effects of cracking in reinforced concrete slabs', *Proc. Engineering Institute of Canada Specialty Conf. on Finite Element Method in Civil Engineering*, Montreal, 1952.
74. O. Buyukozturk and P. V. Marçal, 'Strength of reinforced concrete chambers under external pressure', to be presented at the 2nd National Congress on Pressure Vessel and Piping, San Francisco, California, 23–27 June 1975.
75. B. Saugy, T. Zimmermann, and M. Hussain, 'Three-dimensional rupture analysis of a prestressed concrete pressure vessel including creep effects', *Nucl. Eng. Des.*, **28** (No. 1, July), 97–120, 1974.

76. M. UEDA, M. KAWAHARA, Y. YOSHIOKA, and M. KIKUCHI, 'Non-linear viscoelastic and elasto-plastic finite elements for concrete structures' in *Discrete Methods in Engineering*, C.I.S.E., 1974.
77. D. V. PHILLIPS and O. C. ZIENKIEWICZ, 'Finite element non-linear analysis of concrete structures', *Prov. Instn. Civ. Eng.*, Part 2, **61**, 59–88, March 1976.
78. O. C. ZIENKIEWICZ, D. V. PHILLIPS, and D. R. J. OWEN, 'Finite element analysis of some concrete non-linearities. Theories and examples', *IABSE Symp. on Concrete Structures Subjected to Triaxial Stresses*, Bergamo, 17–19 May 1974.
79. R. MCGEORGE and L. F. SWEC, 'Refined cracked concrete analysis of concrete containment structures subject to loadings', *Operational and Environmental Nucl. Eng. Des.*, **29**, 58–70, 1974.
80. M. SUIDAN and W. C. SCHNOBRICH, 'Finite element analysis of reinforced concrete', *Proc. Am. Soc. Civ. Eng.*, **99**, ST10, 2109–22, 1973.
81. O. C. ZIENKIEWICZ and B. BEST, 'Some non-linear problems in soil and rock mechanics—finite element solution', *Conf. on Rock Mechanics*, Univ. of Queensland, Townsville, June 1969.
82. R. E. GOODMAN, R. L. TAYLOR, and T. BREKKE, 'A model for the mechanics of jointed rock', *Proc. Am. Soc. Civ. Eng.*, **94**, SM3, 637–59, 1968.
83. O. C. ZIENKIEWICZ and G. PANDE, 'Multilaminate models for rock', *Num. Anal. Meth. Geomech.*, **1**, 1977.
84. J. T. ODEN, *Finite Elements of Non-linear Continua*, McGraw-Hill, 1972.
85. F. D. MURNAGHAN, *Finite Deformation of an Elastic Solid*, Wiley, 1951.
86. E. TONTI, 'Variational formulation of non-linear differential equations', *Bull. de l'Acad. Roy. Belg. (Science)*, **55**, 137–278, 1969.
87. M. MUSCAT, *The Flow of Homogenous Fluids through Porous Media*, T. H. Edwards Inc., 1946.
88. R. E. VOLKER, 'Non-linear flow in porous media by finite elements', *Proc. Am. Soc. Civ. Eng.*, **95**, H76, 2093–114, 1969.
89. H. AHMED and D. K. SUNEDA, 'Non-linear flow in porous media', *Proc. Am. Soc. Civ. Eng.*, **95**, H76, 1847–59, 1969.
90. A. M. WINSLOW, 'Numerical solution of the quasi-linear Poisson's equation in a non-uniform triangle mesh', *J. Comp. Phys.*, **1**, 149–72, 1967.
91. M. V. K. CHARI and P. SILVESTER, 'Finite element analysis of magnetically saturated D.C. mechanics, *IEEB Winter Meeting/Power*, New York, 1971.
92. O. C. ZIENKIEWICZ, J. F. LYNESS, and D. R. J. OWEN, 'Three-dimensional magnetic field determination using a scalar potential. A finite element solution', to be published in *Magn. Trans. I.E.E.E.*, 1977.
93. J. F. LYNESS, D. R. J. OWEN, and O. C. ZIENKIEWICZ, 'The finite element analysis of engineering systems governed by a non-linear quasi-harmonic equation', *Comp. Struct.* **5**, 65–79, 1975.
94. D. GELDER, 'Solution of the compressible flow equations', *Int. J. Num. Meth. Eng.*, **3**, 35–43, 1971.
95. C. A. ANDERSON and O. C. ZIENKIEWICZ, 'Spontaneous ignition: finite element solutions for steady and transient conditions', *Am. Soc. Mech. Eng., J. Heat Transfer*, 398–404, August 1974.

19. Geometrically Non-Linear Problems; Large Displacement and Structural Instability

19.1 Introduction

In the previous chapter the question of non-linearities arising from material properties was discussed and methods were developed to allow the standard linear forms to be used in an iterative way to obtain solutions. In this chapter a similar path will be followed in the treatment of geometric non-linearity of structures.

In all problems discussed so far it has been implicitly assumed that both displacements and strains developed in the structure are small. In practical terms this means that geometry of the elements remains basically unchanged during the loading process and that first order, infinitesimal, linear strain approximations can be used.

In practice such assumptions fail frequently even though actual strains may be small and elastic limits of ordinary structural materials not exceeded. If accurate determination of the displacements is needed, geometric non-linearity may have to be considered in some structures. Here, for instance, stresses due to membrane action, usually neglected in plate flexure may cause a considerable decrease of displacements as compared with the linear solution even though displacements are still quite small. Conversely, it may be found that a load is reached where deflections increase more rapidly than predicted by a linear solution and indeed a state may be attained where load-carrying capacity *decreases* with continuing deformation. This classic problem is that of structural stability and obviously has many practical implications. The applications of such analysis are clearly of considerable importance in aerospace engineering, design of radio telescopes, cooling towers, box girder bridges and other relatively slender structures.

In many cases *very large displacements* may occur without causing large strains. Typical in this context is the classical problem of the 'elastica' of which an example is a watch spring.

In this chapter an attempt is made to unify the treatment of all the above problems and to present generally applicable procedures. This is achieved by examining the basic non-linear equilibrium equations together with their solution. Such considerations also lead to the formulation of classical initial stability problems. These concepts are then illustrated by formulating the large deflection and initial stability problems for flat plates. This naturally leads to the general continuum formulation of the large displacement problem. A Lagrangian approach is adopted throughout in which displacements are referred to the *original configuration*. Alternative, Eulcrian forms are not treated here.

One aspect of geometric non-linearity is, however, not discussed in detail. This is the case of large strain such as may occur, even elastically, with such materials as rubber, etc. Here specialized relations between stress and strain have to be introduced[1] and the length of this book prohibits the full discussion. Nevertheless the general processes of the next section are still applicable providing suitable stress and strain laws are introduced.

Geometric non-linearity may often be combined with material non-linearity of the type discussed in the previous chapter, such as small strain plasticity, etc. In principle this does not introduce addditional complexities and the methods of this chapter can easily be extended to deal with this situation.[2]

19.2 General Considerations

19.2.1 *The basic problem.* Whether the displacements (or strains) are large or small, equilibrium conditions between internal and external 'forces' have to be satisfied. Thus, if the displacements are prescribed in the usual manner by a finite number of (nodal) parameters **a**, we can obtain the necessary equilibrium equations using the virtual work principle of Chapter 2. Now, however, different definitions of 'stresses' and 'strains' which are conjugate to each other must be used. We shall discuss some such conjugate quantities in the context of plates, shells, and general elasticity later, but in all cases we find that we can write

$$\boldsymbol{\Psi}(\mathbf{a}) = \int_V \overline{\mathbf{B}}^{\mathrm{T}} \boldsymbol{\sigma} \, \mathrm{d}V - \mathbf{f} = 0 \tag{19.1}$$

where $\boldsymbol{\Psi}$ once again represents the sum of external and internal generalized forces, and in which $\overline{\mathbf{B}}$ is defined from the strain definition as

$$\mathrm{d}\boldsymbol{\varepsilon} = \overline{\mathbf{B}} \, \mathrm{d}\mathbf{a}. \tag{19.2}$$

The bar suffix has now been added as, if displacements are large, the strains depend non-linearly on displacement, and the matrix $\overline{\mathbf{B}}$ is now dependent on **a**. We shall see later that we can conveniently write

$$\overline{\mathbf{B}} = \mathbf{B}_0 + \mathbf{B}_L(\mathbf{a}) \tag{19.3}$$

in which $\mathbf{B}_0$ is the same matrix as in linear infinitesimal strain analysis and only $\mathbf{B}_L$ depends on the displacement. In general $\mathbf{B}_L$ will be found to be a *linear function* of such displacements.

If strains are reasonably small we can still write the general elastic relation

$$\boldsymbol{\sigma} = \mathbf{D}(\boldsymbol{\varepsilon}-\boldsymbol{\varepsilon}_0)+\boldsymbol{\sigma}_0 \tag{19.4}$$

in which $\mathbf{D}$ is the usual set of elastic constants.†

However, any non-linear stress–strain relationship could equally well be written, as the whole process of solution once again reduces to the solution of a non-linear set of equations (19.1).

It is perhaps too obvious to restate that in Eq. (19.1) integrals are in fact carried out element by element and contributions to 'nodal equilibrium' summed in the usual manner.

19.2.2 *Solution processes.* Clearly the solution of Eq. (19.1) will have to be approached iteratively and the general methods described in the previous chapter (in section 18.2) are applicable.

If for instance the Newton–Raphson process is to be adopted we have to find the relation between $\mathrm{d}\mathbf{a}$ and $\mathrm{d}\boldsymbol{\Psi}$, as explained there. Thus taking appropriate variations of Eq. (19.1) with respect to $\mathrm{d}\mathbf{a}$ we have

$$\mathrm{d}\boldsymbol{\Psi} = \int_V \mathrm{d}\overline{\mathbf{B}}^{\mathrm{T}}\boldsymbol{\sigma}\,\mathrm{d}V+\int_V \overline{\mathbf{B}}^{\mathrm{T}}\,\mathrm{d}\boldsymbol{\sigma}\,\mathrm{d}V = \mathbf{K}_{\mathrm{T}}\,\mathrm{d}\mathbf{a} \tag{19.5}$$

and using Eqs. (19.4) and (19.2) we have‡

$$\mathrm{d}\boldsymbol{\sigma} = \mathbf{D}\,\mathrm{d}\boldsymbol{\varepsilon} = \mathbf{D}\overline{\mathbf{B}}\,\mathrm{d}\mathbf{a}$$

and if Eq. (19.3) is valid

$$\mathrm{d}\overline{\mathbf{B}} = \mathrm{d}\mathbf{B}_L.$$

Therefore

$$\mathrm{d}\boldsymbol{\Psi} = \int_V \mathrm{d}\mathbf{B}_L^{\mathrm{T}}\boldsymbol{\sigma}\,\mathrm{d}V+\overline{\mathbf{K}}\,\mathrm{d}\mathbf{a} \tag{19.6}$$

where

$$\overline{\mathbf{K}} = \int_V \overline{\mathbf{B}}^{\mathrm{T}}\mathbf{D}\overline{\mathbf{B}}\,\mathrm{d}V = \mathbf{K}_0+\mathbf{K}_L \tag{19.7}$$

† It is important to remember here that the stress components defined in Eq. (19.4) are those *corresponding* to the strain component used. In some gross displacement problems such strain components are subject to considerable change of direction from original, fixed axes.

‡ Once again if non-linear stress–strain relations are used $\mathbf{D} = \mathbf{D}(\boldsymbol{\sigma})$ is the incremental elasticity matrix as given by Eq. (18.25).

in which $\mathbf{K}_0$ represents the usual, small displacements stiffness matrix, i.e.,

$$\mathbf{K}_0 = \int_V \mathbf{B}_0^{\mathrm{T}} \mathbf{D} \mathbf{B}_0 \, \mathrm{d}V. \tag{19.7a}$$

The matrix $\mathbf{K}_L$ is due to the large displacement and is given by

$$\mathbf{K}_L = \int_V (\mathbf{B}_0^{\mathrm{T}} \mathbf{D} \mathbf{B}_L + \mathbf{B}_L^{\mathrm{T}} \mathbf{D} \mathbf{B}_L + \mathbf{B}_L^{\mathrm{T}} \mathbf{D} \mathbf{B}_0) \, \mathrm{d}V. \tag{19.7b}$$

$\mathbf{K}_L$ is variously known as the *initial displacement* matrix,[3] *large displacement matrix*, etc. and contains only terms which are linear and quadratic in $\mathbf{a}$. It will be found that this is a matrix which would alternatively be obtained by using an infinitesimal strain approach but adjusting element co-ordinates in the computation of the stiffness.

The first term of Eq. (19.6) can generally be written (perhaps less obviously until particular cases are examined) as

$$\int_V \mathrm{d}\mathbf{B}_L^{\mathrm{T}} \boldsymbol{\sigma} \, \mathrm{d}V \equiv \mathbf{K}_\sigma \, \mathrm{d}\mathbf{a} \tag{19.8}$$

where $\mathbf{K}_\sigma$ is a symmetric matrix dependent on the stress level. This matrix is known as *initial stress* matrix[3, 21] or *geometric* matrix.[4, 5] Thus

$$\mathrm{d}\boldsymbol{\Psi} = (\mathbf{K}_0 + \mathbf{K}_\sigma + \mathbf{K}_L) \, \mathrm{d}\mathbf{a} = \mathbf{K}_T \, \mathrm{d}\mathbf{a} \tag{19.9}$$

with $\mathbf{K}_T$ being the total, *tangential stiffness*, matrix. Newton-type iteration can once more be applied precisely in the manner of section 18.2.

To summarize, usually:

(*a*) Elastic linear solution is obtained as a first approximation $\mathbf{a}^0$;

(*b*) $\boldsymbol{\Psi}^0$ is found using Eq. (19.1) with appropriate definition of $\overline{\mathbf{B}}$ and stresses as given by Eq. (19.4) (or any other linear or non-linear law);

(*c*) Matrix $\mathbf{K}_T^0$ is established; and

(*d*) Correction is computed as

$$\Delta \mathbf{a}^0 = -(\mathbf{K}_T^0)^{-1} \boldsymbol{\Psi}^0$$

and processes (*b*), (*c*) and (*d*) repeated until $\boldsymbol{\Psi}_n^n$ becomes sufficiently small.

Again a constant matrix could be used, increasing the number of iterating steps but making use of a semi-inverted, resolve process at smaller computer cost, providing that at each step $\boldsymbol{\Psi}^n$ is calculated by the correct expressions but convergence is sometimes slow by this procedure.

Quite conveniently the tangent stiffness matrix may be made constant after, say, the second iteration of each load increment; this has been used with considerable success.[6]

While all solutions can be accomplished in a one-step operation for a full load, on occasion, as in all non-linear problems, the possibility of a non-unique solution arises and the physically unimportant one may be obtained. In such cases it is wise to proceed by incrementing the load and obtaining the non-linear solution for each increment. This indeed is sometimes computationally cheaper as effects of non-linearity in each step are reduced. Indeed, if load increments of sufficiently small magnitude are taken each incremental solution may be accomplished sufficiently accurately in one step.[4, 5, 7] It is important, however, to check periodically the total equilibrium by using the full Eq. (19.1).

All the solution techniques described in Chapter 18, section 18.2, have been used in the context of geometrically non-linear analysis problems with success. Such techniques have been extensively evaluated by Haisler *et al.*[8]

19.2.3 *Initial stability problem.* It is of interest to note at this stage that $\mathbf{K}_\sigma$ does not explicitly contain the displacements and is proportional to the stress level $\boldsymbol{\sigma}$. Thus, if at the first step of computation we evaluate $\boldsymbol{\sigma}$ by a linear solution we have from Eq. (19.6)

$$d\boldsymbol{\Psi} = (\mathbf{K}_0 + \mathbf{K}_\sigma)\,d\mathbf{a} \tag{19.10}$$

as $\mathbf{K}_L = 0$ at this stage.

If the loads are increased by a factor λ we may find that neutral stability exists, i.e.,

$$d\boldsymbol{\Psi} = (\mathbf{K}_0 + \lambda\mathbf{K}_\sigma)\,d\mathbf{a} \equiv 0 \tag{19.11}$$

From this λ can be obtained by solving the typical *eigenvalue problem* defined above (*vide* Chapter 20 for discussion of eigenvalue problems).

This is the classical, 'initial' stability problem such as occurs in the buckling of struts, plates, shells, etc.

Quite frequently in the literature this type of approach is used beyond its limits of applicability. The 'initial stability' expressed can only give physically significant answers if the elastic ($\mathbf{K}_0$) solution gives such deformations that the *large deformation matrix* $\mathbf{K}_L$ *is identically zero*. This only happens in a very limited number of practical situations (e.g., a perfectly straight strut under axial load, complete sphere under uniform pressure, etc.) The preoccupation of such investigators with the subject of 'initial imperfections' is strictly limited to such situations where a true *bifurcation* can occur. In real engineering situations where the qualitative nature of the behaviour is completely unknown such problems should be investigated using the full tangential stiffness matrix.[7] When $\mathbf{K}_T\,d\mathbf{a}$ is identically zero, neutral equilibrium is obtained. A step-by-step approach is clearly necessary here.

It is possible to achieve a compromise between the classical stability problem and a full non-linear analysis in a number of ways.[9, 10] For example, (i) based on a full non-linear analysis the eigenvalue problem may be considered after each increment of load or, alternatively, (ii) a linear eigenvalue analysis that includes a linearized $\mathbf{K}_L$ as well as the usual $\mathbf{K}_\sigma$ can be used.

19.2.4 *Energy interpretation of the stability criteria.* It was shown in Chapter 2 that the virtual work done during a displacement variation $\mathrm{d}\mathbf{a}$ is in fact equal to the variation of total potential energy Π. Thus for equilibrium

$$\mathrm{d}\Pi = \mathrm{d}\mathbf{a}^{\mathrm{T}}\boldsymbol{\Psi} = 0 \tag{19.12}$$

i.e., *the total potential energy is stationary* (which is equivalent to Eq. (19.1)).

The second variation of Π is (by Eq. (19.9))

$$\mathrm{d}^2\Pi = \mathrm{d}(\mathrm{d}\Pi) = \mathrm{d}\mathbf{a}^{\mathrm{T}}\,\mathrm{d}\boldsymbol{\Psi} = \mathrm{d}\mathbf{a}^{\mathrm{T}}\mathbf{K}_T\,\mathrm{d}\mathbf{a}. \tag{19.13}$$

Stability criterion is given by a positive value of this second variation and conversely instability by a negative value (as in first case energy has to be added to the structure while in second it contains surplus energy). In

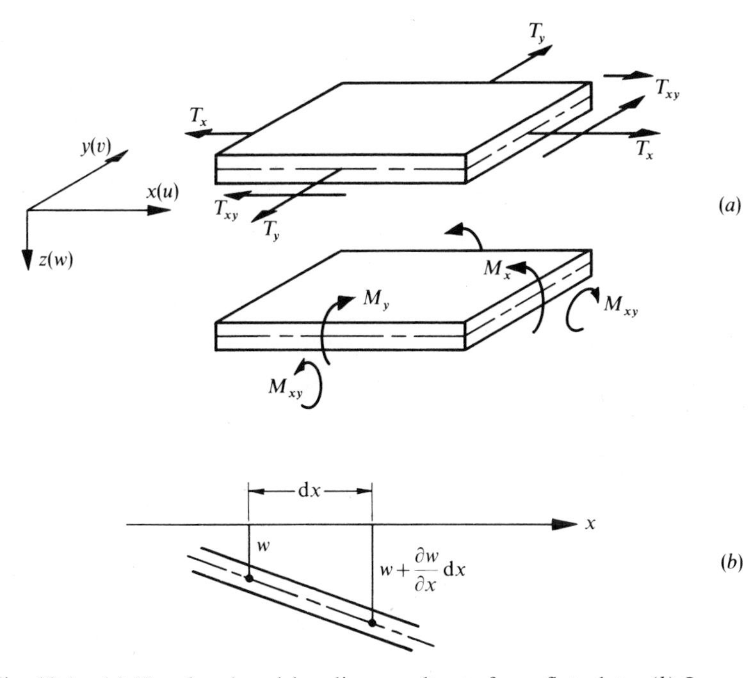

Fig. 19.1 (*a*) 'In plane' and bending resultants for a flat plate. (*b*) Increase of middle surface length due to lateral displacement

other words *if* $\mathbf{K}_T$ *is positive-definite, stability exists.* This criterion is well known[11] and of considerable use when investigating stability during large deformation.†[12–14]

19.2.5 *Forces dependent on deformation.* In the derivation of Eq. (19.5) it was implicitly assumed that the forces **f** are not themselves dependent on the deformation. In some instances this is not true. For instance pressure loads on a grossly deforming structure are in general in this category as indeed are some aerodynamic forces depending on the deformation (flutter).

If forces vary with displacement then in relation to Eq. (19.5) the variation d**f** has to be considered with respect to d**a**. This leads to the introduction of the *load-correction matrix.*[15,16] Stability and large deformation problems under such (non-conservative) loads can be once again studied if proper consideration is given to the above term.

19.3 Large Deflection and 'Initial' Stability of Plates

19.3.1 *Definitions.* As a first example we shall consider the problems associated with the deformation of plates subject to 'in-plane' and 'lateral' forces, when displacements are not infinitesimal but also not excessively large, Fig. 19.1. In this situation the 'change in geometry' effect is less important than the *relative magnitudes* of the linear and non-linear strain–displacement terms, and in fact for 'stiffening' problems the non-linear displacements are always less than the corresponding linear ones, see Fig. 19.2. It is well known that in such situations the lateral displacements will be responsible for development of 'membrane' type strains and now the two problems of 'in-plane' and 'lateral' deformation can no longer be dealt with separately but are *coupled.*

We shall, as before, describe the plate 'strains' in terms of middle surface displacements, i.e., if the x–y plane coincides with the middle surface as in Fig. 19.1(a) we shall have (see Chapters 10 and 13)

$$\boldsymbol{\varepsilon} = \begin{Bmatrix} \varepsilon_x \\ \varepsilon_y \\ \gamma_{xy} \\ -\dfrac{\partial^2 w}{\partial x^2} \\ -\dfrac{\partial^2 w}{\partial y^2} \\ 2\dfrac{\partial^2 w}{\partial x\,\partial y} \end{Bmatrix} = \begin{Bmatrix} \boldsymbol{\varepsilon}^p \\ \boldsymbol{\varepsilon}^b \end{Bmatrix}; \qquad \boldsymbol{\sigma} = \begin{Bmatrix} T_x \\ T_y \\ T_{xy} \\ M_x \\ M_y \\ M_{xy} \end{Bmatrix} = \begin{Bmatrix} \boldsymbol{\sigma}^p \\ \boldsymbol{\sigma}^b \end{Bmatrix}. \tag{19.14}$$

† An alternative test is to investigate the sign of the determinant of $\mathbf{K}_T$.[14]

The 'stresses' are defined in terms of the usual resultants.†

$T_x = \bar{\sigma}_x t$, where $\bar{\sigma}_x$ is the average membrane stress, etc. Now, if the deformed shape is considered as in Fig. 19.1(*b*), we see that displacement w produces some additional extension in the x and y directions of the middle surface and the length dx stretches to

$$\mathrm{d}x' = \mathrm{d}x\sqrt{1+\left(\frac{\partial w}{\partial x}\right)^2} = \mathrm{d}x\left\{1+\frac{1}{2}\left(\frac{\partial w}{\partial x}\right)^2 + \ldots\right\},$$

i.e., in defining the x elongation we can write (to second approximation)

$$\varepsilon_x = \frac{\partial u}{\partial x}+\frac{1}{2}\left(\frac{\partial w}{\partial x}\right)^2.$$

Considering in a similar way the other components[17] we can write as the definition of strain

$$\boldsymbol{\varepsilon} = \left\{\begin{matrix} \dfrac{\partial u}{\partial x} \\ \dfrac{\partial v}{\partial y} \\ \dfrac{\partial u}{\partial y}+\dfrac{\partial v}{\partial x} \\ -\dfrac{\partial^2 w}{\partial x^2} \\ -\dfrac{\partial^2 w}{\partial y^2} \\ 2\dfrac{\partial^2 w}{\partial x\,\partial y} \end{matrix}\right\} + \left\{\begin{matrix} \dfrac{1}{2}\left(\dfrac{\partial w}{\partial x}\right)^2 \\ \dfrac{1}{2}\left(\dfrac{\partial w}{\partial y}\right)^2 \\ \left(\dfrac{\partial w}{\partial x}\right)\left(\dfrac{\partial w}{\partial y}\right) \\ 0 \\ 0 \\ 0 \end{matrix}\right\} = \left\{\begin{matrix} \boldsymbol{\varepsilon}_0^p \\ \boldsymbol{\varepsilon}_0^b \end{matrix}\right\} + \left\{\begin{matrix} \boldsymbol{\varepsilon}_L^p \\ \mathbf{0} \end{matrix}\right\} \tag{19.15}$$

in which the first term is the linear expression already encountered many times and the second gives the non-linear terms. In above u, v, w stand for appropriate displacements of middle surface.

If only linear elastic behaviour is considered, the **D** matrix is composed of plane stress and a bending component (*vide* Chapters 4 and 10).

$$\mathbf{D} = \begin{bmatrix} \mathbf{D}^p & 0 \\ 0 & \mathbf{D}^b \end{bmatrix} \tag{19.16}$$

† In-plane and bending components have here been separated by appropriate superscripts.

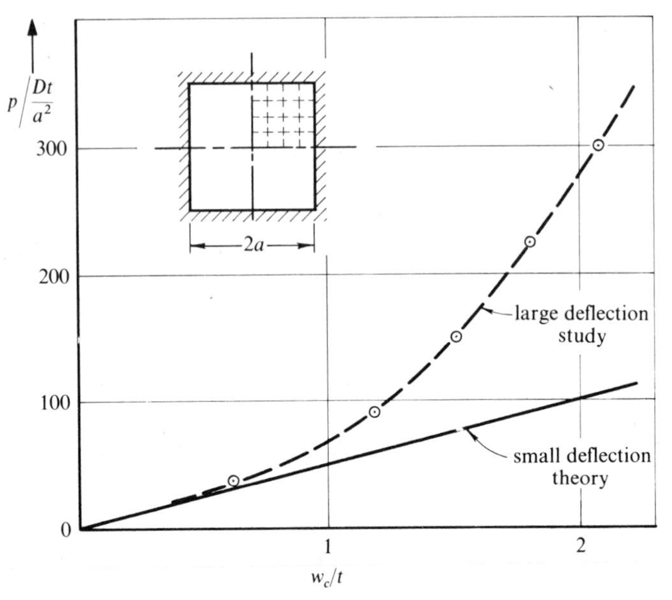

Fig. 19.2 Central deflection, w_c, of a clamped square plate under uniform load p.[14] $u = v = 0$ at edge

Finally the displacements are defined in terms of nodal parameters using the appropriate shape functions.

Thus for instance

$$\begin{Bmatrix} u \\ v \\ w \end{Bmatrix} = \mathbf{N}\mathbf{a}^e \tag{19.17}$$

where a typical set of nodal parameters will be for convenience divided into those which influence in-plane and bending deformation respectively.

$$\mathbf{a}_i = \begin{Bmatrix} \mathbf{a}_i^p \\ \mathbf{a}_i^b \end{Bmatrix} \quad \text{with} \quad \mathbf{a}_i^p = \begin{Bmatrix} u_i \\ v_i \end{Bmatrix} \quad \text{(as in Chapter 4)} \tag{19.18}$$

$$\mathbf{a}_i^b = \begin{Bmatrix} w_i \\ \dfrac{\partial w}{\partial x_i} \\ \dfrac{\partial w}{\partial y_i} \end{Bmatrix} \quad \text{(as in Chapter 10).}$$

Thus the shape function can also be subdivided as

$$\mathbf{N}_i = \begin{bmatrix} \mathbf{N}_i^p & 0 \\ 0 & \mathbf{N}_i^b \end{bmatrix} \tag{19.19}$$

and indeed we shall assume in what follows that the final assembled displacement vector is also subdivided in the manner of Eq. (19.18).

This is convenient as with the exception of the non-linear strain $\boldsymbol{\varepsilon}_L^p$ all the definitions of standard linear analysis apply and therefore do not have to be repeated here.

19.3.2 *Evaluation of* $\overline{\mathbf{B}}$. For further formulation it will be necessary to establish expressions for $\overline{\mathbf{B}}$ and $\mathbf{K}_T$ matrices. First we shall note that

$$\overline{\mathbf{B}} = \mathbf{B}_0 + \mathbf{B}_L \tag{19.20}$$

where

$$\mathbf{B}_0 = \begin{bmatrix} \mathbf{B}_0^p & \mathbf{0} \\ \mathbf{0} & \mathbf{B}_0^b \end{bmatrix}; \qquad \mathbf{B}_L = \begin{bmatrix} \mathbf{0} & \mathbf{B}_L^b \\ \mathbf{0} & \mathbf{0} \end{bmatrix}$$

where $\mathbf{B}_0^{pp}$, $\mathbf{B}_0^b$ are the well-defined, standard matrices of appropriate linear in-plane and bending elements and $\mathbf{B}_L^b$ is found by taking a variation of $\boldsymbol{\varepsilon}_L^p$ with respect to the parameters $\mathbf{a}^b$.

This, non-linear, strain component of Eq. (19.15) can be written conveniently as

$$\boldsymbol{\varepsilon}_L^p = \frac{1}{2} \begin{bmatrix} \dfrac{\partial w}{\partial x} & 0 \\ 0 & \dfrac{\partial w}{\partial y} \\ \dfrac{\partial w}{\partial y} & \dfrac{\partial w}{\partial x} \end{bmatrix} \begin{Bmatrix} \dfrac{\partial w}{\partial x} \\ \dfrac{\partial w}{\partial y} \end{Bmatrix} = \frac{1}{2} \mathbf{A}\boldsymbol{\theta}. \tag{19.21}$$

The derivatives (slopes) of w can be related to the nodal parameters $\mathbf{a}^b$ as

$$\boldsymbol{\theta} = \begin{Bmatrix} \dfrac{\partial w}{\partial x} \\ \dfrac{\partial w}{\partial y} \end{Bmatrix} = \mathbf{G}\mathbf{a}^b \tag{19.22}$$

in which we have

$$\mathbf{G} = \begin{bmatrix} \dfrac{\partial N_1^b}{\partial x}, & \dfrac{\partial N_2^b}{\partial x} & \cdots \\ \dfrac{\partial N_1^b}{\partial y}, & \dfrac{\partial N_2^b}{\partial y} & \cdots \end{bmatrix}. \tag{19.23}$$

Thus $\mathbf{G}$ is a matrix defined purely in terms of the co-ordinates.

Taking the variation of Eq. (19.21) we have†

$$d\boldsymbol{\varepsilon}_L^p = \tfrac{1}{2}\,d\mathbf{A}\boldsymbol{\theta} + \tfrac{1}{2}\mathbf{A}\,d\boldsymbol{\theta} = \mathbf{A}\,d\boldsymbol{\theta} = \mathbf{A}\mathbf{G}\,d\mathbf{a}^b \tag{19.24}$$

and hence immediately, by definition

$$\mathbf{B}_L^b = \mathbf{A}\mathbf{G}. \tag{19.25}$$

19.3.3 *Evaluation of* $\mathbf{K}_T$. The linear, small deformation, matrices are written as

$$\mathbf{K}_0 = \begin{bmatrix} \mathbf{K}_0^p & \mathbf{0} \\ \mathbf{0} & \mathbf{K}_0^b \end{bmatrix} \tag{19.26}$$

with appropriate definitions given in Chapters 4 and 10. Using Eq. (19.7b) the large displacement matrices can be defined on substituting Eq. (19.20). Thus after some manipulation

$$\mathbf{K}_L = \int_V \begin{bmatrix} \mathbf{0}, & \mathbf{B}_0^{pT} & \mathbf{D}^p & \mathbf{B}_L^b \\ \text{Sym.}, & \mathbf{B}_L^{bT} & \mathbf{D}^{b'} & \mathbf{B}_L^b \end{bmatrix} dV. \tag{19.27}$$

† The manipulation of Eq. (19.24) is due to an interesting property of the matrices $\mathbf{A}$ and $\boldsymbol{\theta}$. It is easy to verify that if

$$\mathbf{x} = \begin{Bmatrix} x_1 \\ x_2 \end{Bmatrix}$$

is an arbitrary vector then

$$d\mathbf{A}\mathbf{x} = \begin{bmatrix} d\left(\dfrac{\partial w}{\partial x}\right) & 0 \\ 0 & d\left(\dfrac{\partial w}{\partial y}\right) \\ d\left(\dfrac{\partial w}{\partial y}\right) & d\left(\dfrac{\partial w}{\partial x}\right) \end{bmatrix} \begin{Bmatrix} x_1 \\ x_2 \end{Bmatrix} \equiv \begin{bmatrix} x_1 & 0 \\ 0 & x_2 \\ x_2 & x_1 \end{bmatrix} d\boldsymbol{\theta}$$

thus

$$d\mathbf{A}\boldsymbol{\theta} = \mathbf{A}\,d\boldsymbol{\theta}.$$

Similarly if

$$\mathbf{y} = \begin{Bmatrix} y_1 \\ y_2 \\ y_3 \end{Bmatrix}$$

$$d\mathbf{A}^{\mathrm{T}}\mathbf{y} = \begin{bmatrix} d\left(\dfrac{\partial w}{\partial x}\right) & 0 & d\left(\dfrac{\partial w}{\partial y}\right) \\ 0 & d\left(\dfrac{\partial w}{\partial y}\right) & d\left(\dfrac{\partial w}{\partial x}\right) \end{bmatrix} \begin{Bmatrix} y_1 \\ y_2 \\ y_3 \end{Bmatrix} = \begin{bmatrix} y_1 & y_3 \\ y_3 & y_2 \end{bmatrix} d\boldsymbol{\theta}.$$

Use of this second property will be made later.

Finally $\mathbf{K}_\sigma$ has to be found using the definition of Eq. (19.8). From Eq. (19.20) we have on taking a variation

$$\mathrm{d}\mathbf{B}_L^{\mathrm{T}} = \begin{bmatrix} \mathbf{0} & \mathbf{0} \\ \mathrm{d}\mathbf{B}_L^{b\mathrm{T}} & \mathbf{0} \end{bmatrix} \tag{19.28}$$

which on substitution into Eqs. (19.8) and (19.25) gives

$$\mathbf{K}_\sigma \, \mathrm{d}\mathbf{a} = \int_V \begin{bmatrix} \mathbf{0} & \mathbf{0} \\ \mathbf{G}^{\mathrm{T}} \, \mathrm{d}\mathbf{A}^{\mathrm{T}} & \mathbf{0} \end{bmatrix} \begin{Bmatrix} T_x \\ T_y \\ T_{xy} \\ M_x \\ M_y \\ M_{xy} \end{Bmatrix} \mathrm{d}V. \tag{19.29}$$

But by the special property described in the footnote of page 510, we can write

$$\mathrm{d}\mathbf{A}^{\mathrm{T}} \begin{Bmatrix} T_x \\ T_y \\ T_{xy} \end{Bmatrix} = \begin{bmatrix} T_x & T_{xy} \\ T_{xy} & T_y \end{bmatrix} \mathrm{d}\boldsymbol{\theta} = \begin{bmatrix} T_x & T_{xy} \\ T_{xy} & T_y \end{bmatrix} \mathbf{G} \, \mathrm{d}\mathbf{a}^b$$

and finally we obtain

$$\mathbf{K}_\sigma = \begin{bmatrix} \mathbf{0} & \mathbf{0} \\ \mathbf{0} & \mathbf{K}_\sigma^b \end{bmatrix} \tag{19.30}$$

with

$$\mathbf{K}_\sigma^b = \int_V \mathbf{G}^{\mathrm{T}} \begin{bmatrix} T_x & T_{xy} \\ T_{xy} & T_y \end{bmatrix} \mathbf{G} \, \mathrm{d}V \tag{19.31}$$

a well known *symmetric form of the initial stress*, plate matrix.

19.3.4 *Large deflection problem.* All the ingredients necessary for computing the large deflection plate problem are now available.

As a first step displacements $\mathbf{a}^0$ are found according to the small displacement, uncoupled solution. This determines the actual strains by considering the non-linear contribution defined by Eq. (19.21) together with the appropriate linear contributions. Corresponding stresses can be found by the elastic expressions and $\boldsymbol{\Psi}^0$ determined according to Eq. (19.1). For successive iterations $\mathbf{K}_T^n$ is found from Eqs. (19.26), (19.27), and (19.30).

A typical solution thus obtained,[14] Fig. 19.2, shows the stiffening of the plate with increasing deformation due to development of 'membrane' stresses. At the edges of the plate all in-plate and lateral deformations are restrained. The results show an excellent agreement with an alternative analytical solution.

The element properties were derived using for the in-plane deformation the simplest rectangle function of Chapter 7 and for bending deformation the non-conforming shape function for a rectangle (section 10.4, Chapter 10).

An example of the stress variation with load for a clamped square plate under uniform load is shown in Fig. 19.3.[18] A quarter of the plate is analysed as above with 32 triangular elements, using the 'in-plane' element of Chapter 3 together with a modified[19] version of the non-conforming plate bending element of Chapter 10. Many other examples of large plate deformation obtained by finite element methods are available in the literature.[18, 20–25]

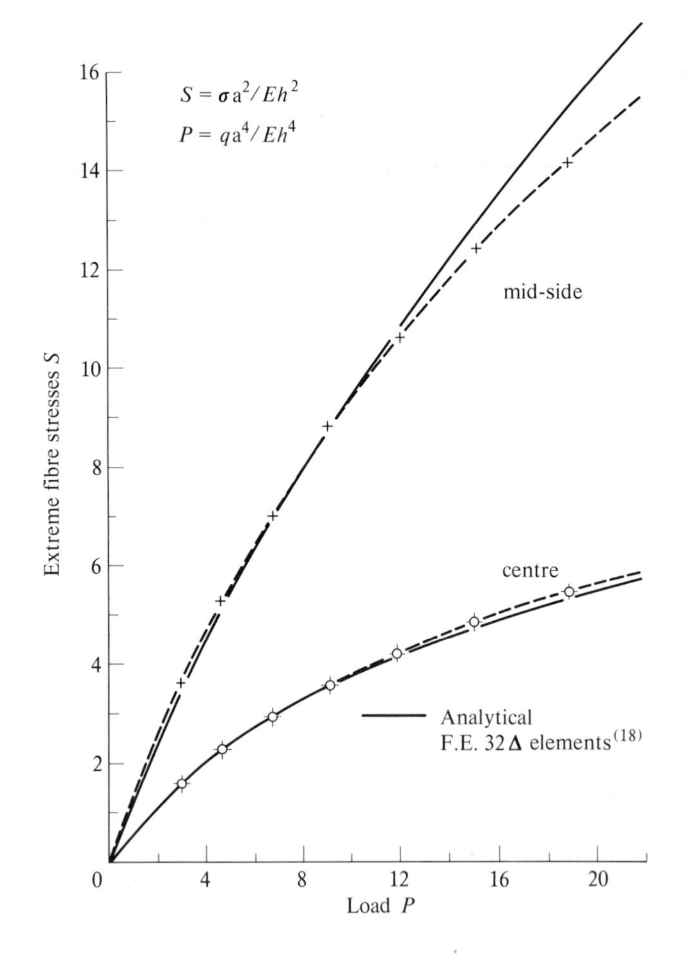

Fig. 19.3 Clamped square plate—stresses

19.3.5 *Bifurcation instability*. In a few practical cases as in the classical Euler problem a bifurcation instability is possible. Consider the situation of a plate loaded purely in its own plane. As lateral deflections, w, are not produced, the small deflection theory gives an exact solution. However, even with zero lateral displacements, the initial stress matrix $\mathbf{K}_\sigma^b$ can be found while $\mathbf{K}_L \equiv 0$. If the in-plane stresses are compressive this matrix will generally be such that real eigenvalues of the bending deformation equation can be found

$$(\mathbf{K}_0^b + \lambda \mathbf{K}_\sigma^b)\mathbf{a}^b = \mathbf{0} \tag{19.32}$$

in which λ denotes the increase factor on in-plane stresses necessary to achieve neutral equilibrium (instability).

At such an increased load incipient buckling occurs and lateral deflections can occur without any lateral load.

The problem is simply formulated by writing only the bending equation with $\mathbf{K}_0^b$ determined as in Chapter 10 and $\mathbf{K}_\sigma^b$ found from Eq. (19.31).

Points of such incipient stability (buckling) for a variety of plate problems have been determined using various element formulations.[26–31] Some comparative results for a simple problem of a square, simply supported plate under a uniform compression T_x in one direction, are given in Table 19.1. In this the buckling parameter is defined as

$$C = T_x a^2/\pi^2 D$$

where a is the side of the plate and D the bending rigidity.

TABLE 19.1
VALUES OF C FOR A SQUARE PLATE SIMPLY SUPPORTED AND COMPRESSED UNIAXIALLY

Elements in quarter plate	Non-compatible		Compatible	
	Rectangle[27] 12 D.O.F.	Triangle[29] 9 D.O.F.	Rectangle[30] 16 D.O.F.	Quadrilateral[31] 16 D.O.F.
2×2		3·22		
4×4	3·77	3·72	4·015	4·029
8×8	3·93	3·90	4·001	4·002

Exact $C = 4{\cdot}00$[17]
D.O.F. = degrees of freedom.

The elements are all of the type described in Chapter 10 and it is of interest to note that all those which are slope compatible always over-estimate the buckling factor. The non-conforming elements in this case under-estimate it although no such bound could be, at present, determined.

Figure 19.4 shows a buckling mode for a geometrically more complex case.[29] Here again the non-conforming triangle was used.

Such incipient stability problems in plates are of limited practical importance. As soon as lateral deflection occurs a stiffening of the plate follows and additional loads can be carried. This stiffening was noted in the example of Fig. 19.2. Post-buckling behaviour thus should be studied by the large deformation process generally described in previous sections.[32–34] To avoid the bifurcation difficulty a slight perturbation (or lateral load) should then be imposed.

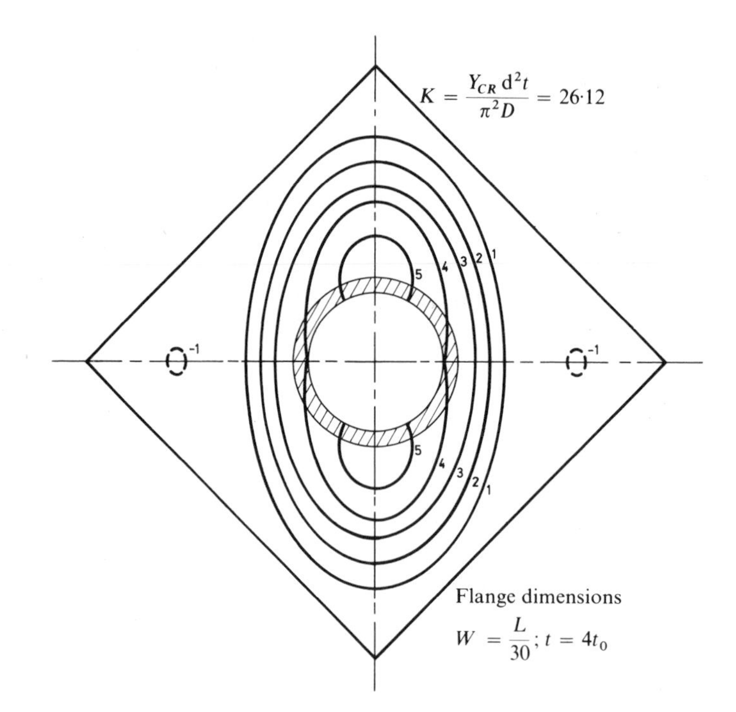

Fig. 19.4 Buckling mode of a square plate under shear—clamped edges, central hole stiffened by flange[29]

19.4 Shells

In shells stability problems are much more relevant than in plates. Here, in general, the problem is one in which the tangential stiffness matrix $\mathbf{K}_T$ should always be determined taking the actual displacements into account, as now the special case of uncoupled membrane and bending effects does not occur under load except in most trivial cases. If the *initial stability* matrix $\mathbf{K}_\sigma$ is determined for the elastic stresses it is, however, sometimes possible to obtain useful results concerning the stability factor

λ, and indeed in the classical work on the subject of shell buckling this initial stability has been almost exclusively considered. The true collapse load may, however, be well *below* the initial stability load and it is important to determine at least approximately the deformation effects.

If shells are assumed to be built up of flat plate elements the same transformations as given in Chapter 13 can be followed with the plate tangential stiffness matrix.[35,36] If curved shell elements are used it is essential to revert to the equation of shell theory and to include in these the non-linear terms.[14,37,38,39] For the complete formulation required the reader is referred to these references.

In the case of shallow shells the transformations of Chapter 13 may conveniently be avoided by adopting a formulation based on Marguerre shallow shell theory.[25,40,41]

It is extremely important to emphasize again, that initial instability calculations are meaningful only in special cases, and that they often overestimate the collapse loads considerably. For correct answers a full non-linear process has to be invoked. The progressive 'softening' of a shell under a load well below the one given by linearized buckling is shown in Fig. 19.5 taken from reference 14. Figure 19.6 shows the progressive collapse

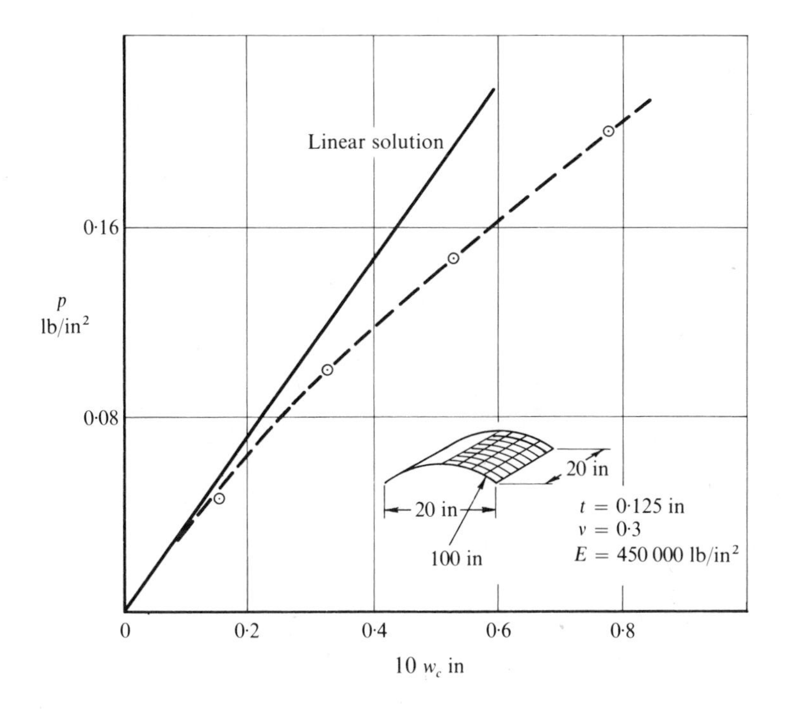

Fig. 19.5 Deflections of cylindrical shell at centre. All edges clamped[14]

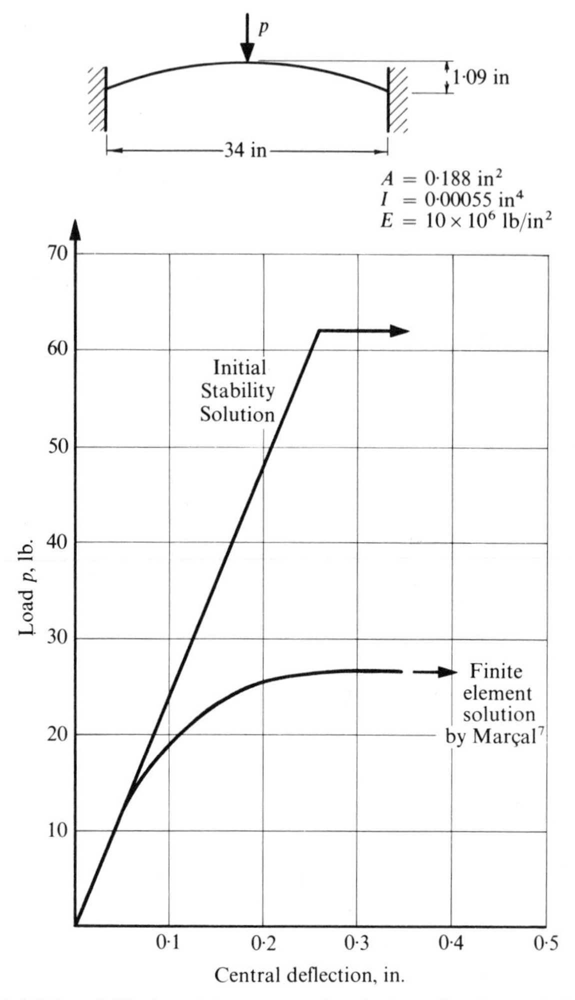

Fig. 19.6 'Initial stability' and incremental solutions for large deformation of an arch under central load p[7]

of an arch at a load much below that given by the linear stability value.[7]

The determination of the actual collapse load of a shell or other slender structure presents obvious difficulties (of a kind already encountered in Chapter 18) as convergence of displacements cannot be obtained when load is increased near the 'peak' carrying capacity.

It is convenient, then, to proceed immediately prescribing displacement increments and computing the corresponding reactions if one concentrated load is considered. By such processes, Argyris[5] and others[39, 42] succeed in following a complete snap through behaviour of an arch.

Pian and Ping Tong[43] show how the process can be simply generalized when a system of proportional loads is considered.

Alternative processes approaching the collapse problem have been described and much work has been accomplished in this important field.[44-48]

19.5 General, Large Strain and Displacement Formulation

The non-linear strain displacement relationship for plates, used in section 19.3 (Eq. (19.15)) was derived on *ad hoc* basis. For shells alternative relationships may be similarly derived but the possibility of diverse approximations arises at all stages. It is, however, possible to use a general definition of strains which is *valid whether displacements or strains are large or small*. Such a definition was introduced by Green and St. Venant and is known as the *Green's strain tensor*. Referred to a fixed Cartesian system of co-ordinates, x, y, z, the displacements u, v, w define strains as[49]

$$\begin{aligned} \varepsilon_x &= \frac{\partial u}{\partial x} + \frac{1}{2}\left[\left(\frac{\partial u}{\partial x}\right)^2 + \left(\frac{\partial v}{\partial x}\right)^2 + \left(\frac{\partial w}{\partial x}\right)^2\right] \\ \gamma_{xy} &= \frac{\partial u}{\partial y} + \frac{\partial v}{\partial x} + \left[\frac{\partial u}{\partial x}\cdot\frac{\partial u}{\partial y} + \frac{\partial v}{\partial x}\cdot\frac{\partial v}{\partial y} + \frac{\partial w}{\partial x}\cdot\frac{\partial w}{\partial y}\right] \end{aligned} \tag{19.33}$$

with other components obtained by suitable permutation.

If displacements are small the general first order linear strain approximation is obtained by neglecting the quadratic terms.

Geometric interpretation of the above strain definitions is not obvious in the general case but it should be noted that they give a measure of elongation and angular distortion of an originally orthogonal element and for small strain result in usual definitions even if the displacements are large.

If the actual strain quantities are small then it is simple to show that ε_x defines the change of length of a unit length *originally oriented parallel to the x axis* while γ_{xy} gives similarly the angle change between two lines *originally* parallel to x and y axes. This is true in the above definition even if large movements have occurred which have rotated or displaced the original axes by gross amounts.

We shall now establish the general non-linear expressions $\overline{\mathbf{B}}$ and $\mathbf{K}_T$ for a complete three-dimensional state of stress. It is a simple matter to specialize from here to one- or two-dimensional forms and such exercises will be left to the reader. Indeed plate and shell problems are conveniently approached via such a general formulation and some terms neglected in the specialized approach to the plate problem of the previous section are now easily included.

19.5.1 *Derivation of the* $\mathbf{B}_L$ *matrix.* The general strain vector in three dimensions can be defined in terms of the infinitesimal and large displacement components

$$\boldsymbol{\varepsilon} = \boldsymbol{\varepsilon}_0 + \boldsymbol{\varepsilon}_L \tag{19.34}$$

where

$$\boldsymbol{\varepsilon}_0 = \begin{Bmatrix} \varepsilon_x \\ \varepsilon_y \\ \varepsilon_z \\ \gamma_{yz} \\ \gamma_{zx} \\ \gamma_{xy} \end{Bmatrix} = \begin{Bmatrix} \dfrac{\partial u}{\partial x} \\ \dfrac{\partial v}{\partial y} \\ \dfrac{\partial w}{\partial z} \\ \dfrac{\partial v}{\partial z} + \dfrac{\partial w}{\partial y} \\ \dfrac{\partial w}{\partial x} + \dfrac{\partial u}{\partial z} \\ \dfrac{\partial u}{\partial y} + \dfrac{\partial v}{\partial x} \end{Bmatrix} \tag{19.35}$$

is the same as defined in Chapter 6. The non-linear terms of Eq. (19.33) can be conveniently rewritten as

$$\boldsymbol{\varepsilon}^L = \tfrac{1}{2} \begin{bmatrix} \boldsymbol{\theta}_x^{\mathrm{T}} & \mathbf{0} & \mathbf{0} \\ \mathbf{0} & \boldsymbol{\theta}_y^{\mathrm{T}} & \mathbf{0} \\ \mathbf{0} & \mathbf{0} & \boldsymbol{\theta}_z^{\mathrm{T}} \\ \mathbf{0} & \boldsymbol{\theta}_z^{\mathrm{T}} & \boldsymbol{\theta}_y^{\mathrm{T}} \\ \boldsymbol{\theta}_z^{\mathrm{T}} & \mathbf{0} & \boldsymbol{\theta}_x^{\mathrm{T}} \\ \boldsymbol{\theta}_y^{\mathrm{T}} & \boldsymbol{\theta}_x^{\mathrm{T}} & \mathbf{0} \end{bmatrix} \begin{Bmatrix} \boldsymbol{\theta}_x \\ \boldsymbol{\theta}_y \\ \boldsymbol{\theta}_z \end{Bmatrix} = \tfrac{1}{2}\mathbf{A}\boldsymbol{\theta} \tag{19.36}$$

in which

$$\boldsymbol{\theta}_x^{\mathrm{T}} = \left[\frac{\partial u}{\partial x}, \frac{\partial v}{\partial x}, \frac{\partial w}{\partial x}\right], \text{ etc.}$$

and $\mathbf{A}$ is a 6×9 matrix.

The reader can readily verify the validity of the above definition and re-establish the properties of the matrices $\mathbf{A}$ and $\boldsymbol{\theta}$ defined in section 19.3.2 (footnote to p. 510). Once again

$$\mathrm{d}\boldsymbol{\varepsilon}_L = \tfrac{1}{2}\,\mathrm{d}\mathbf{A}\boldsymbol{\theta} + \tfrac{1}{2}\mathbf{A}\,\mathrm{d}\boldsymbol{\theta} = \mathbf{A}\,\mathrm{d}\boldsymbol{\theta} \tag{19.37}$$

and as we can determine $\boldsymbol{\theta}$ in terms of the shape function $\mathbf{N}$ and nodal parameters $\mathbf{a}$ we can write

$$\boldsymbol{\theta} = \mathbf{G}\mathbf{a} \tag{19.38}$$

or

$$\mathrm{d}\boldsymbol{\varepsilon}^L = \mathbf{A}\mathbf{G}\,\mathrm{d}\mathbf{a}$$

and

$$\mathbf{B}_L = \mathbf{A}\mathbf{G} \tag{19.39}$$

19.5.2 *Derivation of* $\mathbf{K}_T$ *matrix*. Noting that

$$\overline{\mathbf{B}} = \mathbf{B}_0 + \mathbf{B}_L$$

we can readily form the matrix of Eq. (19.7)

$$\overline{\mathbf{K}} = \mathbf{K}_0 + \mathbf{K}_L = \int_V \overline{\mathbf{B}}^{\mathrm{T}}\mathbf{D}\overline{\mathbf{B}}\,\mathrm{d}V. \tag{19.40}$$

To complete the total tangential stiffness matrix it is necessary only to determine the initial stress matrix $\mathbf{K}_\sigma$. Again by Eq. (19.8) we have

$$\mathbf{K}_\sigma\,\mathrm{d}\mathbf{a} = \int_V \mathrm{d}\mathbf{B}_L^{\mathrm{T}}\boldsymbol{\sigma}\,\mathrm{d}V = \int_V \mathbf{G}^{\mathrm{T}}\,\mathrm{d}\mathbf{A}^{\mathrm{T}}\boldsymbol{\sigma}\,\mathrm{d}V. \tag{19.41}$$

Once again we can verify that we can write

$$\mathrm{d}\mathbf{A}^{\mathrm{T}}\boldsymbol{\sigma} = \begin{bmatrix} \sigma_x\mathbf{I}_3 & \tau_{xy}\mathbf{I}_3 & \tau_{xz}\mathbf{I}_3 \\ & \sigma_y\mathbf{I}_3 & \tau_{yz}\mathbf{I}_3 \\ \text{sym.} & & \sigma_z\mathbf{I}_3 \end{bmatrix} \mathrm{d}\boldsymbol{\theta} = \mathbf{M}\mathbf{G}\,\mathrm{d}\mathbf{a} \tag{19.42}$$

in which $\mathbf{I}_3$ is a 3×3 identity matrix.

Substituting Eq. (19.42) into Eq. (19.41) yields[42]

$$\mathbf{K}_\sigma = \int_V \mathbf{G}^{\mathrm{T}}\mathbf{M}\mathbf{G}\,\mathrm{d}V \tag{19.43}$$

in which $\mathbf{M}$ is a 9×9 matrix of the six stress components arranged as shown in Eq. (19.42). The *symmetric* form of $\mathbf{K}_\sigma$ is once again demonstrated.

Once again we have omitted element superscripts though in fact all of the above matrices would be obtained element by element and added in the standard manner.

The use of the general expressions is a useful starting point in the analysis of plates and shells if consistent approximations are to be made. In the case of the thick shell formulation of Chapter 14 such expressions are essential.

Further if a suitable stress-strain relation can be found they are valid for large strain analysis. Here, however, it is more usual to define directly a strain energy function in terms of the strain components and to obtain generalized forces by direct minimization. Some examples of such large strain analysis have been given by Oden[50–53] who discusses large deformation of rubber membranes and continua.

An example of the application of the above formulation for an axi-symmetric shell under a central point load and ring loads, is shown in Figs. 19.7[42, 54, 55] in which parabolic-linear isoparametric elements were

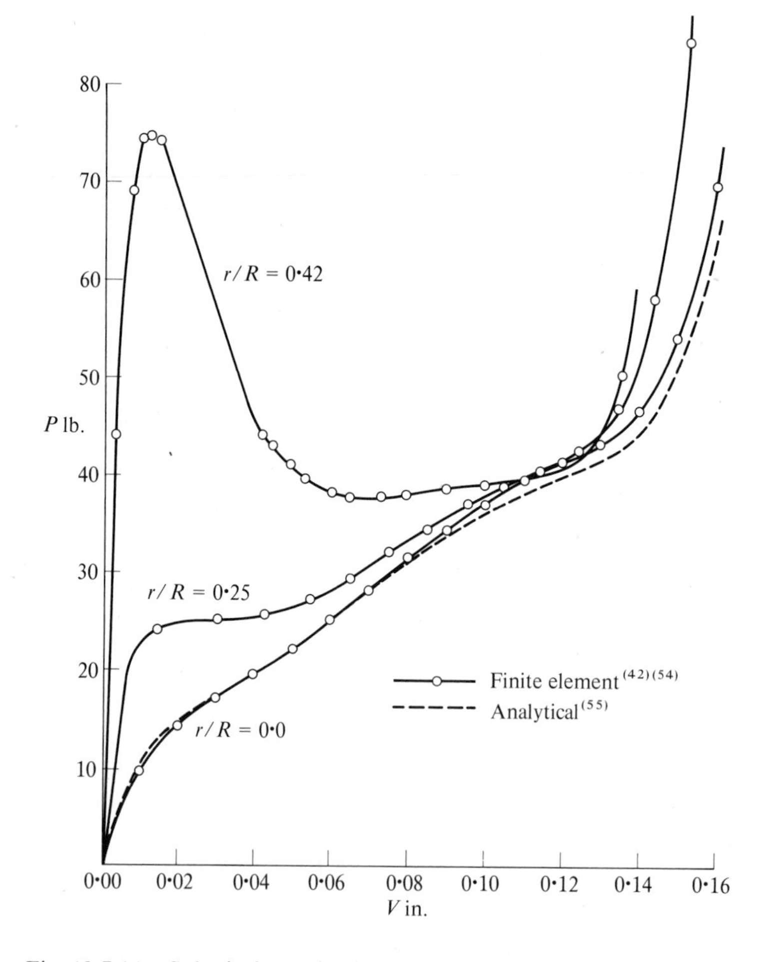

Fig. 19.7 (*a*) Spherical cap: load-deflection curves for various ring loads

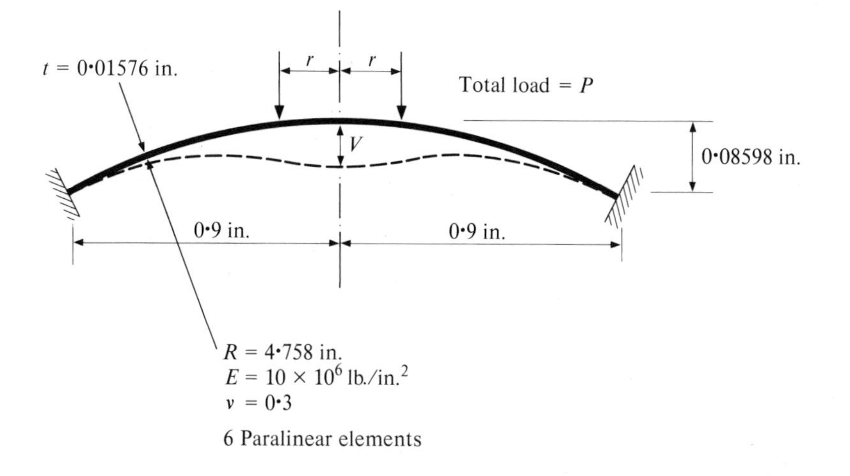

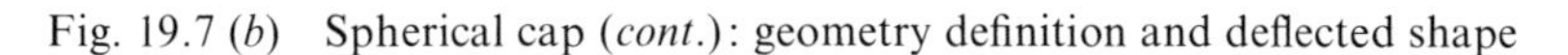

Fig. 19.7 (*b*) Spherical cap (*cont.*): geometry definition and deflected shape

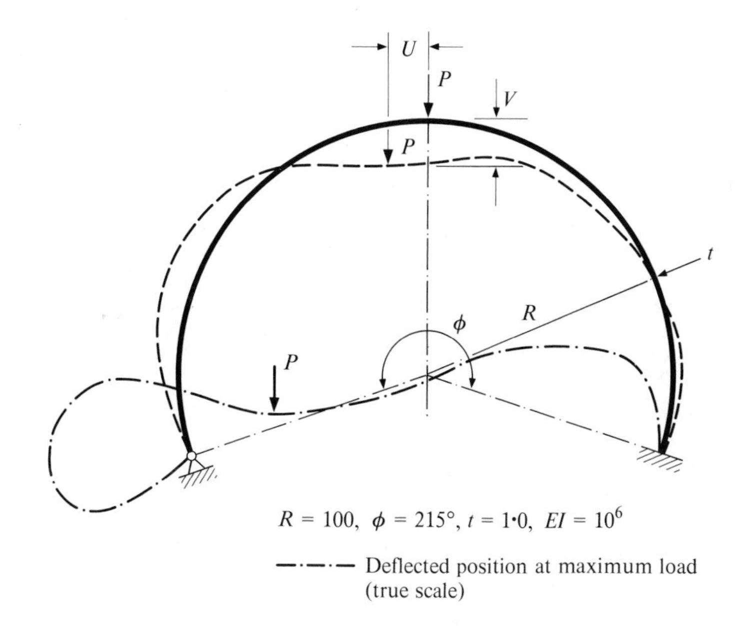

Fig. 19.8 Clamped-hinged arch—load deflection

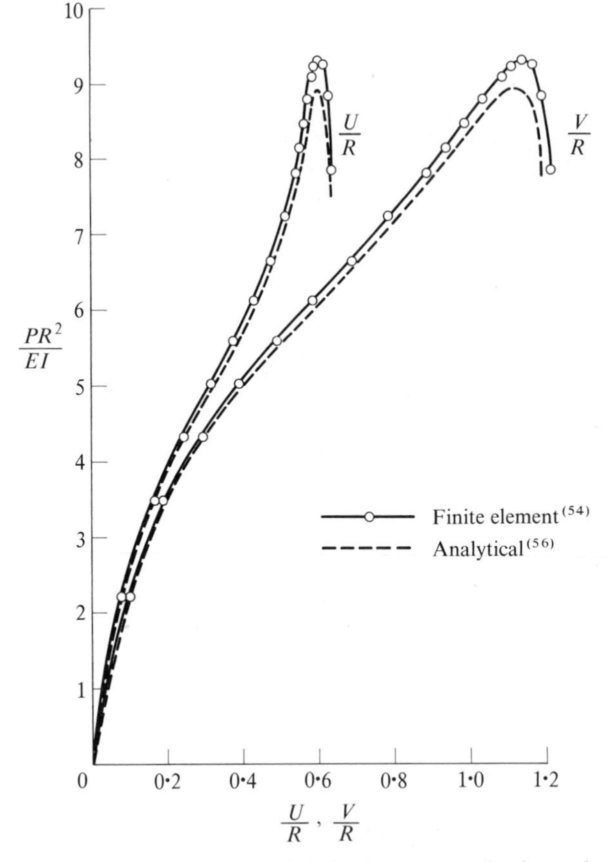

Fig. 19.9 Clamped-hinged arch: load-deflection curves (horizontal and vertical components)

used. Another typical solution using this two-dimensional formulation is shown in Figs. 19.8 and 19.9 for the case of a deep clamped-hinged arch.[54, 56] Here extremely large deflections are followed by the Lagrangian process described.

It is important to note that the use of isoparametric elements in conjunction with the above general formulation leads to particularly concise representations of $\mathbf{K}_0+\mathbf{K}_L$ and $\mathbf{K}_\sigma$. Furthermore, such formulations result in substantial savings in the computation of $\mathbf{K}_\sigma$.[6, 42, 55]

19.6 Concluding Remarks

This chapter attempted to present a unified approach to all large deformation problems. Various methods of solution on the basic non-linear equation system have been outlined and the reader may, rightly, wonder which is preferable. If a single solution of a non-linear large deformation problem is desired the Newton process appears to converge quite rapidly in most instances. For certain cases, however, the constant matrix methods are more economical.

If a full load deformation study is required it has been the common practice to proceed with small load increments and treat for each such increment the problem as a linear elastic one with the tangential stiffness matrix evaluated at the start of the increment.[3,4] Such methods may accumulate error and Brebbia and Connor[14] recommend a complete Newton-type solution to be used every few increments.

Extension of geometrically non-linear problems to dynamic situations is readily accomplished.[57] In Chapter 21 (sect. 21.10) we shall show several examples of this.

Combination of material non-linearity together with geometric non-linearity is particularly simple if an incremental elasticity matrix may be established. Marçal[3] solves a number of such problems in which large deformation is coupled with plasticity. It is interesting to note that the operations required in solution of problems of material and geometric non-linearity are similar and computer systems capable of dealing with both can be developed.

Finally two matters should be noted. One is the apparently lengthy derivation of the initial stress matrix for plates which in previous publications[27,29] is derived in a more direct way. This is due to an attempt at complete generality which hopefully was achieved. The second point is that somewhat involved operations were required in the section dealing with general large strain in order to preserve the convenient matrix formulation used throughout this book. Some simplification would have arisen if tensor notation were used and indeed this could have been used throughout the book as an alternative. No apology is, however, being made for the choice of the more direct and better understood notation.

An alternative to the Lagrangian process described in this chapter is the use of the current (up-dated) element configuration in an Eulerian manner. This is advantageous in large strain situations and a clear description of the procedure is given by McMeeking and Rice.[58]

References

1. C. TRUESDELL (ed.), *Continuum Mechanics IV: Problems of Non-linear Elasticity*, Vol. 8, p. 4, Gordon and Beach, 1965.
2. K. J. BATHE and H. OZDEMIR, 'Elastic-plastic large deformation static and dynamic analysis', *Comp. Struct.*, **6** (No. 2), 81–92, April 1976.
3. P. V. MARÇAL, *Finite element analysis of combined problems of material and geometric behaviour*, Techn. Rep. 1, ONR, Brown University, 1969, also *Proc.*

Am. Soc. Mech. Eng. Conf. on Computational Approaches in Applied Mechanics, 133, June 1969.
4. J. H. ARGYRIS, S. KELSEY and H. KAMEL, *Matrix Methods of Structural Analysis*, AGARD-ograph 72, Pergamon Press, 1963.
5. J. H. ARGYRIS, 'Continua and discontinua', *Proc. Conf. Matrix Methods in Structural Mechanics*, Air Force Inst. Tech., Wright-Patterson A.F. Base, Ohio, Oct. 1965.
6. G. C. NAYAK, *Plasticity and large deformation problems by finite element method*, Ph.D. Thesis, Univ. of Wales, Swansea, 1971 (C/Ph/15/1971).
7. P. V. MARÇAL, *Effect of initial displacement on problem of large deflection and stability*, Tech. Report ARPA E54, Brown Univ., 1967.
8. W. E. HAISLER, J. A. STRICKLIN and F. J. STEBBINS, 'Development and evaluation of solution procedures for geometrically non-linear analysis', *J.A.I.A.A.*, **10**, 264–72, 1972.
9. G. A. DUPUIS, D. D. PFAFFINGER and P. V. MARÇAL, 'Effective use of incremental stiffness matrices in non-linear geometric analysis', *IUTAM Symp. on High Speed Computing of Elastic Structures*, Liége, Aug. 1970.
10. R. H. GALLAGHER and S. T. MAU, *A method of limit point calculation in finite element structural analysis*, NASA CR 12115, Sept. 1972.
11. H. L. LANGHAAR, *Energy Methods in Applied Mechanics*, Wiley, 1962.
12. K. MARGUERRE, 'Über die Anwendung der energetischen Methode auf Stabilitätsprobleme', *Hohrb.*, D.V.L., 252–62, 1938.
13. B. FRAEIJS DE VEUBEKE, 'The second variation test with algebraic and differential contrasts', *Advanced Problems and Methods for Space Flight Optimisation*, Pergamon Press, 1969.
14. C. A. BREBBIA and J. CONNOR, 'Geometrically non-linear finite element analysis', *Proc. Am. Soc. Civ. Eng.*, **95**, EM2, 463–83, 1969.
15. H. D. HIBBITT, P. V. MARÇAL and J. R. RICE, 'A finite element formulation for problems of large strain and large displacement', *Int. Jl. Solids Struct.*, **6**, 1069–86, 1970.
16. J. T. ODEN, Discussion on 'Finite element analysis of non-linear structures', by Mallet and Marçal, *Proc. Am. Soc. Civ. Eng.*, **95**, ST6, 1376–81, 1969.
17. S. P. TIMOSHENKO and J. M. GERE, *Theory of Elastic Stability*, 2nd ed., McGraw-Hill, 1961.
18. R. D. WOOD, *The application of finite element methods to geometrically non-linear finite element analysis*, Ph.D. Thesis, Univ. of Wales, Swansea, 1973 (C/Ph/20/73).
19. A. RAZZAQUE, 'Program for triangular bending elements with derivative smoothing', *Int. J. Num. Meth. Eng.*, **6**, 333–5, 1973.
20. L. A. SCHMIT, F. K. BOGNER and R. L. FOX, 'Finite deflection structural analysis using plate and cylindrical shell discrete elements', *Proc. AIAA/ASME 8th Struct. and Stress Dynamic Conference*, Palm Springs, California, 197–211, March 1967. Also *J.A.I.A.A.*, **5**, 1525–7, 1968.
21. M. J. TURNER, E. H. DILL, H. C. MARTIN and R. J. MELOSH, 'Large deflection of structures subjected to heating and external loads', *J. Aero. Sci.*, **27**, 97–106, 1960.
22. T. KAWAI and N. YOSHIMURA, 'Analysis of large deflection of plates by finite element method', *Int. J. Num. Meth. Eng.*, **1**, 123–33, 1969.
23. R. H. MALLETT and P. V. MARÇAL, 'Finite element analysis of non-linear structures', *Proc. Am. Soc. Civ. Eng.*, **94**, ST9, 2081–105, 1968.
24. D. W. MURRAY and E. L. WILSON, 'Finite element large deflection analysis of plates', *Proc. Am. Soc. Civ. Eng.*, **94**, EM1, 143–165, 1968.

25. P. G. BERGAN and R. W. CLOUGH, 'Large deflection analysis of plates and shallow shells using the finite element method', *Int. J. Num. Meth. Eng.*, **5**, 543–56, 1973.
26. H. C. MARTIN, 'On the derivation of stiffness matrices for the analysis of large deflection and stability problems', *Proc. Conf. Matrix Methods in Structural Mechanics*, Air Force Inst. Tech., Wright-Patterson A.F. Base, Ohio, Oct. 1965.
27. K. K. KAPUR and B. J. HARTZ, 'Stability of thin plates using the finite element method', *Proc. Am. Soc. Civ. Eng.*, EM2, 177–95, 1966.
28. R. H. GALLAGHER and J. PADLOG, 'Discrete element approach to structural instability analysis', *J.A.I.A.A.*, **1**, 1537–9, 1963.
29. R. G. ANDERSON, B. M. IRONS and O. C. ZIENKIEWICZ, 'Vibration and stability of plates using finite elements', *Int. J. Solids Struct.*, **4**, 1031–55, 1968.
30. W. G. CARSON and R. E. NEWTON, 'Plate buckling analysis using a fully compatible finite element', *J.A.I.A.A.*, **8**, 527–9, 1969.
31. Y. K. CHAN and A. P. KABAILA, 'A conforming quadrilateral element for analysis of stiffened plates', *UNICIV Report* R-121, University of New South Wales, 1973.
32. D. W. MURRAY and E. L. WILSON, 'Finite element post buckling analysis of thin elastic plates', *Proc. 2nd Conf. Matrix Methods in Structural Mechanics*, Wright-Patterson A.F. Base, Ohio, 1968.
33. K. C. ROCKEY and D. K. BAGCHI, 'Buckling of plate girder webs under partial edge loadings', *Int. J. Mech. Sci.*, **12**, 61–76, 1970.
34. T. M. ROBERTS and D. G. ASHWELL, *Post-buckling analysis of slightly curved plates by the finite element method*, Report 2, Dept. of Civil and Struct. Engineering, Univ. of Wales, Cardiff, 1969.
35. R. G. ANDERSON, *A finite element eigenvalue solution system*, Ph.D. Thesis, Univ. of Wales, Swansea, 1968.
36. R. H. GALLAGHER, R. A. GELLATLY, R. H. MALLETT and J. PADLOG, 'A discrete element procedure for thin shell instability analysis', *J.A.I.A.A.*, **5**, 138–145, 1967.
37. R. H. GALLAGHER and H. T. Y. YANG, 'Elastic instability predictions for doubly curved shells', *Proc. 2nd Conf. Matrix Methods in Structural Mechanics*, Air Force Inst. Tech., Wright-Patterson A.F. Base, Ohio, 1968.
38. J. L. BATOZ, A. CHATTOPADHYAY and G. DHATT, 'Finite element large deflection analysis of shallow shells', *Int. J. Num. Meth. Eng.*, **10**, 35–8, 1976.
39. T. MATSUI and O. MATSUOKA, 'A new finite element scheme for instability analysis of thin shells', *Int. J. Num. Meth. Eng.*, **10**, 145–70, 1976.
40. T. Y. YANG, 'A finite element procedure for the large deflection analysis of plates with initial imperfections', *J.A.I.A.A.*, **9** (No. 8), 1971.
41. T. M. ROBERTS and D. G. ASHWELL, 'The use of finite element mid-increment stiffness matrices in the post-buckling analysis of imperfect structures', *Int. J. Solids Struct.*, **7**, 805–23, 1971.
42. O. C. ZIENKIEWICZ and G. C. NAYAK, 'A general approach to problems of plasticity and large deformation using isoparametric elements', *Proc. Conf. on Matrix Methods in Structural Mechanics*, Wright-Patterson A.F. Base, Ohio, 1971.
43. T. H. H. PIAN and PING TONG, 'Variational formulation of finite displacement analysis', Symp. Int. Un. Th. Appl. Mech. on 'High speed computing of elastic structures', Liége, 1970.
44. H. C. MARTIN, 'Finite Elements and the analysis of geometrically non-linear

problems', *Recent advances in matrix methods and structural analysis and design*, Univ. of Alabama Press, 1971.

45. A. C. WALKER, 'A non-linear finite element analysis of shallow circular arches', *Int. J. Solids Struct.*, **5**, 97–107, 1969.
46. J. M. T. THOMPSON and A. C. WALKER, 'A non-linear perturbation analysis of discrete structural systems', *Int. J. Solids Struct.*, **4**, 757–767, 1968.
47. J. S. PRZEMIENIECKI, 'Stability analysis of complex structures using discrete element techniques', Symp. on Struct. Stability and optimisation, Loughborough Univ., March 1967.
48. J. CONNOR and N. MORIN, 'Perturbation techniques in the analysis of geometrically non-linear shells', Symp. Int. Un. Th. Appl. Mech. on 'High speed computing of elastic structures', Liége, 1970.
49. Y. C. FUNG, *Foundation of solid mechanics*, Prentice Hall Int., 1965.
50. J. T. ODEN, 'Finite plane strain of incompressible elastic solids by the finite element method', *The Aeronautical Quarterly*, **19**, 254–64, 1967.
51. J. T. ODEN and T. SATO, 'Finite deformation of elastic membranes by the finite element method', *Int. J. Solids Struct.*, **3**, 471–88, 1967.
52. J. T. ODEN, 'Numerical formulation of non-linear elasticity problems', *Proc. Am. Soc. Civ. Eng.*, **93**, ST3, 235–55, 1967.
53. J. T. ODEN, 'Finite element applications in non-linear structural analysis', Proc. Symp. on Application of Finite Element Methods in Civil Engineering, *Am. Soc. Civ. Eng.*, Vanderbilt Univ., 1969.
54. R. D. WOOD and O. C. ZIENKIEWICZ, *Geometrically non-linear finite element analysis of beams– frames– circles and axisymmetric shells*, Dept. Civ. Eng. Report C/R/281/76, Univ. of Wales, Swansea.
55. J. F. MESCALL, 'Large deflections of spherical shells under concentrated loads', *J. Appl. Mech.*, **32**, 936–8, 1965.
56. D. A. DA DEPPO and R. SCHMIDT, 'Instability of clamped-hinged circular arches subjected to a point load', *Trans. Am. Soc. Mech. Eng.*, 894–6, Dec. 1975.
57. J. A. STRICKLIN, 'Non-linear dynamic analysis of shells of revolution', Symp. Int. Un. Th. Appl. Mech. on 'High speed computing of elastic structures', Liège, 1970.
58. R. M. MCMEEKING and J. R. RICE, 'Finite element formulations for problems of large elastic-plastic, deformation. *In. J. Solids Struct.*, **11**, 601–16, 1975.

20. The Time Dimension. Semi-Discretization of Field and Dynamic Problems and Analytical Solution Procedures

20.1 Introduction

In all the problems considered so far in this text conditions which do not vary with time were generally assumed. There is little difficulty in extending the finite element idealization to situations which are time dependent.

The range of practical problems in which the time dimension has to be considered is great. Transient heat conduction, wave transmission in fluids, and dynamic behaviour of structure are typical examples. While it is usual to consider these various problems separately—sometimes classifying them according to the mathematical structure of governing equations as 'parabolic' or 'hyperbolic'[1]—we shall group them in one category to show that the formulation is identical.

In the first part of this chapter we shall formulate, by a simple extension of the methods used so far, matrix differential equations governing such problems for a variety of physical situations. Here a finite element discretization in the space dimension only will be used and a semi-discretization process followed (viz. Chapter 3, p. 60). In the remainder of this chapter various analytical procedures of solution for the resulting ordinary linear differential equation system will be dealt with. These form the basic arsenal of steady state and transient analysis.

Chapter 21 will be devoted to the discretization at the time domain itself.

20.2 Direct Formulation of Time Dependent Problems with Spatial Finite Element Subdivision

20.2.1 *The 'quasi-harmonic' equation with time differentials.* In many physical problems the quasi-harmonic equation takes up a form in which time derivatives of the unknown function ϕ occur. In the three-dimensional case typically we might have

$$\frac{\partial}{\partial x}\left(k_x \frac{\partial \phi}{\partial x}\right)+\frac{\partial}{\partial y}\left(k_y \frac{\partial \phi}{\partial y}\right)+\frac{\partial}{\partial z}\left(k_z \frac{\partial \phi}{\partial z}\right) + \left(\bar{Q}-\mu \frac{\partial \phi}{\partial t}-\rho \frac{\partial^2 \phi}{\partial t^2}\right) = 0. \quad (20.1)$$

In the above, quite generally, all the parameters may be prescribed functions of time, or in non-linear cases of ϕ, i.e.,

$$k_x = k_x(\phi, t); \qquad \bar{Q} = \bar{Q}(\phi, t), \text{ etc.} \quad (20.2)$$

If a situation at a particular instant of time is considered, the time derivatives of ϕ and all the parameters can be treated as *prescribed functions of space co-ordinates*. Thus, at that instant the problem is precisely identified with those treated in a previous chapter (section 17.2) if the whole of the quantity in the last parentheses of Eq. (20.1) is treated as the term Q of Eq. (17.9).

The finite element discretization of this in terms of *space* elements has already been fully discussed and we found that with the prescription

$$\phi = \sum N_i a_i = \mathbf{Na}$$
$$\mathbf{N} = \mathbf{N}(x, y, z) \qquad \mathbf{a} = \mathbf{a}(t) \quad (20.3)$$

for each element, a standard form of assembled equation†

$$\mathbf{Ka}+\bar{\mathbf{f}} = 0 \quad (20.4)$$

was obtained. Element contributions to the above matrices are defined by Eq. 17.17 which need not be repeated here except for that representing the 'load' term due to Q. This gives by

$$\bar{\mathbf{f}}_i^e = -\int_{\Omega^e} Q N_i \, \mathrm{d}\Omega. \quad (20.5)$$

Replacing now Q by the last bracketed term of Eq (20.1) we have

$$\bar{\mathbf{f}}_i^e = -\int_{\Omega^e} N_i \left(\bar{Q}-\mu \frac{\partial \phi}{\partial t}-\rho \frac{\partial^2 \phi}{\partial t^2}\right) \mathrm{d}\Omega \quad (20.6)$$

However, from Eq. (20.3) it is noted that ϕ is approximated in terms of the nodal parameters $\mathbf{a}$. On substitution of this approximation we have

$$\bar{\mathbf{f}} = -\int_{\Omega} \mathbf{N}^{\mathrm{T}} \bar{Q} \, \mathrm{d}\Omega + \left(\int_{\Omega} \mathbf{N}^{\mathrm{T}} \mu \mathbf{N} \, \mathrm{d}\Omega\right) \frac{\mathrm{d}}{\mathrm{d}t} \mathbf{a} + \left(\int_{\Omega} \mathbf{N}^{\mathrm{T}} \rho \mathbf{N} \, \mathrm{d}\Omega\right) \frac{\mathrm{d}^2}{\mathrm{d}t^2} \mathbf{a} \quad (20.7)$$

† We have replaced the matrix **H** of Chapter 17 by **K** to facilitate comparison with other dynamic equations.

and on expanding Eq. (20.4) in its final assembled form we get the following *matrix differential equation*

$$\mathbf{M}\ddot{\mathbf{a}}+\mathbf{C}\dot{\mathbf{a}}+\mathbf{K}\mathbf{a}+\mathbf{f} = 0 \qquad (20.8)$$

$$\dot{\mathbf{a}} \equiv \frac{\mathrm{d}}{\mathrm{d}t}\mathbf{a}; \qquad \ddot{\mathbf{a}} = \frac{\mathrm{d}^2}{\mathrm{d}t^2}\mathbf{a} \qquad (20.9)$$

in which all the matrices are assembled from element submatrices in the standard manner with submatrices $\mathbf{K}^e$ and $\mathbf{f}^e$ still given by relations (17.12) and (17.13),† and

$$C^e_{ij} = \int_{\Omega^e} N_i \mu N_j \, \mathrm{d}\Omega \qquad (20.10)$$

$$M^e_{ij} = \int_{\Omega^e} N_i \rho N_j \, \mathrm{d}\Omega \qquad (20.11)$$

Once again these matrices are symmetric as seen from the above relations.

Boundary conditions imposed at any time instant are treated once again as in the previous chapter.

The variety of physical problems governed by Eq. (20.1) is so large that a comprehensive discussion of them is beyond the scope of this book. A few typical examples will, however, be quoted.

Equation (20.1) *with* $\rho = 0$. This is the standard *transient heat conduction equation*[1,2] which has been discussed in the finite element context by several authors.[3–6] This same equation is applicable in other physical situations—one of these being the *soil consolidation equations*[7] associated with *transient seepage forms*.[8]

Equation (20.1) *with* $\mu = 0$. Now the relationship becomes the famous *Helmholz wave equation* governing a wide range of physical phenomena. Electromagnetic waves[9] fluid surface waves,[10] and compression waves[11] are but a few cases to which the finite element process has been applied.

Equation (20.1) *with* $\mu \neq \rho \neq 0$. This damped wave equation is yet of more general applicability and has particular significance in fluid mechanics (wave) problems.

The reader will recognize that what we have done here is simply an application of the process of partial discretization described in section 3.7 of Chapter 3. It is convenient, however, to perform the operations in the

† In Eq. (20.8) it is implied as usual that the loading term $\bar{\mathbf{f}}$ includes also all prescribed values of ϕ (and hence of $\mathbf{a}$) on the boundaries. A point which seems to be missed by some is that this implies also a prescription of $\dot{\mathbf{a}}$ and $\ddot{\mathbf{a}}$ on such boundaries and appropriate load terms are contributed from *all the matrices* **K**, **C** and **M**. Only when prescribed **a** values are zero, or when the matrices are diagonal, can such terms be neglected.

manner suggested above as all the matrices and discretization expressions obtained from steady-state analysis are immediately available.

20.2.2 *Dynamic behaviour of elastic structures with linear damping.*† While in the previous section we have been concerned with, apparently, a purely mathematical problem, identical reasoning can be applied directly to the wide class of dynamic behaviour of elastic structures following precisely the general lines of Chapter 2.

When displacements of an elastic body vary with time two sets of additional forces are called into play. The first is the inertia force, which for an acceleration characterized by $\ddot{\mathbf{u}}$ can be replaced by its static equivalent

$$-\rho\ddot{\mathbf{u}}$$

using the well-known d'Alembert principle. ($\mathbf{u}$ is here the generalized displacement defined in Chapter 2.)

This is a force with components in directions identical to those of the displacement $\mathbf{u}$ and (generally) given per unit volume.

In this context ρ is simply the mass per unit volume.

The second force is that due to (frictional) resistances opposing the motion. These may be due to microstructure movements, air resistance, etc., and generally are related in a non linear way to the displacement velocity $\dot{\mathbf{u}}$.

For simplicity of treatment, however, only a linear, viscous type, resistance will be permitted resulting again in unit volume forces in an equivalent static problem of magnitude

$$-\boldsymbol{\mu}\dot{\mathbf{u}}$$

In the above $\boldsymbol{\mu}$ is some property which (presumably) can be given numerical values.

The equivalent static problem, at any instant of time, is now discretized precisely in the manner of Chapter 2 but replacing the distributed body force $\mathbf{b}$ by its equivalent

$$\bar{\mathbf{b}}-\rho\ddot{\mathbf{u}}-\boldsymbol{\mu}\dot{\mathbf{u}}$$

The element (nodal) forces given by Eq. (2.13) now become (excluding initial stress–strain contributions)

$$\bar{\mathbf{f}}^e = -\int_{V^e} \mathbf{N}^{\mathrm{T}}\mathbf{b}\,\mathrm{d}V = -\int_{V^e} \mathbf{N}^{\mathrm{T}}\bar{\mathbf{b}}\,\mathrm{d}V + \int_{V^e} \mathbf{N}^{\mathrm{T}}\rho\ddot{\mathbf{u}}\,\mathrm{d}V + \int_{V^e} \mathbf{N}^{\mathrm{T}}\boldsymbol{\mu}\dot{\mathbf{u}}\,\mathrm{d}V \tag{20.12}$$

† For simplicity we shall only consider *distributed* inertia and damping effects—concentrated mass and damping forces being simply a limiting case.

in which the first force is precisely that due to external distributed load of Chapter 2 and need not be further considered.

As the approximation to the displacements is given by Eq. (2.1)

$$\mathbf{u} = \mathbf{N}\mathbf{a}^e \tag{2.1}$$

we can substitute Eq. (20.12) into the general equilibrium equations and obtain finally, on assembly, the following matrix differential equation

$$\mathbf{M}\ddot{\mathbf{a}} + \mathbf{C}\dot{\mathbf{a}} + \mathbf{K}\mathbf{a} + \mathbf{f} = 0 \tag{20.13}$$

in which $\mathbf{K}$ and $\mathbf{f}$ are assembled stiffness and force matrices obtained by usual addition of stiffness coefficients of elements and of element forces due to external, specified loads, initial stresses, etc., in the manner fully described before. The new matrices $\mathbf{C}$ and $\mathbf{M}$ are assembled by the usual rule from element submatrices given by

$$\mathbf{C}_{ij}^e = \int_{V^e} \mathbf{N}^{\mathrm{T}} \mu \mathbf{N} \, \mathrm{d}V \tag{20.14}$$

and

$$\mathbf{M}_{ij}^e = \int_{V^e} \mathbf{N}^{\mathrm{T}} \rho \mathbf{N} \, \mathrm{d}V \tag{20.15}$$

The matrix $\mathbf{M}^e$ is known as the *element mass matrix* and the assembled matrix $\mathbf{M}$ as the system mass matrix.

It is of interest to note that in early attempts to deal with dynamic problems of this nature the mass of the elements was usually arbitrarily 'lumped' at nodes, resulting always in a diagonal matrix even if no actual concentrated masses existed. The fact that such a procedure was, in fact, unnecessary and apparently inconsistent was simultaneously recognized by Archer[12] and independently by Leckie and Lindberg[13] in 1963. The general presentation of the results given in Eq. (20.15) is due to Zienkiewicz and Cheung.[14] The name of 'consistent mass matrix' has been coined for the distributed mass element matrix, a term thought to be unnecessary since it is the logical and natural consequence of the discretization process.

By analogy the matrices $\mathbf{C}^e$ and $\mathbf{C}$ could be called *consistent damping matrices*.

For many computational processes the original lumped mass matrix is, however, more convenient and economical. Many practitioners are today using such matrices exclusively showing often an improvement of accuracy. While with simple elements a physically obvious methodology of lumping is easy to devise, it is not the case with higher order elements and we shall return to the process of 'lumping' later.

The determination of the damping matrix $\mathbf{C}$ is in practice difficult as the knowledge of the viscous matrix $\boldsymbol{\mu}$ is lacking. It is often assumed, therefore, that the damping matrix is a linear combination of stiffness and mass matrices, i.e.

$$\mathbf{C} = \alpha\mathbf{M} + \beta\mathbf{K}. \tag{20.16}$$

Here α and β are determined experimentally.[15]

Such damping is known as 'Rayleigh damping' and has certain mathematical advantages which we shall discuss later. On occasion $\mathbf{C}$ may be more explicitly specified and such approximation devices are not necessary.

It is perhaps worth recognizing that on occasion different shape functions need to be used to describe the inertia forces from those specifying the displacements $\mathbf{u}$. For instance in plates and beams (Chapter 10) the full strain state was prescribed simply by defining w, the lateral displacement, as the additional plate bending assumptions were introduced. When considering, however, the inertia forces it may be desirable not only to include the simple lateral inertia force given by

$$-\rho\frac{\partial^2 w}{\partial t^2}$$

(in which ρ is now the weight per unit area of the plate) but also to consider *rotary inertia couples* of the type

$$\frac{\rho t^2}{12}\frac{\partial^2}{\partial t^2}\left(\frac{\partial w}{\partial x}\right), \text{ etc.}$$

Now it will be simply necessary to describe a more generalized displacement $\bar{\mathbf{u}}$

$$\bar{\mathbf{u}} = \begin{Bmatrix} w \\ \dfrac{\partial w}{\partial x} \\ \dfrac{\partial w}{\partial y} \end{Bmatrix} = \bar{\mathbf{N}}\mathbf{a}^e$$

in which $\bar{\mathbf{N}}$ will follow directly from the definition of $\mathbf{N}$ which specifies only the w component. Relations such as Eq. (20.15) still are valid providing we replace $\mathbf{N}$ by $\bar{\mathbf{N}}$ and put in place of ρ a matrix

$$\begin{bmatrix} \rho & 0 & 0 \\ 0 & \dfrac{\rho t^2}{12} & 0 \\ 0 & 0 & \dfrac{\rho t^2}{12} \end{bmatrix}$$

Such specialized usage is, however, rare.

20.2.3 *'Mass' or 'damping' matrices for some typical elements.* It is impracticable to present in an explicit form all the mass matrices for the various elements discussed in previous chapters. Some selected examples only will be discussed here.

Plane stress and strain. Using triangular elements discussed in Chapter 4, the matrix **N** is defined as

$$\mathbf{N}^e = \mathbf{I}[N_i, N_j, N_k]$$

in which

$$\mathbf{I} = \begin{bmatrix} 1 & 0 \\ 0 & 1 \end{bmatrix}$$

with Eq. (4.8) giving

$$N_i = (a_i + b_i x + c_i y)/2\Delta, \text{ etc.},$$

where Δ is the area of triangle.

If the thickness of the element is t and this is assumed to be constant within the element, we have for the mass matrix Eq. (20.15)

$$\mathbf{M}^e = \rho t \int\int \mathbf{N}^{\mathrm{T}}\mathbf{N}\, \mathrm{d}x\, \mathrm{d}y$$

or

$$\mathbf{M}^e_{rs} = \rho t \mathbf{I} \int\int N_r N_s\, \mathrm{d}x\, \mathrm{d}y. \tag{20.17}$$

If the relationships Eq. (4.8) are substituted, it is easy to show that

$$\int N_r N_s\, \mathrm{d}x\, \mathrm{d}y \begin{matrix} = \frac{1}{12}\Delta & \text{when } r \neq s \\ = \frac{1}{6}\Delta & \text{when } r = s. \end{matrix} \tag{20.18}$$

Thus, taking the mass of the element as

$$\rho t \Delta = W$$

the mass matrix becomes

$$\mathbf{M}^e = \frac{W}{3} \left|\begin{array}{cc:cc:cc} \frac{1}{2} & 0 & \frac{1}{4} & 0 & \frac{1}{4} & 0 \\ 0 & \frac{1}{2} & 0 & \frac{1}{4} & 0 & \frac{1}{4} \\ \hdashline \frac{1}{4} & 0 & \frac{1}{2} & 0 & \frac{1}{4} & 0 \\ 0 & \frac{1}{4} & 0 & \frac{1}{2} & 0 & \frac{1}{4} \\ \hdashline \frac{1}{4} & 0 & \frac{1}{4} & 0 & \frac{1}{2} & 0 \\ 0 & \frac{1}{4} & 0 & \frac{1}{4} & 0 & \frac{1}{2} \end{array}\right| \tag{20.19}$$

If the mass had been lumped at the nodes in three equal parts the mass matrix contributed by the element would have been

$$\mathbf{M}^e = \frac{W}{3}\left[\begin{array}{cc|cc|cc} 1 & 0 & 0 & 0 & 0 & 0 \\ 0 & 1 & 0 & 0 & 0 & 0 \\ \hline 0 & 0 & 1 & 0 & 0 & 0 \\ 0 & 0 & 0 & 1 & 0 & 0 \\ \hline 0 & 0 & 0 & 0 & 1 & 0 \\ 0 & 0 & 0 & 0 & 0 & 1 \end{array}\right] \tag{20.20}$$

Certainly both matrices differ considerably and yet in application the results of analysis are almost identical.

Plate bending. Vibration of plates presents problems of considerable engineering importance. Such practical situations as bridge-deck oscillations, vibration of turbine blades, etc., result in analytically intractable formulations.

The use a consistent mass matrix is illustrated in several references.[15–19]

If the rectangular plate element of section 10.4 is considered, for instance, the displacement function is defined by Eq. (10.19) as

$$\mathbf{N} = \mathbf{P}\mathbf{C}^{-1} \tag{20.21}$$

with notation as defined in Chapter 10.

It will be observed that $\mathbf{C}$ is not dependent on the co-ordinates and that $\mathbf{P}$ is determined as

$$\mathbf{P} = [1, x, y, x^2, xy, y^2, x^3, x^2y, xy^2, y^3, x^3y, xy^3].$$

Thus, the mass matrix for a plate element of constant thickness t becomes from Eq. (20.15)

$$\mathbf{M}^e = \rho t \mathbf{C}^{-1\mathrm{T}}\left(\int\int \mathbf{P}^{\mathrm{T}}\mathbf{P}\,\mathrm{d}x\,\mathrm{d}y\right)\mathbf{C}^{-1}. \tag{20.22}$$

Once again only the central integral needs to be evaluated, thus presenting no difficulty, and the full matrix can be obtained by matrix multiplication. However, an explicit expression has been presented in Dawe[17] and is quoted in Table 20.1.

Similar mass matrices can be obtained for triangular elements discussed in sections 10.6 *et seq.* Explicit formulation is here avoided† and numerical integration procedures are recommended for use with such elements.

Shells. If the mass matrices for the 'in plane' and 'bending' motions of an element are found, then once again the mass matrices referred to a general co-ordinate system can be found. The rules of transformation are,

† Explicit integrals are available in references 20 and 21.

obviously, precisely the same as for forces. The derivation of the mass matrices for each element in general co-ordinates and the final assembly of the mass matrix associated with a node follow the steps for similar operations with stiffness matrices (*see* Chapter 13).

TABLE 20.1

MASS MATRIX OF A RECTANGULAR PLATE ELEMENT

$$\mathbf{M}^e = \mathbf{L}\mathbf{M}\mathbf{L}$$

$$\mathbf{M}^e = \lambda \begin{bmatrix}
3454 & & & & & & & & & & & \\
-461 & 80 & & & & & & & & & & \\
-461 & -63 & 80 & & & & & & & & & \\
1226 & -274 & 199 & 3454 & & & & & & & & \\
274 & -60 & 42 & 461 & 80 & & & & & & & \\
199 & -42 & 40 & 461 & 63 & 80 & & & & & & \\
1226 & -199 & 274 & 394 & 116 & 116 & 3454 & & & & & \\
-199 & 40 & -42 & -116 & -30 & -28 & -461 & 80 & & & & \\
-274 & 42 & -60 & -116 & -28 & -30 & -461 & 63 & 80 & & & \\
394 & -116 & 116 & 1226 & 199 & 274 & 1226 & -274 & -199 & 3454 & & \\
116 & -30 & 28 & 199 & 40 & 42 & 274 & -60 & -42 & 461 & 80 & \\
-116 & 28 & -30 & -274 & -42 & -60 & -199 & 42 & 40 & -461 & -63 & 80
\end{bmatrix}$$

$\mathbf{L}$ is defined in Table 10.1 and $\lambda = \dfrac{\rho tab}{6300}$

In principle, therefore, shell vibration problems present no special difficulties.

Indeed the same may be said of a wide variety of matrices which arise in the field of structural dynamics. Performance of numerical integration allows the mass (or damping) matrices to be evaluated in a direct and simple fashion with procedures described in Chapter 8.

20.2.4 *Mass matrix 'lumping' or 'diagonalization'*. As we have already mentioned, the use of lumped (or diagonal) mass matrices is of considerable computational convenience, and many engineers have persisted with purely physical concepts of lumping for this reason. There is clearly a need for devising a systematic and mathematically acceptable procedure for such lumping exercises. Two basic lines of argument can be here followed. *In the first* we can consider lumping as a process in which a different shape function $\bar{\mathbf{N}}$ is used to describe the unknown function ϕ of Eq. (20.1), or displacement in Eq. (20.13), in the terms from which the mass matrix arises. Thus for instance, in the problem of Eq. (20.1) we could use

$$\phi = \mathbf{N}\mathbf{a} \tag{20.2}$$

for terms arising from the conductivity and which lead to the matrix **K**, and simultaneously use

$$\phi = \hat{\mathbf{N}}\mathbf{a} \tag{20.23}$$

for terms leading to the mass matrix. If the shape functions are piecewise constants, such that $\hat{N}_i = 1$ in a certain part of the element surrounding the node i and zero elsewhere, and if such parts are not overlapping, then clearly the mass matrix of Eq. (20.8) becomes diagonal as

$$\int_\Omega \hat{N}_i c \hat{N}_j \, \mathrm{d}\Omega = 0 \qquad (i \neq j). \tag{20.24}$$

Such an approximation with different shape functions is clearly permissible if the usual finite element criteria of *integrability* and *completeness* are satisfied. Further, it is necessary for the original approximation to be derivable from energy (variational) principles as is the case in this example. As no differentiation is involved in expression (20.8) it suffices for the function itself to be piecewise constant and for the usual condition of summation

$$\sum_{i=1}^{n} \hat{N}_i = 1 \tag{20.25}$$

to be satisfied (see p. 196).

In Fig. 20.1 we show the functions N_i and $\hat{N}_i$ for a triangular element.

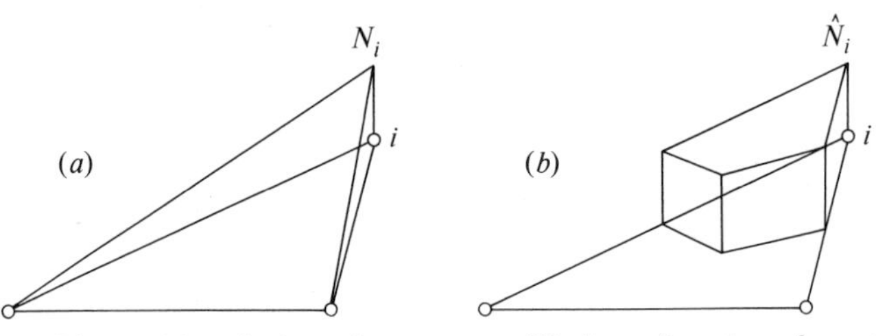

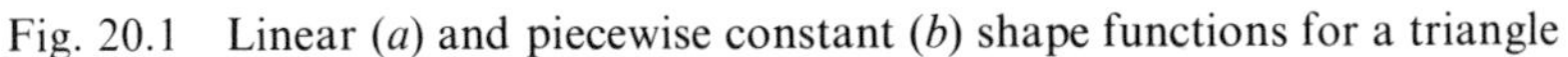

Fig. 20.1 Linear (*a*) and piecewise constant (*b*) shape functions for a triangle

Conditions (20.25) can be interpreted simply as a preservation of total mass as

$$\int_\Omega \sum_{i=1}^{n} \hat{N}_i \rho \, \mathrm{d}\Omega \equiv \int_\Omega \rho \, \mathrm{d}\Omega \tag{20.26}$$

and this simple and obvious rule has been used frequently and explicitly stated by Clough.[22] Thus any lumping which preserves the total mass will lead to convergent results.

Key and Beisinger,[23] Hinton *et al.*,[24] and others have experimented successfully with various procedures which give not only acceptable but often improved results over those attainable with consistent mass matrices. It is, for instance, obvious that in a simple triangle little improvement can be obtained by any other lumping than the simple one in which the total mass is distributed in three equal parts, as shown in Eq. (20.20). For an 8-node two-dimensional isoparametric element no such obvious procedure is available, and Hinton *et al.*[24] show that excellent results can be obtained by *taking the consistent mass matrix and scaling diagonal terms of this so that the total mass is preserved.* This leads for the case of rectangular elements of 8- and 9-node type to lumping schemes shown in Fig. 20.2. P. Tong *et al.*[25] show that use of different interpolations can, in certain cases, lead to a loss of convergence rate and an alternative scheme for lumping has been suggested by Fried.[26]

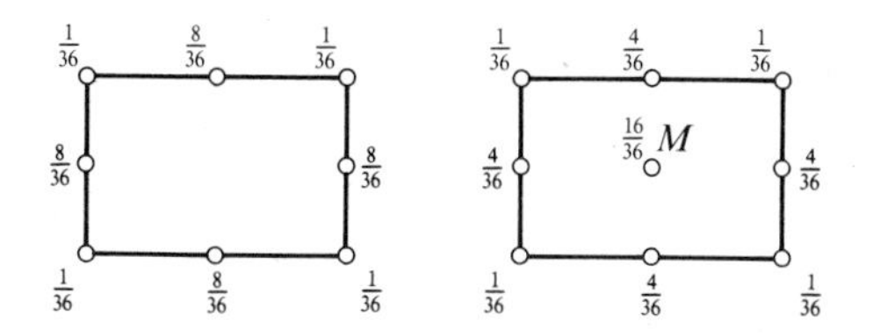

Fig. 20.2 Mass lumping by scaling of diagonal terms in consistent mass matrix (Proportion of total mass shown at each node)

This *second procedure* uses numerical integration to obtain lumping without apparently introducing additional shape functions. Clearly, if numerical integration is used to evaluate the mass matrix of Eq. (20.8) we shall write a typical term as a summation (following Chapter 8)

$$M_{ij} = \int_{\Omega} N_i \rho N_j \, d\Omega = \sum_{q=1}^{m} W_q (N_i \rho N_j)_q \tag{20.27}$$

where q refers to the sampling point at which the integrand has to be found and W_q gives the appropriate weighting.

If the sampling points for the numerical integration are *located at nodes then*, as all except one shape function N_i are zero there,

$$M_{ij} \equiv 0 \qquad (i \neq j)$$

and the matrix becomes diagonal.

In Chapter 8 we have stated an important theorem concerning the order of numerical integration which does not affect the convergence rate. This stated simply that if p was the order of polynomials used and m the order of differentiation present in the variational functional, then any integration which is exact to the order of $2(p-m)$ will not affect the rate of convergence. This order of integration stems from the terms containing derivatives of N and obviously it is unnecessary to use a higher order of integration on terms (like those from which the mass matrix originate) which do not have any differentiation.

If, thus, an integration scheme which uses only nodal points for sampling is devised and which possesses the correct order of integration, then such lumping will not affect the convergence rate. Fried[26] shows such schemes for several elements, and demonstrates not only that convergence order is maintained but that the accuracy *is often improved*. For a simple triangle once again a suitable integration scheme results in a trivial lumping of Eq. (20.20). Let us consider, however, as an example the two-rectangular parabolic elements of Fig. 20.2 in the context of the problem of Eq. (20.1). Here $p = 2$, $m = 1$, and we shall therefore seek an integration formula which integrates exactly polynomials of order 2. (Error order $O(h^3)$.)

Writing a general quadratic polynomial

$$f(\xi, \eta) = \alpha_1 + \alpha_2\xi + \alpha_3\eta + \alpha_4\xi^2 + \alpha_5\xi\eta + \alpha_6\eta^2$$

we desire that an integration formula of the type given by Eq. (20.27) should integrate this exactly. By symmetry the corner nodes must all have the same integrating weight W_1 and the mid-sides W_2. Integrating exactly we find, for an element of Fig. 20.3, that

$$\int_{-1}^{1}\int_{-1}^{1} f(\xi, \eta)\, \mathrm{d}\xi\, \mathrm{d}\eta \equiv 4\alpha_1 + \tfrac{4}{3}\alpha_4 + \tfrac{4}{3}\alpha_6.$$

Writing the above as the sum of Eq. (20.27) we have

$$\iint f(\xi, \eta)\, \mathrm{d}\xi\, \mathrm{d}\eta = \sum_{i=\text{corner}} W_1 f(\xi_i, \eta_i) + \sum_{i=\text{mid-side}} W_2 f(\xi_i, \eta_i) =$$

$$= 4\alpha_1(W_1 + W_2) + \alpha_4(4W_1 + 2W_2) + \alpha_6(4W_1 + 2W_2).$$

For both the above expressions to be identical (i.e., for exact integration of the quadratic polynomial) we must have

$$W_1 + W_2 = 1 \qquad \text{and} \qquad 4W_1 + 2W_2 = 4/3$$

or

$$W_1 = -1/3; \qquad W_2 = 4/3.$$

This results in the total mass being 'lumped' in proportions given in Fig. 20.3(*a*), and we observe that unfortunately some of the masses are negative. While mathematically this is still acceptable, and indeed Fried shows that good results can be obtained with such negative masses, these are numerically inconvenient and seldom used.

For a 9-noded element of Fig. 20.3(*b*) a similar calculation shows that many integration formula are possible and that positive lumped masses can be achieved. With W_3 standing for the weighting of the central node the reader can verify that the previous reasoning leads to

$$W_1 + W_2 + W_3/4 = 1 \qquad \text{and} \qquad 4W_1 + 2W_2 = 4/3$$

or

$$W_1 = -1/3 + W_3/4; \qquad W_2 = 4/3 - W_3/2$$

which for $4/3 < W_3 < 8/3$ gives positive lumped masses shown in Fig. 20.3(*b*). In Fig. 20.3(*c*) we show a particular form which integrates to order $O(h^4)$ and is equivalent to Gauss–Lobatto integration ($W_3 = 16/9$).

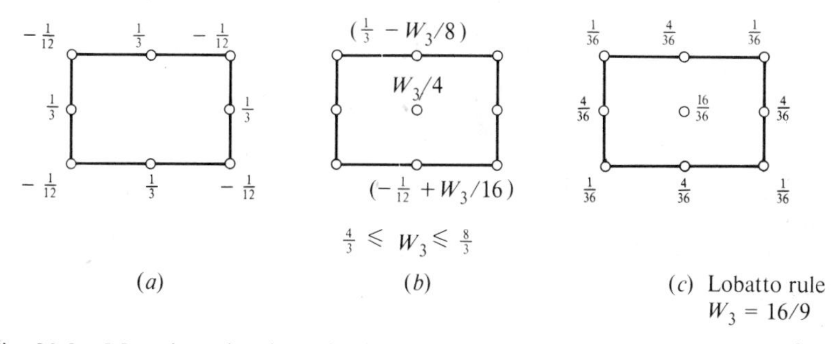

Fig. 20.3 Mass lumping by selective numerical integration of order $O(h^3)$ for (*a*) serendipity and (*b*) Lagrangian rectangles ($b = 2$); (*c*) Shows specific case of (*b*) with integration error $O(h^4)$

(Proportions of total mass shown at each node)

In Fig. 20.4 some lumping formulae derived in reference 26 are shown for linear, parabolic, and cubic triangles.

Whilst the procedures were discussed on the basis of matrices derived for the field problem they are equally applicable for all mass and indeed for damping matrices of this chapter. In the subsequent sections of the next chapter we shall show results which are derived using various types of lumping here described, and these will be seen to show numerical advantage together with an improvement of accuracy over that obtainable with consistent matrices.

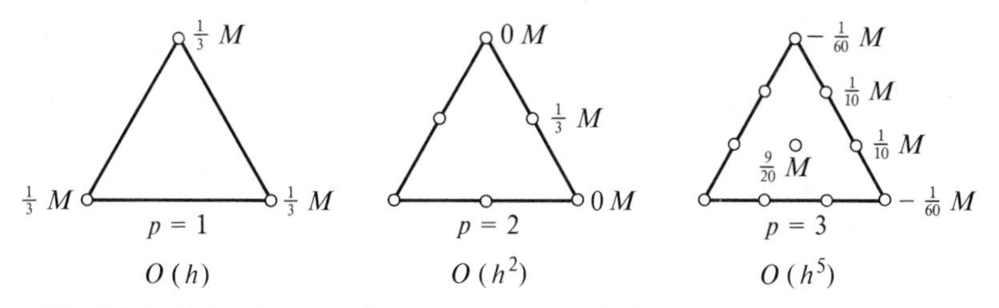

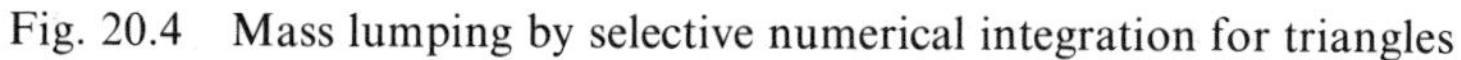
Fig. 20.4 Mass lumping by selective numerical integration for triangles

20.3 'Coupled' Problems

Both sets of problems discussed in the previous section led to the basic identical form of matrix differential equations, characterized by Eq. (20.6) or Eq. (20.13). Other problems with higher order governing equations may be similarly derived. On occasion two separate systems of such equations arise in problems of a coupled kind. To complete the present discussion we shall discuss two such typical cases of some considerable engineering interest.

20.3.1 *Coupled motion of an elastic structure in a fluid. Acoustic problem.*[11,16,27,28] The differential equation governing the pressure distribution, (p), during the small amplitude motions of a compressible fluid is[29]

$$\frac{\partial^2 p}{\partial x^2}+\frac{\partial^2 p}{\partial y^2}+\frac{\partial^2 p}{\partial z^2}+\frac{1}{\bar{c}^2}\frac{\partial^2 p}{\partial t^2} = 0 \tag{20.28}$$

in which $\bar{c}$ stands for the velocity of the acoustic wave, and 'damping' (viscous) terms have been omitted.

On the boundaries either p is specified, or if these are solid and subject to prescribed motion

$$\frac{\partial p}{\partial n} = -\rho\frac{\partial^2}{\partial t^2}(U_n) \tag{20.29}$$

where U_n is the normal component of the displacement. This problem, on finite element subdivision of the fluid region, leads to a discretized equation similar to the form of Eq. (20.8)

$$\mathbf{G}\ddot{\mathbf{p}}+\mathbf{H}\mathbf{p}+\bar{\mathbf{f}}_f = 0 \tag{20.30}$$

in which the matrices $\mathbf{H}$ and $\mathbf{G}$ are obtained in the usual manner. The matrix $\bar{\mathbf{f}}_f$ does not contain any contribution from volume integrals, but is entirely due to boundary integrals corresponding to the prescribed motions on these (see Eq. (17.27)).†

Now, the boundary (interface) motion is prescribed by the movement of the structure. If the structure itself is descretized we can write

$$U_n = \bar{\bar{\mathbf{N}}}\mathbf{a} \tag{20.31}$$

where $\bar{\bar{\mathbf{N}}}$ is determined from the appropriate shape functions and $\mathbf{a}$ are

† In a more general case Eq. (20.30) can be augmented by a term containing the first time derivatives of p. This will arise if viscous terms are included in the equation of fluid motion or if fluid boundaries exist which do not reflect incident pressure waves. Such boundaries are important if the fluid is extending to infinity and for analysis purposes its region is truncated.[22]

the displacement (nodal) parameters. By Eq. (17.13) we have

$$\bar{\mathbf{f}}_f = \mathbf{S}\ddot{\mathbf{a}} \tag{20.32}$$

in which

$$\mathbf{S} = \int_S \mathbf{N}^{\mathrm{T}} \rho \bar{\bar{\mathbf{N}}}\, \mathrm{d}S \tag{20.33}$$

with $\mathbf{N}$ being the shape functions defining the pressure distribution and S the fluid-structure interface.

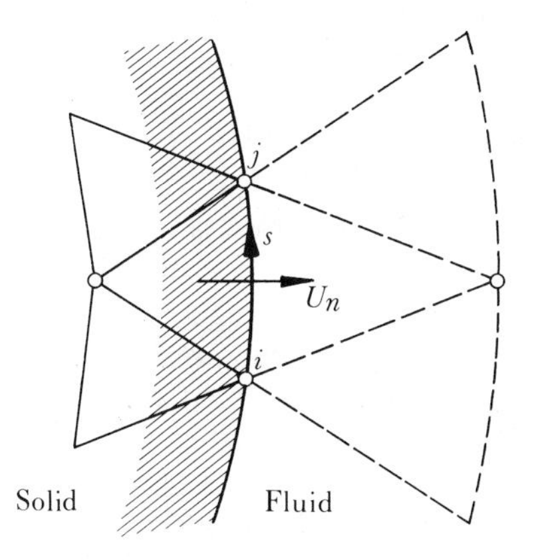

Fig. 20.5 A solid–fluid interface

For the structural problem we have similarly on discretization

$$\mathbf{M}\ddot{\mathbf{a}} + \mathbf{C}\dot{\mathbf{a}} + \mathbf{K}\mathbf{a} + \bar{\mathbf{f}}_s + \mathbf{r} = 0 \tag{20.34}$$

in which the usual terms of Eq. (20.13) are recognized but the forcing terms have been separated into the external part $\mathbf{r}$ assumed independently specified and $\bar{\mathbf{f}}_s$ due to the fluid interface pressures. By virtual work we shall find those to be given by

$$\bar{\mathbf{f}}_s = \int_S \bar{\bar{\mathbf{N}}}^{\mathrm{T}} p\, \mathrm{d}S = \frac{1}{\rho}\mathbf{S}^{\mathrm{T}}\mathbf{p}. \tag{20.35}$$

as

$$p = \mathbf{N}\mathbf{p}$$

Combining Eqs. (20.30), (20.32), (20.34), and (20.35) we have finally a coupled system of matrix differential equations

$$\mathbf{G}\ddot{\mathbf{p}}+\mathbf{H}\mathbf{p}+\mathbf{S}\ddot{\mathbf{a}} = 0 \tag{20.36}$$

and

$$\mathbf{M}\ddot{\mathbf{a}}+\mathbf{C}\dot{\mathbf{a}}+\mathbf{K}\mathbf{a}+\frac{1}{\rho}\mathbf{S}^{\mathrm{T}}\mathbf{p}+\mathbf{r} = 0 \tag{20.37}$$

which governs the problem.

Some aspects of this problem are discussed in references 11 and 16. For the special case of an incompressible fluid ($\bar{c} = \infty$) the first term, **G**, of the first equation becomes zero and this can be solved directly giving

$$\mathbf{p} = -\mathbf{H}^{-1}\mathbf{S}\ddot{\mathbf{a}}. \tag{20.38}$$

Substitution into the second equation gives simply a standard dynamic equation in which the mass matrix is augmented by an '*added mass matrix*'

$$-\frac{1}{\rho}\mathbf{S}^{\mathrm{T}}\mathbf{H}^{-1}\mathbf{S}. \tag{20.39}$$

Derivation of such a (reduced) mass matrix was in fact discussed in section 17.5 and was first suggested by Zienkiewicz *et al.*[4, 16] More recently the same process was used in deriving natural frequencies of arch dams and other submerged or floating structures.[28, 30, 31] Similar problems arise in acoustic engineering where the fluid (air) is simply a compressible medium.[32, 34]

20.3.2 *Elastic behaviour of a porous, saturated material.*[35–41] This second problem is of considerable interest in soil mechanics and a variety of geotechnical situations.

In the porous, elastic medium the fluid pressures existing in pores cause body forces acting in the elastic matrix of magnitude

$$\begin{Bmatrix} X \\ Y \\ Z \end{Bmatrix} = -\begin{Bmatrix} \dfrac{\partial p}{\partial x} \\ \dfrac{\partial p}{\partial y} \\ \dfrac{\partial p}{\partial z} \end{Bmatrix}. \tag{20.40}$$

These have been discussed already in Chapter 4 (pp. 108–12) but the interested reader may find further discussion of this phenomenon in reference 36.

If now the elastic structure is discretized in the finite element manner these body forces will contribute nodal 'forces' of magnitude

$$\bar{\mathbf{f}}_p = \left(\int_V \bar{\mathbf{N}}^{\mathrm{T}} \begin{Bmatrix} \partial/\partial x \\ \partial/\partial y \\ \partial/\partial z \end{Bmatrix} \mathbf{N}\,\mathrm{d}V\right)\mathbf{p} = \mathbf{L}\mathbf{p} \tag{20.41}$$

in which $\bar{\mathbf{N}}$ stands for the shape functions defining the displacements of the elastic body while $\mathbf{N}$ are shape functions defining the pressure distribution.†

Finally for the elastic continuum we have thus

$$\mathbf{Ka}+\mathbf{Lp}+\mathbf{r} = 0 \tag{20.42}$$

the standard discretized equation in which $\mathbf{K}$ is the stiffness matrix and $\mathbf{r}$ represents all specified forces except those due to pore pressure.

Turning our attention to the fluid contained in the pores we shall write the appropriate differential flow equation. This has already been encountered in Chapter 17 as the typical Eq. (17.1) in which k_x, k_y, k_z are now the permeability coefficients and Q represents the rate at which the fluid is generated (or discharged into) a unit volume of space.

Now, with the solid matrix deforming elastically with the displacement components u, v, and w and assuming complete incompressibility of fluid and of the solid skeleton particles we have

$$Q = -\frac{\partial}{\partial t}\left(\frac{\partial u}{\partial x}+\frac{\partial v}{\partial y}+\frac{\partial w}{\partial z}\right) = -\frac{\partial}{\partial t}\begin{Bmatrix}\partial/\partial x\\ \partial/\partial y\\ \partial/\partial z\end{Bmatrix}^{\mathrm{T}} \bar{\mathbf{N}}\mathbf{a}. \tag{20.43}$$

On discretizing Eq. (17.1) with Eq. (20.42) substituted we have

$$\mathbf{Hp}+\mathbf{S}\dot{\mathbf{a}} = 0 \tag{20.44}$$

as the contribution of Q is, by Eq. (17.13),

$$\int_V \mathbf{N}^{\mathrm{T}} Q \,\mathrm{d}V = \left(\int \mathbf{N}^{\mathrm{T}} \begin{Bmatrix}\partial/\partial x\\ \partial/\partial y\\ \partial/\partial z\end{Bmatrix} \bar{\mathbf{N}}\,\mathrm{d}V\right)\dot{\mathbf{a}}. \tag{20.45}$$

Now Eqs. (20.42) and (20.44) form a coupled system of simultaneous matrix differential equations. These are similar to the system derived for the coupled fluid-structure dynamic interaction (Eqs. (20.36) and (20.37)). When fluid compressibility was eliminated in the latter the form is in fact identical.

It should be noted that, formally, only from Eqs. (20.41) and (20.45)

$$\mathbf{S} = \mathbf{L}^{\mathrm{T}}. \tag{20.46}$$

This problem was first discretized with finite elements by Sandhu and Wilson[35] using a somewhat different approach. The physical aspects of the problem are discussed by Crochet and Naghdi[37] and also by Biot.[38]

The conventional consolidation equation which is of the form of Eq. (20.1) (without the second-order time differential) is but a special case of this more general formulation.

† Integrals for whole regions are written to simplify notation as in Chapter 2.

In the above the fluid was assumed to be incompressible. However, if fluid compressibility is also added to the problem an additional term of the form

$$\mathbf{A}\dot{\mathbf{p}}$$

arises simply in Eq. (20.44). With this extension partially saturated soil problems can be dealt with.

Formulations including such compressibility effects and indeed an alternate but equivalent treatment of the pore pressure forces on the solid can be found elsewhere.[38–42] These two examples show that discretized coupled systems now lead to *systems of ordinary differential equations* and solution of these can be obtained in a manner similar to that which will be discussed in the remaining portions of this and in the next chapter.

ANALYTICAL SOLUTION PROCEDURES

20.4 **General Classification**

We have seen that as a result of semi-discretization many time-dependent problems can be reduced to a system of ordinary differential equations of the characteristic form given by Eq. (20.13).

$$\mathbf{M}\ddot{\mathbf{a}} + \mathbf{C}\dot{\mathbf{a}} + \mathbf{K}\mathbf{a} + \mathbf{f} = 0. \tag{20.47}$$

In this, in general, all the matrices are symmetric (although some cases involving non-symmetric matrices will occur in Chapter 22). This second order system often becomes one of the first order if $\mathbf{M} = 0$ as, for instance, in transient heat conduction problems. We shall now discuss some methods of solution of such ordinary differential equation systems. In general, the above equations can be non-linear (if, for instance, stiffness matrices are dependent on non-linear material properties, or if large deformation is involved) but initially we shall concentrate on linear cases only.

Systems of ordinary, linear, differential equations can always in principle be solved analytically without the introduction of additional approximations, and the remainder of this chapter will be concerned with such analytical processes. Whilst such solutions are possible they may be so complex that a further recourse has to be taken to the process of approximation and we shall deal with this matter in the next chapter. The analytical approach provides, however, an insight into the behaviour of the system which the investigator always finds helpful.

Some of the matter in this chapter will be an extension of standard well-known procedures used for the solution of differential equations with constant coefficients which have been encountered by most students of dynamics or mathematics. In the following we shall deal successively

with

(*a*) determination of free response ($\mathbf{f} = 0$)
(*b*) determination of periodic response ($\mathbf{f}(t)$—periodic)
(*c*) determination of transient response ($\mathbf{f}(t)$—arbitrary)

In the first two, initial conditions of the system are of no importance and a general solution is simply sought. The last, most important, phase presents a problem to which considerable attention will be devoted.

20.5 Free Response; Eigenvalues for Second Order Problems and Dynamic Vibrations

20.5.1 *Free dynamic vibration—real eigenvalues.* If no damping or forcing terms exist in the dynamic problem of Eq. (20.47) this is reduced to

$$\mathbf{M}\ddot{\mathbf{a}}+\mathbf{K}\mathbf{a} = 0 \tag{20.48}$$

If a general solution of such an equation is written as

$$\mathbf{a} = \bar{\mathbf{a}}\ e^{i\omega t}$$

(the real part of which represents simply a harmonic response as $e^{i\omega t} \equiv \cos \omega t + i \sin \omega t$) then on substitution we find that ω can be determined from

$$(-\omega^2\mathbf{M}+\mathbf{K})\bar{\mathbf{a}} = 0 \tag{20.49}$$

This is an *eigenvalue or characteristic value problem* and for non-zero solutions the determinant of the above equation must be zero.

$$|-\omega^2\mathbf{M}+\mathbf{K}| = 0. \tag{20.50}$$

Such a determinant will in general give n values of ω^2 (or ω_j, $j = 1-n$) when the size of the matrices $\mathbf{K}$ and $\mathbf{M}$ is $n \times n$. Providing the matrices $\mathbf{K}$ and $\mathbf{M}$ are positive definite—which is the usual case with structural problems—it can be shown that all the roots of the Eq. (20.50) are real positive numbers (for proof of this see reference 1). These are known as the natural frequencies of the system.

Whilst the solution of Eq. (20.50) cannot determine the actual values of $\mathbf{a}$ we can find n vectors $\bar{\mathbf{a}}_j$ which give the proportions of the various terms. Such vectors are known as the *natural modes of the system* and are generally normalized so that

$$\bar{\mathbf{a}}_j^{\mathrm{T}}\mathbf{M}\bar{\mathbf{a}}_j = 1. \tag{20.51}$$

At this stage it is useful to note the property of *modal orthogonality*, i.e., that

$$\left.\begin{aligned}\bar{\mathbf{a}}_j^{\mathrm{T}}\mathbf{M}\bar{\mathbf{a}}_i &\equiv \mathbf{0}\\ \bar{\mathbf{a}}_j^{\mathrm{T}}\mathbf{K}\bar{\mathbf{a}}_i &\equiv \mathbf{0}\end{aligned}\right\}\quad (i \neq j) \tag{20.52}$$

The proof of the above statement is simple. As Eq. (20.49) is valid for any mode we can write

$$\omega_i^2 \mathbf{M}\bar{\mathbf{a}}_i = \mathbf{K}\bar{\mathbf{a}}_i$$

$$\omega_j^2 \mathbf{M}\bar{\mathbf{a}}_j = \mathbf{K}\bar{\mathbf{a}}_j$$

premultiplying the first by $\bar{\mathbf{a}}_j^{\mathrm{T}}$ and the second by $\bar{\mathbf{a}}_i^{\mathrm{T}}$ we have on subtraction (noting the symmetry of the matrix **M**, i.e., $\bar{\mathbf{a}}_i^{\mathrm{T}}\mathbf{M}\bar{\mathbf{a}}_j \equiv \bar{\mathbf{a}}_j^{\mathrm{T}}\mathbf{M}\bar{\mathbf{a}}_i$)

$$(\omega_i^2 - \omega_j^2)\bar{\mathbf{a}}_i^{\mathrm{T}}\mathbf{M}\bar{\mathbf{a}}_j = 0$$

and if $\omega_i \neq \omega_j$ the orthogonality condition for the matrix **M** has been proved. Immediately from this, orthogonality of the matrix **K** follows.

20.5.2 *Determination of eigenvalues.* To find the actual eigenvalues it is seldom practicable to write the polynomial expanding the determinant given in Eq. (20.50) and alternative techniques have to be developed. The discussion of such techniques is best left to specialist texts and indeed many standard computer programs exist today as library routines.

Many new and extremely efficient procedures are being added to the arsenal available. Their description is beyond the scope of this text but the reader can find some interesting matter in references 43–50.

In most processes the starting point is the *special eigenvalue* problem given by

$$\mathbf{H}\mathbf{x} = \lambda\mathbf{x} \tag{20.53}$$

in which **H** is a symmetric, positive definite, matrix. Equation (20.49) can be written as

$$\mathbf{K}^{-1}\mathbf{M}\bar{\mathbf{a}} = \lambda\bar{\mathbf{a}} \tag{20.54}$$

on inverting **K** with $\lambda = 1/\omega^2$ but symmetry is in general lost.

If, however, we write in a triangular form

$$\mathbf{K} = \mathbf{L}\mathbf{L}^{\mathrm{T}} \quad \text{and} \quad \mathbf{K}^{-1} = \mathbf{L}^{\mathrm{T}^{-1}}\mathbf{L}^{-1}$$

in which **L** is a matrix having only zero coefficients above the diagonal we have, on multiplying Eq. (20.54) by $\mathbf{L}^{\mathrm{T}}$

$$\mathbf{L}^{-1}\mathbf{M}\bar{\mathbf{a}} = \lambda\mathbf{L}^{\mathrm{T}}\bar{\mathbf{a}}.$$

Calling

$$\mathbf{L}^{\mathrm{T}}\bar{\mathbf{a}} = \mathbf{x} \tag{20.55}$$

we have finally

$$\mathbf{H}\mathbf{x} = \lambda\mathbf{x} \tag{20.56}$$

in which

$$\mathbf{H} = \mathbf{L}^{-1}\mathbf{M}\mathbf{L}^{\mathrm{T}^{-1}} \tag{20.57}$$

which is of the form of Eq. (20.53), as **H** is now symmetric.

Having determined λ (all, or only few selected largest values corresponding to fundamental periods) the modes of $\mathbf{x}$ are found, and hence by use of Eq. (20.55) the modes of $\bar{\mathbf{a}}$.

If the matrix $\mathbf{M}$ is diagonal—as is the case if the masses have been 'lumped'—the procedure of deriving the standard eigenvalue problem is simplified and here appears the first advantage of such diagonalization which we have discussed in section 20.2.4.

20.5.3 *Free vibration with singular* $\mathbf{K}$ *matrix*. In static problems we have always introduced a suitable number of *support* conditions to allow $\mathbf{K}^{-1}$ to be inverted, or which is equivalent, to solve the static equations uniquely (*vide* Chapter 1). If such 'support' conditions are in fact not specified, as may well be the case with a rocket travelling in space, an arbitrary fixing of a minimum number of support conditions allows a static solution to be obtained without affecting the stresses. In dynamic problems such a fixing is not permissible and frequently one is faced with a problem of a free oscillation for which $\mathbf{K}$ is singular and therefore does not possess an inverse.

To preserve the applicability of the general methods described in the previous section a simple artifice is possible. Equation (20.49) is modified to

$$[(\mathbf{K}+\alpha\mathbf{M})-(\omega^2+\alpha)\mathbf{M}]\bar{\mathbf{a}} = 0 \tag{20.58}$$

in which α is an arbitrary constant of the same order as the typical ω^2 sought.

The new matrix $(\mathbf{K}+\alpha\mathbf{M})$ can be inverted and the standard process maintained to find $(\omega^2+\alpha)$.

This simple but effective sidestepping of otherwise serious difficulties was first suggested by Cox[51] and Jennings.[52] Alternative methods of dealing with above problem are given in references 53 and 54.

20.5.4 *Eigenvalue economizer methods*. Whatever technique is used in the process of determining the eigenvalues and eigenmodes of the system, for a given size of problem the computer effort is larger by an order of magnitude than the solution of an equivalent static situation. Fortunately, reasonably good eigenvalues can be determined with fewer degrees of freedom than needed for a normal static solution.

If a rather fine subdivision is used in the dynamic analysis we can eliminate a number of degrees of freedom and 'lump' the 'mass' and 'damping' effects at a reduced number of nodal parameters. A *consistent* way of doing this has been suggested by Irons[55, 56] and later by Guyan.[57] The similarity with substructure analysis described in Chapter 7, section 7.6, will not escape the notice of the reader.

Let the total vector $\mathbf{a}$ be divided into two parts

$$\mathbf{a} = \begin{Bmatrix} \mathbf{a}^s \\ \mathbf{a}^m \end{Bmatrix} \tag{20.59}$$

and assume that the displacements $\mathbf{a}^s$ depend in some unique way on the displacements $\mathbf{a}^m$. The latter we shall call therefore, 'master' and the former 'slave' variables. Thus, if

$$\mathbf{a}^s = \mathbf{T}\mathbf{a}^m \tag{20.60}$$

we have

$$\mathbf{a} = \begin{bmatrix} \mathbf{I} \\ \mathbf{T} \end{bmatrix} \mathbf{a}^m = \mathbf{T}^*\mathbf{a}^m \tag{20.61}$$

in which $\mathbf{T}$ is the matrix specifying the dependence.

Now the dynamic equation of the whole system

$$\mathbf{M}\ddot{\mathbf{a}} + \mathbf{K}\mathbf{a} = 0 \tag{20.62}$$

can be reduced by applying the constraint on the deformation freedom implied in Eq. (20.61).

Using the principles of transformation derived in Chapter 1 (1.8) (or simply premultiplying Eq. (20.62) by $\mathbf{T}^{*\mathrm{T}}$ after substitution of (20.61)) we can write

$$\mathbf{M}\ddot{\mathbf{a}}^m + \mathbf{K}^*\mathbf{a}^m = 0 \tag{20.63}$$

where

$$\mathbf{K} = \mathbf{T}^{*\mathrm{T}}\mathbf{K}\mathbf{T}; \qquad \mathbf{M}^* = \mathbf{T}^{*\mathrm{T}}\mathbf{M}\mathbf{T}$$

which is a problem involving a smaller number of variables.

The important question is how to determine reasonably the relation between 'slave' and 'master' deflections. A suitable assumption which can be reasonably justified by engineering intuition is to assume that the general pattern of deformation will follow that which would be obtained by imposing displacements $\mathbf{a}^m$ on an otherwise unloaded structure in static condition. Thus partitioning with $\dot{\mathbf{a}} = \ddot{\mathbf{a}} = 0$ we can write Eq. 20.47

$$\mathbf{K}\mathbf{a} = \begin{bmatrix} \mathbf{K}_{ss} & \mathbf{K}_{sm} \\ \mathbf{K}_{sm}^{\mathrm{T}} & \mathbf{K}_{mm} \end{bmatrix} \begin{Bmatrix} \mathbf{a}^s \\ \mathbf{a}^m \end{Bmatrix} = \begin{Bmatrix} 0 \\ \mathbf{f}^m \end{Bmatrix} \tag{20.64}$$

As the 'slave' nodes are unloaded we can write

$$\mathbf{K}_{ss}\mathbf{a}^s + \mathbf{K}_{sm}\mathbf{a}^m = 0$$

or

$$\mathbf{a}^s = -\mathbf{K}_{ss}^{-1}\mathbf{K}_{sm}\mathbf{a}^m$$

thus

$$\mathbf{T} = -\mathbf{K}_{ss}^{-1}\mathbf{K}_{sm} \tag{20.65}$$

Application of such techniques is illustrated well in the literature [58,59] and some examples will be cited later.

One important question remains. This concerns the best choice of the master nodes and obviously here much judgement can be exercised. The obvious choice is to eliminate first the nodes with little or no mass attached. To make the procedure automatic Henshell and Ong[59] suggest that the ratio of the diagonal stiffness and mass terms be first calculated

$$K_{ii}/M_{ii}$$

and the nodes with the largest values of this ratio be first eliminated. This simple procedure appears to be effective and can easily be programmed.

One drawback of the condensation is that in the reduced problem the mass matrix is never diagonal.

20.5.5 *Some examples.* There are a variety of problems for which practical solutions exist, so only a few simple examples will be shown.

Vibration of plates. Figure 20.6 shows the vibration of a rectangular cantilever plate solved using only four triangular elements. The results are compared against an elaborate calculation carried out by Barton.[60] It is seen that the results using the simple non-conforming triangle are here superior to those using the more elaborate formulation and the accuracy is quite remarkable both in frequency and mode shape.

A similar problem is presented in Fig. 20.7 where the effect of using the *eigenvalue economizer* method is examined. It will be seen how very small the changes in the first four frequencies are on restricting the degrees of freedom from 90 through various stages to six.

So many further examples of plate and shell vibration analysis are included in current literature that a list of references is here impracticable and a survey of such literature should be consulted.[61,62]

Shell vibration. Application of the process to any elastic two- or three-dimensional continuum can obviously be made and shell vibrations are a typical problem of much interest. In Fig. 20.8, by contrast to the previous simple examples, the elaborate thick shell elements described in Chapter 16 are used to solve a problem of turbine blade vibration.[63,64]

Some other dynamic analyses of shells are given in references 65–68. Reference 20 also shows some application utilizing full three-dimensional isoparametric elements.

The 'wave' equation. Electromagnetic and fluid problems. The basic dynamic Eq. (20.8) can be derived from a variety of non-structural problems as indeed was shown in the previous chapter. The eigenvalue problem once again occurs with 'stiffness' and 'mass' matrices now having alternate physical meaning.

A particular form of the more general equations discussed earlier is the

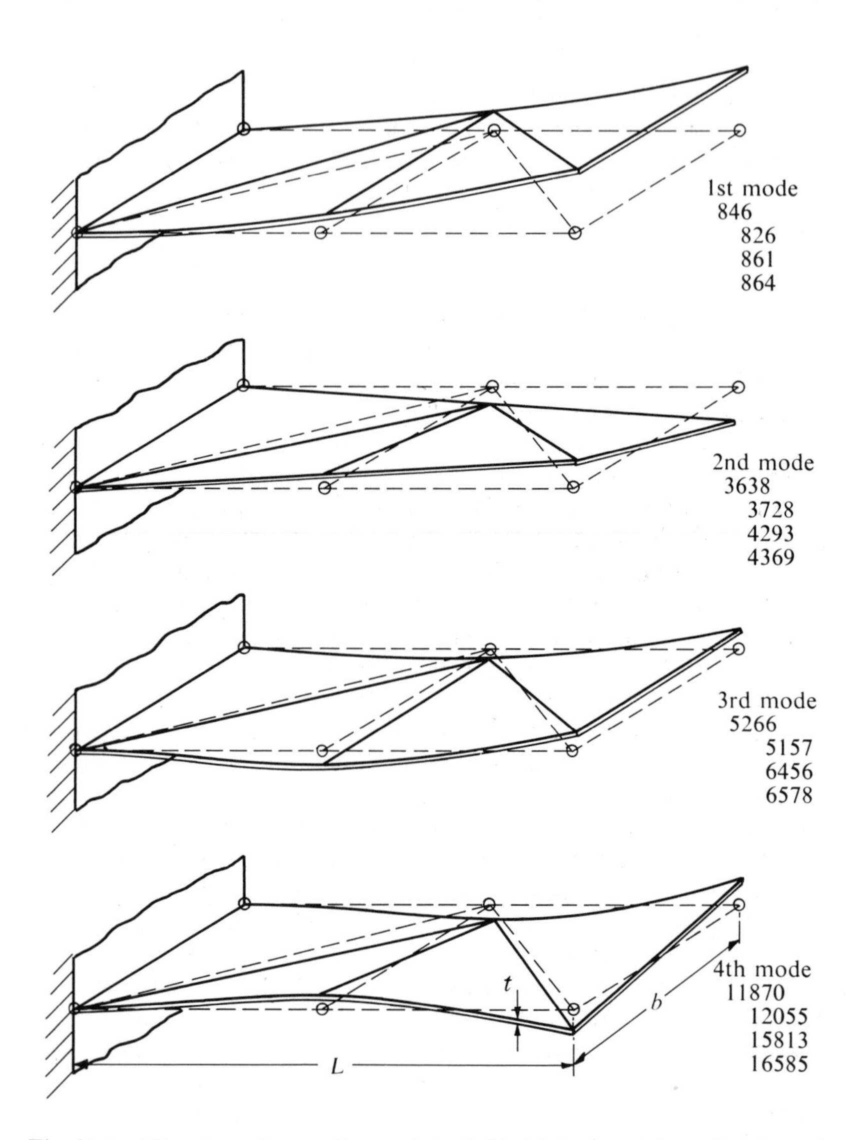

Fig. 20.6 Vibration of a cantilever plate divided into four triangular elements Modal shapes. Data: $E = 30 \times 10^6$ lb/in^2; $t = 0{\cdot}1$ in; $L = 2$ in; $b = 1$ in; $\upsilon = 0{\cdot}3$; density $\rho = 0{\cdot}283$ lb/in^3. The numbers listed show frequencies in cycles/sec for (1) Exact solution (ref. 60); (2) 'Non conforming' triangle; (3) Conforming triangle. Corrective function Eq. (10.31); (4) Conforming triangle. Corrective function Eq. (10.32)

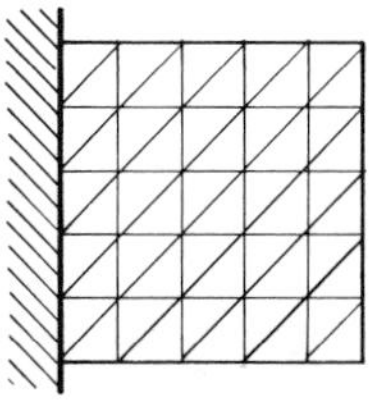

Plate without elimination

Number of degrees of freedom = 90.

Mode	$\omega\sqrt{D/\rho t a^4}$
1	3·469
2	8·535
3	21·450
4	27·059

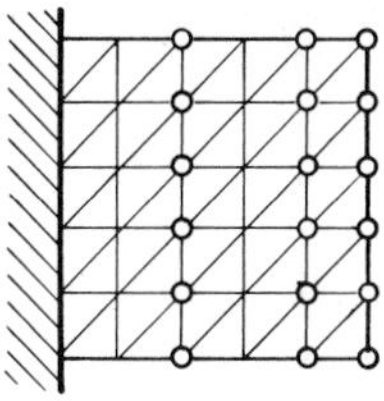

Nodes not ringed are eliminated.

Number of master displacements = 54.

Mode	$\omega\sqrt{D/\rho t a^4}$
1	3·470
2	8·540
3	21·559
4	27·215

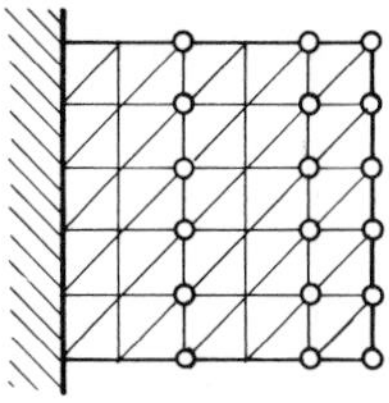

All degrees of freedom eliminated except lateral deflections at ringed nodes.

N.M.D = 18.

Mode	$\omega\sqrt{D/\rho t a^4}$
1	3·470
2	8·543
3	21·645
4	27·296

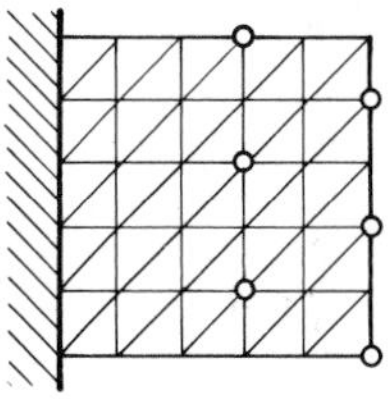

All degrees of freedom eliminated except lateral deflections at ringed nodes.

N.M.D = 6.

Mode	$\omega\sqrt{D/\rho t a^4}$
1	3·473
2	8·604
3	22·690
4	29·490

Fig. 20.7 Use of eigenvalue elimination in vibration of a square cantilever plate (a—size of plate, t—thickness)

well-known Helmholz wave equation which, in two-dimensional form is

$$\frac{\partial^2 \phi}{\partial x^2}+\frac{\partial^2 \phi}{\partial y^2}+\frac{1}{\bar{c}^2}\frac{\partial^2 \phi}{\partial t^2} = 0. \tag{20.66}$$

If the boundary conditions do not force a response, an eigenvalue problem results which has a significance in several fields of physical science.

The first application is to *electromagnetic* fields.[69] Figure 20.9 shows a modal shape of a field for a *waveguide problem*. Simple triangular elements are used here. More complex three-dimensional oscillations are also discussed in reference 69.

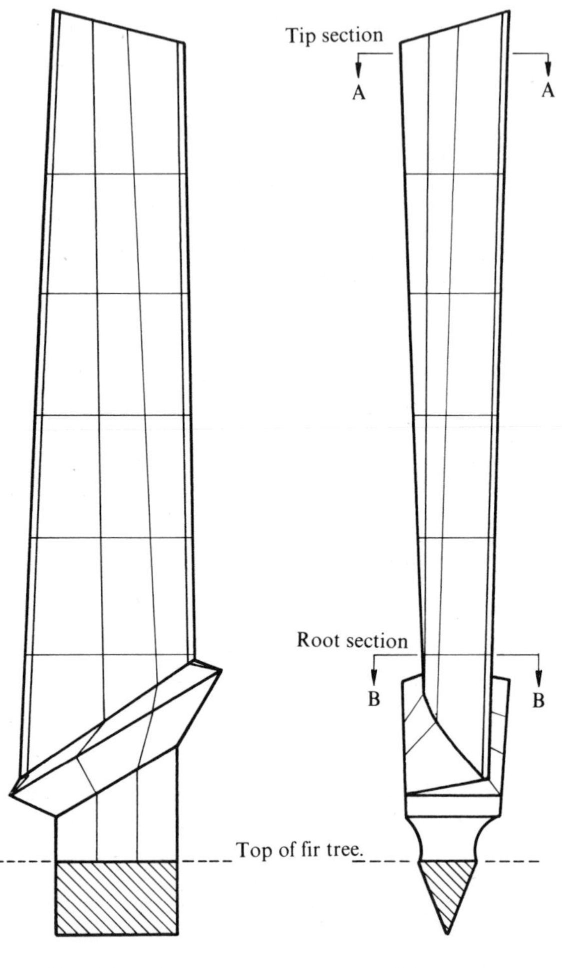

Fig. 20.8 Vibration of turbine blade treated as a thick shell (*a*) Element subdivision (parabolic type), (*b*) modal shapes and frequencies compared with experiment.

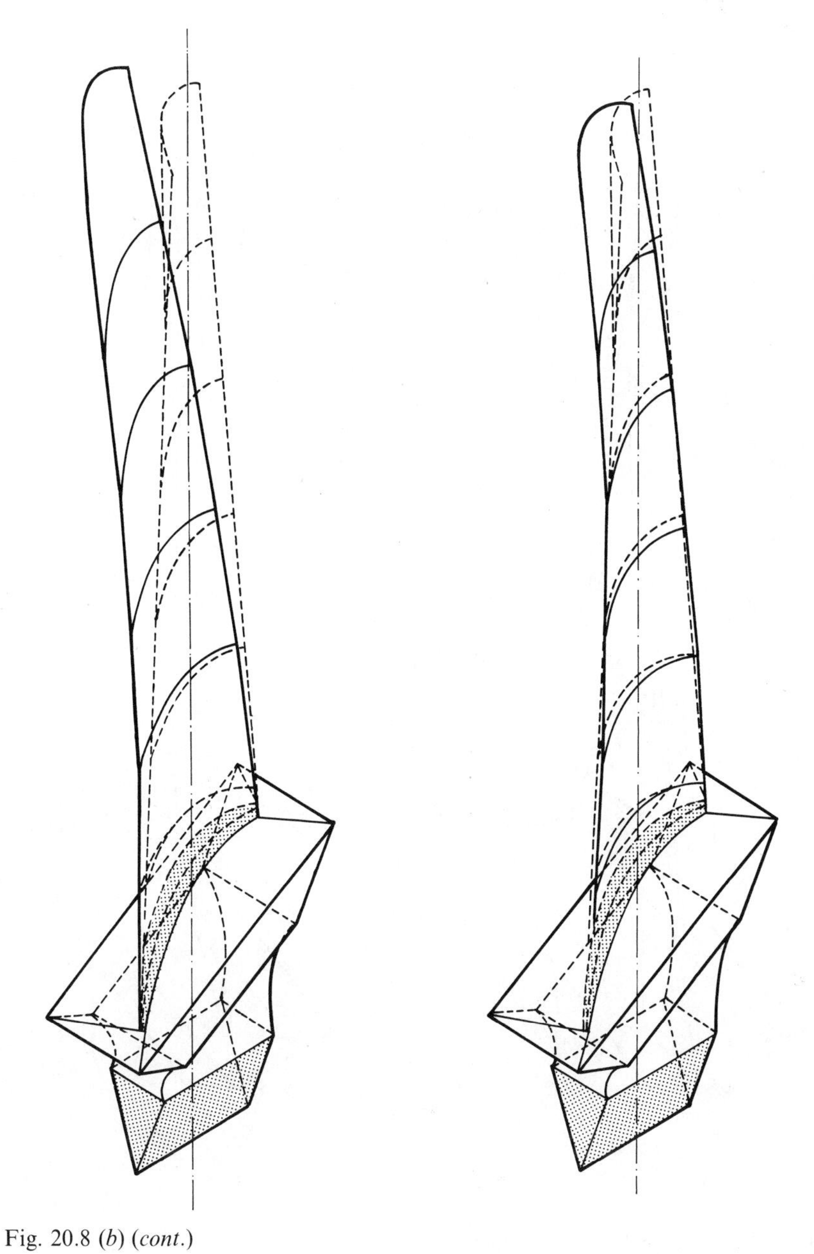

Fig. 20.8 (*b*) (*cont.*)

Mode 1—1st flap.	Measured frequency	= 517 c/s
	Calculated frequency	= 518 c/s
Mode 2—1st edgewise.	Measured frequency	= 1326 c/s
	Calculated frequency	= 1692 c/s

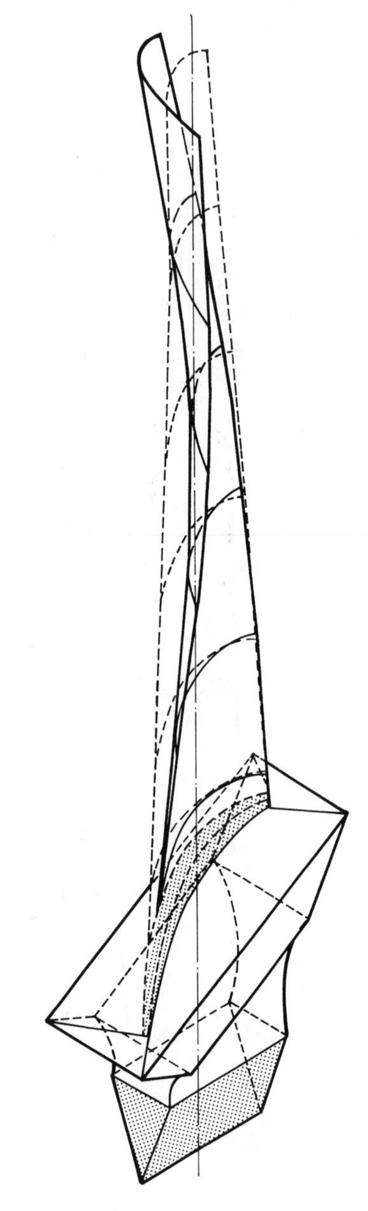

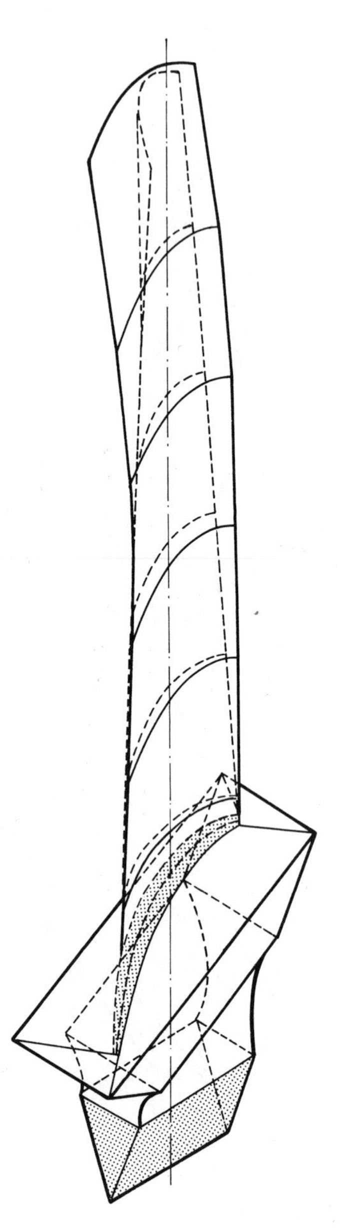

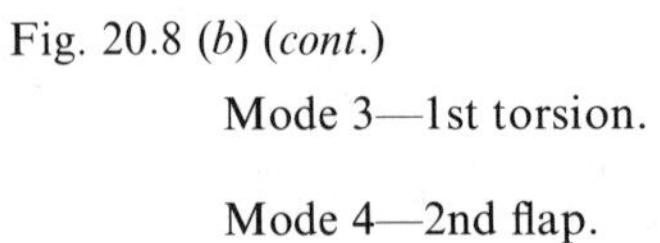

Fig. 20.8 (*b*) (*cont.*)

Mode 3—1st torsion.	Measured frequency	= 2885 c/s
	Calculated frequency	= 2686 c/s
Mode 4—2nd flap.	Measured frequency	= 2510 c/s
	Calculated frequency	= 2794 c/s

A similar equation also describes to a reasonable approximation the behaviour of shallow water waves in a body of water:

$$\frac{\partial}{\partial x}\left(h\frac{\partial \psi}{\partial x}\right)+\frac{\partial}{\partial y}\left(h\frac{\partial \psi}{\partial y}\right)+\frac{1}{g}\frac{\partial^2 \psi}{\partial t^2}=0 \tag{20.67}$$

in which h is the average water depth, ψ the surface elevation above average, and g the gravity acceleration.

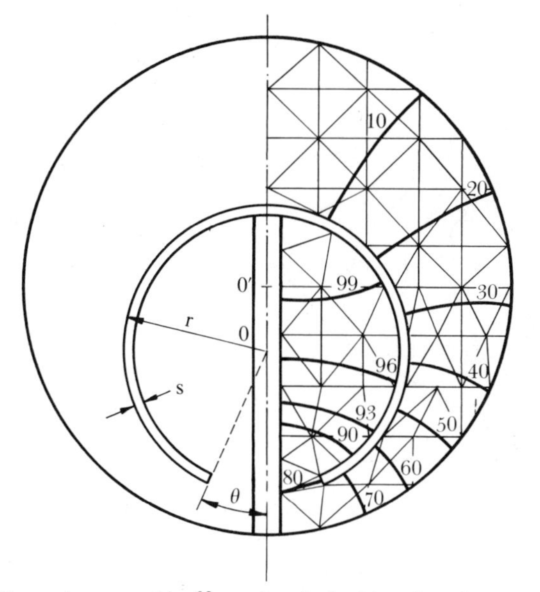

Fig. 20.9 A 'lunar' waveguide;[69] mode of vibration for electro-magnetic field. Outer diameter $= d$, $00' = 1{\cdot}3d$, $r = 0{\cdot}29d$, $S = 0{\cdot}055d$, $\theta = 22°$

Thus natural frequencies of bodies of water contained in harbours of varying depths may easily be found.[10] Figure 20.10 shows the modal shape for a particular harbour.

Coupled Structures—fluid motion. The theory of the problem was outlined in section 20.3 and once again, if damping or forcing is not present the problem leads to an eigenvalue solution.

A simple three-dimensional example showing interaction of an idealized dam and a body of fluid is shown in Fig. 20.11 and indicates the mode shapes of oscillation in which compressibility effects are excluded.[70]

In the actual reduction of the coupled problem to a standard eigenvalue form certain special transformations are helpful. Some of these are outlined in reference 11 and another computation process is presented by Irons.[71]

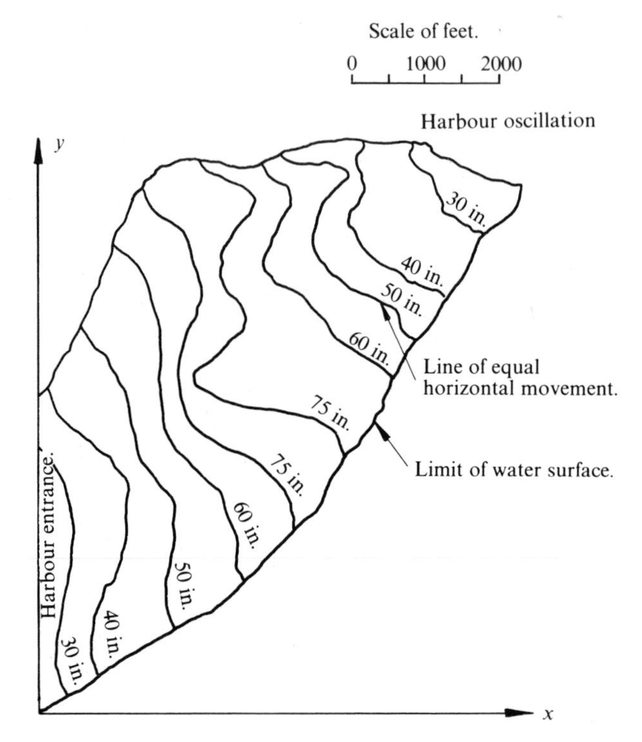

Fig. 20.10 Oscillations of a natural harbour: contours of velocity amplitudes[10]

20.6 Free Response; Eigenvalues for First Order Problems and Heat Conduction, etc.

If in Eq. (20.47) $\mathbf{M} = 0$, we have a form typical of the transient heat conduction equation (see Eq. (20.1)). For free response we seek solution of the homogeneous equation

$$\mathbf{C}\dot{\mathbf{a}} + \mathbf{K}\mathbf{a} = 0. \tag{20.68}$$

Once again an exponential form can be used

$$\mathbf{a} = \bar{\mathbf{a}}\, \mathrm{e}^{-\omega t}$$

Substituting we have

$$(-\omega\mathbf{C} + \mathbf{K})\bar{\mathbf{a}} = 0 \tag{20.69}$$

which gives once again an eigenvalue problem identical to that of Eq. (20.49). As $\mathbf{C}$ and $\mathbf{K}$ are usually positive definite, ω will be positive and real. The solution represents, therefore, simply an exponential decay

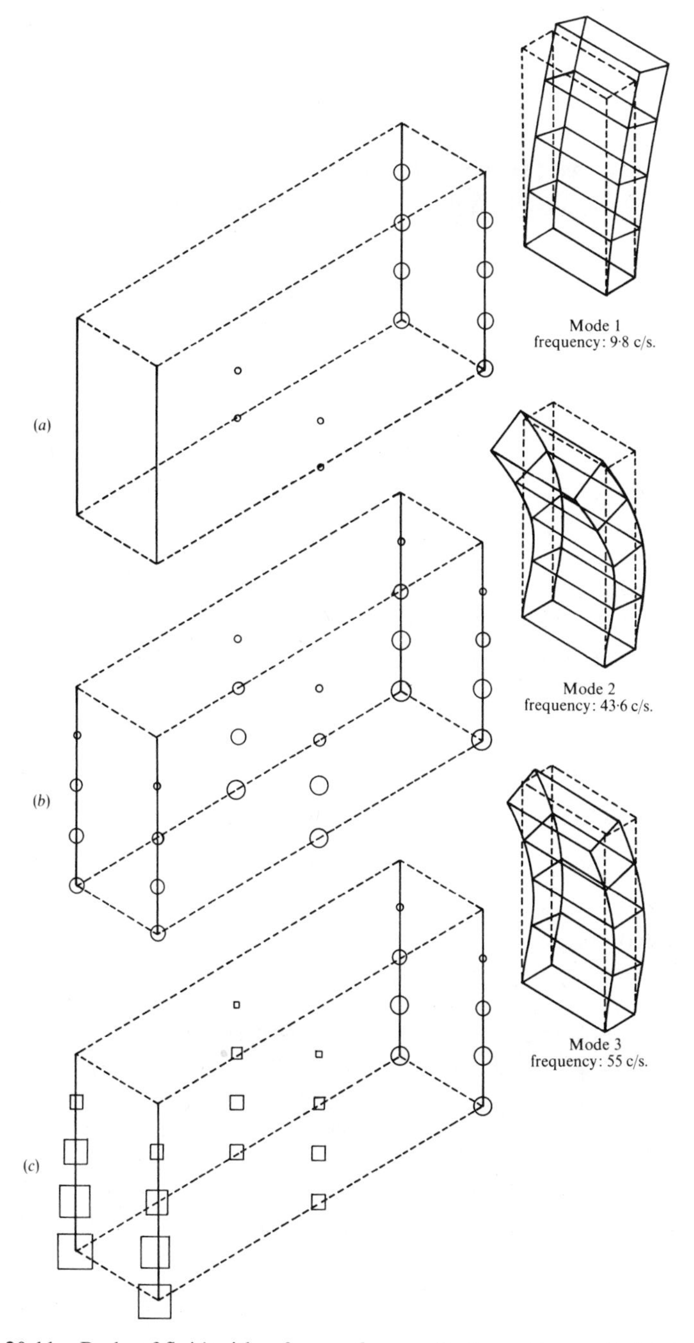

Fig. 20.11 Body of fluid with a free surface oscillating with a wall. Circles show pressure amplitudes and squares indicate opposite signs. A three-dimensional approach using parabolic elements[70]

term and is not really steady state. Combination of such terms, however, can be useful in the solution of initial value transient problems but *per se* is of little value.

20.7 **Free Response; Damped Dynamic Eigenvalues**

We shall now consider the full equation (20.47) for free response conditions. Writing

$$\mathbf{M}\ddot{\mathbf{a}} + \mathbf{C}\dot{\mathbf{a}} + \mathbf{K}\mathbf{a} = 0 \tag{20.70}$$

and substituting

$$\mathbf{a} = \bar{\mathbf{a}}\, e^{\alpha t} \tag{20.71}$$

we have the characteristic equation

$$(\alpha^2\mathbf{M} + \alpha\mathbf{C} + \mathbf{K})\bar{\mathbf{a}} = 0 \tag{20.72}$$

where α and $\bar{\mathbf{a}}$ will in general be found to be complex. The real part of the solution represents a decaying vibration.

The eigenvalue problem involved in Eq. (20.71) is more difficult than that arising in the previous sections. Fortunately seldom this needs to be solved explicitly. The concept of eigenvalues of above kind is of importance in modal analysis, as we shall see later.

20.8 **Forced Periodic Response**

If the forcing term in Eq. (20.47) is periodic or, more generally, if we can express it as

$$\mathbf{f} = \bar{\mathbf{f}}\, e^{\alpha t} \tag{20.73}$$

where α is complex, i.e.,

$$\alpha = \alpha_1 + i\alpha_2 \tag{20.74}$$

then a general solution can be once again written as

$$\mathbf{a} = \bar{\mathbf{a}}\, e^{\alpha t}. \tag{20.75}$$

Substituting the above in Eq. (20.47) gives

$$(\alpha^2\mathbf{M} + \alpha\mathbf{C} + \mathbf{K})\bar{\mathbf{a}} \equiv \mathbf{D}\bar{\mathbf{a}} = -\bar{\mathbf{f}} \tag{20.76}$$

which is no longer an eigenvalue problem but can be solved by inverting the matrix $\mathbf{D}$, i.e., formally

$$\bar{\mathbf{a}} = -\mathbf{D}^{-1}\bar{\mathbf{f}}. \tag{20.77}$$

The solution is thus precisely of the same form as that used for static problems but now, however, has to be determined in terms of complex quantities. Computer programs are available for operation of complex numbers but the computation can be arranged directly, noting that

$$\begin{aligned} e^{\alpha t} &= e^{\alpha_1 t}(\cos \alpha_2 t + \mathrm{i} \sin \alpha_2 t) \\ \bar{\mathbf{f}} &= \bar{\mathbf{f}}_1 + \mathrm{i}\bar{\mathbf{f}}_2 \\ \bar{\mathbf{a}} &= \bar{\mathbf{a}}_1 + \mathrm{i}\bar{\mathbf{a}}_2 \end{aligned} \tag{20.78}$$

in which α_1, α_2, $\mathbf{f}_1$, $\mathbf{f}_2$, $\bar{\mathbf{a}}_1$, and $\bar{\mathbf{a}}_2$ are real quantities. Inserting the above

$$\begin{bmatrix} (\alpha_1^2-\alpha_2^2)\mathbf{M}+\alpha_1\mathbf{C}+\mathbf{K}, & -2\alpha_1\alpha_2\mathbf{M}-\alpha_2\mathbf{C} \\ 2\alpha_1\alpha_2\mathbf{M}+\alpha_2\mathbf{C}, & (\alpha_1^2-\alpha_2^2)\mathbf{M}+\alpha_1\mathbf{C}+\mathbf{K} \end{bmatrix} \begin{Bmatrix} \bar{\mathbf{a}}_1 \\ \bar{\mathbf{a}}_2 \end{Bmatrix} = -\begin{Bmatrix} \bar{\mathbf{f}}_1 \\ \bar{\mathbf{f}}_2 \end{Bmatrix} \tag{20.79}$$

Equations (20.79) form a system in which all quantities are real and from which the response to any periodic input can be determined by direct solution.

The system is no longer positive definite although it is still symmetric.

With periodic input the solution after an initial transient is not sensitive to the initial conditions and this 'guessed' solution represents the finally established response. It is valid for problems of dynamic structural response as well as for the problems typical of heat conduction in which we simply put $\mathbf{M} = 0$.

20.9 Transient Response by Analytical Procedures

20.9.1 *General.* In the previous sections we have been concerned with steady-state general solutions which took no account of the initial conditions of the system or of the non-periodic form of the forcing terms. The response taking these features into account is essential if we consider, for instance, the earthquake behaviour of structures or the transient behaviour of heat conduction problem. The solution of such general cases requires either a full-time discretization which we shall discuss in detail in the next chapter, or the use of special analytical procedures. Here two broad possibilities exist:

(*a*) the frequency response procedure;
(*b*) the modal analysis procedure.

We shall indicate these briefly

20.9.2 *Frequency response procedures.* In section 20.8 we have shown how the response of the system to any forcing terms of the general periodic type or in particular to a periodic forcing function

$$\mathbf{f} = \bar{\mathbf{f}}\, e^{i\omega t} \tag{20.80}$$

can be obtained by solving a simple equation system. As a completely arbitrary forcing function can be represented approximately by a Fourier series or in the limit, exactly, as a Fourier integral, the response to such an input can be obtained by a synthesis of a curve representing the response of any quantity of interest e.g. the displacement at a particular point, etc., to all frequencies ranging from zero to infinity. In fact only a limited number of such forcing frequencies has to be considered and the

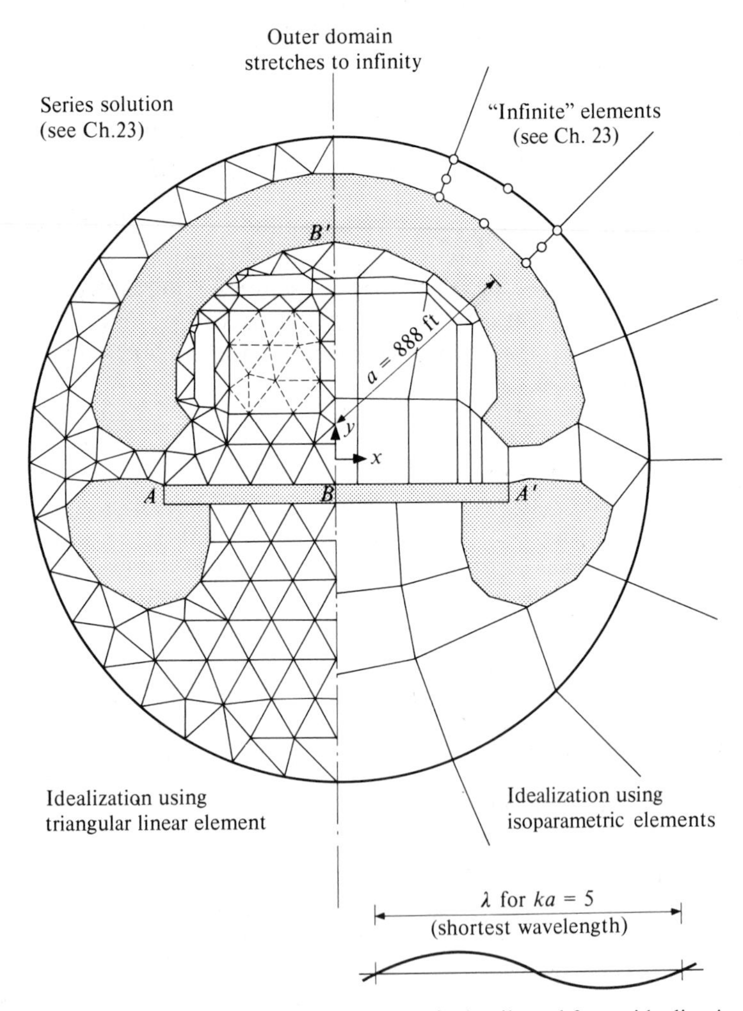

Fig. 20.12 (a) An island harbour geometric details and f.e.m. idealization

results can be synthesized efficiently by Fast Fourier transform techniques.[72] We shall not discuss the mathematical details of such procedures which can be found in standard texts on dynamics.[15]

The technique of frequency response is readily adapted to problems where damping matrix **C** is of an arbitrary specified form. This is not the case with the more widely used modal synthesis procedures which are to be described in the next section.

By way of illustration we show in Fig. 20.12 the frequency response of an artificial harbour (*vide* Eq. (20.67) to an input of waves with different frequencies and damping due to the radiation of reflected waves which imposes a very particular form on the damping matrix. Details of this

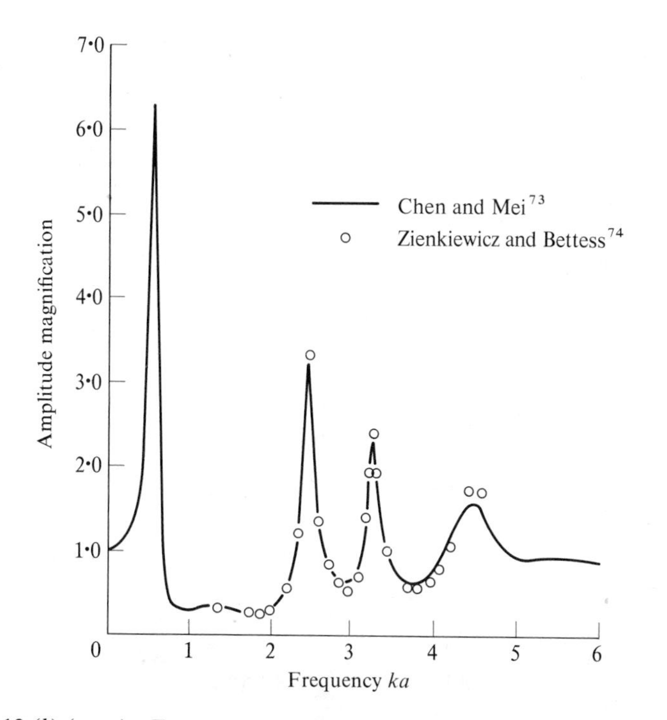

Fig. 20.12 (*b*) (*cont.*) Frequency response of mean depth in harbour related to incident wave height.

problem are given elsewhere.[73, 74] Similar techniques are frequently used in the analysis of foundation response of structures where radiation of energy occurs.[75]

20.9.3 *Modal decomposition analysis*. This procedure is probably the most important and widely used in practice. Further it provides an insight into the behaviour of the whole system which is of value where strictly numerical processes are used and we shall therefore describe it in detail in the context of the general problem of Eq. (20.47), i.e.,

$$\mathbf{M}\ddot{\mathbf{a}}+\mathbf{C}\dot{\mathbf{a}}+\mathbf{K}\mathbf{a}+\mathbf{f} = 0 \tag{20.81}$$

where $\mathbf{f}$ is an arbitrary function of time.

We have seen that the general solution for the free response is of the form

$$\mathbf{a} = \bar{\mathbf{a}}\,\mathrm{e}^{\alpha t} = \sum_{i=1}^{n} \bar{\mathbf{a}}_i\,\mathrm{e}^{\gamma_i t} \tag{20.82}$$

where α_i are the eigenvalues and $\bar{\mathbf{a}}_i$ are the eigenvectors (section 20.7). For forced response we shall assume that the solution can be written in a linear combination of modes as

$$\mathbf{a} = \sum \bar{\mathbf{a}}_i y_i = [\bar{\mathbf{a}}_1, \bar{\mathbf{a}}_2, \ldots]\mathbf{y} \tag{20.83}$$

where the scalar mode participation factor $y_i(t)$ is a function of time. This shows in a clear manner the proportions of each mode occurring. Such a decomposition of an arbitrary vector presents no restriction as all the modes are linearly independent vectors (except for repeated frequencies).

If expression (20.83) is substituted into Eq. (20.81) and the result is premultiplied by $\bar{\mathbf{a}}_i^{\mathrm{T}} (i = 1-n)$, then the result is simply a set of scalar, independent, equations

$$m_i\ddot{y}_i + c_i\dot{y}_i + k_i y_i + f_i = 0 \tag{20.84}$$

where

$$\begin{aligned} m_i &= \bar{\mathbf{a}}_i^{\mathrm{T}}\mathbf{M}\bar{\mathbf{a}}_i = 1 \text{ (if modes normalized)} \\ c_i &= \bar{\mathbf{a}}_i^{\mathrm{T}}\mathbf{C}\bar{\mathbf{a}}_i \\ k_i &= \bar{\mathbf{a}}_i^{\mathrm{T}}\mathbf{K}\bar{\mathbf{a}}_i \\ f_i &= \bar{\mathbf{a}}_i^{\mathrm{T}}\mathbf{f} \end{aligned} \tag{20.85}$$

as for true eigenvectors $\bar{\mathbf{a}}_i$

$$\bar{\mathbf{a}}_i^{\mathrm{T}}\mathbf{M}\bar{\mathbf{a}}_j = \bar{\mathbf{a}}_i^{\mathrm{T}}\mathbf{C}\bar{\mathbf{a}}_j = \bar{\mathbf{a}}_i^{\mathrm{T}}\mathbf{K}\bar{\mathbf{a}}_j = 0. \tag{20.86}$$

when

$$i \neq j$$

(This result was proved in section 20.5 for real eigenvalues but is valid generally for complex eigenvalues and vectors as could be verified by the reader.)

Each scalar equation of (20.84) can be solved by elementary procedures independently and the total vector of response obtained by superposition following Eq. (20.83). In a general case, as we have shown in section 20.7, the eigenvalues and eigenvectors are complex and their determination is not simple.[53] The more usual procedure is to use real eigenvalues corresponding to the solution of Eq. (20.48).

$$(-\omega^2\mathbf{M}+\mathbf{K})\bar{\mathbf{a}} = 0. \tag{20.87}$$

Now repetition of procedures using the process described in Eqs (20.83) to (20.86) leads to decoupled equations with real variables $\mathbf{y}$ only if

$$\bar{\mathbf{a}}_i^{\mathrm{T}}\mathbf{C}\mathbf{a}_j = 0 \tag{20.88}$$

which generally does not occur as the eigenvectors now guarantee only the orthogonality of $\mathbf{M}$ and $\mathbf{K}$ and not of the damping matrix. However, if the damping matrix $\mathbf{C}$ is of the form of Eq. (20.16), i.e. a linear combination of $\mathbf{M}$ and K, such orthogonality will obviously occur. Unless the damping is of a definite form which requires special treatment, an assumption of orthogonality is made and Eq. (20.84) is assumed valid in terms of such eigenvectors.

From Eq. (20.87) we have

$$\omega_i^2\mathbf{M}\bar{\mathbf{a}}_i = \mathbf{K}\bar{\mathbf{a}}_i \tag{20.89}$$

and on pre-multiplying by $\bar{\mathbf{a}}_i^{\mathrm{T}}$ we obtain

$$\omega_i^2 m_i = k_i. \tag{20.90}$$

Writing

$$c_i = 2\omega_i c_i' \tag{20.91}$$

(where c_i' represents the ratio of damping to its critical value) and assuming that the nodes have been normalized so that $m_i = 1$ (*vide* Eq. (20.51)), Eq. (20.84) can be rewritten in a standard form of second degree:

$$\ddot{y}_i + 2\omega_i c_i'\dot{y}_i + \omega_i^2 y_i + f_i = 0 \tag{20.92}$$

The general solution can be then obtained by writing

$$y_i = \int_0^t f_i \, \mathrm{e}^{-c_i'\omega_i(t-\tau)} \sin \omega_i(t-\tau)\, \mathrm{d}\tau. \tag{20.93}$$

Such integration can be carried out numerically and the response obtained. In principle, superposition will result in the full transient required. In practice often a single calculation is carried out for each mode to determine the maximum responses and a suitable addition of these results is used. Such processes are well described in standard texts and are used as standard procedures to calculate the behaviour of structures subject to earthquake shocks.[15,43]

For a first order equation system, such as arises in heat conduction,

$$\mathbf{C}\dot{\mathbf{a}}+\mathbf{K}\mathbf{a}+\mathbf{f} = 0$$

exactly analogous procedures can be used. Using now the real eigenvalues determined in section 20.6, decomposition into scalar equations can be made yielding a set of decoupled equations

$$c_i \dot{y}_i + k_i y_i + f_i = 0 \tag{20.94}$$

and this can be used again to get the general solution analytically.[76] We leave details of such a solution to the reader as an exercise.

20.9.4 *Damping and participation of modes.* The type of calculation implied in modal decomposition apparently necessitates the determination of all modes and eigenvalues; a task of considerable magnitude. In fact only a limited number of modes usually need to be taken into consideration as often the response to higher frequency is critically damped and insignificant.

To show that this is true consider the form of the damping matrices. In section 20.2 (Eq. (20.16)) we have indicated that the damping matrix is often assumed as

$$\mathbf{C} = \alpha\mathbf{M}+\beta\mathbf{K}. \tag{20.95}$$

Indeed a form of this type is necessary for the use of modal decomposition although other generalizations are possible.[77,78] From the definition of c'_i the critical damping ratio in Eq. (20.91) we see that this can now be written as

$$c'_i = \frac{1}{2\omega_i}\bar{\mathbf{a}}_i^{\mathrm{T}}(\alpha\mathbf{M}+\beta\mathbf{K})\bar{\mathbf{a}}_i = \frac{1}{2\omega_i}(\alpha+\beta\omega_i^2). \tag{20.96}$$

Thus if the coefficient β is of larger importance, as is the case with most structural damping, c'_i grows with ω_i and at high frequency an over-damped condition will arise. This is indeed fortunate as, in general, an infinite number of high frequencies exist which are not modelled by any finite element discretizations.

We shall see in the next chapter that in the step-by-step recurrence computation the high frequencies often control the problem, and this effect needs to be filtered out for realistic results.

20.10 **Symmetry and Repeatability**

In concluding this chapter it is worth remarking that in dynamic calculation we have once again encountered all the general principles of assembly, etc., which are applicable to static problems. However some aspects of symmetry and repeatability which were used previously (vide

Chapter 9, section 9.5) need amending. It is obviously possible for a symmetric structure to vibrate in an unsymmetric manner for instance and similarly a repeatable structure contains modes which are themselves non-repeatable. However even here considerable simplification can still be made—and details of this are discussed in references 79–81.

References

1. S. CRANDALL, *Engineering Analysis*, McGraw-Hill, 1956.
2. H. S. CARSLAW and J. C. JAEGER, *Conduction of Heat in Solids*, 2nd ed., Clarendon Press, 1959.
3. W. VISSER, 'A finite element method for the determination of non-stationary temperature distribution and thermal deformation', *Proc. Conf. on Matrix Methods in Structural Mechanics*, Air Force Inst. Tech., Wright-Patterson A.F. Base, Ohio, 1965.
4. O. C. ZIENKIEWICZ and Y. K. CHEUNG, *The Finite Element Method in Structural and Continuum Mechanics*, 1st ed., McGraw-Hill, 1967.
5. E. L. WILSON and R. E. NICKELL, 'Application of finite element method to heat conduction analysis', *Nucl. Eng. Des.*, **4**, 1–11, 1966.
6. O. C. ZIENKIEWICZ and C. J. PAREKH, 'Transient field problems—two and three dimensional analysis by isoparametric finite elements', *Int. J. Num. Meth. Eng.*, **2**, 61–71, 1970.
7. K. TERZHAGI and R. B. PECK, *Soil Mechanics in Engineering Practice*, Wiley, 1948.
8. D. K. TODD, *Ground Water Hydrology*, Wiley, 1959.
9. P. L. ARLETT, A. K. BAHRANI and O. C. ZIENKIEWICZ, 'Application of finite elements to the solution of Helmholz's equation', *Proc. I.E.E.*, **115**, 1962–6, 1968.
10. C. TAYLOR, B. S. PATIL and O. C. ZIENKIEWICZ, 'Harbour oscillation: a numerical treatment for undamped natural modes', *Proc. Inst. Civ. Eng.*, **43**, 141–56, 1969.
11. O. C. ZIENKIEWICZ and R. E. NEWTON, 'Coupled vibrations in a structure submerged in a compressible fluid', *Int. Symp. on Finite Element Techniques*, Stuttgart, 1969.
12. J. S. ARCHER, 'Consistent mass matrix for distributed systems', *Proc. Am. Soc. Civ. Eng.*, **89**, ST4, 161, 1963.
13. F. A. LECKIE and G. M. LINDBERG, 'The effect of lumped parameters on beam frequencies', *Aero Quart.*, **14**, 234, 1963.
14. O. C. ZIENKIEWICZ and Y. K. CHEUNG, 'The finite element method for analysis of elastic isotropic and orthotropic slabs', *Proc. Inst. Civ. Eng.*, **28**, 471, 1964.
15. R. W. CLOUGH and J. PENZIEN, *Dynamics of Structures*, McGraw-Hill, 1975.
16. O. C. ZIENKIEWICZ, B. M. IRONS and B. NATH, 'Natural frequencies of complex free or submerged structures by the finite element method' in *Symp. on Vibration in Civil Engineering*, London, April 1965 (Butterworth, 1966).
17. D. J. DAWE, 'A finite element approach to plate vibration problems', *J. Mech. Eng. Sci.*, **7**, 28, 1965.
18. R. J. GUYAN, 'Distributed mass matrix for plate elements in bending', *J.A.I.A.A.*, **3**, 567, 1965.
19. G. P. BAZELEY, Y. K. CHEUNG, B. M. IRONS and O. C. ZIENKIEWICZ, 'Triangular elements in plate bending—conforming and non-conforming solu-

tions', *Proc. Conf. on Matrix Methods in Structural Mechanics*, Air Force Inst. Tech., Wright-Patterson A.F. Base, Ohio, 1965.

20. R. G. ANDERSON, B. M. IRONS and O. C. ZIENKIEWICZ, 'Vibration and stability of plates using finite elements', *Int. J. Solids Struct.*, **4**, 1031–55, 1968.
21. R. G. ANDERSON, *The application of the non-conforming triangular plate bending element to plate vibration problems*, M.Sc. Thesis, Univ. of Wales, Swansea, 1966.
22. R. W. CLOUGH, 'Analysis of structure vibrations and response', *Recent Advances in Matrix Method of Structure Analysis and Design*, pp. 25–45 (eds. R. H. Gallagher, Y. Yamada and J. T. Oden), First U.S.–Japan Seminar, Alabama Press, 1971.
23. S. W. KEY and Z. E. BEISINGER, 'The transient dynamic analysis of thin shells in the finite element method', *Proc. 3rd Conf. on Matrix Methods and Structural Mechanics*, Wright-Patterson A.F. Base, Ohio, 1971.
24. E. HINTON, A. ROCK and O. C. ZIENKIEWICZ, 'A note on mass lumping in related process in the finite element method', *Int. J. Earthquake Eng. Struct. Dynam.*, **4**, 245–9, 1976.
25. P. TONG, T. H. H. PIAN and L. L. BUCIOVELLI, 'Mode shapes and frequencies by the finite element method using consistent and lump matrices', *J. Comp. Struct.*, **1**, 623–38, 1971.
26. I. FRIED and D. S. MELKUS, 'Finite element mass matrix lumping by numerical integration with the convergence rate loss', *Int. J. Solids Struct.*, **11**, 461–5, 1975.
27. O. C. ZIENKIEWICZ, Discussion of 'Earthquake behaviour of reservoir-dam systems' by A. K. Chopra, *Proc. Am. Soc. Civ. Eng.*, **95**, EM3, 801–3, 1969.
28. P. A. A. BACK, A. C. CASSELL, R. DUNGAR and R. T. SEVERN, 'The seismic study of a double curvature dam', *Proc. Inst. Civ. Eng.*, **43**, 217–48, 1969.
29. SIR H. LAMB, *Hydrodynamics*, 6th ed., Cambridge, 1932.
30. P. CHAKRABARTI and A. K. CHOPRA, 'Earthquake analysis of gravity dams including hydrodynamic interaction', *Int. J. Earthquake Eng. Struct. Dynam.*, **2**, 1973.
31. A. R. CHANDRASEKARAN *et al.*, 'Hydrodynamic pressure on circular cylindrical cantilevered structures surrounded by water', *Proc. 4th Symp. on Earthquake Engineering*, Rourkee, 1970, pp. 161–71.
32. G. M. L. GLADWELL, 'A variational formulation of damped acousto-structural problems', *J. Sound Vib.*, **3**, 233, 1966.
33. H. MORAND and R. OHAYON, 'Variational-formulation for the acoustic vibration problem; finite element results', *Proc. 2nd Int. Symp. on Finite Element Methods in Flow Problems*, St. Margharita, Italy, 1976.
34. A. SOMMERFELD, *Partial Differential Equations in Physics*, Academic Press, 1949.
35. R. S. SANDHU and E. L. WILSON, 'Finite element analysis of seepage in elastic media', *Proc. Am. Soc. Civ. Eng.*, **95**, EM3, 641–51, 1969.
36. J. L. SERAFIM, Chapter 3 of *Rock Mechanics and Engineering Practice* (eds. K. G. Stagg and O. C. Zienkiewicz), Wiley, 1968.
37. J. CROCHET and P. M. NAGHDI, 'On constitutive equations for flow of fluid through an elastic solid', *Int. J. Eng. Sci.*, **4**, 383–401, 1966.
38. M. A. BIOT, 'General theory of three dimensional consolidation', *J. Appl. Phys.*, **12**, 155–64, 1941.
39. O. C. ZIENKIEWICZ, C. HUMPHESON and R. W. LEWIS, 'A unified approach to

soil mechanics problems including plasticity and viscoplasticity', *Conf. on Numerical Methods in Soil and Rock Mechanics*, Karlsruhe Univ., 1975. See chapter 4 of *Finite Elements in Geomechanics*, G. Gudehus, ed., J. Wiley & Son, 1977.)

40. J. GHABOUSSI and E. L. WILSON, 'Flow of compressible fluids in elastic media', *Int. J. Num. Meth. Eng.*, **5**, 419–42, 1973.
41. C. T. HWANG, N. R. MORGENSTERN and D. W. MURRAY, 'On solution of plane strain consolidation problems by finite element methods', *Can. Geotech. J.*, **8**, 109–18, 1971.
42. O. C. ZIENKIEWICZ, R. W. LEWIS, V. A. NORRIS and C. HUMPHESON, 'Numerical analysis for foundation of offshore structures with special reference to programme deformation', *Soc. Petr. Eng. Amsterdam*, April 1976.
43. K. J. BATHE and E. L. WILSON, *Numerical methods in finite element analysis*, Prentice-Hall, 1976.
44. J. H. WILKINSON, *The Algebraic Eigenvalue Problem*, Clarendon Press, Oxford, 1965.
45. I. FRIED, 'Gradient methods for finite element eigen problems', *J.A.I.A.A.*, **7**, 739–41, 1969.
46. O. RENFIELD, 'Higher vibration modes by matrix iteration', *J.A.I.A.A.*, **9**, 505–741, 1971.
47. K. J. BATHE and E. L. WILSON, 'Large eigenvalue problems in dynamic analysis', *Proc. Am. Soc. Civ. Eng.*, **98**, EM6, 1471–85, 1972.
48. K. J. BATHE and E. L. WILSON, 'Solution methods for eigenvalue problems in structural dynamics', *Int. J. Num. Meth. Eng.*, **6**, 213–26, 1973.
49. A. JENNINGS, 'Mass condensation and similarity iterations for vibration problems', *Int. J. Num. Meth. Eng.*, **6**, 543–52, 1973.
50. K. K. GUPTA, 'Solution of eigenvalue problems by Sturm sequence method', *Int. J. Num. Meth, Eng.*, **4**, 379–404, 1972.
51. H. L. COX, 'Vibration of missiles', *Aircraft Eng.*, **33**, 2–7 and 48–55.
52. A. JENNINGS, 'Natural vibration of a free structure', *Aircraft Eng.*, **34**, 8, 1962.
53. W. C. HURTY and M. F. RUBINSTEIN, *Dynamics of Structures*, Prentice-Hall, 1974.
54. A. CRAIG and M. C. C. BAMPTON, 'On the iterative solution of semi definite eigenvalue problems', *Aero. J.*, **75**, 287–90, 1971.
55. B. M. IRONS, 'Eigenvalue economisers in vibration problems', *J. Roy. Aero. Soc.*, **67**, 526, 1963.
56. B. M. IRONS, 'Structural eigenvalue problems: elimination of unwanted variables', *J.A.I.A.A.*, **3**, 961, 1965.
57. R. J. GUYAN, 'Reduction of stiffness and mass matrices', *J.A.I.A.A.*, **3**, 380, 1965.
58. J. N. RAMSDEN and J. R. STOKER, 'Mass condensation; a semi-automatic method for reducing the size of vibration problems', *Int. J. Num. Meth. Eng.*, **1**, 333–49, 1969.
59. R. D. HENSHELL and J. H. ONG, 'Automatic masters for eigenvalue economisation', *Int. J. Earthquake Struct. Dynam.*, **3**, 375–83, 1975.
60. M. V. BARTON, 'Vibration of rectangular and skew cantilever plates', *J. Appl. Mech.*, **18**, 129–34, 1951.
61. G. B. WARBURTON, 'Recent advances in structural dynamics', *Symp. on Dynamic Analysis of Structures*, N.E.L., East Kilbride, Scotland, Oct. 1975.
62. J. C. MACBAIN, 'Vibratory behaviour of twisted cantilever plates', *J. Aircr.*, **12**, 357–359, 1975.

63. S. Ahmad, R. G. Anderson and O. C. Zienkiewicz, 'Vibration of thick, curved, shells with particular reference to turbine blades', *J. Strain Analysis*, **5**, 200–6, 1970.
64. R. G. Anderson, *A finite element eigenvalue system*, Ph.D. Thesis, Univ. of Wales, Swansea, 1968.
65. J. S. Archer and C. P. Rubin, 'Improved linear axi-symmetric shell-fluid model for launch vehicle longitudinal response analysis', *Proc. Conf. on Matrix Methods in Structural Mechanics*, Air Force Inst. Tech. Wright-Patterson A.F. Base, Ohio, 1965.
66. J. H. Argyris, 'Continua and discontinua', *Proc. Conf. on Matrix Methods in Structural Mechanics*, Air Force Inst. Tech., Wright-Patterson A.F. Base, Ohio, 1965.
67. S. Klein and R. J. Sylvester, 'The linear elastic dynamic analysis of shells of revolution by the matrix displacement method', *Proc. Conf. on Matrix Methods in Structural Mechanics*, Air Force Inst. Tech., Wright-Patterson A.F. Base, Ohio, 1965.
68. R. Dungar, R. T. Severn and P. R. Taylor, 'Vibration of plate and shell structures using triangular finite elements', *J. Strain Analysis*, **2**, 73–83, 1967.
69. P. L. Arlett, A. K. Bahrani and O. C. Zienkiewicz, 'Application of finite elements to the solution of Helmholtz's equation', *Proc. I.E.E.*, **115**, 1762–964, 1968.
70. J. Holbeche, Ph.D. Thesis, Univ. of Wales, Swansea, 1971.
71. B. M. Irons, 'Role of part-inversion in fluid structure problems with mixed variables', *J.A.I.A.A.*, **7**, 568, 1970.
72. E. O. Brigham, *The Fast Fourier Transform*, Prentice-Hall, 1974.
73. H. S. Chen and C. C. Mei, 'Hybrid-element method for water waves', *Proc. Modelling Techniques Conf.* (*Modelling 1975*), San Francisco, 3–5 Sept. 1975, Vol. I, pp. 63–81.
74. O. C. Zienkiewicz and P. Bettess, 'Infinite elements in the study of fluid-structure interaction problems', *2nd Int. Symp. on Computing Methods in Applied Science and Engineering*, Versaille, France, Dec. 1975.
75. J. Penzien, 'Frequency domain analysis including radiation damping and water load coupling', in *Numerical Methods in Offshore Engineering* (eds. O. C. Zienkiewicz, R. W. Lewis and K. G. Stagg), J. Wiley (to be published 1977).
76. R. H. Gallagher and R. H. Mallett, *Efficient solution process for finite element analysis of transient heat conduction*, Bell Aero Systems, Buffalo, 1969.
77. E. L. Wilson and J. Penzien, 'Evaluation of orthogonal damping matrices', *Int. J. Num. Meth. Eng.*, **4**, 5–10, 1972.
78. H. T. Thomson, T. Colkins and P. Caravani, 'A numerical study of damping', *Int. J. Earthquake Eng. Struct. Dynam.*, **3**, 97–103, 1974.
79. F. W. Williams, 'Natural frequencies of repetitive structures', *Quart. Jn. Mech. Appl. Math.*, **24**, 285–310, 1971.
80. D. A. Evensen, 'Vibration analysis of multi-symmetric structures', *J.A.I.A.A.*, **14**, 446–53, 1976.
81. D. L. Thomas, 'Standing waves in rotationally periodic structures', *J. Sound Vibr.*, **37**, 288–90, 1974.

21. The Time Dimension; Finite Element Approximation to Initial Value–Transient Problems

21.1 Introduction

In the last chapter we have seen that semi-discretization of dynamic or transient field problems has led to a system of ordinary differential equations of the form

$$\mathbf{M}\ddot{\mathbf{a}}+\mathbf{C}\dot{\mathbf{a}}+\mathbf{K}\mathbf{a}+\mathbf{f} = 0 \tag{21.1}$$

for dynamic situations or

$$\mathbf{C}\dot{\mathbf{a}}+\mathbf{K}\mathbf{a}+\mathbf{f} = 0 \tag{21.2}$$

for heat transfer and similar problems. Whilst various analytical solution possibilities were presented, we noted that actual solution for transient problems was generally difficult and indeed not available for non-linear situations. In this chapter we shall, therefore, revert to a *trial function–finite element* discretization now applied to the time domain in the general manner presented in Chapter 3. As the time dimension is of an infinite extent we shall deal with finite domains of time and repeat the calculation for subsequent domains with new initial conditions. The process will thus lead to a *step-by-step* or *recurrence* calculation and the reader will recognize many procedures of time-stepping generally derived by finite difference or Runge–Kutte procedures. These will, however, turn out to be much more general. Much literature has been devoted to the subject of deriving recurrence relations.[1–5] However the finite element process will be shown once again to embrace most of the conventional procedures and lead to some new possibilities.

At this stage the reader may well ask why, in the original problem, a finite element discretization in the full space–time domain has not been described using functions written in terms of both sets of co-ordinates.

The answer to that is simple. First, the general domain would involve too many simultaneous variables; second, the geometrically simple nature of the time domain offers little incentive to the use of an irregular subdivision into space–time elements. Further, if product type functions were used in the general domain, the answer to successive space and time discretization would prove, indeed, identical to a simultaneous discretization applied to both.

In the following it will be convenient to note that the system of equations (21.1) or (21.2) could be written as a set of modally uncoupled scalar equations (*vide* Chapter 20, section 20.9.3)

$$m_i\ddot{y}_i + c_i\dot{y}_i + k_i y_i + f_i = 0 \tag{21.3}$$

or

$$c_i\dot{y}_i + k_i y_i + f_i = 0 \tag{21.4}$$

in which y_i is the mode participation variable.

Whether the discretization is done on the basis of the full vector variable $\mathbf{a}$ or on the basis of the modal participation variables y_i we shall obtain similar numerical recurrence expressions and indeed identical results.

In general we shall write the recurrence relations for the full original matrix equations. However, for stability considerations it will be convenient to revert to the mode participation scalar equations of the type (21.3) or (21.4) and, indeed, much of the general behaviour pattern can be studied from these relationships.

In sections 21.2 to 21.4 we shall restrict ourselves to the simpler first order (heat conduction) type of problem given by Eq. (21.2) and the simplest types of two-point formulae. This allows the general methodology to be introduced to the reader and the simple inclusion of such considerations as stability, starting point, smoothing, etc. In sections 21.6 and 21.7 a simple extension of the interpolation domain allows us to treat the second order equation of dynamics, (21.1), and establish general multipoint formulae valid for both first and second order equations. Discussion of aspects such as non-linearity of coefficients, etc., is dealt with in the remaining parts of this chapter.

21.2 Two-Point Recurrence Schemes for First Order Equations

21.2.1 *Weighted residual approach.* We shall now consider the first order equation (Eq. (21.2)) and show how a simple kind of recurrence scheme can be devised.

Proceeding in the usual manner of discretization, with time as the independent variable, we can write

$$\mathbf{a} \approx \hat{\mathbf{a}} = \sum N_i \mathbf{a}_i \tag{21.5}$$

where $\mathbf{a}_i$ stands for a *nodal* set of $\mathbf{a}$ at a time i. The shape functions $N_i(t)$

are taken to be the same for each component of the vector **a** and therefore N_i is a scalar.

The lowest order of polynomial necessary for the shape function N_i is as that of first order, as only first order derivatives are involved in Eq. (21.2).

Consider now a typical time 'element' of length Δt with $\mathbf{a}_i$ taking on nodal values $\mathbf{a}_n$ and $\mathbf{a}_{n+1}$, as shown in Fig. 21.1. The interpolation, shape, functions and their first time derivatives can be written in terms of local variables as

$$\begin{aligned} 0 \leqslant \xi \leqslant 1 & \qquad \xi = t/\Delta t \\ N_n = 1-\xi; & \qquad \dot{N}_n = -1/\Delta t \\ N_{n+1} = \xi; & \qquad \dot{N}_{n+1} = 1/\Delta t \end{aligned} \tag{21.6}$$

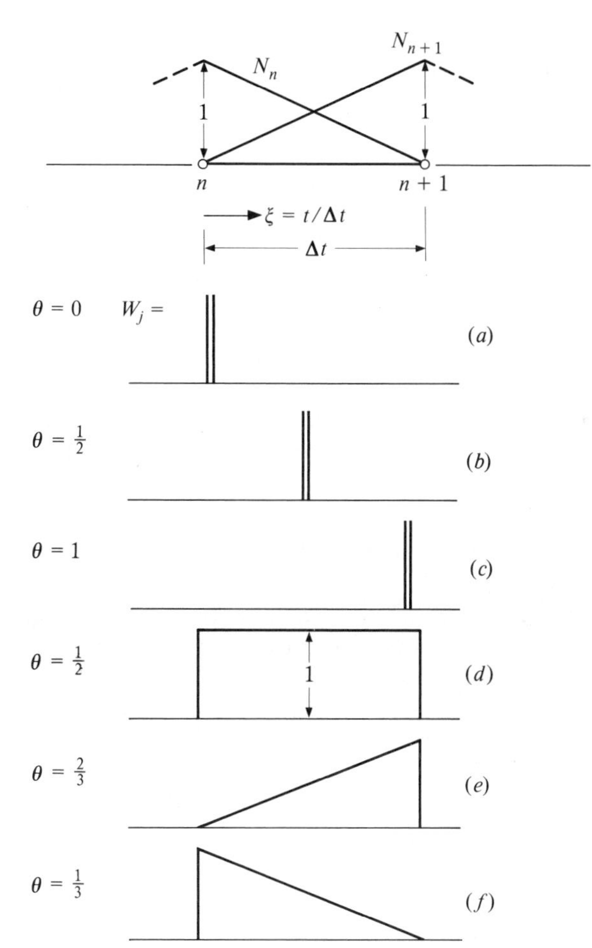

Fig. 21.1 Shape functions and weight functions for two point recurrence formulae

A typical weighted residual equation can now be written *assuming that the full domain of investigation corresponds with that of one element.*

$$\int_0^1 W_j[\mathbf{C}(\mathbf{a}_n\dot{N}_n+\mathbf{a}_{n+1}\dot{N}_{n+1})+\mathbf{K}(\mathbf{a}_nN_n+\mathbf{a}_{n+1}N_{n+1})+\mathbf{f}]\,\mathrm{d}\xi$$

$$= 0 \qquad (j = 1). \qquad (21.7)$$

As the problem is an initial value one, one of the parameter sets $\mathbf{a}_n$ is assumed known and the equation will serve to determine $\mathbf{a}_{n+1}$ approximately. Immediately on inserting Eq. (21.6) a recurrence relation can be written as

$$\left(\mathbf{K}\int_0^1 W_j\xi\,\mathrm{d}\xi+\mathbf{C}\int_0^1 W_j\,\mathrm{d}\xi/\Delta t\right)\mathbf{a}_{n+1}$$

$$+\left(\mathbf{K}\int_0^1 W_j(1-\xi)\,\mathrm{d}\xi-\mathbf{C}\int_0 W_j\,\mathrm{d}\xi/\Delta t\right)\mathbf{a}_n+\int_0^1 W_j\mathbf{f}\,\mathrm{d}\xi = 0 \qquad (21.8)$$

in which various weighting functions can be inserted. In the above, matrices $\mathbf{K}$ and $\mathbf{C}$ have been assumed to be independent of t.
Quite generally expression (21.8) can be rewritten for *any* weighting function as

$$(\mathbf{C}/\Delta t+\mathbf{K}\theta)\,\mathbf{a}_{n+1}+(-\mathbf{C}/\Delta t+\mathbf{K}(1-\theta))\,\mathbf{a}_n+\mathbf{f} = 0 \qquad (21.9)$$

where

$$\theta = \int_0^1 W_j\xi\,\mathrm{d}\xi\bigg/\int_0^1 W_j\,\mathrm{d}\xi$$

and

$$\bar{\mathbf{f}} = \int_0^1 W_j\mathbf{f}\,\mathrm{d}\xi\bigg/\int_0^1 W_j\,\mathrm{d}\xi.$$

From the above $\mathbf{a}_{n+1}$ can be found by a simple solution of equations as $\mathbf{a}_n$ and $\bar{\mathbf{f}}$ are known.

It is logical and convenient to assume that the same interpolation is applied to the function $\mathbf{f}$ as that used for the unknown vector $\mathbf{a}$. If this interpolation is adopted we find that

$$\bar{\mathbf{f}} = \mathbf{f}_{n+1}\theta+\mathbf{f}_n(1-\theta). \qquad (21.10)$$

At this stage we observe that if $\theta = 0$ and $\mathbf{C}$ is a *diagonal* (lumped) matrix, the solution for $\mathbf{a}_{n+1}$ is trivial and each individual value can be computed directly from its precursor. Such schemes are known as *explicit* whilst those in which $\theta \neq 0$, requiring the solution of a non-diagonal system of equations, are known as *implicit*. We shall see later that this computational convenience is accompanied by a serious drawback which requires the use of Δt not exceeding a certain magnitude.

The reader will recognize in Eq. (21.8) a well-known series of finite difference formulae with a modification of using a weighted loading term $\bar{\mathbf{f}}$. In Fig. 21.1 we show a series of weighting functions and the resulting values of θ. The first three, (*a*)–(*c*), represent point collocation weighting applied at n, $n+\frac{1}{2}$, and $n+1$, respectively, and give the well-known forward difference (Euler), mid-difference (Crank-Nicolson) and backward difference formula.

The schemes shown in Fig. 21.1(*e*) and (*f*) represent Galerkin type processes. The scheme (*e*) was first suggested by the author[6] and is equivalent to the normal use of the Galerkin procedure with the weighting corresponding to the unknown function, and this scheme has been shown to lead to considerable computational advantages.[7]

In the foregoing we have assumed the domain of the approximation to correspond to a time Δt and established a *recurrence* relation between two successive values of $\mathbf{a}_{n+1}$ and $\mathbf{a}_n$. The computation can obviously proceed in a step-by-step manner using a sequence of such domains. It is, however, possible to apply the procedure to the *whole domain of time* simultaneously, assuming this domain to be divided into a finite number of elements. Using, for instance, the linear expansions of Eq. (21.6), but integrating over the whole domain, we note that unless the weight functions are confined to a single element, several sets of $\mathbf{a}_i$ will be included simultaneously in a typical equation. For instance, if the Galerkin procedure is used and we assume that

$$W_j = N_j$$

then, if the whole domain is included, three successive values are involved in a typical equation, as shown in Fig. 21.2. For a constant time interval Δt the following expression is obtained, which can be verified by the reader

$$(\mathbf{C}/2\Delta t+\mathbf{K}/6)\,\mathbf{a}_{n+1}+(2\mathbf{K}/3)\,\mathbf{a}_n+(-\mathbf{C}/2\Delta t+\mathbf{K}/6)\,\mathbf{a}_{n-1}+\bar{\mathbf{f}} = 0$$

$$\bar{\mathbf{f}} = \mathbf{f}_{n+1}/6+2\mathbf{f}_n/3+\mathbf{f}_{n-1}/6 \qquad (21.11)$$

Providing *two initial values* are known, such a formula can be used in a recurrent manner for evaluating $\mathbf{a}_{n+1}$ from $\mathbf{a}_n$ and $\mathbf{a}_{n-1}$. We shall discuss such expressions in general in section 21.7 in some detail.

21.2.2 *Variational approaches.* In Chapter 3 we have indicated that as the alternative to the weighted residual approach it is often possible to use a variational functional whose stationarity corresponds to the correct differential equations. For equations discussed in the time domain such functionals can often be obtained. For instance, in the second order dynamic equation Hamiltons principle[8] could provide a starting basis. Gurtin[9, 10] establishes some other useful principles for both first and

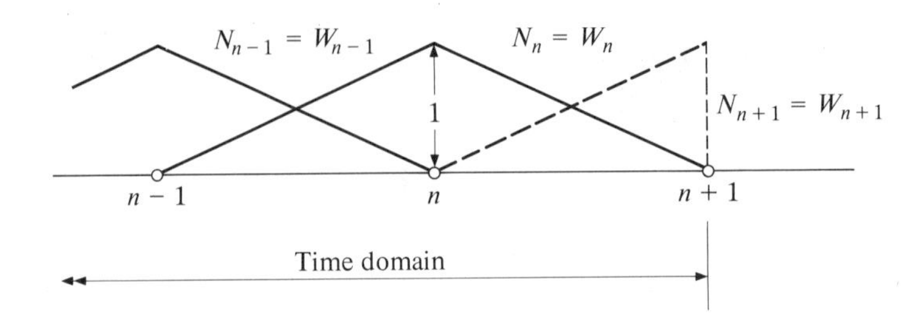

Fig. 21.2 A Galerkin scheme with linear shape functions applied to whole time domain

second order equations. Such variational functionals can be used to establish 'finite elements in time' and have been used as such by Wilson and Nickell,[11] Fried,[12] and Argyris.[13] However, we have shown in Chapter 3 that such variational principles are equivalent to the use of Galerkin procedures and will yield no new alternative numerical schemes. Zienkiewicz and Parekh[14] have used, for the first time, the finite element in time based on weighted residual forms.

One alternative which is not embedded explicitly in standard variational forms is that of using the least square procedure. In this (see Chapter 3) we write a functional minimizing the square of the error. In the context of one element representing the whole domain and a linear interpolation function of Eq. 21.6 we thus minimize

$$\Pi = \int_0^1 [\mathbf{C}(\mathbf{a}_n\dot{N}_n + \mathbf{a}_{n+1}\dot{N}_{n+1}) + \mathbf{K}(\mathbf{a}_n N_n + \mathbf{a}_{n+1}N_{n+1}) + \mathbf{f}]^{\mathrm{T}}[\mathbf{C}(\mathbf{a}_n\dot{N} + \cdots]\,\mathrm{d}\xi \tag{21.12}$$

with respect to the variable $\mathbf{a}_{n+1}$. The reader can verify that the final recurrence scheme now obtained is

$$\begin{aligned}&[\mathbf{C}^{\mathrm{T}}\mathbf{C}/\Delta t + (\mathbf{K}^{\mathrm{T}}\mathbf{C} + \mathbf{C}^{\mathrm{T}}\mathbf{K})/2 + \mathbf{K}^{\mathrm{T}}\mathbf{K}\Delta t/3]\,\mathbf{a}_{n+1} + \\ &[-\mathbf{C}^{\mathrm{T}}\mathbf{C}/\Delta t - (\mathbf{K}^{\mathrm{T}}\mathbf{C} - \mathbf{C}^{\mathrm{T}}\mathbf{K})/2 + \mathbf{K}^{\mathrm{T}}\mathbf{K}\,\Delta t/6]\,\mathbf{a}_n \\ &\qquad + \mathbf{K}^{\mathrm{T}}\int_0^1 \mathbf{f}\xi\,\mathrm{d}\xi/\Delta t^2 + \mathbf{C}^{\mathrm{T}}\int_0^1 \mathbf{f}\,\mathrm{d}\xi/\Delta t = 0.\end{aligned} \tag{21.13}$$

The above scheme obviously involves more computation but results in symmetric equations even if matrices **K** and **C** are not symmetric. Some experiments with such a scheme are given by Zienkiewicz and

Lewis[15] and used successfully by others.[16] To compare the performance of various schemes outlined in this section we shall consider a simple single variable equation with

$$\mathbf{K} = \mathbf{C} = 1; \qquad \mathbf{f} = 0$$

for an initial value of $\mathbf{a}_0 = 1$. In Fig. 21.3 we show results obtained using $\Delta t = 0{\cdot}5$ and $\Delta t = 0{\cdot}9$ for various schemes. The least square process shows

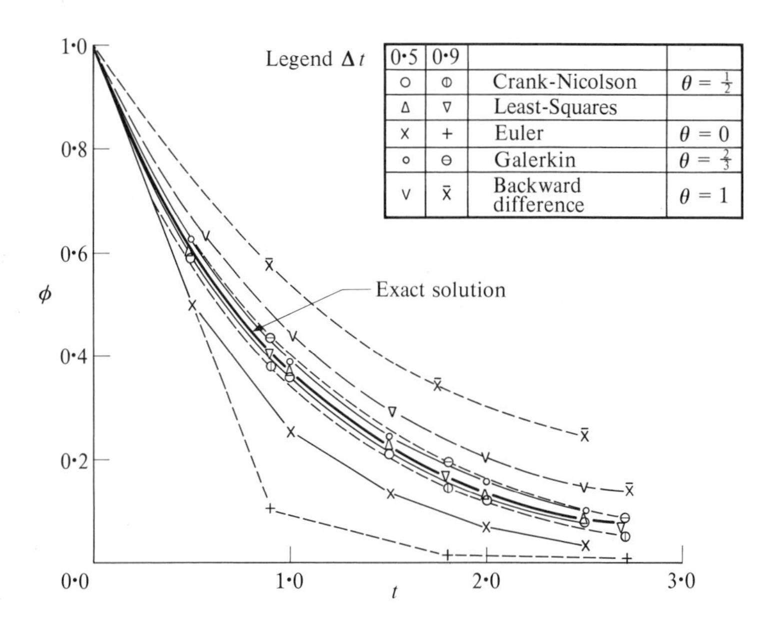

Fig. 21.3 Comparison of various time-stepping schemes on an initial value problem

by far the greatest accuracy, but with associated larger computational cost. Here in the same example the central difference schemes give 'second best' results and this scheme has been used most frequently in practical applications. As we see later, however, it oftens leads to oscillatory solutions which are troublesome, and for that reason the backward difference scheme ($\theta = 1$) is often preferred.

In Fig. 21.4 results for larger time steps $\Delta t = 1{\cdot}5$ and $2{\cdot}5$ are shown for the same problem. Here an interesting phenomenon occurs. The forward integration gives either oscillatory or divergent results while the other schemes still retain an acceptable accuracy. The behaviour shown here is that of instability, which we shall discuss in the next section.

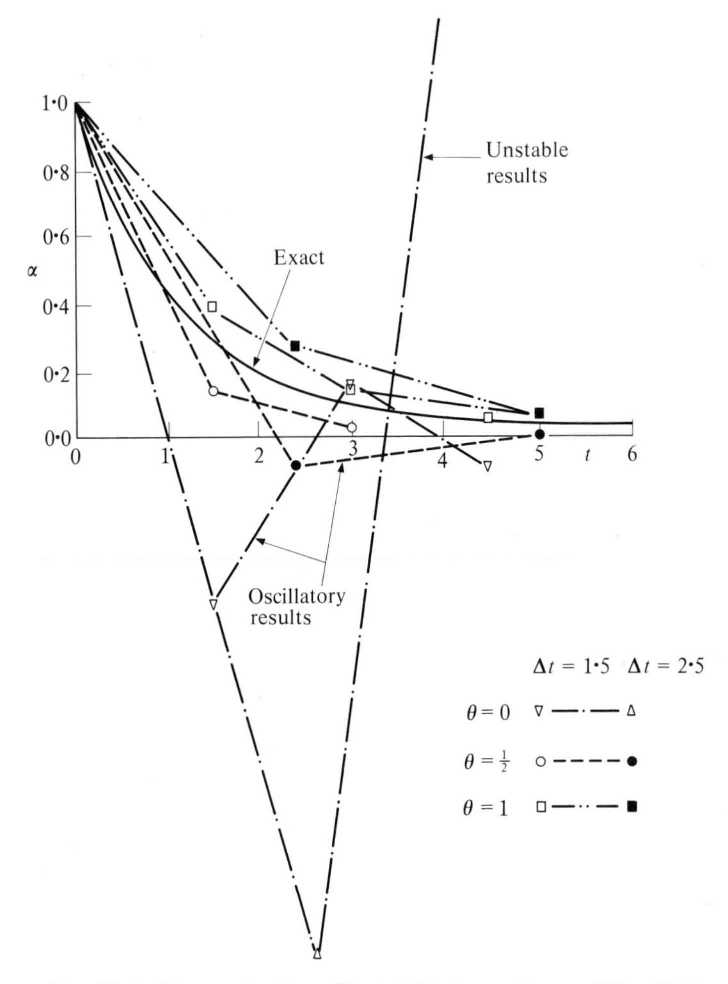

Fig. 21.4 Demonstration of instability in problems of Fig. 20.3

21.3 **Oscillation and Instability for Two Point Reccurence Schemes**

As we have remarked in the introduction, for stability considerations it will be convenient to consider the system of decoupled equations in terms of the modal participation variables y_i, i.e., to deal with a set of independent scalar equations of the form (21.4) for the first order equation

$$c_i \dot{y}_i + k_i y_i + f_i = 0. \tag{21.4}$$

Clearly as any response can be written as a sum of such modes we shall be able to synthesize the total response for a multi-degree-of-freedom system with *eigenvectors* $\bar{\mathbf{a}}_i$ as

$$\mathbf{a} = [\bar{\mathbf{a}}_1, \bar{\mathbf{a}}_2, \ldots] \begin{Bmatrix} y_1 \\ y_2 \\ \vdots \end{Bmatrix}. \tag{21.14}$$

The recurrence relations written previously are obviously applicable here and we shall consider the general recurrence relation of the form (21.9). For *free response*, i.e., with the forcing term f_i set to zero, we can write this recurrence as

$$(c_i/\Delta t + k_i\theta)\,(y_i)_{n+1} + (-c_i/\Delta t + k_i(1-\theta))\,(y_i)_n = 0 \tag{21.15}$$

or writing

$$(y_i)_{n+1} = (y_i)_n\,\lambda \tag{21.16}$$

we have

$$\lambda\,(c_i/\Delta t + k_i\theta) + (-c_i/\Delta t + k_i(1-\theta)) = 0. \tag{21.17}$$

Immediately we see that an 'unbounded' response will arise if $|\lambda| > 1$ and the problem will become unstable, for real solutions a damped behaviour being generally noted. Further, if $\lambda < 0$ an oscillating behaviour will arise which clearly does not represent the required solution well.

Equation (21.16), known as the *characteristic equation of the recurrent scheme*, gives

$$\lambda = \frac{1 - k_i(1-\theta)\,\Delta t/c_i}{1 + k_i\theta\,\Delta t/c_i}. \tag{21.18}$$

We see immediately that if we require

$$|\lambda| < 1$$

the right-hand side of Eq. (21.18) has to be greater than -1. Noting that $k_i/c_i \equiv \omega_i$ (viz Chapter 20, sect. 20.9.3), the eigenvalue corresponding to a mode i, we can write this requirement as

$$1 - \omega_i\,\Delta t(1-\theta) > -1 - \omega_i\,\Delta t\theta$$

or

$$\omega_i\,\Delta t(2\theta - 1) > -2. \tag{21.19}$$

This condition is always satisfied for $\theta \geqslant 1/2$ and schemes of such kind are *unconditionally stable*. When $0 < \theta < 1/2$ stability is conditional, requiring

$$\omega_i\,\Delta t < \frac{2}{1-2\theta}. \tag{21.20}$$

This, in the example of Fig. 21.4, gives the limit of the time step as $\omega_i \Delta t \leqslant 2$ when the forward difference scheme ($\theta = 0$) is used and the violation of this condition has resulted in divergent numbers shown.

The study of the characteristic value λ gives us, however, more information. In Fig. 21.5 we show how λ varies with $\omega_i \Delta t$ for some of the time-stepping schemes discussed. We note that for most of these λ becomes negative with a large time step, and in fact the backward difference scheme with $\theta=1$ is the only one not showing this property.

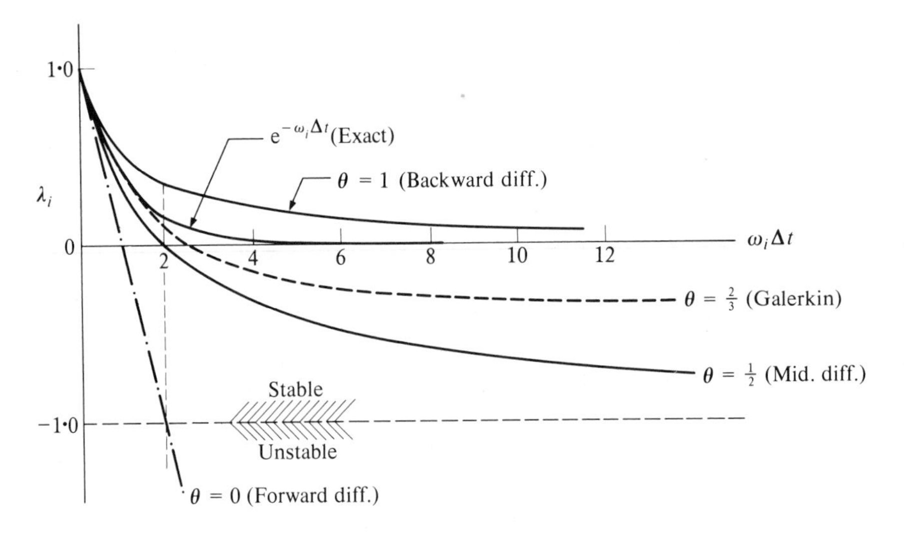

Fig. 21.5 Variation of λ with $W_i \Delta t$ for various recurrence schemes

This means that for any $\theta < 1$ oscillatory results can be given. Although mathematically it can be shown that the error for the mid-difference scheme with $\theta = 1/2$ is of a higher order, this scheme can give less accurate results than the backward scheme for larger time steps, as shown in Fig. 21.4.

The time response of a multi-degree-of-freedom system is a combination of all modal responses. Fortunately, the major part of the response corresponds generally to the modes with low eigenvalues ω_i. When an unconditionally stable scheme is used for a multi-degree-of-freedom system we will usually involve time steps much in excess of those corresponding to good accuracy for higher frequencies ω_i, and with the value of λ for the unconditionally stable schemes having a negative value, oscillations may not damp out as they should in practice. This is particularly noticeable with the Crank–Nicolson $\theta = \frac{1}{2}$ scheme and numerical devices have to be used to eliminate such spurious oscillations.[16, 17] All such devices consist in essence of averaging successive time steps of computation, and we should note that the most severe input starting such oscillations occurs at abrupt changes of the forcing term. Such abrupt changes should therefore be avoided or particular care taken to deal with them. We shall discuss this matter in the next section.

An alternative procedure which helps to iron out the oscillation is the use of $\theta > \frac{1}{2}$. In this context a useful compromise has been found in the Galerkin scheme with $\theta = \frac{2}{3}$ and has been shown to be very useful in practice with almost all oscillation and errors avoided.† In Table 21.1[7] we show the results for a one-dimensional finite element problem where a bar at uniform initial temperature is subjected suddenly to zero temperatures applied at its ends. Here ten linear elements are used in the space dimension with $L = 1$. The oscillational errors occurring with $\theta = \frac{1}{2}$ is much reduced for $\theta = \frac{2}{3}$. The time step used here is much longer than that corresponding to the lowest eigen period and the main cause of the oscillation is in the abrupt discontinuity of temperature change.

TABLE 21.1
PERCENTAGE ERRORS FOR FINITE ELEMENTS IN TIME ($\theta = \frac{2}{3}$) AND CRANK–NICOLSON ($\theta = \frac{1}{2}$) SCHEME; $\Delta t = 0{\cdot}01$.

t	$x = 0{\cdot}1$		$x = 0{\cdot}2$		$x = 0{\cdot}3$		$x = 0{\cdot}4$		$x = 0{\cdot}5$	
	$\theta = \frac{2}{3}$	$\theta = \frac{1}{2}$	$\theta = \frac{2}{3}$	$\theta = \frac{1}{2}$	$\theta = \frac{2}{3}$	$\theta = \frac{1}{2}$	$\theta = \frac{2}{3}$	$\theta = \frac{1}{2}$	$\theta = \frac{2}{3}$	$\theta = \frac{1}{2}$
0·01	10·8	28·2	1·6	3·2	0·5	0·7	0·6	0·1	0·5	0·2
0·02	0·5	3·5	2·1	9·5	0·1	0·0	0·5	0·7	0·7	0·4
0·03	1·3	9·9	0·5	0·7	0·8	3·1	0·5	0·2	0·5	0·6
0·05	0·5	4·5	0·4	0·2	0·5	2·3	0·4	0·8	0·5	1·0
0·10	0·1	1·4	0·1	2·0	0·1	1·5	0·1	1·9	0·1	1·6
0·15	0·3	2·2	0·3	2·1	0·3	2·2	0·3	2·1	0·3	2·2
0·20	0·6	2·6	0·6	2·6	0·6	2·6	0·6	2·6	0·6	2·6
0·30	1·4	3·5	1·4	3·5	1·4	3·5	1·4	3·5	1·4	3·5

Conductivity = specific heat = 1

Conditionally stable schemes, and in particular the explicit scheme, with $0 \leqslant \theta \leqslant \frac{1}{2}$, will always result in instability as long as the time interval used in the calculation exceeds that specified by the highest eigenvalue of the system. This is a serious limitation of the use of explicit time schemes, nevertheless the computation of time saving with no matrix inversion (or repeated solution) being required compensates often for the need of using many small time steps.

† Lambert[2] suggests, by different reasoning, $\theta = 0{\cdot}878 \approx \frac{7}{8}$ as optimal.

The estimate of the critical time step for conditionally stable schemes apparently necessitates the solution of the eigenvalue problem for the whole system. This is not so. The bound on the highest eigenvalue can be simply obtained by the consideration of an individual element. This is established by an important theorem proposed by Irons[18] which proves that the highest system eigenvalue must always be less than the highest eigenvalue of the individual elements. This allows a very easy estimate of critical time steps which (by the above theorem) will err on the safe side.†

We have not discussed the stability conditions for either the three-point scheme given by Eq. (21.11) or the recurrence relation obtained from the least square approximation given by Eq. (21.13), and we leave these as exercises for the reader to perform.

The 'lumping' or diagonalization of the matrix **C** (which is generally banded) follows identical procedures to those described in Chapter 20, section 20.2.4 for mass matrices **M**.

21.4 Accuracy and Initial Conditions

We have stated that in the first order problems discussed here the recurrence starts from some known initial conditions which are specified. However, as in the problem of a rapid change of boundary conditions discussed previously, such initial conditions may cause an abrupt discontinuity to appear at the start and an alternative formulation is physically preferable. In this we start the whole problem from a solution which represents the system at rest, i.e., in which $\mathbf{a}$ = const., and the change of the initial conditions or of any suddenly applied forcing terms can then be considered as a disturbance on this steady system. The initial conditions therefore correspond to those of the steady state and are given *for all times prior to a certain instant.*

If we consider as an example the simple initial value problem on which we have based Figs. 21.3 and 21.4, then the problem can be restated as that requiring a solution of an equation

$$\dot{a}+a+f(t) = 0 \tag{21.21}$$

which is valid in the whole time domain, and where

$$f(t) = -1 \qquad (t<0) \qquad a = 1$$

$$f(t) = 0 \qquad (t \geqslant 0).$$

† It is of importance to observe here that for the first order problems of the type of Eq. (21.2) the element eigenvalues increase as h^2 (h = element size). This means a very rapid decrease of stable Δt with mesh subdivision and explicit schemes are therefore seldom used for such problems. This is not the case with dynamic problems when element eigenvalues are proportioned to h only.

The resulting discontinuity in the new forcing term is shown in Fig. 21.6. Clearly such a discontinuity should be 'smoothed' and this can be done in many ways such as, for instance, placing an interval which straddles this discontinuity as shown. In the same figure we show the numerical results obtained in which such smoothing was applied for the very large time interval $\Delta t = 2{\cdot}5$. The reader will note that the smoothed solution does not show the violent oscillations previously noted, and is more physically acceptable.

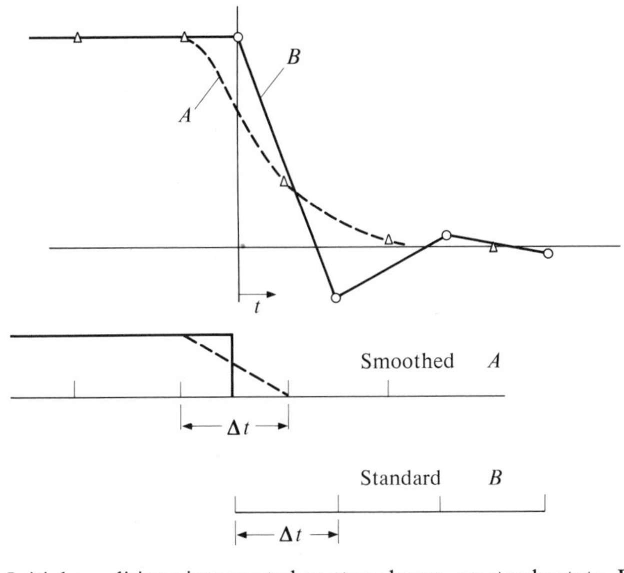

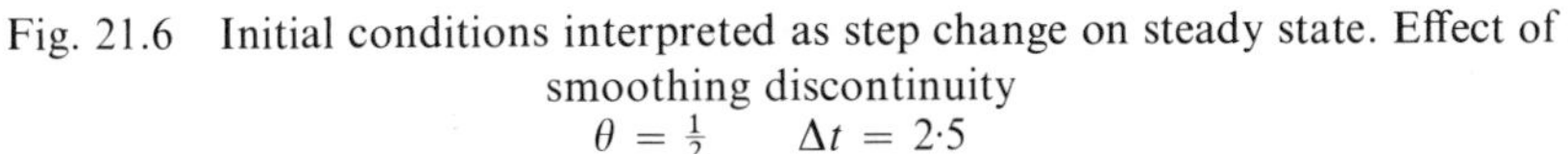

Fig. 21.6 Initial conditions interpreted as step change on steady state. Effect of smoothing discontinuity
$\theta = \frac{1}{2} \qquad \Delta t = 2{\cdot}5$

The process of starting from known steady solutions for $t<0$ and smoothing out any discontinuities will simplify the computation for multi-step schemes as it *implies the knowledge of all the vectors* $\mathbf{a}$ *at times prior to the initial instant*. We have now shown that this simplification does, in fact, lead to more realistic answers and we shall invariably apply it both in first and second order equations.

21.5 **Three-point Recurrence Schemes for Second Order Equations**

The procedures of section 21.2 can be applied directly to the second order equation (21.1) but the shape functions $N_i(t)$ have now to be at least of second order (parabolic) in t as a second order derivative has to be represented. We need now, thus, a minimum of three sets of $\mathbf{a}_i$ to describe approximately the variation of this function. Once again we can write

$$\mathbf{a} \approx \hat{\mathbf{a}} = \sum N_i \mathbf{a}_i \tag{21.22}$$

and use for $i = n, n+1, n-1$.

The shape functions normalized to an interval $2\Delta t$, i.e., with $-1<\xi<1$, are the standard parabolic ones shown in Fig. 21.7, and can be written as (see Chapter 7)

$$\begin{aligned} \xi &= t/\Delta t \\ N_{n+1} &= \xi(1+\xi)/2 \\ N_n &= (1-\xi)(1+\xi) \\ N_{n-1} &= -\xi(1-\xi)/2. \end{aligned} \tag{21.23}$$

The time derivatives of these functions are given as

$$\begin{aligned} \dot{N}_{n+1} &= (\tfrac{1}{2}+\xi)/\Delta t & \ddot{N}_{n+1} &= 1/\Delta t^2 \\ \dot{N}_n &= -2\xi/\Delta t & \ddot{N}_n &= -2/\Delta t^2 \\ \dot{N}_{n-1} &= (-\tfrac{1}{2}+\xi)/\Delta t & \ddot{N}_{n-1} &= 1/\Delta t^2 \end{aligned} \tag{21.24}$$

We shall now use as a 'domain' of time the period of the size of the element $2\Delta t$ and assume that $\mathbf{a}_n$ and $\mathbf{a}_{n-1}$ are prescribed, thus leaving $\mathbf{a}_{n+1}$ to be determined. The assumption of two known values in fact corresponds precisely to the requirement that two initial conditions be known in order to solve the second order equation system. Thus we write a single weighted residual equation of the form

$$\int_{-1}^{1} W_j[\mathbf{M}(\mathbf{a}_{n-1}\ddot{N}_{n-1}+\mathbf{a}_n\ddot{N}_n+\mathbf{a}_{n+1}\ddot{N}_{n+1})+\mathbf{C}(\mathbf{a}_{n-1}\dot{N}_{n-1}+\mathbf{a}_n\dot{N}_n$$
$$+\mathbf{a}_{n+1}\dot{N}_{n+1})+\mathbf{K}(\mathbf{a}_{n-1}N_{n-1}+\mathbf{a}_nN_n+\mathbf{a}_{n+1}N_{n+1})+\mathbf{f}]\,d\xi = 0 \qquad (j=1). \tag{21.25}$$

In a similar analysis to that used in deriving Eq. (21.8) we find that the above can be written as

$$\begin{aligned} &[\mathbf{M}+\gamma\,\Delta t\mathbf{C}+\beta\,\Delta t^2\mathbf{K}]\,\mathbf{a}_{n+1} \\ &\quad +[-2\mathbf{M}+(1-2\gamma)\,\Delta t\mathbf{C}+(\tfrac{1}{2}-2\beta+\gamma)\,\Delta t^2\mathbf{K}]\,\mathbf{a}_n \\ &\quad +[\mathbf{M}-(1-\gamma)\,\Delta t\mathbf{C}+(\tfrac{1}{2}+\beta-\gamma)\,\Delta t^2\mathbf{K}]\mathbf{Q}_{n-1}+\bar{\mathbf{f}}\,\Delta t^2 = 0 \end{aligned} \tag{21.26}$$

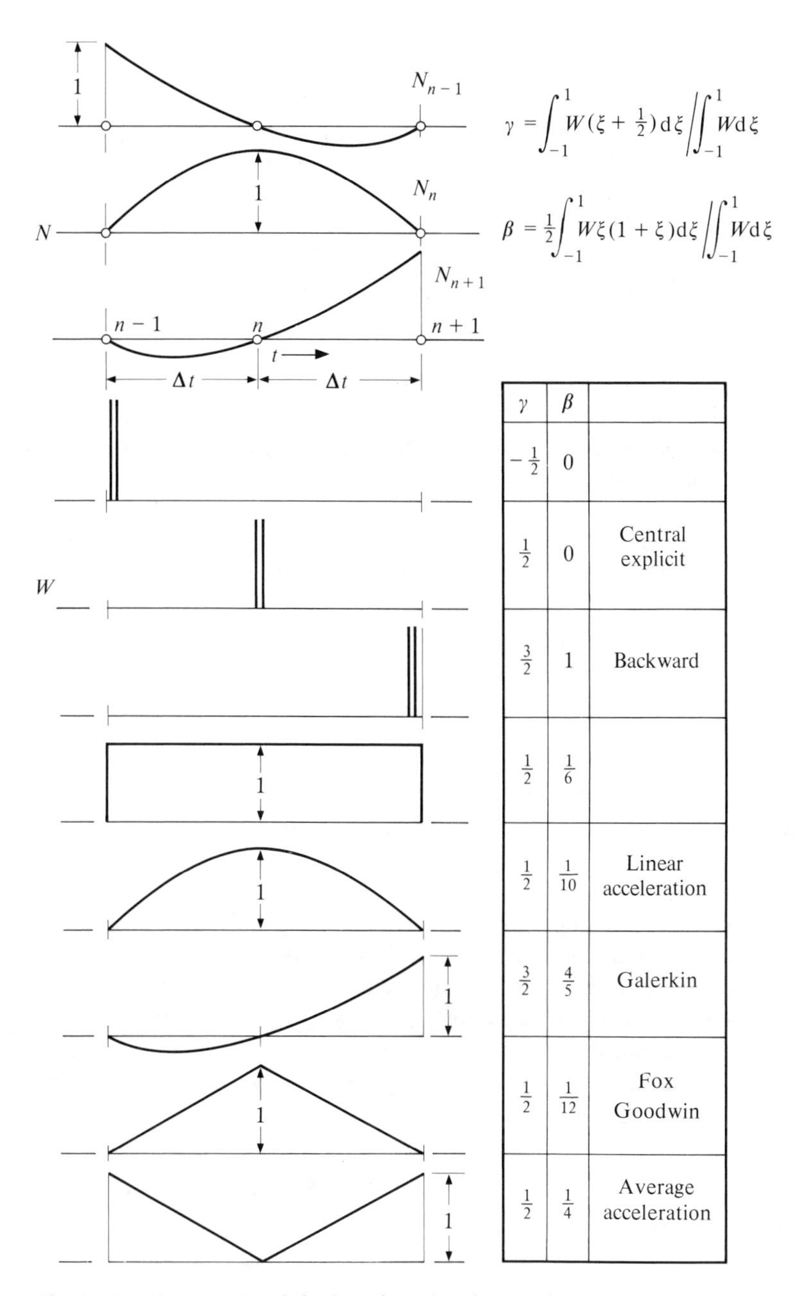

γ	β	
$-\frac{1}{2}$	0	
$\frac{1}{2}$	0	Central explicit
$\frac{3}{2}$	1	Backward
$\frac{1}{2}$	$\frac{1}{6}$	
$\frac{1}{2}$	$\frac{1}{10}$	Linear acceleration
$\frac{3}{2}$	$\frac{4}{5}$	Galerkin
$\frac{1}{2}$	$\frac{1}{12}$	Fox Goodwin
$\frac{1}{2}$	$\frac{1}{4}$	Average acceleration

Fig. 21.7 Shape and weight functions for three point reccurence formulae

in which

$$\gamma = \int_{-1}^{1} W_j(\xi + \tfrac{1}{2})\, \mathrm{d}\xi \Big/ \int_{-1}^{1} W_j\, \mathrm{d}\xi$$

$$\beta = \int_{-1}^{1} W_i\, \tfrac{1}{2}\xi(1+\xi)\, \mathrm{d}\xi \Big/ \int_{-1}^{1} W_j\, \mathrm{d}\xi \qquad (21.27\text{a})$$

$$\bar{\mathbf{f}} = \int_{-1}^{1} W_j \mathbf{f}\, \mathrm{d}\xi \Big/ \int_{-1}^{1} W_j\, \mathrm{d}\xi$$

If the same interpolation is used for **f** as implied for **a**, then

$$\bar{\mathbf{f}} = \mathbf{f}_{n+1}\beta + \mathbf{f}_n(\tfrac{1}{2} - 2\beta + \gamma) + \mathbf{f}_{n-1}(\tfrac{1}{2} + \beta - \gamma). \qquad (21.27\text{b})$$

Expression (21.26) is in fact a general algorithm corresponding to that derived by Newmark[19] in a completely different manner and is one of the best known recurrence relationships† (albeit not usually stated explicitly in the above form).[20]

Once again various forms of weighting function can be used ranging from point collocation to Galerkin forms. In Fig. 21.7 we show the values of β and γ corresponding to a series of such weightings. Newmark recommends that $\gamma = \frac{1}{2}$ be generally taken. This corresponds to symmetric weighting functions of all forms.

Again we note that if $\beta = 0$ and the matrices **M** (and **C**) are diagonal (lumped), then no inversion is necessary to determine $\mathbf{a}_{n+1}$ and the scheme is *explicit*. One such scheme with $\gamma = \frac{1}{2}$ corresponds to the central difference formula which is extremely economical. However, just as in first order equations we shall find that stability here is conditional and the time interval Δt has to be suitably limited. Such schemes have proved extremely efficient for many linear and non-linear problems in practice and are discussed in references.[21–24]

As we have already remarked, two starting values $\mathbf{a}_n$ and $\mathbf{a}_{n-1}$ are necessary to commence calculation. Frequently these are specified by the initial displacements and the initial velocity. This is inconvenient and as recommended in the linear systems it is always an easy matter to translate a problem to that of an initially steady system to which a perturbation is applied. Alternatively, various starting schemes have to be used to predict two successive values of the vector $\mathbf{a}_i$.

21.6 Stability of Three-point Recurrence Schemes for Second Order Equations[25–29]

The stability of the schemes developed in the previous section can now be investigated using the procedures identical to those described in section

† β and γ correspond to precisely the same symbols introduced by Newmark.

21.3. First we note that the response can be studied for a series of uncoupled equations of the general form

$$m_i\ddot{y}_i + c_i\dot{y}_i + k_i y_i + f_i = 0. \tag{21.28}$$

We have remarked in the previous chapter that in general to achieve such uncoupling it is necessary to consider complex natural frequencies but if the **C** matrix is of a suitable form the mode decomposition can be accomplished using real eigenvalues of the free vibration

$$\omega_i^2 = k_i/m_i. \tag{21.29}$$

It is mainly with this situation that we shall be concerned although the arguments given below are quite general.

Once again we shall study the free response in which $f_i = 0$ and *assume a solution in the form of*

$$\begin{aligned}(y_i)_{n+1} &= \lambda(y_i)_n \\ (y_i)_n &= \lambda(y_i)_{n-1}.\end{aligned} \tag{21.30}$$

Substituting the above into the general recurrence relation (21.26) (now written for the single-degree-of-freedom system of Eq. (21.28)) we have a *characteristic equation*

$$\lambda^2[m_i + \gamma\,\Delta t c_i + \beta\,\Delta t^2 k_i] + \lambda[-2m_i + (1-2\gamma)\,\Delta t c_i + (\tfrac{1}{2} - 2\beta + \gamma)\,\Delta t^2 k_i] + [m_i - (1-\gamma)\,\Delta t c_i + (\tfrac{1}{2} + \beta - \gamma)\,\Delta t^2 k_i] = 0. \tag{21.31}$$

The roots of this equation can be found and will indicate the behaviour of the numerical solution and, in particular, show whether this grows in an unbounded manner which obviously is a physically unacceptable situation. One difference from the solutions discussed in section 21.3 is apparent. The roots of the quadratic will generally be complex; this indeed would be expected from the fact that the solution corresponds to one of a (damped) oscillation. However, we shall still require that the absolute value (modulus) of the complex root satisfies

$$|\lambda| \leqslant 1 \tag{21.32}$$

for the oscillation to remain bounded.

The general case of a damped oscillation can be studied in detail by the reader and here we shall consider answers only for the undamped case, i.e., putting the damping matrix $c_i = 0$. We know that for such cases the true solution must be of the form given below

$$y_i = \bar{y}_i\, e^{i\omega t} \tag{21.33}$$

thus establishing the exact value of λ which is

$$\lambda = \frac{(y_i)_{n+1}}{(y_i)_n} = \frac{e^{i\omega(t+\Delta t)}}{e^{i\omega t}} = e^{i\omega\,\Delta t} \tag{21.34}$$

This has an absolute value

$$|\lambda| = 1 \tag{21.35}$$

thus indicating an undamped persisting oscillation.

Any numerical scheme which produces $|\lambda|<1$ will give a stable but *artificially damped* solution. Considering now the solution of the characteristic equation, (Eq. (21.31)), in which

$$c_i = 0$$

$$p_i = \frac{k_i}{m_i}\Delta t^2 = \omega_i^2\,\Delta t^2$$

we have

$$\lambda^2[1+\beta p_i]+\lambda[-2+(\tfrac{1}{2}-2\beta+\gamma)\,p_i]+[1+(\tfrac{1}{2}+\beta-\gamma)\,p_i] = 0. \tag{21.36}$$

Writing

$$g = \frac{(\frac{1}{2}+\gamma)\,p_i}{1+\beta p_i}; \qquad h = \frac{(\frac{1}{2}-\gamma)\,p_i}{1+\beta p_i}$$

we can obtain the roots of Eq. (21.36) as

$$\lambda_{1,2} = \frac{(2-g)\pm\sqrt{(2-g)^2-4(1+h)}}{2}. \tag{21.37}$$

These roots will be complex if the quantity under the square root is negative, i.e., if

$$4(1+h)>(2-g)^2$$

or

$$p_i[4\beta-(\tfrac{1}{2}+\gamma)^2]>-4. \tag{21.38}$$

The modulus of λ is then

$$|\lambda| = \sqrt{1+h} \leqslant 1$$

which imposes the requirement

$$-1<h = \frac{(\frac{1}{2}-\gamma)\,p_i}{1+\beta p_i}<0 \tag{21.39}$$

Summarizing the results of Eqs. (21.38) and (21.39) we require therefore that

$$\begin{aligned} \beta &\geqslant \tfrac{1}{4}(\tfrac{1}{2}+\gamma)^2 \\ \gamma &\geqslant \tfrac{1}{2} \\ \tfrac{1}{2}-\gamma+\beta &\geqslant 0 \end{aligned} \tag{21.40}$$

to achieve stability, i.e., for the stabillty to be unconditional for all values of p_i.

If the above conditions are not satisfied, stability can still be achieved providing

$$p_i < p_{\text{crit}}$$

$$\Delta t < \Delta t_{\text{crit}}$$

where the critical value can be again found from Eq. (21.38).

Only two of the schemes shown in Fig. 21.7 are unconditionally stable ($\gamma = \frac{3}{2}$, $\beta = 1$ and $\gamma = \frac{1}{2}$, $\beta = \frac{1}{4}$) and for all the others the stability is conditional. As remarked by Newmark, all schemes for which

$$\gamma = \tfrac{1}{2}, \qquad \beta \geqslant \tfrac{1}{4}$$

are unconditionally stable and indeed show no artificial damping, i.e., $|\lambda| = 1$. For conditionally stable schemes such as the central difference expression a simple calculation from Eq. (21.38) shows that for $\gamma = \frac{1}{2}$, $\beta = 0$ we require for stability

$$p_i < 4$$

or

$$\omega_i \, \Delta t < 2$$

putting $\omega_i = 2\pi/T_i$, where T_i is the period we can write the stability condition as

$$\Delta t \leqslant \frac{1}{\pi} T_i. \tag{21.41}$$

An alternative but popular scheme with

$$\gamma = \tfrac{1}{2}, \qquad \beta = \tfrac{1}{6}$$

requires for stability

$$\Delta t \leqslant \frac{\sqrt{3}}{\pi} T_i$$

as can easily be verified by the reader.

All schemes with $\gamma > \frac{1}{2}$ show an appreciable numerical damping leading to an inaccuracy of results. In Fig. 21.8 we show a plot of $|\lambda|$ against $\Delta t/T_i$ for various such schemes, showing how large this damping can be for larger values of Δt.

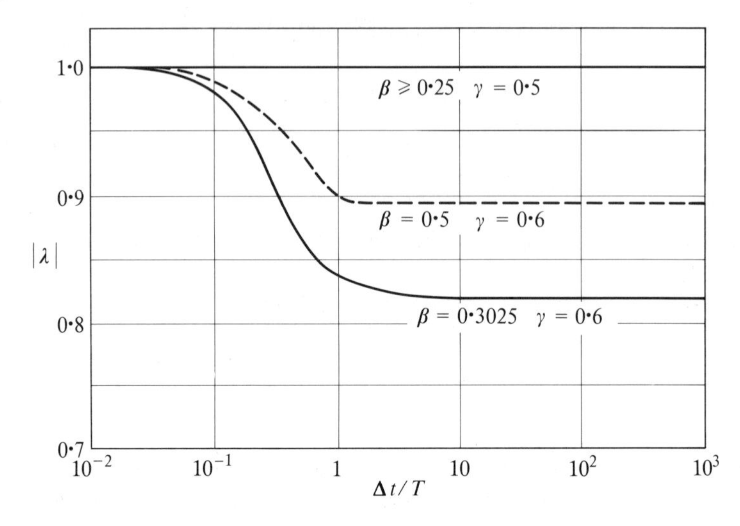

Fig. 21.8 Three level (Newmark) expression. 'Spectral' radius, $|\lambda|$ versus $\Delta t/T$

In practice some degree of numerical damping is desirable as invariably Δt used in unconditionally stable schemes will be much larger than that corresponding to the highest frequencies of the system which cannot be reproduced exactly and only lead to numerical noise. Ideally, therefore, we will seek schemes which have $|\lambda| = 1$ for $\Delta t/T < 1$ and show as much damping as possible when $\Delta t/T > 1$.

One way of obtaining such performance is to include, deliberately, a damping matrix of a form

$$\mathbf{C}_a = \varepsilon \mathbf{K}\, \Delta t. \tag{21.42}$$

Obviously as $\Delta t \to 0$ this damping will disappear and when $\Delta t \to \infty$ it will have a large value.

Such artificial 'adjustment' of numerical performance is of some interest and was used by Grant[30] to adjust the performance of an explicit $\gamma = \beta = 0$ Newmark scheme. A more sophisticated use is developed by Hilber *et al.*[31] (and we shall return to these results later).

In addition to numerical damping, the value of λ obtained from the characteristic equation (21.31) will, by comparison with the exact λ of Eq. (21.34), give information on the *relative period error*.

Results are shown in Fig. 21.9 for the various schemes of second order.

For a very full discussion of stability and performance of various schemes the reader can consult references 25–31.

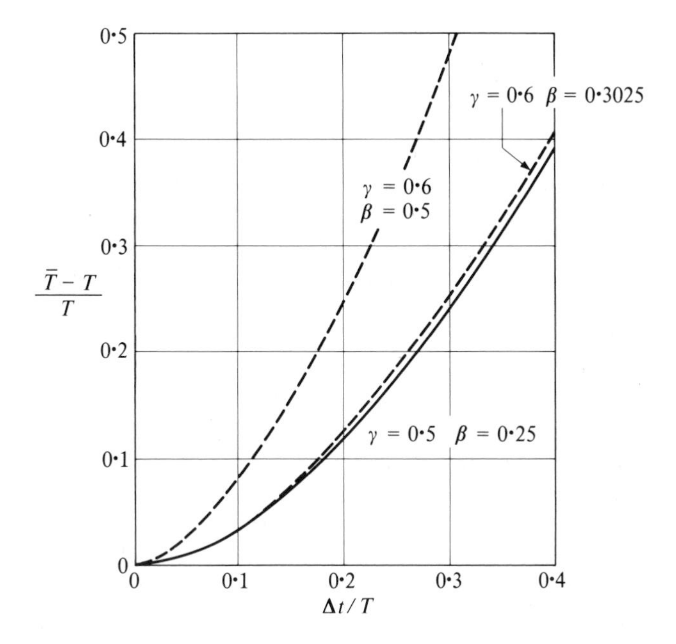

Fig. 21.9 Three level (Newmark) expression. Relative period error versus $\Delta t/T$

21.7 **Multipoint Recurrence Schemes**

21.7.1 *General.* The procedures for deriving recurrence schemes by the weighting process can be extended to domains of $n\ \Delta t$ using higher order interpolation, and indeed to the whole domain using a set of elements as we have shown in section 21.2 when formula (21.11) was derived. In this section we shall consider domains of three points for the first order equation (presuming the determination of $\mathbf{a}_{n+1}$ from known values of $\mathbf{a}_n$ and $\mathbf{a}_{n-1}$) in which parabolic interpolation is used and four-point recurrence schemes for the second order equation (determining $\mathbf{a}_{n+1}$ from $\mathbf{a}_n$, $\mathbf{a}_{n-1}$, and $\mathbf{a}_{n-2}$) in which a cubic shape functipn determines the variation over the time domain.

21.7.2 *Three-point schemes—first order equation.* The first set of expressions is in fact already given by the derivation of section 21.5. Indeed, if we put the second order term $\mathbf{M} = 0$, Eq. (21.26) will immediately result in

$$[\gamma\mathbf{C}+\beta\ \Delta t\mathbf{K}]\ \mathbf{a}_{n+1}+[(1-2\gamma)\mathbf{C}+(\tfrac{1}{2}-2\beta+\gamma)\Delta t\mathbf{K}]\ \mathbf{a}_n$$
$$+[-(1-\gamma)\mathbf{C}+(\tfrac{1}{2}+\beta-\gamma)\Delta t\mathbf{K}]\ \mathbf{a}_{n-1}+\bar{\mathbf{f}} = 0 \qquad (21.43)$$

in which $\bar{\mathbf{f}}$ is given by previous interpolation expression (21.27b) with β and γ having the same meaning as before. $\gamma = \frac{1}{2}$ and $\beta = \frac{1}{3}$ give an algorithm due to Lees[32] much used in practice for non-linear problems.[33] $\gamma = \frac{1}{2}$ and $\beta = \frac{1}{6}$ result from subdomain collocation (uniform $W_i = 1$) of Fig. 21.7. (This incidentally corresponds exactly to the recurrence formula (21.10) derived by considering linear interpolation over the whole time domain.) Once again the stability can be investigated as in section 21.6. Dahlquist[34] discusses the problem and establishes the following conditions for unconditional stability

$$\gamma \geqslant \tfrac{1}{2} \qquad \beta \geqslant \tfrac{1}{2}\gamma. \tag{21.44}$$

Conditional stability can once again be achieved in certain circumstances. However, it is, interesting to note that the central difference scheme with $\gamma = \frac{1}{2}$ and $\beta = 0$ is now unconditionally unstable and cannot be used for any time interval.

21.7.3 *Four-point Schemes for second and first order equations.* Here the extension of the weighting procedure is obvious and can be done as an exercise by the reader. Using the time domain of $3\Delta t$, as shown in Fig. 21.10, and normalizing the co-ordinate as $\xi = 2t/3\Delta t$ the standard cubic shape functions can be used to obtain the expansion of

$$\mathbf{a} \approx \hat{\mathbf{a}} = \sum N_i a_i \qquad (i = n-2, n-1, n, n+1) \tag{21.45}$$

and a weighted residual form obtained. The reader is now invited to derive the explicit general form now given in terms of three independent parameters α, β, and γ, by using the previous established procedures.

This form is given as:

$$\begin{aligned}
&[\mathbf{M}(\gamma-1)+(\tfrac{1}{2}\beta-\gamma+\tfrac{1}{3})\mathbf{C}\,\Delta t+(\tfrac{1}{6}\alpha-\tfrac{1}{2}\beta+\tfrac{1}{3}\gamma)\mathbf{K}\,\Delta t^2]\mathbf{a}_{n+1}\\
&+[(-3\gamma+4)\mathbf{M}+(-\tfrac{3}{2}\beta+4\gamma-\tfrac{3}{2})\mathbf{C}\,\Delta t+(-\tfrac{1}{2}\alpha+2\beta-\tfrac{3}{2}\gamma)\mathbf{K}\,\Delta t^2]\mathbf{a}_n\\
&+[(3\gamma-5)\mathbf{M}+(\tfrac{3}{2}\beta-5\gamma+3)\mathbf{C}\,\Delta t+(\tfrac{1}{2}\alpha-\tfrac{5}{2}\beta+3\gamma)\mathbf{K}\,\Delta t^2]\mathbf{a}_{n-1}\\
&+[(-\gamma+2)\mathbf{M}+(-\tfrac{1}{2}\beta+2\gamma-\tfrac{11}{6})\mathbf{C}\,\Delta t+(-\tfrac{1}{6}\alpha+\beta-\tfrac{11}{6}\gamma+1)\mathbf{K}\,\Delta t^2]\mathbf{a}_{n-2}\\
&+(\tfrac{1}{6}\alpha-\tfrac{1}{2}\beta+\tfrac{1}{3}\gamma)\mathbf{f}_{n+1}\,\Delta t^2+(-\tfrac{1}{2}\alpha+2\beta-\tfrac{3}{2}\gamma)\mathbf{f}_n\,\Delta t^2\\
&+(\tfrac{1}{2}\alpha-\tfrac{5}{2}\beta+3\gamma)\mathbf{f}_{n-1}\,\Delta t^2+(-\tfrac{1}{6}\alpha+\beta-\tfrac{11}{6}\gamma+1)\mathbf{f}_{n-2}\,\Delta t^2 = 0
\end{aligned} \tag{21.46}$$

$$\begin{aligned}
\alpha &= \int_0^3 W\xi^3\,\mathrm{d}\xi \Big/ \int_0^3 W\,\mathrm{d}\xi\\
\beta &= \int_0^3 W\xi^2\,\mathrm{d}\xi \Big/ \int_0^3 W\,\mathrm{d}\xi\\
\gamma &= \int_0^3 W\xi\,\mathrm{d}\xi \Big/ \int_0^3 W\,\mathrm{d}\xi
\end{aligned} \tag{21.47}$$

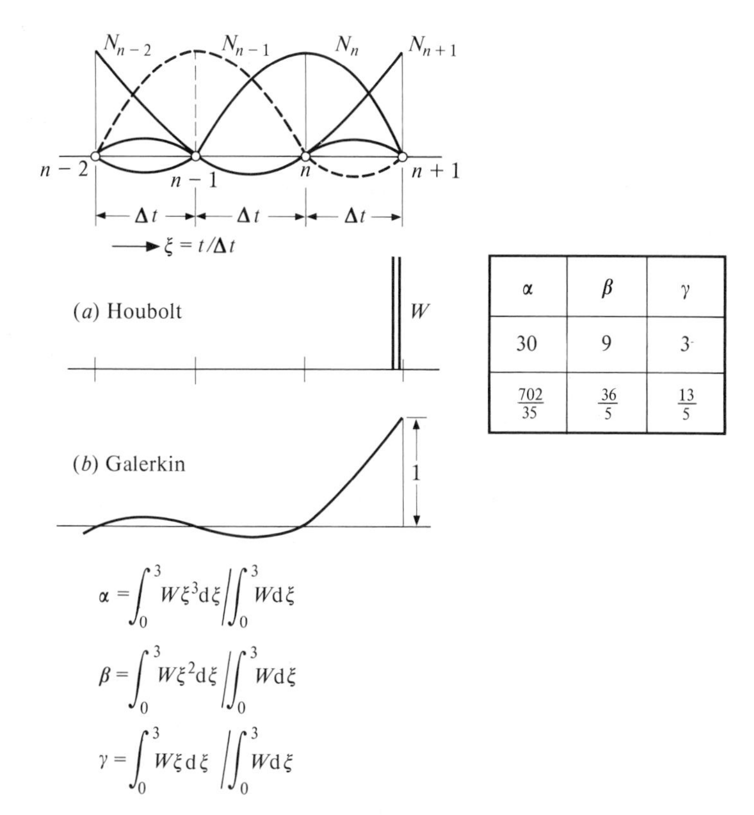

α	β	γ
30	9	3
$\frac{702}{35}$	$\frac{36}{5}$	$\frac{13}{5}$

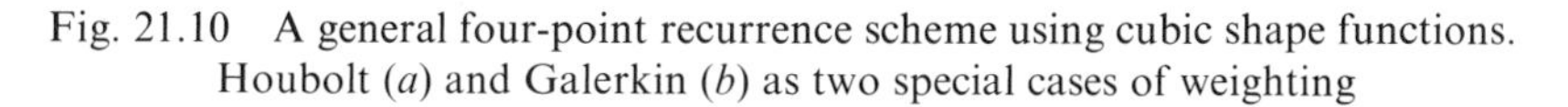
Fig. 21.10 A general four-point recurrence scheme using cubic shape functions. Houbolt (a) and Galerkin (b) as two special cases of weighting

The general expression given here was derived by the author[35] and its general stability properties are discussed in detail by Wood.[36] Just as is the case of the previous three-point generalized Newmark expression given by Eq. (21.26), some of the permutations of the three constants result in already known and widely used expressions.

With $\alpha = 30$, $\beta = 9$, and $\gamma = 3$ the Houbolt[37] formula is recognized. This is unconditionally stable and has characteristics shown in Fig. 21.11 and 21.12 which the reader can compare with Figs. 21.8 and 21.9 for the Newmark schemes.

With

$$\begin{aligned} \alpha &= 2+4\theta+3\theta^2+\theta^3 \\ \beta &= \tfrac{4}{3}+2\theta+\theta^2 \\ \gamma &= 1+\theta \end{aligned} \tag{21.48}$$

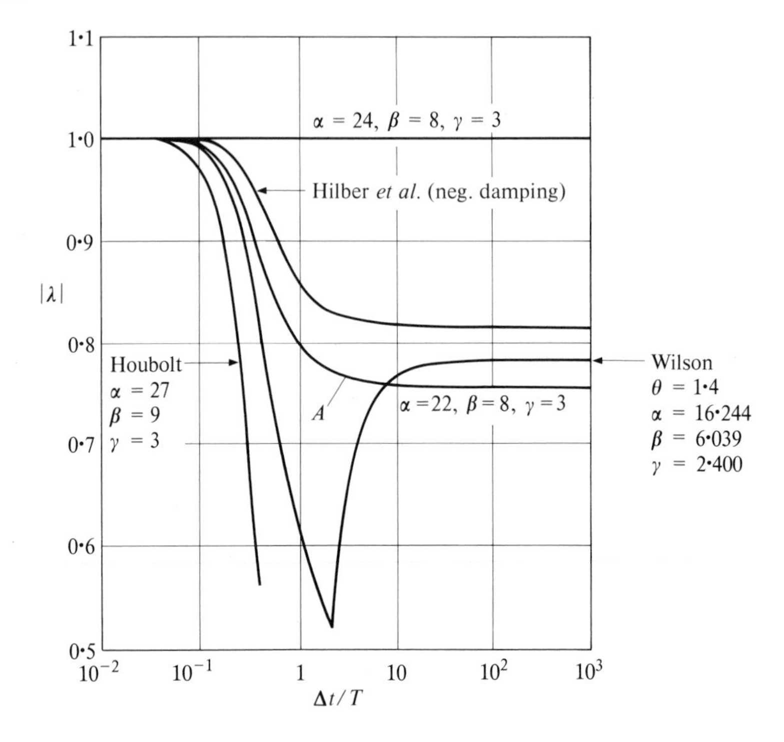

Fig. 21.11 Four level time scheme. Spectral radius $|\lambda|$ for some combinations of α, β and γ.

the so-called Wilson-θ procedure[24, 38] is obtained. This was introduced to obtain improved high frequency mode elimination. $\theta = 1{\cdot}4$–$2{\cdot}0$ have frequently been used here and it can be shown that for unconditional stability $\theta > 1{\cdot}366$ is required.[36] It is of interest to remark that $\theta = 1$ reduces the four-point algorithm in the undamped case to that of Newmark ($\gamma = \frac{1}{2}$; $\beta = \frac{1}{6}$).

The algorithms proposed by Hilber *et al.*[31] once again can be identified with the general four-level schemes with the introduction of artificial (negative or positive) damping of expression (21.42).

Clearly the general expression (21.46) possesses many other possibilities not yet fully tested in practice. For suitable choices of α. β, γ explicit expressions can be derived.

The general stability conditions for the undamped equations are not as simple as with the three-point expression and can be summarized as a requirement that[36]

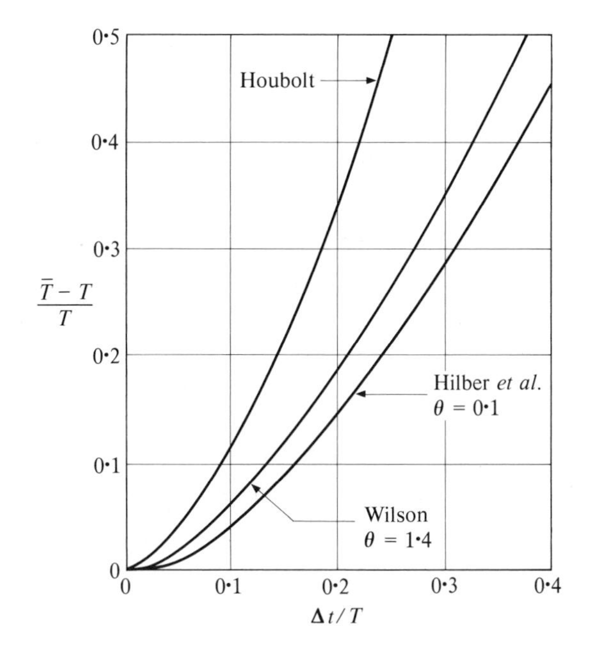

Fig. 21.12 Four level time scheme. Relative period error for some combinations α, β and γ

$$
\begin{aligned}
\tfrac{3}{2} < \gamma &\leqslant \beta/3 + \tfrac{1}{2} \\
\tfrac{3}{4} + 9\beta/2 - 5\gamma < \alpha &\leqslant -9\gamma^2 + 3\beta\gamma + 13\gamma - 6
\end{aligned}
\qquad (21.49)
$$

When equality exists in the last relation then the algorithm shows no artificial damping. The performance of one, previously unnamed, scheme with $\alpha = 22, \beta = 8$ and $\gamma = 3$ is shown to be very satisfactory in Fig. 21.11.

Putting $M = 0$ a series of four-point expressions for first order problems can be obtained. Formulae of a somewhat similar type have been discussed by Cryer.[39]

21.8 **Alternative Derivation of Recurrence Formulae**

In this chapter we have shown that many of the widely available finite difference expressions for recurrence schemes are but particular cases of a weighted residual finite element approach in the time domain. However, alternative possibilities for derivation of recurrence schemes are discussed in many texts dealing with the subject. Some of these are derived by completely different reasoning. In particular use can be made of the fact that for single-degree-of-freedom linear systems the exact solution is known. For instance, in the first order scalar equation (Eq. (21.4)) we know that

$$\lambda = e^{-\omega_i \Delta t} \qquad \omega_i = k_i/c_i \tag{20.50}$$

and in the second order equation with no damping we have already established, as shown in Eq. (21.34), that

$$\lambda = e^{i\omega \Delta t}. \tag{21.51}$$

Clearly the use of such exact values will be only applicable to decoupled modes (see, for instance, Fig. 21.5 in which the exact λ is contrasted with approximations derived previously). In a general case we can, however, approximate λ by rational matrix expansions. For instance, in the general case of the first order of equations we can write

$$\lambda = e^{-\mathbf{C}^{-1}\mathbf{K}\,\Delta t} \approx [1+\beta_1\mathbf{C}^{-1}\mathbf{K}+\beta_2(\mathbf{C}^{-1}\mathbf{K})^2+\cdots]^{-1} \times [1+\alpha_1\mathbf{C}^{-1}\mathbf{K}+\alpha_2(\mathbf{C}^{-1}\mathbf{K})^2+\cdots] \tag{21.52}$$

and seek α and β coefficients which will give the best approximation to a series expansion. Some of the formulae thus obtained will be found identical to those already established, but other possibilities are now offered.

Such approximations are known by the name of Padé and have some merit in linear problems. A particular form of interest is that given by

$$\lambda = [(\mathbf{I}+\beta\mathbf{C}^{-1}\mathbf{K})^n]^{-1}[\mathbf{I}+\alpha_1\mathbf{C}^{-1}\mathbf{K}+\alpha_2(\mathbf{C}^{-1}\mathbf{K})^2+\cdots] \tag{21.53}$$

where only one inverse has to be obtained and can be used repeatedly. Approximations of this type have been derived by Nørsett[40] and have recently been found to be of some practical value.[41–42]

21.9 Non-linear Time Marching Schemes

The major field of application of time marching procedures is to non-linear problems, where one or more of the matrices **M**, **C**, **K**, or **f** depend on the unknown vector **a**. Here the alternatives of analytical solution present in linear situations do not exist. Clearly the weighted residual expressions, such as given by Eqs. (21.7) and (21.25) for first and second order systems respectively, remain valid but their simplified, integrated, forms (21.9) and (21.26) are not. Indeed the full integration can only be carried out numerically and, *in general, iteration within each time step appears to be needed.* Further, at each time step a different set of equations will require solution, thus making procedures costly.

In practice, with such non-linear problems point collocation procedures have often been used (finite differences) avoiding the difficulties of integration in the time domain which are now trivial. This loses some of the accuracy of the formulation and we strongly recommend that an integration of the appropriate quantities be carried out numerically within an interval. The alternative to an iteration, however, within a time interval is the extrapolation of the various values **M**, **C**, **K**, or **f** from the previously established steps. Such extrapolation leads to more stable solutions[43–45] as indeed within each interval the usual linearized stability criteria obtain. Further, instability which may sometimes arise solely due to iteration is avoided.

A very simple extrapolation uses, in a particular interval, the values of the matrices corresponding to those at the last known value, i.e.,

$$\mathbf{M}_{n+1} = \mathbf{M}_n, \quad \mathbf{C}_{n+1} = \mathbf{C}_n, \quad \text{etc.} \tag{21.54}$$

and for first order problems any of the two- or three-point formulae previously developed can be used directly.

Culham and Varga[43] use such a scheme for the solution of a first order equation with a variable **K** and combine it with a backward difference ($\theta = 1$) numerical scheme. Their results show a better accuracy than that obtainable with other schemes where iteration has been used.

In explicit schemes with lumped matrices the use of approximation 21.55 is natural. Now the term

$$\mathbf{K}_n \mathbf{a}_n = \mathbf{p}_n \tag{21.55}$$

is evaluated as a *vector* such as may be caused, for instance, by a non-linear material or geometrical behaviour of a structure, and the matrix **K** is, in fact, not explicitly formed. Such schemes appear extremely useful in many structural short-period investigations[21–24] and although the time interval is still subject to stability, the simplicity of computation makes this scheme very commendable. We shall discuss it more extensively in the example section following.

Another use of the assumption (21.54) is in the three-point schemes applied to first order equations in which all the matrices are subject to non-linearity. This has been demonstrated as an effective procedure by Comini *et al.*[33] for a heat conduction problem in which conductivity of specific heat and heat generation all depend on temperature. Again we shall discuss these examples later.

21.10 **Examples**

The reader will find numerous examples of time-stepping applied to both linear and non-linear problems in the current literature.

Transient heat conduction—linear. Two examples taken from reference 14 illustrate the application of a two-point $\theta = \frac{1}{2}$ algorithm to a heat conduction problem governed by the equation

$$\frac{\partial}{\partial x}\left(k\frac{\partial T}{\partial x}\right)+\frac{\partial}{\partial y}\left(k\frac{\partial T}{\partial y}\right)+\frac{\partial}{\partial z}\left(k\frac{\partial T}{\partial z}\right)+\rho c\frac{\partial T}{\partial t}-Q = 0 \qquad (21.56)$$

with a linearized radiation boundary condition

$$k\frac{\partial T}{\partial n} = -\alpha(T-T_a) \qquad (21.57)$$

where α is known as the heat transfer coefficient.

In both examples k, the thermal conductivity, c the specific heat, and ρ the density are constants not depending on temperature. In the first example transient conditions in a cooled rotor blade are studied (Fig. 21.13) using two-dimensional isoparametric, cubic, elements. In the second example the accuracy of the time-stepping solution is demonstrated on a three-dimensional problem by comparison with an analytical solution[46] (Fig. 21.14).

Transient heat conduction—non-linear. The non-linearities in practical applications of heat conduction problems may vary from those which are very mild to those which are very severe. We shall illustrate the possibilities by quoting two examples of the latter kind. The first concerns the *freezing of ground* in which the latent heat of freezing is represented by varying the specific heat with temperature in a narrow zone, as shown in Fig. 21.15. Further, in the transition from the fluid to the frozen state a variation in conductivity occurs. We now thus have a problem in which both matrices **C** and **K** are variable and the solution of Fig. 21.16 illustrates the progression of a freezing front which was derived using the three-point (Lees) algorithm[32, 33] with $\mathbf{C} = \mathbf{C}_n$ and $\mathbf{K} = \mathbf{K}_n$.

A computational feature of some significance arises in this problem as values of the specific heat become very high in the transition zone, and in time-stepping can be missed if the temperature step *straddles* the freezing point. To avoid this difficulty and keep the heat balance correct the concept of enthalpy is introduced, defining

$$H = \int_0^T \rho c \, \mathrm{d}T. \qquad (21.58)$$

Now, whenever a change of temperature is considered an appropriate value of ρc is calculated which gives the correct change of H.

The heat conduction problem involving phase change is of some importance in welding and casting technology. Some very useful finite element solutions of these problems have been obtained in reference 47.

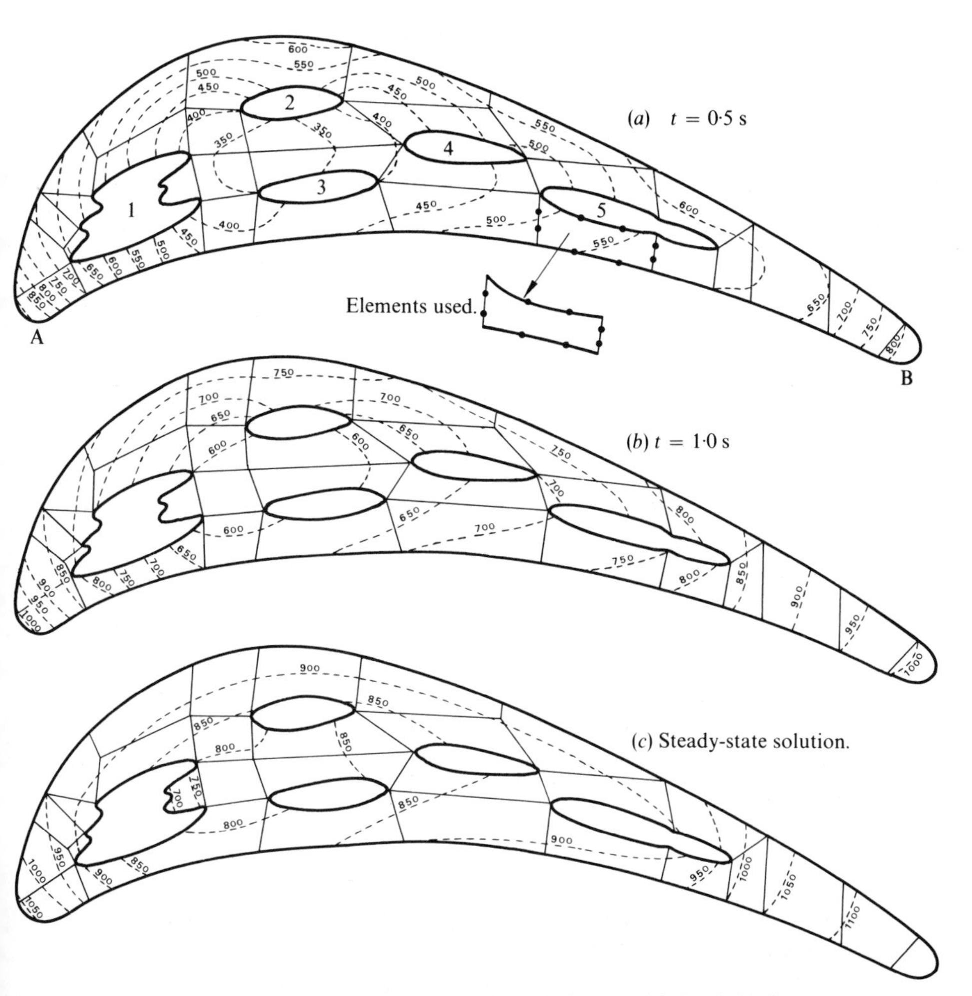

Fig. 21.13 Temperature distribution in a cooled rotor blade, initially at zero temperature

specific heat $c = 0{\cdot}11$ cal/gm °C
density $\rho = 7{\cdot}99$ gm/cm^3
conductivity $k = 0{\cdot}05$ cal/sec cm °C
gas temperature around blade = 1145°C

Heat transfer coefficient α varies from 0·390 to 0·056 on the outside surfaces of the blade (A–B)

Hole Number	Cooling Hole Temperature	α around perimeter of each hole
1	545°C	0·0980
2	587°C	0·0871

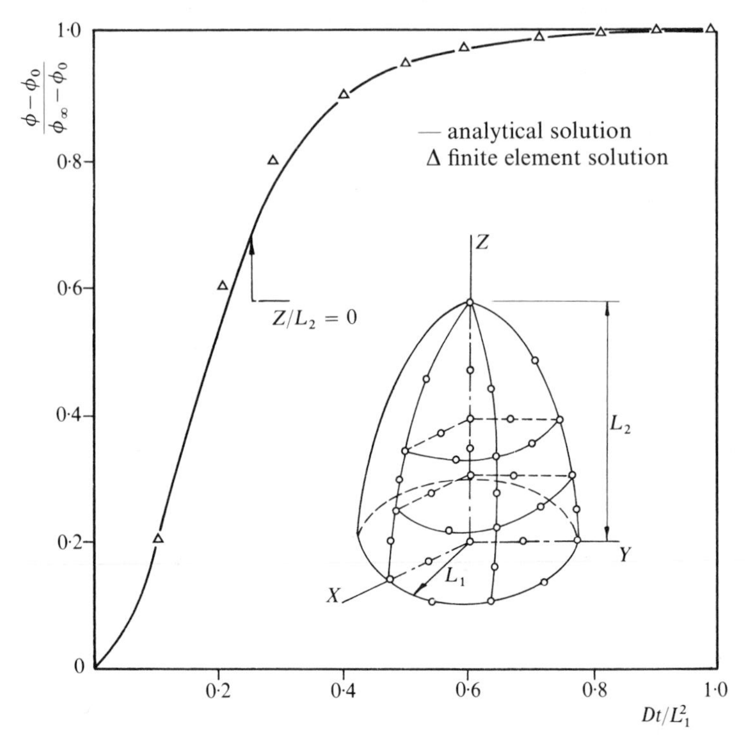

Fig. 21.14 Temperature variation with time at $x = y = z = 0$ in a prolate spheroidal solid

$$\frac{L_2}{L_1} = 2 \qquad \frac{\alpha L_1}{k} = \infty$$

The second non-linear example concerns the problem of *spontaneous ignition*.[48] (See also Chapter 18, p. 493.) Here the heat generated depends on the temperature

$$Q = \delta \, e^T \tag{21.59}$$

and the situation can become *physically unstable* with temperatures rising continuously to extreme values. In Fig. 21.17 we show a transient solution for a sphere at an initial temperature of $T = 290$ °K immersed in a bath of 500 °K. The solution is given for two values of the parameter δ with $k = \rho c = 1$, and the non-linearities are now so severe that an iterative solution in each time increment was necessary. For the larger value of δ the temperature increases to an infinite value in *a finite time* and the time interval for the computation had to be changed continuously to account

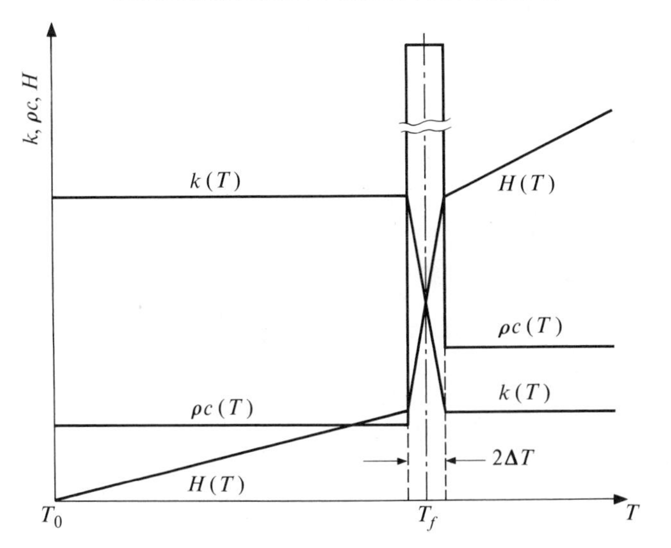

Fig. 21.15 Estimation of thermophysical properties in phase change problems. Latent heat effect is approximated by a large capacity over a small temperature interval $2\Delta T$

for this. The finite time for this point to be reached is known as the *induction time* and is shown in Fig. 21.17 for various values of δ.

The question of changing the time interval during the computation has not been discussed in detail, but clearly this must be done quite frequently to avoid large changes of the unknown function which will result in inaccuracies.

The final example concerns *dynamic structural transient with material and geometric non-linearity.*

This last example concerns a highly non-linear dynamic structural problem in which both geometrical and material non-linearity occur. Neglecting damping forces, we can write the equation of motion (Eq. (21.1)) in the discretized system as

$$\mathbf{M}\ddot{\mathbf{a}}+\mathbf{P}(\mathbf{a})+\mathbf{f} = 0 \tag{21.60}$$

where

$$\mathbf{P}(\mathbf{a}) \equiv \mathbf{K}\mathbf{a} \tag{21.61}$$

is a vector of resisting internal forces. This, as has been shown in the Chapters 18 and 19 can be computed from the knowledge of the stresses at any stage as

$$\mathbf{P}(\mathbf{a}) = \int_{\Omega} \overline{\mathbf{B}}^{\mathrm{T}}\boldsymbol{\sigma}\, \mathrm{d}\Omega \tag{21.62}$$

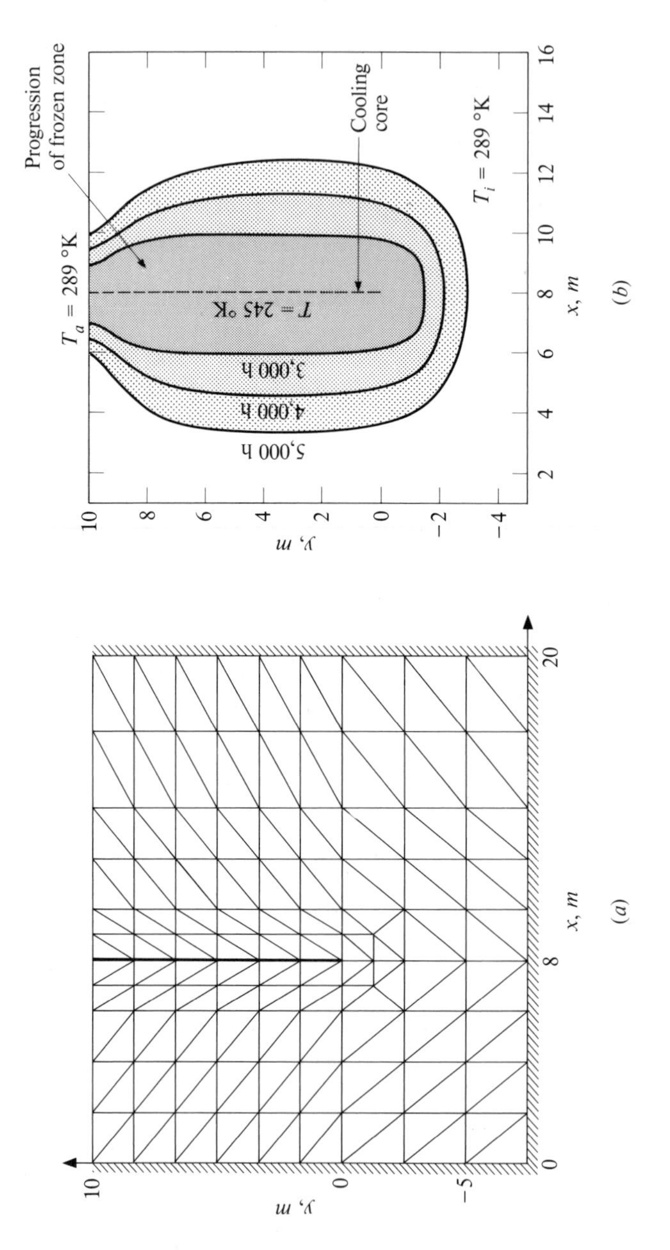

Fig. 21.16 Freezing of a moist soil (sand), (*a*) and (*b*)

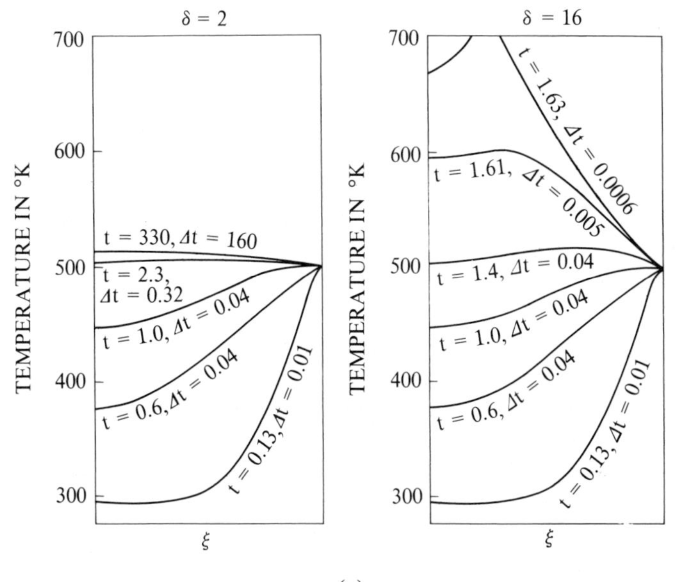

(*a*)

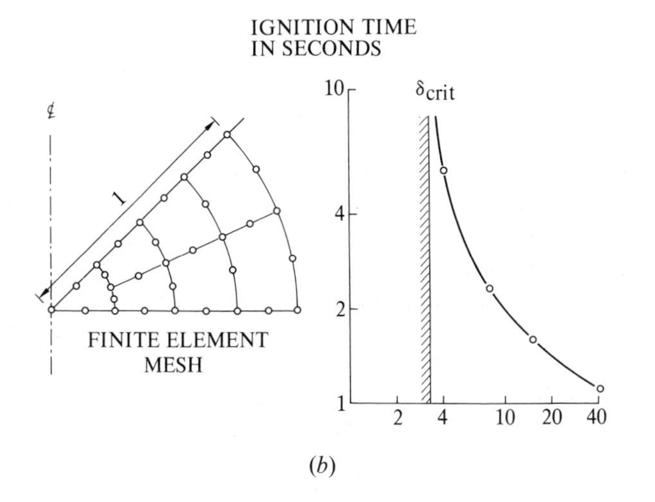

(*b*)

Fig. 21.17 Reactive sphere (*a*) induction time versus Frank–Kamenetskii parameter. (*b*) Temperature profiles for ignition ($\delta = 16$) and non-ignition ($\delta = 2$) transient behaviour of a reactive sphere

where $\overline{\mathbf{B}}$ is the non-linear strain matrix and $\boldsymbol{\sigma}$ is the stress state. If the three-point, explicit, weighting is used corresponding to that of $\gamma = \frac{1}{2}$, $\beta = 0$ in Eq. (21.26) we can write the recurrence scheme as

$$\mathbf{M}(\mathbf{a}_{n+1} - 2\mathbf{a}_n + \mathbf{a}_n) + \mathbf{P}_n + \mathbf{f}_n = 0 \tag{21.63}$$

or

$$\mathbf{a}_{n+1} = -\mathbf{M}^{-1}(\mathbf{P}_n + \mathbf{f}) + 2\mathbf{a}_n - \mathbf{a}_{n-1}. \tag{21.64}$$

With a diagonal *lumped* matrix $\mathbf{M}$ the inversion is trivial and, further, the computation only has to evaluate $\mathbf{P}_n$ or $\mathbf{f}$ node by node avoiding large storage requirements. Stability considerations limit the time step but, the computation being simple, the total cost of it remains reasonable even with the large number of time steps used.

In Fig. 21.18 results of such a computation for a pressure vessel subject to a sudden pressure pulse are shown.[22]

It is of interest to note that here an identical formulation was used for both solid and fluid portions. In the latter, simply the shear rigidity was taken as zero thus modelling the essential difference between the behaviour of solid and fluid. This device is of considerable practical interest and has been used with other integration schemes without apparent deleterious effect.[49]

21.11 **Concluding Remarks**

This chapter has introduced many classical and some non-classical procedures of recurrent schemes for solving time-dependent problems. The whole field is so wide that details of application methodology to specific problems have had to be scant. In particular we have not discussed the solution of creep problems referred to in Chapter 18, which essentially involve the time-stepping scheme and thus are subject to all the stability considerations given here.[50] Further, a range of problems in which the time-stepping involves geometrical boundary changes have not been referred to. Amongst this class of application is the solution of the free surface seepage problem.[51, 52]

The reader will have noticed the simplicity of solution in dynamic non-linear problems, apparent in both programming and storage requirement, when the explicit algorithm is used. One corollary of such a formulation is the possibility of using the time-stepping procedure simply as a device to obtain a solution of a non-linear set of equations

$$\mathbf{P}(\mathbf{a}) = 0. \tag{21.65}$$

If this is augmented by including an artificial mass and damping matrix to a form

$$\mathbf{P}(\mathbf{a}) + \mathbf{M}\ddot{\mathbf{a}} + \mathbf{C}\dot{\mathbf{a}} = 0 \tag{21.66}$$

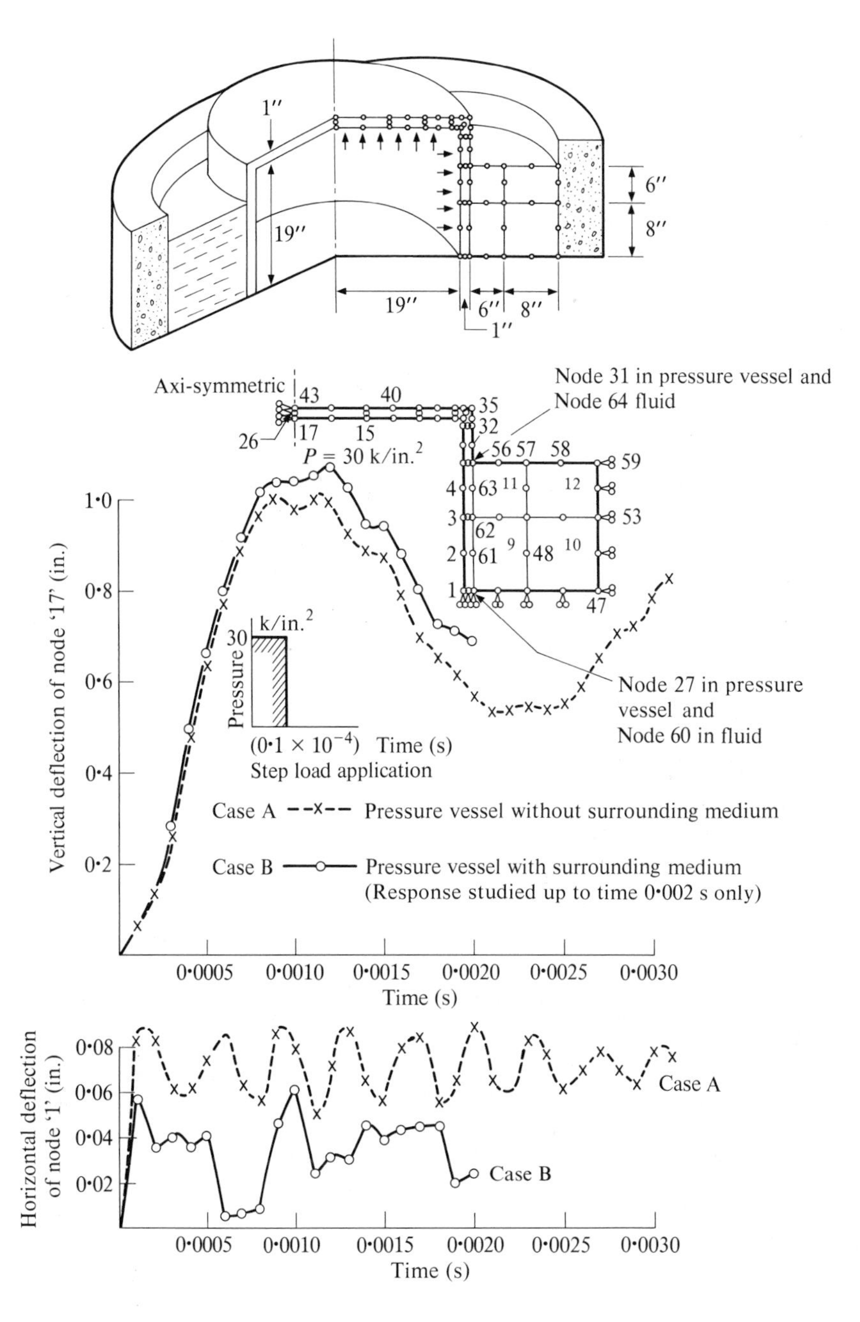

Fig. 21.18 Dynamic large elastic plastic response of a pressure vessel under a short pulse

and an explicit transient solution followed until steady-state conditions are reached, i.e., until

$$\ddot{\mathbf{a}} = \dot{\mathbf{a}} = 0$$

the solution to the original non-linear problem is obtained. This type of technique is quite efficient and has been applied successfully in the context of finite differences under the name of 'dynamic relaxation'.[53] Its possibilities remain yet to be explored in the general finite element context.

References

1. R. D. RICHTMYER and K. W. MORTON, *Difference Methods for Initial Value Problems*, Wiley (Interscience), 1967.
2. T. D. LAMBERT, *Computational Methods in Ordinary Differential Equations*, Wiley, 1973.
3. L. FOX and E. T. GOODWIN, 'Some new methods for the numerical integration of ordinary differential equations', *Proc. Camb. Phil. Soc.*, **49**, 373, 1953.
4. F. B. HILDEBRAND, *Finite Difference Equations and Simulations*, Prentice-Hall, 1968.
5. P. HENRICI, *Discrete Variable Methods in Ordinary Differential Equations*, Wiley, 1962.
6. O. C. ZIENKIEWICZ, *The Finite Element Method in Engineering Science*, pp. 335–9, McGraw-Hill, 1971.
7. M. ZLAMAL, 'Finite element methods in heat conduction problems', pp. 85–104, 'The mathematics of finite elements and applications II', ed. J. Whiteman, Academic Press, 1977.
8. K. WASHIZU, *Variational Methods in Elasticity and Plasticity*, 2nd ed., Pergamon Press, 1975.
9. M. GURTIN, 'Variational principles for linear elastodynamics', *Arch. Nat. Mech. Anal.*, **16**, 34–50, 1969.
10. M. GURTIN, 'Variational principles for linear initial-value problems', *Quart. Appl. Math.*, **22**, 252–6, 1964.
11. E. L. WILSON and R. E. NICKELL, 'Application of finite element method to heat conduction analysis', *Nucl. Eng. Des.*, **4**, 1–11, 1966.

12a. I. FRIED, 'Finite element analysis of time-dependent phenomena', *J.A.I.A.A.*, **7**, 1170–3, 1969.

12b. J. T. ODEN, 'A general theory of finite elements—I: Topological considerations', pp. 205–21, and 'II: Applications', pp. 247–60. *Int. J. Num. Meth. Eng.*, **1**, 1969.

13. J. H. ARGYRIS and D. W. SCHARPF, 'Finite elements in time and space', *Nucl. Eng. Des.*, **10**, 456–69, 1969.
14. O. C. ZIENKIEWICZ and C. J. PAREKH, 'Transient field problems—two and three dimensional analysis by isoparametric finite elements', *Int. J. Num. Meth. Eng.*, **2**, 61–71, 1970.
15. O. C. ZIENKIEWICZ and R. H. LEWIS, 'An analysis of various time stepping schemes for initial value problems', *Int. J. Earthquake Eng. Struct. Dynam.*, **1**, 407–8, 1973.
16. W. L. WOOD and R. H. LEWIS, 'A comparison of time marching schemes for the transient heat conduction equation', *Int. J. Num. Meth. Eng.*, **9**, 679–89, 1975.

17. D. G. JONES and R. D. HENSHELL, 'Oscillations in transient thermal calculations' (to be published).
18. B. M. IRONS, *Applications of a theorem on eigenvalues to finite element problems*, Univ. of Wales, Dept. of Civil Eng., Swansea, 1970 (CR/132/70).
19. N. M. NEWMARK, 'A method for computation of structural dynamics', *Proc. Am. Soc. Civ. Eng.*, **85**, EM3, 67–94, 1959.
20. S. P. CHAN, H. L. COX and O. BENFIELD, 'Transient analysis of forced vibrations of complex structural mechanical systems', *J. Roy. Aero. Soc.*, **66**, 457–60, 1962.
21. T. BELYTSCHKO, R. L. CHIAPETTA and H. D. BARTEL, 'Efficient large scale non-linear transient analysis by finite elements', *Int. J. Num. Meth. in Eng.*, **10**, 579–96, 1976.
22. D. SHANTARAM, D. R. J. OWEN and O. C. ZIENKIEWICZ, 'Dynamic transient behaviour of two and three dimensional structures including plasticity, large deformation and fluid interaction', *Int. J. Earthquake Eng. Struct. Dynam.*, **4**, 561–78, 1976.
23. R. D. KREIG and S. W. KEY, 'Transient shock response by numerical time integration', *Int. J. Num. Meth. Eng.*, **7**, 273–86, 1973.
24. C. C. FU, 'On the stability of explicit methods for numerical integration of the equations of matrices in finite element methods', *Int. J. Num. Meth. Eng.*, **4**, 95–107, 1972.
25. K. J. BATHE and E. L. WILSON, 'Stability and accuracy analysis of direct integration methods', *Int. J. Earthquake Eng. Struct. Dynam.*, **1**, 283–91, 1973.
26. R. E. NICKELL, 'On the stability of approximation operators in problems of structural dynamics', *Int. J. Solids Struct.*, **7**, 301–19, 1971.
27. R. S. DUNHAM *et al.*, 'Integration operators for transient structural response', *J. Comp. Struct.*, **2**, 1972.
28. G. L. GOUDREAU and R. L. TAYLOR, 'Evaluation of numerical integration methods in elasto dynamics', *Comp. Meth. Appl. Mech. Eng.*, **2**, 69–97, 1972.
29. L. FOX and E. T. GOODWIN, 'Some new methods for the numerical integration of ordinary differential equations', *Proc. Camb. Phil. Soc.*, **49**, 373–88, 1949.
30. J. E. GRANT, 'Response computation using Taylor series', *Proc. A.S.C.E.*, **97**, EM2, 295–303, 1971.
31. H. M. HILBER, T. J. R. HUGHES and R. L. TAYLOR, 'Improved numerical dissipation for time integration algorithms in structural mechanics', *Int. J. Earthquake Eng. Struct. Dynam.*, **5**, 283–92, 1977.
32. M. LEES, 'A linear three level difference scheme for quasilinear parabolic equations', *Maths. Comp.*, **20**, 516–622, 1966.
33. G. COMINI, S. DEL GUIDICE, R. H. LEWIS and O. C. ZIENKIEWICZ, 'Finite element solution of non-linear heat conduction problems with special reference to phase change', *Int. J. Num. Meth. Eng.*, **8**., 613–24, 1974.
34. G. DAHLQUIST, 'A special stability problem for linear multistep methods', *B.I.T.*, **3**, 27–43, 1963.
35. O. C. ZIENKIEWICZ, 'A new look at the Newmark, Houbolt and other time stepping schemes. A weighted residual approach', *Int. J. Earthquake Struct. Dynam.*, **5**, 413–18, 1977.
36. W. L. WOOD, 'On the Zienkiewicz four time level scheme for the numerical integration of vibration problems', *Int. J. Num. Meth. Eng.*, **11**, 1519–28, 1977.
37. J. C. HOUBOLT, 'A recurrence matrix solution for the dynamic response of elastic aircraft', *J. Aero. Sci.*, **17**, 540–50, 1950.
38. I. FARHOOMAND, *Non-linear dynamic stress analysis of two dimensional solids*, Ph.D. Dissertation, Univ. of California, Berkeley, 1970.

39. C. H. Cryer, 'A new class of highly stable methods: A_0-stable methods', *B.I.T.*, **13**, 153–9, 1973.
40. S. P. Norsett, 'One step methods of Hermite type for numerical integration of stiff systems', *B.I.T.*, **14**, 63–77, 1974.
41. D. M. Trujillo, 'The direct numerical integration of linear matrix differential equations using Padé approximations, *Int. J. Num. Meth. Eng.*, **9**, 259, 1975.
42. I. M. Smith, J. L. Siemieniuch, and I. Gladwell, *A comparison of old and new methods for large systems of ordinary differential equations*, Univ. of Manchester, Dept. of Mathematics, NAR 13, 1975.
43. W. E. Culham and R. S. Varga, 'Numerical methods for time dependent non linear boundary value problems', *J. Soc. Petrol. Eng.*, 374–87, Dec. 1971.
44. P. T., Boggs, 'The solution of non-linear systems of equations by *A*-stable integration techniques', *SIAM Jl.*, **8**, 767–85, 1975.
45. J. A. Stricklin *et al.*, 'Non-linear dynamic analysis of shells of revolution by matrix displacement methods', *Proc. Conf. on Structural Dynamics*, A.S.M.E., Denver, 1971.
46. A. Haji-Sheikh and E. M. Sparrow, 'Transient heat conduction in a prolate spheroidal solid', *J. Heat Transfer, Trans. Am. Soc. Mech. Eng.*, **88**, 331–3, 1966.
47. H. D. Hibbitt and P. V. Marcal, 'Numerical thermo-mechanical model for the welding and subsequent loading of a fabricated structure', *Comp. Struct.*, **3**, 1145–74, 1973.
48. C. A. Anderson and O. C. Zienkiewicz, 'Spontaneous ignition: finite element solutions for steady and transient conditions', *J. Heat Transfer, Trans. Am. Soc. Mech. Eng.*, 398–404, 1974.
49. E. L. Wilson, 'Finite elements for foundations, joints and fluids', *Conf. on Numerical Methods in Soil and Rock Mechanics*, Karlsruhe Univ., 1975.
50. I. C. Cormeau, 'Numerical stability in quasi-static elasto and visco plasticity', *Int. J. Num. Meth. Eng.*, **9**, 109–27, 1975.
51. P. W. France, C. J. Parekh, J. C. Peters and C. J. Taylor, 'Numerical analysis of linear free surface seepage problems', *Proc. Am. Soc. Civ. Eng.*, **97**, IR1, 165–79, 1971.
52. C. J. Taylor, P. W. France and O. C. Zienkiewicz, 'Some free surface transient flow problems of seepage and irrotational flow', pp. 313–26. *The Mathematics of Finite Elements and Applications* (ed. J. Whiteman), Academic Press, 1973.
53. J. R. M. Otter, 'Dynamic relaxation', *Proc. Inst. Civ. Eng.*, **35**, 633–56, 1966.

22. Flow of Viscous Fluids; Some Special Problems of Convective Transport

22.1 **Introduction**

Throughout this book we have endeavoured to present the reader with a systematic approach to a variety of problems of the physical world which, once posed in mathematical terms, could be discretized and hence solved numerically. Problems of solid mechanics have, however, been predominant and to redress the balance this chapter is devoted to fluid mechanics.

Although we could start by writing the appropriate governing differential equations—and then solve these by applying the general principles of Chapter 3—we prefer to approach the mechanics of viscous fluid flow via their analogy to solid mechanics, and the first sections of this chapter are devoted to such an approach. This will permit the reader to utilize directly, or with minor modification, some of the programs developed for solids to solve certain fluid problems.

The major difference from the formulations already encountered lies in the convective terms which enter the equations of fluid mechanics problems. These lead to non-symmetric matrices if the conventional, Galerkin, approach is used in their discretization. Further, instability of computation can occur and this necessitates special discretization procedures so far not encountered in this text. We shall outline these in section 22.8.

Space limitations will not allow an exhaustive treatment to be presented here. In particular high-speed compressible (trans- or super-sonic) flow will not be considered. For supplementation the reader is referred to a series of conference proceedings and texts.[1–6] However, it is hoped that the contents of this chapter, together with those of Chapters 17, 20, and 23 in which some special fluid flow cases are treated, will give the reader a reasonably full picture of the possibilities open in this field. Some prior knowledge of fluid mechanics is naturally assumed—and for more detailed treatment of the essentials the reader should consult some of the well-known texts.[7,8]

22.2 Basic Concepts of Viscous, Slightly Compressible Flow

22.2.1 *Equilibrium.* If an isolated volume of fluid is considered at some instance of time (Fig. 22.1) then, just as in a solid, in its interior the stresses $\boldsymbol{\sigma}$ must be in equilibrium with the body forces $\mathbf{b}$ which include the appropriate acceleration forces. Further, on its external surfaces the stresses $\boldsymbol{\sigma}$ must be in balance with the applied traction $\bar{\mathbf{t}}$. Thus, both the internal equilibrium equations and those on the boundary are *identical* to those pertaining to the solid. Using the nomenclature of Chapter 3, Eq. 3.40, and of Chapter 12, Eq. 12.14, we can write

$$\mathbf{L}^{\mathrm{T}}\boldsymbol{\sigma}+\mathbf{b} = 0 \quad \text{in } \Omega \tag{22.1a}$$

and

$$\mathbf{G}\boldsymbol{\sigma} = \bar{\mathbf{t}} \quad \text{on } \Gamma_t \tag{22.1b}$$

where Ω is the problem domain and Γ_t its boundary on which tractions are prescribed.

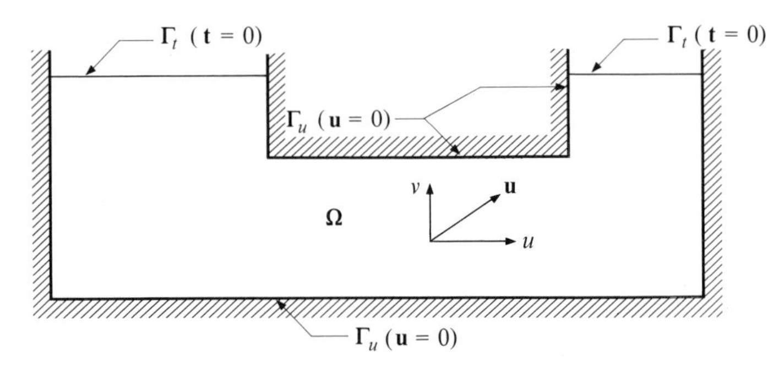

Fig. 22.1 A two-dimensional fluid flow domain

Thus the virtual work relationships used in Chapter 2 and discussed in Chapter 3 can once again be invoked. It is convenient now to apply virtual velocities $\delta\mathbf{u}$ in place of virtual displacements and we can write in place of Eqs. 22.1 the equivalent statement

$$\int_{\Omega} \delta\boldsymbol{\varepsilon}^{\mathrm{T}}\boldsymbol{\sigma}\,\mathrm{d}\Omega - \int_{\Omega} \delta\mathbf{u}^{\mathrm{T}}\mathbf{b}\,\mathrm{d}\Omega - \int_{\Gamma_t} \delta\mathbf{u}^{\mathrm{T}}\,\bar{\mathbf{t}}\,\mathrm{d}\Gamma = 0 \tag{22.2}$$

where Γ_t stands for the part of the boundary on which tractions are specified and $\delta\mathbf{u} \neq 0$ there.

In the above,

$$\delta\dot{\boldsymbol{\varepsilon}} = \mathbf{L}\,\delta\mathbf{u} \quad \text{and} \quad \dot{\boldsymbol{\varepsilon}} = \mathbf{L}\mathbf{u} \tag{22.3}$$

defines the virtual *strain rate* by an identical expression to that used previously to define virtual strains (viz. Eq. 6.9. Chapter 6 for such a definition in three dimensions).

In fluid mechanics, due to the continually changing *displacements*, it is natural that we concentrate our attention on *velocities* and these at a fixed point of space will be denoted by **u**—an identical symbol to that previously used for displacements. The body forces **b** per unit volume can be written, as in solid mechanics (*vide* Chapter 20), invoking d'Alembert's principle, as

$$\mathbf{b} = \mathbf{b}_0 - \rho\mathbf{c} \tag{22.4}$$

where **c** is the acceleration vector acting on each particle and ρ is the density. As we have defined velocity **u** at a point in space rather than with reference to a particle, the simple differentiation of the latter with respect to time does not suffice to define the acceleration. This is now given by the *total* (or particle) derivative of **u**, e.g. for the x component

$$c_x = \frac{\mathrm{D}u}{\mathrm{d}t} = \frac{\partial u}{\partial t} + \frac{\partial u}{\partial x}\frac{\partial x}{\partial t} + \frac{\partial u}{\partial y}\frac{\partial y}{\partial t} + \frac{\partial u}{\partial z}\frac{\partial z}{\partial t}, \quad \text{etc.} \tag{22.5a}$$

As $\partial x/\partial t \equiv u$, etc., we can write the total acceleration vector as

$$\mathbf{c} = \frac{\partial \mathbf{u}}{\partial t} + (\boldsymbol{\nabla}\cdot\mathbf{u}^{\mathrm{T}})^{\mathrm{T}}\mathbf{u} \tag{22.5b}$$

where $\boldsymbol{\nabla}^{\mathrm{T}} = [\partial/\partial x, \partial/\partial y, \partial/\partial z]$ and $(\boldsymbol{\nabla}\cdot\mathbf{u}^{\mathrm{T}}) \equiv \mathbf{J}(\mathbf{u})$ is a Jacobian matrix.

Now even if the flow is steady, i.e., $\partial\mathbf{u}/\partial t = 0$, acceleration exists and here lies the principal difference from the solid mechanics formulation. Further, the expression for acceleration is non-linear in **u** and the problem is immediately of a non-linear nature.

22.2.2 *Constitutive relations*. In a fluid, by definition, no deviatoric stresses can be supported unless motion occurs. We can thus state quite generally that the deviatoric stresses are a function of the *strain rates* $\dot{\boldsymbol{\varepsilon}}$.

If we define the pressure p as

$$p = -\sigma_m = (\sigma_x + \sigma_y + \sigma_z)/3 \tag{22.6}$$

we can write a very general linear relationship between the deviatoric stress $\boldsymbol{\sigma}'$ and strain rate as

$$\boldsymbol{\sigma}' \equiv \boldsymbol{\sigma} + \mathbf{m}p = \mathbf{D}'\dot{\boldsymbol{\varepsilon}} \tag{22.7}$$

with

$$\mathbf{m}^{\mathrm{T}} = [1, 1, 1, 0, 0, 0]$$

For an isotropic incompressible fluid, by analogy with solid mechanics, one constant μ, known as viscosity, defines completely the $\mathbf{D}'$ matrix.

$$\mathbf{D}' = \mu \begin{bmatrix} 2 & & & & & \\ & 2 & & & 0 & \\ & & 2 & & & \\ & & & 1 & & \\ & 0 & & & 1 & \\ & & & & & 1 \end{bmatrix} \tag{22.8}$$

Clearly μ plays here an identical role to that of the shear modulus G in elasticity (*vide* Chapter 11, Eq. (11.22)).

The constitutive relationship (22.7) is thus of an identical form to that pertaining to incompressible solid mechanics with the strain rates now playing the role of strains, and additional constraint is thus necessary before the solution can be attempted.

22.2.3 *Continuity equation.* If an infinitesimal volume of space is considered then, quite generally, we can state that the nett rate of mass inflow is equal to the rate of mass accumulation. Thus, if ρ is the density we can write

$$\frac{\partial}{\partial x}(\rho u)+\frac{\partial}{\partial y}(\rho v)+\frac{\partial}{\partial z}(\rho w)-\frac{\partial \rho}{\partial t} \equiv \mathbf{\nabla}^{\mathrm{T}}(\rho \mathbf{u})-\frac{\partial \rho}{\partial t} = 0. \tag{22.9}$$

Quite generally the pressure p and the density ρ are related by a suitable state relation

$$\rho = \rho(p). \tag{22.10}$$

If, however, the density changes are small, the continuity relationship can be simplified to

$$\nabla^{\mathrm{T}}\mathbf{u} \equiv \dot{\varepsilon}_{\upsilon} = 0 \tag{22.11}$$

stating simply that the rate of volumetric straining is identically zero. This is analogous to the constraint used in incompressible solid mechanics (Chapters 11 and 12) and we shall in the main be concerned only with problems where this incompressibility is enforced.

22.2.4 *Summary.* We have noted that a completely formal analogy exists between the elasticity and viscous fluid mechanical problems.

Indeed, if we disregard the difference which occurs in the acceleration forces and consider a purely incompressible flow the analogy is exact. *Thus all the methodology developed for the solution of incompressible elastic solids is immediately available for the solution of viscous incompressible flow under steady-state conditions, omitting acceleration terms.*

Following identification of terms is necessary

Elasticity		$\longleftrightarrow$	*Viscous flow*	
displacement	$\mathbf{u}$	$\longleftrightarrow$	velocity	$\mathbf{u}$
strain	ε	$\longleftrightarrow$	strain rate	$\dot{\varepsilon}$
stress	$\boldsymbol{\sigma}$	$\longleftrightarrow$	stress	$\boldsymbol{\sigma}$
shear modulus	G	$\longleftrightarrow$	viscosity	μ

The flow in which acceleration effects are negligible is generally known as creeping—and clearly for its solution any of the techniques already described for the solution of incompressible elasticity are immediately available.

Amongst these we have already encountered (and obviously more alternatives are possible):

1. The use of $\mathbf{u}$ and p as variables—with p entering the variational form as a Lagrangian multiplier (Chapter 12, p. 323).
2. The use of $\mathbf{u}$ as the only variable with the incompressibility constraint entering by use of a penalty function (Chapter 11, p. 286).
3. The use of equilibrating formulations (Chapter 12, p. 306).
4. The use of stream functions (Chapter 12, p. 324).

If acceleration effects are not negligible—their insertion into the discretization process (if this is achieved by Galerkin procedures) is simple and follows the lines used in structural dynamic effects in Chapter 20. However, the use of variational principles even in steady-state cases is no longer possible as true variational principles no longer exist.[9] In the next section we shall perform the discretization of the various types explicitly.

When formulating elasticity problems in Chapter 2 and elsewhere, we started from the virtual work principle as a basis and did not state the full governing equations explicitly. Such equations could be readily derived for elasticity if displacement formulation were to be used by eliminating the stresses and strains from the equilibrium equations and bear the name of Navier. Indeed these equations could (less conveniently) have been used for obtaining the first finite element discretization.

In fluid mechanics the conventional starting point of discretization is often based on similar equations.[10–21] Although we shall not pursue this line, which obviously leads to the same results as those obtainable by direct use of virtual work,[19] it is of interest to state explicitly the governing equations, which are known as those of Navier–Stokes.

Thus, if we eliminate $\boldsymbol{\sigma}$ from Eq. (22.1a) using relationships (22.3–5) and (22.7) we obtain a general Navier–Stokes equation

$$\rho\left[\frac{\partial \mathbf{u}}{\partial t}+(\nabla \mathbf{u}^{\mathrm{T}})^{\mathrm{T}}\mathbf{u}\right] = -\mathbf{L}^{\mathrm{T}}\mathbf{m}p+\mathbf{L}^{\mathrm{T}}\mathbf{D}'\mathbf{L}\mathbf{u}+\mathbf{b}_0. \tag{22.12}$$

With the form of $\mathbf{D}'$ given by Eq. (22.8) and $\mathbf{L}$ as defined previously, the above can be simplified to a more standard form. Thus in x direction

$$\rho\left(\frac{\partial u}{\partial t}+u\frac{\partial u}{\partial x}+v\frac{\partial u}{\partial y}+w\frac{\partial u}{\partial z}\right) = -\frac{\partial p}{\partial x}+2\frac{\partial}{\partial x}\left(\mu\frac{\partial u}{\partial x}\right)+\frac{\partial}{\partial y}\left(\mu\frac{\partial u}{\partial y}+\mu\frac{\partial v}{\partial x}\right) + \frac{\partial}{\partial z}\left(\mu\frac{\partial u}{\partial z}+\mu\frac{\partial w}{\partial x}\right) \tag{22.13}$$

with similar equations in the y and z directions. For solution this equation has to be combined with that of continuity (Eq. (22.9)).

Alternative formulations can be devised, however, in which the stresses $\boldsymbol{\sigma}$ may be used as explicit variables.

22.3 Discretization of Viscous Flow Equations

Obviously with the analogy stated in the previous section, the discretization details could be omitted as these follow precisely the lines used in equivalent solid mechanics sections. For completion, however, we still give here brief details of the three useful forms.

22.3.1 *Velocity and pressure as variables.* In this we shall discretize the velocity and pressure in terms of independent parameters

$$\mathbf{u} = \mathbf{N}^u\mathbf{a}^u; \qquad p = \mathbf{N}^p\mathbf{a}^p \tag{22.14}$$

Using the virtual work statement of Eq. (22.2), with

$$\delta\mathbf{u} = \mathbf{N}^u\,\delta\mathbf{a}^u \qquad \text{and} \qquad \delta\dot{\boldsymbol{\varepsilon}} = (\mathbf{L}\mathbf{N}^u)\,\delta\mathbf{a}^u \equiv \mathbf{B}\,\delta\mathbf{a}^u \tag{22.15}$$

we can write

$$\delta\mathbf{a}^{u\mathrm{T}}\left[\int_\Omega \mathbf{B}^\mathrm{T}\boldsymbol{\sigma}\,\mathrm{d}\Omega-\int_\Omega \mathbf{N}^{u\mathrm{T}}\mathbf{b}\,\mathrm{d}\Omega-\int_{\Gamma_t}\mathbf{N}^{u\mathrm{T}}\bar{\mathbf{t}}\,\mathrm{d}\Gamma\right] = 0 \tag{22.16}$$

Noting that this is true for all variations $\delta\mathbf{a}^u$ we have, on inserting Eqs. (22.3–5) and (22.7)

$$\mathbf{K}\mathbf{a}^u+\bar{\mathbf{K}}\mathbf{a}^u+\mathbf{K}^p\mathbf{a}^p+\mathbf{M}\frac{\mathrm{d}\mathbf{a}^u}{\mathrm{d}t}+\mathbf{f}^u = 0 \tag{22.17}$$

with coefficients given by

$$\mathbf{K}_{ij} = \int_\Omega \mathbf{B}_i^\mathrm{T}\mathbf{D}'\mathbf{B}_j\,\mathrm{d}\Omega \tag{22.18a}$$

$$\bar{\mathbf{K}}_{ij} = \int_\Omega \rho(\mathbf{N}_i^u)^\mathrm{T}\,(\boldsymbol{\nabla}.(\mathbf{N}\mathbf{a}^u)^\mathrm{T})^\mathrm{T}\mathbf{N}_j\,\mathrm{d}\Omega \tag{22.18b}$$

$$\mathbf{K}_{ij}^p = -\int_\Omega \mathbf{B}_i^u\mathbf{m}\mathbf{N}_j^p\,\mathrm{d}\Omega \tag{22.18c}$$

$$\mathbf{M} = \int_{\Omega} (\mathbf{N}_i^u)^{\mathrm{T}} \rho \mathbf{N}_j^u \, \mathrm{d}\Omega \tag{22.18d}$$

$$\mathbf{f} = -\int_{\Omega} (\mathbf{N}_i^u)^{\mathrm{T}} \mathbf{b}_0 \, \mathrm{d}\Omega - \int_{\Gamma_t} (\mathbf{N}_i^u)^{\mathrm{T}} \bar{\mathbf{t}} \, \mathrm{d}\Gamma. \tag{22.18e}$$

We note immediately that, with the exception of the second matrix, as expected, the standard forms of elastic analysis are rediscovered with appropriate stiffness, mass, and force matrices.

To obtain the second equation necessary in view of the constraint equations we shall use the Galerkin process and simply pre-multiply the continuity equation (22.11) by $(\delta p)^{\mathrm{T}}$ and integrate for the case of complete incompressibility.

Thus we have

$$(\delta \mathbf{a}^p)^{\mathrm{T}} \int_{\Omega} (\mathbf{N}^p)^{\mathrm{T}} \varepsilon_v \, \mathrm{d}\Omega = 0 \tag{22.19}$$

or, noting that this is true for all $\delta \mathbf{a}^p$ and writing

$$\varepsilon_v \equiv \mathbf{m}^{\mathrm{T}} \mathbf{L} \mathbf{u} \equiv \mathbf{m}^{\mathrm{T}} \mathbf{L} \mathbf{N}^u \mathbf{a}^u \equiv \mathbf{m}^{\mathrm{T}} \mathbf{B} \mathbf{a}^u \tag{22.20}$$

this results in an equation

$$(\mathbf{K}^p)^{\mathrm{T}} \mathbf{a}^u = 0 \tag{22.21}$$

with $\mathbf{K}^p$ taking on the form already given in Eq. (22.18c).

Equation systems (22.17) and (22.21) can be written as

$$\begin{bmatrix} \mathbf{K} + \bar{\mathbf{K}}, & \mathbf{K}^p \\ \mathbf{K}^{p\mathrm{T}}, & 0 \end{bmatrix} \begin{Bmatrix} \mathbf{a}^u \\ \mathbf{a}^p \end{Bmatrix} + \begin{bmatrix} \mathbf{M} & 0 \\ 0 & 0 \end{bmatrix} \frac{\mathrm{d}}{\mathrm{d}t} \begin{Bmatrix} \mathbf{a}^u \\ \mathbf{a}^p \end{Bmatrix} + \begin{Bmatrix} \mathbf{f} \\ 0 \end{Bmatrix} = 0 \tag{22.22}$$

and can be used for the solution of transient viscous flow problems.

The reader will note that

(*a*) the equations are non-symmetric and non-linear if the velocities are large enough for $\bar{\mathbf{K}}$ to be significant;

(*b*) when the problem is one of steady state and $\bar{\mathbf{K}}$ is neglected, the Lagrangian form of incompressible elasticity equations of Chapter 12, p. 323, has been re-derived (now without use of a variational statement).

The formulation just presented is one of the most popular in the context of fluid mechanics and has been used frequently.[11, 14, 21] All the remarks made previously about over-constraint are once again applicable and practitioners find that, generally, a lower order of interpolation of p compared with that of $\mathbf{u}$ is desirable only to avoid over-constraints. Arguments of 'consistency' have, however, been used in the above context but we believe that these are not the correct reasons for improved performance with such mixed interpolations.[20, 21]

22.3.2 *Stream function formulation.* It is possible to define the velocity field **u** by means of an auxiliary set of functions so that the continuity (incompressibility) condition is automatically satisfied.

Although vector stream functions can be obtained in three dimensions these have not proved successful and the approach is restricted generally to two dimensions. Writing thus

$$\mathbf{u} = \hat{\mathbf{L}}\psi; \qquad \hat{\mathbf{L}}^{\mathrm{T}} = \left[\frac{\partial}{\partial y}, -\frac{\partial}{\partial x}\right] \tag{22.23}$$

the velocity field automatically satisfies Eq. (22.11) written in two space dimensions.

If now we discretize the stream functions by writing

$$\psi = \hat{\mathbf{N}}\mathbf{a} \tag{22.24}$$

we note that the virtual work equation, (22.2), can be written again as

$$\delta\mathbf{a}^{\mathrm{T}}\left[\int_{\Omega} \hat{\mathbf{B}}^{\mathrm{T}}\boldsymbol{\sigma}\,\mathrm{d}\Omega - \int_{\Omega} (\hat{\mathbf{L}}\hat{\mathbf{N}})^{\mathrm{T}}\mathbf{b}\,\mathrm{d}\Omega - \int_{\Gamma_t} (\hat{\mathbf{L}}\hat{\mathbf{N}})^{\mathrm{T}}\bar{\mathbf{t}}\,\mathrm{d}\Gamma\right] = 0 \tag{22.25}$$

with

$$\hat{\mathbf{B}} = \mathbf{L}\hat{\mathbf{L}}\hat{\mathbf{N}}. \tag{22.26}$$

On insertion of Eqs. (22.5) and (22.7) we shall find that the coefficients of p disappear and a set of equations of the standard form

$$(\hat{\mathbf{K}} + \hat{\hat{\mathbf{K}}})\mathbf{a} + \hat{\mathbf{M}}\frac{\mathrm{d}\mathbf{a}}{\mathrm{d}t} + \mathbf{f} = 0 \tag{22.27}$$

can be written with

$$\begin{aligned}
\hat{\mathbf{K}}_{ij} &= \int_{\Omega} \hat{\mathbf{B}}^{\mathrm{T}}\mathbf{D}'\hat{\mathbf{B}}\,\mathrm{d}\Omega \qquad \hat{\mathbf{M}}_{ij} = \int_{\Omega} (\hat{\mathbf{L}}\hat{\mathbf{N}}_i)^{\mathrm{T}}\rho\hat{\mathbf{L}}\hat{\mathbf{N}}_j\,\mathrm{d}\Omega \\
\hat{\hat{\mathbf{K}}}_{ij} &= \int_{\Omega} (\hat{\mathbf{L}}\hat{\mathbf{N}}_i)^{\mathrm{T}}\rho(\boldsymbol{\nabla}(\hat{\mathbf{L}}\hat{\mathbf{N}}\mathbf{a})^{\mathrm{T}})^{\mathrm{T}}\hat{\mathbf{L}}\hat{\mathbf{N}}\,\mathrm{d}\Omega \\
\mathbf{f}_i &= -\int_{\Omega} (\hat{\mathbf{L}}\hat{\mathbf{N}}_i)\mathbf{b}_0\,\mathrm{d}\Omega - \int_{\Gamma i} (\hat{\mathbf{L}}\hat{\mathbf{N}}_i)^{\mathrm{T}}\bar{\mathbf{t}}\,\mathrm{d}\Gamma.
\end{aligned} \tag{22.28}$$

Two points are worth noting beyond the existence of the non-linear and non-symmetric matrix $\hat{\hat{\mathbf{K}}}$ as in the previous formulation. These are, first, that the shape function $\hat{\mathbf{N}}$ now needs to possess C_1 continuity as second order derivatives exist in $\hat{\mathbf{B}}$ and, second, that the formulation is (almost) identical to that of plate bending problems. Indeed, solutions with this procedure have invariably utilized this analogy using many of the elements formerly noted in Chapter 10.[18, 22]

22.3.3 *'Penalty' function formulation.* This has been introduced as an effective procedure for incompressible elasticity in Chapter 11 and, hence, in the present context the method should be applicable. We shall, however, approach it without stating a variational principle.

To eliminate the variable p let us write in the constitutive relation (22.7)

$$p = \alpha \dot{\varepsilon}_v \tag{22.27}$$

where α is a large number. As $\dot{\varepsilon}_v \to 0$ by the constraint equation, Eq. (22.11), p will thus be a finite quantity.

With this substitution the need for discretizing p is eliminated, and for a discretized velocity $\mathbf{u} = \mathbf{Na}$ we have

$$\dot{\varepsilon}_v = \mathbf{m}^{\mathrm{T}}\mathbf{LNa}. \tag{22.30}$$

Pursuing the discretization of Eqs. (22.16) and (22.17) we arrive now at

$$\mathbf{Ka} + \bar{\mathbf{K}}\mathbf{a} + \bar{\bar{\mathbf{K}}}\mathbf{a} + \mathbf{M}\frac{\mathrm{d}}{\mathrm{d}t}\mathbf{a} + \mathbf{f} = 0 \tag{22.31}$$

where all, but one, of the matrices are defined in Eq. (22.18) and

$$\bar{\bar{\mathbf{K}}}_{ij} = \int_{\Omega} (\mathbf{m}^{\mathrm{T}}\mathbf{B}_i)^{\mathrm{T}}\, \alpha(\mathbf{m}^{\mathrm{T}}\mathbf{B}_j)\, \mathrm{d}\Omega. \tag{23.32}$$

Again α can be recognized as analogous to the bulk modulus of elasticity, and indeed the standard form of nearly incompressible elasticity used in Chapter 11 is obtained for slow flow.

The procedure was first formulated in Reference 19 and used subsequently for creeping flow with 'reduced' integration elements. The first effective use of solutions for the full equations of viscous flow was made by Hughes *et al.*[23] using a bi-linear quadrilateral with a single point integration for the volumetric strain rate terms.

22.4 Some Applications of Viscous Flow Forms and Solution Techniques

22.4.1 *Steady-state creeping Newtonian flow.* By 'Newtonian' we mean that the problem is linear with a constant viscosity. With all acceleration terms rejected the formulation gives linear equations and little has to be said about this solution.

Entry flow. In this first example[19] of Fig. 22.2 a solution of entry flow in an axi-symmetric case is obtained

(*a*) by a stream function form in which the Hermitian rectangles of Chapter 10 are used;

(*b*) by a standard elasticity program with isoparametric, 8-node elements utilizing 2×2 Gauss point, 'reduced' integration (near-

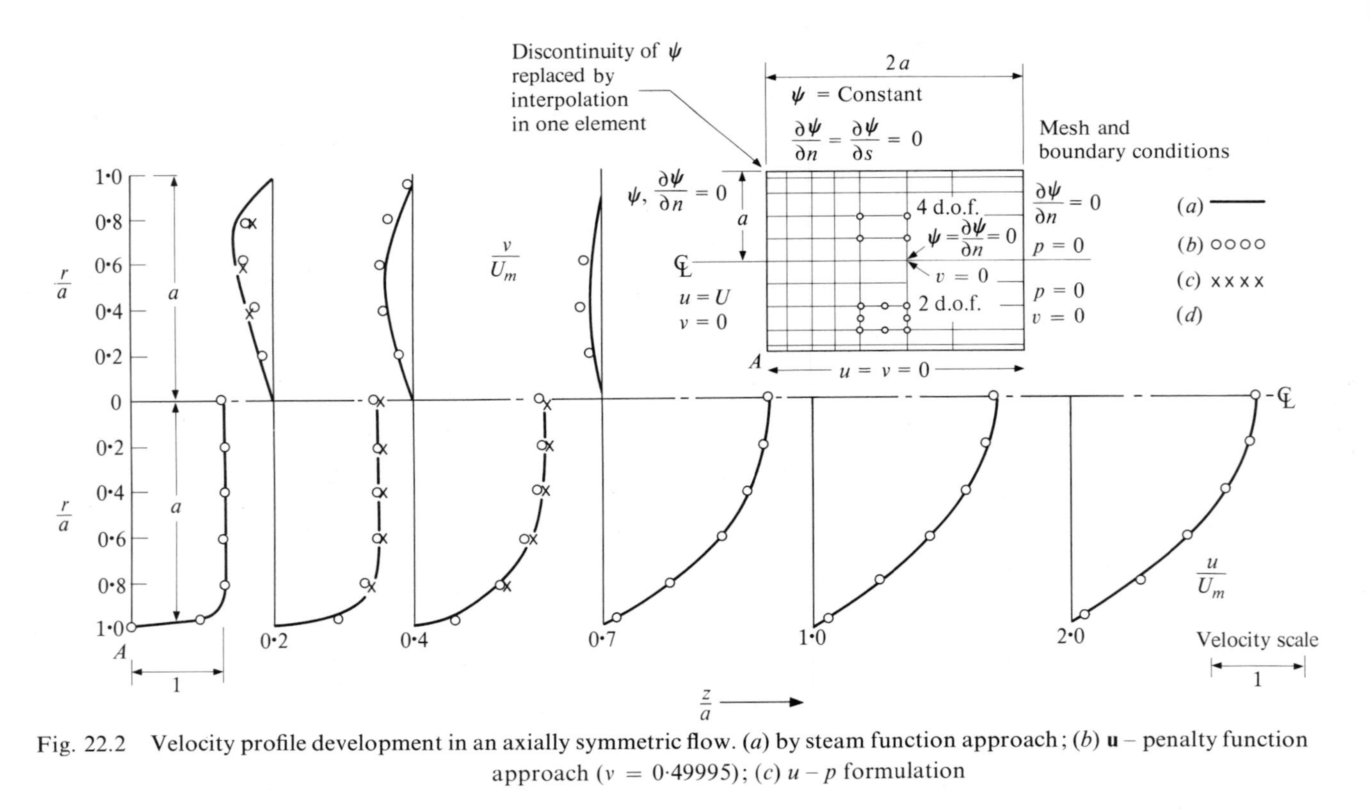

Fig. 22.2 Velocity profile development in an axially symmetric flow. (*a*) by steam function approach; (*b*) **u** – penalty function approach ($\nu = 0{\cdot}49995$); (*c*) $u - p$ formulation

incompressibility is here achieved by setting the equivalent of Poisson's ratio as 0·49995); and

(*c*) by velocity-pressure formulation using a parabolic interpolation for velocities and linear for pressures.[24]

All solutions give almost identical results at a comparable cost and compare well with a finite difference solution of the same problem carried out with a very fine subdivision.[25]

Although there is no apparent difference in the solution technique the last two procedures are easily generalized to three dimensions.

Solution for such three-dimensional flows have been presented in references 19 and 26.

Flow past an obstacle. This example, illustrated in Fig. 22.3 in which solution by both stream function and by penalty-velocity forms were obtained, brings out a further point of difficulty encountered with the former type of discretization. As the distribution of flow is not known initially, the value of the stream function on the obstacle is not known *a priori*—except that it is constant.

An additional requirement has now to be introduced.[27] This states that the rate of work done by boundary tractions on the stationary object must be zero, i.e., that

$$\int_{\Gamma} \delta \mathbf{u}^{\mathrm{T}} \bar{\mathbf{t}} \, \mathrm{d}\Gamma \equiv \int_{\Gamma} (\hat{\mathbf{L}}\, \delta\psi)^{\mathrm{T}} \bar{\mathbf{t}} \, \mathrm{d}\Gamma = 0 \tag{22.33}$$

where Γ is the surface of the obstacle.

Imposition of this condition on the stream function parameters can be made if two independent solutions are carried out—a procedure which is clearly inconvenient and computationally expensive.

22.4.2 *Steady-state, creeping non-Newtonian flow. Visco-plastic metal flow.* In many fluids for which slow rates of flow are of interest the viscosity is a function of the strain rate $\dot{\boldsymbol{\varepsilon}}$. Such fluids comprise many oils, chemicals, and indeed metals.

If the constitutive relation of Eqs. (22.7) and (22.8) is examined we find that it is convenient, in isotropic materials, to write it in terms of second stress and strain rate invariants, $\bar{\sigma}$ and $\dot{\bar{\varepsilon}}$ (for definitions see Chapter 18).

The relation can be written simply as

$$\bar{\sigma} = \mu \dot{\bar{\varepsilon}} \tag{22.34}$$

which, for a Newtonian fluid, is of a form shown in Fig. 22.4. For non-Newtonian fluids the stress strain–rate invariant relationship may take various forms—a typical one being written as

$$\bar{\sigma} = \beta \dot{\bar{\varepsilon}}^{n}. \tag{22.35}$$

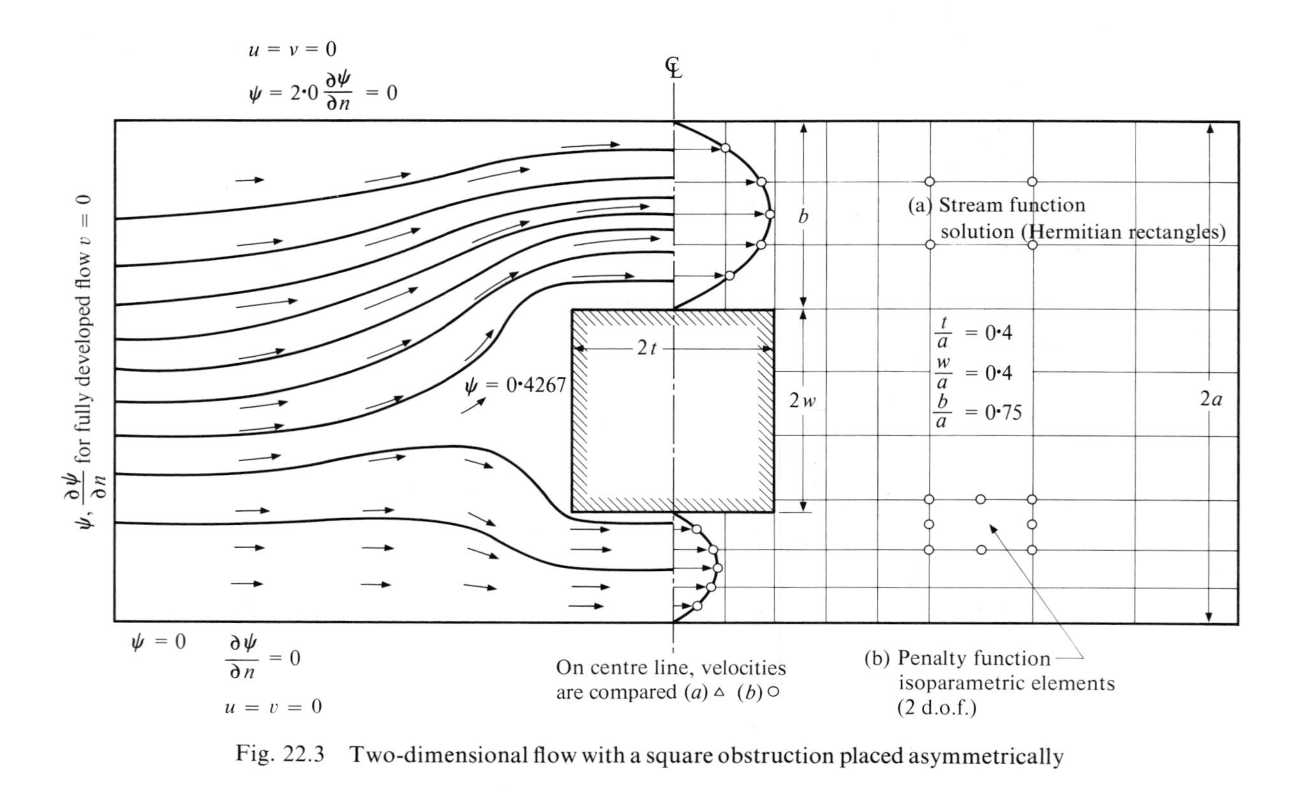

Fig. 22.3 Two-dimensional flow with a square obstruction placed asymmetrically

Clearly this can be represented as a behaviour of the standard form with

$$\mu \equiv \beta \dot{\bar{\varepsilon}}^{n-1}. \tag{22.36}$$

A very characteristic form of behaviour is known as that of Bingham fluid in which a yield stress is exhibited.

This can be written as a particular case of viscoplasticity, giving (see Fig. 22.4)

$$\dot{\bar{\varepsilon}} = \gamma(\bar{\sigma} - \bar{\sigma}_y). \tag{22.37}$$

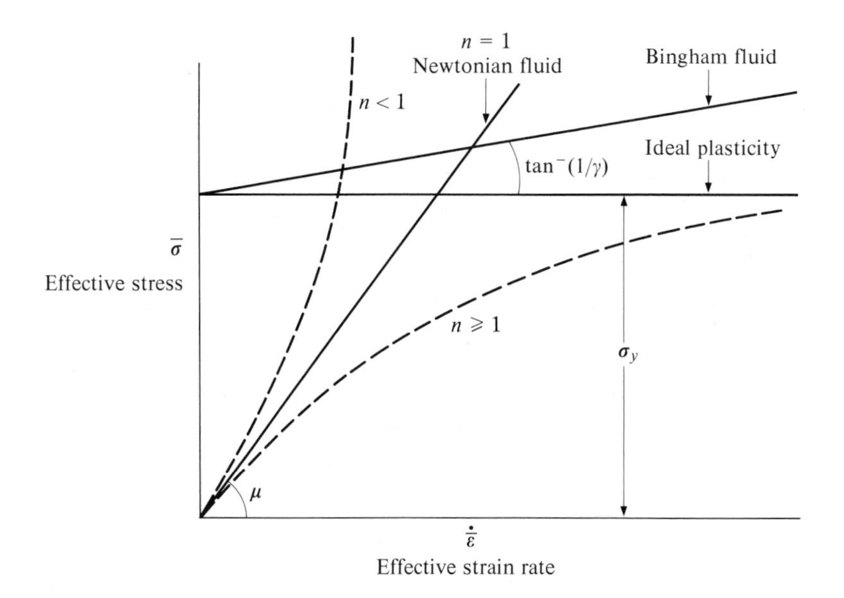

Fig. 22.4 Newtonian and non-Newtonian effective stress–strain relations

Again we can interpret this in terms of variable viscosity of expression (22.34), with

$$\mu = \frac{\dot{\bar{\varepsilon}}/\gamma + \bar{\sigma}_y}{\dot{\bar{\varepsilon}}} \tag{22.38}$$

reducing for an ideally plastic behaviour to

$$\mu = \frac{\bar{\sigma}_y}{\dot{\bar{\varepsilon}}}. \tag{22.39}$$

In Chapter 18 we have discussed the elasto-plastic and elasto-visco-plastic behaviour of many materials. If the deformations are such that elastic strains can be neglected, all such solids behave in effect as non-Newtonian fluids and solutions for their behaviour are easily attainable.

The creeping flow formulations of all types discussed in previous sections have resulted in a general form

$$\mathbf{Ka}+\mathbf{f} = 0 \tag{22.40}$$

Now $\mathbf{K} = \mathbf{K}(\mathbf{a})$ and is a symmetric matrix dependent on the viscosity and hence on the velocity parameters which define the effective strain rate $\dot{\bar{\varepsilon}}$.

Various non-linear solution techniques can be adopted for this problem but the simplest in which the matrix $\mathbf{K}$ is recalculated iteratively with

$$\mathbf{a}^{m+1} = -\mathbf{K}^{m-1}\mathbf{f} \tag{22.41}$$

gives very rapid convergence, even if quite severe non-linearity of the type given by Eq. (22.31) is encountered, providing the forcing function is one of specified boundary velocities.

Many solutions of non-Newtonian flow are available in the literature[28, 29] but the plastic flow situations are of greatest interest.[19, 24, 27, 30]

In Fig. 22.5 we show, for instance, a problem of steady-state extrusion solved[30] as a case of non-Newtonian flow. Comparison with solutions available for the same problems by classical slip line solution confirm the accuracy attainable in modelling such a flow.

22.4.3 *Steady-state viscous flow with inclusion of convective acceleration terms.* We have already remarked that, in the case of steady-state viscous flow in which the convective acceleration terms are retained, all formulations give non-linear equation systems (even if viscosity is Newtonian) of the form

$$(\mathbf{K}+\overline{\mathbf{K}}(\mathbf{a}))\mathbf{a}+\mathbf{f} = 0 \tag{22.42}$$

in which $\overline{\mathbf{K}}(\mathbf{a})$ is a non-symmetric matrix dependent on the solution parameters (velocities).

In fluid mechanics it is usual to characterize the flow by a non-dimensional parameter known as the Reynolds number R_n and defined as

$$R_n = \frac{\rho U d}{\mu} \tag{22.43}$$

where U and d are a characteristic velocity and dimension, respectively. The creeping form which we have previously discussed is thus the limiting case of $R_n \to 0$. Now we shall consider increasing the Reynolds number.

The solution techniques for the non-linear equation system depends evidently on the value of R_n.

For small values of R_n a modified Newton–Raphson technique is effective, using only the constant and symmetric matrix in solution.

At higher Reynolds numbers a full Newton–Raphson iteration is necessary and this involves a repeated solution of a *non-symmetric* equa-

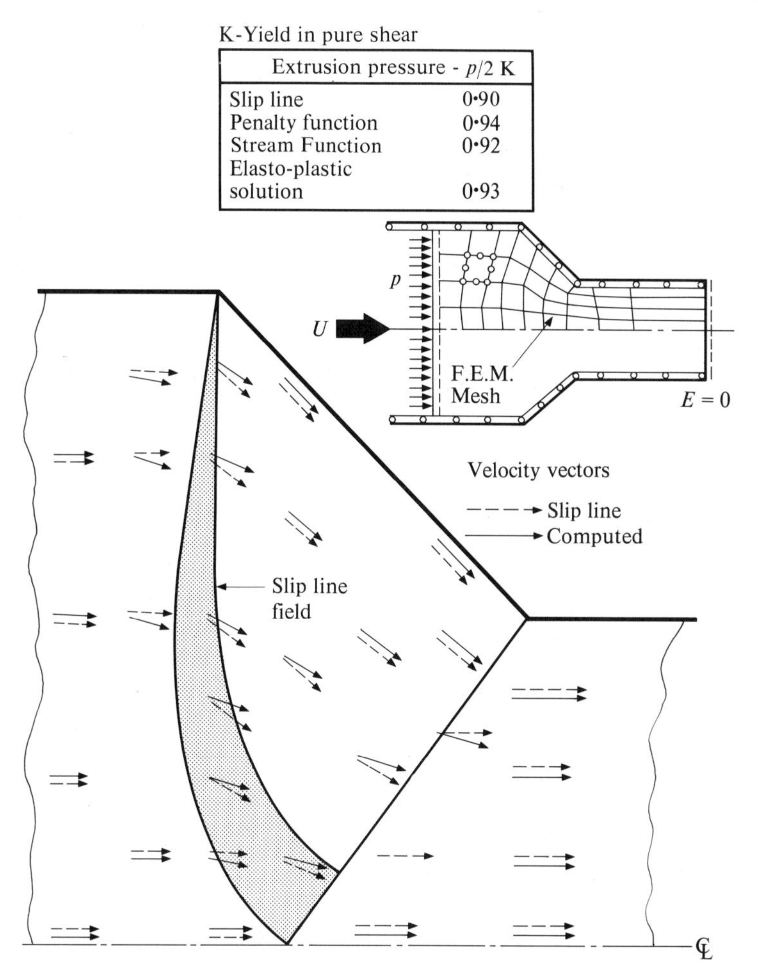

K-Yield in pure shear

	Extrusion pressure - $p/2$ K
Slip line	0·90
Penalty function	0·94
Stream Function	0·92
Elasto-plastic solution	0·93

Fig. 22.5 Plane strain extrusion; ideal plasticity; frictionless walls; penalty function solution

tions system. Other techniques, such as perturbation methods, etc., have also been used with success.[31]

When the Reynolds number becomes very large, and the convective terms predominate, convergence generally ceases. This occurs due to two causes: first, at some value of R_n the flow becomes physically highly unstable and turbulence sets in; second, an instability may be induced by the special character of the approximation to the convective term when the standard Galerkin form is used. This numerical instability will be discussed in depth later.

Purely as an example of such higher R_n computation, we show in Fig. 22.6 some solutions obtained using the velocity–pressure techniques for a flow around a two-dimensional obstacle.[20]

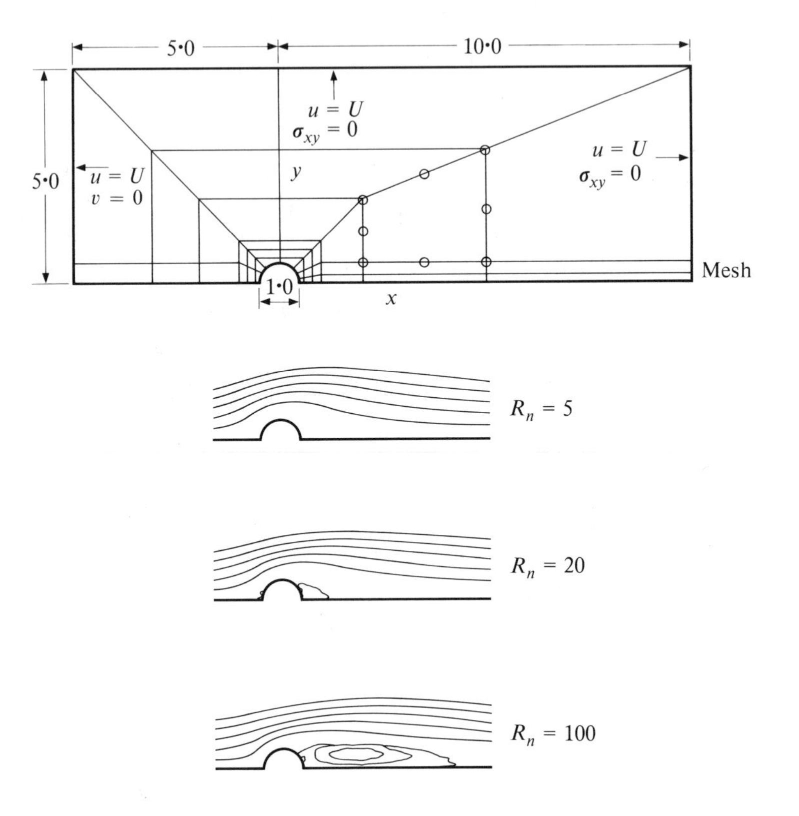

Fig. 22.6 Flow round a cylinder

22.5 Turbulent Flow

As the value of R_n increases, the turbulence which starts with large isolated eddies increases until it becomes widely distributed through the fluid. If *average velocities* are considered then the effect of the turbulence is analogous to that of viscosity and the flow can be represented by the standard viscous equations with the viscosity coefficient now replaced by an *eddy viscosity*, $\bar{\mu}$, which is dependent on the whole velocity field and its gradients. Indeed this behaviour may well be anisotropic, i.e., specified by several such coefficients. In principle, thus, turbulent flow approached in this way presents no more difficulties than those associated with non-Newtonian situations. In practice, unfortunately, no general explicit expressions for determining the eddy viscosity coefficients exist and, at best, very rough solutions are attainable. We shall, however, make use of such turbulence concepts in some aspect of shallow water flow.

22.6 **Transient, Time-dependent Flow and Free Surface Problems**

In principle the transient flow Equations 22.22, 22.27 or 22.31 can be integrated using one or other of the time-stepping processes discussed in

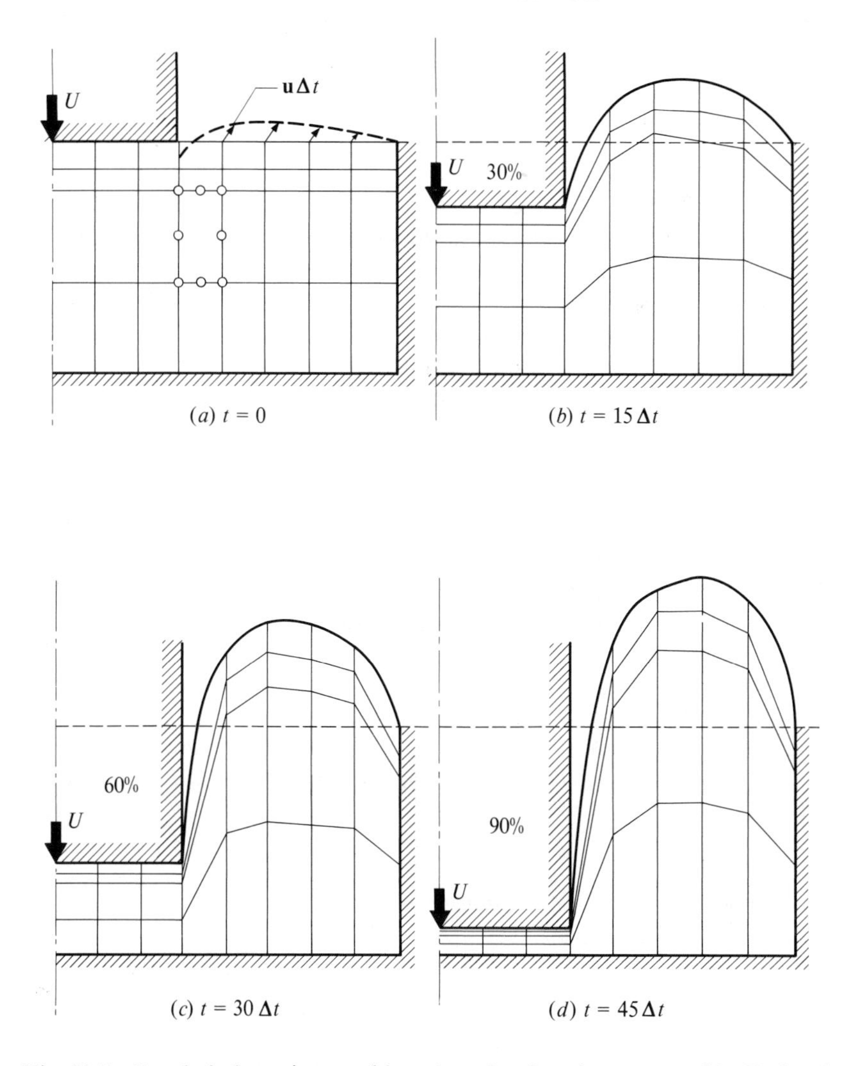

Fig. 22.7 Punch indentation problem (penalty function approach). Updated mesh and surface profile. 24 isoparametric elements. Ideally plastic material; (*a*), (*b*), (*c*) and (*d*) show various depths of indentation.

the previous chapter. Indeed it appears that such a procedure may present a useful technique for obtaining steady-state solutions if these exist. Littie work has as yet been done on this aspect, but some solutions to simple problems have been obtained.[32, 33]

Time-stepping techniques can readily be adopted to follow the development of the form of the free surface. With the initial position of this known, the velocities at the start of a time step determine the position of particles on the free surface at a later time. Iteration can be used here, but if the time interval is not large, a single forward integration giving a change of position, $\mathbf{x}$, as

$$\Delta \mathbf{x}^{m+1} = \mathbf{u}^m \Delta t \tag{22.44}$$

is effective.

This technique, when used in the context of slow (creeping) flow in which all acceleration effects are ignored, necessitates simply an updating of the free surface and a successive resolution of the problem with a new configuration. Indeed the whole mesh can be updated in this way, but if this is found to produce badly shaped elements a new mesh can be generated to fit the new surface at each stage.

Techniques of this kind are extremely useful in a variety of metal forming and rolling processes.[24, 27, 34, 35, 36] In Fig. 22.7 we show successive stages of deformation caused in an ideally plastic metal by a punch.[27]

Steady-state free surface problems are in a sense more difficult. Here we have to ensure that the traction-free surface develops velocities which are *strictly tangential to this surface*,[34, 37] In such cases it is convenient to specify the original surface, obtain a velocity solution, and recompute a new surface by integrating from a known point, noting that the slope is given by the direction of the velocity vector at all points. Three or four repetitions of this process frequently suffice. In Fig. 22.8 we show an axi-symmetric drawing problem of a creeping Newtonian fluid emerging from a tube. The problem is of some importance in glass fibre drawing.

22.7 **Shallow Water Flow: Estuaries and Lakes**

22.7.1 *General equations.* In many problems of practical engineering importance the concern is with flow in bodies of water whose plan dimension is much larger than the depth. Lakes, estuaries, and indeed the oceans provide such examples for which a study of currents caused by wind action, periodic tidal forces, or wave drag is of interest. In contrast to the corresponding plane stress problems, the distribution of velocities across the depth is not uniform and often the changes of depth

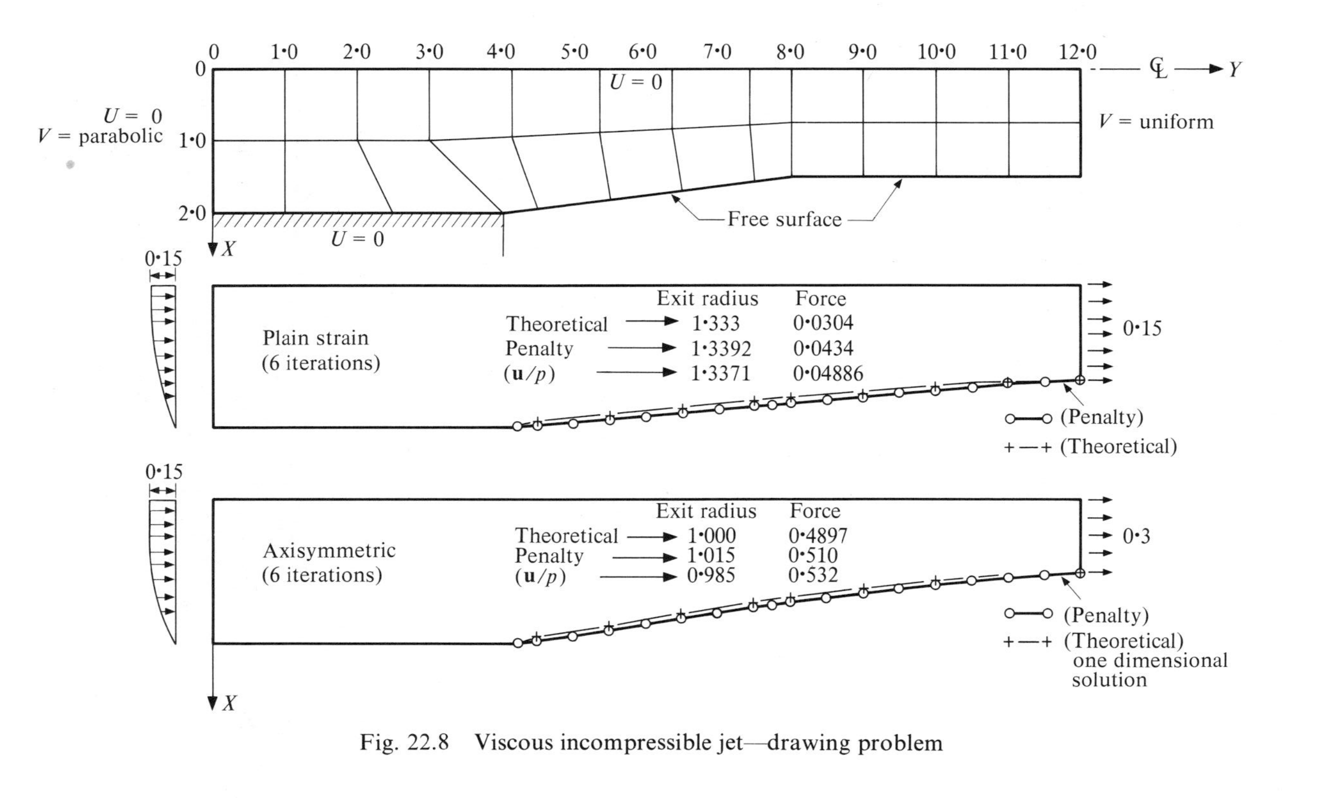

Fig. 22.8 Viscous incompressible jet—drawing problem

provide the main driving forces. Nevertheless, complete three-dimensional analysis of such flows is not practicable and two-dimensional approximations have to be made. Various forms of such approximations are available,[38, 39, 40] and here we shall present a derivation of a set of quite general equations which are of a form not dissimilar to those of Navier–Stokes' equations already derived, but now involve some additional terms.

In subsequent parts of this section we shall show particular simplified forms of the equations which are of some practical interest.

The derivation of basic shallow water equations uses the assumptions that the vertical accelerations are negligible and that the pressure distribution in the vertical directions is hydrostatic, i.e. (see Fig. 22.9),

$$p = \rho g(\eta - z) + p_a \tag{22.45}$$

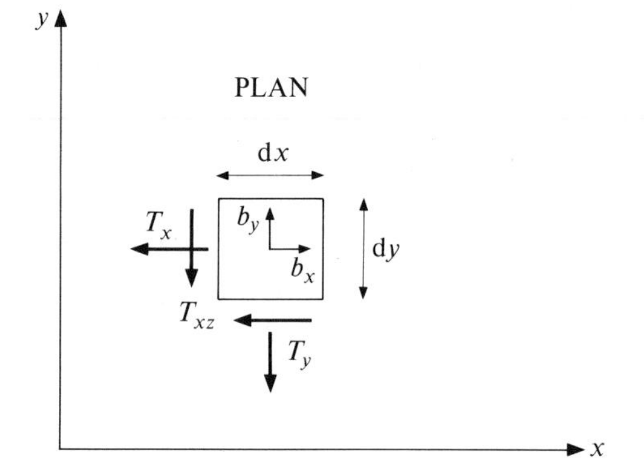

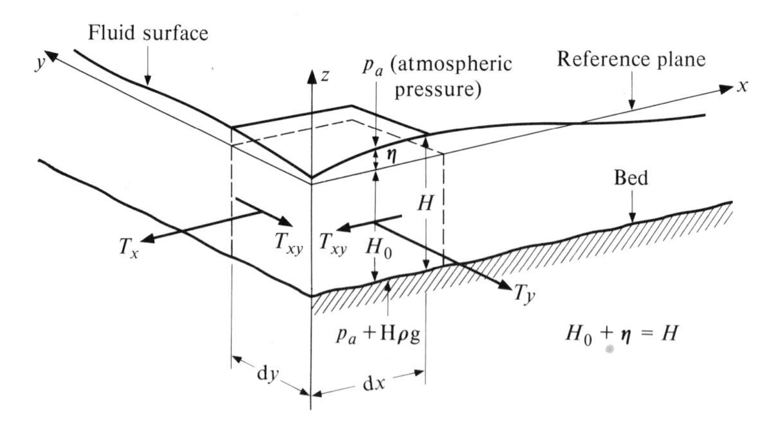

Fig. 22.9 Definitions for shallow water flow problem

where p_a is the atmospheric pressure. Further, we shall be concerned only with the average velocities in the plan direction, i.e., U or V.

$$U = \frac{1}{H}\int_{-H_0+\eta}^{\eta} u \, dz; \qquad V = \frac{1}{H}\int_{-H_0+\eta}^{\eta} v \, dz. \tag{22.46}$$

With these two assumptions the overall continuity and equilibrium relations can be written.

For complete generality we assume that the density ρ can vary with position in the plan (of importance when density currents are considered) and we can write the continuity condition for a prism of unit plan area shown in Fig. 22.9 as

$$\frac{\partial}{\partial x}(\rho HU)+\frac{\partial}{\partial y}(\rho HV)-\rho\frac{\partial \eta}{\partial t} = 0 \tag{22.47}$$

where the last term gives the rate of fluid accumulation due to the rising surface.

To examine the equilibrium in the plan directions we shall proceed in a manner completely analogous to that used in deriving the general viscous equilibrium equations.

First we observe that on the faces of an elementary prism we have tractions T_x, T_y, and T_{xy} which are due partly to the pressures and partly to turbulent mass transfer in which $\bar{\mu}$ is the eddy viscosity. Thus

$$T_x = -\int_{-H_0+\eta}^{\eta} p \, dz + 2\bar{\mu}H\frac{\partial U}{\partial x} = -\rho g\frac{H^2}{2} - p_a H + 2\bar{\mu}H\frac{\partial U}{\partial x} \tag{22.48a}$$

Similarly,

$$T_y = -\rho g\frac{H^2}{2} - p_a H + 2\bar{\mu}H\frac{\partial V}{\partial y} \tag{22.48b}$$

and

$$T_{xy} = \bar{\mu}H\left(\frac{\partial U}{\partial y}+\frac{\partial V}{\partial x}\right). \tag{22.48c}$$

We note that the tractions due to turbulent mass transfer are of precisely the same form as those associated with the viscosity coefficients in Eq. (22.7) but that the 'pressures' are now defined in terms of the depth H which plays, in shallow water equations, the same role as the density in compressible flow (in fact we shall find that the equations of shallow water flow bear a striking resemblance to compressible flow equations).

The tractions given by Eq. (22.48) must be in equilibrium with the appropriate body force vector **b** and the equilibrium equations (Eqs.

(22.1a) and (22.1b)) can again be written with the operator **L** appropriate to the two-dimensional problems as

$$\mathbf{L}^{\mathrm{T}}\mathbf{T}+\mathbf{b} = 0 \qquad \mathbf{T}^{\mathrm{T}} = [T_x, T_y, T_{xy}]. \tag{22.49}$$

For discretization it is convenient to use the virtual work equation corresponding to Eq. (22.2), i.e.,

$$\int_{\Omega} \delta\dot{\boldsymbol{\varepsilon}}^{\mathrm{T}}\mathbf{T}\,\mathrm{d}\Omega - \int_{\Omega} \delta\mathbf{U}^{\mathrm{T}}\mathbf{b} - \int_{\Gamma} \delta\mathbf{U}^{\mathrm{T}}\overline{\mathbf{T}}\,\mathrm{d}\Gamma = 0 \tag{22.50}$$

where $\overline{\mathbf{T}}$ stands for prescribed boundary 'tractions' (which depend on the depths H) on suitable boundaries.

Once the body force vector is available the discretization can be written in the standard manner which we shall not pursue here in detail. However, it is essential to define the body force vector, as several terms not previously encountered now enter the problem. As before we can write (remembering that a depth of fluid H is considered)

$$\mathbf{b} = \mathbf{b}_0 - \rho\mathbf{c}H \tag{22.51}$$

where the acceleration **c** is

$$\mathbf{c} = \frac{\partial \mathbf{U}}{\partial t} + (\nabla\mathbf{U}^{\mathrm{T}})^{\mathrm{T}}U$$
$$\mathbf{U}^{\mathrm{T}} = [U, V]. \tag{22.52}$$

Further, the vector $\mathbf{b}_0$ is now specifically divided into several causes—and can be written as the sum of the following:

(*a*) Coriolis effects: if rotation of the earth is important due to extent of problems $-\rho f\begin{bmatrix} 0 & 1 \\ -1 & 0 \end{bmatrix}\mathbf{U}$, where f is the Coriolis parameter (f = 2 × angular velocity of frame of reference rotation).

(*b*) Surface traction due to wind (or waves), $\boldsymbol{\tau}$.

(*c*) Bottom traction resisting motion, $-\beta\mathbf{U}$, with β a coefficient dependent on the absolute value $|\mathbf{U}|$ if turbulent conditions exist.

(*d*) Horizontal component of surface pressure, $p_a \nabla\eta$.

(*e*) Horizontal component of bottom pressure, $(p_a + \rho g H)\nabla H_0$.

Noting that the essential variables of the problem are the velocity (mean) vector **U** and the surface elevation η as

$$H = \eta + H_0 \tag{22.53}$$

where H_0 is a known depth of the mean water surface, the full discretized equations can be written in a manner analogous to that described in section 22.3.1 by using the virtual work statement and a weighted form of the continuity equation, Eq. (22.47). We shall spare the reader the details which he can readily fill in, but before proceeding further we shall give an explicit form of the equilibrium equation (in a manner equivalent to that of (Eq. 22.13)) as such equations have been used as the starting point of the discretization by many investigators. With an explicit form of the operator **L**, which we remind the reader is given by

$$\mathbf{L}^{\mathrm{T}} = \begin{bmatrix} \dfrac{\partial}{\partial x} & 0 & \dfrac{\partial}{\partial y} \\ 0 & \dfrac{\partial}{\partial y} & \dfrac{\partial}{\partial x} \end{bmatrix} \tag{22.54}$$

the substitution of **b** and **T** into Eq. (22.49) results in two differential equations, i.e.,

$$\begin{aligned} H\rho\left(\frac{\partial U}{\partial t}+U\frac{\partial U}{\partial x}+V\frac{\partial U}{\partial y}-fV\right) \\ = -\rho g H\frac{\partial \eta}{\partial x} - g\frac{H^2}{2}\frac{\partial \rho}{\partial x} - \beta U + H\frac{\partial p_a}{\partial x} + 2\frac{\partial}{\partial x}\left(\bar{\mu}\frac{\partial}{\partial x}(HU)\right) \\ + \frac{\partial}{\partial y}\left(\bar{\mu}\frac{\partial}{\partial y}(HU)+\bar{\mu}\frac{\partial}{\partial x}(HV)\right)+\tau_x = 0 \end{aligned} \tag{22.55}$$

with a similar equation for the y direction.

As the surface elevation η is generally small compared with the depth H it simplifies matters to assume that

$$H \approx H_0.$$

With this simplification the reader will note that Eq. (22.55) together with the Eq. (22.47) are essentially similar to the Navier–Stokes equation in two dimension Eq. (22.13) and Eq. (22.9), but now contain some additional terms. Techniques of solution for both steady-state and transient situations for both problems will be essentially identical and therefore it is convenient to write programs capable of incorporating both classes of problems simultaneously.

In Fig. 22.10 we show some results of computations carried out for determination of tidal currents in Tokyo harbour.[41] Here the periodic nature of the tides was used to simplify the time response by a harmonic analysis applied despite the inherent non-linearity of the equations.

22.7.2 *Simplified shallow water flow equations.* Very few investigators have so far used the full set of equations of shallow flow even if certain forces or effects are absent.[42,43]

First, the horizontal eddy viscosity terms are frequently dropped.[44,45] When this is done it should be remembered that only one velocity component can be prescribed as the boundaries (the second order of the

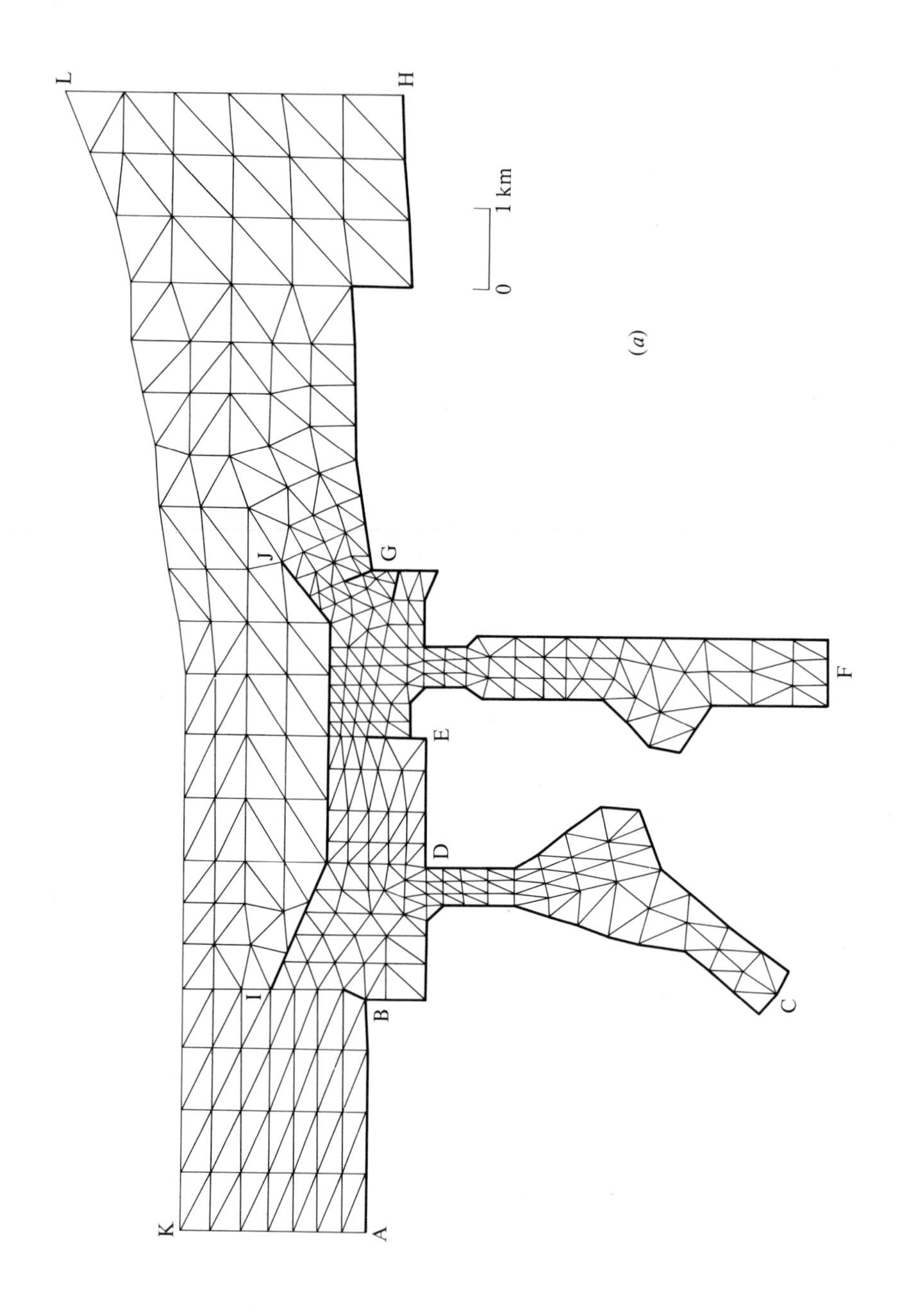

(a)

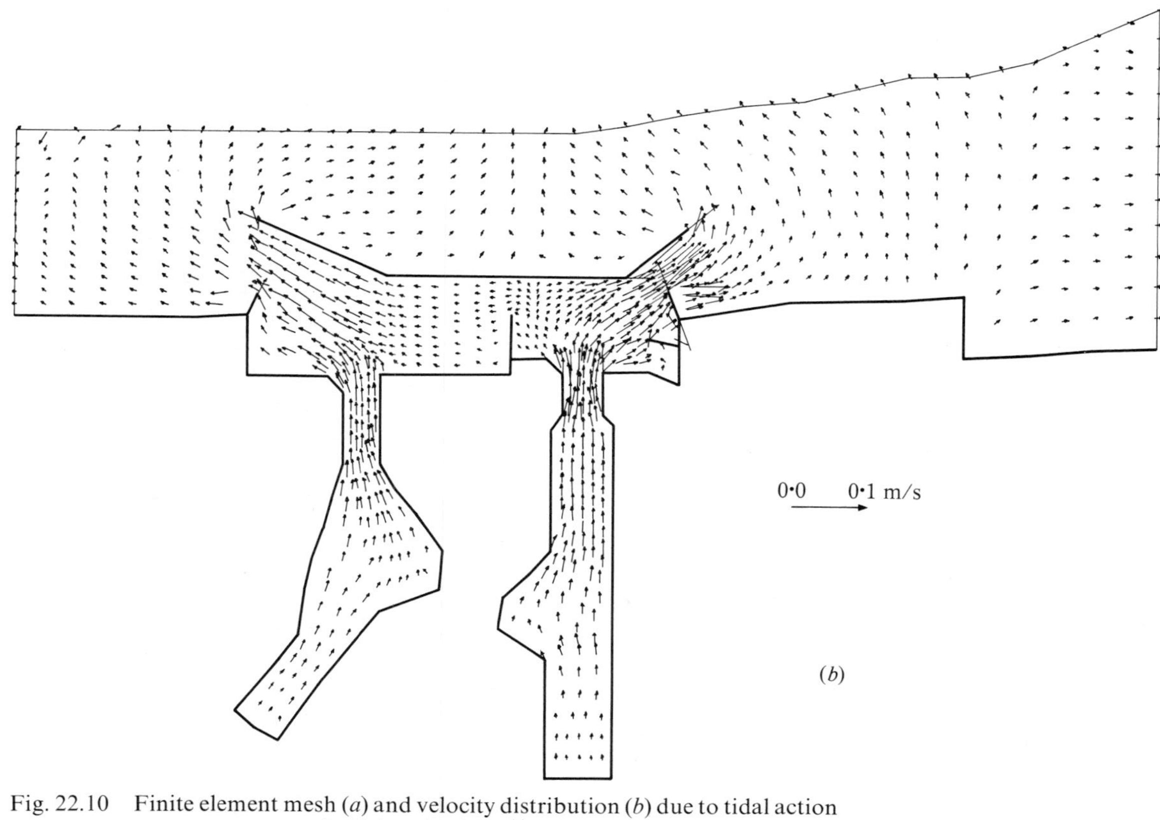

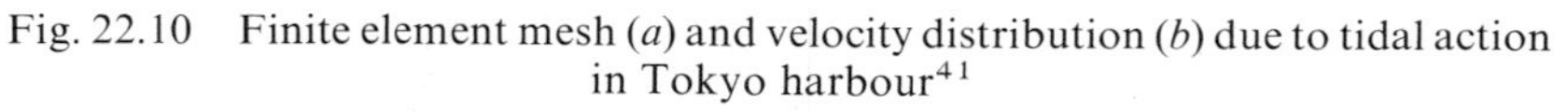
Fig. 22.10 Finite element mesh (*a*) and velocity distribution (*b*) due to tidal action in Tokyo harbour[41]

equations being now reduced to first). Indeed the 'no slip' condition on boundaries can no longer be imposed.

Second, the convective terms are frequently omitted thus linearizing the equation system (22.47) and (22.55) to a much simpler form (for constant density ρ and taking $H = H_0$).[46]

$$
\begin{aligned}
&\frac{\partial}{\partial x}(H_0 U)+\frac{\partial}{\partial y}(H_0 V)-\frac{\partial \eta}{\partial t} = 0 \\
&\rho H_0\left(\frac{\partial U}{\partial t}-fV\right) = -\rho g H_0 \frac{\partial \eta}{\partial x}-H_0\frac{\partial p_a}{\partial x}-\beta U+\tau_x \\
&\rho H_0\left(\frac{\partial V}{\partial t}+fU\right) = -\rho g H_0 \frac{\partial \eta}{\partial y}-H_0\frac{\partial p_a}{\partial x}-\beta V+\tau_y .
\end{aligned}
\tag{22.56}
$$

For steady-state conditions it is convenient to introduce once again the notion of a stream function. Defining

$$
H_0 U = \frac{\partial \psi}{\partial y}; \qquad H_0 V = -\frac{\partial \psi}{\partial x} \tag{22.57}
$$

the first of Eqs. (22.56) (with time differentiation omitted) is identically satisfied.

On elimination of η between the second pair we find that the governing equation reduces simply to the quasi-harmonic form (discussed in Chapter 17):

$$
\frac{\partial}{\partial x}\left(\frac{\beta}{H_0}\frac{\partial \psi}{\partial x}\right)+\frac{\partial}{\partial y}\left(\frac{\beta}{H_0}\frac{\partial \psi}{\partial y}\right) = \frac{\partial \tau_x}{\partial y}-\frac{\partial \tau_y}{\partial x} \tag{22.58}
$$

It is of interest to see that the Coriolis and pressure gradient forces do not now affect the solution. Despite this drastic simplification apparently reasonable predictions of wind (or wave) induced velocities can be made.[47,48] It seems, however, desirable to ascertain in all cases the errors due to the approximation; this can always be done by computing the contributions due to the omitted terms.

22.7.3 *Long wave equations.* If the time derivative terms of Eqs. (22.56) are not omitted we find that these equations are typical of wave problems.

For instance, if the drag, Coriolis, and pressure force terms are omitted we can eliminate U and V (by differentiation of the first equation with respect of time and the second and third by x and y, respectively) and obtain the classical wave equation of the form

$$
\frac{\partial}{\partial x}\left(H_0\frac{\partial \eta}{\partial x}\right)+\frac{\partial}{\partial y}\left(H_0\frac{\partial \eta}{\partial y}\right)-\frac{1}{\rho g}\frac{\partial^2 \eta}{\partial t^2} = 0. \tag{22.59}
$$

We have discussed this type of equation in Chapter 20 and some aspects of it are dealt with in Chapter 23.

With the drag terms included it is possible, by the introduction of some further mathematical approximations, to derive an equation of type (22.59) with a damping term included.

22.8 **Convective Transport Equation and some Special Finite Element Problems. 'Upwind' Weighting**

22.8.1 *The convective transport problem.* In the basic fluid problem discussed in this chapter we have encountered a new type of term, i.e., that of convective acceleration (*vide* Eq. (22.5))

$$\rho(\nabla \mathbf{u}^{\mathrm{T}})^{\mathrm{T}}\mathbf{u} \tag{22.60}$$

which has caused a major difficulty by introducing a *non-symmetric matrix* into the final equation.

Terms of this kind arise in an Eulerian formulation (i.e., one in which a fixed space element is considered) when a certain quantity is *transported* by a velocity field $\mathbf{u}$. In Eq. (22.60) the quantity is the *momentum* but in many problems it may be, say, the amount of a chemical dissolved in the fluid or, of heat carried by it, etc.

If, for instance, we consider the heat transfer in a moving fluid in which the velocity $\mathbf{u}$ is known, then the heat balance equation derived by precisely the same reasoning as that concerning heat diffusion in Chapter 17 and Chapter 20 (Eq. 17.6 and Eq. 20.1) now contains an additional term. The balance equation now takes the following form:

$$\nabla^{\mathrm{T}}(k\,\nabla\phi)+Q-c\frac{\partial\phi}{\partial t}-\nabla^{\mathrm{T}}(c\phi\mathbf{u}) = 0 \tag{22.61}$$

where the last term is due to the transport of the heat content $c\phi$ by the moving fluid.

In steady flow, if the velocities obey the incompressibility condition, i.e. if

$$\nabla^{\mathrm{T}}\mathbf{u} = 0 \tag{22.62}$$

this equation can be written as

$$\nabla^{\mathrm{T}}(k\,\nabla\phi)-\nabla^{\mathrm{T}}(c\phi)\mathbf{u}+Q = 0 \tag{22.63}$$

or if c is independent of position and the diffusivity is isotropic we have

$$\nabla^{\mathrm{T}}(k'\,\nabla\phi)-(\nabla^{\mathrm{T}}\phi)\mathbf{u}+Q' = 0 \tag{22.64}$$

with $k' = k/c$ and $Q' = Q/c$.

This type of convective problem is of extreme importance in all branches of physics and engineering. Heat transfer in fluid machinery, dispersion of pollutant in shallow water, etc., are but a few examples.

In Chapter 3 (section 3.5, p. 56) we have already treated this specific example by the Galerkin process and, beyond remarking that now non-symmetric matrices arise (just as in the corresponding fluid mechanics case), the impression was given that no special difficulties arise. However, in the context of high velocity fluid flows we have already remarked on page 621 that numerical instability has sometimes been noted. We shall now investigate the problem further—in the context of the simple form of Eq. (22.64). All the remarks which will be made are applicable to the more complex forms and, indeed, to the basic fluid flow problem itself.

In passing it should be noted that in equations of the type (22.64) the relative importance of the two terms will be obviously of crucial importance on the very nature of the problem. If, for instance, the conductivity (diffusivity) terms were to become zero, then we would have only a first order equation which clearly would not allow the specification of the same number of boundary conditions and would be one of initial value-propagation type.[49] We would indeed find that the temperature conditions at entry would govern entirely the solution and *downstream* conditions could not be imposed. Clearly, if k' has a small, but non-zero, value the solution will still have to be of the same nature, and the downstream effects will be highly localized. It is here that the essentials problems lie.

22.8.1 *General discretized form and the Galerkin approximation.* In section 3.5, p. 56 of Chapter 3 we have discretized the problem of Eq. (22.64) *a priori*, specifying the Galerkin form of weighting. Proceeding with an arbitrary weighting function set W_i we can similarly derive a more general discretization (which the reader can check as an exercise).

Writing

$$\phi = \sum N_i a_i = \mathbf{N}\mathbf{a} \tag{22.65}$$

a system of linear equations of standard form

$$\mathbf{K}\mathbf{a}+\mathbf{f} = 0 \tag{22.66}$$

is obtained, in which (omitting all boundary contributions for simplicity)

$$\mathbf{K}_{ij} = \int_\Omega (\nabla W_i)^{\mathrm{T}} k' (\nabla N_j)\, \mathrm{d}\Omega + \int_\Omega W_i \mathbf{u}^{\mathrm{T}} \nabla N_j \, \mathrm{d}\Omega \tag{22.67}$$

$$\mathbf{f}_i = \int_\Omega W_i Q \, \mathrm{d}\Omega$$

Now the formulations and solutions can be obtained using any desired element form or weighting function.

To illustrate the difficulty which is encountered we shall consider a simple entry flow region, Fig. 22.11, where the velocity solution is assumed known (or derived as it was in this example by processes of viscous flow solution). With the diffusivity k' assumed constant, the solutions will be characterized by a non-dimensional (Péclet) number

$$P_e = Ud/k' \tag{22.68}$$

where U is the entry velocity and d a typical problem dimension (in our case half the duct width).

Using $W_i = N_i$, i.e., the standard Galerkin procedure, and a bilinear isoparametric element mesh shown in Fig. 22.11, we find that reasonable results are obtained for $P_e = 3{\cdot}75$ but that on increasing this the results deteriorate until at $P_e = 37{\cdot}5$ a meaningless oscillation is obtained. Clearly this situation is not acceptable and some corrective measure has to be taken.

22.8.2 *One- and two-dimensional problems—'upwind' weighting functions.* The difficulty just mentioned has been noted repeatedly in finite difference context[50–52] and in the finite element field the remedy was derived very recently.[53–56]

To appreciate the problem it is convenient to consider it first in one dimension, using standard linear interpolation functions shown in Fig. 22.12 and elements of a constant size h with a constant velocity u throughout. Without loss of generality the homogenous problem ($Q = 0$) is used and the weighting function will be assumed within each element to have a form

$$W_i = N_i(x) + \alpha F(x) \tag{22.68}$$

where α is positive when u is directed towards node; and

$$F(x) = -3x(x-h)/h^2 \tag{22.69}$$

is chosen so as to have a positive value in each element and zero values at the nodes to preserve C_0 continuity. Clearly $\alpha = 0$ will reproduce once again the classic Galerkin process.

On discretization and assembly of equations for a typical node the following (difference) equation is found

$$\left[1+\frac{\gamma}{2}(|\alpha|+1)\right]\phi_{i-1}-(2+\gamma\,|\alpha|)\phi_i+\left[1+\frac{\gamma}{2}(|\alpha|-1)\phi_{i+1}\right] = 0 \tag{22.70}$$

where the parameter γ is defined as

$$\gamma = uh/k' \tag{22.71}$$

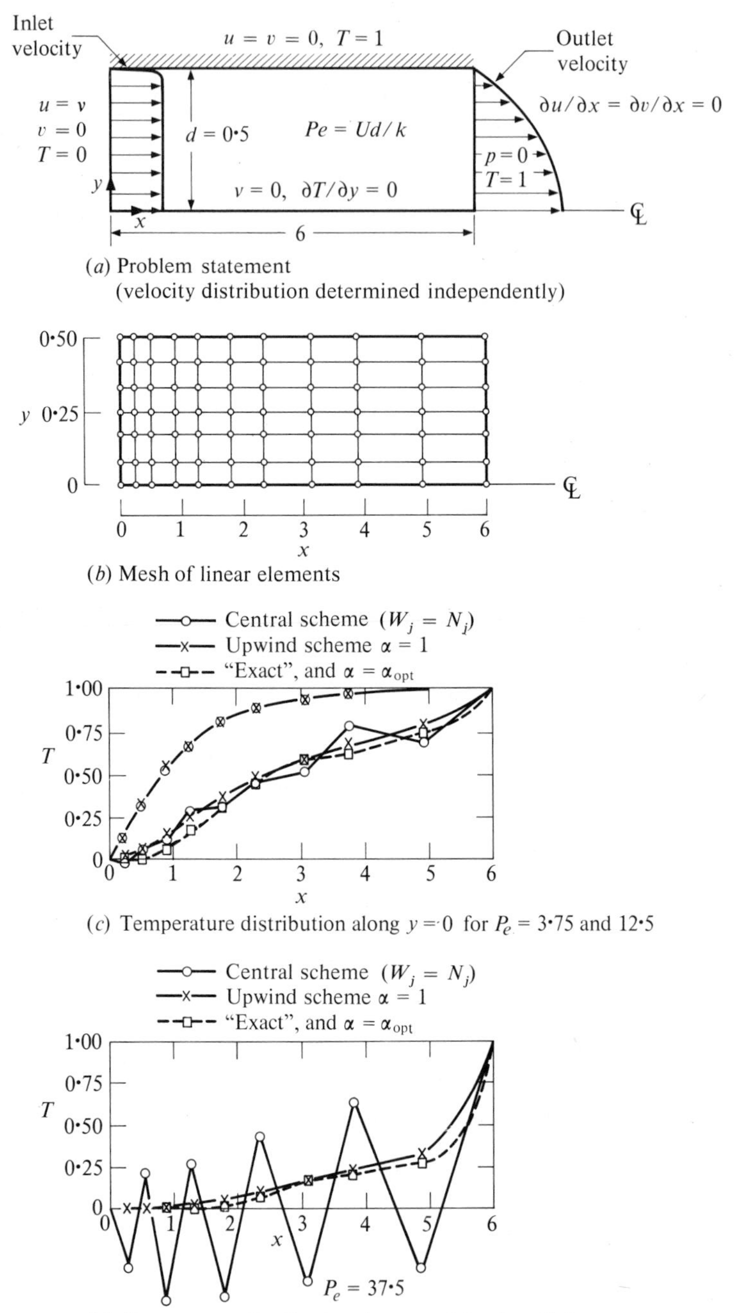

Fig. 22.11 Thermal convection—diffusion in entry flow. (*a*) Problem statement; (*b*) Mesh 2—linear elements; (*c*); and (*d*) temperature distribution for various P_e numbers and discretization procedures.

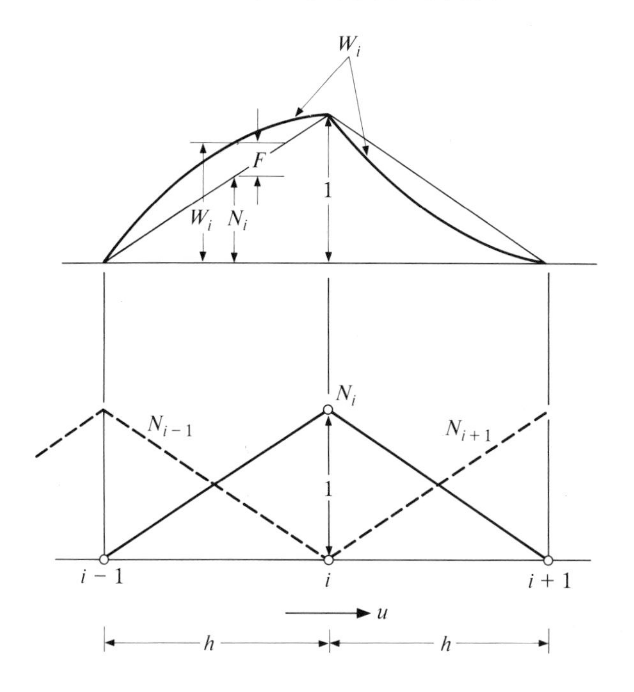

Fig. 22.12 One-dimensional problem. Shape functions (N_i) and weighting functions (W_i). Constant velocity u

The 'exact' solution of this difference equation is obtained in reference 54 and is given by

$$\phi_i = A + B\left[\frac{1+(|\alpha|+1)\,\gamma/2}{1+(|\alpha|-1)\,\gamma/2}\right]^i \tag{22.72}$$

where A and B are constants determined from the boundary conditions. The solution will be oscillatory unless

$$|\alpha| > \alpha_c = 1 - 2/\gamma; \qquad (\text{or } \gamma \leqslant 2). \tag{22.73}$$

Further it can be shown[54] that the *exact* solution to the original differential equation will be obtained at nodal points if

$$|\alpha| = \alpha_0 = \coth \gamma/2 - 2/\gamma. \tag{22.74}$$

In Chapter 3 we have indeed noted that in a simple diffusion problem of Fig. 3.4, in which $\gamma = 0$, such exact solutions were obtained with all meshes at the nodal points. The above result generalizes this observation, indicating the *best choice of a weighting function*.

In Fig. 22.13 we show that the stable and optimal values of α differ but little for higher values of γ and due to simplicity of computation the simpler expression (22.73) may be preferred. The process gives results very similar to those achieved in finite difference approaches by using 'upwind differences'. In a recent publication by Barrett[57] the possibility of using such optimal values of α is indeed also suggested.

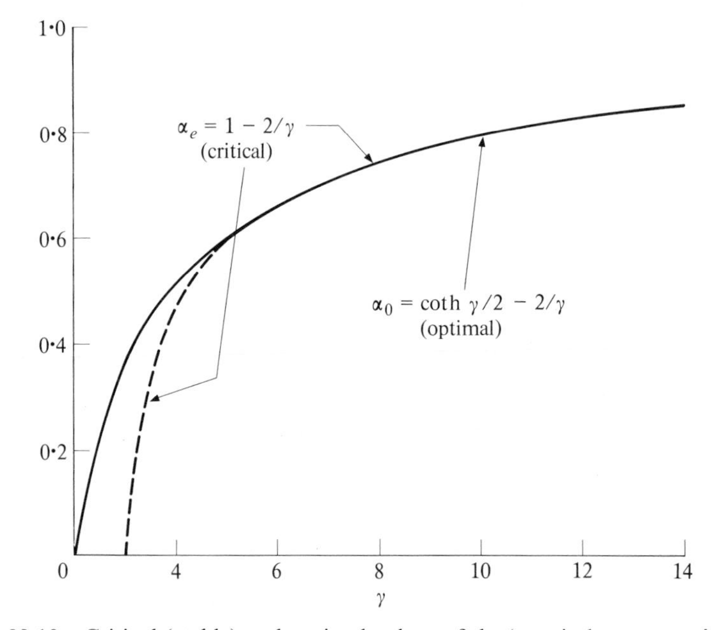

Fig. 22.13 Critical (stable) and optimal values of the 'upwind parameter' α for different values of $\gamma = uh/k'$

To generalize the result to a two- (or three-) dimensional field appears at first sight to be difficult, as simple 'exact' solutions of difference equations are not available. However, proceeding pragmatically the two-dimensional weighting functions can be derived by simply using the appropriate products of such one-dimensional functions as shown in Fig. 22.14. After all this was the basic process of deriving the two-dimensional shape functions discussed in Chapter 7.

To take account of the generally varying velocity field, the optimal α value is chosen in accordance with expression (22.74), depending on the flow *velocity component along the side* (e.g., u_{ij} from node i to node j). Figure 22.14 illustrates how such weighting functions may vary from node to node of the bilinear element. The success and stability of using $\boldsymbol{\alpha} = \boldsymbol{\alpha}_0$ is shown in example of Fig. 22.12.

Processes similar to the one described above can be developed, albeit

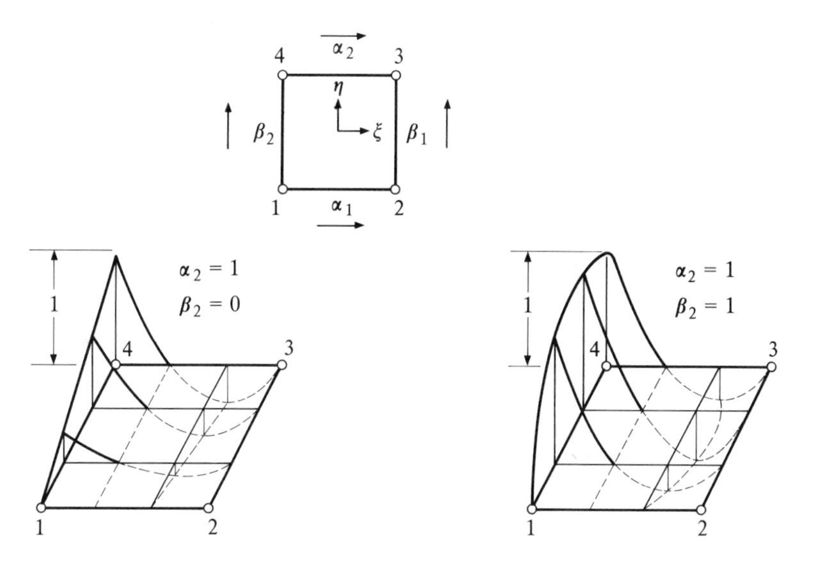

Fig. 22.14 Typical weighting functions for a bi-linear two-dimensional element (parent co-ordinates). Velocity sign convention

with more difficult mathematics, for higher order elements. This has been done in reference 56 for quadratic elements where now two parameters have to be determined along each side. The necessity for doing this is still there as the oscillations develop even to a greater extent with such higher order elements. In Fig. 22.15 we show, for instance, the meaningless results obtained with curved isoparametric quadratic elements for a diffusion problem to a hypothetic estuary and the improvement consequent on the use of 'upwinding'.

22.9 **Some Further Problems of Fluid Mechanics and Concluding Remarks**

The scope of this chapter has permitted only a brief mention of some typical fluid flow problems. Compressible subsonic and supersonic flow have not been touched upon despite quite an extensive literature already appearing on finite element approximations in those areas.[58–60]

Even with incompressible situations many interesting coupled flow problems could not be discussed. For instance, density changes owing to temperature variation are frequently important contributors to the velocity field development. As the temperature field is in turn influenced by such velocity variations, an iterative complex formulation is necessary. Figure 22.16 from reference 53 shows such a coupled solution in a ventilating duct. Other similar coupled problems of heat generation, and the consequent temperature changes (which in turn affect the viscosities), are discussed in the context of plastic non-Newtonian flows of metals in references 24, 34, and 56—the latter showing the importance of considering the convective terms by processes discussed in the last section.

While the applications of the finite element method in the field of classical fluid mechanics is fairly obvious, the flow of rocks on a geological

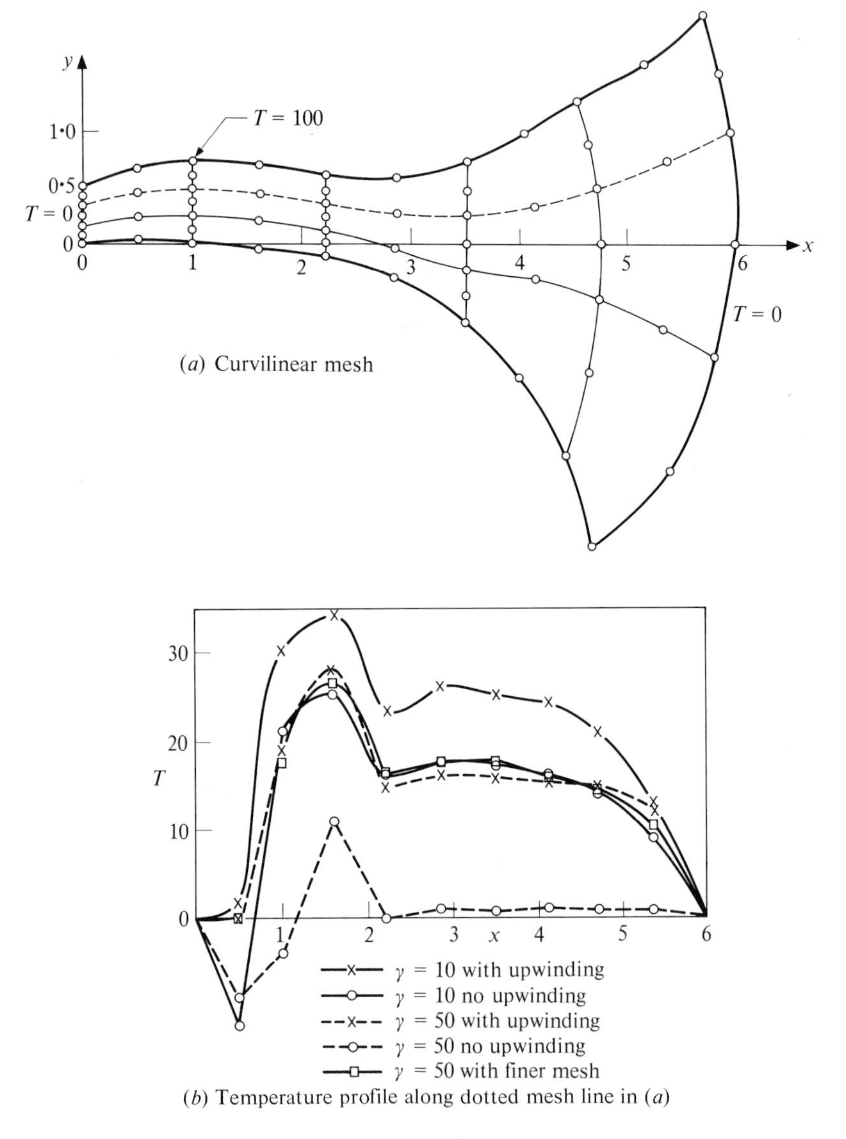

Fig. 22.15 Steady state convective diffusion of a pollutant in an estuary

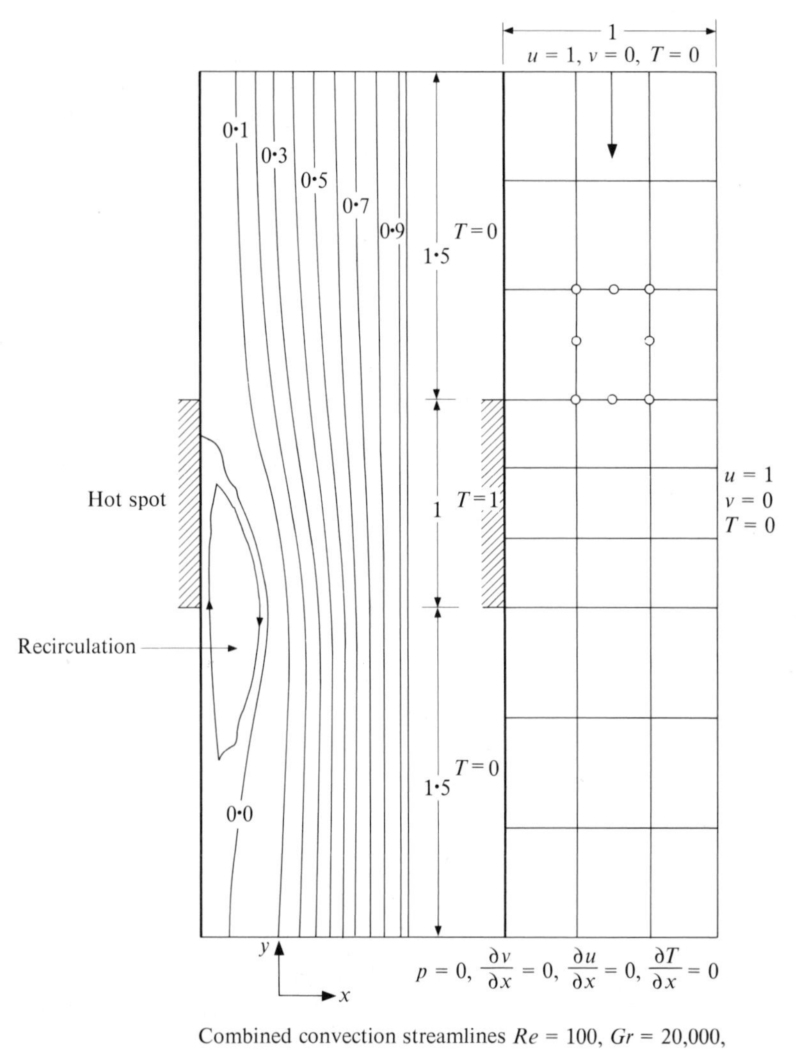

Combined convection streamlines $Re = 100$, $Gr = 20{,}000$, $Pr = 1{\cdot}0$ heated duct

Fig. 22.16 Recirculation caused by a hot spot in a duct with uniform velocity

time scale is but one of the new areas in which the principles of this chapter are directly applicable. A survey of possible application is given in reference 61.

References

1. J. T. ODEN, O. C. ZIENKIEWICZ, R. H. GALLAGHER, and C. TAYLOR (eds.), *Finite Element Methods in Flow Problems* (Proceedings 1st Symp., Swansea, 1974), Univ. of Alabama Press, 1974.
2. J. T. ODEN, O. C. ZIENKIEWICZ, R. H. GALLAGHER, and C. TAYLOR (eds.), *Finite Elements in Fluids*, Vols. I and II, J. Wiley and Sons, 1975.
3. *Proc. 2nd Int. Symp. on Finite Elements in Fluid Problems ICCAD*, St. Margharita Ligure, Italy, 1976.
4. *Finite Elements in Fluids*, Vol. III. Survey lectures presented at 2nd Int. Symp. at St. Margharita Ligure. To be published by J. Wiley & Sons, 1977.
5. B. L. HEWITT, C. R. ILLINGWORTH, G. C. LOCK, K. W. MANGLER, T. H. MCDONELL, C. RICHARDSON and F. WALKDEN (eds.), *Computational Methods and Problems on Aeronautical Fluid Dynamics*, Proc. of Conf. at Univ. of Manchester, Academic Press, 1976.
6. J. J. CONNOR and C. A. BREBBIA, *Finite Element Techniques for Fluid Flow*, Newnes–Butterworths, 1976.
7. H. LAMB, *Hydrodynamics*, 6th ed., Cambridge Univ. Press, 1932.
8. G. K. BATCHELOR, *An Introduction to Fluid Dynamics*, Cambridge Univ. Press, 1967.
9. B. A. FINLAYSON, *The Method of Weighted Residuals and Variational Principles*, Academic Press, 1972.
10. J. T. ODEN and D. SOMOGYI, 'Finite element applications in fluid dynamics', *J. Eng. Mech. Div., Proc. Am. Soc. Civ. Eng.*, **95**, EM4, 821–6, 1969.
11. O. C. ZIENKIEWICZ and C. TAYLOR, 'Weighted residual processes in finite elements with particular reference to some transient and coupled problems', pp. 415–58, *Lectures on Finite Element Method in Continuum Mechanics*, 1970, Lisbon (eds. J. T. Oden and E. R. A. Oliveira), Univ. Alabama Press, Huntsville, 1973.
12. J. H. ARGYRIS and G. MARECZEK, 'Finite element analysis of slow incompressible viscous fluid motion', *Ingenieur Archiv.*, **43**, 92–109, 1974.
13. (a) J. T. ODEN, 'A finite element analog of the Navier–Stokes equations', *Proc. Am. Soc. Civ. Eng.*, **96**, EM4, 529–34, 1970.
13. (b) J. T. ODEN, 'The finite element in fluid mechanics', pp. 151–86, *Lectures on Finite Element Method in Continuum Mechanics*, 1970, Lisbon (eds. J. T. Oden and E. R. A. Oliveira), Univ. Alabama Press, Hunstville, 1973.
14. C. TAYLOR and P. HOOD, 'A numerical solution of the Navier–Stokes equations using the finite element techniques', *Comp. Fluids*, **1**, 73–100, 1973.
15. M. KAWAHARA, N. YOSHIMURA, and K. NAKAGAWA, 'Analysis of steady incompressible viscous flow', *Finite Element Methods in Flow Problems*, pp. 107–20, (eds. J. T. Oden, O. C. Zienkiewicz, R. H. Gallagher, and C. Taylor), Univ. Albama Press, Huntsville, 1974.
16. J. T. ODEN and L. C. WELLFORD Jr., 'Analysis of viscous flow by the finite element method', *J.A.I.A.A.*, **10**, 1590–9, 1972.
17. A. J. BAKER, 'Finite element solution algorithm for viscous incompressible fluid dynamics', *Int. J. Num. Meth. Eng.*, **6**, 89–101, 1973.
18. M. D. OLSON, 'Variational finite element methods for two dimensional and Navier–Stokes equations', Ref. 2, Vol. 1, pp. 57–72, J. Wiley and Sons, 1974.
19. O. C. ZIENKIEWICZ and P. N. GODBOLE, 'Viscous incompressible flow with special reference to non-Newtonian (plastic) flow', Ref. 2, Vol. 1, Ch. 2, pp. 25–71, J. Wiley and Sons, 1975.
20. P. HOOD and C. TAYLOR, 'Navier–Stokes equations using mixed inter-

polation', *Finite Element Method in Flow Problems*, pp. 121–32 (eds. J. T. Oden, O. C. Zienkiewicz, R. H. Gallagher, and C. Taylor), Univ. Alabama Press, Huntsville, 1974.

21. M. D. OLSON and S. Y. TUENN, 'Primitive variables versus stream function finite element solutions of the Navier–Stokes equation', pp. 55–68 of ref. 3.
22. B. ATKINSON, C. C. M. CARD, and B. M. IRONS, 'Application of the finite element method to creeping flow problems', *Trans. Inst. Chem. Eng.*, **48**, 276–84, 1970.
23. T. J. R. HUGHES, R. L. TAYLOR, and J. F. LEVY, 'A finite element method for incompressible viscous flows', pp. 1–16 of ref. 3.
24. P. C. JAIN, *Plastic flow in solids. Static, quasistatic and dynamic situations including temperature effects*, Ph.D. Thesis, Univ. of Wales, Swansea, to be submitted, 1976.
25. H. S. LEW and Y. C. FUNG, 'On low Reynolds number entry flow into a circular tube', *J. Bio-mech.*, **2**, 105–19, 1969.
26. P. N. GODBOLE, 'Creeping flow in rectangular ducts by the finite element method', *Int. J. Num. Meth. Eng.*, **9**, 727–30, 1975.
27. O. C. ZIENKIEWICZ and P. N. GODBOLE, 'Flow of plastic and visco-plastic solids with special reference to extrusion and forming processes', *Int. J. Num. Meth. Eng.*, **8**, 3–16, 1974.
28. K. PALIT and R. T. FENNER, 'Finite element analysis of two dimensional slow non-Newtonian flows', *A.I.Ch.E. Jl*, **18**, 1163–9, 1972.
29. K. PALIT and R. T. FENNER, 'Finite element analysis of slow non-Newtonian channel flow', *A.I.Ch.E. Jl*, **18**, 628–33, 1972.
30. O. C. ZIENKIEWICZ and P. N. GODBOLE, 'Penalty function approach to problems of plastic flow of metals with large surface deformations', *J. Strain Analysis*, **10**, 180–3, 1975.
31. M. KAWAHARA, N. YOSHIMURA, K. NAKAGAWA and H. OHSAKA, 'Steady and unsteady finite element analysis of incompressible viscous flow', *Int. J. Num. Meth. Eng.*, **10**, 437–56, 1976.
32. S. L. SMITH and C. A. BREBBIA, 'Finite element solution of Navier–Stokes equations for transient 2-dimensional incompressible flow', *J. Comp. Phys.*, **17**, 235–45, 1975.
33. C. H. LEE, 'Finite element method for transient linear viscous flow problems', *Proc. Int. Conf. on Numerical Methods in Fluid Dynamics*, 1973.
34. O. C. ZIENKIEWICZ, P. C. JAIN and E. OÑATE, *Flow of solids during forming and extrusion. Some aspects of numerical solutions*, Univ. College of Swansea, Report No. C/R/283/76.
35. G. C. CORNFIELD and R. H. JOHNSON, 'Theoretical prediction of plastic flow in hot rolling including the effect of various temperature distributions', *J. Iron Steel Inst.*, **211**, 567–73, 1973.
36. J. W. H. PRICE and J. M. ALEXANDER, 'The finite element analysis of two high temperature metal deformation processes', pp. 715–20 of ref. 3.
37. R. E. NICKELL, R. I. TANNER, and B. CASWELL, 'The solution of viscous incompressible jet and free surface flows using finite element methods', *J. Fluid Mech.*, **65**, Part 1, 189–206, 1974.
38. J. J. DRONKERS, 'Tidal computation for rivers, coastal waters and seas', *Proc. Am. Soc. Civ. Eng.*, **95**, H71, 29–77, 1969.
39. P. WELANDER, 'Wind action on shallow sea, some generalisations of Eckmann's theory', *Tellus*, **9**, 47–52, 1957.
40. R. T. CHENG, T. M. POWELL and T. M. DILLON, 'Numerical models of wind driven circulation in lakes', *Appl. Math. Modelling*, **1**, 141–59, 1976.

41. M. Kawahara and K. Hasegawa, 'Periodic Galerkin finite element method of tidal flow', *Int. J. Num. Meth. Eng.*, **12**, 115–27, 1978.
42. T. Tenaka, T. Hirai and T. Katayama, 'Finite element applications to lake circulations on diffusion problems in Lake Drive', *Proc. 2nd Int. Symp. on Finite Elements in Fluid Problems*, St. Margharita, Italy, 1976.
43. J. Connor and J. Wang, 'Finite element modelling of hydrodynamic circulation' from *Numerical Methods in Fluid Dynamics* (eds. C. Brebbia and J. Connor), Pentech Press, 1974.
44. C. Taylor and J. M. Davis, 'Tidal propagation and dispersion in estuaries', Ref. 2, Vol. 1, Ch. 5, pp. 95–118, J. Wiley, 1975.
45. P. F. Hamblin, 'Finite element methods approach to the modelling of circulation, seiches, tides and storm surges in large lakes', (see ref. 3).
46. R. T. Cheng, 'Numerical investigation of Lake circulation around islands by the finite element method', *Int. J. Num. Meth. Eng.*, **5**, 103–12, 1972.
47. F. Arrizebalaya, G. M. Kovadi, and R. J. Krizek, 'Variational model for lake circulation' (see ref. 3).
48. R. T. Cheng and C. Tung, 'Wind driven lake circulation by the finite element method', *Proc. 13th Conf. on Great Lakes Research*, 1970.

49. S. Crandall, *Engineering Analysis*, McGraw-Hill, 1956.
50. R. Courant, E. Isaacson, and M. Rees, 'On the solution of non-linear hyperbolic differential equations by finite differences', *Comm. Pure Appl. Math.*, **V**, 243–55, 1952.
51. A. K. Runchal and M. Wolfstein, 'Numerical integration procedure for the steady state Navier–Stokes equations', *J. Mech. Eng. Sci.*, **11**, 445–53, 1969.
52. D. B. Spalding, 'A novel finite difference formulation for differential equations involving both first and second derivatives', *Int. J. Num. Meth. Eng.*, **4**, 551–9, 1972.
53. O. C. Zienkiewicz, R. H. Gallagher, and P. Hood, 'Newtonian and non-Newtonian viscous incompressible flow. Temperature induced flows. Finite element solutions', *The Mathematics of finite elements and applications II*, ed. J. Whiteman, Academic Press, 1977.
54. I. Christie, D. F. Griffiths, A. R. Mitchell and O. C. Zienkiewicz, 'Finite element methods for second order differential equations with significant first derivatives', *Int. J. Num. Meth. Eng.*, **10**, 1389–96, 1976.
55. O. C. Zienkiewicz, J. C. Heinrich, P. S. Huyakorn and A. R. Mitchell, 'An upwind finite element scheme for two dimensional convective transport equations', *Int. J. Num. Meth. Eng.*, **11**, 131–44, 1977.
56. J. C. Heinrich and O. C. Zienkiewicz, 'Quadratic finite element schemes for two dimensional convective-transport problems', *Int. J. Num. Meth. Eng.*, **1**, 1831–44, 1977.
57. K. E. Barrett, 'The numerical solution of singular perturbation boundary value problems', *Q. J. Mech. Appl. Math.*, **27**, 57–68, 1974.
58. J. Periaux, 'Three dimensional analysis of compressible potential flow', *Int. J. Num. Mech. Eng.*, **9**, 775–83, 1975.
59. T. E. Laskaris, 'Finite element analysis of compressible and incompressible viscous flow and heat transfer problems', *Physics of Fluids*, **18**, 1639–48, 1975.
60. S. G. Margolis, 'Finite element methods for compressible gas dynamic steam review', **17**, 385, 1975.
61. O. C. Zienkiewicz, 'The finite element method and the solution of some geophysical problems', *Phil. Trans. R. Soc. Lond. A.*, **283**, 139–51, 1976.

23. 'Boundary Solution' Processes and the Finite Element Method. Infinite Domains; Singularity in Fracture Mechanics

23.1 **Introduction**

Despite its versatility and enormous field of application the finite element process implemented with locally defined polynomial expansions suffers certain drawbacks. In particular, such difficulties are manifested in two definite areas, i.e., (*a*) when the domain becomes infinite and (*b*) when singularities (at which some, or all, of the derivatives are infinite) occur.

The first is clearly unattainable with *finite* elements; the second is poorly approximated by polynomial expansions and indeed convergence theorems cease to be applicable here as Taylor's expansion no longer converges near such singularities.

The pragmatic, engineering approach quite correctly often ignores both difficulties as they are, in practice, mathematical fictions—and this allows useful results to be obtained by using *large* though not infinite regions and *nearly* singular points. However, these mathematical 'fictions' are often useful as substantial economies of computation can be obtained by their use. Indeed, it is well known that many *extremely simple*, *exact solutions* are available for dealing with both 'infinity' and 'singularity' and it would be worthwhile to make use of these whenever possible. We therefore pose as an aim of this chapter the demonstration of the manner in which use of such analytical solutions can be made in the process of numerical discretizations. While many alternatives are possible for the incorporation (or simply an amendment) of the problem so that infinite regions or singularities are avoided,[1–5] the most powerful technique is presented by the so-called 'boundary solution' or Trefftz processes.[6] We shall, therefore, discuss first in some detail similarities and differences of this process from the finite element method and will show that if properly conceived and processed, all its merits can be preserved in the finite element context. We shall find that some of the procedures used are identical to those used in the derivation of various hybrid elements of Chapter 12.

The Boundary Solution Procedure consists in its essence of choosing a set of trial functions for the unknown function in a standard form, i.e.,

$$\hat{\mathbf{u}} = \sum \mathbf{N}_i \mathbf{a}_i = \mathbf{N}\mathbf{a} \tag{23.1}$$

but now, as distinct from the standard finite element process, these trial functions are so chosen that the governing equation—invariably of a linear kind (*vide* Chapter 3, Eq. (3.6))

$$\mathbf{A}(\mathbf{u}) \equiv \bar{\mathbf{L}}\mathbf{u} + \mathbf{p} = 0 \tag{23.2}$$

in Ω is automatically satisfied, at least in its homogeneous form, i.e.,

$$\bar{\mathbf{L}}\mathbf{N} \equiv 0 \tag{23.3}$$

giving thus, by superposition, a trial function set $\mathbf{N}$ which satisfies all the requirements within the domain.

As the non-homogeneous part $\mathbf{p}$ can usually be removed from the consideration in the domain Ω by a simple addition of a particular solution and an appropriate change of boundary conditions, the latter remain now as the only constraints and the parameters $\mathbf{a}$ are determined by an approximate satisfaction of these, i.e., ensuring that in some mean sense

$$\mathbf{B}(\hat{\mathbf{u}}) = 0 \qquad \text{on } \Gamma. \tag{23.4}$$

Clearly a variety of different procedures can here be used and all the general approaches outlined in Chapter 3 are again applicable. Conceived in this manner the boundary solution processes differ from the more conventional finite element applications in

(1) choice of shape functions satisfying Eq. (23.3)
(2) forming of the approximation on the boundary of the problem alone.

As the discretization process now only involves the boundary a much reduced number of parameters can be used as compared with the standard finite element process, and this has in some cases proved an economic advantage and accounted for the intense development process which, since the early sixties, has paralleled that of the finite element method. A further advantage of the boundary solution procedure is that now, obviously, the analytical trial function dealing with singularities or infinity can be used thus avoiding the difficulties of F.E.M. already mentioned.

On the debit side the difficulty of dealing with non-linear or non-homogeneous situations and the complete equation coupling (as against the narrow band structure of the F.E.M. matrices) are evident. Evidently a 'marriage' of the two processes is desirable to secure their appropriate merits.

At this point it is of interest to mention briefly the history and the development of boundary solution procedures.

The most important classification concerns the nature of the trial functions assumed. Here two alternatives exist:

(*a*) superposition of series of functions with arbitrary parameters $\mathbf{a}$;
(*b*) development of *Boundary Integral Equations* which represent the exact solution and a subsequent discretization of these in terms of parameters $\mathbf{a}$, which often can be identified with the unknown function values on the boundary.

The latter process which, incidentally, usually guarantees *completeness* of expansion, has today become predominant and a recent survey[7] is recommended for basic study dealing with problems of elasticity, potential, etc. Just as in the history of the F.E.M. it is difficult to trace its origins; the source distributions of von Karman[8] in solution of air flow problems in 1930 contain some of its essence which is elaborated in the context of potential theory further by Jaswon[9–10] and Symm[11] and in the context of elasticity by Massonet[12] and Oliveira,[13] Cruse,[14] Rizzo,[15,16] and others.[17–20] Since then much development and application of the procedure in other fields has taken place and a comprehensive bibliography is given in references 7, 21 and 22.

The series solution approaches have developed in parallel, and here the work of Hess,[22] Quinlan,[23] and others,[24] is noteworthy. The particular advantage of such processes lies in the free choice of functions available to represent singularity or other special behaviour but it is difficult in general to satisfy completeness requirements.

23.2 Boundary Solution 'Elements' Coupled with a Finite Element Field

Consider Fig. 23.1 in which a field of conventional finite elements surrounds a region Ω^I in which a 'boundary type' solution of a series type is prescribed. To define the terms we shall use the terms of linear elasticity but, obviously, the analogous terms are available for other elliptic self-adjoint, mathematical field problems.

In the region Ω^I a field of displacements $\mathbf{u}$ or stresses $\boldsymbol{\sigma}$ is defined in terms of a finite set of parameters $\mathbf{b}$ so that both compatability and equilibrium are preserved, i.e., the appropriate differential equation (Eq. (23.2)) is satisfied. On the interface I, where other elements join, we can write thus

$$\mathbf{u} = \mathbf{N}(s)\mathbf{b} \tag{23.5}$$

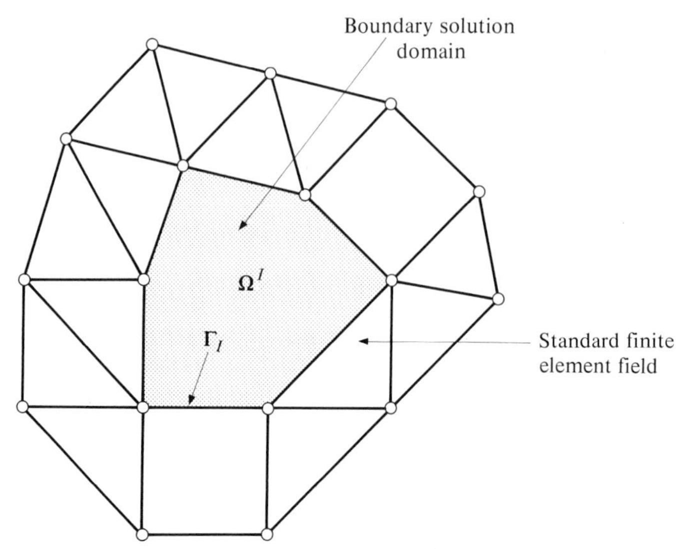

Fig. 23.1 Boundary solution 'element' Ω^I in a standard finite element field

for the displacements and

$$\mathbf{t} = \mathbf{M}(s)\mathbf{b} \tag{23.6}$$

to define the 'traction'. In the above, s defines the position of Γ_I. Here the reader will observe that $\mathbf{N}(s)$ and $\mathbf{M}(s)$ are related. If, for the whole region

$$\mathbf{u} = \mathbf{N}\mathbf{b} \tag{23.7}$$

then, using the appropriate operator for strains $\mathbf{L}$, and introducing the elasticity matrix $\mathbf{D}$, the stresses $\boldsymbol{\sigma}$ can be found as

$$\boldsymbol{\sigma} = \mathbf{DLu} = \mathbf{DLNb} \tag{23.8}$$

and the tractions determined on s are

$$\mathbf{t} = \mathbf{GDLNb} \equiv \mathbf{M}(s)\mathbf{b}. \tag{23.9}$$

thus defining $\mathbf{M}(s)$.

In the finite element field we are minimizing the total potential energy defined in terms of displacements at nodes given by parameters $\mathbf{a}$. On the interface the displacements are thus prescribed as

$$\bar{\mathbf{u}} = \overline{\mathbf{N}}\mathbf{a} \tag{23.10}$$

$\overline{\mathbf{N}}$ being the appropriate finite element shape function.

The potential energy of the region Ω^I can therefore be written as

$$\Pi^I = \int_{\Omega^I} \tfrac{1}{2}\boldsymbol{\sigma}^{\mathrm{T}}\boldsymbol{\varepsilon}\, \mathrm{d}\Omega - \int_{\Gamma_I} \bar{\mathbf{u}}^{\mathrm{T}}\mathbf{t}\, \mathrm{d}\Gamma \tag{23.11}$$

where the last term 'couples' the internal $\boldsymbol{\sigma}$–$\mathbf{u}$ field with the finite element displacements $\bar{\mathbf{u}}$.

As the assumed field satisfies all the governing equations it is easy to show that the first, volume, integral can be replaced by a boundary integral as

$$\int_{\Omega^I} \tfrac{1}{2}\boldsymbol{\sigma}^{\mathrm{T}}\boldsymbol{\varepsilon}\; \mathrm{d}\Omega \equiv \tfrac{1}{2}\int_{\Gamma_I} \mathbf{u}^{\mathrm{T}}\mathbf{t}\; \mathrm{d}\Gamma \tag{23.12}$$

and on insertion of the discretized expression (23.5) and (25.6) we can write

$$\Pi^I = \mathbf{b}^{\mathrm{T}}\left(\tfrac{1}{2}\int_{\Gamma_I} \mathbf{N}^{\mathrm{T}}\mathbf{M}\; \mathrm{d}\Gamma\right)\mathbf{b} - \mathbf{a}^{\mathrm{T}}\left(\int_{\Gamma_I} \bar{\mathbf{N}}^{\mathrm{T}}\mathbf{M}\; \mathrm{d}\Gamma\right)\mathbf{b}. \tag{23.13}$$

Immediately within the 'element' defining the boundary solution domain

$$\frac{\partial \Pi^I}{\partial \mathbf{b}} \equiv 0 = \mathbf{K}_b^I \mathbf{b} + \mathbf{K}_{ba}^I \mathbf{a} \tag{23.14}$$

$$\mathbf{K}_b^I = \tfrac{1}{2}\left[\int_{\Gamma_I} \mathbf{N}^{\mathrm{T}}\mathbf{M}\; \mathrm{d}\Gamma + \int_{\Gamma_I} \mathbf{M}^{\mathrm{T}}\mathbf{N}\; \mathrm{d}\Gamma\right] \qquad \mathbf{K}_{ba}^I = -\int_{\Gamma_I} \mathbf{M}^{\mathrm{T}}\bar{\mathbf{N}}\; \mathrm{d}\Gamma$$

and $\mathbf{b}$ can be eliminated leaving a standard type 'stiffness type' expression

$$\frac{\partial \Pi^I}{\partial \mathbf{a}} = \mathbf{K}^I \mathbf{a}. \tag{23.15}$$

This can be coupled to any of the other elements and being variationally obtained we find, as usual, a symmetrical stiffness matrix $\mathbf{K}^I$. The reader can verify easily that

$$\mathbf{K}^I = (\mathbf{K}_{ba}^I)^{\mathrm{T}}\,(\mathbf{K}_b^I)^{-1}\,\mathbf{K}_{ba}^I. \tag{23.16}$$

The procedure adopted above is, in fact, identical to that used in the derivation of hybrid and equilibrating elements of Chapter 12 (section 12.3) except that the choice of exact fields has permitted the substitution of boundary in place of volume integrals.

The choice of fields satisfying both equilibrium and compatibility is by no means simple, especially as for convergence certain completeness conditions have to be satisfied. For these obvious reasons the material properties within the region Ω^I will need to be assumed as homogeneous—a restriction of a kind not encountered in the previous finite element methodology. The Boundary Integral Equation process to be described in the next section will introduce a relatively general procedure for determining generally suitable functions.

The domain Ω^I can be treated for all purposes as a standard 'finite element' but, as we shall show later, it is easy to include in it most complex shapes. Further, the new element need not be 'finite', if the shape functions

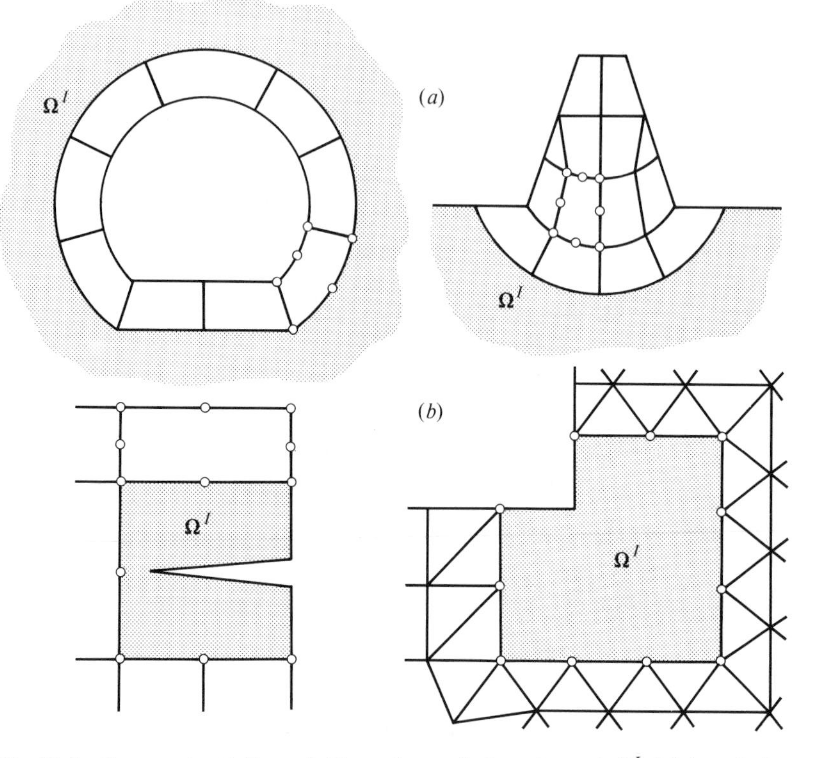

Fig. 23.2 Some potentially useful boundary solution elements Ω^I. (*a*) As exterior, infinite element. (*b*) As singularity elements

N are so chosen as to satisfy asymptotically the infinity boundary condition, Fig. 23.2(*a*). Similarly, special elements for singularities present near corners or crack tips can readily be developed.[25, 26] We shall return to such examples later.

At this stage it is interesting to remark that we have already encountered at an early stage of this book an element which, though derived in a direct fashion, falls into the category of boundary solution elements here described. This is the simple plane triangle (or, in three dimensions, the tetrahedron) where the linear displacement and constant stress field automatically satisfy the internal equilibrium equations.

23.3 'Boundary Integral Equation' Elements

23.3.1 *General formulation in elasticity.* We have already remarked that the generation of shape functions which satisfy all the governing equations is fraught with difficulties.

A general procedure is available by synthesis of continuously distributed, unit singular solutions (Green's functions) which, with a homogeneous system, lead to an integral equation which has to be satisfied only on the boundary. A very convenient form of such integral equations uses the value of the variable itself (say the displacement $\mathbf{u}$) and of its conjugate (say $\mathbf{t}$) on the boundary as the basic unknown. This is known as the *direct* formulation.

Thus, for instance, in the context of linear elasticity we can write for any point on the boundary Γ_I of Fig. 23.2 which surrounds the domain Ω^I

$$\mathbf{u}(s)/2 = \int_{\Gamma_I} \mathbf{T}(s, \Gamma)\mathbf{u}(\Gamma)\, d\Gamma + \int_{\Gamma_I} \mathbf{U}(s, \Gamma)\mathbf{t}(\Gamma)\, d\Gamma \qquad (23.17)$$

where

$$\mathbf{T} = \mathbf{T}(s, \Gamma) \quad \text{and} \quad \mathbf{U} = \mathbf{U}(s, \Gamma)$$

are matrices depending on stresses and displacements obtained from an exact solution for a point load in an infinite medium. The above equation is known as the Somigliana identity—and for its derivation the reader should consult appropriate references[15, 27, 28] although in the next section we shall derive a simple equivalent for potential problems.

Now the *exact* relationship is available between the displacements and tractions defined on the boundary of Ω^I and before proceeding further a discretization of the integral equation (Eq. (23.17)) has to be made. Again we can use an interpolation of the form given by Eqs. (23.5) and (23.6) but now the shape functions $\mathbf{N}(s)$ and $\mathbf{M}(s)$ are only defined on the interface and, therefore, may be fixed quite independently. It is obviously convenient to identify $\mathbf{N}(s)$ with $\overline{\mathbf{N}}$, the finite element representation, on the interface and a direct coupling is then available.

Thus we can write

$$\mathbf{u} = \bar{\mathbf{u}} = \overline{\mathbf{N}}\mathbf{a} \qquad \overline{\mathbf{N}} = \overline{\mathbf{N}}(\Gamma) \qquad (23.18a)$$

and

$$\mathbf{t} = \mathbf{M}\mathbf{c} \qquad \mathbf{M} = \mathbf{M}(\Gamma) \qquad (23.18b)$$

where $\mathbf{c}$ are parameters defining the traction distribution (e.g., nodal values of $\mathbf{t}$).

Equation (23.17) can now be discretized by suitable collocation say writing for each node, $\mathbf{u}(s_i) \equiv \mathbf{a}_i$, and we can represent the set in a matrix form as

$$\mathbf{A}\mathbf{a} = \mathbf{B}\mathbf{c} \qquad (23.19)$$

with the cofficients obtained by successive contour integrations for all s_i values.

Equation (23.19) is the basis of the Boundary Integral Equation solutions *per se* and leads to fully coupled matrices. Certain difficulties arise in integration due to presence of singularities in the expressions for $\mathbf{t}$ and $\mathbf{u}$ at the point s. Further, the matrices $\mathbf{A}$ and $\mathbf{B}$ are not symmetric. In the solution a certain number of *tractions/displacements* are specified and the remainder found by solving the equation.

As the integrals in Eq. (23.17) do not demand continuity of $\mathbf{u}$ or $\mathbf{t}$ in the earlier methods using Boundary Integral Equations, the boundary contour was generally divided into straight segments and in each constant values of each of these quantities were taken representing the parameters $\mathbf{a}$ and $\mathbf{c}$. However, more recently isoparametric types of interpolation became popular using continuous distributions of the same form for both quantities.[19] In formulating our problems here we assume that $\overline{\mathbf{N}}$ is of such a form but that the distribution of $\mathbf{t}$, i.e., the shape functions, $\mathbf{M}$ may be discontinuous. The best forms of the two interpolations are still being studied.[28]

To couple the solution of the region Ω^I with the finite element domain we have to determine, as in the preceding section, the potential energy of the domain. This now can be done directly as parameters $\mathbf{a}$ represent the common nodal variable of both regions and the continuity of $\mathbf{u}$ is assured.

Inverting thus Eq. (23.19) we can write

$$\mathbf{c} = \mathbf{B}^{-1}\mathbf{A}\mathbf{a} \tag{23.20}$$

and, from (23.18b)

$$\mathbf{t} = \mathbf{M}\mathbf{B}^{-1}\mathbf{A}\mathbf{a}. \tag{23.21}$$

The potential energy of the boundary solution element is thus given as

$$\Pi^I = \frac{1}{2}\int_{\Gamma_I} \mathbf{u}^{\mathrm{T}}\mathbf{t}\,d\Gamma = \frac{1}{2}\mathbf{a}^{\mathrm{T}}\left(\int_{\Gamma_I} \overline{\mathbf{N}}^{\mathrm{T}}\mathbf{M}\,d\Gamma\right)\mathbf{B}^{-1}\mathbf{A}\mathbf{a} \tag{23.22}$$

from which the stiffness matrix $\mathbf{K}^I$ is directly available as

$$\mathbf{K}^I = \frac{1}{2}\left[\int_{\Gamma} \overline{\mathbf{N}}^{\mathrm{T}}\mathbf{M}\,d\Gamma\mathbf{B}^{-1}\mathbf{A} + \left(\int_{\Gamma} \overline{\mathbf{N}}^{\mathrm{T}}\mathbf{M}\,d\Gamma\mathbf{B}^{-1}\mathbf{A}\right)^{\mathrm{T}}\right] \tag{23.23}$$

The outline of the formulation given here follows the procedures used in conventional direct Boundary Integral approaches. We note, however, that the independent interpolation of $\mathbf{u}$ and $\mathbf{t}$ is not consistent with the full satisfaction of the governing equations and that 'stiffness' derived above do not satisfy complete equilibrium conditions unlike their finite element counterparts. To improve the performance a direct 'imposition of equilibrium constraints' is suggested in ref. 18 and this is used to modify Eq. (23.19). This results in an improvement of performance.

As an alternative of the above Boundary Integral Equations an *indirect formulation* can be used avoiding the difficulty. This follows, in principle, a series type expansion of Eq. (23.5) but replaces a finite number of terms by a distribution of singularities. Details of such formulations are available elsewhere and will not be discussed here.

23.3.2 *Some examples of potential field type.* The applications of the B.I.E. process to field problems of various kinds follow precisely the same lines as that of the elasticity formulation given in the preceding section. Let us, for instance, consider a problem of the same type as shown in Fig. 23.1, but in which the quasi-harmonic equation (see Chapter 17) has to be solved, i.e.,

$$\frac{\partial}{\partial x}\left(k\frac{\partial \phi}{\partial x}\right)+\frac{\partial}{\partial y}\left(k\frac{\partial \phi}{\partial y}\right) = 0. \tag{23.24}$$

Further, in Ω^I the material is homogeneous so the above reduces to

$$k\left(\frac{\partial^2 \phi}{\partial x^2}+\frac{\partial^2 \phi}{\partial y^2}\right) \equiv k\,\nabla^2\phi = 0. \tag{23.25}$$

We have observed (in Chapter 17) that on boundaries

$$q \equiv k\frac{\partial \phi}{\partial n} \tag{23.26}$$

plays the same role as that of the tractions in the corresponding elasticity problem.

The 'potential energy' of the finite element domain can be written in terms of standard expressions (see Chapter 17, p. 426) and we shall therefore now, only consider the domain Ω^I. Here, integrating by parts, we note that (*vide* Appendix 3)

$$\Pi^I = \tfrac{1}{2}\int_{\Omega^I}\left(k\left(\frac{\partial \phi}{\partial x}\right)^2+k\left(\frac{\partial \phi}{\partial y}\right)^2\right)\mathrm{d}\Omega \equiv -\int_{\Omega^I} k\,\nabla^2\phi\,\mathrm{d}\Omega+\tfrac{1}{2}\int_{\Gamma_I}\frac{\partial \phi}{\partial n}\phi\,\mathrm{d}\Gamma \tag{23.27a}$$

and if ϕ is a solution of Eq. (23.25) we can write analogously to Eq. (23.12)

$$\Pi^I = \tfrac{1}{2}\int_{\Gamma_I} q\phi\,\mathrm{d}\Gamma. \tag{23.27b}$$

To obtain the Boundary Integral Equation we proceed as follows. First we note that for *any two* functions ψ and ϕ we can write the following identity by using the principles of integration by parts (Green's identity):

$$\int_{\Omega^I}\phi\,\nabla^2\psi\,\mathrm{d}\Omega-\int_{\Omega^I}\psi\,\nabla^2\phi\,\mathrm{d}\Omega = \int_{\Gamma_I}\phi\frac{\partial \psi}{\partial n}\mathrm{d}\Gamma-\int_{\Gamma_I}\frac{\partial \phi}{\partial n}\psi\,\mathrm{d}\Gamma \tag{23.28}$$

If now we choose for ψ a singularity function which satisfies Eq. (23.25) (except at the pole) such as, for instance, a source function in two dimensions

$$\psi = \log_e r \tag{23.29}$$

and if ϕ is the exact solution of Eq. (23.25), then we can write for points p in the interior of the domain†

$$2\pi\phi_p = \int \phi \frac{\partial \psi}{\partial n} \, d\Gamma - \int \frac{\partial \phi}{\partial n} \psi \, d\Gamma \tag{23.30}$$

and for points P on the boundary

$$\pi\phi_P = \int \phi \frac{\partial \psi}{\partial n} \, d\Gamma - \int \frac{\partial \phi}{\partial n} \psi \, d\Gamma. \tag{23.31}$$

This supplies the necessary integral equation replacing that of the previous section (Eq. (23.17)).

From here onwards the procedure of formulation of the element matrix is almost identical to that described in the previous section and we leave the details to the reader.

At this point it is essential to note, however, that the integrals contain always a singularity at point p and that r measures the distance between that point and the current integration co-ordinate, as shown in Fig. 23.3 Numerical integration with standard Gaussian quadrature can be used except in the vicinity of the singularities where either more refined numerical integration or locally exact integrations are used.

Figures 23.4 to 23.6 illustrate various solutions in which simple linear finite elements are used with an appropriate B.I.E. element in which the function ϕ is linearly interpolated while $\partial\phi/\partial n$ is taken as piecewise constant on each segment of the boundary. In Fig. 23.4, heat conduction on a radial segment is calculated using an arbitrary division of the domain into linear triangles and B.I.E. elements.

In Fig. 23.5 an 'exterior' problem with an infinite element is dealt with. This shows the distribution of electro-static potential between two charged plates. Here the same standard element was used, and the fact that the singular solution ψ tends to zero or infinity provides the necessary boundary conditions there for ϕ which now tends to zero there. This problem is also discussed by Silvester[29] who, however, uses a different form of 'exterior' representation.

In Fig. 23.6 we turn our attention to an aeronautical problem of describing the flow of an ideal fluid around an obstacle. Here, in general, 'circulation' can occur and a flow potential is multivalued. The problem

† Note that the only term which does not disappear on the left-hand side of Eq. (23.28) is the integral over the singularity of p. For an internal point this gives

$$\phi_p \cdot \int \nabla^2 \psi \, d\Omega = \phi_p \cdot \oint \frac{\partial \psi}{\partial n} \, d\Gamma = \phi_p 2\pi$$

and half of this value at boundary points. For a detailed discussion see references 27 and 28.

is thus solved by a stream function formulation. This function ϕ defines the flow velocities as

$$u = \frac{\partial \phi}{\partial x} \quad \text{and} \quad v = -\frac{\partial \phi}{\partial y} \tag{23.32}$$

and ψ has to satisfy[30]

$$\nabla^2 \phi = 0. \tag{23.33}$$

At infinity a uniform velocity $u = U$ has to exist and thus it is convenient to separate the stream function into two parts:

$$\phi = \phi_1 + \phi_2, \qquad \phi_1 = Uy. \tag{23.34}$$

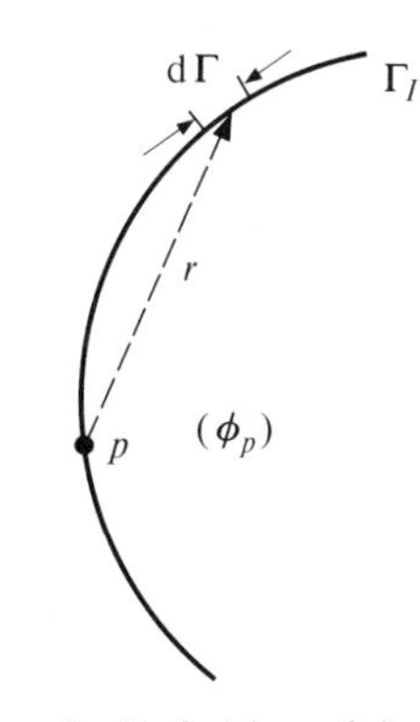

Fig. 23.3 Singularity at integrals. Definition of singular pole p and line element dr

We note that the definition of ϕ_1 satisfies the infinity condition and ϕ_2 therefore tends to zero there. Further, we still have

$$\nabla^2 \phi_2 = 0 \tag{23.25}$$

by virtue of (23.33) and (23.34), so that now the problem is simply posed in terms of ϕ_2. As on the boundary of the obstacle $\phi = $ const., the boundary condition for ϕ_2 is

$$\phi_2 = \text{const.} - Uy \text{ on } \Gamma_I. \tag{23.36}$$

In the context of an arbitrary section around which circulation occurs the value of the constant has to be determined so that the velocity at the trailing edge is finite (Kutta–Joukowsky condition). This necessitates superposition of two separate computations. In the symmetric section

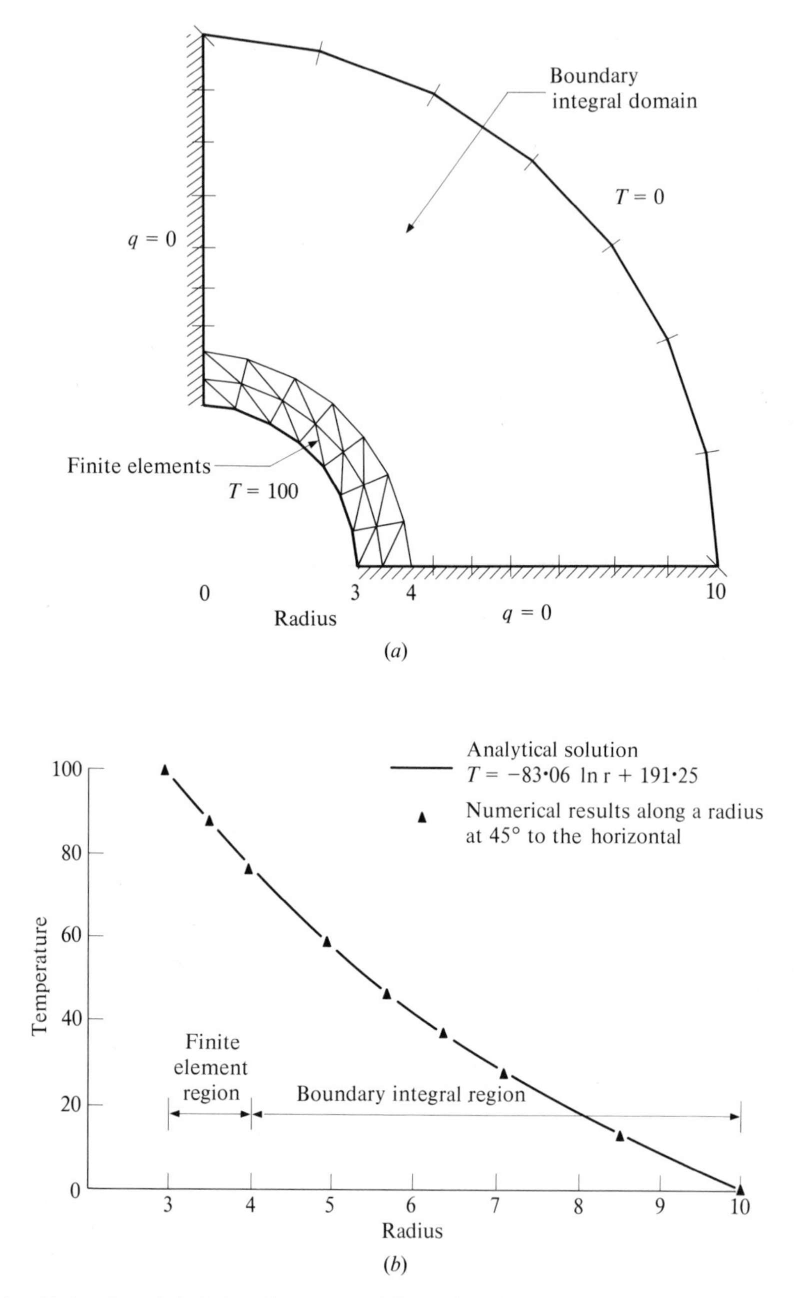

Fig. 23.4 Coupled finite element and boundary integral equation methods—heat conduction on a disc. (*a*) element subdivision. (*b*) Variation of temperature along a radius

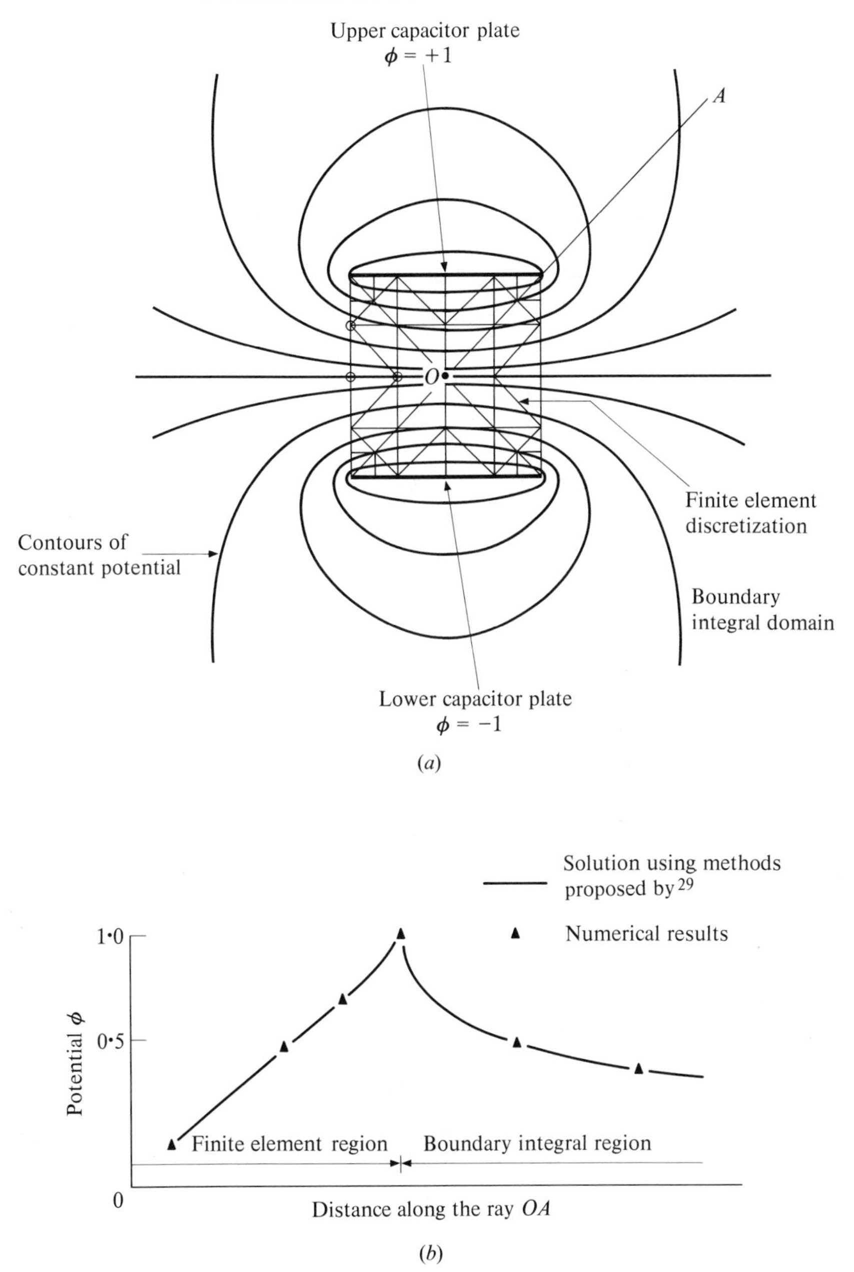

Fig. 23.5 Coupled finite element and boundary integral equation methods—the electric field for a parallel plate capacitor. (*a*) element subdivision and potential contours. (*b*) Potential distribution along *OA*

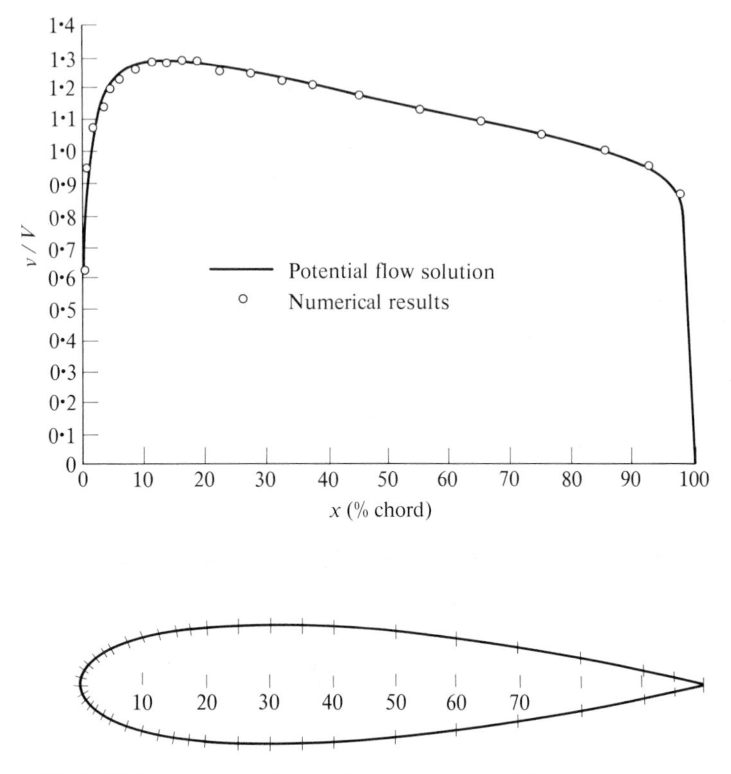

NACA 0018 wing section showing boundary segmentation used (piecewise flat segments)

Fig. 23.6 Flow around a symmetric obstacle in an irrotational flow field

shown on Fig. 23.6 a single solution leads immediately to a velocity distribution. In the case considered only one infinite B.I.E. element was used.

Solutions of this type have been discussed extensively in the literature and are a very standard part of aeronautical calculation.[8, 30–32]

Obviously the exterior solution problem has many applications in many problems of diverse origin.

The second field problem concerns the Helmholz equation which governs propagation of waves in shallow water, compressible media and in electromagnetic problems (see Chapter 20, p. 529 and p. 551). For steady-state response to forcing input proportional to $e^{i\omega t}$ the equation becomes

$$\frac{\partial}{\partial x}\left(k\,\frac{\partial \phi}{\partial x}\right)+\frac{\partial}{\partial y}\left(k\,\frac{\partial \phi}{\partial y}\right)+\omega^2\phi = 0 \tag{23.37}$$

and, in general, ϕ is a complex quantity (see Chapter 20). In the homogeneous domain this equation becomes simply

$$k\,\nabla^2\phi+\omega^2\phi = 0. \tag{23.38}$$

The reader can verify that the boundary integral equation (Eq. (23.30–31)) is still valid providing the singular solution satisfies Eq. (23.38). Such a solution is provided for two-dimensional problems by a Hankel function giving

$$\phi_P = \frac{i}{2}\int_{\Gamma_I}\left[H_0(\alpha r)\frac{\partial\phi}{\partial n}-\phi\frac{\partial H_0(\alpha r)}{\partial n}\right]\mathrm{d}\Gamma \qquad (\alpha = \omega/\sqrt{k}) \tag{23.39}$$

In a typical situation we are concerned here with the interaction of an incident wave with an obstacle, and a separation of the type used in Eq. (23.34) is necessary. The reflected wave ϕ_2 is such that it becomes zero at infinity and does *not* return (no reflection). The Boundary Integral

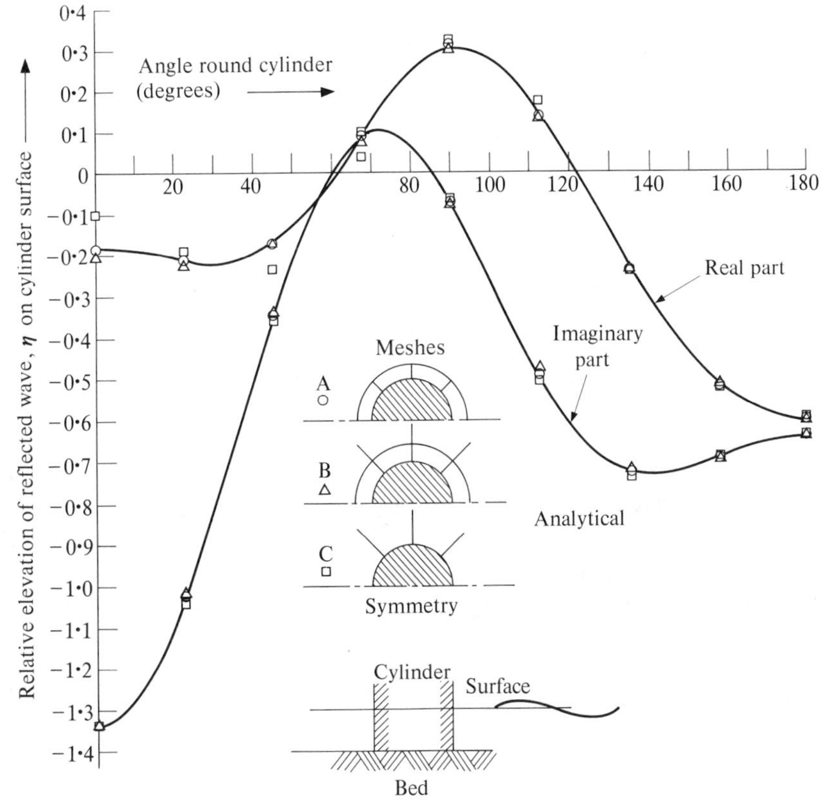

Fig. 23.7 Elevation of reflected waves around cylinder. (A) B.I.E. type solution in exterior with one row of 8 node isoparametric elements. (B) Infinite elements in exterior but with one row of 8 node isoparametric elements. (C) as (B) but with infinite elements only

Wave number = 2; Radius = 1

Equation is formed so that it automatically satisfies this condition. In Fig. 23.7 we show an exterior problem of wave diffraction caused by a cylinder placed in a field of uniform waves. As an exact solution is available, a comparison with this is made and shows an excellent accuracy. Although not particularly advantageous here the field is again divided into a set of conventional (now isoparametric parabolic) elements and a single exterior B.I.E. element.

Similar solutions have been obtained using boundary solution elements of the type discussed in section 23.2 with an independent expansion of non-singular form. The accuracy of both procedures appears to be the same.[24, 33, 34, 35]

A similar formulation can be used effectively for dealing with problems of electromagnetic oscillation (wave guide) and in providing simple elements for coupling structural and fluid oscillations.

23.3.3 *Simultaneous use of Boundary Integral Equations and other expansions*. The two alternative approaches to generating boundary solution 'elements' can, on occasion, be usefully combined. This could be advantageous if we knew *a priori* that a certain type of solution form must be predominant. The process can be used globally, extending such a solution over all elements—whether of boundary or conventional type—and thus concentrating on a less significant part of the total variable. This kind of approach is presented in references 1 and 2, for instance, when concentrated load effects are so eliminated in the context of elasticity. Mote[36] introduces this concept quite generally in the context of stress analysis. However, it is possible to combine such solutions at the element level alone, and here the idea is particularly useful in the context of boundary type elements. Possibilities of such approaches have not yet been fully explored and the reader may find some useful applications here.

23.4 Alternative Approaches to Infinity and Singularity Problems (Special Shape Functions in Domain Integrals)

In the previous sections we have explored the possibilities offered by a direct use of exact analytical solution in certain elements (or problem subdomains). This has enabled us to avoid integration over these and transfer all such integrations to the boundary alone. Another use of analytical solutions is, however, possible. *In this we use these only to obtain an indication of the shape function form which should be adopted, usually in addition to the standard polynomial terms.* The integration is still carried out in the standard manner over the whole element domain although, on occasion, additional interface terms have to be used to ensure inter-element compatibility.

If, in addition to the simple polynomial terms in each element, we impose such functions, convergence in the sense referred to previously in this book is maintained (and, indeed, improved if singularities are so modelled). On the other hand, we may simply use such solutions to approximate the behaviour of a subdomain without enquiring about convergence in the sense of decreasing element size.

Typical here are so-called *infinite elements* introduced by Bettess and Zienkiewicz[37–39] and others.[40] Here the infinite elements are defined as radiating strips in the exterior region, and numerical integration is used taking advantage of the fact that shape functions decay to zero at infinity.

We shall illustrate the procedure on the type of problems presented by Eq. (23.38) of the previous section. In these the form of the elements is shown in Fig. 23.8 and a standard Lagrangian mapping is used to define the co-ordinates, specifying reference points at a finite distance. The shape functions are defined, however, in co-ordinates η and r where the latter is a *scaled* distance in the ξ direction. These are essentially again Lagrangian polynomials but are multiplied by

$$e^{-r/L}\, e^{ikr} \tag{23.40}$$

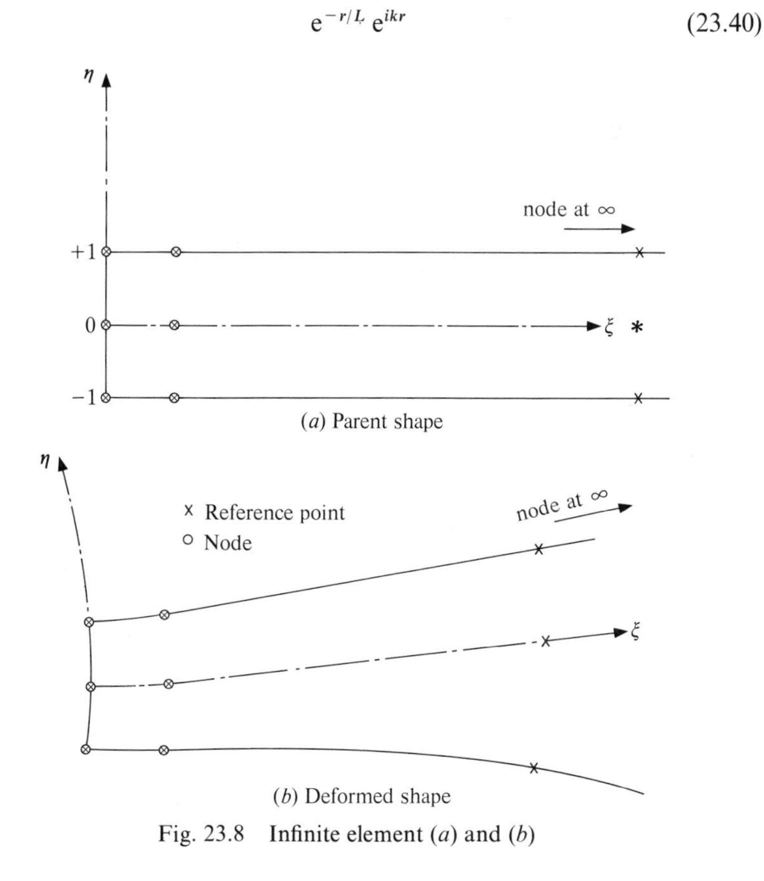

Fig. 23.8 Infinite element (*a*) and (*b*)

where k is the wave number ($= 2\pi$/wavelength), and L is a measure of the severity of the decay. The first term ensures decay of the shape function for large r, and the second a basically periodic behaviour. For the *sth* node row the shape function can thus be written using a Lagrangian expansion in r

$$e^{(r_s - r)/L} \, e^{ikr} \, l_s^n(r) \tag{23.41}$$

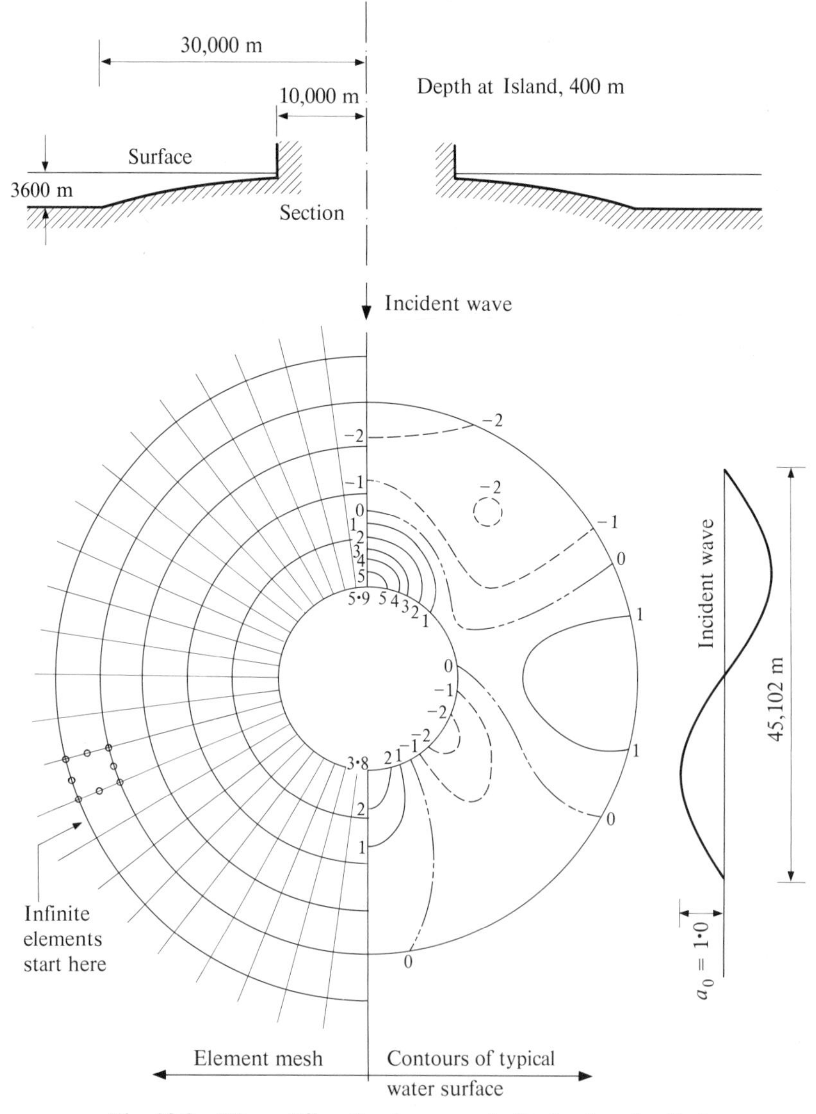

Fig. 23.9 Wave diffraction by a parabolic shoal and cylinder

where there are n rows of nodes, the last being infinitely distant. These shape functions are combined with conventional Lagrange polynomials in the circumferential direction η (vide Chapter 7, p. 155).

This formulation ensures compatibility of ϕ with interior elements and a complete flexibility of the number of nodal points on the side of the element which stretches to infinity.

Numerical integration of element matrices—which now preserve their standard, finite, form—needs some elaboration, the details of which are given in reference 39. Although here we cannot discuss convergence in the sense of decreasing element size, this can be achieved by simply extending the finite/infinite element interface to progressively larger distances. The approximation is, however, remarkable; with eight infinite and eight finite elements surrounding the cylinder problem of Fig. 23.7(b), the results obtained are indistinguishable from the exact ones. Without the zone of finite elements the values show a small discrepancy which again is indicated in Fig. 23.7(c).

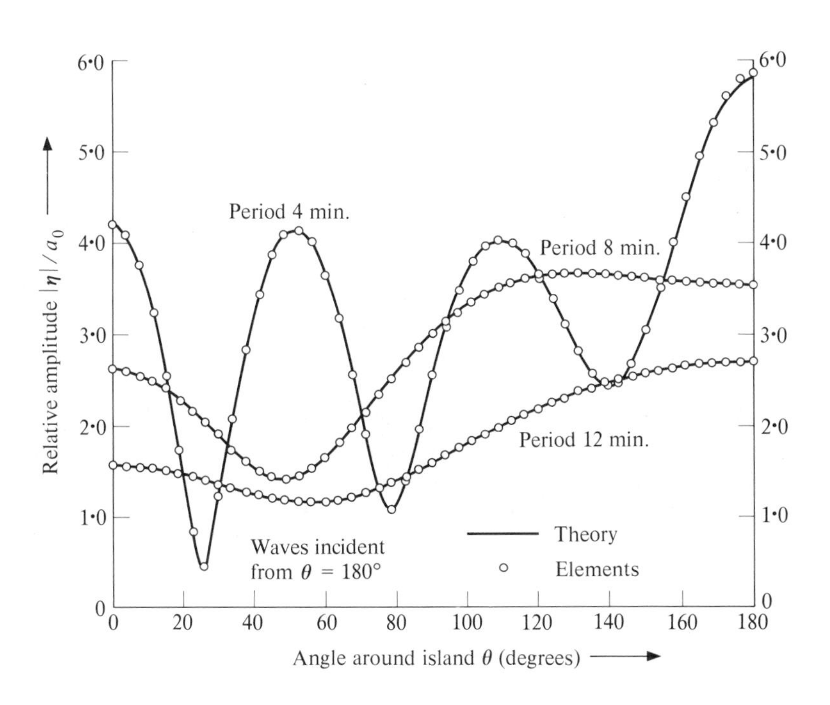

Fig. 23.10 Relative amplitudes on cylinder Problem of Fig. 23.9

This remarkable accuracy, achieved with a very simple formulation, is demonstrated again in Figs. 23.9 and 23.10 where the problem of wave diffraction/refraction in a shoaling region is dealt with. Here the varying depth region is modelled by conventional elements with infinite elements dealing with the exterior.

Clearly the narrow banding of the final equations as compared with the full exterior B.I.E. element form is an attractive feature of the approach outlined.

Similar exterior infinite elements for steady-state viscous flow and elasticity are discussed in reference 38 and 40.

We shall see in the next section that special element forms using an analytically 'suspected' result can be achieved on occasions by a simple element curvilinear mapping.

23.5 Some Aspects of Fracture Mechanics

23.5.1 *Basic ideas.* The vast amount of general literature and facts dealing with the subject, and the number of publications on finite element approaches, justifies inclusion of this section which is designed to introduce the reader to this subject and indicate very selectively some of the current approaches. For a comprehensive survey the reader should consult references 41–51.

The subject falls, as far as the finite elements are concerned, into this chapter as its essence is the problem of a stress singularity initiating a crack progression. This starts at re-entrant corners of irregularities and cavities found in every solid. At such singularities it is well known that theoretically infinite stresses will arise, and the question of whether propagation will occur depends on the manner in which such stresses decay at some distance from a crack tip. If the stress drops off in a very short distance compared with that of the atomic structure of the material, then it may not be sufficient to break the bonds and fracture will be arrested.

As early as 1920 Griffiths[52] suggested the propagation of a crack depends on a criterion which relates *the energy release of the structure to the surface area of a propagating crack*.

For example, if we consider a crack shown in Fig. 23.11 in a uniform tensile stress field and we note *the work done by external forces* (or change of strain energy per unit thickness) when the crack extends by a distance a, then for propagation we must have

$$\mathrm{d}\Pi = \text{Energy release} \geqslant \mathrm{d}a\mathscr{G} \quad \text{or} \quad \frac{\mathrm{d}\Pi}{\mathrm{d}a} \geqslant \mathscr{G} \tag{23.42}$$

where the value of $\mathscr{G}$ is associated with characteristics of a given material.

If we examine the detail of elastic stress distribution in the vicinity

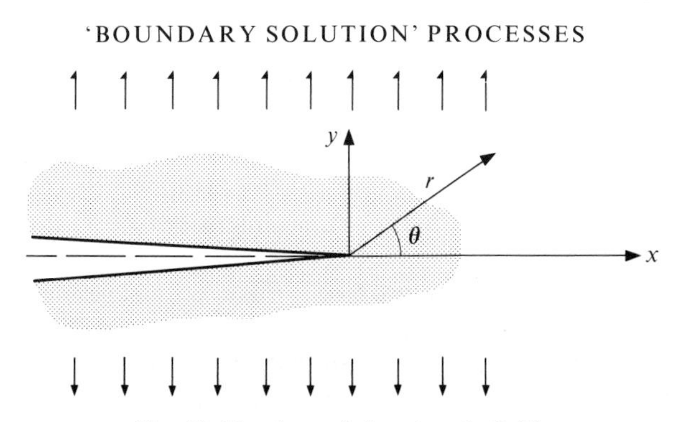

Fig. 23.11 A crack in a tensile field

of the crack tip we find that the *predominant* terms giving the stress and displacement distribution there can be written as

$$
\begin{aligned}
\sigma_x &= \frac{K_\mathrm{I}}{\sqrt{2\pi r}} \cos\frac{\theta}{2}\left(1-\sin\frac{\theta}{2}\sin\frac{3\theta}{2}\right)\\
\sigma_y &= \frac{K_\mathrm{I}}{\sqrt{2\pi r}} \cos\frac{\theta}{2}\left(1+\sin\frac{\theta}{2}\sin\frac{3\theta}{2}\right)\\
\tau_{xy} &= \frac{K_\mathrm{I}}{\sqrt{2\pi r}} \sin\frac{\theta}{2}\cos\frac{\theta}{2}\cos\frac{3\theta}{2}\\
u &= \frac{K_\mathrm{I}}{4G}\sqrt{\frac{r}{2\pi}}\left[(2\kappa-1)\cos\frac{\theta}{2}-\cos\frac{3\theta}{2}\right]\\
v &= \frac{K_\mathrm{I}}{4G}\sqrt{\frac{r}{2\pi}}\left[(2\kappa+1)\sin\frac{\theta}{2}-\sin\frac{3\theta}{2}\right]
\end{aligned}
\tag{23.43}
$$

where

$$
\begin{aligned}
\kappa &= (3-4\nu) \text{ for plane strain}\\
&= (3-\nu)/(1+\nu) \text{ for plane stress}
\end{aligned}
$$

and K_I is a constant depending on the magnitude of the externally applied stress.

It is possible to show that the energy release rate can be related to this constant as

$$\frac{\mathrm{d}\Pi}{\mathrm{d}a} = K_\mathrm{I}^2(\kappa+1)/8G \tag{23.44}$$

and Irwin[41] suggests that this quantity, known as the *stress intensity factor* K_I, should be used as a criterion in preference to the critical energy release rate $\mathscr{G}$. However, the obvious equivalence of the two makes no difference in practice if only one cracking mode described were considered, and as the strain energy is a quantity more readily determined in finite element analysis than the stresses, the former concept is more convenient to use (although usually critical values of stress intensity factors rather than $\mathscr{G}$ are reported).

In fully three-dimensional problems three possible modes of crack extension illustrated in Fig. 23.12 exist. In each case the local stress distribution is similar in form to that given in Eq. (23.43) and quite generally three stress intensity factors K_{I}, K_{II}, or K_{III} are often reported. These critical values are related to the critical energy release rate as

$$\mathscr{G} = \frac{1}{E}(K_{\mathrm{I}}^2 + K_{\mathrm{II}}^2 + K_{\mathrm{III}}^2). \tag{23.45}$$

For combined modes this single energy criterion appears to be just as effective as a separate specification of the individual stress intensity factors.

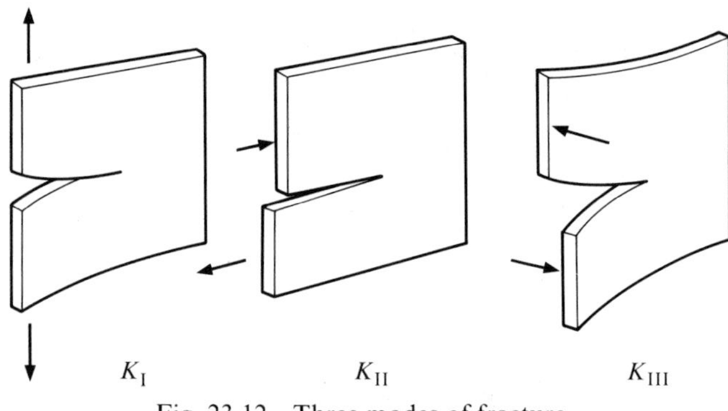

Fig. 23.12 Three modes of fracture

While all considerations so far assumed purely elastic behaviour, the conclusions are reasonably valid when the extent of plasticity developing near a crack tip is small. For larger plastic regions the concepts of crack propagation are by no means yet fully agreed and the reader could well consult some recent references on the subject.[50, 53–57] A very comprehensive picture of the special aspects of elastic–plastic fracture mechanism is given by Rice and Tracey.[50] However, the finite element treatment of the problem still presents the same singularity difficulties and it is with these that we are mainly concerned.

23.5.2 *Determination of the critical energy release rate (or of stress intensity factors).* Irrespective of the finite element formulation used, the problem of the best method of determining the critical energy release rate or of equivalent stress intensity factors has occupied much attention in the literature. In some early work the stress (or displacement) distributions in the vicinity of the crack tip were plotted and, by comparison with expression (23.43), the stress intensity factors were found. Much more effective are, however, calculations based on the direct evaluation of changes in the potential energy content as the crack progresses. For instance, the potential energy could be found for two different positions of the crack tip. In Fig. 23.13 we show, somewhat crudely, a finite element idealization of a structure including a crack. The energy is now evaluated for two different positions of the crack separated by Δa and we have approximately

$$\frac{d\Pi}{da} \approx \frac{\Pi_1 - \Pi_2}{\Delta a}. \tag{23.46}$$

Such approaches were first suggested by Dixon and Pook[58] and followed by others.[59–61]

The two separate solutions implicit in such processes are, however, uneconomic and the direct determination of K factors from a single analysis would seem preferable. However, with a simple modification suggested by Parks[62] and Hellen,[63] it is possible to avoid such 'double work'.

To describe this process, now extensively used, we shall once again consider the system illustrated in Fig. 23.13. Let $\mathbf{K}$, $\mathbf{a}$, and $\mathbf{f}$ correspond to the stiffness matrix, displacement parameters, and loads on the structure

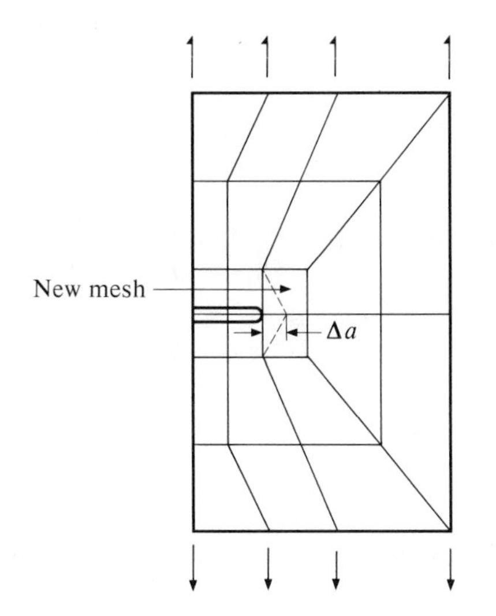

Fig. 23.13 Crack extension Δa and mesh re-adjustment

with original position of the crack. Further, let $\Delta\mathbf{K}$ and $\Delta\mathbf{a}$ be the changes in these quantities due to crack extension Δa ($\mathbf{f}$ does not change).

With the original crack position we have

$$\mathbf{K}\mathbf{a}+\mathbf{f} = 0 \tag{23.47}$$

and the potential energy is

$$\Pi = \tfrac{1}{2}\mathbf{a}^{\mathrm{T}}\mathbf{K}\mathbf{a}+\mathbf{a}^{\mathrm{T}}\mathbf{f}. \tag{23.48}$$

We can now write the change of potential energy as

$$\Delta\Pi = \tfrac{1}{2}(\mathbf{a}+\Delta\mathbf{a})^{\mathrm{T}}(\mathbf{K}+\Delta\mathbf{K})(\mathbf{a}+\Delta\mathbf{a})+(\mathbf{a}+\Delta\mathbf{a})^{\mathrm{T}}\mathbf{f}-\tfrac{1}{2}\mathbf{a}^{\mathrm{T}}\mathbf{K}\mathbf{a}-\mathbf{a}^{\mathrm{T}}\mathbf{f}. \tag{23.49}$$

Neglecting second order terms and using (23.47), we can write the above as

$$\Delta\Pi = \tfrac{1}{2}\mathbf{a}^{\mathrm{T}}\Delta\mathbf{K}\mathbf{a} \tag{23.50}$$

and it is evident that to determine $\Delta\Pi/\Delta a$ it is only necessary

(*a*) to evaluate $\mathbf{a}$ by a single solution, and
(*b*) to determine the change $\Delta\mathbf{K}$ by a calculation of appropriate stiffness changes when the crack configuration alters by Δa.

The additional computations involved in the process implicit in Eq. (23.50) involve only a recalculation of element stiffnesses in the immediate vicinity of the crack tip where the geometric change has occurred. This localized calculation is very economical and can be used easily in both two- and three-dimensional contexts.

A yet different alternative of a direct evaluation of energy release by involving only a single analysis has been proposed by Rice.[64] This involves a calculation of an integral on any path surrounding the crack tip. This integral known as the J integral is path independent and is of the form

$$J = \int_{\Gamma}\left(U-\sigma_x\frac{\partial u}{\partial x}-\tau_{xy}\frac{\partial v}{\partial x}\right)\mathrm{d}y+\int\left(\tau_{xy}\frac{\partial u}{\partial x}+\sigma_y\frac{\partial v}{\partial x}\right)\mathrm{d}x \tag{23.51}$$

in two-dimensional stress fields. Here U is the strain energy density, given in linear elasticity by

$$U \equiv \tfrac{1}{2}\boldsymbol{\sigma}^{\mathrm{T}}\boldsymbol{\varepsilon} \tag{23.52}$$

and x is the direction of the propagating crack.

A special advantage of the above integral form is that it is also applicable in non-linear elastic cases—and indeed with suitable modifications can be used for elastic–plastic investigation.[50]

As

$$J \equiv \frac{\partial \Pi}{\partial a} \tag{23.53}$$

the full information on the fracture criteria is again available.

Clearly, whatever the procedure of determining the energy release rate, an adequate modelling of the singularity is essential.

The simple refinement of a mesh using standard finite elements can be used, but it is found that the convergence rate is low—indeed, this is now almost independent of the polynomial order used in the approximation and the use of higher order elements is not advantageous.[65] Obviously, therefore, special approaches of the type discussed in sections 23.1–23.3 of this chapter are necessary, and in the next section we shall give a description of some possibilities used.

23.5.3 *Elements for fracture mechanics.* As the form of the predominant singularity present near the crack tip is known (*vide* Eq. (23.43)) for elastic analysis, it is natural to attempt to include this in the finite element model. Two main lines of attack are possible, following the general procedures we have discussed at the outset of this chapter. These are:

(*a*) Use of exact solution series in the interior of a special element which incorporates the crack, and an interface/boundary matching of such elements with adjacent, standard ones. Here the integrals would involve only the element interface as shown in sections 23.2 and 23.3.

(*b*) Use of (displacement) assumptions for shape functions which include or approximate to the exact singularity form but which are given as ordinary shape functions necessitating the formation of element matrices by domain integration. This, indeed, is the form of attack outlined in section 23.4.

In category (*a*) an element developed by Tong and Pian[25,26] is a prime example. Here a series expansion satisfying the boundary conditions on the sides of the crack is used in an element shown in Fig. 23.14. This element, using up to 31 terms in the series expansion, yields extremely accurate results. Its formulation follows precisely the pattern described in section 23.2.

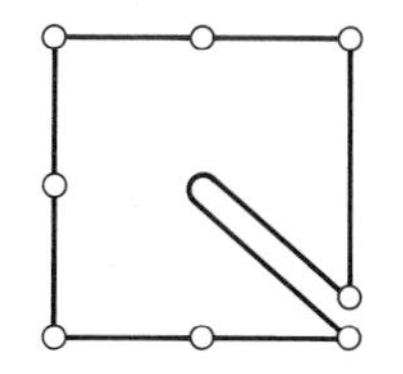

Fig. 23.14 A crack element

Boundary Integral Equations forms of section 23.3 can clearly be used here *without any modification*, and results obtained by Cruse and van Buren[66] show that excellent accuracies can be obtained although the crack singularities are not explicitly contained in the expressions. The solution in reference 66 does not couple the field with finite elements, but obviously could do so if tackled in the manner of section 23.3 and herein lies its attractiveness as special elements would no longer be needed.

In category (*b*) a much wider range of possibilities exist, and indeed here lie some of the costliest attempts at solution. The first of these is due to Byskov[67] who uses an element embodying the same singularities as those used by Tong and Pian[25] but combines these with a standard finite element expansion in a triangle and integrates over its domain with certain incompatibilities remaining on the boundary.

Similar lines have been followed by others[68–77] deriving more or less successful elements. Two of the approaches are of special interest due to their simplicity and will be discussed in some more detail.

In the first of these, suggested by Benzley,[77] a standard C_0 element shape function (*vide* Chapter 7) is used with a supplementation yielding displacement singularity functions of the predominant kind typified by Eq. (23.43). For such a singularity situated, say, at the corner of a typical element shown in Fig. 23.15, we can write (in plane analysis)

$$
\begin{aligned}
u' &= K_{\mathrm{I}} Q_{u\mathrm{I}} + K_{\mathrm{II}} Q_{u\mathrm{II}} \\
v' &= K_{\mathrm{I}} Q_{\mathrm{VI}} + K_{\mathrm{II}} Q_{\mathrm{VII}}
\end{aligned}
\tag{23.54}
$$

where K_{I} and K_{II} are the stress intensities and the functions $Q(r, \theta)$ correspond to the predominant modes

$$
\begin{aligned}
Q_{u\mathrm{I}} &= \frac{1}{G\sqrt{2\pi}} \cos\frac{\theta}{2}\left[\frac{\kappa-1}{2}+\sin^2\frac{\theta}{2}\right]\sqrt{r} \\
Q_{u\mathrm{II}} &= \frac{1}{G\sqrt{2\pi}} \sin\frac{\theta}{2}\left[\frac{\kappa+1}{2}+\cos^2\frac{\theta}{2}\right]\sqrt{r} \\
Q_{v\mathrm{I}} &= \frac{1}{G2\pi} \sin\frac{\theta}{2}\left[\frac{\kappa+1}{2}-\cos^2\frac{\theta}{2}\right]\sqrt{r} \\
Q_{v\mathrm{II}} &= \frac{1}{G2\pi} \cos\frac{\theta}{2}\left[-\frac{\kappa-1}{2}+\cos^2\frac{\theta}{2}\right]\sqrt{r}
\end{aligned}
\tag{23.55}
$$

To incorporate such fields in the analysis we can *either* proceed in a conventional manner, identifying nodal parameters with displacements and write

$$
\mathbf{u} = \alpha_1 + \alpha_2 x + \cdots + K_{\mathrm{I}} Q_{u\mathrm{I}} + K_{\mathrm{II}} Q_{u\mathrm{II}} \tag{23.56}
$$

and obtain element shape functions in terms of nodal variables $\mathbf{u}_i$ and the parameters K_I and K_{II}, *or* use the standard expressions with nodal parameters not corresponding to the total displacement and use terms of Eqs. (23.54) over *all elements*.

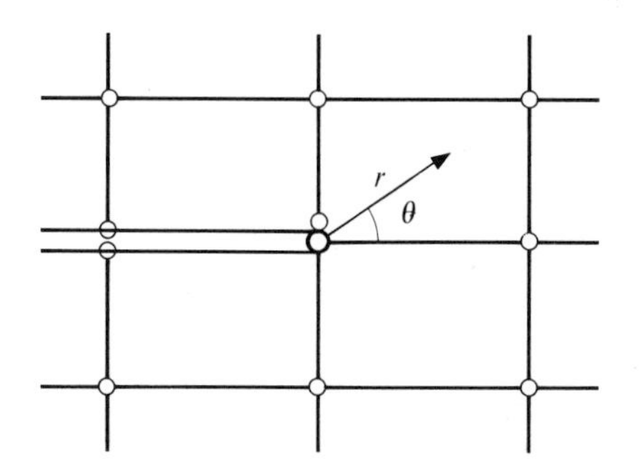

Fig. 23.15 Elements of linear type with supplementary singularity functions

The former process limits the singularity affects to the immediate vicinity of it and can, by suitable modification given by Benzley,[77] yield fully conforming shape functions. The latter includes all elements—and now parameters K_I and K_{II}, for each singularity appear in all terms of the final assembled matrix thus destroying its banded structure. However, with the use of front type solution techniques[78] this is of little consequence and the technique shows some promise.[79]

The second, even simpler, approach to the singularity element relies on the properties of a co-ordinate mapping introduced in isoparametric elements. We have already noted (Chapter 8) that in such elements a singularity can develop if the Jacobian determinant becomes zero. This, in general, was inconvenient but now use of such properties can be made to *ensure* interpolations which contain *suitable singularities*.

An element of this kind, shown in Fig. 23.16(*a*), was introduced almost simultaneously by Henshell[73] and Barsoum[74, 75] for quadrilaterals by a simple shift of the mid-side node in quadratic, isoparametric elements to the quarter point.

It can now be shown (and we leave this exercise to the curious reader) that along the element edges the derivatives $\partial u/\partial x$ (or strains) vary as $1/\sqrt{r}$ where r is the distance from the corner node at which the singularity develops. Although good results are achievable with such elements the singularity is, in fact, not well modelled on lines other than element edges. A more recent development suggested by Hibbitt[76] achieves a better result by using triangular second order elements for this purpose (Fig. 23.16(*b*)).

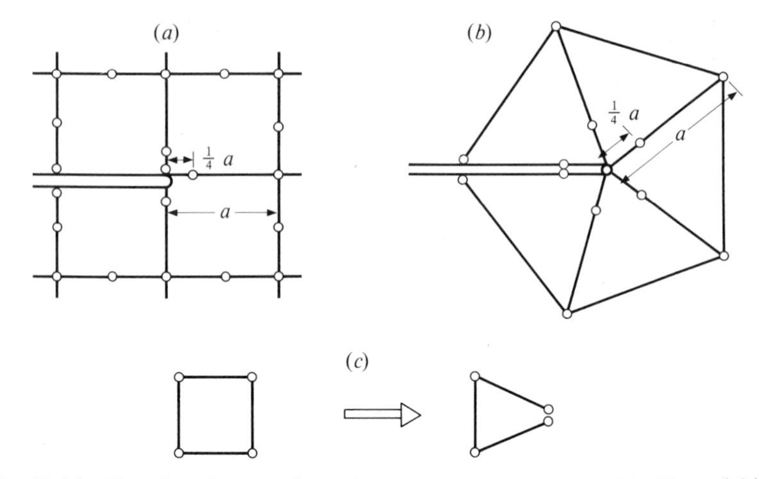

Fig. 23.16 Singular elements from degenerate isoparameters. (*a*), (*b*), and (*c*)

Indeed, the use of distorted or degenerate isoparametrics is not confined to elastic singularities. Rice[64] shows that in the case of plasticity a shear strain singularity of $1/r$ type develops and Levy *et al.*[57] use an isoparametric, linear quadrilateral to generate such a singularity by the simple device of coalescing two nodes but treating these displacements independently. A variant of this is developed by Rice and Tracey.[50]

The elements just described are evidently simple to implement without any changes in a standard finite element program.

23.6 **Concluding Remarks**

Although this chapter has dealt with a variety of problems, its main thread of thought was the *use of analytically available solutions to supplement numerical computation.*

We have also noted how another powerful technique of numerical analysis, i.e., the boundary solution method, can in fact be interpreted as a special case of a trial function/finite element process and thus be incorporated in the finite element process.

These concepts are, perhaps, most fitting ones for the closure of this text in which we have, it is hoped, indicated the power and width of application inherent in finite element approaches. However, if the reader becomes obsessed with the potential of the techniques offered and neglects to make use of the parallel body of knowledge available from analytical methods or, indeed, from the proper physical understanding of the problems dealt with, then he fails as an engineer or a physicist and wastes valuable computer time in computations which could be avoided, or at least substantially reduced. It is in the combination of all such knowledge that the finite element methods can be made use of most effectively.

References

1. O. C. ZIENKIEWICZ and R. W. GERSTNER, 'The method of interface stress adjustment and its uses in some plane elasticity problems', *Int. J. Mech. Sci.*, **2**, 267–76, 1961.
2. O. C. ZIENKIEWICZ and R. W. GERSTNER, 'Stress analysis and special problems of prestressed dams', *Proc. Am. Soc. Civ. Eng.*, **87**, PO1, 7–43, 1961.
3. M. J. M. BERNAL and J. R. WHITEMAN, 'Numerical treatment of biharmonic boundary value problems with re-entrant boundaries', *Comp. J.*, **13**, 87–91, 1970.
4. R. WAIT and A. R. MITCHELL, 'Corner singularities in elliptic problems by finite element methods', *J. Comp. Phys.*, **8**, 45–52, 1971.
5. Y. YAMAMOTO, 'Finite element approaches with the aid of analytical solutions', *Recent Advances in Matrix Method of Structural Analysis and Design*, pp. 85–103 (eds. R. H. Gallagher, Y. Yamada and T. J. Oden), Univ. of Alabama Press, Huntsville, 1970.
6. E. TREFFTZ, 'Gegenstück zom Ritz'schen Verfahren', *Proc. Sec. Int. Congress Applied Mechanics*, Zurich, 1926.
7. T. A. CRUSE and F. J. RIZZO (eds.), 'Boundary-integral equation method: computational applications in applied mechanics', *Proc. Am. Soc. Mech. Eng.*, Sp. Publ., **11**, 1975.
8. T. VON KARMAN, *Calculation of pressure distribution on airship hulls*, NACA TM 574, 1930.
9. M. A. JASWON, 'Integral equation methods in potential theory: I', *Proc. Roy. Soc.*, *A*, **275**, 23–32, 1963.
10. M. A. JASWON and A. R. PONTER, 'An integral equation solution of the torsion problem', *Proc. Roy. Soc.*, **A273**, 237–46, 1963.
11. G. T. SYMM, 'Integral equation methods in potential theory: II', *Proc. Roy. Soc.*, **A275**, 33–46, 1963.
12. C. E. MASSONNET, 'Numerical use of integral procedures', Chapter 10 of *Stress Analysis* (eds. O. C. Zienkiewicz and G. S. Holister), Wiley, 1965.
13. E. R. A. OLIVEIRA, 'Plane stress analysis by a general integral method', *J. Eng. Mech. Div., Proc. Am. Soc. Civ. Eng.*, pp. 79–101, Feb. 1968.
14. T. A. CRUSE, 'Application of the boundary-integral equation method to 3D stress analysis', *J. Comp. Struct.*, **3**, 509–27, 1973.
15. F. J. RIZZO. 'An integral equation approach to boundary value problems of classical elastostatics', *Q. J. Appl. Math.*, **25**, 83–95, 1967.
16. F. J. RIZZO and D. J. SHIPPY, 'A formulation and solution procedure for the general non-homogeneous elastic inclusion problem', *Int. J. Solids Struct.*, **5**, 1161–73, 1968.
17. R. BUTTERFIELD and P. K. BANNERJEE, 'The elastic analysis of compressible piles and pile groups', *Geotechnique*, **21** (No. 1), 43–60, 1971.
18. J. M. BOISSENOT, J. C. LACHAT and J. WATSON, 'Étude par équations integrales d'une éprouvette C.T.15', *Rev. Phys. Appliq.*, **9**, 611–15, Dept. D.T.E.-CETIM, Senlis, France, July 1974.
19. J. C. LACHAT and J. O. WATSON, 'Effective numerical treatment of boundary integral equations; A formulation for three-dimensional elastostatics', *Int. J. Num. Meth. Eng.*, **10**, 991–1006, 1976.

20. W. VANBUREN, *The indirect potential method for three-dimensional boundary value problems of classical elastic equilibrium*, Research Report 68-ID7-MEKMA-R2, Westinghouse Research Laboratories, Pitsburg, 1968.
21. O. C. ZIENKIEWICZ, D. W. KELLY and P. BETTESS, 'The coupling of the finite element method and boundary solution procedures', *Int. J. Num. Meth. Eng.*, **11**, 355–76, 1977.
22. J. L. HESS, 'Review of integral-equation techniques for solving potential-flow problems with emphasis on the surface-source method', *Comp. Meth. Appl. Mech. Eng.*, **5**, 145–96, 1975.
23. P. M. QUINLAN, 'The edge-function method in elasto-statics' from *Studies in Numerical Analysis*, Academic Press, 1974.
24. H. S. CHEN and C. C. MEI, 'Oscillations and wave forces in a man-made harbor in open sea'. (Paper presented at 10th Naval Hydrodynamics Symposium, June 1974.) Dept. of Civil Eng., M.I.T., Cambridge, Mass.
25. P. TONG and T. H. H. PIAN, 'A hybrid-element approach to crack problems in plane elasticity', *Int. J. Num. Meth. Eng.*, **7**, 297–308, 1973.
26. P. TONG and T. H. H. PIAN, 'On the convergence of the finite element method for problems with singularity', *Int. J. Solids Struct.*, **9**, 313–21, 1972.
27. C. SOMIGLIANA, 'Sopra l'equilibrio di un corpo elastico isotropo', *Il Nuovo Ciemento*, t. 17–19, 1885–86.
28. O. C. ZIENKIEWICZ, D. W. KELLY and P. BETTESS, '"Marriage a la mode", Finite elements and boundary integrals', in *Proc. Conf. Innovative Numerical Analysis in Engineering Science* CETIM, Paris, 1977 (to be published).
29. P. SILVESTER and M. S. HSIEH, 'Finite-element solution of 2D exterior field problems', *Proc. I.E.E.*, **118** (No. 12), 1943–7, Dec. 1971.
30. G. K. BATCHELOR, *An Introduction to Fluid Dynamics*, Cambridge Univ. Press, 1967.
31. I. H. ABBOTT and A. E. VON DOENHOFF, *Theory of wing sections*, Dover, 1959.
32. I. FRIED, *Finite element analysis of problems formulated by an integral equation; application to potential flow*, Institut fur Statik & Dynamik der Luft- und Raumfahrtkonstruktionene, Universität Stuttgart, Oct. 1968.
33. S. HOMMA, 'On the behaviour of seismic sea waves round circular island', *Geophys. Mag.*, **XXI**, 199–208, 1950.
34. A. KARAIOSSIFIDIS, *A comparison of solution methods for exterior surface wave problems*, M.Sc. Thesis, Univ. College of Wales, Swansea, 1976.
35. O. C. ZIENKIEWICZ, P. BETTESS and D. W. KELLY, 'The finite element method for determining fluid loadings on rigid structures', Ch. 4 of *Numerical Methods in Offshore Engineering*, ed. O. C. Zienkiewicz, R. Lewis and K. G. Stagg, J. Wiley and Son, 1977.
36. C. D. MOTE, Jr., 'Global-local finite element', *Int. J. Num. Meth. Eng.*, **3**, 565–74, 1971.
37. O. C. ZIENKIEWICZ and P. BETTESS, 'Infinite elements in the study of fluid-structure interaction problems'. *Second Int. Symp. in Computing Methods in Applied Science and Engineering*, IRIA, Versailles, France, 1975.
38. P. BETTESS, 'Infinite elements', *Int. J. Num. Meth. Eng*, **11**, 53–64, 1977.
39. P. BETTESS and O. C. ZIENKIEWICZ, 'Diffraction and refraction of surface waves using finite and infinite elements', *Int. J. Num. Meth. Eng.*, **11**, 1271–90, 1977.
40. D. K. GARTLING and E. B. BECKER, 'Finite element analysis of viscous incompressible fluid flow', *Comp. Meth. Appl. Mech. Eng.*, **8**, 51–60, 1976.
41. G. R. IRWIN, 'Fracture', *Handbuch der Physik*, Vol. 6, 551–90, Springer, Berlin, 1958.

42. G. R. Irwin, 'Fracture mechanics' in *Structural Mechanics*, pp. 557–94, *Proc. 1st Symp. on Naval Structural Mechanics* (eds. J. N. Goodier and N. J. Hoff), Pergamon Press, 1960.
43. G. C. Sih (ed.), *Mechanics of Fracture—Vol. 1: Methods of Analysis and Solutions of Crack Problems*, Noordhoff, 1973.
44. Y. Tada, P. C. Paris and G. R. Irwin, *The Stress Analysis of Cracks Handbook*, Del Research Corp., Hellertown, Penn, 1973.
45. J. F. Knott, *Fundamentals of Fracture Mechanics*, Butterworths, 1973.
46. R. H. Gallagher, 'Survey and evaluation of the finite element method in fracture mechanics analysis', *Proc. 1st Int. Conf. on Structural Mechanics in Reactor Technology*, Vol. **6**, Part L, 637–53, Berlin, 1971.
47. N. Levy, P. V. Marçal and J. R. Rice, 'Progress in three-dimensional elastic–plastic stress analysis for fracture mechanics', *Nucl. Eng. Des.*, **17**, 64–75, 1971.
48. J. J. Oglesby and O. Lomacky, 'An evaluation of finite element methods for the computation of elastic stress intensity factors', *J. Eng. Ind.*, **95**, 177–83, 1973.
49. T. H. H. Pian, 'Crack elements' in *Proc. World Congress on Finite Element Methods in Structural Mechanics*, Vol. 1, pp. F1–F39, Bournemouth, 1975.
50. J. R. Rice and D. M. Tracey, 'Computational fracture mechanics' in *Numerical and Computer Methods in Structured Mechanics*, pp. 555–624 (eds. S. J. Fenves *et al.*), Academic Press, 1973.
51. E. F. Rybicki and S. E. Benzley (eds.), *Computational fracture mechanics*, A.S.M.E. Special Publication, 1975.
52. A. A. Griffiths, 'The phenomena of flow and rupture in solids', *Phil. Trans. Roy. Soc. (London)*, **A221**, 163–98, Oct. 1920.
53. B. Aamodt and P. G. Bergan, 'Propagation of elliptical surface cracks and nonlinear fracture mechanics by the finite element method', *5th Conf. on Dimensioning and Strength Calculations*, Budapest, Oct. 1974.
54. P. G. Bergan and B. Aamodt, 'Finite element analysis of crack propagation in three-dimensional solids under cyclic loading' in *Proc. of 2nd Int. Conf. on Structural Mechanics in Reactor Technology*, Vol. **III**, Part G-H.
55. J. L. Swedlow, 'Elasto-plastic cracked plates in plane strain', *Int. J. Fract. Mech.*, **5**, 33–44, March 1969.
56. T. Yokobori and A. Kamei, 'The size of the plastic zone at the tip of a crack in plane strain state by the finite element method', *Int. J. Fract. Mech.*, **9**, 98–100, 1973.
57. N. Levy, P. V. Marçal, W. J. Ostergren and J. R. Rice, 'Small scale yielding near a crack in plane strain: a finite element analysis', *Int. J. Fract. Mech.*, **7**, 143–57, 1967.
58. J. R. Dixon and L. P. Pook, 'Stress intensity factors calculated generally by the finite element technique', *Nature*, **224**, 166, 1969.
59. J. R. Dixon and J. S. Strannigan, 'Determination of energy release rates and stress-intensity factors by the finite element method', *J. Strain Analysis*, **7**, 125–31, 1972.
60. V. B. Watwood, 'Finite element method for prediction of crack behavior', *Nucl. Eng. Des.*, **II** (No. 2), 323–32, March 1970.
61. D. F. Mowbray, 'A note on the finite element method in linear fracture mechanics', *Eng. Fract. Mech.*, **2**, 173–6, 1970.
62. D. M. Parks, 'A stiffness derivative finite element technique for determination of elastic crack tip stress intensity factors', *Int. J. Fract*, **10**, 487–502, 1974.

63. T. K. Hellen, 'On the method of virtual crack extensions', *Int. J. Num. Meth. Eng.*, **9** (No. 1), 187–208, 1975.
64. J. R. Rice, 'A path-independent integral and the approximate analysis of strain concentration by notches and cracks', *J. Appl. Mech., Trans. Am. Soc. Mech. Eng.*, **35**, 379–86, 1968.
65. P. Tong and T. H. H. Pian, 'On the convergence of the finite element method for problems with singularity', *Int. J. Solids Struct.*, **9**, 313–21, 1972.
66. T. A. Cruse and W. Vanburen, 'Three dimensional elastic stress analysis of a fracture specimen with edge crack', *Int. J. Fract. Mech.*, **7**, 1–15, 1971.
67. E. Byskov, 'The calculation of stress intensity factors using the finite element method with cracked elements', *Int. J. Fract. Mech.*, **6**, 159–67, 1970.
68. P. F. Walsh, 'Numerical analysis in orthotropic linear fracture mechanics', *Inst. Eng. Australia, Civ. Eng. Trans.*, **15**, 115–19, 1973.
69. P. F. Walsh, 'The computation of stress intensity factors by a special finite element technique', *Int. J. Solids Struct.*, **7**, 1333–42, Oct. 1971.
70. A. K. Rao, I. S. Raju and A. Murthy Krishna, 'A powerful hybrid method in finite element analysis', *Int. J. Num. Meth. Eng.*, **3**, 389–403, 1971.
71. W. S. Blackburn, 'Calculation of stress intensity factors at crack tips using special finite elements', in *The Mathematics of Finite Elements*, pp. 327–36 (ed. J. R. Whiteman), Academic Press, 1973.
72. D. M. Tracey, 'Finite elements for determination of crack tip elastic stress intensity factors', *Eng. Fract. Mech.*, **3**, 255–65, 1971.
73. R. D. Henshell and K. G. Shaw, 'Crack tip elements are unnecessary', *Int. J. Num. Meth. Eng.*, **9**, 495–509, 1975.
74. R. S. Barsoum, 'On the use of isoparametric finite elements in linear fracture mechanics', *Int. J. Num. Meth. Eng.*, **10**, 25–38, 1976.
75. R. S. Barsoum, 'Triangular quarter point elements as elastic and perfectly elastic crack tip elements', *Int. J. Num. Meth. Eng.*, **11**, 85–98, 1977.
76. H. D. Hibbitt, 'Some properties of singular isoparametric elements', *Int. J. Num. Meth. Eng.*, **11**, 180–4, 1977.
77. S. E. Benzley, 'Representation of singularities with isoparametric finite elements', *Int. J. Num. Meth. Eng.*, **8** (No. 3), 537–45, 1974.
78. B. M. Irons, 'A frontal solution program for finite element analysis', *Int. J. Num. Meth. Eng.*, **2**, 5–32, 1970.
79. A. J. Fawkes, D. R. J. Owen and A. R. Luxmoore, *Finite elements applied to crack tip singularities—an assessment of current models*, Internal Civil Engineering Report, Univ. College, Swansea. (To be published.)

24. Computer Procedures for Finite Element Analysis

[R. L. TAYLOR]

24.1 **Introduction**

In this chapter we shall consider some of the steps which are involved in the development of a finite element computer program to carry out analyses for the theory presented in previous chapters. The computer program which is discussed here can be utilized to solve one-, two- and three-dimensional problems which may be linear or non-linear, steady-state or transient.

Finite element programs can be separated into two basic parts:

(*a*) data input module and preprocessor, and
(*b*) solution and output modules to carry out the actual analysis (see Fig. 24.1 for a simplified schematic).

Each of the modules can in practice be very complex. In the subsequent sections we shall discuss in some detail the programming aspects for each of the modules. It will be assumed that the reader is familiar with programming, and in particular with FORTRAN IV. Readers who merely intend to use the program listed at the end of the chapter may wish to skip the following sections and go to the user instructions included in sections 24.3 and 24.4.

The chapter is broken down into 8 sections. Sections 24.2 and 24.3 discuss the procedure adopted for data input, describing a finite element problem and the instructions for data card preparation, respectively. Basically the data consists of nodal quantities (e.g., co-ordinates, boundary condition data, loading, etc.) and element quantities (e.g., connection data, material properties, etc.).

Section 24.4 discusses solution algorithms for various classes of finite element analyses. In order to have a computer program which can solve many types of finite element problems a *macro* programming language is introduced. The macro programming language is associated with a set of compact subprograms each designed to compute one or at most a few basic steps in a finite element solution process. Examples in the macro language are commands to form the global stiffness matrix, as well as commands to solve equations, print results, etc. The macro programming concept permits inclusion of a wide class of solution algorithms in the computer program presented here.

In section 24.5 we discuss a methodology commonly used to develop element matrices. In particular, numerical integration is used to derive the element 'stiffness', 'mass', and 'load' matrices for problems in linear heat transfer and elasticity. The concept of using basic shape function routines is exploited in these developments (see Chapter 8).

In section 24.6 methods for solving the large set of algebraic equations resulting from finite element formulations is presented. The method adopted for the computer program is an active column profile (skyline) method which is based upon the Crout method of Gauss elimination. The method consists of a factorization of the stiffness matrix into the product of a lower triangular matrix and an upper triangular matrix. The use of this scheme with active column profile storage leads to a very compact program and allows for inclusion of a resolve capability (i.e., new load cases) without any significant additional programming effort. Use of the resolve capability can substantially reduce costs for analysing subsequent load cases. Included are solution subprograms for both symmetric and non-symmetric coefficient matrices, thus permitting the consideration of finite element formulation which do not lead to a symmetric stiffness matrix, e.g., fluid problems (Chapter 22). In connection with the solution of equations this section concludes with a discussion of the assembly of the global stiffness matrix from the element stiffness matrices.

In present-day practice there are many complex and efficient system programs for finite element analysis capable of dealing with very large numbers of variables and formulations. The very complexity of such systems means that it is difficult to update them in order to introduce new developments of technology. The program presented here is written specifically as a research and educational tool in which the various 'modules' can be changed or added to as desired. Indeed, quite different combinations of the subroutines for purposes which may even today not be obviously needed are possible.

For simplicity, all equations have been retained in core; thus, the capacity of the program is limited by the computer used. With the core size available in some modern computers, the program can handle realistic engineering problems with several hundred unknowns. If larger problems are to be considered the program can (at the expense of some complexity and efficiency) be extended, and we shall deal with this matter in section 24.7.

Finally, section 24.8 contains a complete listing of the program discussed in this chapter. Included also are element routines to carry out analyses for two-dimensional problems of linear elasticity, heat transfer, and fluid mechanics (Navier–Stokes equations). Using the format of these routines the reader should, after mastering this chapter, be able to program additional routines for other problems and greatly extend the capabilities of the program.

24.2 **Data Input Module**

The data input module shown in Fig. 24.1 must transmit sufficient information to the other modules so that each problem can be solved. In the program discussed here the data input module is used to read from cards the necessary geometric, material, and loading data so that all subsequent finite element arrays can be established. In the program a set of dimensioned arrays are established which store nodal co-ordinates, element connections, material properties, boundary restraint codes, prescribed nodal forces and displacements, nodal temperatures, etc. Table 24.1 lists the array names (and their dimensions) which are used to store these quantities.

The notation used for the arrays is at variance with that used in the text. For example, on the text it was found convenient to refer to nodal co-ordinates as x_i, y_i, z_i, whereas in the program these are called X(1, i),

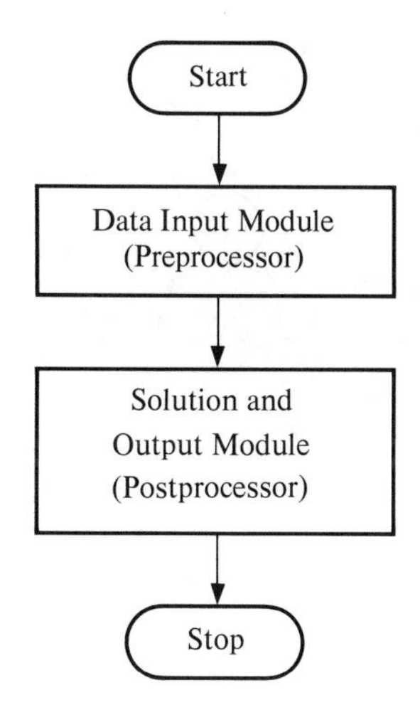

Fig. 24.1 Simplified schematic of finite element program

X(2, i), X(3, i), respectively. This change was made so that all arrays used in the program could be dynamically dimensioned; thus, if a two-dimensional problem is analysed, space will not be reserved for the X(3, i) co-ordinates, and likewise for X(2, i) in one-dimensional problems.

In addition, the nodal displacement in the text were called a_i; in the program these are called either U(i) or U(1, i), U(2, i), etc., where the first subscript refers to the degrees of freedom at a node (from 1 to NDF).

24.2.1 *Storage allocation*. A single array **M** is partitioned to store all

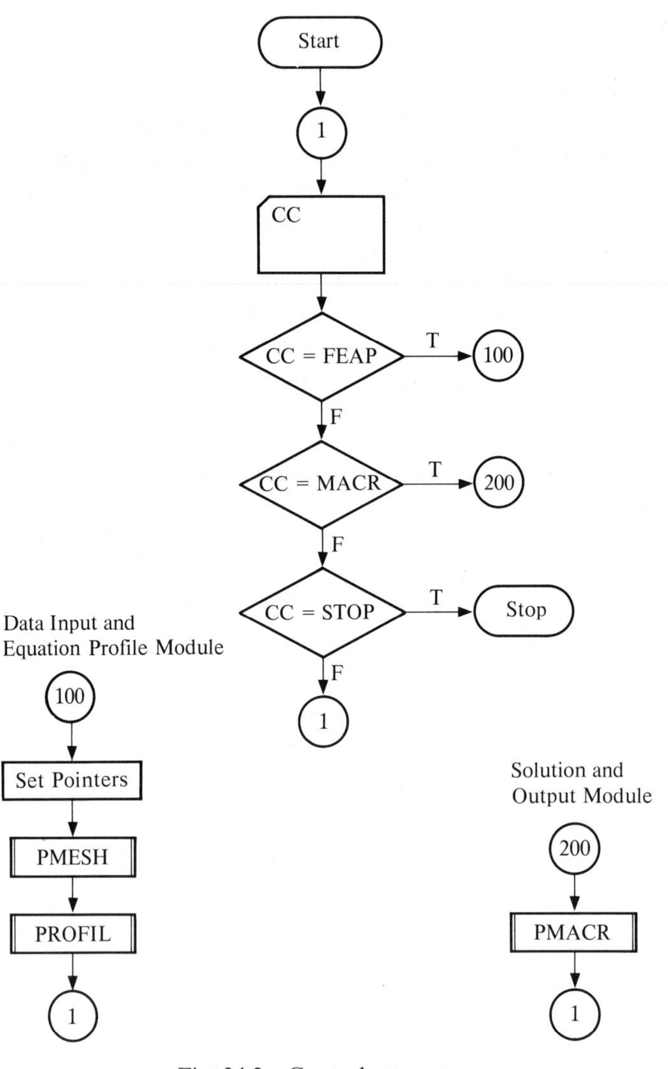

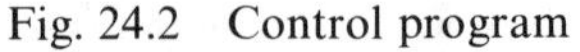
Fig. 24.2 Control program

the data arrays, as well as, the global arrays, e.g., stiffness, load etc. Each array given in Table 24.1 is variably dimensioned to the exact size required for each problem by using a set of pointers established in the control program (see Fig. 24.2). In this way no space is wasted in data storage and a maximum amount of space is reserved to store the global arrays. Since this automatic dimensioning method is used it is not possible to establish absolute values for maximum number of material sets, nodes, or elements.

TABLE 24.1
FORTRAN VARIABLE NAMES USED FOR DATA STORAGE

Variable name (dimension)	Description
D(10,NUMMAT)	Material property data sets, limited to 10 words per set.
F(NDF,NUMNP)	Nodal forces and displacements.
ID(NDF,NUMNP)	Boundary restraint conditions after input of data changed to equation numbers in global arrays.
IE(NUMMAT)	Element type for each material set.
IX(NEN1,NUMEL)	Element nodal connections and material set numbers.
T(NUMNP)	Nodal temperatures.
X(NDM,NUMNP)	Nodal co-ordinates.
NDF	Maximum number of degrees of freedom at any node (maximum of 6).
NDM	Spatial dimension of problem (maximum is 3).
NEN	Maximum number of nodes connected to any element.
NEN1	NEN+1
NUMEL	Number of elements.
NUMMAT	Number of material sets.
NUMNP	Number of nodes.

The storage required in M for the mesh data is

$$\begin{aligned}\text{NE} = {} & \text{IPR*[NEN*(NDM}+1)+\text{NST*(NST}+3)\\ & +10\text{*NUMMAT}+\text{NUMNP*(NDM}+\text{NDF}+1)\\ & +2\text{*NEQ]}+[\text{NST}+\text{NUMMAT}+\text{NDF*NUMNP}\\ & +\text{NUMEL*(NEN}+1)+\text{NEQ]} \qquad (24.1)\end{aligned}$$

where, in addition to variables defined in Table 24.1, IPR is the precision of real variables and NEQ is the number of active equations (which is NDF* NUMNP minus the number of boundary restraints). The program will check that sufficient space exists to solve each problem and, if not, an error message will be printed. The total capacity of the program is controlled by the dimension of the array M in the blank common of the main program and the corresponding value of MAX.

24.2.2 *Element and co-ordinate data.* Once a meshing for a problem has been established data can be prepared for the computer program (see section 24.3 for card formats). As an example, consider the speci-

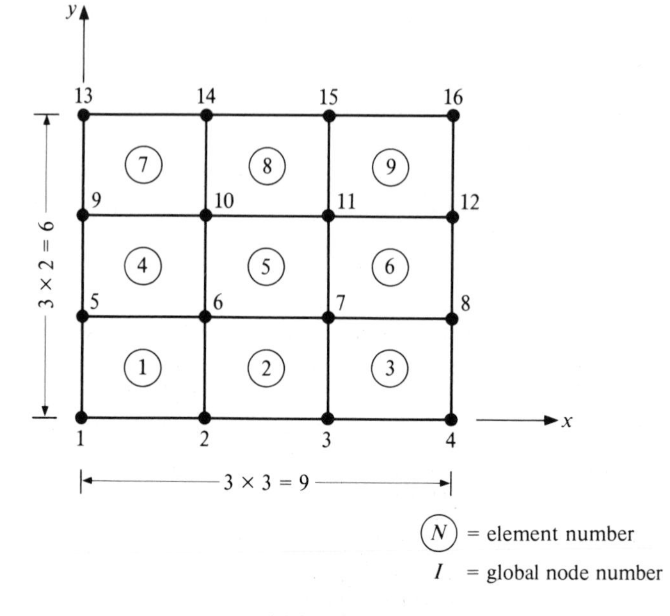

Fig. 24.3 Simple mesh

fication of the nodal co-ordinate and element connection data for the sample two-dimensional (i.e., NDM = 2) rectangular region shown in Fig. 24.3, where a mesh of nine 4-node rectangular elements (NUMEL = 9 and NEN = 4) and 16 nodes (NUMNP) has been established. To describe the nodal and element data, values must be assigned to each X(i, j) for $i = 1, 2$ and $j = 1$ to 16 and for each IX(k, n) for $k = 1$ to 4 and $n = 1$ to 9. In the definition of the co-ordinate array X, the 'i' subscript indicates co-ordinate direction and the 'j' subscript defines the node number. Thus the value of X(1, 3) is the x co-ordinate for node 3 and the value of X(2, 3) is the y co-ordinate for node 3. In a similar way for the element connection array IX the 'k' subscript is the local node number of the element and 'n' is the element number. The value of any IX(k, n) is the number of a global node. The convention for the first local node number is somewhat arbitrary. The local node number 1 for element 3 in Fig. 24.3 could be associated with the global node 3, 4, 7, or 8. Once the first local node is established the others follow according to the convention adopted for each particular element type. For example the 4-node quadrilateral can be numbered according to Fig. 24.4. If we consider once again element 3 we have four possibilities for specifying the

IX(k, 3) array. These are

Option no.	Local node number 1	2	3	4
a	3	4	8	7
b	4	8	7	3
c	8	7	3	4
d	7	3	4	8

The computation of the element matrices from any of the above descriptions must produce the same coefficients for the global arrays.

For a very large mesh the preparation of each piece of mesh data would be very tedious; consequently, a program should provide capabilities to generate much of the data. A simple scheme for nodal generation is to input end points of any line and generate by some scheme the interior points, e.g., linear interpolation is used in the program given here. Thus for the mesh in Fig. 24.3 one could input co-ordinates of nodes 1 and 4 and generate co-ordinates for nodes 2 and 3. Even for this simple problem nodal co-ordinate data preparation is reduced by half.

For elements a pattern usually exists from which elements can be generated. Again consider the mesh in Fig. 24.3. The nodal values for element 2 are those of element 1 incremented by 1; and nodes of element 3 are those of element 2 incremented by 1. Thus, one could input the nodal connections for elements 1, 4, and 7 and generate the rest using a specified increment.

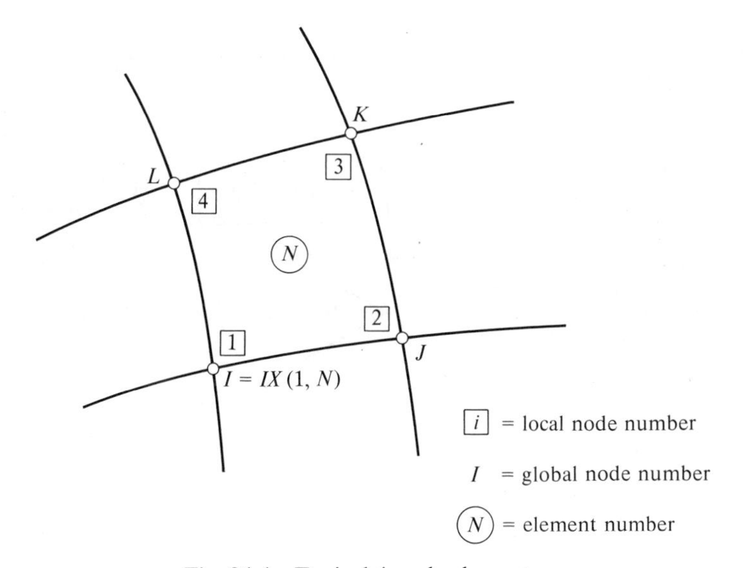

Fig. 24.4 Typical 4-node element

More sophisticated generation schemes can be developed, e.g., see reference 1. The procedure adopted to input the mesh data must be consistent with the particular class of problems the user wishes to analyse, as well as the facilities which are available. The data input scheme included in the program given here is simple and should suffice for most analyses within the capacity of the program. If a user wishes to prepare his own input generation scheme, it can easily be interfaced with subroutine PMESH since each input segment does not interact with any of the others. A flow chart for PMESH is given in Fig. 24.5.

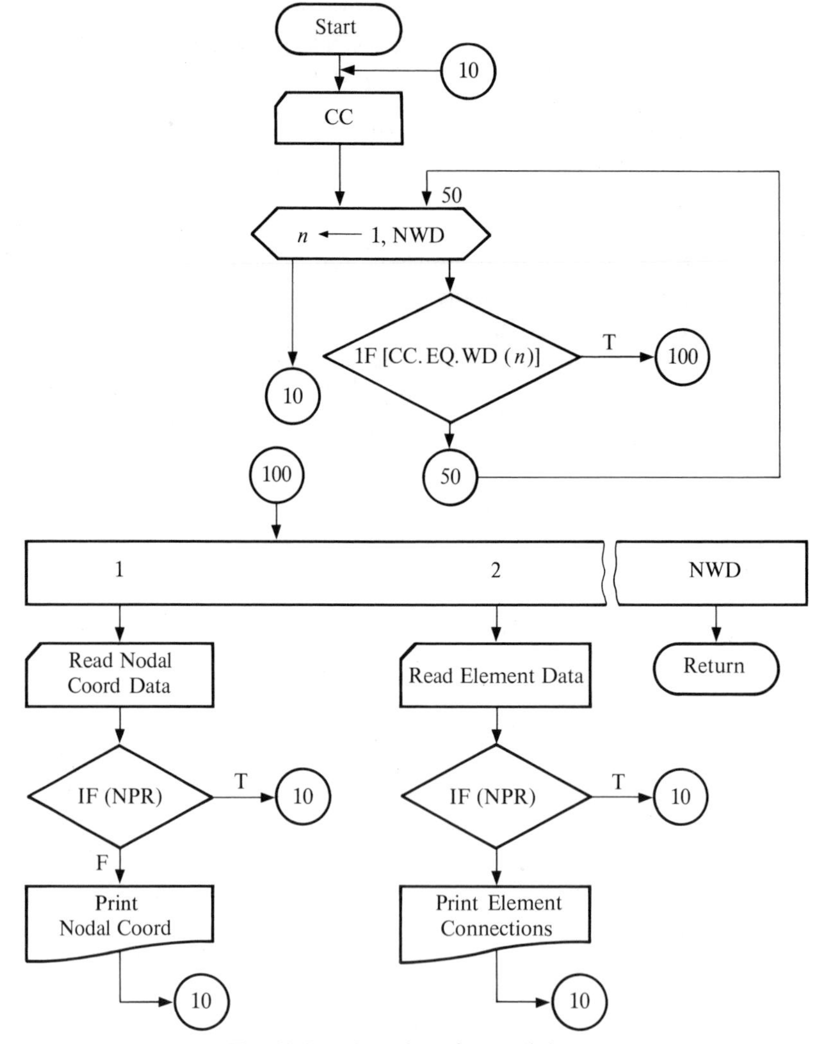

Fig. 24.5 Flow chart for mesh input

24.2.3 *Material property specification—different element routines.* The above discussion has focussed only on the data arrays for nodal co-ordinates and element connections. It is also necessary to specify the material properties associated with each element, loadings, and the restraints for each node.

Each element has material properties associated with it, e.g., for linear isotropic elastic materials Young's modulus E and Poisson's ratio ν describe the material constitution for an isothermal state. In most situations several elements have the same material properties and it is unnecessary to specify properties for each element individually. In this case an element can be 'keyed' to a material property set by a single number on the element card and the properties can be specified once. For example, if the region shown in Fig. 24.3 is all the same material, only one material property set is required and each element would reference this set.

A more complicated example is that shown in Chapter 1, Fig. 1.4 where elements 1, 2, 4, and 5 might be plane elements while element 3 is a truss element. (In realistic engineering problems several elements may be needed at the same time.) In this case at least two different types of element stiffness formulations must be computed. In the computer program given here facilities exist for using up to four different types of element routines in any analysis. The program has been designed so that all computations associated with any element are contained in one element subroutine called ELMTnn where nn is between 01 and 04 (see section 24.5.3 for a discussion on organization of ELMTnn). Each element type to be used is specified as part of the material property data. Thus if element type 1, e.g., computed using subroutine ELMT01 is a plane linear elastic 3- or 4-node element and element type 2 is a truss element, e.g., computed using subroutine ELMT02, the data given for the example in Fig. 1.4 might be:

(*a*) *Material properties*

Material set no.	Element type	Material property data
1	2	E_1, A_1
2	1	E_2, ν_2, NG_2

(*b*) *Element connections*

Element	Material property set	Connection
1	2	1 3 4
2	2	1 4 2
3	1	2 5
4	2	3 6 7 4
5	2	4 7 8 5

where E is Young's modulus, ν is Poisson's ratio, A is area, and NG is the number of integration points used to compute element stiffness (see Chapter 8 and section 24.5). Thus elements 1, 2, 4, and 5 have material property set 2 which is associated with element type 1—that is, it is a plane linear elastic element as Fig. 1.4 shows. In a similar way, element 3 has material property set 1 which is associated with element type 2, the truss element. It will be seen later that the above scheme leads to an organization of an element routine which inputs material data sets and computes all necessary arrays for finite element analyses.

24.2.4 *Boundary conditions—equation numbers.* The process of specifying the boundary conditions and the procedure for modification for specified displacements is tied to the method adopted to store the global arrays, e.g., stiffness and mass matrices. In the program only those coefficients within a non-zero profile in the global arrays are stored.

The storage of the non-zero profile of the equations leads to considerable savings over the more traditional banded solution storage.† In addition, it is usually more efficient to delete the rows and columns for the equations corresponding to specified boundary displacements. As an example consider the stiffness matrix corresponding to the problem given in Fig. 1.1, storing all terms within the upper profile requires 54 words, whereas if the equations corresponding to the restrained nodes 1 and 6 are deleted only 32 words are required to store the compacted stiffness (see Fig. 24.6). This is a saving of over 40 per cent for the stiffness matrix alone. The effort (as measured by computer time) to solve equations by a profile method is approximately proportional to the sum of the column heights squared. For the example in Fig. 24.6 compacted storage also leads to savings of over 40 per cent in equation solution.

To facilitate the compact storage operation a boundary condition array is input for each node. The array is called ID and is dimensioned as shown in Table 24.1. During data input the value of the ID array for any nodal degree of freedom for which either the value is specified *a priori* or for which no unknown exists (i.e., different nodes can have different numbers of associated unknowns) is set to a *non-zero* value; all other degrees of freedom have a *zero* value. Table 24.2a shows the table of values ID for the example shown in Chapter 1, Fig. 1.1 where it is evident that nodes 1 and 6 are fully restrained.

The numbers for the compacted equations are constructed from Table 24.2a by replacing each non-zero value with a zero and each zero value by the appropriate equation number. In the program this step is performed in subroutine PROFIL starting with the degrees of freedom associated

† It will be shown in section 24.6 that the profile storage by columns leads to a very efficient direct equation solution method.

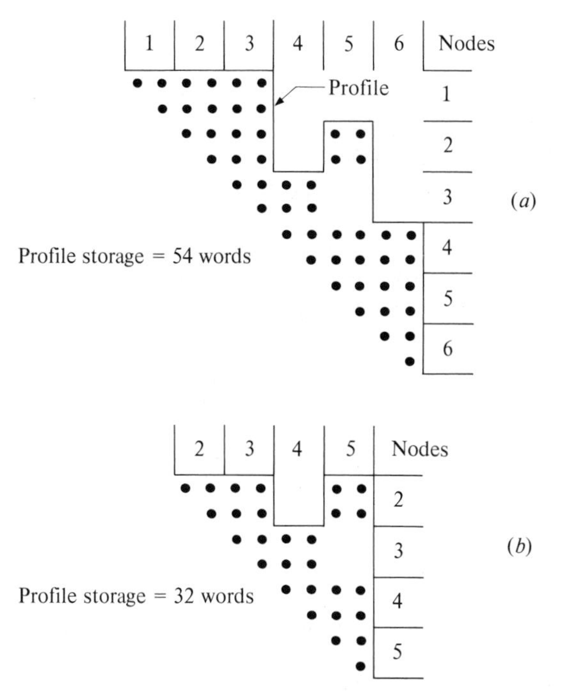

Fig. 24.6 Stiffness matrix. (*a*) Total stiffness storage. (*b*) Storage after deletion of boundary condition

TABLE 24.2a
BOUNDARY RESTRAINT CODES VALUES
AFTER DATA INPUT OF PROBLEM IN FIG. 1.1.

Node	Degree of freedom 1	2
1	1	1
2	0	0
3	0	0
4	0	0
5	0	0
6	−1	1

with node 1, etc. The result for the example leads to values shown in Table 24.2b, which contains the boundary restraint information but, in addition, tells how to assemble the compacted equations. Accordingly the first equation of the compacted equations shown in Fig. 24.6 is associated with the first degree of freedom of node 2, etc.

The equation number scheme can be further exploited to handle repeating boundaries (see Chapter 9, section 9.5) where nodes on two boundaries are required to have the same displacement but its value is unknown. This is accomplished by setting the equation numbers to the same value (discard the unused one). The equation profile is then adjusted to accommodate the connection (see section 24.2.7).

TABLE 24.2b
COMPACTED EQUATION NUMBERS
FOR PROBLEM IN FIG. 1.1.

Node	Degree of freedom 1	2
1	0	0
2	1	2
3	3	4
4	5	6
5	7	8
6	0	0

24.2.5 *Loading*. The non-zero nodal forces or displacements associated with each degree of freedom must be specified. In the program these are both stored in the array F and the distinction between load and displacement is made by comparing the corresponding value of the boundary restraint condition (from the equation number table) for each degree of freedom. For the example of Fig. 1.1, if F(1, 1) was set to 0·01 it would signify that the displacement of the first degree of freedom (i.e., u) is specified to be ·01 units, whereas F(2, 3) set to 5 would indicate that the force for the second degree of freedom has a value of 5 units. (It should be emphasized that a finite element program such as the one given here rarely has built-in units; these are established by the user using any consistent set of units.)

In many problems the loading may be distributed and in these cases the loading must first be converted to nodal forces. This can be facilitated in the data input by adding a new macro command DIST which makes a call to subroutine SLDnn to perform the computation of the equivalent nodal loads. Once any distributed load is converted to the generalized forces it is treated in the same way as for the loads input by FORC. A word of caution—the use of the DIST macro must follow specification of FORC or BOUN macro commands if concentrated nodal forces act at the same node as a distributed load, or if the node which has distributed loading applied to it is restrained.

The program also provides facilities to specify nodal temperatures. (The association here is for thermal loading of structural problems; for other classes of problems other interpretations can be inserted by the user.) Temperatures are input for each node in the same way as for forces and co-ordinates.

Specific instructions for the preparation of data cards for each of the data items discussed above are given in section 24.3.

24.2.6 *Mesh data checking.* Once all data for the geometric, material, and loading are supplied the program is ready to initiate execution of the solution module; however, prior to this step it is usually preferable to perform some checks on the input data. The simplest such check would be a review of the input data (and generated values) as given on a printed output from the program. For large problems any checking by this procedure will leave considerable doubt as to the accuracy of the data—it is easy to mispunch data such that visual checks reveal nothing! It is advisable to use some automatic plot to scale of the mesh as an alternative check on accuracy. In addition, checks on the value of the Jacobians in isoparametric elements, as suggested in Chapter 8, can be used. The program given here provides for the latter check. A program for plotting two-dimensional meshes was given in the previous book[2] and can easily be interfaced. For three-dimensional problems mesh plot routines can also be prepared; however, without capabilities of erasing hidden lines, rotating, slicing etc., the usual plot is unintelligible for any but the simplest problem. The general topic of automatic plotting is outside the scope of this text and the reader is referred to reference 3 for further information on this very important aspect of any practical finite element solution package.

24.2.7 *Profile determination.* As discussed above, the global arrays for stiffness and mass are to be stored in a compact profile form as shown in Fig. 24.6. The storage mode will be by columns above the principle diagonal, and for problems with unsymmetric arrays by rows below the diagonal. It will be shown in section 24.6 that this mode of storage leads to a very efficient algorithm for direct solution of equations. In unsymmetric problems it is assumed that the profile is symmetric, which will certainly be the case for finite element formulations based upon variational or Galerkin methods. It is necessary to know beforehand the profile of the equations, and this is determined by first numbering the active equations as described above and then using the element connection array, IX, together with the equation number and boundary condition array, ID, to determine the maximum column height of each equation. Finally the equations are compacted into a vector and the column heights are used to construct the address of the diagonal elements in the storage vector. The programming steps for the profile determination are given in subroutine PROFIL. The total number of equations is determined by the maximum value of the ID array and is called NEQ. The total storage requirement for either the upper or lower half of the matrix profile is given by the address for the diagonal of NEQ (i.e. JDIAG(NEQ)). Thus the storage requirements over that given in the introduction must be increased by this amount for each profile matrix required, e.g., by JDIAG(NEQ) for linear symmetric steady-state problems.

24.3 **User Instructions for Computer Program**

The solution of a finite element problem using the program given at the end of this chapter begins with a sketch of a mesh covering the region to be analysed. If boundaries are curved the mesh will only approximate the shape of the region (e.g., see Fig. 24.7). In sketching the mesh the type and order (linear, quadratic, etc.) of elements must be taken into consideration: for triangular elements in two dimensions the mesh is described by a net of triangles, whereas for quadrilateral isoparametric 4-node elements the region can be described by a net of quadrilaterals. The user may wish to use both triangles and quadrilaterals. In this case two element routines may be necessary, one for triangles and one for quadrilaterals. The shape function routine, SHAPE, for quadrilaterals given in this chapter includes the 3-node triangle by coalescing the shape functions of two nodes, hence in this case only one element routine need be used; if, however, a quadratic element is employed, the simple trick of coalescing nodes no longer works and additional programming effort is required.[4]

After a sketch of the mesh has been made the elements and nodes are numbered in consecutive order. The order of numbering the elements is not crucial. However, the order of numbering the nodes will strongly influence the profile of non-zero coefficients. As a general rule, the numbering should be such as to minimize the nodal difference for each element (maximum node number minus minimum node number). A numbering can usually be improved using an automatic nodal renumbering scheme for bandwidth or, preferably, profile minimization, e.g., see references 5 and 6.

Once the sketch and numbering of the mesh is completed the user can proceed to the preparation of data for the program. The first step consists of specifying problem title and control information given in Table 24.3, which is used during subsequent data input and also is used to allocate memory in the program.

In addition to the input data card formats, Table 24.3 gives the variable names used in the program. The variables NDF, NEN, and NAD are used to calculate the size of the element arrays, NST. Normally for displacement formulations NDF*NEN is the size of the element array;

TABLE 24.3

TITLE AND CONTROL INFORMATION FORMATS

TITLE CARD—FORMAT (20A4)
The title card also serves as a start of problem card. The first four (4) columns must contain the start word FEAP.

Columns	*Description*	*Variable*
1 to 4	Must contain FEAP	TITL(1)
5 to 80	Alphanumeric information to be printed with output as page header.	TITL(I). I = 2, 20

CONTROL CARD—FORMAT (7I5)

Columns	*Description*	*Variable*
1 to 5	Number of nodes	NUMNP
6 to 10	Number of elements	NUMEL
11 to 15	Number of material sets	NUMMAT
16 to 20	Spatial dimension ($\leqslant 3$)	NDM
21 to 25	Number of unknowns per node ($\leqslant 6$)	NDF
26 to 30	Number of nodes/element	NEN
31 to 35	Added size to element matrices, in excess of NDF*NEN	NAD

however, if nodeless variables or hybrid methods are used it may be necessary to increase the size of the element array by NAD.

Once the control data is supplied the program expects the data cards for the mesh description, e.g., nodal co-ordinates, element connections, etc. Each problem or problem class may require different types and amounts of data; consequently, the flow of data to the program is controlled by a set of *macro* commands. The available macro commands are given in Table 24.4; others may be added by suitably modifying the data list WD in subroutine PMESH. The macro commands PRINt† and NOPRint allow the user to print and suppress printing, respectively, of any data which is input subsequently. Thus, once a mesh has been fully checked and subsequent analyses are desired it is not necessary to reprint all the mesh data.

An analysis will require at least:

(*a*) co-ordinate data which follows the macro command COOR and is prepared as described in Table 24.5;

(*b*) element data which follows the macro command ELEM and is prepared according to Table 24.6; and

(*c*) material data which follows the macro command MATE and is prepared according to Table 24.7 and the data required for each particular element (see section 24.8.3).

In addition, most analyses will require specification of nodal boundary restraint conditions, macro BOUN, and the corresponding nodal force

† Only capital characters are data.

or displacement value, macro FORC, which are specified according to Tables 24.8 and 24.9, respectively. Some analyses may have auxiliary nodal quantities which specify a loading. For example, in the analysis of elasticity problems, temperature may provide loading. The program provides a capability of specifying a temperature (or corresponding nodal loading) using the TEMP macro followed by data prepared according to Table 24.10.

The end of any mesh data is indicated by use of an END macro card. The use of the macro command cards allows the user to specify only those data items actually needed for each analysis. The END macro card signifying end of mesh data input. The use of macro cards also reduces the chance of data input errors due to extraneous blank cards. Strict card sequencing is necessary only within each macro segment. Several blank cards may exist after actual data without affecting the execution of the program.

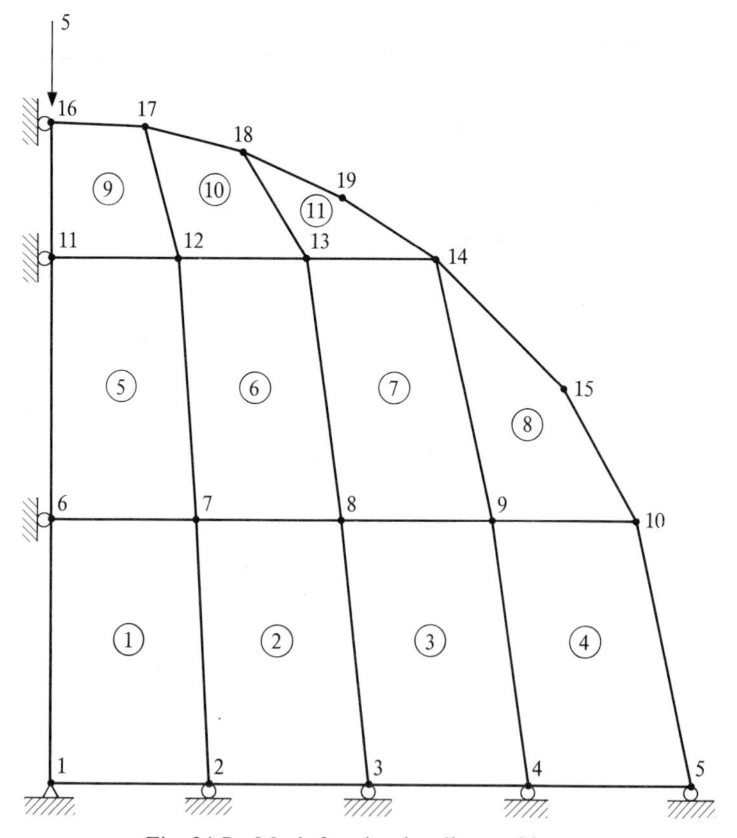

Fig. 24.7 Mesh for circular disc problem

Once all necessary mesh data is supplied the user can select to solve the problem or not. If only a check of the mesh is desired, insert either a STOP macro command to stop execution or start a new problem as described in Table 24.1. If a solution to the problem is desired, additional data is required as discussed in the next section.

As an example of the data input required to describe a mesh consider the mesh shown in Fig. 24.7 for the quadrant of a circular disc. The input data card images for this problem are shown in Table 24.11.

TABLE 24.4
DATA INPUT: MACRO CONTROL STATEMENTS

INPUT MACRO CONTROL CARDS—FORMAT (A4)

The input of each data segment is controlled by the value assigned to CC. The following values are admissible and each CC card must be immediately followed by the appropriate data (described in Tables 24.5 to 24.9)

CC Value	*Data to be input*
COOR	Co-ordinate data
ELEM	Element data
MATE	Material data
BOUN	Boundary condition data
FORC	Prescribed nodal force data
TEMP	Temperature data
PRIN	Print subsequent mesh data (default mode)
NOPR	Do not print subsequent mesh data
END	Must be last card in mesh data, terminates mesh input

Except for the END card the data segments can be in any order. If the values of BOUN, FORC, or TEMP are zero, no input data is required.

TABLE 24.5
CO-ORDINATE DATA

CO-ORDINATE DATA—FORMAT (2I5, 7F10.0)—must immediately follow a COOR macro card

The co-ordinate data card contains the node number N and the value of the co-ordinates for the node. Only the values of (XL(I), I = 1,NDM) are used, where NDM is the value input on the control card.

Nodal co-ordinates can be generated along a straight line described by the values input on two successive cards. The value of the node number is computed using the N and NG on the first card to compute the sequence N, N+NG, N+2NG, etc. NG may be input as a negative number, if it has incorrect sign the sign will be changed. Nodes need not be in order.

Columns	*Description*	*Variable*
1 to 5	Node number	N
6 to 10	Generator increment	NG
11 to 20	X1-co-ordinate	XL(1)→X(1,N)
21 to 30	X2-co-ordinate	XL(2)→X(2,N)
31 to 40	X3-co-ordinate	XL(3)→X(3,N)

N.B. Terminate with blank card(s).

TABLE 24.6
ELEMENT DATA

ELEMENT DATA—FORMAT (16I5)—must immediately follow an ELEM card
The element data card contains the element number, material set number (which also selects the element type, see Table 24.7), and the sequence of nodes connected to the element. If there are less than NEN nodes (see Table 24.1 for input of NEN) either leave the appropriate fields blank or punch zeros.

Elements must be in order. If element cards are omitted the element data will be generated from the previous element with the same material number and the nodes all incremented by the LX on the previous element. Generation to the maximum element number occurs when a blank card is encountered.

Columns	*Description*	*Variable*
1 to 5	Element number	L
6 to 10	Material set number	IX(NEN1,L)
11 to 15	Node 1 number	IX(1,L)
16 to 20	Node 2 number	IX(2,L)
etc.	⋮	⋮
etc.	Node NEN number	IX(NEN,L)
etc.	Generation increment	LX

TABLE 24.7
MATERIAL PROPERTY DATA

MATERIAL DATA SETS—must immediately follow a MATE macro card
Each material property set also selects the element type which will be used for the material property data.
Card 1). FORMAT (I5, 4X, I1, 17A4)

Columns	*Description*	*Variable*
1 to 5	Material set number	MA
6 to 9	Not used	
10	Element type number (1 to 4)	IEL
11 to 78	Alphanumeric information to be output	XHED

Each material Card 1.) must be followed immediately by the material property data required for the element type IEL being used, e.g., see section 24.8.3 (ISW = 1)

TABLE 24.8
BOUNDARY RESTRAINT DATA

BOUNDARY CONDITION CARDS—FORMAT (16I5)—must immediately follow a BOUN macro card
For each node which has at least one degree of freedom with a specified displacement, a boundary condition card must be input. The convention used for boundary restraints is

$= 0$ no restraint, force specified
$\neq 0$ restrained, displacement specified

Values of force or displacement input in FORC (Table 24.9)

Columns	*Description*	*Variable*
1 to 5	Node number	N
6 to 10	Generation increment	NX
11 to 15	DOF 1 boundary code	IDL(1)→ID(1,N)
16 to 20	DOF 2 boundary code	IDL(2)→ID(2,N)
etc.	⋮	⋮
etc.	DOF NDF boundary code	IDL(NDF)→ID(NDF,N)

N.B. When generating boundary condition codes for subsequent nodes IDL is set to zero if it was input $\geqslant 0$, and is set to -1 if input negative. All degrees of freedom with non-zero codes are assumed fixed. Terminate with blank card(s).

TABLE 24.9

NODAL FORCED BOUNDARY VALUE DATA

FORCE CARDS—FORMAT (2I5, 7F10.0)—must immediately follow a FORC macro card
For each node which has a non-zero nodal force or displacement a force card must be input or generated. Generation is the same as for co-ordinate data (see Table 24.5). The value specified is a force if the corresponding restraint code is zero and a displacement if the corresponding restraint code is non-zero.

Columns	*Description*	*Variable*
1 to 5	Node number	N
6 to 10	Generation increment	NG
11 to 20	DOF 1 Force (Displ.)	XL(1)→F(1,N)
21 to 30	DOF 2 Force (Displ.)	XL(2)→F(2,N)
etc.	⋮	
etc.	DOF NDF Force (Displ.)	XL(NDF)→F(NDF,N)

N.B. Terminate with a blank card.

TABLE 24.10

NODAL TEMPERATURE DATA

TEMPERATURE DATA CARD—FORMAT (2I5, F10.0)—must immediately follow a TEMP macro card
For each node which has a non-zero temperature the value must be input. Generation of values can be performed as described for co-ordinates (see Table 24.5).

Columns	*Description*	*Variable*
1 to 5	Node number	N
6 to 10	Generation increment	NG
11 to 20	Nodal temperature	XL(1)→T(N)

N.B. Terminate with blank card.

24.4 **Solution of Finite Element Problems—the Macro Programming Language**

At the completion of data input and mesh checking we are prepared to initiate a problem solution. It is at this stage that the particular type of solution mode must be available to the user in the program. In many existing programs only a small number of fixed algorithm solution modes are available to the user. For example, the program may only be able to solve linear steady-state problems. Or, in addition, it may be able to solve linear transient problems. In practical engineering problems fixed algorithm programs are often too restrictive and the user must continually modify the program to solve his problem—often at the expense of another user! For this reason it is desirable to have a program which has modules for variable algorithm capabilities and which, if necessary, can be modified without interrupting other users' capabilities. The program which we discuss here is indeed elementary and the reader can undoubtedly see many ways to improve and extend the program for other classes of problems. One important extension would be to include a matrix interpretive language so that individual terms or equations can be modified for specific needs.

TABLE 24.11

DATA INPUT CARD IMAGES FOR DISC PROBLEM

```
FEAP * * QUADRANT OF A CIRCULAR DISK  (EXAMPLE PROBLEM)                    1
   19    11     1     2     2     4                                          2
COORD                                                                        3
    1     1   0.          0.                                                 4
    5     0   5.          0.                                                 5
    6     1   0.          2.                                                 6
   10     0   4.5828      2.                                                 7
   11     1   0.          4.                                                 8
   14     0   3.0         4.                                                 9
   15     0   4.0         3.                                                10
   16     0   0.          5.                                                11
   17     0   0.75        4.9434                                            12
   18     0   1.5         4.7697                                            13
   19     0   2.25        4.4651                                            14
                                                                            15
ELEM                                                                        16
    1     1     1     2     7     6     1                                   17
    5     1     6     7    12    11     1                                   18
    9     1    11    12    17    16     1                                   19
                                                                            20
BOUN                                                                        21
    1     1     1    -1                                                     22
    5     0     0     1                                                     23
    6     5    -1     0                                                     24
   16     0     1     0                                                     25
                                                                            26
FORC                                                                        27
   16     0      0.          -5.                                            28
                                                                            29
MATE                                                                        30
    1     1                                                                 31
   100.       0.3         0.0          2     1     1                        32
                                                                            33
END                                                                         34
```

24.4.1 *Linear steady-state problems.* The basic aspect of the variable algorithm program is a macro instruction language which can be used to construct modules for specific algorithms as needed. The user only needs to learn the mnemonics of the language to use it. For example, if one wishes to form the global stiffness matrix the program instruction TANG is used (TANG is the mnemonic for a symmetric tangent stiffness matrix and for non-linear elements would form and assemble into the global stiffness matrix the element tangent stiffness computed about the current displacement state; for linear elements this is just the linear stiffness matrix). For a problem with an unsymmetric tangent stiffness the macro command UTAN is used. If one wishes to form the right-hand side of the equations modified for specified displacements one uses the program instruction, FORM. The resulting equations are solved using the instruction SOLV. Printed output can be obtained using the instructions DISP for displacements and STRE for element variables such as strains and stresses. The above instructions are sufficient to solve linear steady-state problems, that is, the macro instructions

```
TANG (or UTAN)
FORM
SOLV
DISP
STRE
```

are precisely the required instruction to solve any linear steady-state problem. The reader will undoubtedly observe at this time that ordering can sometimes be changed without affecting the algorithm. For example, use of the macro instruction

```
FORM
TANG (or UTAN)
SOLV
STRE
DISP
```

produces the same algorithm except element quantities are printed before the nodal displacements.

The variable algorithm program as described by the macro programming language can be extended as necessary. For example, when multiple load problems are analysed the global stiffness matrix is always the same and need only be formed once. The right-hand side changes and the new displacements need to be computed. The procedure to solve two load cases requires changing nodal loads and/or specified displacements. The macro instruction MESH causes the program to enter the data input module again, and at this stage loads can be changed. Data appears *after* the macro program instructions which terminate with the END statement. Thus the macro program and data for two load cases could be

```
TANG (or UTAN)
MESH ⎫
FORM ⎪
SOLV ⎬ instructions for problem 1
DISP ⎪
STRE ⎭
MESH ⎫
FORM ⎪
SOLV ⎬ instructions for problem 2
DISP ⎪
STRE ⎭
END    end macro program
FORC
```

```
        loads for problem 1
END        end mesh data input
FORC
        loads for problem 2
END        end mesh data input
```

The reader should notice that the same block of instructions is repeated twice and that if ten load cases were desired, considerable effort is wasted in preparing the macro instruction data cards. To rectify this, looping commands are introduced as the instruction pair

```
LOOP    n
  ⋮
NEXT
```

which indicates that looping over all instructions between LOOP and NEXT will occur n times, hence the macro program for two load cases is now

```
TANG (or UTAN)
LOOP      2
MESH
FORM
SOLV
DISP
STRE
NEXT
END
FORC
      loads for problem 1
END
FORC
      loads for problem 2
END
```

For this program ten load cases are as simple as two (except for the FORC data).

The reader will note that the TANG instruction is executed only once while the SOLV instruction is executed twice. The program will automatically recognize that the second execution uses a stiffness matrix for which the triangular decomposition has already been performed and will select a *resolve* mode of operation (see equation solution section).

Many other classes of problems can be solved using the simple macro instruction list given in Table 24.12. We summarize a few algorithms in the subsequent paragraphs.

TABLE 24.12

LIST OF MACRO PROGRAMMING COMMANDS

Following is a list of macro instruction commands which can be used to construct solution algorithms. The first instruction must be a card with MACR in columns 1 to 4.

	Columns 1 to 4	*Columns* 6 to 9	*Columns* 11 to 15	*Description*
	CHEC			Perform Check of Mesh (ISW = 2)†
	CMAS			Consistent Mass Formulation (ISW = 5)
	CONV			Displacement Convergence Test.
	DATA	**		Read Data **Macro commands (a ** Macro card is inserted as data following the Macro program)
	DISP		N	Print Nodal Displacements every N steps in loop
**	DT		V	Set time increment to value V
	EIGE			Compute dominant eigen value and vector of current mass and symmetric stiffness
	EXCD			Explicit centered difference integration of equations of motion using lumped mass. First call reserves memory only.
	FORM			Form right hand side of equations (ISW = 6)
	LMAS			Lumped mass formulation (ISW = 5)
	LOOP		N	Loop N times between all instructions between matching NEXT instruction
	MESH			Input mesh changes (must not change boundary conditions) Data follows Macro program.
	NEXT			End of loop instruction.
	PROP		1	Input proportional load table (data follows macro program).
	REAC			Compute nodal reactions (ISW = 6)
	SOLV			Solve tangent equations. Update nodal displacements.
	STRE		N	Print element variables (e.g., Stresses) every N steps in loop (ISW = 4)
	TANG			Symmetric tangent stiffness formulation (ISW = 3)
	TIME			Advance time by DT value
**	TOL		V	Set solution convergence tolerance to value V.
	UTAN			Unsymmetric tangent stiffness formulation (ISW = 3)
	END			End of macro program instructions. Data for program follows in order of use.

24.4.2 *Non-linear steady state.*

(*a*) *Newton–Raphson iteration.* The Newton–Raphson iteration algorithm is defined by (viz. Chapter 18, section 18.2.3)

$$\begin{aligned} \mathbf{K}_T(\mathbf{a}^i)\mathbf{v}^i &= -\mathbf{f} - \mathbf{P}(\mathbf{a}^i) = -\boldsymbol{\Psi}^i \\ \mathbf{a}^{i+1} &= \mathbf{a}^i + \mathbf{v}^i \end{aligned} \tag{24.2}$$

Usually we begin with zero as the $\mathbf{a}^0$; however, some other vector may be used. If we compare the above algorithm with the algorithm for the linear steady-state problem we note that the only difference is that we must solve the linear problem several times, consequently the algorithm is

```
LOOP    10
TANG (or UTAN)
FORM
```

† The indicated ISW value is used by each element routine to perform the appropriate computations. See section 24.5.3.

```
SOLV
DISP     2
NEXT
STRE
DISP
END
```

The above algorithm indicates that 10 iterations are to be performed. In addition, the 2 on the DISP command indicates that displacements are to be printed every two iterations.

The program has an internal check on the value of $|\Psi^i|$, where:

$$|\Psi^i| = [\boldsymbol{\Psi}^i . \boldsymbol{\Psi}^i]^{1/2}$$

Whenever

$$|\Psi^i| < \text{TOL} \max_j |\Psi^j| \qquad (j = 1, 2, \ldots, i)$$

where TOL is a predefined tolerance (10^{-9} is the default value), the iteration ceases and a skip to the macro command immediately following the first NEXT occurs. Thus, if convergence occurred in seven iterations the program would transfer to the STRE command.

The DISP instruction in the loop allows the user to observe the displacements as they converge (or to observe any tendencies not to converge!).

(*b*) *Modified Newton–Raphson iteration.* The modified Newton–Raphson algorithm is given by (viz. Chapter 18, section 18.2.4)

$$\begin{aligned} \mathbf{K}_T(\mathbf{a}^j)\mathbf{v}^i &= -\mathbf{f} - \mathbf{P}(\mathbf{a}^i) = -\boldsymbol{\Psi}^i \qquad (j \leqslant i) \\ \mathbf{a}^{i+1} &= \mathbf{a}^i + \mathbf{v}^i \end{aligned} \tag{24.3}$$

The corresponding macro program for $j = 0$ is:

```
TANG
LOOP     10
FORM
SOLV
DISP     2
NEXT
STRE
DISP
END
```

The only difference from (*a*) is that the tangent stiffness is formed (and factored) only once and an alternative is:

```
LOOP    2
TANG
LOOP    5
FORM
SOLV
DISP
NEXT
NEXT
STRE
DISP
END
```

In this algorithm the tangent stiffness is reformed every five iterations. Many other alternatives are possible. In particular, the user may wish to check on the displacement convergence

$$|v^i| < \text{TOL} \max_j |v^j| \qquad j = 1, 2, \ldots, i$$

This can be accomplished by inserting a CONV macro card in the loop.

24.4.3 *Incremental load method.* The macro instruction program to solve a steady-state non-linear problem by incrementing the load is given below (viz. Chapter 18, section 18.2.5). The program does not iterate at each step to achieve convergence but will perform a load correction, the programs given before could be combined if desired, to give an incremental load, Newton–Raphson solution.

DT	·1	$\Delta t = \cdot 1$
PROP	1	$\mathbf{f} = t\mathbf{f}_0$
LOOP	10	increment loop
TIME		$t = t + \Delta t$
TANG		compute $\mathbf{K}_T$
FORM		compute $-\mathbf{f} - \mathbf{P}(\mathbf{a})$
SOLV		compute $\mathbf{v}$ set $\mathbf{a} = \mathbf{a} + \mathbf{v}$
DISP		print $\mathbf{a}$
STRE		print stress
NEXT		
END		

A simplified proportional loading is permitted with

$$\text{PROP} = A_1 + A_2 t + A_3 (\sin (A_4 t + A_5))^L$$

where the coefficients are input on a data card following the END macro card according to the following table.

PROPORTIONAL LOAD CARD—FORMAT (5X,I5, 7F10.0)

Column	*Description*
6 to 10	L
11 to 20	t_{min}, minimum time for which PROP is computed
21 to 30	t_{max}, maximum time for which PROP is computed
31 to 40	A_1
41 to 50	A_2
51 to 60	A_3
61 to 70	A_4
71 to 80	A_5

24.4.4 *Evaluation of fundamental eigenvalue.* The determination of the dominant (smallest) eigenvalue of the problem

$$\mathbf{Ka} = \lambda\mathbf{Ma}$$

by inverse iteration is given by the algorithm

(1) Set $\mathbf{a}^0 = \mathbf{v}$ a starting vector and λ^0 to zero. A superscript denotes iteration number.
(2) Compute $\mathbf{y}^i = \mathbf{Ma}^i$ for each $i = 0, 1, 2, \ldots$
(3) Construct Rayleigh quotient to get estimate for eigenvalue

$$\lambda^i = \frac{\mathbf{a}^{iT}\mathbf{Ka}^i}{\mathbf{a}^{iT}\mathbf{y}^i}$$

(4) If $|\lambda^i - \lambda^{i-1}| < \mathrm{TOL}\ |\lambda^i|$; stop iteration
(5) Set $\mathbf{z}^i = \mathbf{y}^i/|y^i|$, where $|y^i| = (\mathbf{y}^{iT}\mathbf{y}^i)^{1/2}$
(6) Solve $\mathbf{Ka}^{i+1} = \mathbf{z}^i$ for $\mathbf{a}^{i+1}$
(7) Repeat (2) to (6) until either (4) is satisfied or a maximum specified number of iterations is exceeded.

The macro program instructions for the inverse iteration using a consistent mass matrix are:

TANG
CMAS
EIGE (uses subroutine PEIGS)

It should be noted that the numerator to the Rayleigh quotient is computed during the solution of the equations in step (6), as described in the equation solution section.

More general eigenvalue programs could also be included. The reader is referred to references 7 and 8 for more detailed information on this subject.

24.4.5 *Explicit integration of the equations of motion.* The explicit integration of the equations of motion for structural systems can be simply treated for those cases where a lumped mass is used. The equations are given by

$$\mathbf{M}\ddot{\mathbf{a}} + \mathbf{P}(\mathbf{a}, \dot{\mathbf{a}}) = -\mathbf{f}$$

where it is assumed that the problem is non-linear. An explicit consistent algorithm is given using centered differences, as

$$\ddot{\mathbf{a}}_{n+1} = \mathbf{M}^{-1}(-\mathbf{f} - \mathbf{P}(\mathbf{a}_{n+1}, \dot{\mathbf{a}}_{n+1/2}))$$

where

$$\mathbf{a}_{n+1} = \mathbf{a}_n + \Delta t_n \dot{\mathbf{a}}_{n+1/2}$$
$$\dot{\mathbf{a}}_{n+1/2} = \dot{\mathbf{a}}_{n-1/2} + \overline{\Delta t}_n \ddot{\mathbf{a}}_n$$

with

$$\overline{\Delta t}_n = \tfrac{1}{2}(\Delta t_n + \Delta t_{n-1})$$

If the internal force vector is computed using the new displacement and velocity (which is localized into UL(NDF+1,i), etc.) the new acceleration can be computed for a lumped matrix by merely dividing the force by the appropriate mass term. A macro command EXCD (explicit centered difference) has been added to perform these steps. The macro program

```
EXCD
PROP     1
LMAS
DT       Δt
LOOP     n
TIME
FORM
EXCD
DISP
STRESS
NEXT
END
```

will perform n steps of explicit integration at constant Δt and a proportional loading specified on a card following the END macro card.

The reader should by this time be familiar with the types of problems which can be formulated with the macro instruction program. In the next section we discuss some of the aspects which go into constructing a macro instruction module for the program. The macro program concept forces the program to be modular, and the reader can observe in the subroutine PMACR how the modular aspect leads to compact statements to carry out each instruction. New modules should not interfere at all with existing ones so that the needs of one user do not destroy the capabilities of another.

24.4.6 *Programming for macro instructions.* The macro instruction module for the program discussed here is named subroutine PMACR. This subprogram reads and prints all macro instructions, compiles, and then executes the macro program. The possible macro instructions are contained in the array WD. Thus, to add an instruction the array WD must first be extended and the new upper limit assigned to the integer variable NWD. Secondly, the GO TO list must be similarly extended to include transfers to the new macro instruction modules. And, finally, the appropriate statement added. The reader is advised to carefully study the modules already included before attempting to add a new one. In particular, if more space is required, new pointers in array M must be added. This is facilitated by use of the subroutine PSETM with a logical variable preset to .TRUE. being used to ensure that space is added only once. Subroutine PSETM automatically checks that sufficient space is available in M to add the new array.

24.5 Computation of Finite Element Solution Modules

In the establishment of the macro instruction for element computations (e.g., TANG and FORM) many of the operations are the same. In this section we discuss aspects which enter into the computation of the element arrays for finite element computation. The first step is to localize all the geometric, material, and displacement data for the element array to be computed. The particular element quantities to be computed include stiffness, mass, internal forces, stresses, strains, etc. We discuss aspects for computing these for the problems of linear elasticity and heat transfer in section 24.5.2. The organization of the element routines for the program is given in section 24.5.3.

24.5.1 *Localization of element data.* When we want to compute an element array, for example an element stiffness matrix, $\mathbf{S}$, or a load or internal force vector, $\mathbf{P}$, we only need those quantities associated with the element in question—all the other values are superfluous. The nodal and material quantities which are required can be determined from the node and material numbers stored in the IX array for each element. In the program the necessary values are moved from the global arrays to a set of local arrays before the appropriate element routine, i.e., subroutine ELMTnn, is called. This process will be called *localization*. The quantities which are localized are:

(1) nodal co-ordinates which are stored in the array XL,
(2) nodal displacements which are stored in the array UL,
(3) nodal temperatures which are stored in the array TL, and
(4) equation numbers which are stored in the destination array LD.

The LD array described in step (4) is used to map the element stiffness (or mass) matrix and element load or internal force vector to the global stiffness (or mass) matrix and load vector, respectively. Accordingly, for the following element arrays

$$
\begin{array}{cccc} [\mathrm{LD}(1) & \mathrm{LD}(2) & \mathrm{LD}(3) & \ldots] \end{array}
$$

$$
\begin{bmatrix} \mathrm{S}(1,1) & \mathrm{S}(1,2) & \mathrm{S}(1,3) & \\ \mathrm{S}(2,1) & \mathrm{S}(2,2) & & \\ \vdots & & & \end{bmatrix} \quad \begin{bmatrix} \mathrm{P}(1) \\ \mathrm{P}(2) \\ \vdots \end{bmatrix}
$$

the term S(i, j) would be assembled into the global stiffness (or mass) matrix in the position corresponding to row LD(i) and column LD(j), i.e., the LD array contains the equation numbers of the global matrices. Similarly, P(i) would be assembled into the position corresponding to the LD(i) value.

The localization process is the same for every type of finite element and is thus centralized into subroutine PFORM, which organizes all computations associated with elements including looping over the list of elements as described by the IX array. The element properties are stored in the square array S and the vector P. The LD array is a destination vector of element to global equation numbers and is used to map S and P onto the global arrays.

During the localization process the number of nodes actually connected to each element (i.e., NEL which may be less than NEN) is determined by finding the largest non-zero entry in the IX array of that element number. Intermediate zeros are interpreted as no node connected. In this way the program permits mixing of elements with different numbers of connected nodes, e.g., 3-node triangles can be mixed with 4-node quadrilaterals.

Since the current value of the nodal displacements is localized for all element computations, the program can be used to solve non-linear problems. This is, in fact, the only additional information over that for linear problems which is required to construct tangent stiffness matrices, etc., for the solution to non-linear problems as discussed in Chapters 18 and 19.

It should be pointed out before proceeding to the computation of element arrays that the localization step (except for current value of displacements) could be done once and for all and the global data arrays could then be destroyed. This would involve more programming steps and also efficient use of buffering and backing storage to maintain adequate working space in core for subsequent global array determination.

24.5.2 *Element array computations.* The efficient computation of element arrays (in both programmer and computer time) is a crucial aspect of any finite element development. The development of routines to evaluate element stiffness (or tangent stiffness) and load arrays can be accomplished by a combination of appropriate numerical methods. In order to explicate the development, a statement of the essential steps is first given and then some details are shown for the plane stress/strain case.

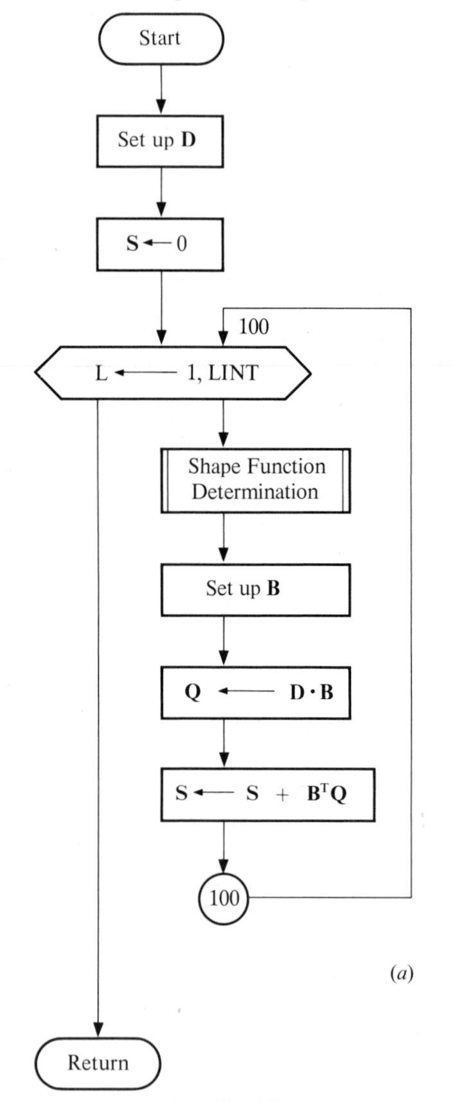

Fig. 24.8 Procedure to compute element stiffness matrix. (*a*) Using shape function routine and numerical integration. (*b*) For elements with constant material properties

The essential steps to compute an element stiffness matrix, S, are summarized in Fig. 24.8. The key ingredients are the numerical integration, the use of shape function routines (which are the same for all problems with the same required continuity) and efficient formulation of the matrix triple product.

Usually Gauss quadrature formulae are utilized to compute all element integrals since they give highest accuracy for effort expended (see Chapter 8). In some instances it is desirable to use other formulae. For example, if one employs a quadrature formula which samples only at the nodes,

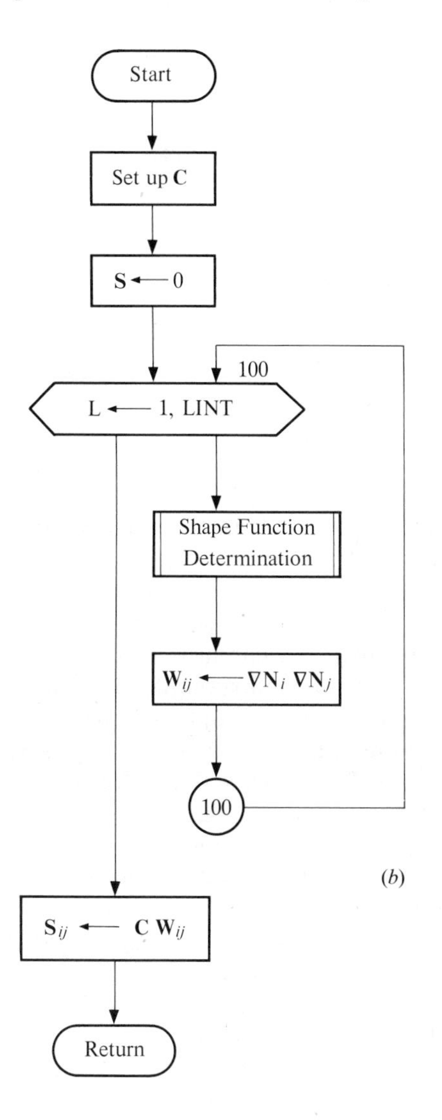

Fig. 24.8 (*cont.*)

then evaluation of the mass term leads to a diagonal mass matrix which is often more advantageous for dynamic problems.

Shape function subprograms allow the programmer to develop elements for many problems quickly and reliably. The shape function subprogram evaluates not only the shape function but also its derivatives with respect to the global co-ordinate frame. As an example, consider the two-dimensional C_0 problem where we need only first derivatives of each shape function N_i. For the 4-node isoparametric quadrilateral we have

$$N_i = \tfrac{1}{4}(1+\xi_i\xi)(1+\eta_i\eta) \tag{24.4}$$

where ξ_i, η_i are the ξ, η co-ordinates of the nodes.

Using the isoparametric concept we have

$$\begin{aligned} x &= N_i x_i \\ y &= N_i y_i \end{aligned} \tag{24.5}$$

and derivatives given by

$$\begin{Bmatrix} N_{i,\xi} \\ N_{i,\eta} \end{Bmatrix} = \begin{bmatrix} x_{,\xi} & y_{,\xi} \\ x_{,\eta} & y_{,\eta} \end{bmatrix} \begin{Bmatrix} N_{i,x} \\ N_{i,y} \end{Bmatrix} \tag{24.6}$$

$$\begin{Bmatrix} N_{i,x} \\ N_{i,y} \end{Bmatrix} = \frac{1}{J} \begin{bmatrix} y_{,\eta} & -y_{,\xi} \\ -x_{,\eta} & x_{,\xi} \end{bmatrix} \begin{Bmatrix} N_{i,\xi} \\ N_{i,\eta} \end{Bmatrix} \tag{24.7}$$

where J is the Jacobian determinant and $(\;)_{,x}$ denotes the partial derivative $\partial(\;)/\partial x$, etc. The above relations define steps for the shape function subprogram given in Fig. 24.9 where it is assumed that the nodal co-ordinates have been transferred to the local co-ordinate array XL.

This shape function routine can be used for all two-dimensional C_0 problems which use the 4-node element (e.g., two-dimensional plane and axi-symmetric elasticity, heat conduction, flow in porous media, fluid dynamics, etc.). Shape function subprograms can also be used for the generation of mesh data.[1] It is a simple task to extend the shape function routine to higher order elements. As an example the routines SHAPE and SHAP2 (see 24.8.3) give shape functions for 3-node triangles up to 8-node serendipity and 9-node Lagrange quadrilaterals. The elements may even have edges with linear expansion while others have quadratic expansion by simply omitting the mid-side node number for a linear edge.

The generation of the matrix triple product also deserves special attention since many zeros often exist in the **B** and **D** matrices. Several methods can be used to reduce the number of operations. The first is to form explicitly the matrix triple product. While this at first appears to involve too many hand computations, it in fact is elementary if it is done nodewise. For example, consider the two-dimensional plane linear

```
      SUBROUTINE SHAPEF(S,T,XL,XSJ,SHP)                                   SHA   1
C                                                                         SHA   2
C.... SHAPE FUNCTION ROUTINE FOR FOUR NODE ISOPARAMETRIC QUADRILATERAL     SHA   3
C                                                                         SHA   4
C           S,T          NATURAL COORDINATES WHERE SHAPE FUNCTIONS        SHA   5
C                        ARE EVALUATED.                                   SHA   6
C           SHP(1,I)     X DERIVATIVE OF SHAPE FUNCTIONS                  SHA   7
C           SHP(2,I)     Y DERIVATIVE OF SHAPE FUNCTIONS                  SHA   8
C           SHP(3,I)     SHAPE FUNCTIONS                                  SHA   9
C           XS           JACOBIAN ARRAY                                   SHA  10
C           XSJ          JACOBIAN DETERMINANT                             SHA  11
C           XL(1,I)      X NODAL COORDINATES                              SHA  12
C           XL(2,I)      Y NODAL COORDINATES                              SHA  13
C                                                                         SHA  14
      DIMENSION SHP(3,4),XL(2,4),XS(2,2),SI(4),TI(4)                      SHA  15
      DATA SI,TI/-.5,.5,.5,-.5,-.5,-.5,.5,.5/                             SHA  16
C                                                                         SHA  17
C.... COMPUTE SHAPE FUNCTIONS AND DERIVATIVES IN NATURAL COORDINATES      SHA  18
C                                                                         SHA  19
      DO 100 I = 1,4                                                      SHA  20
      SHP(3,I) = (0.5+SI(I)*S)*(0.5+TI(I)*T)                              SHA  21
      SHP(1,I) = SI(I)*(0.5+TI(I)*T)                                      SHA  22
100   SHP(2,I) = TI(I)*(0.5+SI(I)*S)                                      SHA  23
C                                                                         SHA  24
C.... COMPUTE JACOBIAN TRANSFORMATION FROM X,Y TO S,T                     SHA  25
C                                                                         SHA  26
      DO 200 I = 1,2                                                      SHA  27
      DO 200 J = 1,2                                                      SHA  28
      XS(I,J) = 0.0                                                       SHA  29
      DO 200 K = 1,4                                                      SHA  30
200   XS(I,J) = XS(I,J) + XL(I,K)*SHP(J,K)                                SHA  31
C                                                                         SHA  32
C.... COMPUTE JACOBIAN DETERMINANT                                        SHA  33
C                                                                         SHA  34
      XSJ = XS(1,1)*XS(2,2)-XS(1,2)*XS(2,1)                               SHA  35
C                                                                         SHA  36
C.... TRANSFORM NATURAL DERIVATIVES TO X,Y DERIVATIVES                    SHA  37
C                                                                         SHA  38
      DO 300 I = 1,4                                                      SHA  39
      TEMP     = ( XS(2,2)*SHP(1,I)-XS(2,1)*SHP(2,I))/XSJ                 SHA  40
      SHP(2,I) = (-XS(1,2)*SHP(1,I)+XS(1,1)*SHP(2,I))/XSJ                 SHA  41
300   SHP(1,I) = TEMP                                                     SHA  42
      RETURN                                                              SHA  43
      END                                                                 SHA  44
```

Fig. 24.9

elasticity problem, where

$$\mathbf{B}_i = \begin{bmatrix} N_{i,x} & 0 \\ 0 & N_{i,y} \\ N_{i,y} & N_{i,x} \end{bmatrix} \tag{24.8}$$

and for isotropy

$$\mathbf{D} = \begin{bmatrix} D_{11} & D_{12} & 0 \\ D_{12} & D_{11} & 0 \\ 0 & 0 & D_{33} \end{bmatrix}. \tag{24.9}$$

where D_{33} is an appropriate combination of D_{11} and D_{12} (viz. Chapter 4, section 4.2.4).

Thus for typical nodal pairs i and j the element stiffness $\mathbf{k}_{ij}$ is given by

$$\mathbf{k}_{ij} = \mathbf{B}_i^{\mathrm{T}}\mathbf{Q}_j \tag{24.10a}$$

where

$$\mathbf{Q}_j = \mathbf{D}\mathbf{B}_j \tag{24.10b}$$

Thus

$$\mathbf{Q}_j = \begin{bmatrix} D_{11}N_{j,x} & D_{12}N_{j,y} \\ D_{12}N_{j,x} & D_{11}N_{j,y} \\ D_{33}N_{j,y} & D_{33}N_{j,x} \end{bmatrix} \tag{24.11}$$

and

$$\mathbf{k}_{ij} = \begin{bmatrix} (N_{i,x}Q_{11}+N_{i,y}Q_{31}) & (N_{i,x}Q_{12}+N_{i,y}Q_{32}) \\ (N_{i,y}Q_{21}+N_{i,x}Q_{31}) & (N_{i,y}Q_{22}+N_{i,x}Q_{32}) \end{bmatrix}. \tag{24.12}$$

Thus for each nodal pair it is required to compute 14 multiplications to form the $\mathbf{k}_{ij}$, whereas, formal multiplication of $\mathbf{B}^{\mathrm{T}}\mathbf{D}\mathbf{B}$ would give 30 multiplications. Also, when the element matrix is symmetric only one half would be formed during numerical integration (the other half would be formed by reflection). A typical routine for the stiffness computation is given in Fig. 24.10, where it is assumed that the Gauss quadrature points and weights are stored in the arrays SG, TG and WG.

An extension to anisotropic problems can also be made by replacing the isotropic $\mathbf{D}$ matrix with the appropriate anisotropic one and then recomputing the $\mathbf{Q}_j$ and $\mathbf{k}_{ij}$ matrices.

The computation of element stiffness matrices for problems which have material properties constant within an element can be made more efficient by noting that the internal energy can be written, using index notation, as

$$(u_k)_i C_{klmn} \int_{V^e} (N_{i,l})(N_{j,n})\, \mathrm{d}V (u_m)_j$$

where C_{klmn} are the elastic constants; k, l, m, n, range over the dimension of the problem; and i, j are nodal indices which in an element range from 1 to NEL. The element stiffness matrix for the nodal pair i, j is thus given by

$$(K_{km})_{ij} = (W_{ln})_{ij} C_{klmn}$$

where

$$(W_{ln})_{ij} = \int_{V^e} (N_{i,l})(N_{j,n})\, \mathrm{d}V.$$

For isotropic materials the elastic constants are given by

$$C_{klmn} = \delta_{kl}\,\delta_{mn}\lambda + (\delta_{km}\,\delta_{ln} + \delta_{lm}\,\delta_{kn})\mu$$

where λ and μ are the Lamé constants which are related to the usual elastic constants E and ν as $\lambda = \nu E/(1+\nu)(1-2\nu)$, $\mu = E/2(1+\nu)$.

```
C.... ISOPARAMETRIC ELEMENT STIFFNESS COMPUTATION FOR ISOTROPIC LINEAR      RIC    1
C.... ELASTICITY * * PLANE STRESS AND PLANE STRAIN DIFFER ONLY IN VALUES    RIC    2
C.... OF THE CONSTANTS D(1),D(2),D(3) SUPPLIED                              RIC    3
C                                                                           RIC    4
C          D(1)   IS A MATERIAL PROPERTY                                    RIC    5
C                 FOR PLANE STRESS                                          RIC    6
C                   D(1) = E/(1.-NU*NU)                                     RIC    7
C                 FOR PLANE STRAIN                                          RIC    8
C                   D(1) = E*(1.-NU)/((1.+NU)*(1.-2.*NU))                   RIC    9
C          D(2)   IS A MATERIAL PROPERTY                                    RIC   10
C                 FOR PLANE STRESS                                          RIC   11
C                   D(2) = NU*D(1)                                          RIC   12
C                 FOR PLANE STRAIN                                          RIC   13
C                   D(2) = NU*D(1)/(1.-NU)                                  RIC   14
C          D(3)   IS A MATERIAL PROPERTY                                    RIC   15
C                   D(3) = E/(2.*(1.+NU))                                   RIC   16
C          DV     IS AN AREA WEIGHTING                                      RIC   17
C          LINT   IS THE NUMBER OF INTEGRATION POINTS                       RIC   18
C          NEL    IS THE NUMBER OF NODES ON THE ELEMENT                     RIC   19
C          S      IS THE ELEMENT STIFFNESS                                  RIC   20
C          SG,TG  ARE INTEGRATION POINTS IN NATURAL COORDINATES             RIC   21
C          SHP    CONTAINS VALUES OF X,Y DERIVATIVES OF SHAPE FUNCTIONS     RIC   22
C          XL     IS THE ARRAY OF NODAL COORDINATES                         RIC   23
C          XSJ    IS THE JACOBIAN DETERMINANT                               RIC   24
C          WG     IS THE INTEGRATION WEIGHT                                 RIC   25
C                                                                           RIC   26
C.... FOR EACH INTEGRATION POINT COMPUTE CONTRIBUTION TO STIFFNESS          RIC   27
      DO 100 L = 1,LINT                                                     RIC   28
      CALL SHAPEF(SG(L),TG(L),XL,XSJ,SHP)                                   RIC   29
      DV = XSJ*WG(L)                                                        RIC   30
      D11 = D(1)*DV                                                         RIC   31
      D12 = D(2)*DV                                                         RIC   32
      D33 = D(3)*DV                                                         RIC   33
C                                                                           RIC   34
C.... FOR EACH J NODE COMPUTE DB = D*B                                      RIC   35
C                                                                           RIC   36
      DO 100 J = 1,NEL                                                      RIC   37
      DB11 = D11*SHP(1,J)                                                   RIC   38
      DB12 = D12*SHP(2,J)                                                   RIC   39
      DB21 = D12*SHP(1,J)                                                   RIC   40
      DB22 = D11*SHP(2,J)                                                   RIC   41
      DB31 = D33*SHP(2,J)                                                   RIC   42
      DB32 = D33*SHP(1,J)                                                   RIC   43
C                                                                           RIC   44
C.... FOR EACH I NODE COMPUTE S = BT*DB                                     RIC   45
C                                                                           RIC   46
      DO 100 I = 1,J                                                        RIC   47
      S(I+I-1,J+J-1) = S(I+I-1,J+J-1) + SHP(1,I)*DB11+SHP(2,I)*DB31        RIC   48
      S(I+I-1,J+J  ) = S(I+I-1,J+J  ) + SHP(1,I)*DB12+SHP(2,I)*DB32        RIC   49
      S(I+I  ,J+J-1) = S(I+I  ,J+J-1) + SHP(1,I)*DB31+SHP(2,I)*DB21        RIC   50
  100 S(I+I  ,J+J  ) = S(I+I  ,J+J  ) + SHP(1,I)*DB32+SHP(2,I)*DB22        RIC   51
C                                                                           RIC   52
C.... COMPUTE LOWER TRIANGULAR PART BY SYMMETRY                             RIC   53
C                                                                           RIC   54
      NL = NEL+NEL                                                          RIC   55
      DO 200 I = 2,NL                                                       RIC   56
      DO 200 J = 1,I                                                        RIC   57
  200 S(I,J) = S(J,I)                                                       RIC   58
      RETURN                                                                RIC   59
      END                                                                   RIC   60
```

Fig. 24.10

Thus the stiffness matrix is computed from

$$(K_{km})_{ij} = \lambda(W_{km})_{ij} + \mu[(W_{mk})_{ij} + \delta_{km}(W_{nn})_{ij}].$$

Using this procedure the steps to compute the element stiffness matrix for plane elasticity are given in Fig. 24.8(b). This procedure for computing stiffness matrices was noted in reference 9 and results in about a 25 per cent reduction over the procedure of Fig. 24.8(a). In three dimensions the savings are even greater.

The above procedure is incorporated into the subroutine for linear elasticity included in section 24.8.3, where the programming steps described by Fig. 24.8(b) are given.

The computation of other element arrays can also be effected using a shape function routine. For example, the computation of the element mass matrix for transient or eigenvalue computations can be performed. The mass matrix for the two-dimensional plane problem is obtained from

$$\mathbf{M}_{ij}^e = \mathbf{I} \int_{V^e} N_i \rho N_j \, \mathrm{d}V \tag{24.13}$$

where $\mathbf{I}$ is a 2×2 identity matrix and ρ is the mass density. The mass matrix is always symmetric, hence only the upper or lower triangular part need be computed during numerical integration, also the mass coefficient for each degree of freedom is the same and a further reduction in numerical integration is possible if only one term of each nodal pair is computed and the other merely copied. A set of statements to compute the mass matrix for the plane problem is shown in Fig. 24.11, where the element consistent mass matrix is stored in the array S, if needed the lumped mass matrix is stored in the array P.

```
C.... ISOPARAMETRIC ELEMENT MASS MATRICES FOR PLANE ELASTICITY          RIC   1
C                                                                       RIC   2
C             P        IS THE LUMPED MASS MATRIX                        RIC   3
C             S        IS THE CONSISTENT MASS MATRIX                    RIC   4
C             OTHER ARRAYS AND VARIABLES ARE DEFINED IN FIG 24.10       RIC   5
C                                                                       RIC   6
C.... COMPUTE MASS MATRICES AT EACH INTEGRATION POINT                   RIC   7
      DO 500 L = 1,LINT                                                 RIC   8
C.... COMPUTE SHAPE FUNCTIONS                                           RIC   9
      CALL SHAPEF(SG(L),TG(L),XL,XSJ,SHP)                               RIC  10
      DV = WG(L)*XSJ*D(4)                                               RIC  11
C.... FOR EACH NODE J COMPUTE DB = RHO*SHAPE*DV                         RIC  12
      DO 500 J = 1,NEL                                                  RIC  13
      DB = SHP(3,J)*DV                                                  RIC  14
      P(J+J) = P(J+J) + DB                                              RIC  15
C.... FOR EACH NODE I COMPUTE MASS MATRIX (UPPER TRIANGULAR PART)       RIC  16
      DO 500 I = 1,J                                                    RIC  17
500   S(I+I,J+J) = S(I+I,J+J) + SHP(3,I)*DB                             RIC  18
C.... COMPUTE MISSING PARTS AND LOWER TRIANGULAR PART BY SYMMETRIES     RIC  19
      NL = NEL + NEL                                                    RIC  20
      P(1) = P(2)                                                       RIC  21
      S(1,1) = S(2,2)                                                   RIC  22
      DO 510 I = 4,NL,2                                                 RIC  23
      P(I-1) = P(I)                                                     RIC  24
      DO 510 J = 2,I,2                                                  RIC  25
      S(I,J) = S(J,I)                                                   RIC  26
      S(I-1,J-1) = S(J,I)                                               RIC  27
510   S(J-1,I-1) = S(J,I)                                               RIC  28
      RETURN                                                            RIC  29
      END                                                               RIC  30
```

Fig. 24.11 Consistent mass matrix for 4-node quadrilateral

The shape function routine may also be used to compute the element strains, stresses, and internal forces. For the two-dimensional plane problem the strains are computed from

$$\boldsymbol{\varepsilon} = \mathbf{B}_i \mathbf{a}_i \tag{24.14}$$

and stresses from

$$\boldsymbol{\sigma} = \mathbf{D}\boldsymbol{\varepsilon} = \mathbf{Q}_i \mathbf{a}_i. \tag{24.15}$$

The $\mathbf{B}_i$ matrix is given above for the two-dimensional problem and depends only on the derivatives of the shape functions. It has previously been computed at the Gauss points when the element stiffness matrix was evaluated. If stress output points are also at the Gauss points the **B** matrix could be saved on tape or disc and recalled at the time stresses and strains are to be output. (In fact **Q** could be saved also.) This procedure was adopted in the program given in reference 10. Often, however, the points where the stresses and strains are to be determined do not coincide with the stiffness Gauss points. In these cases the **B** matrix must be re-computed. In non-linear problems the computation of strains and stresses

```
C.... ISOPARAMETRIC ELEMENT STRESSES, STRAINS, AND INTERNAL FORCES        RIC   1
C                                                                         RIC   2
C           P      IS THE INTERNAL FORE VECTOR                            RIC   3
C           UL     IS THE ARRAY OF NODAL DISPLACEMENTS                    RIC   4
C           OTHER ARRAYS AND VARIABLES DEFINED IN FIGS 24.10 AND 24.11    RIC   5
C                                                                         RIC   6
C.... COMPUTE ELEMENT STRESSES, STRAINS, AND FORCES                       RIC   7
      DO 600 L = 1,LINT                                                   RIC   8
C.... COMPUTE ELEMENT SHAPE FUNCTIONS                                     RIC   9
      CALL SHAPEF(SG(L),TG(L),XL,XSJ,SHP)                                 RIC  10
C.... COMPUTE STRAINS AND COORDINATES                                     RIC  11
      DO 410 I = 1,3                                                      RIC  12
410   EPS(I) = 0.0                                                        RIC  13
      XX = 0.0                                                            RIC  14
      YY = 0.0                                                            RIC  15
      DO 420 J = 1,NEL                                                    RIC  16
      XX = XX + SHP(3,J)*XL(1,J)                                          RIC  17
      YY = YY + SHP(3,J)*XL(2,J)                                          RIC  18
      EPS(1) = EPS(1) + SHP(1,J)*UL(1,J)                                  RIC  19
      EPS(2) = EPS(2) + SHP(2,J)*UL(2,J)                                  RIC  20
420   EPS(3) = EPS(3) + SHP(1,J)*UL(2,J) + SHP(2,J)*UL(1,J)               RIC  21
C.... COMPUTE STRESSES                                                    RIC  22
      SIG(1) = D(1)*EPS(1) + D(2)*EPS(2)                                  RIC  23
      SIG(2) = D(2)*EPS(1) + D(1)*EPS(2)                                  RIC  24
      SIG(3) = D(3)*EPS(3)                                                RIC  25
C.... OUTPUT STRESSES AND STRAINS                                         RIC  26
      IF(MCT.GT.0) GO TO 430                                              RIC  27
      WRITE(6,2001) O,HEAD                                                RIC  28
      MCT = 50                                                            RIC  29
430   WRITE(6,2003) N,MA,XX,YY,SIG,EPS                                    RIC  30
C.... COMPUTE INTERNAL FORCES                                             RIC  31
      DV = XSJ*WG(L)                                                      RIC  32
      DO 610 J = 1,4                                                      RIC  33
      P(J+J-1) = P(J+J-1) + (SHP(1,J)*SIG(1) + SHP(2,J)*SIG(3))*DV        RIC  34
610   P(J+J  ) = P(J+J  ) + (SHP(1,J)*SIG(3) + SHP(2,J)*SIG(2))*DV        RIC  35
600   CONTINUE                                                            RIC  36
      RETURN                                                              RIC  37
      END                                                                 RIC  38
```

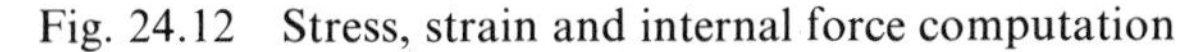

Fig. 24.12 Stress, strain and internal force computation

must be performed directly; hence, it seems desirable for a computer program to be able to compute strains and stresses as necessary. In addition, in the macro instruction FORM the internal forces given by

$$\mathbf{P}_i = -\int_V \mathbf{B}_i^{\mathrm{T}} \boldsymbol{\sigma} \, \mathrm{d}V \tag{24.16}$$

are computed using current displacements (including all specified boundary displacements). In this way the displacement boundary conditions are properly handled. The programming steps to compute strains, stresses, and internal forces for the two-dimensional problem are given in Fig. 24.12. The local co-ordinates ξ, η, where stresses and strains are evaluated, are called SG, TG. The user could read these as data or preset them to specify particular output points; on the other hand, they could be specified as Gauss points.

If stresses are computed at Gauss points they can be extrapolated to the nodes and smoothed.[11,12] Shape function subprograms can be used for this purpose also.

```
C.... ISOPARAMETRIC ELEMENT CONDUCTIVITY COMPUTATION FOR LINEAR HEAT     RIC    1
C.... TRANSFER ANALYSIS                                                  RIC    2
C                                                                        RIC    3
C             D(1)     CONDUCTIVITY IN ELEMENT                           RIC    4
C             DV       AREA WEIGHTING                                    RIC    5
C             LINT     NUMBER OF INTEGRATION POINTS                      RIC    6
C             NEL      NUMBER OF NODES CONNECTED TO ELEMENT              RIC    7
C             S        ELEMENT CONDUCTIVITY MATRIX                       RIC    8
C             SG,TG    INTEGRATION POINTS IN NATURAL COORDINATES         RIC    9
C             SHP      X,Y DERIVATIVES AND SHAPE FUNCTIONS               RIC   10
C             XL       NODAL COORDINATES OF ELEMENT                      RIC   11
C             XSJ      JACOBIAN DETERMINANT                              RIC   12
C             WG       INTEGRATION WEIGHT                                RIC   13
C                                                                        RIC   14
C.... FOR EACH INTEGRATION POINT COMPUTE CONTRIBUTION TO CONDUCTIVITY     RIC   15
C                                                                        RIC   16
      DO 100 L = 1,LINT                                                  RIC   17
      CALL SHAPEF(SG(L),TG(L),XL,XSJ,SHP)                                RIC   18
      DV = XSJ*WG(L)                                                     RIC   19
      D11 = D(1)*DV                                                      RIC   20
C                                                                        RIC   21
C.... FOR EACH J NODE COMPUTE DB = D*B                                   RIC   22
C                                                                        RIC   23
      DO 100 J = 1,NEL                                                   RIC   24
      DB11 = D11*SHP(1,J)                                                RIC   25
      DB21 = D11*SHP(2,J)                                                RIC   26
C                                                                        RIC   27
C.... FOR EACH I NODE COMPUTE S = BT*DB                                  RIC   28
C                                                                        RIC   29
      DO 100 I = 1,J                                                     RIC   30
100   S(I,J) = S(I,J) + SHP(1,I)*DB11 + SHP(2,I)*DB21                    RIC   31
C                                                                        RIC   32
C.... COMPUTE LOWER TRIANGULAR PART BY SYMMETRY                          RIC   33
C                                                                        RIC   34
      DO 200 I = 2,NEL                                                   RIC   35
      DO 200 J = 1,I                                                     RIC   36
200   S(I,J) = S(J,I)                                                    RIC   37
C                                                                        RIC   38
      RETURN                                                             RIC   39
      END                                                                RIC   40
```

Fig. 24.13 Stiffness for linear heat transfer

The generality of an isoparametric C_0 shape function routine can be exploited to program element routines for other problems. For example, Fig. 24.13 gives the necessary program instructions to compute the 'stiffness' matrix for the problems of linear heat conduction discussed in Chapter 17.

24.5.3 *Organization of element routines.* The previous discussion has focused on procedures for determining element arrays. The reader will note that all the element square matrices have been stored in the array S

```
      SUBROUTINE ELMTNN(D,UL,XL,IX,TL,S,P,NDF,NDM,NST,ISW)              ELM    1
C                                                                       ELM    2
C.... MOCK ELEMENT ROUTINE                                              ELM    3
C                                                                       ELM    4
      DIMENSION D(10),UL(NDF,1),XL(NDM,1),IX(1),TL(1),S(NST,1),P(1)     ELM    5
      COMMON /CDATA/ O,HEAD(20),NUMNP,NUMEL,NUMMAT,NEN,NEQ,IPR          ELM    6
      COMMON /ELDATA/ DM,N,MA,MCT,IEL,NEL                               ELM    7
C                                                                       ELM    8
C.... TRANSFER TO APPROPRIATE SEGMENT                                   ELM    9
C                                                                       ELM   10
      GO TO (1,2,3,4,5,6), ISW                                          ELM   11
C                                                                       ELM   12
C.... READ ANS PRINT MATERIAL PROPERTY DATA                             ELM   13
C                                                                       ELM   14
1         THE ARRAY D(10) IS USED TO STORE UP TO 10 WORDS OF INFORMATION   ELM   15
          FOR EACH MATERIAL SET.  AFTER COMPLETION THEN                 ELM   16
C                                                                       ELM   17
      RETURN                                                            ELM   18
C                                                                       ELM   19
C.... COMPUTE JACOBIANS, ETC. TO CHECK AN ELEMENT FOR ERRORS            ELM   20
C                                                                       ELM   21
2         COMPUTE AND REQUIRED ERROR CHECKS AND PRINT ERROR MESSAGE     ELM   22
          THEN                                                          ELM   23
C                                                                       ELM   24
      RETURN                                                            ELM   25
C                                                                       ELM   26
C.... COMPUTE ELEMENT (TANGENT) STIFFNESS MATRIX                        ELM   27
C                                                                       ELM   28
3         THE S(NST,NST) ARRAY IS USED TO STORE THE ELEMENT STIFFNESS,  ELM   29
          THEN                                                          ELM   30
C                                                                       ELM   31
      RETURN                                                            ELM   32
C                                                                       ELM   33
C.... COMPUTE AND OUTPUT ELEMENT QUATITIES (E.G., STRESSES)             ELM   34
C                                                                       ELM   35
4         MCT IS USED AS ALINE COUNTER AND IS SET TO ZERO IN *PFORM* EACH  ELM   36
          TIME A NEW ELEMENT TYPE IS ENCOUNTERED, AFTER OUTPUT THEN      ELM   37
C                                                                       ELM   38
      RETURN                                                            ELM   39
C                                                                       ELM   40
C.... COMPUTE MASS MATRIX                                               ELM   41
C                                                                       ELM   42
5         THE S(NST,NST) ARRAY IS USED TO SDTRE A CONSISTENT MASS MATRIX.  ELM   43
          THE P(NST) ARRAY IS USED TO STORE A LUMPED MASS MATRIX, THEN  ELM   44
C                                                                       ELM   45
      RETURN                                                            ELM   46
C                                                                       ELM   47
C.... COMPUTE INTERNAL FORCES                                           ELM   48
C                                                                       ELM   49
6         THE P(NST) ARRAY IS USED TO STORE THE INTERNAL FORCES         ELM   50
          (I.E.,  P(A) IN CHAPTER 18), THEN                             ELM   51
C                                                                       ELM   52
      RETURN                                                            ELM   53
C                                                                       ELM   54
      END                                                               ELM   55
```

Fig. 24.14 Mock element routine layout

while element vectors are stored in array P. This was intentional since all aspects of computing element arrays for the program are to be consolidated into a single subprogram called the 'Element Routine'. An element routine is called by the subprogram ELMLIB, which is the element library generation routine. As given here, the element library provides space for four element subprograms at any one time, where the names of the element routines are ELMT01, ELMT02, ELMT03, and ELMT04. This can easily be increased by adding more element routine names in ELMLIB. The subroutine ELMLIB is, in turn, called by subroutine PFORM which, as mentioned previously, is the subroutine to loop through all elements, set up local arrays for co-ordinates (XL), displacements (UL), and destinations in the global arrays (LD). The subroutine PFORM also calls subroutine ADDSTF to assemble element arrays into global arrays. When an element routine is accessed the value of the parameter ISW is specified between 1 and 6. The parameter specifies what action is to be taken in the element routine. Each element routine must provide appropriate transfers for each value of ISW. A mock element routine is shown in Fig. 24.14.

24.6 Solution of Simultaneous, Linear Algebraic Equation

In solving problems by the finite element method we are eventually faced with the task of solving a large set of simultaneous, linear algebraic equations. For example, in the analysis of linear steady-state problems the direct assembly of the element stiffness matrices leads to the set of linear algebraic equations, whereas, for non-linear problems it is a linearization or iteration which produces the large set of algebraic equations. In this section methods are considered which solve the algebraic equation by a direct procedure, where an *a priori* calculation on the number of numerical operations can be made, and by an indirect or iterative method where no *a priori* estimate can be made.

24.6.1 *Direct solution.* Consider first the general problem of direct solution of the set of algebraic equations given by

$$\mathbf{K}\mathbf{a} = \mathbf{r} \tag{24.17}$$

where $\mathbf{K}$ is a square coefficient matrix, $\mathbf{a}$ is a vector of unknowns, and $\mathbf{r}$ is a vector of specified quantities. The reader can associate these to the quantities described previously: namely, the stiffness matrix, the nodal unknowns, and the specified forces.

In the discussion to follow it is assumed that the coefficient matrix has nice properties such that interchanges of rows and/or columns is never necessary in order to solve the equations. This is true in cases where

K is symmetric, positive (or negative) definite. It may or may not be true when the equations are unsymmetric, or indefinite, conditions which may occur when the finite element formulation is based on a mixed variational principle or some weighted residual method. In these cases some checks or additional analysis are necessary to ensure that the equations can be solved.[8] If interchanges are necessary, modification must be made to the equation solving subprograms given in this chapter; e.g., see references 8, 13 or 14.

For the moment consider that the coefficient matrix can be written as the product of a lower triangular matrix with unit diagonals and an upper triangular matrix, i.e.,

$$\mathbf{K} = \mathbf{L}\mathbf{U} \tag{24.18}$$

where

$$\mathbf{L} = \begin{bmatrix} 1 & 0 & \dots & 0 \\ L_{21} & 1 & \dots & 0 \\ \vdots & & \ddots & \vdots \\ L_{n1} & L_{n2} & \dots & 1 \end{bmatrix} \tag{24.19}$$

and

$$\mathbf{U} = \begin{bmatrix} U_{11} & U_{12} & \dots & U_{1n} \\ 0 & U_{22} & \dots & U_{2n} \\ \vdots & & \ddots & \vdots \\ 0 & 0 & \dots & U_{nn} \end{bmatrix} \tag{24.20}$$

This step is called the triangular decomposition of **K**. The solution to the equations can now be obtained by solving the pair of equations

$$\mathbf{L}\mathbf{y} = \mathbf{r}$$

and (24.21)

$$\mathbf{U}\mathbf{a} = \mathbf{y}$$

where $\mathbf{y}$ is introduced to facilitate the separation, e.g., see references 13 or 14.

The reader can easily observe that the solution to these equations is trivial. In terms of the elements of the equations the solution is

$$\begin{aligned} y_1 &= r_1 \\ y_i &= r_i - \sum_{j=1}^{i-1} L_{ij} y_j \qquad (i = 2, 3, \dots, n) \end{aligned} \tag{24.22}$$

and

$$
\begin{aligned}
a_n &= y_n / U_{nn} \\
a_i &= \left(y_i - \sum_{j=i+1}^{n} U_{ij}\, a_j \right) \Big/ U_{ii} \qquad (i = n-1, n-2, \ldots, 1)
\end{aligned}
\tag{24.23}
$$

Equation (24.22) is called 'forward elimination' while Eq. (24.23) is called 'back substitution',

The problem remains to construct the triangular decomposition of the coefficient matrix. This step is accomplished using a compact Crout method which is a variation on Gauss elimination. In practice, the operations necessary for the triangular decomposition are performed directly in the coefficient array; however, to make the steps clear, the basic steps are shown in Table 24.13 using separate arrays. The decomposition is performed in the same way that the two subprograms UACTCL and ACTCOL operate, thus, the reader can easily grasp the details of these routines once the steps in Table 24.13 are mastered.

The Crout variation of Gauss elimination is used to successively reduce

TABLE 24.13
TRIANGULAR DECOMPOSITION OF **K**

Active zone

$$
\begin{bmatrix} K_{11} & K_{12} & K_{13} \\ K_{21} & K_{22} & K_{23} \\ K_{31} & K_{32} & K_{33} \end{bmatrix}
\qquad
\begin{bmatrix} L_{11} = 1 & & \\ & & \\ & & \end{bmatrix}
\qquad
\begin{bmatrix} U_{11} = K_{11} & & \\ & & \\ & & \end{bmatrix}
$$

Step 1. Active Zone. First row and column to principal diagonal.

Reduced zone
Active zone

$$
\begin{bmatrix} & K_{12} & K_{13} \\ K_{21} & K_{22} & K_{23} \\ K_{31} & K_{32} & K_{33} \end{bmatrix}
\qquad
\begin{bmatrix} 1 & 0 \\ L_{21} = K_{21}/U_{11} & L_{22} = 1 \\ & \end{bmatrix}
\qquad
\begin{bmatrix} U_{11} & U_{12} = K_{12} \\ 0 & U_{22} = K_{22} - L_{21}U_{12} \\ & \end{bmatrix}
$$

Step 2. Active zone. Second row and column to principal diagonal. Use first row of **K** to eliminate $L_{21}U_{11}$. The active zone uses only values of **K** from the active zone and values of **L** and **U** which have already been computed in Steps 1 and 2.

Reduced zone
Active zone

$$
\begin{bmatrix} & & K_{13} \\ & & K_{23} \\ K_{31} & K_{32} & K_{33} \end{bmatrix}
\qquad
\begin{bmatrix} 1 & 0 & 0 \\ L_{21} & 1 & 0 \\ L_{31} & L_{32} & L_{33} = 1 \end{bmatrix}
\qquad
\begin{bmatrix} U_{11} & U_{12} & U_{13} = K_{13} \\ 0 & U_{22} & U_{23} = K_{23} - L_{21}U_{13} \\ 0 & 0 & U_{33} = K_{33} - L_{31}U_{13} - L_{32}U_{23} \end{bmatrix}
$$

$$
\begin{aligned}
L_{31} &= K_{31}/U_{11} \\
L_{32} &= (K_{32} - L_{31}U_{12}) / U_{22}
\end{aligned}
$$

Step 3. Active zone. Third row and column to principal diagonal. Use first row to eliminate $L_{31}U_{11}$; use second row of reduced terms to eliminate $L_{32}U_{22}$ (reduced coefficient K_{32}). Reduce column 3 to reflect eliminations below diagonal.

the original coefficient array to upper triangular form. The lower triangular portion is not actually set to zero but is used to construct **L**, as shown in Table 24.13. As mentioned above, the upper and lower triangular matrices will replace the original coefficient matrix; consequently, it is not possible to retain the principal diagonal elements of both **L** and **U**. Those of **L** are understood since it is known by definition that they are all unity.

Based upon the organization of Table 24.12, it is convenient to consider the coefficient array to be divided into three parts: part one being the region that is fully reduced, part two the region which is currently being reduced (called the active zone), and part three the region which contains the original unreduced coefficients. These regions are shown in Fig. 24.15 where the jth column above the diagonal and the jth row below the diagonal constitute the active zone. The algorithm for the triangular decomposition of an $n \times n$ square matrix can be deduced from Table 24.12 and Fig. 24.16 as follows:

$$\begin{aligned} U_{11} &= K_{11} \\ L_{11} &= 1. \end{aligned} \tag{24.24a}$$

For each active zone j from 2 to n

$$\begin{aligned} L_{j1} &= K_{j1}/U_{11} \\ U_{1j} &= K_{1j} \end{aligned} \tag{24.24b}$$

then

$$\begin{aligned} L_{ji} &= \left(K_{ji} - \sum_{m=1}^{i-1} L_{jm} U_{mi}\right) \Big/ U_{ii} \\ U_{ij} &= K_{ij} - \sum_{m=1}^{i-1} L_{im} U_{mj} \end{aligned} \qquad (i = 1, 2, \ldots, j-1) \tag{24.24c}$$

and finally

$$\begin{aligned} L_{jj} &= 1 \\ U_{jj} &= K_{jj} - \sum_{m-1}^{j-1} L_{jm} U_{mj}. \end{aligned} \tag{24.24d}$$

The ordering of the reduction process and the terms used are shown in Fig. 24.16. The results from Table 24.13 and Eqs. (24.24) can be verified by the reader using the matrix given in the example shown in Table 24.14.

Once the triangular decomposition of a matrix is computed, several solutions for different right-hand sides **r** can be computed using Eqs. (24.22) and (24.23). This process is often called resolution since it is not necessary to recompute **L** and **U**. For very large size coefficient matrices

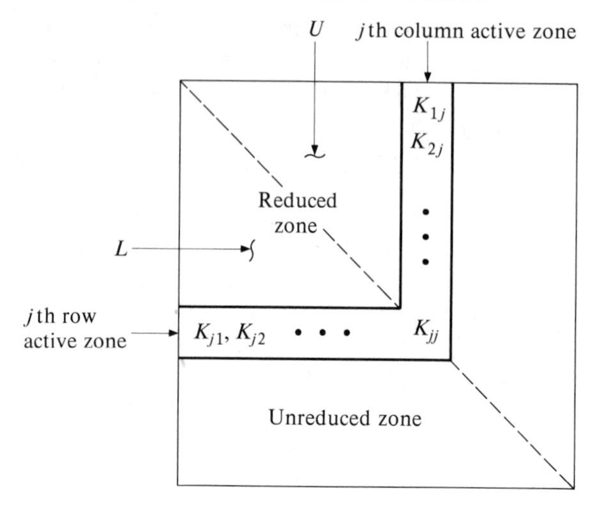

Fig. 24.15

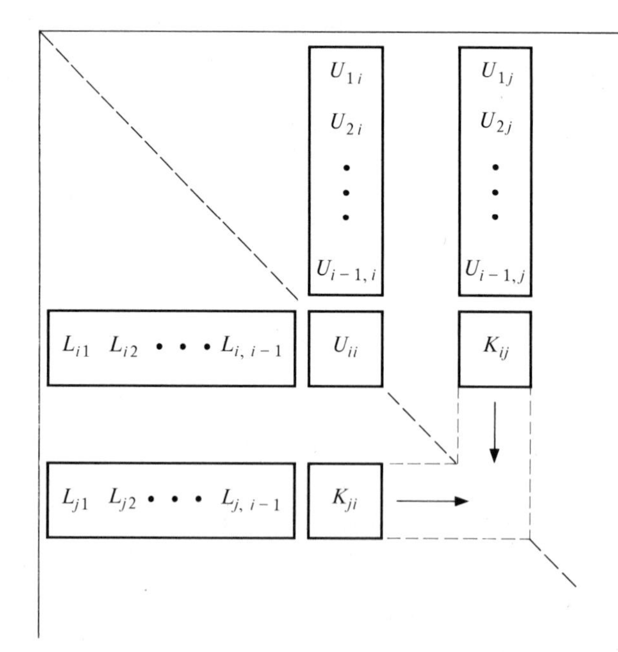

Fig. 24.16

the decomposition process is very costly while a resolution is relatively cheap, consequently, a resolution capability is necessary in any finite element solution system.

The above discussion considered the general case of equation solving (without interchanges). In coeffcient matrices resulting from a finite element problem some special properties are usually present. Often the stiffness matrix is symmetric ($K_{ij} = K_{ji}$) and it is easy to verify in this case that

$$U_{ij} = L_{ji} U_{ii}. \tag{24.25}$$

For this problem class it is not necessary to store the entire coefficient array. It is sufficient to store only those coefficients above (or below) the principal diagonal and use Eq. (24.25) to construct the missing part.

TABLE 24.14

EXAMPLE: TRIANGULAR DECOMPOSITION OF 3 × 3 MATRIX

$$\mathbf{K} \qquad \mathbf{L} \qquad \mathbf{U}$$

$$\begin{bmatrix} 4 & 2 & 1 \\ 2 & 4 & 2 \\ 1 & 2 & 4 \end{bmatrix} \quad \begin{bmatrix} 1 & & \\ & & \\ & & \end{bmatrix} \quad \begin{bmatrix} 4 & & \\ & & \\ & & \end{bmatrix}$$

Step 1: $L_{11} = 1$, $U_{11} = 4$

$$\begin{bmatrix} & 2 & 1 \\ 2 & 4 & 2 \\ 1 & 2 & 4 \end{bmatrix} \quad \begin{bmatrix} 1 & & \\ 0{\cdot}5 & 1 & \\ & & \end{bmatrix} \quad \begin{bmatrix} 4 & 2 & \\ & 3 & \\ & & \end{bmatrix}$$

Step 2: $L_{21} = \frac{2}{4} = 0{\cdot}5$, $U_{12} = 2$, $L_{22} = 1$, $U_{22} = 4-0{\cdot}5\times 2 = 3$

$$\begin{bmatrix} & & 1 \\ & & 2 \\ 1 & 2 & 4 \end{bmatrix} \quad \begin{bmatrix} 1 & & \\ 0{\cdot}5 & 1 & \\ 0{\cdot}25 & 0{\cdot}5 & 1 \end{bmatrix} \quad \begin{bmatrix} 4 & 2 & 1 \\ & 3 & 1{\cdot}5 \\ & & 3 \end{bmatrix}$$

Step 3: $L_{31} = \frac{1}{4} = 0{\cdot}25$; $U_{13} = 1$, $L_{32} = \dfrac{2-0{\cdot}25\times 2}{3} = \dfrac{1{\cdot}5}{3} = 0{\cdot}5$

$U_{23} = 2-0{\cdot}5\times 1 = 1{\cdot}5$ $L_{33} = 1$ $U_{33} = 4-0{\cdot}25\times 1-0{\cdot}5\times 1{\cdot}5 = 3$

$$\begin{bmatrix} 1 & & \\ 0{\cdot}5 & 1 & \\ 0{\cdot}25 & 0{\cdot}5 & 1 \end{bmatrix} \begin{bmatrix} 4 & 2 & 1 \\ & 3 & 1{\cdot}5 \\ & & 3 \end{bmatrix} = \begin{bmatrix} 4 & 2 & 1 \\ 2 & 4 & 2 \\ 1 & 2 & 4 \end{bmatrix}$$

Step 4: Check

This reduces by almost half the required storage for the coefficient array. Still larger savings in storage can be achieved if only the terms within a non-zero *band* are stored. In finite element formulations the maximum 'bandwidth' of non-zero coefficients can usually be made small compared with the number of unknowns—often 10–20 per cent, which reduces the storage from $n(n+1)/2$ to $(0{\cdot}1$ to $0{\cdot}2)n^2$ for symmetric problems. This method was used in the program included in the previous book.[2] A typical storage method for symmetric banded equations is shown in Fig. 24.17. A discussion on band solutions is given by Meyer[15, 16] together with an extensive bibliography.

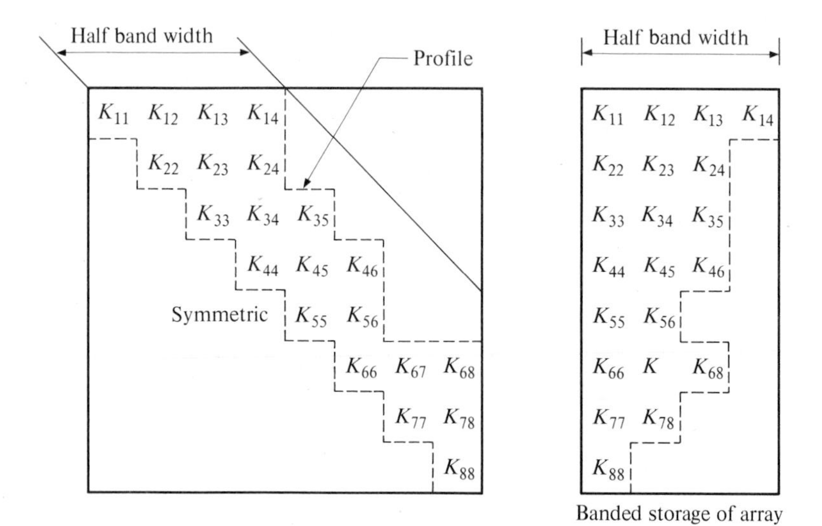

Fig. 24.17

It is possible to reduce the required storage and computational effort still further by storing the necessary parts of the upper triangular portion of the stiffness matrix by columns and the lower triangular portion by rows, as shown in Fig. 24.18. This was noted for symmetric matrices in references 7, 17, 18 and 19. Now it is necessary to store and compute only within the non-zero *profile* of the equations. This method of storage has definite advantages over a banded storage. First, it always requires less storage (unless the matrix is diagonal!); second, the storage requirements are not severely affected by a few very long columns, as shown in Fig. 24.18, and, last, it is very easy to use vector dot product routines to effect the triangular decomposition and forward reduction.[19] This last fact is extremely important to modern machines which are vector oriented.

Two active zone, profile equation-solving subprograms are included for use in the finite element solution package. These are called ACTCOL for symmetric equations and UACTCL for unsymmetric equations. The profile of both cases must be symmetric. The columns above the principal diagonal or the rows below the diagonal are stored in a single subscript array as shown in Fig. 24.18. A pointer array is used to locate the diagonal elements. Table 24.15 defines the labels used in the solution subprograms ACTCOL and UACTCL.

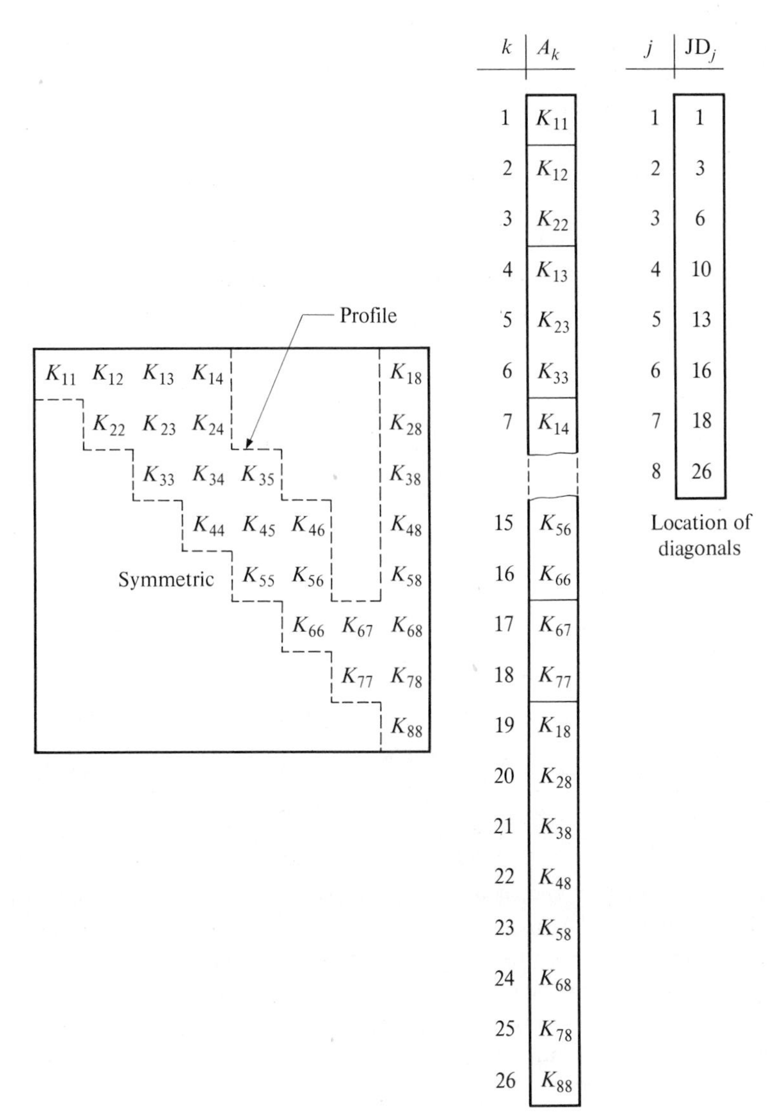

Fig. 24.18

TABLE 24.15
VARIABLES USED IN EQUATION SOLUTION SUBPROGRAMS
ACTCOL AND UACTCL

A(NAD)	Upper triangular coefficients when called, replaced by **U** on return.
B(NEQ)	Right-hand side vector at call, solution vector upon return to calling program.
C(NAD)	Lower triangular coefficients at call, **L** at return (used in UACTCL only).
JDIAG(NEQ)	Pointer array to determine location in A or C of diagonal pivots.
NEQ	Number of equations to be solved.
NAD	Length of A or C arrays: equal to JDIAG(NEQ).
AFAC	Logical variable: if true, triangular decomposition performed.
BACK	Logical variable: if true, forward reduction and back substitution performed.

The procedure used to assemble the profile storage of the stiffness matrix is discussed in section 24.6.4. Before continuing with this discussion, however, a brief presentation of an indirect solution procedure is given. The method included here is the Gauss–Seidel iteration method with successive over-relaxations.

24.6.2 *Iterative solution.* To carry out the Gauss–Seidel iterations we first write an additive decomposition of the coefficient matrix[14]

$$\mathbf{K} = \mathbf{L}+\mathbf{U} \qquad (24.26)$$

where **L** is lower triangular with

$$L_{ij} = K_{ij} \qquad (i = 1, 2, 3, \ldots, n; \quad j = 1, 2, \ldots, i) \qquad (24.27)$$

U is upper triangular with

$$U_{ij} = K_{ij}(i = 1, 2, \ldots, n-1; \quad j = i+1, i+2, \ldots, n) \qquad (24.28)$$

and all other elements of the **L** and **U** arrays are zero.

The basic Gauss–Seidel iteration scheme is given by the algorithm

$$\begin{aligned} \mathbf{a}^0 &= \mathbf{v} \\ \mathbf{L}\mathbf{a}^{n+1} &= \mathbf{r}-\mathbf{U}\mathbf{a}^n \end{aligned} \qquad (24.29)$$

where $\mathbf{v}$ is a starting vector and a superscript refers to iteration number. If the coefficient matrix is symmetric and positive definite the Gauss–Seidel method is known to converge (e.g., see reference 14); however, the rate of convergence may be unacceptably slow. In these cases the computational effort can usually be significantly reduced by using an over-relaxation factor. To facilitate the use of over-relaxation the term $\mathbf{L}\mathbf{a}^n$ is substracted from both sides of Eq. (24.29), giving

$$\mathbf{L}\,\Delta\mathbf{a} = \mathbf{r}-\mathbf{K}\mathbf{a}^n \qquad (24.30)$$

and the solution is advanced using

$$\mathbf{a}^{n+1} = \mathbf{a}^n + \omega\, \Delta\mathbf{a} \tag{24.31}$$

where ω is a problem dependent over-relaxation factor with a value between 0 and 2. The above process is called 'successive over-relaxation' or SOR. The main advantages of iteration are the reduced central memory storage demands and the elimination of the triangular decomposition which is the most costly part of a direct solution. The disadvantages are: the lack of knowledge on how many iterations are necessary to achieve an acceptable solution (often hundreds or thousands of iterations are required); also the value of ω can change significantly the convergence (many people continuously change ω during the solution to achieve an optimal value); the method fails on indefinite or unsymmetric problems; and finally in non-linear problems, or multiple right-hand sides, no advantage (except perhaps the optimum ω value) can be taken of a previous solution process as the whole iteration process must be repeated. In direct solutions once the triangular decomposition is performed a resolution is relatively cheap. The disadvantages usually far outweigh the advantages for SOR; consequently, most finite element solution programs today use direct solution methods to solve the algebraic equations.

24.6.3 *Computation of energy.* When solving finite element problems which result from minimum (or maximum) principles it is often desirable to compute the minimum (or maximum) value of the functional. In the discrete problem this is equivalent to computing the energy given by

$$-2\Pi(\mathbf{a}) = \mathbf{a}^{\mathrm{T}}\mathbf{K}\mathbf{a} = \mathbf{a}^{\mathrm{T}}\mathbf{r} \tag{24.32}$$

where $\mathbf{K}$ is symmetric since we now consider only minimum (or maximum) principles. Often it is not convenient to have the right-hand side and the solution in the central memory at the same time. In this case it is possible to compute the value while solving the equations. Using the triangular decomposition and symmetry conditions for K_{ij}, Eq. (24.32) can be written as

$$\sum_{m=1}^{n} \sum_{i=1}^{n} \sum_{j=1}^{n} a_i U_{mi} U_{mm}^{-1} U_{mj} a_j = \sum_{i=1}^{n} a_i r_i \tag{24.33}$$

which becomes

$$\sum_{m=1}^{n} y_m^2\, U_{mm}^{-1} = \sum_{i=1}^{n} a_i r_i \tag{24.34}$$

when Eq. (24.21) is used. Thus it is possible to compute the value of the energy during the forward reduction of the right-hand side without having both the solution and right-hand side at the same time.

The value of the computed discrete energy can be used in the Rayleigh quotient and to assess rate of convergence in energy since the energy of the error is equal to the error of the energy, e.g., see reference 20, where it is shown that

$$\Pi(a-a^h) = \Pi(a)-\Pi(a^h) \tag{24.35}$$

with Π being the energy. Rate of convergence in energy can then be assessed by plotting a curve of log $(\Pi(a)-\Pi(a^h))$ versus log h, where h is a measure of mesh size and a, a^h are the exact and approximate solutions.

24.6.4 *Assembly of global arrays from element arrays.* In the previous sections a procedure for computing element arrays, e.g., element stiffness matrices, etc., and the procedure for directly solving algebraic equations representing the global arrays has been presented. The missing link is the assembly of the element arrays to form the global arrays.

The assembly of the global stiffness matrix stored in profile form is accomplished using the pointer array JDIAG. To facilitate the assembly a destination vector was established for each element. In the solution system discussed here the vector was called LD and established in subprogram PFORM from the ID and IX arrays during the localization of data to the element level. Each element of LD contains the equation number for a row and/or column in the global arrays for a corresponding row and/or column of the element arrays. The assembly of the global arrays for stiffness, mass, forces is then performed by subroutine ADDSTF. Logical variables are set to define which array or arrays are to be assembled.

24.7 **Extensions and Modification to the Computer Program**

The previous sections describe the program listed in the next section. The capabilities of the program, while quite significant, can still be improved. Improvements could include increased capacity to handle large problems, increased power to the macro programming language and, finally, added postprocessors to prepare graphic output of solution features.

In performing finite element analyses on many engineering problems the capacity of the program discussed here will be inadequate. The inadequacy appears first in the number of unknowns which can be treated—primarily the size of the stiffness and mass matrices is limiting the capacity. The capacity can be extended by blocking the stiffness and mass arrays as shown in Fig. 24.19.[21] It would then only be necessary to have two of these blocks in core at any one time instead of the entire set of equations. When large computers are used this single change would extend the capacity of the program to handle a few thousand unknowns. Efficiency is enhanced when blocking the equations by writing mesh data onto backing core during the equation solving steps so that maximum core area is available to the global arrays. In addition, the assembly of the global equations will have to be modified since the whole set of equations is not available at any one time. The element arrays will now have to be saved on backing store and the equations assembled block by block (e.g., see reference 10). Efficiency will be greatly improved if the elements are grouped in an array, storing together as many as core space will permit. This buffering of the elements will greatly decrease the I/O costs over that of writing one element at a time.

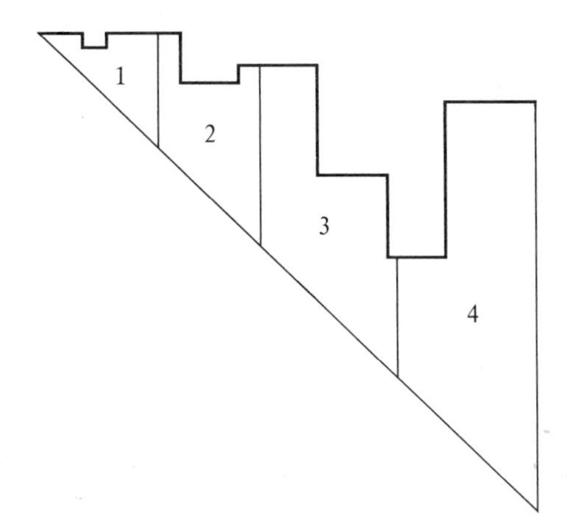

Fig. 24.19 A block storage scheme for profile solution

If it is desired to increase the capacity still further, a radical modification for the storage of the data arrays must be made. The data arrays must also be blocked, stored on backing store, and retrieved as needed. In this case it will be necessary to write special software packages to efficiently handle the extensive amounts of I/O and it will be most advantageous in this case to perform the localization step only once.

An alternative to the blocked profile equation solution schemes is the 'frontal method' which is described in references 22, 23, 24 and 25. This method has been successfully employed in finite element solution packages and is a direct solution method which is closely connected to the finite element method. A frontal solution package can easily be interfaced with the program discussed here. The necessary modifications will include a replacement of the PROFIL subprogram by a pre-front package to determine the order of elimination of the equations, replacing ADDSTF by a taping of the element arrays, and replacement of ACTCOL and UACTCL by frontal solution packages.

It has been stated[23] that the frontal method will be more efficient than band solution methods because the front width is smaller than the band width (when mid-side nodes are present). If nodes are numbered in frontal elimination order the active column method has the same number of numerical operations as the frontal method. There is, however, more than a simple measure of front versus band or profile width that enters into the cost of solving equations. For the discussion here we restrict our comments to the case of symmetric equations. The profile and band solution methods perform the triangular decomposition of the 'stiffness' matrix working from the first equation to the last. On the other hand, the frontal scheme works element by element (solution efficiency is a function of element order, not node order), forming only that part of the stiffness matrix belonging to the front (see Fig. 24.20). After an element is introduced the equations which are completed, e.g. x-equation in Fig. 24.20, are eliminated by Gauss elimination. These equations may be anywhere in the front stiffness matrix—and are seldom the first equations, e.g., x-equation in Fig 24.20. Consequently, the elimination must be performed

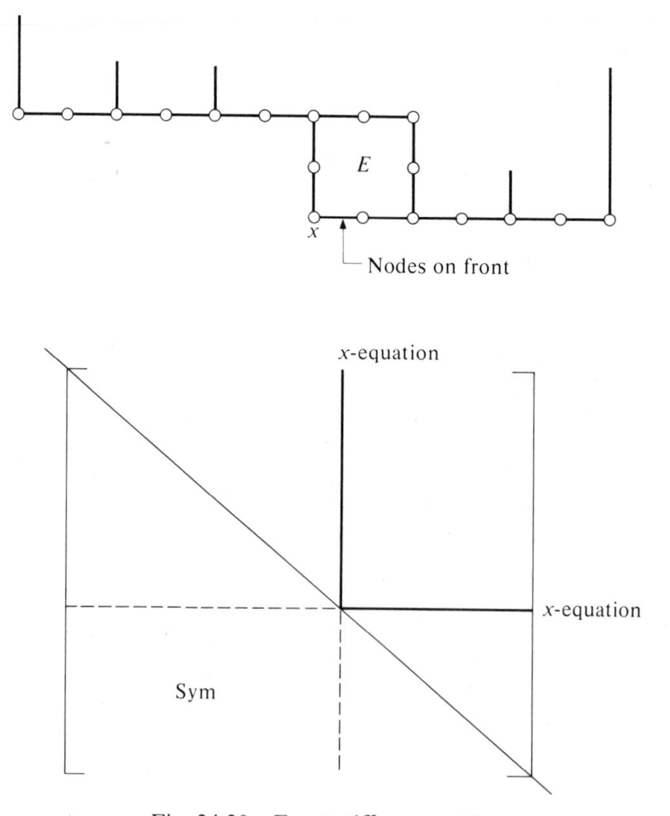

Fig. 24.20 Front stiffness matrix

on both the preceding and succeeding equations. The storage mode of the front stiffness is usually by columns above the diagonal. In this way columns can be added without changing storage mode of former columns. The elimination of the x-equation cannot take advantage of the compact Crout elimination aspects, and this leads to more coding details for the critical inner loop of the triangular decomposition (i.e., the dot product calls in ACTCOL and UACTCL). In reference 24 the inner loop involves one index not in registers, one extra statement involving two-look-ups (one addition and one store), one multiplication, and one division.† This is more than the computations in the inner loop of ACTCOL. This is the primary disadvantage of a frontal method. The other limitations are that the front stiffness must fit into core and usually there is an extensive amount of moving of coefficients before elimination. On the other hand, the active column profile solution method will require considerable amount of effort to assemble the global stiffness matrix when many blocks are involved to store the total array. The issue is not clear as to which is the better method and thus individual users must choose between the two. There is no question that the active column profile method is superior to conventional band solution systems.

In addition to extending the capacity of the program to deal with larger numbers of unknowns, it will be necessary to add macro instructions which extend the problem types which can be treated. In section 24.4 we included but a few possible commands and the program we provide cannot, for example, solve general time-stepping algorithms. (Space is the primary limitation on the types of commands we have included.) In general time-stepping schemes, two important types of macro commands should be added. The first is a command to add linear combinations of the mass and stiffness matrices to form the coefficient matrix for the time-stepping scheme (see Chapter 17 for specific schemes and their coefficient matrices); and secondly, a command to update the solution vectors is needed at the conclusion of each time step. Each of these macro commands will be different for each time-stepping scheme which is included.

For first order problems (i.e., only first derivatives are present) the above additions are unnecessary if special elements are developed which will create the linear combination of the matrices as the tangent element matrix (ISW = 3 in the element routine) and the correct right-hand side as the internal element force vector, ISW = 6 in the element routine. The updating is then done correctly since only the solution at t_n enters into the equations (see Chapter 17).

Many other macro commands could be added to interface other algorithms or processors with the program.

Finally, we have alluded to the problem of checking meshes and stated

† With minor coding changes the division can be removed from the inner loop.

that a viable solution package needs a graphics package. While this is true for data checking it is doubly true for interpretation of output. For large analyses, especially those which are time dependent, it is not possible to interpret very much of a printed output. In these cases a graphics package is certainly a necessity. The graphics package should be capable

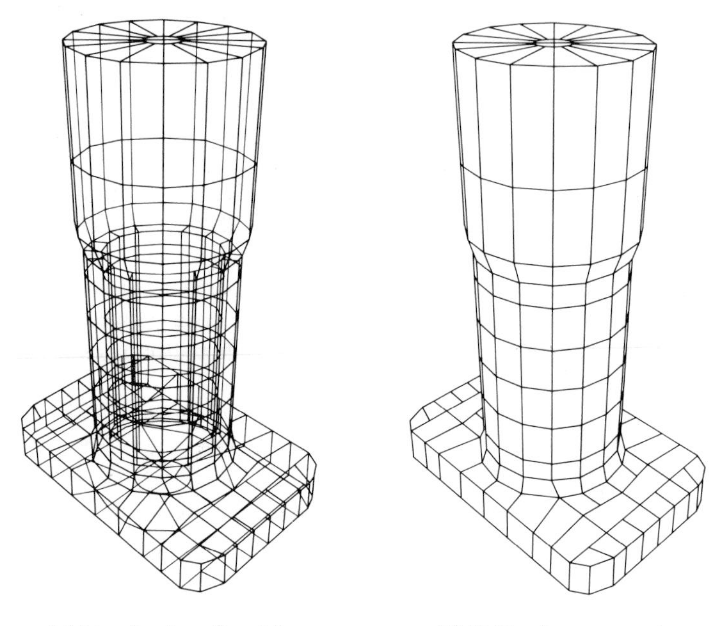

(*a*) Line drawing of model (*b*) Hidden lines removed

Fig. 24.21 Three-dimensional mesh plots with hidden line capability (courtesy of Prof. H. N. Christiansen, Brigham Young University, Provo, Utah)

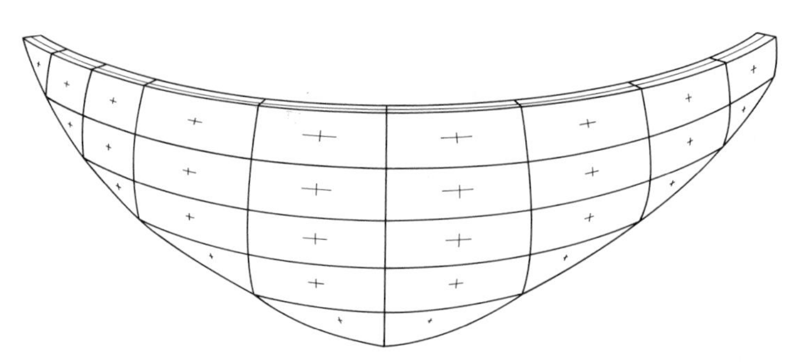

Fig. 24.22 Stress intensity measures by vectors—(give magnitude and direction)[26]

of plotting deformed grids (magnifying the displacements if necessary), producing stress and strain plots, and producing time history plots.[3]

A few examples of plots are shown in Figs. 24.21 to 24.24 and suggest some of the kinds of plotting facilities which should be available.

General plot packages can also be added to the program using the macro language to control the creation of the plot data files.

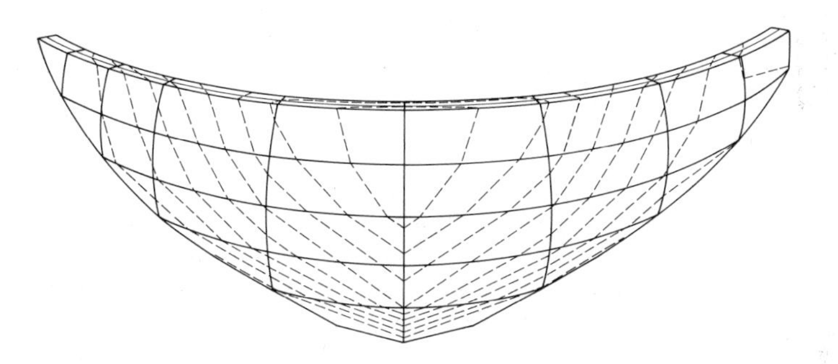

Fig. 24.23 Contour lines for stress levels[26]

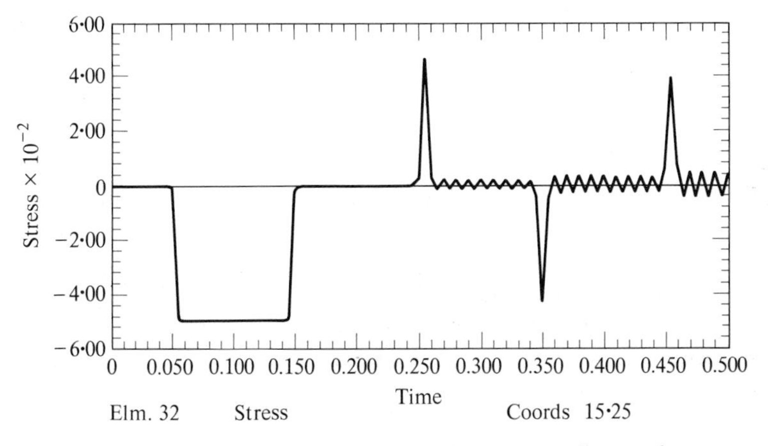

Fig. 24.24 Time history plots—quickly show anomalies such as growing oscillations

24.8 **Listing of Finite Element Computer Program**

We include here the complete FORTRAN listing for the computer program discussed in this chapter. The section is broken down into three parts: the first part consists of the routines for the control and data input modules, the second the routine for the solution and output modules, and the third contains element routines for linear elasticity, linear heat transfer, and a steady-state fluid element.

24.8.1 *Control and data input modules.* The control and data input modules for the program consist of the subroutines PCONTR, PMESH, GENVEC, SETMEM, and the main program. The control of a problem solution is performed by subroutine PCONTR and data input of the mesh parameters and arrays occurs in PCONTR, PMESH, and GENVEC. The subprogram GENVEC is used to generate by linear interpolation missing data for real arrays. Finally, SETMEM is used to monitor available memory in the blank common array M. For machines which permit changing the length of blank common during execution (e.g., CDC 6000 and 7000 series computers), the blank common dimension could be changed to M(1), MAX to the maximum central memory size permissible, and SETMEM modified by replacing the instructions

```
K = J
IF(K·LE·MAX) RETURN
```

by the instructions

```
      K = LOCF(M)+J
      IF(K·GT·MAX) GO TO 100
  appropriate card to reset memory length
      RETURN
  100 CONTINUE
```

Now the field length will be adjusted as necessary and only that actually necessary for each problem will be used.

It is possible to suppress all page ejects by setting the variable O in BLOCK DATA to /1H0/. Also the precision on real variables is controlled by the variable IPR in BLOCK DATA which should be set to /2/ for double precision and /1/ for single precision.

```
      MASTER MINIFEM                                                    IFEM   1
C.... SET PROGRAM CAPACITY * MAX MUST AGREE WITH DIMENSION OF M         IFEM   2
      COMMON M(2000)                                                    IFEM   3
      COMMON/PSIZE/ MAX                                                 IFEM   4
      MAX = 2000                                                        IFEM   5
      CALL PCONTR                                                       IFEM   6
      STOP                                                              IFEM   7
      END                                                               IFEM   8
      SUBROUTINE PCONTR                                                  PCO   1
C                                                                        PCO   2
C.... FINITE ELEMENT ANALYSIS PROGRAM (FEAP)  FOR SOLUTION OF GENERAL     PCO   3
C.... PROBLEM CLASSES USING THE FINITE ELEMENT METHOD.  PROBLEM SIZE      PCO   4
C.... IS CONTROLLED BY THE DIMENSION OF BLANK COMMON AND VALUE OF MAX     PCO   5
C.... AS SET IN MAIN PROGRAM.  ALL ARRAYS MUST RESIDE IN CENTRAL MEMORY.  PCO   6
C                                                                        PCO   7
C.... PROGRAMMED BY PROF. R.L. TAYLOR, DEPARTMENT OF CIVIL ENGINEERING,   PCO   8
C.... UNIVERSITY OF CALIFORNIA, BERKELEY, CALIFORNIA 94720, U.S.A.       PCO   9
C                                                                        PCO  10
      LOGICAL PCOMP                                                      PCO  11
      COMMON/CDATA/ O,HEAD(20),NUMNP,NUMEL,NUMMAT,NEN,NEQ,IPR            PCO  12
      COMMON/LABEL/ PDIS(6),A(6),BC(2),DI(6),CD(3),TE(3),FD(3)           PCO  13
      COMMON M(1)                                                        PCO  14
      DIMENSION TITL(20),WD(3)                                           PCO  15
      DATA WD/4HFEAP,4HMACR,4HSTOP/                                      PCO  16
C.... READ A CARD AND COMPARE FIRST 4 COLUMNS WITH MACRO LIST             PCO  17
1     READ(5,1000) TITL                                                  PCO  18
      IF(PCOMP(TITL(1),WD(1))) GO TO 100                                 PCO  19
      IF(PCOMP(TITL(1),WD(2))) GO TO 200                                 PCO  20
      IF(PCOMP(TITL(1),WD(3))) RETURN                                    PCO  21
      GO TO 1                                                            PCO  22
C.... READ AND PRINT CONTROL INFORMATION                                  PCO  23
100   DO 101 I = 1,20                                                    PCO  24
101   HEAD(I) = TITL(I)                                                  PCO  25
      READ(5,1001) NUMNP,NUMEL,NUMMAT,NDM,NDF,NEN,NAD                    PCO  26
      WRITE(6,2000) HEAD,NUMNP,NUMEL,NUMMAT,NDM,NDF,NEN,NAD              PCO  27
C.... SET POINTERS FOR ALLOCATION OF DATA ARRAYS                          PCO  28
      PDIS(2) = A(NDM)                                                   PCO  29
      NEN1 = NEN + 1                                                     PCO  30
      NST = NEN*NDF + NAD                                                PCO  31
      NO =  1 + NST*2*IPR                                                PCO  32
      N1 = NO + NEN*NDM*IPR                                              PCO  33
      N2 = N1 + NEN*IPR                                                  PCO  34
      N3 = N2 + NST                                                      PCO  35
      N4 = N3 + NST*IPR                                                  PCO  36
      N5 = N4 + NST*NST*IPR                                              PCO  37
      N6 = N5 + NUMMAT                                                   PCO  38
      N7 = N6 + 10*NUMMAT*IPR                                            PCO  39
      N8 = N7 + NDF*NUMNP                                                PCO  40
      N9 = N8 + NDM*NUMNP*IPR                                            PCO  41
      N10 = N9  + NEN1*NUMEL                                             PCO  42
      N11 = N10 + NDF*NUMNP*IPR                                          PCO  43
      N12 = N11 + NUMNP*IPR                                              PCO  44
      N13 = N12 + NDF*NUMNP                                              PCO  45
C.... CHECK THAT SUFFICIENT MEMORY EXISTS                                 PCO  46
      CALL SETMEM(N13)                                                   PCO  47
      CALL PZERO(M,N12)                                                  PCO  48
C.... CALL MESH INPUT SUBROUTINE TO READ AND PRINT ALL MESH DATA          PCO  49
      III = 0                                                            PCO  50
      CALL PMESH(M(N2),M(N5),M(N6),M(N7),M(N8),M(N9),M(N10),M(N11),NDF,  PCO  51
     1   NDM,NEN1,III)                                                   PCO  52
C.... ESTABLISH PROFILE OF RESULTING EQUATIONS FOR STIFFNESS, MASS, ETC   PCO  53
      CALL PROFIL(M(N12),M(N7),M(N9),NDF,NEN1,NAD)                       PCO  54
```

```
C.... SET POINTERS FOR SOLUTION ARRAYS * CHECK FOR SUFFICIENT MEMORY      PCO  55
      N13 = N12 + NEQ                                                    PCO  56
      N14 = N13 + NEQ*IPR                                                PCO  57
      NE  = N14 + NUMNP*NDF*IPR                                          PCO  58
      CALL SETMEM(NE)                                                    PCO  59
      CALL PZERO(M(N13),NEQ)                                             PCO  60
      GO TO 1                                                            PCO  61
C.... CALL MACRO SOLUTION MODULE FOR ESTABLISHING SOLUTION ALGOTITHM      PCO  62
200   CALL PMACR(M,M(NO),M(N1),M(N2),M(N3),M(N4),M(N5),M(N6),M(N7),M(N8) PCO  63
     1    ,M(N9),M(N10),M(N11),M(N12),M(N13),M(N14),M(NE),NDF,NDM,NEN1,  PCO  64
     2    NST,NE)                                                        PCO  65
      GO TO 1                                                            PCO  66
```

```
C.... INPUT/OUTPUT FORMATS                                                       PCO  67
1000  FORMAT(20A4)                                                               PCO  68
1001  FORMAT(16I5)                                                               PCO  69
2000  FORMAT(1H1,20A4//5X,30HNUMBER OF NODAL POINTS       =,I6/5X,30HNUM        PCO  70
     1BER OF ELEMENTS           =,I6/5X,30HNUMBER OF MATERIAL SETS              PCO  71
     2=,I6/5X,30HDIMENSION OF COORDINATE SPACE=,I6/5X,30HDEGREE OF FREED        PCO  72
     3OMS/NODE      =,I6/5X,30HNODES PER ELEMENT (MAXIMUM)  =,I6/5X,30HE         PCO  73
     4XTRA D.O.F. TO ELEMENT      =,I6)                                         PCO  74
      END                                                                        PCO  75


      BLOCK DATA                                                                       1
      COMMON /CDATA/ O,HEAD(20),NUMNP,NUMEL,NUMMAT,NEN,NEQ,IPR                         2
      COMMON /LABEL/ PDIS(6),A(6),BC(2),DI(6),CD(3),TE(3),FD(3)                        3
      DATA A/2H,1,2H,2,2H,3,2H,4,2H,5,2H,6/,CD/4H COO,4HRDIN,4HATES/                   4
      DATA TE/4H TEM,4HPERA,4HTURE/,FD/4H FOR,4HCE/D,4HISPL/                           5
      DATA PDIS/4H(I10,2H, ,4HF13.,4H4,  ,4H6E13,4H.4) /                               6
      DATA BC/4H B.C,2H. /,DI/4H DIS,2HPL,4H VEL,2HOC,4H ACC,2HEL/                     7
      DATA O/1HO/,IPR/2/                                                               8
      END                                                                              9


      SUBROUTINE GENVEC(NDM,X,CD,PRT,ERR)                                        GEN   1
C                                                                                GEN   2
C.... GENERATE REAL DATA ARRAYS BY LINEAR INTERPOLATION                          GEN   3
C                                                                                GEN   4


      LOGICAL PRT,ERR,PCOMP                                                      GEN   5
      COMMON /CDATA/ O,HEAD(20),NUMNP,NUMEL,NUMMAT,NEN,NEQ,IPR                   GEN   6
      DIMENSION X(NDM,1),XL(7),CD(2)                                             GEN   7
      DATA BL/4HBLAN/                                                            GEN   8
      N = 0                                                                      GEN   9
      NG = 0                                                                     GEN  10
102   L = N                                                                      GEN  11
      LG = NG                                                                    GEN  12
      READ(5,1000) N,NG,XL                                                       GEN  13
      IF(N.LE.0.OR.N.GT.NUMNP) GO TO 108                                         GEN  14
      DO 103 I = 1,NDM                                                           GEN  15
103   X(I,N) = XL(I)                                                             GEN  16
      IF(LG) 104,102,104                                                         GEN  17
104   LG = ISIGN(LG,N-L)                                                         GEN  18
      LI =(IABS(N-L+LG)-1)/IABS(LG)                                              GEN  19
      DO 105 I = 1,NDM                                                           GEN  20
105   XL(I) = (X(I,N)-X(I,L))/LI                                                 GEN  21
106   L = L + LG                                                                 GEN  22
      IF((N-L)*LG.LE.0) GO TO 102                                                GEN  23
      IF(L.LE.0.OR.L.GT.NUMNP) GO TO 110                                         GEN  24
      DO 107 I = 1,NDM                                                           GEN  25
107   X(I,L) = X(I,L-LG) + XL(I)                                                 GEN  26
      GO TO 106                                                                  GEN  27
110   WRITE(6,3000) L,(CD(I),I=1,3)                                              GEN  28
      ERR = .TRUE.                                                               GEN  29
      GO TO 102                                                                  GEN  30
108   DO 109 I = 1,NUMNP,50                                                      GEN  31
      IF(PRT) WRITE(6,2000) O,HEAD,(CD(L),L=1,3),(L,CD(1),CD(2),L=1,NDM)         GEN  32
      N = MIN0(NUMNP,I+49)                                                       GEN  33
      DO 109 J = I,N                                                             GEN  34
      IF(PCOMP(X(1,J),BL).AND.PRT) WRITE(6,2008) N                               GEN  35
109   IF(.NOT.PCOMP(X(1,J),BL).AND.PRT) WRITE(6,2009) J,(X(L,J),L=1,NDM)         GEN  36
      RETURN                                                                     GEN  37
1000  FORMAT(2I5,7F10.0)                                                         GEN  38
2000  FORMAT(A1,20A4//5X, 5HNODAL,3A4//6X,4HNODE,9(I7,A4,A2))                    GEN  39
2008  FORMAT(I10,32H HAS NOT BEEN INPUT OR GENERATED)                            GEN  40
2009  FORMAT(I10,9F13.4)                                                         GEN  41
3000  FORMAT(5X,43H**FATAL ERROR 02** ATTEMPT TO GENERATE NODE,I5,3H IN          GEN  42
     1  ,3A4)                                                                    GEN  43
      END                                                                        GEN  44
```

```
      SUBROUTINE PMESH(IDL,IE,D,ID,X,IX,F,T,NDF,NDM,NEN1,III)           PME   1
C                                                                       PME   2
C.... DATA INPUT ROUTINE FOR MESH DESCRIPTION                           PME   3
C                                                                       PME   4
      LOGICAL PRT,ERR,PCOMP                                             PME   5
      COMMON /CDATA/ O,HEAD(20),NUMNP,NUMEL,NUMMAT,NEN,NEQ,IPR          PME   6
      COMMON /ELDATA/ DM,N,MA,MCT,IEL,NEL                               PME   7
      COMMON /LABEL/ PDIS(6),A(6),BC(2),DI(6),CD(3),TE(3),FD(3)         PME   8
      DIMENSION IE(1),D(10,1),ID(NDF,1),X(NDM,1),IX(NEN1,1),XHED(7)     PME   9
     1  ,IDL(6),XL(3),F(NDF,1),FL(6),T(1),WD(10),VA(2)                  PME  10
      DATA WD/4HCOOR,4HELEM,4HMATE,4HBOUN,4HFORC,4HTEMP,4HEND ,4HPRIN,  PME  11
     1  4HNOPR,4HPAGE/,BL/4HBLAN/,VA/4H VAL,2HUE/,LIST/10/,PRT/.TRUE./  PME  12
C.... INITIALIZE ARRAYS                                                 PME  13
      ERR = .FALSE.                                                     PME  14
      IF(III.LT.0) GO TO 10                                             PME  15
      DO 101 N = 1,NUMNP                                                PME  16
      DO 100 I = 1,NDF                                                  PME  17
      ID(I,N) = 0                                                       PME  18
      X(I,N) = BL                                                       PME  19
100   F(I,N) = 0.0                                                      PME  20
101   T(N) = 0.0                                                        PME  21
10    READ(5,1000)CC                                                    PME  22
      DO 20 I = 1,LIST                                                  PME  23
20    IF(PCOMP(CC,WD(I))) GO TO 30                                      PME  24
      GO TO 10                                                          PME  25
30    GO TO (1,2,3,4,5,6,7,8,9,11),I                                    PME  26
C.... NODAL COORDINATE DATA INPUT                                       PME  27
1     CALL GENVEC(NDM,X,CD,PRT,ERR)                                     PME  28
      GO TO 10                                                          PME  29
C.... ELEMENT DATA INPUT                                                PME  30
2     L = 0                                                             PME  31
      DO 206 I = 1,NUMEL,50                                             PME  32
      IF(PRT) WRITE(6,2001) O,HEAD,(K,K=1,NEN)                          PME  33
      J = MIN0(NUMEL,I+49)                                              PME  34
      DO 206 N = I,J                                                    PME  35
      IF(L-N) 200,202,203                                               PME  36
200   READ(5,1001) L,LK,(IDL(K),K=1,NEN),LX                             PME  37
      IF(L.EQ.0) L = NUMEL+1                                            PME  38
      IF(LX.EQ.0) LX=1                                                  PME  39
      IF(L-N) 201,202,203                                               PME  40
201   WRITE(6,3001) L,N                                                 PME  41
      ERR = .TRUE.                                                      PME  42
      GO TO 206                                                         PME  43
202   NX = LX                                                           PME  44
      DO 207 K = 1,NEN1                                                 PME  45
207   IX(K,L) = IDL(K)                                                  PME  46
      IX(NEN1,L) = LK                                                   PME  47
      GO TO 205                                                         PME  48
203   IX(NEN1,N) = IX(NEN1,N-1)                                         PME  49
      DO 204 K = 1,NEN                                                  PME  50
      IX(K,N) = IX(K,N-1) + NX                                          PME  51
204   IF(IX(K,N-1).EQ.0) IX(K,N) = 0                                    PME  52
205   IF(PRT) WRITE(6,2002) N,IX(NEN1,N),(IX(K,N),K=1,NEN)              PME  53
206   CONTINUE                                                          PME  54
      GO TO 10                                                          PME  55
C.... MATERIAL DATA INPUT                                               PME  56
3     WRITE(6,2004) O,HEAD                                              PME  57
      DO 300 N = 1,NUMMAT                                               PME  58
      READ(5,1002) MA,IEL,XHED                                          PME  59
      WRITE(6,2003)MA,IEL,XHED                                          PME  60
      IE(MA) = IEL                                                      PME  61
300   CALL ELMLIB(D(1,MA),DUM,X,IX,T,S,P,NDF,NDM,NST,1)                 PME  62
      GO TO 10                                                          PME  63
C.... READ IN THE RESTRAINT CONDITIONS FOR EACH NODE                    PME  64
4     IF(PRT) WRITE(6,2000) O,HEAD,(I,BC,I=1,NDF)                       PME  65
      III = 1                                                           PME  66
      N = 0                                                             PME  67
      NG = 0                                                            PME  68
402   L = N                                                             PME  69
      LG = NG                                                           PME  70
      READ(5,1001) N,NG,IDL                                             PME  71
```

```
      IF(N.LE.0.OR.N.GT.NUMNP) GO TO 60                                  PME  72
      DO 51 I = 1,NDF                                                    PME  73
      ID(I,N) = IDL(I)                                                   PME  74
51    IF(L.NE.0.AND.IDL(I).EQ.0.AND.ID(I,L).LT.0) ID(I,N) = -1           PME  75
      LG = ISIGN(LG,N-L)                                                 PME  76
52    L = L + LG                                                         PME  77
      IF((N-L)*LG.LE.0) GO TO 402                                        PME  78
      DO 53 I = 1,NDF                                                    PME  79
53    IF(ID(I,L-LG).LT.0) ID(I,L) = -1                                   PME  80
      GO TO 52                                                           PME  81
60    DO 58 N = 1,NUMNP                                                  PME  82
      DO 56 I = 1,NDF                                                    PME  83
56    IF(ID(I,N).NE.0) GO TO 57                                          PME  84
      GO TO 58                                                           PME  85
57    IF(PRT) WRITE(6,2007) N,(ID(I,N),I=1,NDF)                          PME  86
58    CONTINUE                                                           PME  87
      GO TO 10                                                           PME  88
C.... FORCE/DISPL DATA INPUT                                              PME  89
5     CALL GENVEC(NDF,F,FD,PRT,ERR)                                      PME  90
      GO TO 10                                                           PME  91
C.... TEMPERATURE DATA INPUT                                              PME  92
6     CALL GENVEC(1,T,TE,PRT,ERR)                                        PME  93
      GO TO 10                                                           PME  94
7     IF(ERR) STOP                                                       PME  95
      RETURN                                                             PME  96
8     PRT = .TRUE.                                                       PME  97
      GO TO 10                                                           PME  98
9     PRT = .FALSE.                                                      PME  99
      GO TO 10                                                           PME 100
11    READ(5,1000) O                                                     PME 101
      GO TO 10                                                           PME 102
1000  FORMAT(A4,75X,A1)                                                  PME 103
1001  FORMAT(16I5)                                                       PME 104
1002  FORMAT(I5,4X,I1,17A4)                                              PME 105
2000  FORMAT(A1,20A4//5X,17HNODAL B.C.       //6X,4HNODE,9(I7,A4,A2)/1X) PME 106
2001  FORMAT(A1,20A4//5X,8HELEMENTS//3X,7HELEMENT,2X,8HMATERIAL,         PME 107
     1   14(I3,5H NODE)/(20X,14(I3,5H NODE)))                            PME 108
2002  FORMAT(2I10,14I8/(20X,14I8))                                       PME 109
2003  FORMAT(/5X,12HMATERIAL SET,I3,17H FOR ELEMENT TYPE,I2,5X,17A4/1X)  PME 110
2004  FORMAT(A1,20A4//5X,19HMATERIAL PROPERTIES)                         PME 111
2005  FORMAT(A1,20A4//5X,17HNODAL FORCE/DISPL//6X,4HNODE,9(I7,A4,A2))    PME 112
2006  FORMAT(I10,9E13.3)                                                 PME 113
2007  FORMAT(I10,9I13)                                                   PME 114
3001  FORMAT(5X,20H**ERROR 03** ELEMENT,I5,22H APPEARS AFTER ELEMENT,I5) PME 115
      END                                                                PME 116

      SUBROUTINE SETMEM(J)                                               SET   1

C                                                                        SET   2
C.... MONITOR AVAILABLE MEMORY IN BLANK COMMON                            SET   3
C                                                                        SET   4


      COMMON M(1)                                                        SET   5
      COMMON /PSIZE/ MAX                                                 SET   6
      K = J                                                              SET   7
      IF(K.LE.MAX) RETURN                                                SET   8
      WRITE(6,1000) K,MAX                                                SET   9
      STOP                                                               SET  10
1000  FORMAT(5X,49H**ERROR 01** INSUFFICIENT STORAGE IN BLANK COMMON/    SET  11
     1   17X,11HREQUIRED  =,I8/17X,11HAVAILABLE =,I8/)                   SET  12
      END                                                                SET  13
```

24.8.2 *Solution and output modules.* The solution and output for each problem is controlled by subroutine PMACR. Depending on the macro instructions which are used for each analysis, the appropriate set of other routines listed in this section will also be employed.

Except for PMACR, subprograms are included in the listing alphabetically by their name.

```
      SUBROUTINE PMACR (UL,XL,TL,LD,P,S,IE,D,ID,X,IX,F,T,JDIAG,B,DR,CT   PMA   1
     1,NDF,NDM,NEN1,NST,NEND)                                            PMA   2
C                                                                        PMA   3
C.... MACRO INSTRUCTION SUBPROGRAM                                       PMA   4
C                                                                        PMA   5
C.... CONTROLS PROBLEM SOLUTION AND OUTPUT ALGORITHMS BY                 PMA   6
C.... ORDER OF SPECIFYING MACRO COMMANDS IN ARRAY WD.                    PMA   7
C                                                                        PMA   8
      LOGICAL AFR,BFR,CFR,AFL,BFL,CFL,DFL,EFL,FFL,GFL,PCOMP             PMA   9
      COMMON M(1)                                                        PMA  10
      COMMON /CDATA/ O,HEAD(20),NUMNP,NUMEL,NUMMAT,NEN,NEQ,IPR           PMA  11
      COMMON /LABEL/ PDIS(6),Z(6),BC(2),DI(6),CD(3),TE(3),FD(3)          PMA  12
      COMMON /PRLOD/ PROP                                                PMA  13
      COMMON /TDATA/ TIME,DT,C1,C2,C3,C4,C5                              PMA  14
      DIMENSION WD(21),CT(4,1),CTL(4),LVS(9),LVE(9),JDIAG(1),            PMA  15
     1   UL(1),XL(1),TL(1),LD(1),P(1),S(1),IE(1),D(1),ID(1),X(1),        PMA  16
     2   IX(1),F(1),T(1),B(1),DR(1)                                      PMA  17
      DATA WD/4HTOL ,4HDT  ,4HSTRE,4HDISP,4HTANG,4HFORM,4HLOOP,4HNEXT,   PMA  18
     1        4HPROP,4HDATA,4HTIME,4HCONV,4HSOLV,4HLMAS,4HCMAS,4HMESH,   PMA  19
     2        4HEIGE,4HEXCD,4HUTAN,4HREAC,4HCHEC/                        PMA  20
      DATA NWD/21/,ENDM/4HEND /,NV,NC/1,1/                               PMA  21
C.... SET INITIAL VALUES OF PARAMETERS                                   PMA  22
      DT = 0.0                                                           PMA  23
      PROP = 1.0                                                         PMA  24
      RNMAX = 0.0                                                        PMA  25
      TIME = 0.0                                                         PMA  26
      TOL = 1.E-9                                                        PMA  27
      UN = 0.0                                                           PMA  28
      AFL = .TRUE.                                                       PMA  29
      AFR = .FALSE.                                                      PMA  30
      BFL = .TRUE.                                                       PMA  31
      BFR = .FALSE.                                                      PMA  32
      CFL = .TRUE.                                                       PMA  33
      CFR = .FALSE.                                                      PMA  34
      DFL = .TRUE.                                                       PMA  35
      EFL = .TRUE.                                                       PMA  36
      FFL = .FALSE.                                                      PMA  37
      GFL = .TRUE.                                                       PMA  38
      NE = NEND                                                          PMA  39
      NNEQ = NDF*NUMNP                                                   PMA  40
      NPLD = 0                                                           PMA  41
      WRITE(6,2001) O,HEAD                                               PMA  42
C.... READ MACRO CARDS                                                   PMA  43
      LL = 1                                                             PMA  44
      LMAX = 16                                                          PMA  45
      CALL SETMEM(NE+LMAX*4*IPR)                                         PMA  46
      CT(1,1) = WD(7)                                                    PMA  47
      CT(3,1) = 1.0                                                      PMA  48
100   LL = LL + 1                                                        PMA  49
      IF(LL.LT.LMAX) GO TO 110                                           PMA  50
      LMAX = LMAX + 16                                                   PMA  51
      CALL SETMEM(NE+LMAX*4*IPR)                                         PMA  52
110   READ(5,1000)  (CT(J,LL),J=1,4)                                     PMA  53
      WRITE(6,2000) (CT(J,LL),J=1,4)                                     PMA  54
      IF(.NOT.PCOMP(CT(1,LL),ENDM)) GO TO 100                            PMA  55
200   CT(1,LL)= WD(8)                                                    PMA  56
C.... SET LOOP MARKERS                                                   PMA  57
      NE = NE + LMAX*4*IPR                                               PMA  58
      LX = LL - 1                                                        PMA  59
      DO 230 L = 1,LX                                                    PMA  60
```

```
      IF(.NOT.PCOMP(CT(1,L),WD(7))) GO TO 230                          PMA  61
      J = 1                                                             PMA  62
      K = L + 1                                                         PMA  63
      DO 210 I = K,LL                                                   PMA  64
      IF(PCOMP(CT(1,I),WD(7))) J = J + 1                                PMA  65
      IF(J.GT.9) GO TO 401                                              PMA  66
      IF(PCOMP(CT(1,I),WD(8))) J = J - 1                                PMA  67
210   IF(J.EQ.0) GO TO 220                                              PMA  68
      GO TO 400                                                         PMA  69
220   CT(4,I) = L                                                       PMA  70
      CT(4,L) = I                                                       PMA  71
230   CONTINUE                                                          PMA  72
      J = 0                                                             PMA  73
      DO 240 L = 1,LL                                                   PMA  74
      IF(PCOMP(CT(1,L),WD(7))) J = J + 1                                PMA  75
240   IF(PCOMP(CT(1,L),WD(8))) J = J - 1                                PMA  76
      IF(J.NE.0) GO TO 400                                              PMA  77
C.... EXECUTE MACRO INSTRUCTION PROGRAM                                  PMA  78
      LV = 0                                                            PMA  79
      L = 1                                                             PMA  80
299   DO 300 J = 1,NWD                                                  PMA  81
300   IF(PCOMP(CT(1,L),WD(J))) GO TO 310                                PMA  82
      GO TO 330                                                         PMA  83
310   I = L - 1                                                         PMA  84
      IF(L.NE.1.AND.L.NE.LL)                                            PMA  85
     1WRITE(6,2010) I,(CT(K,L),K = 1,4)                                 PMA  86
      GO TO (1,2,3,4,5,6,7,8,9,10,11,12,13,14,15,16,17,18,19,20,21),J   PMA  87
C.... SET SOLUTION TOLERANCE                                             PMA  88
1     TOL = CT(3,L)                                                     PMA  89
      GO TO 330                                                         PMA  90
C.... SET TIME INCREMENT                                                 PMA  91
2     DT = CT(3,L)                                                      PMA  92
      GO TO 330                                                         PMA  93
C.... PRINT STRESS VALUES                                                PMA  94
3     LX = LVE(LV)                                                      PMA  95
      IF(AMOD(CT(3,LX),AMAX1(CT(3,L),1.)).EQ.0.0)                       PMA  96
     1 CALL PFORM(UL,XL,TL,LD,P,S,IE,D,ID,X,IX,F,T,JDIAG,DR,DR,DR,      PMA  97
     2   NDF,NDM,NEN1,NST,4,B,M(NV),.FALSE.,.FALSE.,.FALSE.,.FALSE.)    PMA  98
      GO TO 330                                                         PMA  99
C.... PRINT DISPLACEMENTS                                                PMA 100
4     LX = LVE(LV)                                                      PMA 101
      IF(AMOD(CT(3,LX),AMAX1(CT(3,L),1.)).NE.0.0) GO TO 330             PMA 102
      WRITE(6,2003) O,HEAD,TIME,PROP                                    PMA 103
      CALL PRTDIS(ID,X,B,F,NDM,NDF)                                     PMA 104
      GO TO 330                                                         PMA 105
C.... FORM TANGENT STIFFNESS                                             PMA 106
19    IF(CFL) CALL PSETM(NC,NE,JDIAG(NEQ)*IPR,CFL)                      PMA 107
      CALL PZERO(M(NC),JDIAG(NEQ))                                      PMA 108
      CFR = .TRUE.                                                      PMA 109
5     IF(J.EQ.5) CFR = .FALSE.                                          PMA 110
      IF(GFL) CALL PSETM(NA,NE,JDIAG(NEQ)*IPR,GFL)                      PMA 111
      IF(NPLD.GT.0) PROP = PROPLD(TIME,0)                               PMA 112
      CALL PZERO(M(NA),JDIAG(NEQ))                                      PMA 113
      CALL PFORM(UL,XL,TL,LD,P,S,IE,D,ID,X,IX,F,T,JDIAG,DR,M(NA),M(NC), PMA 114
     2   NDF,NDM,NEN1,NST,3,B,M(NV),.TRUE.,.FALSE.,CFR,.FALSE.)         PMA 115
      AFR = .TRUE.                                                      PMA 116
      GO TO 330                                                         PMA 117
C.... FORM OUT OF BALANCE FORCE FOR TIME STEP/ITERATION                  PMA 118
6     IF(NPLD.GT.0) PROP = PROPLD(TIME,0)                               PMA 119
      CALL PLOAD(ID,F,DR,NNEQ,PROP)                                     PMA 120
      CALL PFORM(UL,XL,TL,LD,P,S,IE,D,ID,X,IX,F,T,JDIAG,DR,DR,DR,       PMA 121
     2   NDF,NDM,NEN1,NST,6,B,M(NV),.FALSE.,.TRUE.,.FALSE.,.FALSE.)     PMA 122
      BFR = .TRUE.                                                      PMA 123
      RN = 0.                                                           PMA 124
      DO 61 N = 1,NEQ                                                   PMA 125
61    RN = RN + DR(N)**2                                                PMA 126
      RN = SQRT(RN)                                                     PMA 127
      RNMAX = AMAX1(RNMAX,RN)                                           PMA 128
      WRITE(6,2005) RNMAX,RN,TOL                                        PMA 129
      IF(RN.GE.RNMAX*TOL) GO TO 330                                     PMA 130
      LX = LVE(LV)                                                      PMA 131
      LO = LVS(LV)                                                      PMA 132
```

```
      CT(3,LX) = CT(3,IO)                                              PMA 133
      L = LX - 1                                                       PMA 134
      GO TO 330                                                        PMA 135
C.... SET LOOP START INDICATORS                                        PMA 136
7     LV = LV + 1                                                      PMA 137
      LX = CT(4,L)                                                     PMA 138
      LVS(LV) = L                                                      PMA 139
      LVE(LV) = LX                                                     PMA 140
      CT(3,LX) = 1.                                                    PMA 141
      GO TO 330                                                        PMA 142
C.... LOOP TERMINATOR CONTROL                                           PMA 143
8     N = CT(4,L)                                                      PMA 144
      CT(3,L) = CT(3,L) + 1.0                                          PMA 145
      IF(CT(3,L).GT.CT(3,N)) LV = LV - 1                               PMA 146
      IF(CT(3,L).LE.CT(3,N)) L = N                                     PMA 147
      GO TO 330                                                        PMA 148
C.... INPUT PROPORTIONAL LOAD TABLE                                     PMA 149
9     NPLD = CT(3,L)                                                   PMA 150
      PROP = PROPLD(0.,NPLD)                                           PMA 151
      GO TO 330                                                        PMA 152
C.... READ COMMAND                                                      PMA 153
10    READ(5,1000) (CTL(I),I=1,4)                                      PMA 154
      IF(.NOT.PCOMP(CT(2,L),CTL(1))) GO TO 402                         PMA 155
      IF(PCOMP(CTL(1),WD(1))) TOL = CTL(3)                             PMA 156
      IF(PCOMP(CTL(1),WD(2))) DT = CTL(3)                              PMA 157
      GO TO 330                                                        PMA 158
C.... INCREMENT TIME                                                    PMA 159
11    TIME = TIME + DT                                                 PMA 160
      RNMAX = 0.0                                                      PMA 161
      UN = 0.0                                                         PMA 162
      GO TO 330                                                        PMA 163
C.... COMPUTE CONVERGENCE TEST                                          PMA 164
12    RN = 0.0                                                         PMA 165
      DO 121 N = 1,NEQ                                                 PMA 166
      UN = UN + B(N)**2                                                PMA 167
121   RN = RN + DR(N)**2                                               PMA 168
      UN = AMAX1(UN,RN)                                                PMA 169
      CN = SQRT(UN)                                                    PMA 170
      RN = SQRT(RN)                                                    PMA 171
      WRITE(6,2002) CN,RN,TOL                                          PMA 172
      LX = LVE(LV)                                                     PMA 173
      LO = LVS(LV)                                                     PMA 174
      IF(RN.LT.CN*TOL) CT(3,LX) = CT(3,LO)                             PMA 175
      GO TO 330                                                        PMA 176
C.... SOLVE THE EQUATIONS                                               PMA 177
13    IF(CFR) GO TO 131                                                PMA 178
      CALL ACTCOL(M(NA),DR,JDIAG,NEQ,AFR,BFR)                          PMA 179
      GO TO 132                                                        PMA 180
131   CALL UACTCL(M(NA),M(NC),DR,JDIAG,NEQ,AFR,BFR)                    PMA 181
132   AFR = .FALSE.                                                    PMA 182
      IF(.NOT.BFR) GO TO 330                                           PMA 183
      BFR = .FALSE.                                                    PMA 184
      DO 133 N = 1,NEQ                                                 PMA 185
133   B(N) = B(N) + DR(N)                                              PMA 186
      GO TO 330                                                        PMA 187
C.... FORM A LUMPED MASS APPROXIMATION                                  PMA 188
14    AFL = .FALSE.                                                    PMA 189
      BFL = .TRUE.                                                     PMA 190
      IF(EFL) CALL PSETM(NN,NE,NEQ*IPR,EFL)                            PMA 191
139   CALL PZERO(M(NN),NEQ)                                            PMA 192
      GO TO 140                                                        PMA 193
C.... FORM A CONSISTENT MASS APPROXIMATION                              PMA 194
15    AFL = .TRUE.                                                     PMA 195
      BFL = .FALSE.                                                    PMA 196
      IF(DFL) CALL PSETM(NM,NE,JDIAG(NEQ)*IPR,DFL)                     PMA 197
152   CALL PZERO (M(NM),JDIAG(NEQ))                                    PMA 198
140   CALL PFORM(UL,XL,TL,LD,P,S,IE,D,ID,X,IX,F,T,JDIAG,M(NN),M(NM),   PMA 199
     1  M(NM),NDF,NDM,NEN1,NST,5,B,M(NV),AFL,BFL,.FALSE.,.FALSE.)      PMA 200
      GO TO 330                                                        PMA 201
16    I = -1                                                           PMA 202
      CALL PMESH(LD,IE,D,ID,X,IX,F,T,NDF,NDM,NEN1,I)                   PMA 203
      IF(I.GT.0) GO TO 404                                             PMA 204
```

```
      GO TO 330                                                          PMA 205
17    J = NM                                                             PMA 206
      IF(DFL) J = NN                                                     PMA 207
      CALL PEIGS(M(NA),M(J),F,X,B,DR,ID,IX,JDIAG,NDF,NDM,NEN1,DFL)       PMA 208
      GO TO 330                                                          PMA 209
18    IF(FFL) GO TO 181                                                  PMA 210
C.... MACRO *EXCD* EXPLICIT INTEGRATION OF EQUATIONS OF MOTION           PMA 211
      NQ = NE                                                            PMA 212
      NV = NQ + NEQ*IPR                                                  PMA 213
      NR = NV + NDF*NUMNP*IPR                                            PMA 214
      NE = NR + NEQ*IPR                                                  PMA 215
      CALL SETMEM(NE+1)                                                  PMA 216
      CALL PZERO(M(NQ),NE-NQ)                                            PMA 217
      FFL = .TRUE.                                                       PMA 218
      GO TO 330                                                          PMA 219
181   IF(.NOT.BFR.OR.EFL) GO TO 403                                      PMA 220
      CALL EUPDAT(DR,B,M(NQ),M(NR),M(NN),DT,NEQ)                         PMA 221
      GO TO 330                                                          PMA 222
C.... COMPUTE REACTIONS AND PRINT                                         PMA 223
20    CALL PZERO(DR,NNEQ)                                                PMA 224
      CALL PFORM(UL,XL,TL,LD,P,S,IE,D,ID,X,IX,F,T,JDIAG,DR,DR,DR,        PMA 225
     1   NDF,NDM,NEN1,NST,6,B,M(NV),.FALSE.,.TRUE.,.FALSE.,.TRUE.)       PMA 226
      CALL PRTREA(DR,NDF)                                                PMA 227
      GO TO 330                                                          PMA 228
21    CALL PFORM(UL,XL,TL,LD,P,S,IE,D,ID,X,IX,F,T,JDIAG,DR,DR,DR,        PMA 229
     1   NDF,NDM,NEN1,NST,2,B,F,.FALSE.,.FALSE.,.FALSE.,.FALSE.)         PMA 230
330   L = L + 1                                                          PMA 231
      IF(L.GT.LL) RETURN                                                 PMA 232
      GO TO 299                                                          PMA 233
400   WRITE(6,4000)                                                      PMA 234
      RETURN                                                             PMA 235
401   WRITE(6,4001)                                                      PMA 236
      RETURN                                                             PMA 237
402   WRITE(6,4002)                                                      PMA 238
      RETURN                                                             PMA 239
403   WRITE(6,4003)                                                      PMA 240
404   WRITE(6,4004)                                                      PMA 241
      RETURN                                                             PMA 242
1000  FORMAT(A4,1X,A4,1X,2F5.0)                                          PMA 243
2000  FORMAT(10X,A4,1X,A4,1X,2G15.5)                                     PMA 244
2001  FORMAT(A1,20A4//5X,18HMACRO INSTRUCTIONS//5X,15HMACRO STATEMENT,5X PMA 245
     1,10HVARIABLE 1,5X,10HVARIABLE 2)                                   PMA 246
2002  FORMAT(5X,29HDISPLACEMENT CONVERGENCE TEST/10X,7HUNMAX =,G15.5,5X, PMA 247
     1   7HUN    =,G15.5,5X,7HTOL   =,G15.5)                             PMA 248
2003  FORMAT(A1,20A4,10X,4HTIME,G13.5//5X,17HPROPORTIONAL LOAD,G13.5)    PMA 249
2004  FORMAT(5X,4HCN =,G12.5,5X,4HDN =,G12.5,5X,4HUN =,G12.5,5X,4HAG =   PMA 250
     1   ,G12.5,5X,4HAC =,G12.5)                                         PMA 251
2005  FORMAT(5X,22HFORCE CONVERGENCE TEST/10X,7HRNMAX =,G15.5,5X,        PMA 252
     1   7HRN    =,G15.5,5X,7HTOL   =,G15.5)                             PMA 253
2010  FORMAT(2X,19H**MACRO INSTRUCTION,I4,13H EXECUTED**  ,2(A4,2X), 6H  PMA 254
     1V1 = ,G13.4, 8H  , V2 = ,G13.4)                                    PMA 255
4000  FORMAT(5X,46H**FATAL ERROR 10** UNBALANCED LOOP/NEXT MACROS )      PMA 256
4001  FORMAT(5X,45H**FATAL ERROR 11** LOOPS NESTED DEEPER THAN 8)        PMA 257
4002  FORMAT(5X,57H**FATAL ERROR 12** MACRO LABEL MISMATCH ON A READ COM PMA 258
     1MAND)                                                              PMA 259
4003  FORMAT(5X,63H**FATAL ERROR 13** MACRO EXCD MUST BE PRECEDED BY LMA PMA 260
     1S AND FORM)                                                        PMA 261
4004  FORMAT(5X,84H**FATAL ERROR 14** ATTEMPT TO CHANGE BOUNDARY RESTRAI PMA 262
     1NT CODES DURING MACRO EXECUTION )                                  PMA 263
      END                                                                PMA 264

      SUBROUTINE ACTCOL(A,B,JDIAG,NEQ,AFAC,BACK)                         ACT   1
      LOGICAL AFAC,BACK                                                  ACT   2
      COMMON/ENGYS/ AENGY                                                ACT   3
      DIMENSION A(1),B(1),JDIAG(1)                                       ACT   4
C                                                                        ACT   5
C.... ACTIVE COLUMN PROFILE SYMMETRIC EQUATION SOLVER                    ACT   6
C                                                                        ACT   7
C.... FACTOR A TO UT*D*U, REDUCE B                                        ACT   8
      AENGY = 0.0                                                        ACT   9
      JR = 0                                                             ACT  10
      DO 600 J = 1,NEQ                                                   ACT  11
```

```
      JD = JDIAG(J)                                                     ACT  12
      JH = JD - JR                                                      ACT  13
      IS = J - JH + 2                                                   ACT  14
      IF(JH-2) 600,300,100                                              ACT  15
100   IF(.NOT.AFAC) GO TO 500                                           ACT  16
      IE = J - 1                                                        ACT  17
      K = JR + 2                                                        ACT  18
      ID = JDIAG(IS - 1)                                                ACT  19
C.... REDUCE ALL EQUATIONS EXCEPT DIAGONAL                              ACT  20
      DO 200 I = IS,IE                                                  ACT  21
      IR = ID                                                           ACT  22
      ID = JDIAG(I)                                                     ACT  23
      IH = MIN0(ID-IR-1,I-IS+1)                                         ACT  24
      IF(IH.GT.0) A(K) = A(K) - DOT(A(K-IH),A(ID-IH),IH)                ACT  25
200   K = K + 1                                                         ACT  26
C.... REDUCE DIAGONAL TERM                                              ACT  27
300   IF(.NOT.AFAC) GO TO 500                                           ACT  28
      IR = JR + 1                                                       ACT  29
      IE = JD - 1                                                       ACT  30
      K = J - JD                                                        ACT  31
      DO 400 I = IR,IE                                                  ACT  32
      ID = JDIAG(K+I)                                                   ACT  33
      IF(A(ID).EQ.0.0) GO TO 400                                        ACT  34
      D = A(I)                                                          ACT  35
      A(I) = A(I)/A(ID)                                                 ACT  36
      A(JD) = A(JD) - D*A(I)                                            ACT  37
400   CONTINUE                                                          ACT  38
C.... REDUCE RHS                                                        ACT  39
500   IF(BACK) B(J) = B(J) - DOT(A(JR+1),B(IS-1),JH-1)                  ACT  40
600   JR = JD                                                           ACT  41
      IF(.NOT.BACK) RETURN                                              ACT  42
C.... DIVIDE BY DIAGONAL PIVOTS                                         ACT  43
      DO 700 I = 1,NEQ                                                  ACT  44
      ID = JDIAG(I)                                                     ACT  45
      IF(A(ID).NE.0.0) B(I) = B(I)/A(ID)                                ACT  46
700   AENGY = AENGY + B(I)*B(I)*A(ID)                                   ACT  47
C.... BACKSUBSTITUTE                                                    ACT  48
      J = NEQ                                                           ACT  49
      JD = JDIAG(J)                                                     ACT  50
800   D = B(J)                                                          ACT  51
      J = J - 1                                                         ACT  52
      IF(J.LE.0) RETURN                                                 ACT  53
      JR = JDIAG(J)                                                     ACT  54
      IF(JD-JR.LE.1) GO TO 1000                                         ACT  55
      IS = J - JD + JR + 2                                              ACT  56
      K = JR - IS + 1                                                   ACT  57
      DO 900 I = IS,J                                                   ACT  58
900   B(I) = B(I) - A(I+K)*D                                            ACT  59
1000  JD = JR                                                           ACT  60
      GO TO 800                                                         ACT  61
      END                                                               ACT  62

      SUBROUTINE ADDSTF(A,B,C,S,P,JDIAG,LD,NST,NEL,AFL,BFL,CFL)         ADD   1
C                                                                       ADD   2
C.... ASSEMBLE GLOBAL ARRAYS                                            ADD   3
C                                                                       ADD   4
      LOGICAL AFL,BFL,CFL                                               ADD   5
      DIMENSION A(1),B(1),JDIAG(1),P(1),S(NST,1),LD(1) ,C(1)            ADD   6
      DO 200 J = 1,NEL                                                  ADD   7
      K = LD(J)                                                         ADD   8
      IF(K.EQ.0) GO TO 200                                              ADD   9
      IF(BFL) B(K) = B(K) + P(J)                                        ADD  10
      IF(.NOT.AFL.AND..NOT.CFL) GO TO 200                               ADD  11
      L = JDIAG(K) - K                                                  ADD  12
      DO 100 I = 1,NEL                                                  ADD  13
      M = LD(I)                                                         ADD  14
      IF(M.GT.K.OR.M.EQ.0) GO TO 100                                    ADD  15
      M = L + M                                                         ADD  16
      IF(AFL) A(M) = A(M) + S(I,J)                                      ADD  17
      IF(CFL) C(M) = C(M) + S(J,I)                                      ADD  18
100   CONTINUE                                                          ADD  19
200   CONTINUE                                                          ADD  20
      RETURN                                                            ADD  21
      END                                                               ADD  22
```

```
      FUNCTION   DOT(A,B,N)                                         DOT   1
C                                                                   DOT   2
C.... VECTOR DOT PRODUCT                                             DOT   3
C                                                                   DOT   4
      DIMENSION A(1),B(1)                                           DOT   5
      DOT = 0.0                                                     DOT   6
      DO 100 I = 1,N                                                DOT   7
100   DOT = DOT + A(I)*B(I)                                         DOT   8
      RETURN                                                        DOT   9
      END                                                           DOT  10

      SUBROUTINE ELMLIB(D,U,X,IX,T,S,P,I,J,K,ISW)                   ELM   1
C                                                                   ELM   2
C.... ELEMENT LIBRARY                                               ELM   3
C                                                                   ELM   4
      COMMON /ELDATA/ DM,N,MA,MCT,IEL,NEL                           ELM   5
      DIMENSION P(K),S(K,K),D(1),U(1),X(1),IX(1),T(1)               ELM   6
      IF(IEL.LE.0.OR.IEL.GT.4) GO TO 400                            ELM   7
      IF(ISW.LT.3) GO TO 30                                         ELM   8
      DO 20 L = 1,K                                                 ELM   9
      P(L) = 0.0                                                    ELM  10
      DO 20 M = 1,K                                                 ELM  11
20    S(L,M) = 0.0                                                  ELM  12
30    GO TO (1,2,3,4),IEL                                           ELM  13
1     CALL ELMT01(D,U,X,IX,T,S,P,I,J,K,ISW)                         ELM  14
      GO TO 10                                                      ELM  15
2     CALL ELMT02(D,U,X,IX,T,S,P,I,J,K,ISW)                         ELM  16
      GO TO 10                                                      ELM  17
3     CALL ELMT03(D,U,X,IX,T,S,P,I,J,K,ISW)                         ELM  18
      GO TO 10                                                      ELM  19
4     CALL ELMT04(D,U,X,IX,T,S,P,I,J,K,ISW)                         ELM  20
10    RETURN                                                        ELM  21
400   WRITE(6,4000) IEL                                             ELM  22
      STOP                                                          ELM  23
4000  FORMAT(5X,39H**FATAL ERROR 04** ELEMENT CLASS NUMBER,I3,6H INPUT)  ELM  24
      END                                                           ELM  25

      SUBROUTINE EUPDAT(DR,U,V,A,XM,DT,NEQ)                         EUP   1
      DIMENSION U(1),V(1),A(1),DR(1),XM(1)                          EUP   2
C.... UPDATE SOLUTION USING EXPLICIT CENTRAL DIFFERENCES            EUP   3
      DATA DTHP/0.0/                                                EUP   4
      DTH = DT/2.                                                   EUP   5
      DTAV = DTH + DTHP                                             EUP   6
      DTHP = DTH                                                    EUP   7
      DO 100 N = 1,NEQ                                              EUP   8
      A(N) = DR(N)/XM(N)                                            EUP   9
      V(N) = V(N) + DTAV*A(N)                                       EUP  10
100   U(N) = U(N) + DT*V(N)                                         EUP  11
      RETURN                                                        EUP  12
      END                                                           EUP  13

      SUBROUTINE NORM(X,Y,N)                                        NOR   1
C                                                                   NOR   2
C.... NORMALIZE VECTOR Y TO UNIT VECTOR X                           NOR   3
C                                                                   NOR   4
      DIMENSION X(1),Y(1)                                           NOR   5
      SCALE = SQRT(DOT(Y,Y,N))                                      NOR   6
      DO 100 I = 1,N                                                NOR   7
100   X(I) = Y(I)/SCALE                                             NOR   8
      RETURN                                                        NOR   9
      END                                                           NOR  10

      LOGICAL FUNCTION PCOMP(A,B)                                   NCTI  1
      PCOMP = .FALSE.                                               NCTI  2
C.... IT MAY BE NECESSARY TO REPLACE THE FOLLOWING ALPHANUMERIC     NCTI  3
C.... COMPARISON STATEMENT IF COMPUTER PRODUCES AN OVERFLOW         NCTI  4
      IF(A.EQ.B) PCOMP = .TRUE.                                     NCTI  5
      RETURN                                                        NCTI  6
      END                                                           NCTI  7
```

```
      SUBROUTINE PEIGS(A,B,F,X,Y,Z,ID,IX,JDIAG,NDF,NDM,NEN1,DFL)         PEI   1
C                                                                        PEI   2
C.... COMPUTE DOMINANT EIGENVALUE BY INVERSE ITERATION                   PEI   3
C                                                                        PEI   4
      LOGICAL DFL                                                        PEI   5
      COMMON /CDATA/ O,HEAD(20),NUMNP,NUMEL,NUMMAT,NEN,NEQ,IPR           PEI   6
      COMMON/ENGYS/ AENGY                                                PEI   7
      DIMENSION A(1),B(1),F(1),X(1),Y(1),Z(1),ID(1),IX(1),JDIAG(1)       PEI   8
      DATA ITS/100/,TOL/1.E-9/                                           PEI   9
C.... GET START VECTOR FROM DIAGONAL OF MASS MATRIX                      PEI  10
      DO 100 I = 1,NEQ                                                   PEI  11
      J = JDIAG(I)                                                       PEI  12
      IF(DFL) J = I                                                      PEI  13
100   Y(I) = B(J)                                                        PEI  14
      EIGP = 0.                                                          PEI  15
      CALL ACTCOL(A,Z,JDIAG,NEQ,.TRUE.,.FALSE.)                          PEI  16
      DO 200 I = 1,ITS                                                   PEI  17
      CALL PZERO(Z,NEQ)                                                  PEI  18
      CALL PROMUL(B,Y,Z,JDIAG,NEQ)                                       PEI  19
C.... RAYLEIGH QUOTIENT                                                  PEI  20
      EIG = AENGY/DOT(Y,Z,NEQ)                                           PEI  21
      IF(ABS(EIG-EIGP).LT.TOL*ABS(EIG)) GO TO 300                        PEI  22
      CALL NORM(Y,Z,NEQ)                                                 PEI  23
      EIGP = EIG                                                         PEI  24
C.... INVERSE ITERATION                                                  PEI  25
200   CALL ACTCOL(A,Y,JDIAG,NEQ,.FALSE.,.TRUE.)                          PEI  26
      WRITE(6,2001) ITS                                                  PEI  27
      RETURN                                                             PEI  28
300   WRITE(6,2000) O,HEAD,EIG,I                                         PEI  29
      CALL NORM(Z,Y,NEQ)                                                 PEI  30
      CALL PRTDIS(ID,X,Z,F,NDM,NDF)                                      PEI  31
      RETURN                                                             PEI  32
2000  FORMAT(A1,20A4//5X,14HEIGENVALUE =  ,G13.4/5X,14HITERATIONS =   ,  PEI  33
     1  I9/)                                                             PEI  34
2001  FORMAT(5X,57H**FATAL ERROR 09** NO CONVERGENCE IN EIGENVALUES, ITS PEI  35
     1 =  ,I5)                                                           PEI  36
      END                                                                PEI  37

      SUBROUTINE PFORM(UL,XL,TL,LD,P,S,IE,D,ID,X,IX,F,T,JDIAG,B,A,C,NDF, PFO   1
     1   NDM,NEN1,NST,ISW,U,UD,AFL,BFL,CFL,DFL)                          PFO   2
C                                                                        PFO   3
C.... COMPUTE ELEMENT ARRAYS AND ASSEMBLE GLOBAL ARRAYS                  PFO   4
C                                                                        PFO   5
      LOGICAL AFL,BFL,CFL,DFL                                            PFO   6
      COMMON /CDATA/ O,HEAD(20),NUMNP,NUMEL,NUMMAT,NEN,NEQ,IPR           PFO   7
      COMMON /ELDATA/ DM,N,MA,MCT,IEL,NEL                                PFO   8
      COMMON/PRLOD/ PROP                                                 PFO   9
      DIMENSION XL(NDM,1),LD(NDF,1),P(1),S(NST,1),IE(1),D(10,1),ID(NDF,1 PFO  10
     1),X(NDM,1),IX(NEN1,1),F(NDF,1),JDIAG(1),B(1),A(1),C(1),UL(NDF,1)   PFO  11
     2   ,TL(1),T(1),U(1),UD(NDF,1)                                      PFO  12
C.... LOOP ON ELEMENTS                                                   PFO  13
      IEL = 0                                                            PFO  14
      DO 110 N = 1,NUMEL                                                 PFO  15
C.... SET UP LOCAL ARRAYS                                                PFO  16
      DO 108 I = 1,NEN                                                   PFO  17
      II = IX(I,N)                                                       PFO  18
      IF(II.NE.0) GO TO 105                                              PFO  19
      TL(I) = 0.                                                         PFO  20
      DO 103 J = 1,NDM                                                   PFO  21
103   XL(J,I) = 0.                                                       PFO  22
      DO 104 J = 1,NDF                                                   PFO  23
      UL(J,I) = 0.                                                       PFO  24
      UL(J,I+NEN)=0.                                                     PFO  25
104   LD(J,I) = 0                                                        PFO  26
      GO TO 108                                                          PFO  27
105   IID = II*NDF - NDF                                                 PFO  28
      NEL = I                                                            PFO  29
      TL(I) = T(II)                                                      PFO  30
      DO 106 J = 1,NDM                                                   PFO  31
106   XL(J,I) = X(J,II)                                                  PFO  32
      DO 107 J = 1,NDF                                                   PFO  33
      K = IABS(ID(J,II))                                                 PFO  34
```

```
      UL(J,I) = F(J,II)*PROP                                          PFO  35
      UL(J,I+NEN)=UD(J,II)                                            PFO  36
      IF(K.GT.0) UL(J,I) = U(K)                                       PFO  37
      IF(DFL) K = IID + J                                             PFO  38
107   LD(J,I) = K                                                     PFO  39
108   CONTINUE                                                        PFO  40
C.... FORM ELEMENT ARRAY                                               PFO  41
      MA = IX(NEN1,N)                                                 PFO  42
      IF(IE(MA).NE.IEL) MCT = 0                                       PFO  43
      IEL = IE(MA)                                                    PFO  44
      CALL ELMLIB(D(1,MA),UL,XL,IX(1,N),TL,S,P,NDF,NDM,NST,ISW)       PFO  45
C.... ADD TO TOTAL ARRAY                                               PFO  46
      IF(AFL.OR.BFL.OR.CFL) CALL ADDSTF(A,B,C,S,P,JDIAG,LD,NST,NEL*NDF, PFO  47
     1   AFL,BFL,CFL)                                                 PFO  48
110   CONTINUE                                                        PFO  49
      RETURN                                                          PFO  50
      END                                                             PFO  51

      SUBROUTINE PLOAD(ID,F,B,NN,P)                                   PLO   1
C                                                                     PLO   2
C.... FOM LOAD VECTOR IN COMPACT FORM                                  PLO   3
C                                                                     PLO   4
      DIMENSION ID(1),F(1),B(1)                                       PLO   5
      DO 100 N = 1,NN                                                 PLO   6
      J = ID(N)                                                       PLO   7
100   IF(J.GT.0) B(J) = F(N)*P                                        PLO   8
      RETURN                                                          PLO   9
      END                                                             PLO  10

      SUBROUTINE PROFIL (JDIAG,ID,IX,NDF,NEN1,NAD)                    PRO   1
C                                                                     PRO   2
C.... COMPUTE PROFILE OF GLOBAL ARRAYS                                 PRO   3
C                                                                     PRO   4
      COMMON /CDATA/ O,HEAD(20),NUMNP,NUMEL,NUMMAT,NEN,NEQ,IPR        PRO   5
      DIMENSION JDIAG(1),ID(NDF,1),IX(NEN1,1)                         PRO   6
C.... SET UP THE EQUATION NUMBERS                                      PRO   7
      NEQ = 0                                                         PRO   8
      DO 50 N = 1,NUMNP                                               PRO   9
      DO 40 I = 1,NDF                                                 PRO  10
      J = ID(I,N)                                                     PRO  11
      IF(J) 30,20,30                                                  PRO  12
20    NEQ = NEQ + 1                                                   PRO  13
      ID(I,N) = NEQ                                                   PRO  14
      JDIAG(NEQ) = 0                                                  PRO  15
      GO TO 40                                                        PRO  16
30    ID(I,N) = 0                                                     PRO  17
40    CONTINUE                                                        PRO  18
50    CONTINUE                                                        PRO  19
C.... COMPUTE COLUMN HEIGHTS                                           PRO  20
      DO 500 N = 1,NUMEL                                              PRO  21
      DO 400 I = 1,NEN                                                PRO  22
      II = IX(I,N)                                                    PRO  23
      IF(II.EQ.0) GO TO 400                                           PRO  24
      DO 300 K = 1,NDF                                                PRO  25
      KK = ID(K,II)                                                   PRO  26
      IF(KK.EQ.0) GO TO 300                                           PRO  27
      DO 200 J = I,NEN                                                PRO  28
      JJ = IX(J,N)                                                    PRO  29
      IF(JJ.EQ.0) GO TO 200                                           PRO  30
      DO 100 L = 1,NDF                                                PRO  31
      LL = ID(L,JJ)                                                   PRO  32
      IF(LL.EQ.0) GO TO 100                                           PRO  33
      M = MAX0(KK,LL)                                                 PRO  34
      JDIAG(M) = MAX0(JDIAG(M),IABS(KK-LL))                           PRO  35
100   CONTINUE                                                        PRO  36
200   CONTINUE                                                        PRO  37
300   CONTINUE                                                        PRO  38
400   CONTINUE                                                        PRO  39
500   CONTINUE                                                        PRO  40
C.... COMPUTE DIAGONAL POINTERS FOR PROFILE                            PRO  41
      NAD = 1                                                         PRO  42
      JDIAG(1) = 1                                                    PRO  43
```

```
      IF(NEQ.EQ.1) RETURN                                              PRO  44
      DO 600 N = 2,NEQ                                                 PRO  45
600   JDIAG(N) = JDIAG(N) + JDIAG(N-1) + 1                             PRO  46
      NAD = JDIAG(NEQ)                                                 PRO  47
      RETURN                                                           PRO  48
      END                                                              PRO  49

      SUBROUTINE PROMUL(A,B,C,JDIAG,NEQ)                               PRO   1
      DIMENSION A(1),B(1),C(1),JDIAG(1)                                PRO   2
C                                                                      PRO   3
C.... ROUTINE TO FORM C = C + A*B WHERE A IS A SYMMETRIC SQUARE MATRIX  PRO   4
C.... STORED IN PROFILE FORM, B,C ARE VECTORS, AND JDIAG LOCATES THE   PRO   5
C.... DIAGONALS IN A.                                                  PRO   6
C                                                                      PRO   7
      JS = 1                                                           PRO   8
      DO 200 J = 1,NEQ                                                 PRO   9
      JD = JDIAG(J)                                                    PRO  10
      IF(JS.GT.JD) GO TO 200                                           PRO  11
      BJ = B(J)                                                        PRO  12
      AB = A(JD)*BJ                                                    PRO  13
      IF(JS.EQ.JD) GO TO 150                                           PRO  14
      JB = J - JD                                                      PRO  15
      JE = JD - 1                                                      PRO  16
      DO 100 JJ = JS,JE                                                PRO  17
      AB = AB + A(JJ)*B(JJ+JB)                                         PRO  18
100   C(JJ+JB) = C(JJ+JB) + A(JJ)*BJ                                   PRO  19
150   C(J) = C(J) + AB                                                 PRO  20
200   JS = JD + 1                                                      PRO  21
      RETURN                                                           PRO  22
      END                                                              PRO  23

      FUNCTION PROPLD(T,J)                                            ROPL   1
C                                                                     ROPL   2
C.... PROPORTIONAL LOAD TABLE (ONE LOAD CARD ONLY)                    ROPL   3
C                                                                     ROPL   4
      DIMENSION A(5)                                                  ROPL   5
      IF(J.GT.0) GO TO 200                                            ROPL   6
C.... COMPUTE VALUE AT TIME T                                         ROPL   7
      PROPLD = 0.0                                                    ROPL   8
      IF(T.LT.TMIN.OR.T.GT.TMAX) RETURN                               ROPL   9
      L = MAX0(L,1)                                                   ROPL  10
      PROPLD = A(1) + A(2)*T + A(3)*(SIN(A(4)*T+A(5)))**L             ROPL  11
      RETURN                                                          ROPL  12
C.... INPUT TABLE OF PROPORTIONAL LOADS                               ROPL  13
200   I = 1                                                           ROPL  14
      READ(5,1000)      K,L,TMIN,TMAX,A                               ROPL  15
      WRITE(6,2000) I,K,L,TMIN,TMAX,A                                 ROPL  16
      RETURN                                                          ROPL  17
1000  FORMAT(2I5,7F10.0)                                              ROPL  18
2000  FORMAT(5X,23HPROPORTIONAL LOAD TABLE//24H  NUMBER     TYPE     EXP.,  ROPL  19
     1 14H  MINIMUM TIME,15H   MAXIMUM TIME, 5X,2HA1,13X,2HA2,13X,2HA3,  ROPL  20
     2   13X,2HA4,13X,2HA5/(3I8,7G15.5))                              ROPL  21
      END                                                             ROPL  22

      SUBROUTINE PRTDIS(ID,X,B,F,NDM,NDF)                              PRT   1
C                                                                      PRT   2
C.... OUTPUT NODAL VALUES                                              PRT   3
C                                                                      PRT   4
      LOGICAL PCOMP                                                    PRT   5
      COMMON/PRLOD/ PROP                                               PRT   6
      COMMON /CDATA/ O,HEAD(20),NUMNP,NUMEL,NUMMAT,NEN,NEQ,IPR         PRT   7
      COMMON /LABEL/ PDIS(6),A(6),BC(2),DI(6),CD(3),TE(3),FD(3)        PRT   8
      COMMON /TDATA/ TIME,DT,C1,C2,C3,C4,C5                            PRT   9
      DIMENSION X(NDM,1),B(1),UL(6),ID(NDF,1),F(NDF,1)                 PRT  10
      DATA BL/4HBLAN/                                                  PRT  11
      DO 102 II = 1,NUMNP,50                                           PRT  12
      WRITE(6,2000) O,HEAD,TIME,(I,CD(1),CD(2),I=1,NDM),(I,DI(1)       PRT  13
     1    ,DI(2),I=1,NDF)                                              PRT  14
      JJ = MIN0(NUMNP,II+49)                                           PRT  15
      DO 102 N = II,JJ                                                 PRT  16
      IF(PCOMP(X(1,N),BL)) GO TO 101                                   PRT  17
      DO 100 I = 1,NDF                                                 PRT  18
```

```
      UL(I) = F(I,N)*PROP                                               PRT  19
      K = IABS(ID(I,N))                                                 PRT  20
100   IF(K.GT.0) UL(I) = B(K)                                           PRT  21
      WRITE(6,PDIS) N,(X(I,N),I=1,NDM),(UL(I),I=1,NDF)                  PRT  22
101   CONTINUE                                                          PRT  23
102   CONTINUE                                                          PRT  24
      RETURN                                                            PRT  25
2000  FORMAT(A1,20A4//5X,19HNODAL DISPLACEMENTS,5X,4HTIME,E13.5//       PRT  26
     1   6X,4HNODE,9(I7,A4,A2))                                         PRT  27
      END                                                               PRT  28

      SUBROUTINE PRTREA(R,NDF)                                          PRT   1
                                                                        PRT   2
C.... PRINT NODAL REACTIONS                                              PRT   3
C                                                                       PRT   4
      DIMENSION R(NDF,1),RSUM(6),ASUM(6)                                PRT   5
      COMMON /CDATA/ O,HEAD(20),NUMNP,NUMEL,NUMMAT,NEN,NEQ,IPR          PRT   6
      DO 50 K = 1,NDF                                                   PRT   7
      RSUM(K) = 0.                                                      PRT   8
50    ASUM(K) = 0.                                                      PRT   9
      DO 100 N = 1,NUMNP,50                                             PRT  10
      J = MIN0(NUMNP,N+49)                                              PRT  11
      WRITE(6,2000) O,HEAD,(K,K=1,NDF)                                  PRT  12
      DO 100 I = N,J                                                    PRT  13
      DO 75 K = 1,NDF                                                   PRT  14
      R(K,I) = -R(K,I)                                                  PRT  15
      RSUM(K) = RSUM(K) + R(K,I)                                        PRT  16
75    ASUM(K) = ASUM(K) + ABS(R(K,I))                                   PRT  17
100   WRITE(6,2001) I,(R(K,I),K=1,NDF)                                  PRT  18
C.... PRINT STATICS CHECK                                                PRT  19
      WRITE(6,2002) (RSUM(K),K=1,NDF)                                   PRT  20
      WRITE(6,2003) (ASUM(K),K=1,NDF)                                   PRT  21
      RETURN                                                            PRT  22
2000  FORMAT(A1,20A4//5X,15HNODAL REACTIONS//6X,4HNODE,                 PRT  23
     1  6(I9,4H DOF))                                                   PRT  24
2001  FORMAT(I10,6E13.4)                                                PRT  25
2002  FORMAT(/7X,3HSUM,6E13.4)                                          PRT  26
2003  FORMAT(/3X,7HABS SUM,6E13.4)                                      PRT  27
      END                                                               PRT  28

      SUBROUTINE PSETM(NA,NE,NJ,AFL)                                    PSE   1
C                                                                       PSE   2
C.... SET POINTER FOR ARRAYS                                             PSE   3
C                                                                       PSE   4
      LOGICAL AFL                                                       PSE   5
      NA = NE                                                           PSE   6
      NE = NE + NJ                                                      PSE   7
      AFL = .FALSE.                                                     PSE   8
      CALL SETMEM(NE)                                                   PSE   9
      RETURN                                                            PSE  10
      END                                                               PSE  11

      SUBROUTINE PZERO(V,NN)                                            PZE   1
C                                                                       PZE   2
C.... ZERO REAL ARRAY                                                    PZE   3
C                                                                       PZE   4
      DIMENSION V(NN)                                                   PZE   5
      DO 100 N = 1,NN                                                   PZE   6
100   V(N) = 0.0                                                        PZE   7
      RETURN                                                            PZE   8
      END                                                               PZE   9

      SUBROUTINE UACTCL(A,C,B,JDIAG,NEQ,AFAC,BACK)                      UAC   1
      LOGICAL AFAC,BACK                                                 UAC   2
      DIMENSION A(1),B(1),JDIAG(1),C(1)                                 UAC   3
C                                                                       UAC   4
C.... UNSYMMETRIC, ACTIVE COLUMN PROFILE EQUATION SOLVER                 UAC   5
C                                                                       UAC   6
C.... FACTOR A TO UT*D*U, REDUCE B TO Y                                  UAC   7
      JR = 0                                                            UAC   8
      DO 300 J = 1,NEQ                                                  UAC   9
      JD = JDIAG(J)                                                     UAC  10
```

```
      JH = JD - JR                                              UAC  11
      IF(JH.LE.1) GO TO 300                                     UAC  12
      IS = J + 1 - JH                                           UAC  13
      IE = J - 1                                                UAC  14
      IF(.NOT.AFAC) GO TO 250                                   UAC  15
      K = JR + 1                                                UAC  16
      ID = 0                                                    UAC  17
C.... REDUCE ALL EQUATIONS EXCEPT DIAGONAL                       UAC  18
      DO 200 I = IS,IE                                          UAC  19
      IR = ID                                                   UAC  20
      ID = JDIAG(I)                                             UAC  21
      IH = MINO(ID - IR - 1,I - IS)                             UAC  22
      IF(IH.EQ.0) GO TO 150                                     UAC  23
      A(K) = A(K) - DOT(A(K-IH),C(ID-IH),IH)                    UAC  24
      C(K) = C(K) - DOT(C(K-IH),A(ID-IH),IH)                    UAC  25
150   IF(A(ID).NE.0.0) C(K) = C(K)/A(ID)                        UAC  26
200   K = K + 1                                                 UAC  27
C.... REDUCE DIAGONAL TERM                                      UAC  28
      A(JD) = A(JD) - DOT(A(JR+1),C(JR+1),JH-1)                 UAC  29
C.... FORWARD REDUCE THE R.H.S.                                  UAC  30
250   IF(BACK) B(J) = B(J) - DOT(C(JR+1),B(IS),JH-1)            UAC  31
300   JR = JD                                                   UAC  32
      IF(.NOT.BACK) RETURN                                      UAC  33
C.... BACKSUBSTITUTION                                           UAC  34
      J = NEQ                                                   UAC  35
      JD = JDIAG(J)                                             UAC  36
500   IF(A(JD).NE.0.0) B(J) = B(J)/A(JD)                        UAC  37
      D = B(J)                                                  UAC  38
      J = J - 1                                                 UAC  39
      IF(J.LE.0) RETURN                                         UAC  40
      JR = JDIAG(J)                                             UAC  41
      IF(JD - JR.LE.1) GO TO 700                                UAC  42
      IS = J -JD + JR + 2                                       UAC  43
      K = JR - IS + 1                                           UAC  44
      DO 600 I = IS,J                                           UAC  45
600   B(I) = B(I) - A(I+K)*D                                    UAC  46
700   JD = JR                                                   UAC  47
      GO TO 500                                                 UAC  48
      END                                                       UAC  49
```

24.8.3 *Element modules.* In this section we include listings for element modules to compute the arrays necessary to carry out analyses in

(*a*) two-dimensional, isotropic, linear elasticity,
(*b*) two-dimensional linear heat transfer, and
(*c*) non-linear, steady-state, Navier–Stokes flows.

All of the elements employ the same shape function routines, SHAPE and SHAP2, which are a generalization on the routine given in section 24.5. The routines are capable of constructing two-dimensional shape functions for a 3-node triangle, a 4-node linear quadrilateral, an 8-node quadratic serendipity quadrilateral, a 9-node quadratic Lagrangian quadrilateral, or any combination between. The local node numbering for the 9-node Lagrangian shape functions is shown in Fig. 24.25. The 8-node serendipity shape functions occur if the ninth node number is omitted (blank or zero on a data card). The 4-node quadrilateral shape functions are computed if only the first four nodal connections are non-zero and the 3-node triangle if the first three nodal connections are non-zero. Finally, if a mid-side nodal connection is omitted the edge is linear;

thus, in this way the shape functions can be constructed to give a transition from linear elements to quadratic elements.

The control information is similar for all elements and this is discussed here. Each element is two-dimensional, hence, the spatial dimension of the problem, NDM, *must* be 2. The maximum number of nodes per element will depend on the type of elements to be used in each analysis, and may be three to nine.

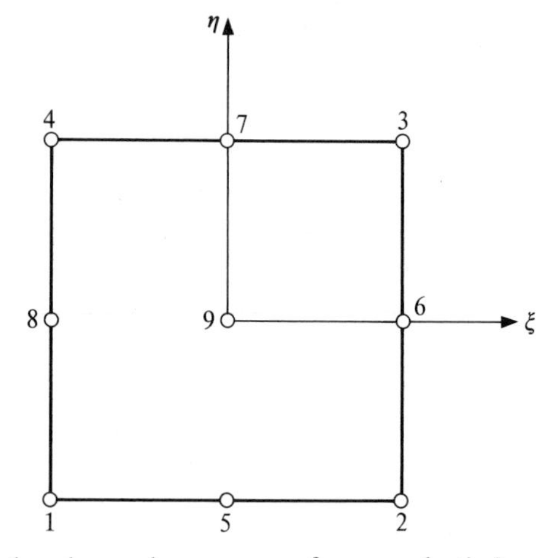

Fig. 24.25 Local node number sequence for a quadratic Lagrangian element

The elasticity and fluid elements have two degrees of freedom per node (i.e., u, v), thus, the user must specify the number of degrees of freedom per node, NDF, to be $\geqslant 2$. Both elements work if NDF > 2 but produce zero displacements for the added degrees of freedom and waste central memory. Additional degrees of freedom should be used only if the element is mixed with some other element which has more degrees of freedom, e.g., a typical frame element with three degrees of freedom per node where the plane element is, for example, then used as a shear panel.

The heat transfer element requires only one degree of freedom per node, although again more could be used if necessary.

The remaining control information depends on each particular problem and must be set accordingly.

Isotropic linear elastic element (ELMTØ1). The elasticity element is called ELMTØ1 (name of subroutine in element library) and thus the data in column 10 of each material number card must be 1 (one) when this element is requested. The second data card of each material set provides the material property and quadrature information and is prepared as follows.

Column		*Description*
1 to 10	E	Young's modulus
11 to 20	ν	Poisson's ratio
21 to 30	ρ	mass density
31 to 35	L	order of Gauss quadrature in each direction (1, 2, or 3)
36 to 40	K	order of Gauss quadrature for (stress) outputs
41 to 45	0	for plane strain, non-zero for plane stress

The total number of Gauss points will be equal to L squared; the co-ordinates will be stored in the arrays SG and TG and the weights in WG.

The elasticity element can be employed for either plane stress or plane strain analysis by specifying the data as shown above. The stiffness matrix is computed according to Fig. 24.8(*b*), assuming the material properties are constant in each element.

The element can be used in

(*a*) any steady-state analysis (linear or non-linear),
(*b*) for computation of the dominant eigenvalue, and
(*c*) for transient solution by explicit step-by-step integration.

A user could easily add macro instructions to carry out the implicit step-by-step integration schemes discussed in Chapter 21.

```
      SUBROUTINE ELMT01(D,UL,XL,IX,TL,S,P,NDF,NDM,NST,ISW)                ELM   1
C                                                                         ELM   2
C.... PLANE LINEAR ELASTIC ELEMENT ROUTINE                                ELM   3
C                                                                         ELM   4
      COMMON /CDATA/ O,HEAD(20),NUMNP,NUMEL,NUMMAT,NEN,NEQ,IPR            ELM   5
      COMMON /ELDATA/ DM,N,MA,MCT,IEL,NEL                                 ELM   6
      DIMENSION D(1),UL(NDF,1),XL(NDM,1),IX(1),TL(1),S(NST,1),P(1)        ELM   7
     1  ,SHP(3,9),SG(9),TG(9),WG(9),SIG(6),EPS(3),WD(2)                   ELM   8
      DATA WD/4HRESS,4HRAIN/                                              ELM   9
C.... GO TO CORRECT ARRAY PROCESSOR                                       ELM  10
      GO TO(1,2,3,4,5,4), ISW                                             ELM  11
C.... INPUT MATERIAL PROPERTIES                                           ELM  12
1     READ(5,1000) E,XNU,D(4),L,K,I                                       ELM  13
      IF(I.NE.0) I = 1                                                    ELM  14
      IF(I.EQ.0) I = 2                                                    ELM  15
      D(1) = E*(1.+(1-I)*XNU)/(1.+XNU)/(1.-I*XNU)                         ELM  16
      D(2) = XNU*D(1)/(1.+(1-I)*XNU)                                      ELM  17
      D(3) = E/2./(1.+XNU)                                                ELM  18
      L = MINO(3,MAXO(1,L))                                               ELM  19
      D(5) = L                                                            ELM  20
      K = MINO(3,MAXO(1,K))                                               ELM  21
      D(6) = K                                                            ELM  22
      LINT = 0                                                            ELM  23
      WRITE(6,2000) WD(I),E,XNU,D(4),L,K                                  ELM  24
      RETURN                                                              ELM  25
2     RETURN                                                              ELM  26
3     L = D(5)                                                            ELM  27
      IF(L*L.NE.LINT) CALL PGAUSS(L,LINT,SG,TG,WG)                        ELM  28
C.... FAST STIFFNESS COMPUTATION,  COMPUTE INTEGRALS OF SHAPE FUNCTIONS   ELM  29
      DO 320 L = 1,LINT                                                   ELM  30
```

```
      CALL SHAPE(SG(L),TG(L),XL,SHP,XSJ,NDM,NEL,IX,.FALSE.)             ELM  31
      XSJ = XSJ*WG(L)                                                   ELM  32
C.... LOOP OVER ROWS                                                    ELM  33
      J1 = 1                                                            ELM  34
      DO 320 J = 1,NEL                                                  ELM  35
      W11 = SHP(1,J)*XSJ                                                ELM  36
      W12 = SHP(2,J)*XSJ                                                ELM  37
C.... LOOP OVER COLUMNS (SYMMETRY NOTED)                                 ELM  38
      K1 = J1                                                           ELM  39
      DO 310 K = J,NEL                                                  ELM  40
      S(J1  ,K1  ) = S(J1  ,K1  ) + W11*SHP(1,K)                        ELM  41
      S(J1  ,K1+1) = S(J1  ,K1+1) + W11*SHP(2,K)                        ELM  42
      S(J1+1,K1  ) = S(J1+1,K1  ) + W12*SHP(1,K)                        ELM  43
      S(J1+1,K1+1) = S(J1+1,K1+1) + W12*SHP(2,K)                        ELM  44
310   K1 = K1 + NDF                                                     ELM  45
320   J1 = J1 + NDF                                                     ELM  46
C.... ASSEMBLE THE STIFFNESS MATRIX FROM INTEGRALS AND MATERIAL PROPS.   ELM  47
      NSL = NEL*NDF                                                     ELM  48
      DO 330 J = 1,NSL,NDF                                              ELM  49
      DO 330 K = J,NSL,NDF                                              ELM  50
      W11 = S(J,K)                                                      ELM  51
      W12 = S(J,K+1)                                                    ELM  52
      W21 = S(J+1,K)                                                    ELM  53
      W22 = S(J+1,K+1)                                                  ELM  54
      S(J  ,K  ) = D(1)*W11 + D(3)*W22                                  ELM  55
      S(J  ,K+1) = D(2)*W12 + D(3)*W21                                  ELM  56
      S(J+1,K  ) = D(2)*W21 + D(3)*W12                                  ELM  57
      S(J+1,K+1) = D(1)*W22 + D(3)*W11                                  ELM  58
C.... FORM LOWER PART BY SYMMETRY                                        ELM  59
      S(K,J) = S(J,K)                                                   ELM  60
      S(K,J+1) = S(J+1,K)                                               ELM  61
      S(K+1,J) = S(J,K+1)                                               ELM  62
330   S(K+1,J+1) = S(J+1,K+1)                                           ELM  63
      RETURN                                                            ELM  64
4     L = D(5)                                                          ELM  65
      IF(ISW.EQ.4) L = D(6)                                             ELM  66
      IF(L*L.NE.LINT) CALL PGAUSS(L,LINT,SG,TG,WG)                      ELM  67
C.... COMPUTE ELEMENT STRESSES, STRAINS, AND FORCES                      ELM  68
      DO 600 L = 1,LINT                                                 ELM  69
C.... COMPUTE ELEMENT SHAPE FUNCTIONS                                    ELM  70
      CALL SHAPE(SG(L),TG(L),XL,SHP,XSJ,NDM,NEL,IX,.FALSE.)             ELM  71
C.... COMPUTE STRAINS AND COORDINATES                                    ELM  72
      DO 410 I = 1,3                                                    ELM  73
410   EPS(I) = 0.0                                                      ELM  74
      XX = 0.0                                                          ELM  75
      YY = 0.0                                                          ELM  76
      DO 420 J = 1,NEL                                                  ELM  77
      XX = XX + SHP(3,J)*XL(1,J)                                        ELM  78
      YY = YY + SHP(3,J)*XL(2,J)                                        ELM  79
      EPS(1) = EPS(1) + SHP(1,J)*UL(1,J)                                ELM  80
      EPS(3) = EPS(3) + SHP(2,J)*UL(2,J)                                ELM  81
420   EPS(2) = EPS(2) + SHP(1,J)*UL(2,J) + SHP(2,J)*UL(1,J)             ELM  82
C.... COMPUTE STRESSES                                                   ELM  83
      SIG(1) = D(1)*EPS(1) + D(2)*EPS(3)                                ELM  84
      SIG(3) = D(2)*EPS(1) + D(1)*EPS(3)                                ELM  85
      SIG(2) = D(3)*EPS(2)                                              ELM  86
      IF(ISW.EQ.6) GO TO 620                                            ELM  87
      CALL PSTRES(SIG,SIG(4),SIG(5),SIG(6))                             ELM  88
C.... OUTPUT STRESSES AND STRAINS                                        ELM  89
      MCT = MCT - 2                                                     ELM  90
      IF(MCT.GT.0) GO TO 430                                            ELM  91
      WRITE(6,2001) O,HEAD                                              ELM  92
      MCT = 50                                                          ELM  93
430   WRITE(6,2002) N,MA,XX,YY,SIG,EPS                                  ELM  94
      GO TO 600                                                         ELM  95
C.... COMPUTE INTERNAL FORCES                                            ELM  96
620   DV = XSJ*WG(L)                                                    ELM  97
      J1 = 1                                                            ELM  98
      DO 610 J = 1,NEL                                                  ELM  99
      P(J1) = P(J1) - (SHP(1,J)*SIG(1) + SHP(2,J)*SIG(2))*DV            ELM 100
      P(J1+1) = P(J1+1) - (SHP(1,J)*SIG(2) + SHP(2,J)*SIG(3))*DV        ELM 101
610   J1 = J1 + NDF                                                     ELM 102
```

```
600    CONTINUE                                                         ELM 103
       RETURN                                                           ELM 104
C....  COMPUTE CONSISTENT MASS MATRIX                                   ELM 105
5      L = D(5)                                                         ELM 106
       IF(L*L.NE.LINT) CALL PGAUSS(L,LINT,SG,TG,WG)                     ELM 107
       DO 500 L = 1,LINT                                                ELM 108
C....  COMPUTE SHAPE FUNCTIONS                                           ELM 109
       CALL SHAPE(SG(L),TG(L),XL,SHP,XSJ,NDM,NEL,IX,.FALSE.)            ELM 110
       DV = WG(L)*XSJ*D(4)                                              ELM 111
C....  FOR EACH NODE J COMPUTE DB = RHO*SHAPE*DV                         ELM 112
       J1 = 1                                                           ELM 113
       DO 500 J = 1,NEL                                                 ELM 114
       W11 = SHP(3,J)*DV                                                ELM 115
C....  FOR EACH NODE K COMPUTE MASS MATRIX (UPPER TRIANGULAR PART)       ELM 116
       K1 = J1                                                          ELM 117
       DO 510 K = J,NEL                                                 ELM 118
       S(J1,K1) = S(J1,K1) + SHP(3,K)*W11                               ELM 119
510    K1 = K1 + NDF                                                    ELM 120
500    J1 = J1 + NDF                                                    ELM 121
C....  COMPUTE MISSING PARTS AND LOWER PART BY SYMMETRIES                ELM 122
       NSL = NEL*NDF                                                    ELM 123
       DO 520 J = 1,NSL,NDF                                             ELM 124
       DO 520 K = J,NSL,NDF                                             ELM 125
       S(J+1,K+1) = S(J,K)                                              ELM 126
       S(K,J) = S(J,K)                                                  ELM 127
520    S(K+1,J+1) = S(J,K)                                              ELM 128
C....  FORMATS FOR INPUT-OUTPUT                                          ELM 129
       RETURN                                                           ELM 130
1000   FORMAT(3F10.0,3I5)                                               ELM 131
2000   FORMAT(/5X,8HPLANE ST,A4,23H LINEAR ELASTIC ELEMENT//            ELM 132
     1 10X,7HMODULUS,E18.5/10X,13HPOISSON RATIO,F8.5/10X,               ELM 133
     2 7HDENSITY,E18.5/10X,13HGAUSS PTS/DIR,I3/10X,10HSTRESS PTS,I6)    ELM 134
2001   FORMAT(A1,20A4//5X,16HELEMENT STRESSES//20H   ELEMENT  MATERIAL  ELM 135
     1   ,6X,7H1-COORD,6X,7H2-COORD,4X,9H11-STRESS,4X,9H12-STRESS,4X,   ELM 136
     2   9H22-STRESS,5X,8H1-STRESS,5X,8H2-STRESS,3X,5HANGLE/50X,        ELM 137
     3  9H11-STRAIN,4X,9H12-STRAIN,4X,9H22-STRAIN)                      ELM 138
2002   FORMAT(2I10,2F13.4,5E13.4,F8.2/46X,3E13.4)                       ELM 139
       END                                                              ELM 140
       SUBROUTINE PGAUSS(L,LINT,R,Z,W)                                  PGA   1
C                                                                       PGA   2
C....  GAUSS POINTS AND WEIGHTS FOR TWO DIMENSIONS                       PGA   3
C                                                                       PGA   4
       DIMENSION LR(9),LZ(9),LW(9),R(1),Z(1),W(1)                       PGA   5
       DATA LR/-1,1,1,-1,0,1,0,-1,0/,LZ/-1,-1,1,1,-1,0,1,0,0/           PGA   6
       DATA LW/4*25,4*40,64/                                            PGA   7
       LINT = L*L                                                       PGA   8
       GO TO (1,2,3),L                                                  PGA   9
C....  1X1 INTEGRATION                                                   PGA  10
1      R(1) = 0.                                                        PGA  11
       Z(1) = 0.                                                        PGA  12
       W(1) = 4.                                                        PGA  13
       RETURN                                                           PGA  14
C....  2X2 INTEGRATION                                                   PGA  15
2      G = 1./SQRT(3.)                                                  PGA  16
       DO 21 I = 1,4                                                    PGA  17
       R(I) = G*LR(I)                                                   PGA  18
       Z(I) = G*LZ(I)                                                   PGA  19
21     W(I) = 1.                                                        PGA  20
       RETURN                                                           PGA  21
C....  3X3 INTEGRATION                                                   PGA  22
3      G = SQRT(0.6)                                                    PGA  23
       H = 1./81.                                                       PGA  24
       DO 31 I = 1,9                                                    PGA  25
       R(I) = G*LR(I)                                                   PGA  26
       Z(I) = G*LZ(I)                                                   PGA  27
31     W(I) = H*LW(I)                                                   PGA  28
       RETURN                                                           PGA  29
       END                                                              PGA  30
```

```
      SUBROUTINE PSTRES(SIG,P1,P2,P3)                                   PST   1
C                                                                       PST   2
C.... COMPUTE PRINCIPAL STRESSES (2 DIMENSIONS)                         PST   3
C                                                                       PST   4
      DIMENSION SIG(3)                                                  PST   5
C.... STRESSES MUST BE STORED IN ARRAY SIG(3) IN THE ORDER              PST   6
C           TAU-XX,TAU-XY,TAU-YY                                        PST   7
      XI1 = (SIG(1) + SIG(3))/2.                                        PST   8
      XI2 = (SIG(1) - SIG(3))/2.                                        PST   9
      RHO = SQRT(XI2*XI2 + SIG(2)*SIG(2))                               PST  10
      P1 = XI1 + RHO                                                    PST  11
      P2 = XI1 - RHO                                                    PST  12
      P3 = 45.0                                                         PST  13
      IF(XI2.NE.0.0) P3 = 22.5*ATAN2(SIG(2),XI2)/ATAN(1.0)              PST  14
      RETURN                                                            PST  15
      END                                                               PST  16

      SUBROUTINE SHAPE(SS,TT,X,SHP,XSJ,NDM,NEL,IX,FLG)                  SHA   1
C                                                                       SHA   2


C.... SHAPE FUNCTION ROUTINE FOR TWO DIMENSIONAL ELEMENTS                SHA   3
C                                                                       SHA   4

      LOGICAL FLG                                                       SHA   5
      DIMENSION SHP(3,1),X(NDM,1),S(4),T(4),XS(2,2),SX(2,2),IX(1)       SHA   6
      DATA S/-0.5,0.5,0.5,-0.5/,T/-0.5,-0.5,0.5,0.5/                    SHA   7
C.... FORM 4-NODE QUADRILATERAL SHAPE FUNCTIONS                         SHA   8
      DO 100 I = 1,4                                                    SHA   9
      SHP(3,I) = (0.5+S(I)*SS)*(0.5+T(I)*TT)                            SHA  10
      SHP(1,I) = S(I)*(0.5+T(I)*TT)                                     SHA  11
100   SHP(2,I) = T(I)*(0.5+S(I)*SS)                                     SHA  12
      IF(NEL.GE.4) GO TO 120                                            SHA  13
C.... FORM TRIANGLE BY ADDING THIRD AND FOURTH TOGETHER                 SHA  14
      DO 110 I = 1,3                                                    SHA  15
110   SHP(I,3) = SHP(I,3)+SHP(I,4)                                      SHA  16
C.... ADD QUADRATIC TERMS IF NECESSARY                                  SHA  17
120   IF(NEL.GT.4) CALL SHAP2(SS,TT,SHP,IX,NEL)                         SHA  18
C.... CONSTRUCT JACOBIAN AND ITS INVERSE                                SHA  19
      DO 130 I = 1,NDM                                                  SHA  20
      DO 130 J = 1,2                                                    SHA  21
      XS(I,J) = 0.0                                                     SHA  22
      DO 130 K = 1,NEL                                                  SHA  23
130   XS(I,J) = XS(I,J) + X(I,K)*SHP(J,K)                               SHA  24
      XSJ = XS(1,1)*XS(2,2)-XS(1,2)*XS(2,1)                             SHA  25
      IF(FLG) RETURN                                                    SHA  26
      SX(1,1) = XS(2,2)/XSJ                                             SHA  27
      SX(2,2) = XS(1,1)/XSJ                                             SHA  28
      SX(1,2) =-XS(1,2)/XSJ                                             SHA  29
      SX(2,1) =-XS(2,1)/XSJ                                             SHA  30
C.... FORM GLOBAL DERIVATIVES                                           SHA  31
      DO 140 I = 1,NEL                                                  SHA  32
      TP       = SHP(1,I)*SX(1,1)+SHP(2,I)*SX(2,1)                      SHA  33
      SHP(2,I) = SHP(1,I)*SX(1,2)+SHP(2,I)*SX(2,2)                      SHA  34
140   SHP(1,I) = TP                                                     SHA  35
      RETURN                                                            SHA  36
      END                                                               SHA  37
      SUBROUTINE SHAP2(S,T,SHP,IX,NEL)                                  SHA   1
C                                                                       SHA   2
C.... ADD QUADRATIC FUNCTIONS AS NECESSARY                              SHA   3
C                                                                       SHA   4
      DIMENSION IX(1),SHP(3,1)                                          SHA   5
      S2 = (1.-S*S)/2.                                                  SHA   6
      T2 = (1.-T*T)/2.                                                  SHA   7
      DO 100 I = 5,9                                                    SHA   8
      DO 100 J = 1,3                                                    SHA   9
100   SHP(J,I) = 0.0                                                    SHA  10
C.... MIDSIDE NODES (SERENDIPITY)                                       SHA  11
      IF(IX(5).EQ.0) GO TO 101                                          SHA  12
      SHP(1,5) = -S*(1.-T)                                              SHA  13
      SHP(2,5) = -S2                                                    SHA  14
      SHP(3,5) = S2*(1.-T)                                              SHA  15
101   IF(NEL.LT.6) GO TO 107                                            SHA  16
      IF(IX(6).EQ.0) GO TO 102                                          SHA  17
```

```
      SHP(1,6) = T2                                                SHA  18
      SHP(2,6) = -T*(1.+S)                                         SHA  19
      SHP(3,6) = T2*(1.+S)                                         SHA  20
102   IF(NEL.LT.7) GO TO 107                                       SHA  21
      IF(IX(7).EQ.0) GO TO 103                                     SHA  22
      SHP(1,7) = -S*(1.+T)                                         SHA  23
      SHP(2,7) = S2                                                SHA  24
      SHP(3,7) = S2*(1.+T)                                         SHA  25
103   IF(NEL.LT.8) GO TO 107                                       SHA  26
      IF(IX(8).EQ.0) GO TO 104                                     SHA  27
      SHP(1,8) = -T2                                               SHA  28
      SHP(2,8) = -T*(1.-S)                                         SHA  29
      SHP(3,8) = T2*(1.-S)                                         SHA  30
C.... INTERIOR NODE (LAGRANGIAN)                                    SHA  31
104   IF(NEL.LT.9) GO TO 107                                       SHA  32
      IF(IX(9).EQ.0) GO TO 107                                     SHA  33
      SHP(1,9) = -4.0*S*T2                                         SHA  34
      SHP(2,9) = -4.0*T*S2                                         SHA  35
      SHP(3,9) = 4.*S2*T2                                          SHA  36
C.... CORRECT EDGE NODES FOR INTERIOR NODE (LAGRANGIAN)             SHA  37
      DO 106 J= 1,3                                                SHA  38
      DO 105 I = 1,4                                               SHA  39
105   SHP(J,I) = SHP(J,I) - 0.25*SHP(J,9)                          SHA  40
      DO 106 I = 5,8                                               SHA  41
106   IF(IX(I).NE.0) SHP(J,I) = SHP(J,I) - .5*SHP(J,9)             SHA  42
C.... CORRECT CORNER NODES FOR PRESENSE OF MIDSIDE NODES            SHA  43
107   K = 8                                                        SHA  44
      DO 109 I = 1,4                                               SHA  45
      L = I + 4                                                    SHA  46
      DO 108 J = 1,3                                               SHA  47
108   SHP(J,I) = SHP(J,I) - 0.5*(SHP(J,K)+SHP(J,L))                SHA  48
109   K = L                                                        SHA  49
      RETURN                                                       SHA  50
      END                                                          SHA  51
```

Linear heat transfer element (*ELMT02*). The heat transfer element is called ELMT02 and thus the data in column 10 of each material number card must be 2 when this element is requested. The second data card of each material set gives the material property and geometry type and is prepared as follows:

Column	*Description*
1 to 10	k conductivity
11 to 20	c heat capacity
21 to 30	ρ mass density
31 to 35	KAT, $=2$ for axisymmetry
	$\neq 2$ for plane geometry

A consistent set of units must be used to specify the data.

```
      SUBROUTINE ELMT02(D,UL,XL,IX,TL,S,P,NDF,NDM,NST,ISW)               ELM   1
C                                                                        ELM   2
C.... TWO DIMENSIONAL HEAT TRANSFER ELEMENT                              ELM   3
C                                                                        ELM   4
      COMMON /CDATA/ O,HEAD(20),NUMNP,NUMEL,NUMMAT,NEN,NEQ,IPR           ELM   5
      COMMON/ELDATA/ DM,N,MA,MCT,IEL,NEL                                 ELM   6
      DIMENSION D(1),UL(1) ,XL(NDM,1),IX(1),TL(1),S(NST,1),P(1),         ELM   7
     1SHP(3,9),SG(4),TG(4) ,WLAB(2)                                      ELM   8
      DATA  SG/1.,1.,-1.,-1./,TG/-1.,1.,1.,-1./                          ELM   9
      DATA WLAB/6H PLANE,6HAXISYM/                                       ELM  10
C.... TRANSFER TO CORRECT PROCESSOR                                      ELM  11
      GO TO (1,2,3,2,5,3),ISW                                            ELM  12
C.... INPUT MATERIAL PROPERTIES                                          ELM  13
1     READ(5,1000) D(1),D(2),D(3),KAT                                    ELM  14
      WRITE(6,2000) D(1),D(2),D(3)                                       ELM  15
      G=1./SQRT(3.)                                                      ELM  16
      D(2)=D(2)*D(3)                                                     ELM  17
      IF(KAT.NE.2) KAT=1                                                 ELM  18
      WRITE(6,2001) WLAB(KAT)                                            ELM  19
      RETURN                                                             ELM  20
C.... INSERT CHECK OF MESH IF DESIRED                                    ELM  21
2     RETURN                                                             ELM  22
C.... COMPUTE CONDUCTIVITY (STIFFNESS) MATRIX                            ELM  23
3     DO 102 L=1,4                                                       ELM  24
      CALL SHAPE(SG(L)*G,TG(L)*G,XL,SHP,XSJ,NDM,NEL,IX,.FALSE.)          ELM  25
      IF(KAT.NE.2) GO TO 101                                             ELM  26
      RR=0.                                                              ELM  27
      DO 100 I=1,NEL                                                     ELM  28
100   RR=RR+SHP(3,I)*XL(1,I)                                             ELM  29
      XSJ=XSJ*RR                                                         ELM  30
101   DO 102 J=1,NEL                                                     ELM  31
      SHJ=SHP(3,J)*XSJ                                                   ELM  32
      A1=D(1)*SHP(1,J)*XSJ                                               ELM  33
      A2=D(1)*SHP(2,J)*XSJ                                               ELM  34
      DO 102 I=1,NEL                                                     ELM  35
102   S(I,J)=S(I,J)+A1*SHP(1,I)+A2*SHP(2,I)                              ELM  36
      DO 106 I = 1,NEL                                                   ELM  37
      DO 106 J = 1,NEL                                                   ELM  38
106   P(I) = P(I) - S(I,J)*UL(J)                                         ELM  39
      RETURN                                                             ELM  40
C.... COMPUTE HEAT CAPACITY (MASS) MATRIX                                ELM  41
5     DO 105 L=1,4                                                       ELM  42
      CALL SHAPE(SG(L)*G,TG(L)*G,X,SHP,XSJ,NDM,NEL,IX,.FALSE.)           ELM  43
      IF(KAT.NE.2) GO TO 104                                             ELM  44
      RR=0.                                                              ELM  45
      DO 103 I=1,NEL                                                     ELM  46
103   RR=RR+SHP(3,I)*XL(1,I)                                             ELM  47
      XSJ=XSJ*RR                                                         ELM  48
104   DO 105 J=1,NEL                                                     ELM  49
      SHJ=D(2)*SHP(3,J)*XSJ                                              ELM  50
      P(J) = P(J) + SHJ                                                  ELM  51
      DO 105 I=1,NEL                                                     ELM  52
105   S(I,J)=S(I,J)+SHJ*SHP(3,I)                                         ELM  53
      RETURN                                                             ELM  54
C.... FORMATS                                                            ELM  55
1000  FORMAT(3F10.0,I5)                                                  ELM  56
2000  FORMAT(5X,30HLINEAR HEAT CONDUCTION ELEMENT   // 5X,               ELM  57
     1  12HCONDUCTIVITY ,E12.5, 5X,15HSPECIFIC HEAT   ,E12.5, 5X,        ELM  58
     2  12H  DENSITY        ,E12.5  )                                    ELM  59
2001  FORMAT(10X,A6,9H ANALYSIS )                                        ELM  60
      END                                                                ELM  61
```

Fluid element for steady-state Navier–Stokes flow (ELMTØ3). The fluid element is called ELMTØ3 and thus data in column 10 of each material number card must be 3 when this element is requested. The material properties and quadrature information is prepared as follows:

Column	*Description*	
1 to 10	μ	viscosity
11 to 20	λ	penalty coefficient
21 to 30	ρ	mass density
31 to 35	L	order of Gaussian quadrature in each direction (1, 2, or 3)

The nodal degrees of freedom are the velocities in the co-ordinate directions (x, y), thus, NDF should be set to 2. It is recommended that only 8- or 9-node elements be used with 2×2 Gaussian quadrature (reduced integration is discussed in Chapter 11, section 11.6).

For non-zero density this element produces an unsymmetric tangent stiffness matrix for the steady-state, non-linear, Navier–Stokes equations; consequently, the macro program should use UTAN and loops to obtain convergence. When the density is zero the problem is linear (Stokes flow) and the stiffness matrix is symmetric; hence, in this case, considerable efficiency results by using the TANG macro command instead of UTAN.

```
      SUBROUTINE ELMT03(D,UL,XL,IX,TL,S,P,NDF,NDM,NST,ISW)             ELM   1
C                                                                      ELM   2
C.... TWO DIMENSIONAL FLUID ELEMENT FOR NAVIER-STOKES EQUATIONS         ELM   3
C                                                                      ELM   4
      DIMENSION XL(NDM,1),UL(NDF,1),P(NDF,1),S(NST,1),D(1),V(2),DV(2,2)  ELM   5
     1,SHP(3,9),IX(1)                                                  ELM   6
      GO TO (1,2,3,4,5,3), ISW                                         ELM   7
C.... INPUT/OUTPUT FLUID PROPERTIES                                    ELM   8
1     READ(5,1000) D(1),D(2),D(3),L                                    ELM   9
      WRITE(6,2000) D(1),D(2),D(3),L                                   ELM  10
      D(4) = L                                                         ELM  11
      LINT = 0                                                         ELM  12
      RETURN                                                           ELM  13
2     RETURN                                                           ELM  14
C.... COMPUTE UNSYMMETRIC TANGENT STIFFNESS OR OUT OF BALANCE FORCES   ELM  15
3     L = D(4)                                                         ELM  16
      IF(L*L.NE.LINT) CALL PGAUSS(L,LINT,SG,TG,WG)                     ELM  17
      DO 65 L = 1,LINT                                                 ELM  18
      CALL SHAPE(SG(L),TG(L),XL,SHP,XSJ,NDM,NEL,IX,.FALSE.)            ELM  19
      XLAM = D(2)*XSJ*WG(L)                                            ELM  20
      XMU  = D(1)*XSJ*WG(L)                                            ELM  21
      XRHO = D(3)*XSJ*WG(L)                                            ELM  22
C.... COMPUTE VELOCITIES AND GRADIENTS                                 ELM  23
      DO 32 I = 1,2                                                    ELM  24
      V(I) = 0.                                                        ELM  25
      DO 31 K = 1,NEL                                                  ELM  26
31    V(I) = V(I) + SHP(3,K)*UL(I,K)                                   ELM  27
      DO 32 J = 1,2                                                    ELM  28
      DV(I,J) = 0.0                                                    ELM  29
      DO 32 K = 1,NEL                                                  ELM  30
32    DV(I,J) + DV(I,J) + SHP(J,K)*UL(I,K)                             ELM  31
      IF(ISW.EQ.6) GO TO 60                                            ELM  32
```

```
C.... COMPUTE TANGENT, LOOP OVER COLUMNS OF S                          ELM  33
      K1 = 1                                                           ELM  34
      DO 34 K = 1,NEL                                                  ELM  35
      A1 = XMU*SHP(1,K)                                                ELM  36
      A2 = XMU*SHP(2,K)                                                ELM  37
      A3 = XRHO*(DV(1,1)*SHP(3,K)+V(1)*SHP(1,K)+V(2)*SHP(2,K))         ELM  38
      A4 = XRHO*(DV(2,2)*SHP(3,K)+V(1)*SHP(1,K)+V(2)*SHP(2,K))         ELM  39
      A5 = XRHO*DV(1,2)*SHP(3,K)                                       ELM  40
      A6 = XRHO*DV(2,1)*SHP(3,K)                                       ELM  41
      B1 = XLAM*SHP(1,K)                                               ELM  42
      B2 = XLAM*SHP(2,K)                                               ELM  43
C.... LOOP OVER ROWS OF S                                              ELM  44
      J1 = 1                                                           ELM  45
      DO 33 J = 1,NEL                                                  ELM  46
      S(J1  ,K1  ) + S(J1  ,K1  ) + SHP(1,J)*(A1+A1+B1)+SHP(2,J)*A2    ELM  47
      S(J1  ,K1+1) + S(J1  ,K1+1) + SHP(1,J)*B2+SHP(2,J)*A1            ELM  48
      S(J1+1,K1  ) + S(J1+1,K1  ) + SHP(1,J)*A2+SHP(2,J)*B1            ELM  49
      S(J1+1,K1+1) + S(J1+1,K1+1) + SHP(1,J)*A1+SHP(2,J)*(A2+A2+B2)    ELM  50
33    J1 = J1 + NDF                                                    ELM  51
34    K1 = K1 + NDF                                                    ELM  52
      GO TO 65                                                         ELM  53
C.... COMPUTE DIVERGENCE TERM                                           ELM  54
60    XDIV = (DV(1,1)+DV(2,2))*XLAM                                    ELM  55
C.... COMPUTE INTERNAL FORCES                                           ELM  56
      DO 64 K = 1,NEL                                                  ELM  57
      DO 64 J = 1,2                                                    ELM  58
      SUM = XDIV*SHP(J,K)                                              ELM  59
      DO 63 I = 1,2                                                    ELM  60
63    SUM = SUM + XMU*(DV(J,I)+DV(I,J))*SHP(I,K) +                     ELM  61
     1  XRHO*V(I)*DV(J,I)*SHP(3,K)                                     ELM  62
64    P(J,K) = P(J,K) - SUM                                            ELM  63
65    CONTINUE                                                         ELM  64
      RETURN                                                           ELM  65
C.... COMPUTE STRESSES AND VELOCITY GRADIENTS                           ELM  66
4     RETURN                                                           ELM  67
C.... COMPUTE MASS MATRIX                                               ELM  68
5      RETURN                                                          ELM  69
C.... FORMATS                                                           ELM  70
1000  FORMAT(3F10.0,I5)                                                ELM  71
2000  FORMAT(5X,29HTWO DIMENSIONAL FLUID ELEMENT//10X,12HVISCOSITY  =, ELM  72
     1 E12.5/10X,12HCONSTRAINT =,E12.5/10X,12HDENSITY    =,E12.5/      ELM  73
     2  10X,12HGAUSS PT/DIR,I5/)                                       ELM  74
      END                                                              ELM  75
```

References

1. O. C. ZIENKIEWICZ and D. V. PHILLIPS, 'An automatic mesh generation scheme for plane and curved surfaces by isoparametric coordinates', *Int. J. Num. Meth. Eng.*, **3**, 519–28, 1971.
2. O. C. ZIENKIEWICZ, *The Finite Element Method in Engineering Science*, McGraw-Hill, 1971.
3. W. PILKEY, K. SACZALSKI and H. SCHAEFFER (eds.), *Structural Mechanics Computer Programs*, Univ. Press of Virginia, Charlottesville, 1974.
4. B. M. IRONS, 'A technique for degenerating brick type isoparametric elements using hierarchical midside nodes', *Int. J. Num. Meth. Eng.*, **8**, 209–11, 1973.
5. N. E. GIBBS, W. G. POOLE, Jr., and P. K. STOCKMEYER, 'An algorithm for reducing the bandwidth and profile of a sparse matrix', *SIAM J. Num. Anal.*, **13**, 236–50, 1976.
6. W.-H. LIU and A. H. SHERMAN, 'Comparative analysis of the Cuthill–McKee and the reversed Cuthill–McKee ordering algorithms for sparse matrices', *SIAM J. Num. Anal.*, **13**, 198–213, 1976.

7. K. J. BATHE and E. L. WILSON, *Numerical Methods in Finite Element Analyses*, Prentice-Hall, 1976.
8. J. H. WILKINSON and C. REINSCH, *Linear Algebra. Handbook for Automatic Computation, II*, Springer-Verlag, 1971.
9. A. K. GUPTA and B. MOHRAZ, 'A method of computing numerically integrated stiffness matrices', *Int. J. Num. Meth. Eng.*, **5**, 83–9, 1972.
10. E. L. WILSON, 'SAP—A general structural analysis program for linear systems', *Nucl. Engr. Des.*, **25**, 257–74, 1973.
11. E. HINTON and J. S. CAMPBELL, 'Local and global smoothing of discontinuous element functions using a least square method', *Int. J. Num. Meth. Eng.*, **8**, 461–80, 1974.
12. E. HINTON, F. C. SCOTT and R. E. RICKETTS, 'Local least square stress smoothing for parabolic isoparametric elements', *Int. J. Num. Meth. Eng.*, **9**, 235–56, 1975.
13. A. RALSTON, *A First Course in Numerical Analyses*, McGraw-Hill, 1965.
14. L. FOX, *An Introduction to Numerical Linear Algebra*, Oxford Univ. Press, 1965.
15. C. MEYER, 'Solution of equations; state-of-the-art', *J. Struct. Div. A.S.C.E.*, **99** (7), 1507–26, 1973.
16. C. MEYER, 'Special problems related to linear equation solvers', *J. Struct. Div. A.S.C.E.*, **101** (4), 869–90, 1975.
17. A. JENNINGS, 'A compact storage scheme for the solution of symmetric simultaneous equations', *Comp. J.*, **9**, 281–5, 1966.
18. D. P. MONDKAR and G. H. POWELL, 'Towards optimal in-core equation solving', *Comp. Struct.*, **4**, 531–48, 1974.
19. C. A. FELIPPA, 'Solution of linear equations with skyline-stored symmetric matrix', *Comp. Struct.*, **5**, 13–30, 1975.
20. G. STRANG and G. J. FIX, *An Analysis of the Finite Element Method*, Prentice-Hall, 1973.
21. D. P. MONDKAR and G. H. POWELL, 'Large capacity equation solver for structural analyses', *Comp. Struct.*, **4**, 699–728, 1974.
22. R. J. MELOSH and R. M. BAMFORD, 'Efficient solution of load-deflection equations', *J. Struct. Div. A.S.C.E.*, **95**, 661–76, 1969.
23. B. M. IRONS, 'A frontal solution program', *Int. J. Num. Meth. Eng.*, **2**, 5–32, 1970.
24. E. HINTON and D. R. J. OWEN, *Finite Element Programming*, Academic Press, 1977.
25. P. HOOD, 'Frontal solution program for unsymmetric matrices', *Int. J. Num. Meth. Eng.*, **10**, 379–400, 1976.
26. A. FRANCAVILLA and O. C. ZIENKIEWICZ, 'Presentation of three dimensional stress for dam analysis', *Proc. Symp. on Numerical Analysis of Dams*, Swansea, 1975.

APPENDIX 1

Matrix Algebra

The mystique surrounding matrix algebra is perhaps due to the texts on the subject requiring the student to 'swallow too much' in one operation. It will be found that in order to follow the present text and carry out the necessary computation only a limited knowledge of a few basic definitions is required.

Definition of a matrix

The linear relationship between a set of variables x and b

$$\begin{aligned} a_{11}x_1+a_{12}x_2+a_{13}x_3+a_{14}x_4 &= b_1 \\ a_{21}x_1+a_{22}x_2+a_{23}x_3+a_{24}x_4 &= b_2 \\ a_{31}x_1+a_{32}x_2+a_{33}x_3+a_{34}x_4 &= b_3 \end{aligned} \tag{A.1.1}$$

can be written, in a shorthand way, as

$$[A]\{x\} = \{b\}$$

or (A.1.1a)

$$\mathbf{Ax} = \mathbf{b}$$

where

$$\mathbf{A} \equiv [A] = \begin{bmatrix} a_{11}, a_{12}, a_{13}, a_{14} \\ a_{21}, a_{22}, a_{23}, a_{24} \\ a_{31}, a_{32}, a_{33}, a_{34} \end{bmatrix} \tag{A.1.2}$$

$$\mathbf{x} \equiv \{x\} = \begin{Bmatrix} x_1 \\ x_2 \\ x_3 \\ x_4 \end{Bmatrix} \quad \mathbf{b} \equiv \{b\} = \begin{Bmatrix} b_1 \\ b_2 \\ b_3 \end{Bmatrix}.$$

The above notation contains within it both the definition of a matrix and of the process of multiplication. Matrices are *defined* as 'arrays of numbers' of the type shown in (A.1.2). The particular form listing a single column of numbers is often referred to as a vector or column matrix. The multiplication of a matrix by a column vector is *defined* by the equivalence of the left sides of Eqs (A.1.1) and (A.1.1a).

The use of bold characters to define both vectors and matrices will be followed throughout the text—generally lower case letters denoting vectors and capitals matrices.

If another relationship, using the same constants, but a different set of x and b, exists, and is written as

$$\begin{aligned} a_{11}x'_1 + a_{12}x'_2 + a_{13}x'_3 + a_{14}x'_4 &= b'_1 \\ a_{21}x'_1 + a_{22}x'_2 + a_{23}x'_3 + a_{24}x'_4 &= b'_2 \\ a_{31}x'_1 + a_{32}x'_2 + a_{33}x'_3 + a_{34}x'_4 &= b'_3 \end{aligned} \tag{A.1.3}$$

then we could write

$$[A][X] = [B]$$

or (A.1.4)

$$\mathbf{AX} = \mathbf{B}$$

in which

$$\mathbf{X} \equiv [X] = \begin{bmatrix} x_1, & x'_1 \\ x_2, & x'_2 \\ x_3, & x'_3 \\ x_4, & x'_4 \end{bmatrix} \qquad \mathbf{B} \equiv [B] = \begin{bmatrix} b_1, & b'_1 \\ b_2, & b'_2 \\ b_3, & b'_3 \end{bmatrix} \tag{A.1.5}$$

implying both the statements (A.1.1) and (A.1.3) arranged simultaneously as

$$\begin{bmatrix} a_{11}x_1 + \cdots, & a_{11}x'_1 + \cdots \\ a_{21}x_1 + \cdots, & a_{21}x'_1 + \cdots \\ a_{31}x_1 + \cdots, & a_{31}x'_1 + \cdots \end{bmatrix} = \begin{bmatrix} b_1, & b'_1 \\ b_2, & b'_2 \\ b_3, & b'_3 \end{bmatrix}. \tag{A.1.4a}$$

It is seen, incidentally, that matrices can be equal only if each of the individual terms is equal.

The multiplication of full matrices is defined above, and it is obvious that it has a meaning only if the number of columns in **A** is equal to the number of rows in **X** for a relation of type (A.1.4). One property which distinguishes matrix multiplication is that, in general,

$$\mathbf{AX} \neq \mathbf{XA}$$

i.e., multiplication of matrices is not commutative as in ordinary algebra.

Matrix addition or subtraction

If relations of form from (A.1.1) and (A.1.3) are added then we have

$$\begin{aligned} a_{11}(x_1+x_1')+a_{12}(x_2+x_2')+a_{13}(x_3+x_3')+a_{14}(x_4+x_4') &= b_1+b_1' \\ a_{21}(x_1+x_1')+a_{22}(x_2+x_2')+a_{23}(x_3+x_3')+a_{24}(x_4+x_4') &= b_2+b_2' \quad \text{(A.1.6)} \\ a_{31}(x_1+x_1')+a_{32}(x_2+x_2')+a_{33}(x_3+x_3')+a_{34}(x_4+x_4') &= b_3+b_3' \end{aligned}$$

which will also follow from

$$\mathbf{Ax}+\mathbf{Ax}' = \mathbf{A}(\mathbf{x}+\mathbf{x}') = \mathbf{b}+\mathbf{b}' = \mathbf{b}'+\mathbf{b}$$

if we define the addition of matrices by a simple addition of the individual terms of the array. Clearly this can be done only if the size of the matrices is identical, i.e., for example

$$\begin{bmatrix} a_{11}, a_{12}, a_{13} \\ \\ a_{21}, a_{22}, a_{23} \end{bmatrix} + \begin{bmatrix} b_{11}, b_{12}, b_{13} \\ \\ b_{21}, b_{22}, b_{23} \end{bmatrix} = \begin{bmatrix} a_{11}+b_{11}, a_{12}+b_{12}, a_{13}+b_{13} \\ \\ a_{21}+b_{21}, a_{22}+b_{22}, a_{23}+b_{23} \end{bmatrix}$$

or

$$\mathbf{A}+\mathbf{B} = \mathbf{C} \tag{A.1.7}$$

implies that every term of **C** is equal to the sum of the appropriate terms of **A** and **B**.

Subtraction obviously follows similar rules.

Transpose of matrix

This is simply a definition for re-ordering of the number of an array in the following manner

$$\begin{bmatrix} a_{11} & a_{12} & a_{13} \\ a_{21} & a_{22} & a_{23} \\ a_{31} & a_{32} & a_{33} \end{bmatrix}^{\mathrm{T}} = \begin{bmatrix} a_{11} & a_{21} & a_{31} \\ a_{12} & a_{22} & a_{32} \\ a_{13} & a_{23} & a_{33} \end{bmatrix} \tag{A.1.8}$$

and will be indicated by the symbol T as shown.

Its use is not immediately obvious but will be indicated later and can be treated here as a simple prescribed operation.

Inverse of a matrix

If in the relationship (A.1.1a) the matrix **A** is 'square', i.e., it represents the coefficients of simultaneous equations of type (A.1.1) equal in number to the number of unknowns **x**, then in general it is possible to solve for the unknowns **x** in terms of the known coefficients **b**. This solution can be written as

$$\mathbf{x} = \mathbf{A}^{-1}\mathbf{b} \tag{A.1.9}$$

in which the matrix $\mathbf{A}^{-1}$ is known as the 'inverse' of the square matrix $\mathbf{A}$. Clearly $\mathbf{A}^{-1}$ is also square and of the same size as $\mathbf{A}$.

We could obtain (A.1.9) by multiplying both sides of (A.1.1a) by $\mathbf{A}^{-1}$ and hence

$$\mathbf{A}\mathbf{A}^{-1} = \mathbf{A}^{-1}\mathbf{A} = \mathbf{I} \qquad \text{(A.1.10)}$$

where $\mathbf{I}$ is an identity matrix having zero on all 'off diagonal' positions and unity on each of the diagonal positions.

If the equations are singular and have no solution then clearly an inverse does not exist.

A sum of products

In problems of mechanics we often encounter a number of quantities such as forces which can be listed as a matrix 'vector'

$$\mathbf{f} = \begin{Bmatrix} f_1 \\ f_2 \\ \vdots \\ f_n \end{Bmatrix}. \qquad \text{(A.1.11)}$$

These, in turn, are often associated with the same number of displacements given by another vector, say,

$$\mathbf{u} = \begin{Bmatrix} u_1 \\ u_2 \\ \vdots \\ u_n \end{Bmatrix}. \qquad \text{(A.1.12)}$$

It is known that the work is represented as a sum of products of force and displacement

$$W = \sum f_n u_n.$$

Clearly the transpose becomes useful here as we can write, by the first rule of matrix multiplication

$$W = [f_1, f_2, \dots f_n] \begin{Bmatrix} a_1 \\ a_2 \\ \vdots \\ a_n \end{Bmatrix} = \mathbf{f}^{\mathrm{T}}\mathbf{a} \equiv \mathbf{a}^{\mathrm{T}}\mathbf{f}. \qquad \text{(A.1.13)}$$

Use of this fact is made frequently in this book.

Transpose of a product

An operation which sometimes occurs is that of taking the transpose of a matrix product. It can be left to the reader to prove from previous definitions that

$$(\mathbf{AB})^{\mathrm{T}} = \mathbf{B}^{\mathrm{T}}\mathbf{A}^{\mathrm{T}}. \tag{A.1.14}$$

Symmetric matrices

In structural problems symmetric matrices are often encountered. If a term of a matrix $\mathbf{A}$ is defined as a_{ij}, then for a symmetric matrix

$$a_{ij} = a_{ji}.$$

It can be shown that the inverse of a symmetric matrix is always symmetric.

Partitioning

It is easy to verify that a matrix product

$$\mathbf{AB}$$

in which, for example,

$$\mathbf{A} = \left[\begin{array}{ccc:cc} a_{11} & a_{12} & a_{13} & a_{14} & a_{15} \\ a_{21} & a_{22} & a_{23} & a_{24} & a_{25} \\ \hdashline a_{31} & a_{32} & a_{33} & a_{34} & a_{35} \end{array}\right]$$

$$\mathbf{B} = \left[\begin{array}{cc} b_{11} & b_{12} \\ b_{21} & b_{22} \\ b_{31} & b_{32} \\ \hdashline b_{41} & b_{42} \\ b_{51} & b_{52} \end{array}\right]$$

could be obtained by dividing each matrix into submatrices, indicated by the dotted lines, and applying the rules of matrix multiplication first to each of such submatrix as if it were a scalar number and then carrying out further multiplication in the usual way. Thus, if we write

$$\mathbf{A} = \begin{bmatrix} \mathbf{A}_{11} & \mathbf{A}_{12} \\ \mathbf{A}_{21} & \mathbf{A}_{22} \end{bmatrix} \qquad \mathbf{B} = \begin{bmatrix} \mathbf{B}_1 \\ \mathbf{B}_2 \end{bmatrix}$$

Then

$$\mathbf{AB} = \begin{bmatrix} \mathbf{A}_{11}\mathbf{B}_1 + \mathbf{A}_{12}\mathbf{B}_2 \\ \mathbf{A}_{21}\mathbf{B}_1 + \mathbf{A}_{22}\mathbf{B}_2 \end{bmatrix}$$

can be verified as representing the complete product by further multiplication.

The essential feature of partitioning is that the size of subdivisions has to be such as to make the products of type $\mathbf{A}_{11}\mathbf{B}_1$ meaningful, i.e., the number of columns in $\mathbf{A}_{11}$ must be equal to the number of rows in $\mathbf{B}_1$ etc. If the above definition holds, then all further operations can be conducted on partitioned matrices treating each partition as if it were a scalar.

It should be noted that any matrix can be multiplied by a scalar (number). Here, obviously, the requirements of equality of appropriate rows and columns no longer apply.

If a symmetric matrix is divided into an equal number of submatrices $\mathbf{A}_{ij}$ in rows and columns then

$$\mathbf{A}_{ij} = \mathbf{A}_{ji}^{\mathrm{T}}.$$

APPENDIX 2

Basic Equations of Displacement Analysis (Chapter 2)

Displacement

(2.1) $\mathbf{u} \approx \hat{\mathbf{u}} = \sum \mathbf{N}_i \mathbf{a}_i = \mathbf{N}\mathbf{a}$

Strain

(2.2–2.3) $\boldsymbol{\varepsilon} = \mathbf{L}\mathbf{u} \approx \sum \mathbf{B}_i \mathbf{a}_i = \mathbf{B}\mathbf{a}$

(2.4) $\mathbf{B}_i = \mathbf{L}\mathbf{N}_i;\ \mathbf{B} = \mathbf{L}\mathbf{N}$

Stress–strain—constitutive relation of linear elasticity

(2.5) $\boldsymbol{\sigma} = \mathbf{D}(\boldsymbol{\varepsilon} - \boldsymbol{\varepsilon}_0) + \boldsymbol{\sigma}_0$

Approximate equilibrium equations

(2.23) $\mathbf{K}\mathbf{a} + \mathbf{f} = \mathbf{r}$

(2.24)
$$\mathbf{K}_{ij} = \int_V \mathbf{B}_i^{\mathrm{T}} \mathbf{D} \mathbf{B}_j \, \mathrm{d}V$$

$$\mathbf{f}_i = -\int_V \mathbf{N}_i^{\mathrm{T}} \mathbf{b} \, \mathrm{d}V - \int_A \mathbf{N}_i^{\mathrm{T}} \bar{\mathbf{t}} \, \mathrm{d}A - \int_V \mathbf{B}_i^{\mathrm{T}} \mathbf{D} \boldsymbol{\varepsilon}_0 \, \mathrm{d}V - \int_V \mathbf{B}_i^{\mathrm{T}} \mathbf{D}_0 \, \mathrm{d}V$$

APPENDIX 3

Integration by Parts in Two or Three Dimensions (Green's Theorem)

Consider the integration by parts of the following two-dimensional expression

$$\int\int_{\Omega} \phi \frac{\partial \psi}{\partial x} \, \mathrm{d}x \, \mathrm{d}y. \tag{A.3.1}$$

Integrating first with respect to x and using the well-known relation for integration by parts

$$\int_{x_L}^{x_R} u \, \mathrm{d}v = -\int_{x_L}^{x_R} v \, \mathrm{d}u + (uv)_{x=x_R} - (uv)_{x=x_L} \tag{A.3.2}$$

we have, with symbols of Fig. A.3.1,

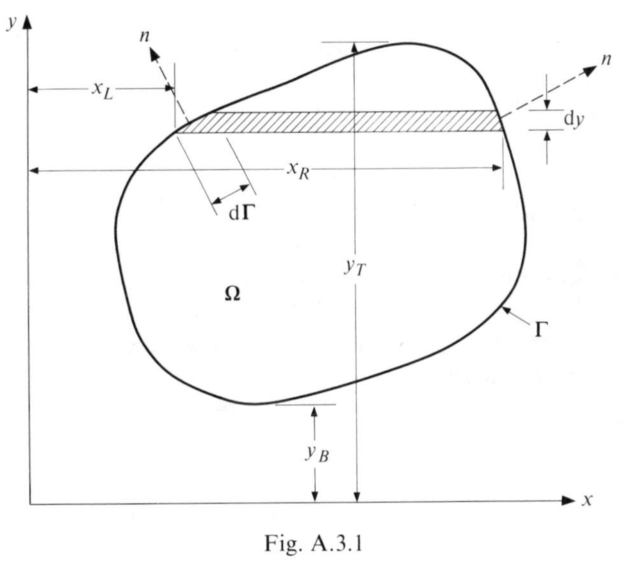

Fig. A.3.1

$$\int\int_{\Omega} \phi \frac{\partial \psi}{\partial x} \, \mathrm{d}x \, \mathrm{d}y = -\int\int_{\Omega} \frac{\partial \phi}{\partial x} \psi \, \mathrm{d}x \, \mathrm{d}y + \\ + \int_{y=y_B}^{y=y_T} [(\phi\psi)_{x=x_R} - (\phi\psi)_{x=x_L}] \, \mathrm{d}y. \quad \text{(A.3.3)}$$

If now we consider a direct segment of the boundary $\mathrm{d}\Gamma$ on the right-hand boundary we note that

$$\mathrm{d}y = \mathrm{d}\Gamma n_x \quad \text{(A.3.4)}$$

where n_x is the direction cosine between the normal and the x direction. Similarly, on the left-hand section we have

$$\mathrm{d}y = -\mathrm{d}\Gamma n_x. \quad \text{(A.3.5)}$$

The final term of Eq. (A3.2) can thus be expressed as the integral taken around an anticlockwise direction

$$\oint_{\Gamma} \phi\psi \, \mathrm{d}\Gamma n_x. \quad \text{(A.3.6)}$$

If several closed contours are encountered this integration has to be taken around each such contour. The general expression is in all cases

$$\int\int_{\Omega} \phi \frac{\partial \psi}{\partial x} \, \mathrm{d}x \, \mathrm{d}y \equiv -\int\int_{\Omega} \frac{\partial \phi}{\partial x} \psi \, \mathrm{d}x \, \mathrm{d}y + \oint_{\Gamma} \phi\psi n_x \, \mathrm{d}\Gamma. \quad \text{(A.3.7)}$$

Similarly if differentiation in the y direction arises we can write

$$\int\int_{\Omega} \phi \frac{\partial \psi}{\partial y} \, \mathrm{d}x \, \mathrm{d}y \equiv -\int\int_{\Omega} \frac{\partial \phi}{\partial y} \psi \, \mathrm{d}x \, \mathrm{d}y + \oint_{\Gamma} \phi\psi n_y \, \mathrm{d}\Gamma \quad \text{(A.3.8)}$$

where l_y is the cosine between the outward normal and the y axis.

In three dimensions by identical procedure we can write

$$\int\int\int_{\Omega} \phi \frac{\partial \psi}{\partial x} \, \mathrm{d}x \, \mathrm{d}y \, \mathrm{d}z = -\int\int\int_{\Omega} \frac{\partial \phi}{\partial x} \psi \, \mathrm{d}x \, \mathrm{d}y \, \mathrm{d}z + \oint_{\Gamma} \phi\psi n_x \, \mathrm{d}\Gamma \quad \text{(A.3.9)}$$

where Γ becomes the element of the surface area and the last integral is taken over the whole surface.

APPENDIX 4

Some Integration Formulae for a Triangle

(Fig. 4.1)

Let a triangle be defined in the x–y plane by three points (x_i, y_i), (x_j, y_j), (x_m, y_m) with the origin at the co-ordinates taken at the centroid, i.e.,

$$\frac{x_i+x_j+x_m}{3} = \frac{y_i+y_j+y_m}{3} = 0.$$

Then integrating over the triangle area

$$\int x \, \mathrm{d}x \, \mathrm{d}y = \int y \, \mathrm{d}x \, \mathrm{d}y = 0$$

$$\int \mathrm{d}x \, \mathrm{d}y = \frac{1}{2} \begin{vmatrix} 1 & x_i & y_i \\ 1 & x_j & y_j \\ 1 & x_m & y_m \end{vmatrix} = \Delta = \text{area of triangle}$$

$$\int x^2 \, \mathrm{d}x \, \mathrm{d}y = \frac{\Delta}{12} (x_i^2 + x_j^2 + x_m^2)$$

$$\int y^2 \, \mathrm{d}x \, \mathrm{d}y = \frac{\Delta}{12} (y_i^2 + y_j^2 + y_m^2)$$

$$\int xy \, \mathrm{d}x \, \mathrm{d}y = \frac{\Delta}{12} (x_i y_i + x_j y_j + x_m y_m)$$

APPENDIX 5

Some Integration Formulae for a Tetrahedron (Fig. 6.1)

Let a tetrahedron be defined in the co-ordinate system (x, y, z) by four points (x_i, y_i, z_i), (x_j, y_j, z_j), (x_m, y_m, z_m), (x_p, y_p, z_p) with the origin at the co-ordinates taken at the centroid, i.e.,

$$\frac{x_i+x_j+x_m+x_p}{4} = \frac{y_i+y_j+y_m+y_p}{4} = \frac{z_i+z_j+z_m+z_p}{4} = 0.$$

Then integrating over the tetrahedron volume

$$\int \mathrm{d}x\,\mathrm{d}y\,\mathrm{d}z = \frac{1}{6}\begin{vmatrix} 1 & x_i & y_i & z_i \\ 1 & x_j & y_j & z_j \\ 1 & x_m & y_m & z_m \\ 1 & x_p & y_p & z_p \end{vmatrix} = V = \text{volume of tetrahedron.}$$

Provided the order of numbering is as indicated on Fig. 6.1 then also:

$$\int x\,\mathrm{d}x\,\mathrm{d}y\,\mathrm{d}z = \int y\,\mathrm{d}x\,\mathrm{d}y\,\mathrm{d}z = \int z\,\mathrm{d}x\,\mathrm{d}y\,\mathrm{d}z = 0$$

$$\int x^2\,\mathrm{d}x\,\mathrm{d}y\,\mathrm{d}z = \frac{V}{20}(x_i^2+x_j^2+x_m^2+x_p^2)$$

$$\int y^2\,\mathrm{d}x\,\mathrm{d}y\,\mathrm{d}z = \frac{V}{20}(y_i^2+y_j^2+y_m^2+y_p^2)$$

$$\int z^2\,\mathrm{d}x\,\mathrm{d}y\,\mathrm{d}z = \frac{V}{20}(z_i^2+z_j^2+z_m^2+z_p^2)$$

$$\int xy\,\mathrm{d}x\,\mathrm{d}y\,\mathrm{d}z = \frac{V}{20}(x_iy_i+x_jy_j+x_my_m+x_py_p)$$

$$\int xz\,\mathrm{d}x\,\mathrm{d}y\,\mathrm{d}z = \frac{V}{20}(x_iz_i+x_jz_j+x_mz_m+x_pz_p)$$

$$\int yz\,\mathrm{d}x\,\mathrm{d}y\,\mathrm{d}z = \frac{V}{20}(y_iz_i+y_jz_j+y_mz_m+y_pz_p).$$

Some Vector Algebra

Some knowledge and understanding of basic vector algebra is needed in dealing with complexities of elements oriented in space such as occur in shells, etc. Some of the operations are here summarized.

Vectors (in the geometric sense) can be described by their components along the directions of the x, y, z axis.

Thus the vector $\mathbf{V}_{01}$ shown in Fig. A.6.1 can be written as

$$\mathbf{V}_{01} = \mathbf{i}x_1 + \mathbf{j}y_1 + \mathbf{k}z_1 \tag{A.6.1}$$

in which **i**, **j**, **k** are unit vectors in directions of the axes x, y, z.

Alternatively the same vector could be written as

$$\mathbf{V}_{01} = \begin{Bmatrix} x_1 \\ y_1 \\ z_1 \end{Bmatrix} \tag{A.6.2}$$

(now a 'vector' in the matrix sense) in which the components are distinguished by positions in the column.

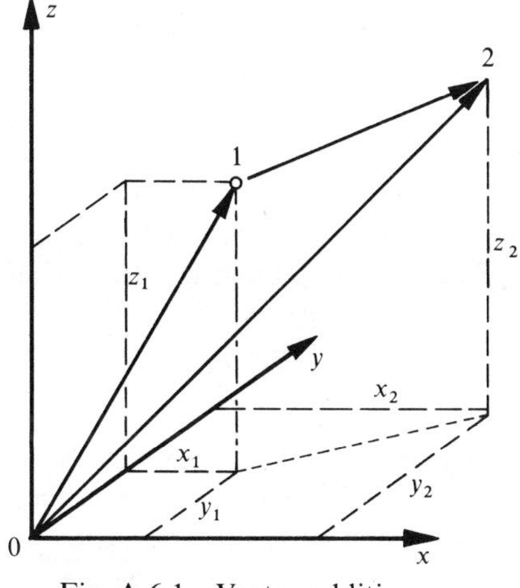

Fig. A.6.1 Vector addition

Addition and subtraction. Addition and subtraction is defined by addition and subtraction of components. Thus for example

$$\mathbf{V}_{02}-\mathbf{V}_{01} = \mathbf{V}_{21} = \mathbf{i}(x_2-x_1)+\mathbf{j}(y_2-y_1)+\mathbf{k}(z_2-z_1). \quad \text{(A.6.3)}$$

The same result is achieved by the definitions of matrix algebra, thus

$$\mathbf{V}_{02}-\mathbf{V}_{01} = \mathbf{V}_{21} = \begin{Bmatrix} x_2-x_1 \\ y_2-y_1 \\ z_2-z_1 \end{Bmatrix}. \quad \text{(A.6.4)}$$

Length of vector. The length of the vector $\mathbf{V}_{21}$ is given, purely geometrically, as

$$l_{21} = \sqrt{(x_2-x_1)^2+(y_2-y_1)^2+(z_2-z_1)^2} \quad \text{(A.6.5)}$$

or in terms of matrix algebra

$$l_{12} = \sqrt{\mathbf{V}_{12}^{\mathrm{T}}\mathbf{V}_{12}} \quad \text{(A.6.6)}$$

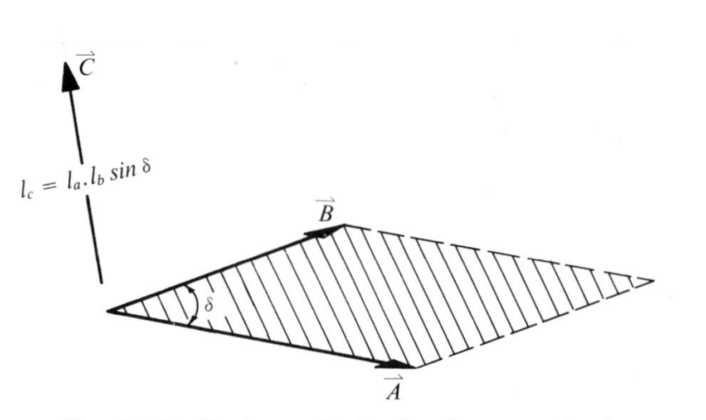

Fig. A.6.2 Vector multiplication (cross product)

Direction cosines. Direction cosines of a vector are simply, from the definition of the projected component lengths, given as

$$\cos \alpha_x = \lambda_{vx} = \frac{x_2-x_1}{l_{12}}, \text{ etc.} \quad \text{(A.6.7)}$$

where α_x is the angle between the vector and x axis.

'*Scalar*' *products.* A scalar product of two vectors is *defined* as the product of the length of one vector by the scalar projection of the other vector on it. Or if γ is the angle between two vectors **A** and **B** and their length l_a and l_b respectively

$$\mathbf{A}.\mathbf{B} = l_a l_b \cos \gamma = \mathbf{B}.\mathbf{A}. \quad \text{(A.6.8)}$$

If

$$\mathbf{A} = \mathbf{i}a_x + \mathbf{j}a_y + \mathbf{k}a_z$$

and (A.6.9)

$$\mathbf{B} = \mathbf{i}b_x + \mathbf{j}b_y + \mathbf{k}b_z$$

$$\mathbf{A}.\mathbf{B} = a_x b_x + a_y b_y + a_z b_z \tag{A.6.10}$$

if we note that, by above definition

$$\mathbf{i}.\mathbf{i} = \mathbf{j}.\mathbf{j} = \mathbf{k}.\mathbf{k} = 1$$

$$\mathbf{i}.\mathbf{j} = \mathbf{j}.\mathbf{k} = \mathbf{k}.\mathbf{i} = 0, \text{ etc.}$$

Using the matrix notation

$$\mathbf{A} = \begin{Bmatrix} a_x \\ a_y \\ a_z \end{Bmatrix}, \qquad \mathbf{B}. = \begin{Bmatrix} b_x \\ b_y \\ b_z \end{Bmatrix} \tag{A.6.11}$$

$$\mathbf{A}.\mathbf{B} = \mathbf{A}^{\mathrm{T}}\mathbf{B} = \mathbf{B}^{\mathrm{T}}\mathbf{A}. \tag{A.6.12}$$

'*Vector' or cross product*. Another product of the vector is *defined* as a vector oriented normally to the plane given by the two vectors and equal in magnitude to the product of the length of the two vectors, multiplied by the sine of the angle between them. Further, its direction follows the right-hand rule as shown in Fig. A.6.2 in which

$$\mathbf{A} \times \mathbf{B} = \mathbf{C}. \tag{A.6.13}$$

is shown.

Thus

$$\mathbf{A} \times \mathbf{B} = -\mathbf{B} \times \mathbf{A}. \tag{A.6.14}$$

It is worth noting that the magnitude (*or length*) of **C** is equal to the area of the parallelogram shown in Fig. A.6.2.

Using definition of Eq. (A.6.9) and noting that

$$\mathbf{i} \times \mathbf{i} = \mathbf{j} \times \mathbf{j} = \mathbf{k} \times \mathbf{k} = 0$$

$$\mathbf{i} \times \mathbf{j} = \mathbf{k}, \qquad \mathbf{j} \times \mathbf{k} = \mathbf{i}, \qquad \mathbf{k} \times \mathbf{i} = \mathbf{j} \tag{A.6.15}$$

we have

$$\mathbf{A} \times \mathbf{B} = \det \begin{vmatrix} \mathbf{i} & \mathbf{j} & \mathbf{k} \\ a_x & a_y & a_z \\ b_x & b_y & b_z \end{vmatrix}$$

$$= (a_y b_z - a_z b_y)\mathbf{i} + (a_z b_x - a_x b_z)\mathbf{j} + (a_x b_y - a_y b_x)\mathbf{k}. \tag{A.6.16}$$

In matrix algebra this does not find a simple counterpart but we can use the above to define the vector **C**†

$$\mathbf{C} = \mathbf{A} \times \mathbf{B} = \begin{Bmatrix} a_y b_z - a_z b_y \\ a_z b_x - a_x b_z \\ a_x b_y - a_y b_x \end{Bmatrix}. \tag{A.6.17}$$

The vector product will be found particularly useful when the problem of erecting a normal direction to a surface (*vide* Chapter 11) is considered.

Elements of area and volume. If ξ and η are some curvilinear co-ordinates then vectors in two dimensional plane

$$\mathrm{d}\boldsymbol{\xi} = \begin{Bmatrix} \dfrac{\partial x}{\partial \xi} \\ \dfrac{\partial y}{\partial \xi} \end{Bmatrix} \mathrm{d}\xi; \qquad \mathrm{d}\boldsymbol{\eta} = \begin{Bmatrix} \dfrac{\partial x}{\partial \eta} \\ \dfrac{\partial y}{\partial \eta} \end{Bmatrix} \mathrm{d}\eta \tag{A.6.18}$$

defined from the relationship between the Cartesian and curvilinear co-ordinates, are vectors directed tangentially to the ξ = const. and η = const. contours respectively. As the *length* of the vector resulting from a cross product of $\mathrm{d}\xi \times \mathrm{d}\eta$ is equal to the area of the elementary parallelogram we can write

$$\mathrm{d(Area)} = \det \begin{vmatrix} \dfrac{\partial x}{\partial \xi} & \dfrac{\partial x}{\partial \eta} \\ \dfrac{\partial y}{\partial \xi} & \dfrac{\partial y}{\partial \eta} \end{vmatrix} \partial\xi\, \partial\eta \tag{A.6.19}$$

by Eq. (A.6.17).

Similarly if we have three curvilinear co-ordinates ξ, η, ζ in the Cartesian space the 'triple scalar' or box product defines a unit volume

$$\mathrm{d(Vol)} = \mathrm{d}\boldsymbol{\xi} \cdot (\mathrm{d}\boldsymbol{\eta} \times \mathrm{d}\boldsymbol{\zeta}) = \det \begin{vmatrix} \dfrac{\partial x}{\partial \xi} & \dfrac{\partial x}{\partial \eta} & \dfrac{\partial x}{\partial \zeta} \\ \dfrac{\partial y}{\partial \xi} & \dfrac{\partial y}{\partial \eta} & \dfrac{\partial y}{\partial \zeta} \\ \dfrac{\partial z}{\partial \xi} & \dfrac{\partial z}{\partial \eta} & \dfrac{\partial z}{\partial \zeta} \end{vmatrix} . \ \mathrm{d}\xi\, \mathrm{d}\eta\, \mathrm{d}\zeta \tag{A.6.20}$$

† If we rewrite **A** as a skew symmetric matrix

$$\hat{\mathbf{A}} \equiv \begin{bmatrix} 0, & -a_z, & a_y \\ a_z, & 0, & -a_x \\ -a_y, & a_x, & 0 \end{bmatrix}$$

then the reader can verify that an alternative representation of the vector product in matrix form is (T. Crouch, private communication)

$$\mathbf{C} = \hat{\mathbf{A}}\mathbf{B}$$

This follows simply from the geometry. The bracketed product, by definition, forms a vector whose length is equal to the parallelogram area with sides tangent to two of the co-ordinates. The second scalar multiplication by a length and cosine of the angle between that length and the normal to the parallelogram establishes an elementary volume.

Author Index

Numbers in bold type refer to the list of references at the end of each chapter.

Subject Index